Glossary of Notation

$C(X)$	[illegible]
$C^n(X)$	Set [illegible] having n continuous derivatives on X *3*
$C^\infty(X)$	Set of all functions having derivatives of all orders on X *3*
$0.\overline{3}$	A decimal in which the numeral 3 repeats indefinitely *3*
$\mathbb{R}$	Set of real numbers *9*
$fl(y)$	Floating-point form of the real number y *16*
$O(\cdot)$	Order of convergence *23*
Δ	Forward difference *51*
$f[\cdot]$	Divided difference of the function f *74*
$\binom{n}{k}$	The kth binomial coefficient of order n *76*
∇	Backward difference *77*
$\rightarrow$	Equation replacement *238*
$\leftrightarrow$	Equation interchange *238*
(a_{ij})	Matrix with a_{ij} as the entry in the ith row and jth column *239*
$\mathbf{x}$	Column vector or element of $\mathbb{R}^n$ *240*
$[A, \mathbf{b}]$	Augmented matrix *240*
δ_{ij}	Kronecker delta, 1 if $i = j$, 0 if $i \neq j$ *258*
I_n	$n \times n$ identity matrix *258*
A^{-1}	Inverse matrix of the matrix A *258*
A^t	Transpose matrix of the matrix A *261*
M_{ij}	Minor of a matrix *261*
$\det A$	Determinant of the matrix A *261*
$\mathbf{0}$	Vector with all zero entries *264*
$\mathbb{R}^n$	Set of ordered n-tuples of real numbers *288*
$\lVert\mathbf{x}\rVert$	Arbitrary norm of the vector $\mathbf{x}$ *288*
$\lVert\mathbf{x}\rVert_2$	The l_2 norm of the vector $\mathbf{x}$ *288*
$\lVert\mathbf{x}\rVert_\infty$	The l_∞ norm of the vector $\mathbf{x}$ *288*
$\lVert A\rVert$	Arbitrary norm of the matrix A *292*
$\lVert A\rVert_\infty$	The l_∞ norm of the matrix A *292*
$\lVert A\rVert_2$	The l_2 norm of the matrix A *293*
$\rho(A)$	The spectral radius of the matrix A *300*
$K(A)$	The condition number of the matrix A *316*
Π_n	Set of all polynomials of degree n or less *334*
$\tilde{\Pi}_n$	Set of all monic polynomials of degree n *343*
$\mathcal{T}_n$	Set of all trigonometric polynomials of degree n or less *352*
C	Set of complex numbers *370*
$\mathbf{F}$	Function mapping $\mathbb{R}^n$ into $\mathbb{R}^n$ *400*
$J(\mathbf{x})$	Jacobian matrix *403*
∇g	Gradient of the function g *418*

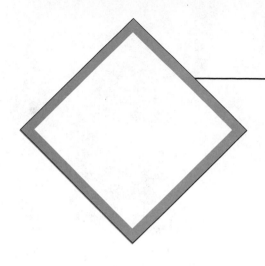

Numerical Methods

SECOND EDITION

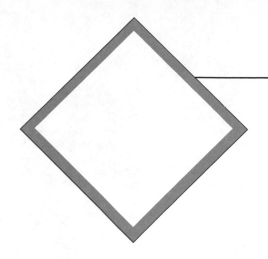

Numerical Methods

SECOND EDITION

J. Douglas Faires
Youngstown State University

Richard Burden
Youngstown State University

Brooks/Cole Publishing Company
I(T)P® An International Thomson Publishing Company

Pacific Grove • Albany • Belmont • Bonn • Boston • Cincinnati • Detroit • Johannesburg
London • Madrid • Melbourne • Mexico City • New York • Paris • Singapore • Tokyo
Toronto • Washington

GWO A GARY W. OSTEDT BOOK

Publisher: *Gary W. Ostedt*
Marketing Team: *Jean V. Thompson, Caroline Croley, Michele Mootz*
Editorial Associate: *Carol Ann Benedict*
Production Coordinator: *Marjorie Z. Sanders*
Production: *Integre Technical Publishing Co., Inc.*
Manuscript Editor: *Linda Thompson*
Interior Design: *Merry Obrecht Sawdey*
Interior Illustration: *Scientific Illustrators*
Cover Design: *Vernon T. Boes*
Cover Photo: *H. Sakuramoto, Photonica*
Typesetting: *Integre Technical Publishing Co., Inc.*
Cover Printing: *Phoenix Color Corporation, Inc.*
Printing and Binding: *R.R. Donnelley & Sons, Crawfordsville*

For more information, contact:

BROOKS/COLE PUBLISHING COMPANY
511 Forest Lodge Road
Pacific Grove, CA 93950
USA

International Thomson Publishing Europe
Berkshire House 168–173
High Holborn
London WC1V 7AA
England

Thomas Nelson Australia
102 Dodds Street
South Melbourne, 3205
Victoria, Australia

Nelson Canada
1120 Birchmount Road
Scarborough, Ontario
Canada M1K 5G4

International Thomson Editores
Seneca 53
Col. Polanco
11560 México, D.F., México

International Thomson Publishing GmbH
Königswinterer Strasse 418
53227 Bonn
Germany

International Thomson Publishing Asia
221 Henderson Road
#05–10 Henderson Building
Singapore 0315

International Thomson Publishing Japan
Hirakawacho Kyowa Building, 3F
2-2-1 Hirakawacho
Chiyoda-ku, Tokyo 102
Japan

Printed in the United States of America

10 9 8 7 6 5 4 3 2 1

Library of Congress Cataloging-in-Publication Data

Faires, J. Douglas.
Numerical methods / J. Douglas Faires, Richard Burden. — 2nd ed.
p. cm.
Includes index.
ISBN 0-534-35187-5 (alk. paper)
1. Numerical analysis. I. Burden, Richard L. II. Title.
QA297.F35 1998
519.4—dc21 97-47340
CIP

THIS BOOK IS PRINTED ON ACID-FREE RECYCLED PAPER

Contents

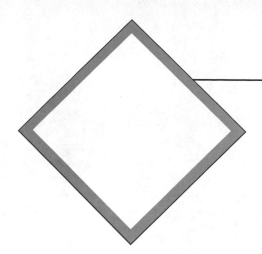

Chapter 11

Boundary-Value Problems for Ordinary Differential Equations 425

Chapter 12

Numerical Methods for Partial-Differential Equations 467

Preface

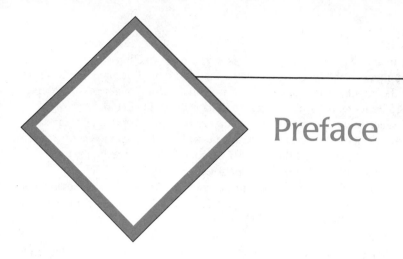

The teaching of numerical approximation techniques to undergraduates is done in a variety of ways. The traditional Numerical Analysis course emphasizes both the approximation methods and the mathematical analysis that produces them. A Numerical Methods course is generally more concerned with the choice and application of techniques to solve problems in engineering and the physical sciences than with the derivation of the methods.

The books used in the Numerical Methods courses differ widely in both content and intent. Sometimes a book written for Numerical Analysis is adapted for a Numerical Methods course by deleting the more theoretical topics and derivations. The advantage of this approach is that the leading Numerical Analysis books are mature; they have been through a number of editions, and they have a wealth of proven examples and exercises. They are also written for a full year coverage of the subject, so they have methods that can be used for reference, even when there is not sufficient time for discussing these methods in the course. The weakness of using a Numerical Analysis book for a Numerical Methods course is that material will need to be omitted, and students have difficulty distinguishing what is important from what is tangential.

The second type of book used for a Numerical Methods course is one that is specifically written for a service course. These books follow the established line of service-oriented mathematics books, similar to those written for business, life sciences, and technical calculus and the statistics books designed for students in economics, psychology, and business. However, the engineering and science students for whom the Numerical Methods course is designed have a much stronger mathematical background than their counterparts in other disciplines. They are quite capable of mastering the material in a Numerical Analysis course, but they do not have the time for, nor often the interest in, the theoretical aspects of such a course. What they need is a sophisticated introduction to the approximation techniques that are used to solve the problems that arise in science and engineering. They also need to know why the methods work, what type of error to expect, and when an application might lead to difficulties. Finally, they need information, with recommendations, regarding the availability of high quality software for numerical approximation routines. In such a course the mathematical analysis is reduced due to a lack of time, not because of the mathematical abilities of the students.

The emphasis in this Numerical Methods book is on the intelligent application of approximation techniques to the types of problems that commonly occur in engineering and the physical and computer sciences. The book is designed for a one-semester course, but contains at least 50% more material, so instructors have flexibility in topic coverage and students have a reference for future work. The techniques covered are essentially the same as those included in our book designed for the Numerical Analysis course (see Burden and Faires, *Numerical Analysis*, Sixth Edition, 1997, Brooks/Cole). However, the emphasis in the two books is quite different. In *Numerical Analysis*, a book with about 800 text pages, each technique is given a mathematical justification before the implementation of the method is discussed. If some portion of the justification is beyond the mathematical level of the book, then it is referenced, but the book is, for the most part, mathematically self-contained. In *Numerical Methods*, each technique is motivated and described from an implementation standpoint. The aim of the motivation is to convince the student that the method is reasonable both mathematically and computationally. A full mathematical justification is included only if it is concise and adds to the understanding of the method.

In the past decade a number of software packages have been developed to produce symbolic mathematical computations. Predominant among them are *MACSYMA*, *DERIVE*, Maple, *Mathematica*, and MATLAB. There are versions of the software packages for most common computer systems and student versions of some are available at reasonable prices. Although there are significant differences among the packages, both in performance and price, they all can perform standard algebra and calculus operations. Having a symbolic computation package available can be very useful in the study of approximation techniques. The results in most of our examples and exercises have been generated using problems for which exact values can be determined, since this permits the performance of the approximation method to be monitored. Exact solutions can often be obtained quite easily using symbolic computation.

We have chosen Maple as our standard package, and have added examples and exercises whenever we felt that a computer algebra system would be of significant benefit. In addition, we have discussed the approximation methods that Maple employs when it is unable to solve a problem exactly. These generally parallel those that are given in the text.

Software is included with and is an integral part of *Numerical Methods*, and a program disk is included with the books. For each method discussed in the text there is a program in C, FORTRAN, Pascal, and a worksheet in Maple, *Mathematica*, and MATLAB. The programs and worksheets permit students to generate all the results that are included in the examples and to modify the programs to generate solutions to problems of their choice. The intent of the software is to provide students with programs that will solve most of the problems that they are likely to encounter in their studies.

Occasionally, exercises in the text contain problems for which the programs do not give satisfactory solutions. These are included to illustrate the difficulties that can arise in the application of approximation techniques and to show the need for the flexibility provided by the standard general purpose software packages that are available for scientific computation. Information about the standard general purposes software packages is discussed in the text. Included are those in packages distributed by the International Mathematical and Statistical Library (IMSL), those produced by the National Algorithms Group (NAG), and specialized techniques in LAPACK, and the routines in MATLAB.

We have completely redone the exercise sets for this edition, including, wherever possible, problems designed to be solved in five to ten minutes, and to get to the heart of the technique being considered. We have also added more application problems, many of which require multiple steps for their solution, and some call for the multiple approximation techniques.

A *Student Study Guide* is available with this edition. It includes solutions to representative exercises, particularly those that extend the theory in the text. The publisher can also provide instructors with a complete *Instructor's Manual* that provides solutions to all the exercises in the book. All the results in this *Instructor's Manual* were regenerated for this edition using the programs on the disk.

The following chart shows the chapter dependencies in the book. We have tried to keep the prerequisite material to a minimum to allow greater flexibility.

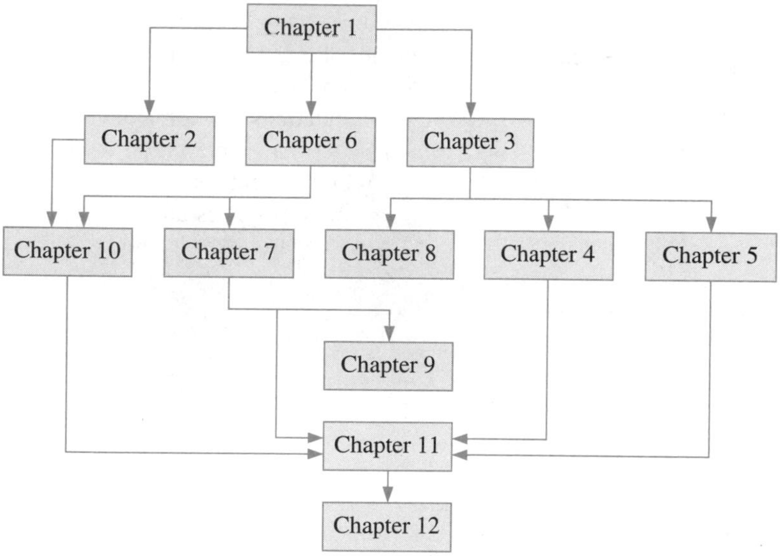

In addition to the many unsolicited comments on the previous edition, we have had the benefit of receiving reviews from the following, whose efforts we greatly appreciate: James Baglama, Texas Tech University; Virginia McGlone, College of St. Elizabeth; Ossawa Salah, University of Tennessee; and Barton Willis, University of Nebraska at Kearney.

Christy Conn, a senior mathematics major at Youngstown State University, has helped us prepare manuscripts since her freshman year, and has done far more than would be expected of a student assistant. She will be leaving us in June, and we want to take this opportunity to thank her and wish her well.

J. Douglas Faires
faires@math.ysu.edu

Richard L. Burden
burden@math.ysu.edu

CHAPTER 1

Mathematical Preliminaries and Error Analysis

1.1 Introduction

This book examines problems that can be solved by methods of approximation, techniques we call *numerical methods*. We begin by considering some of the mathematical and computational topics that arise when approximating a solution to a problem.

Nearly all the problems whose solutions can be approximated involve continuous functions, so calculus is the principal tool to use for deriving numerical methods and verifying that they solve the problems. The calculus definitions and results included in the next section provide a handy reference when these concepts are needed later in the book.

There are two things to consider when applying a numerical technique to solve a problem. The first and most obvious is to obtain the approximation. The equally important second objective is to determine a safety factor for the approximation: some assurance, or at least a sense, of the accuracy of the approximation. Sections 1.3 and 1.4 deal with a standard difficulty that occurs when applying techniques to approximate the solution to a problem: Where and why is computational error produced and how can it be controlled?

The final section in this chapter describes various types and sources of mathematical software for implementing numerical methods.

1.2 Review of Calculus

Fundamental to the study of calculus are the concepts of limit and continuity of a function. A function f defined on a set X of real numbers has the **limit** L at x_0 in X, written $\lim_{x\to x_0} f(x) = L$, if, given any real number $\epsilon > 0$, there exists a real number $\delta > 0$ such that $|f(x) - L| < \epsilon$ whenever $0 < |x - x_0| < \delta$. This definition ensures that values of the function will be close to L whenever x is sufficiently close to x_0. (See Figure 1.1.)

Figure 1.1

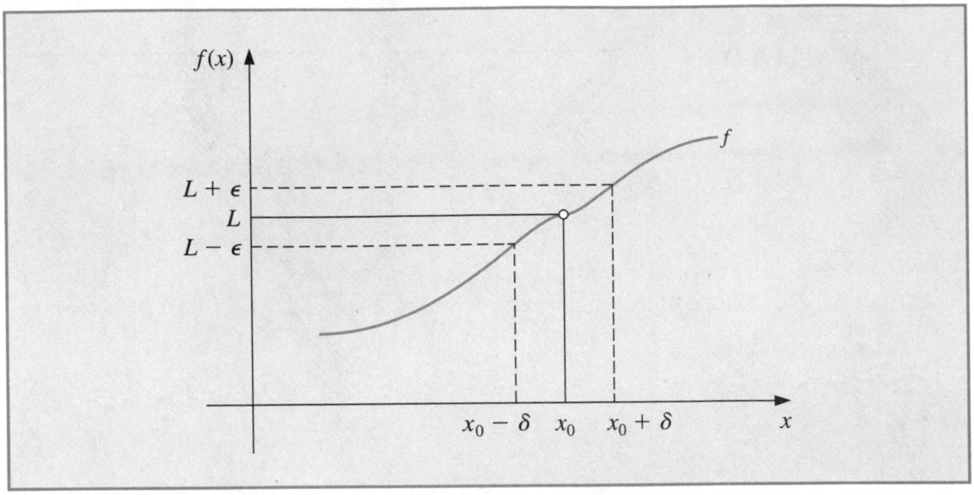

The function f is **continuous** at x_0 if $\lim_{x \to x_0} f(x) = f(x_0)$, and f is **continuous on the set** X if it is continuous at each number in X. We use $C(X)$ to denote the set of all functions that are continuous on X. When X is an interval of the real line, the parentheses in this notation are omitted. For example, the set of all functions that are continuous on the closed interval $[a, b]$ is denoted $C[a, b]$.

The limit of a sequence of real or complex numbers is defined in a similar manner. An infinite sequence $\{x_n\}_{n=1}^{\infty}$ **converges** to a number x if, given any $\epsilon > 0$, there exists a positive integer $N(\epsilon)$ such that $|x_n - x| < \epsilon$ whenever $n > N(\epsilon)$. The notation $\lim_{n \to \infty} x_n = x$, or $x_n \to x$ as $n \to \infty$, means that the sequence $\{x_n\}_{n=1}^{\infty}$ converges to x.

Continuity and Sequence Convergence

If f is a function defined on a set X of real numbers and $x_0 \in X$, then the following are equivalent:

a. f is continuous at x_0;
b. If $\{x_n\}_{n=1}^{\infty}$ is any sequence in X converging to x_0, then

$$\lim_{n \to \infty} f(x_n) = f(x_0).$$

The functions we consider when discussing numerical methods are continuous since this is a minimal requirement for predictable behavior. Functions that are not continuous can skip over points of interest, which can cause difficulties when we attempt to approximate a solution to a problem. More sophisticated assumptions about a function generally lead to better approximation results. For example, a function with a smooth graph would normally behave more predictably than one with numerous jagged features. Smoothness relies on the concept of the derivative.

If f is a function defined in an open interval containing x_0, then f is **differentiable** at x_0 when

$$f'(x_0) = \lim_{x \to x_0} \frac{f(x) - f(x_0)}{x - x_0}$$

exists. The number $f'(x_0)$ is called the **derivative** of f at x_0. The derivative of f at x_0 is the slope of the tangent line to the graph of f at $(x_0, f(x_0))$, as shown in Figure 1.2.

Figure 1.2

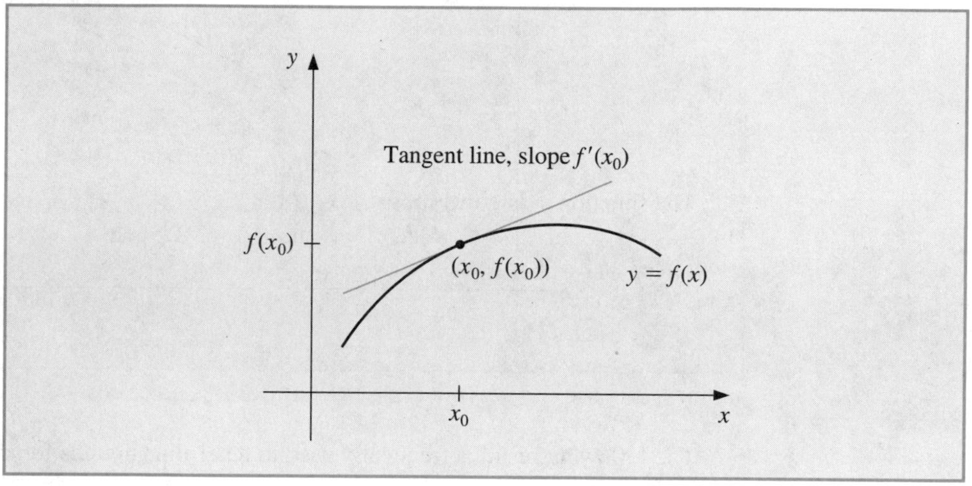

A function that has a derivative at each number in a set X is **differentiable** on X. Differentiability is a stronger condition on a function than continuity in the following sense.

Differentiability Implies Continuity

If the function f is differentiable at x_0, then f is continuous at x_0.

The set of all functions that have n continuous derivatives on X is denoted $C^n(X)$, and the set of functions that have derivatives of all orders on X is denoted $C^\infty(X)$. Polynomial, rational, trigonometric, exponential, and logarithmic functions are in $C^\infty(X)$, where X consists of all numbers at which the function is defined.

The next results are of fundamental importance in deriving methods for error estimation. The proofs of most of these can be found in any standard calculus text.

Mean Value Theorem

If $f \in C[a, b]$ and f is differentiable on (a, b), then a number c in (a, b) exists such that (see Figure 1.3)

$$f'(c) = \frac{f(b) - f(a)}{b - a}.$$

Figure 1.3

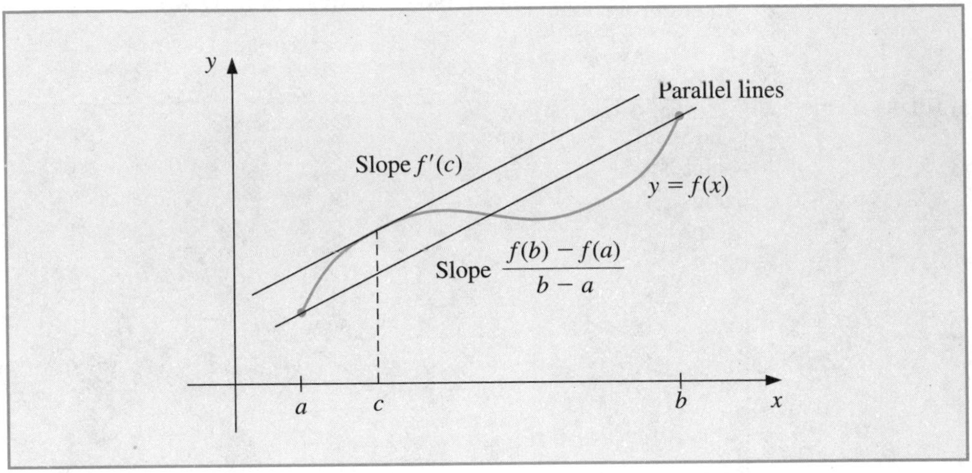

The following result is frequently used to determine bounds for error formulas.

Extreme Value Theorem

If $f \in C[a, b]$, then c_1 and c_2 in $[a, b]$ exist with $f(c_1) \le f(x) \le f(c_2)$ for all x in $[a, b]$. If, in addition, f is differentiable on (a, b), then the numbers c_1 and c_2 occur either at endpoints of $[a, b]$ or where f' is zero.

As mentioned in the preface, we will use the computer algebra system Maple whenever appropriate. We have found this package to be particularly useful for symbolic differentiation and plotting graphs. Both techniques are illustrated in Example 1.

EXAMPLE 1 Find $\max_{a \le x \le b} |f(x)|$ for

$$f(x) = 5\cos 2x - 2x \sin 2x$$

on the intervals $[1, 2]$ and $[0.5, 1]$.

We first illustrate the graphing capabilities of Maple. To access the graphing package, enter the command

```
>with(plots);
```

The commands within the package are then displayed. We define f by entering

```
>f:= 5*cos(2*x)-2*x*sin(2*x);
```

The response is

$$f := 5\cos(2x) - 2x\sin(2x)$$

To graph f on the interval $[0.5, 2]$ requires the command

```
>plot(f,x=0.5..2);
```

The graph appears as shown in Figure 1.4, and we can determine the coordinates of any point of the graph by moving the mouse pointer to the desired point and clicking the left mouse button. This technique can by used to estimate extrema of functions and where a graph crosses the axes.

Figure 1.4

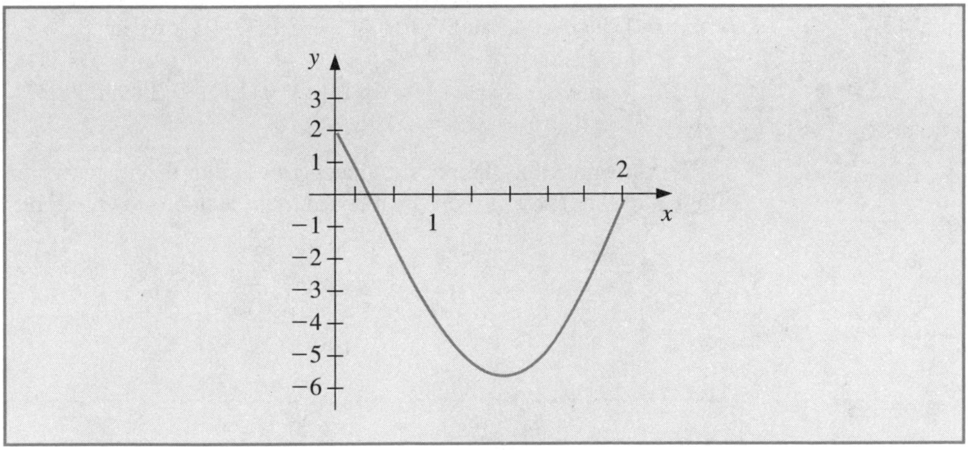

We complete the example using the Extreme Value Theorem. First, consider the interval $[1, 2]$. To obtain the first derivative $g = f'$, enter

```
>g:=diff(f,x);
```

Maple gives

$$g := -12\sin(2x) - 4x\cos(2x)$$

We can solve $g(x) = 0$ for $1 \le x \le 2$ with the command

```
>fsolve(g,x,1..2);
```

obtaining 1.358229874, and compute $f(1.358229874) = -5.675301338$ using

```
>evalf(subs(x=1.358229874,f));
```

Since $f(1) = -3.899329037$ and $f(2) = -0.241008124$, we have

$$\max_{1\le x\le 2} |5\cos 2x - 2x\sin 2x| = |f(1.358229874)| = 5.675301338.$$

If we try to solve $g(x) = 0$ for $0.5 \le x \le 1$, we find that when we enter

```
>fsolve(g,x,0.5..1);
```

Maple responds with

$$\text{fsolve}(-12\sin(2x) - 4x\cos(2x), x, .5\,..\,1)$$

This indicates that Maple could not find a solution in $[0.5, 1]$. If you graph g, you will see that there is no solution in this interval. Hence, f' is never zero in $[0.5, 1]$, and since $f(0.5) = 1.860040545$ and $f(1.0) = -3.899329037$, we have

$$\max_{0.5\le x\le 1} |5\cos 2x - 2x\sin 2x| = |f(1.0)| = 3.899329037.$$ ◆ ◆ ◆

The integral is the other basic concept of calculus that is used extensively. The **Riemann integral** of the function f on the interval $[a, b]$ is the following limit, provided it exists.

$$\int_a^b f(x)\,dx = \lim_{\max \Delta x_i \to 0} \sum_{i=1}^{n} f(z_i)\,\Delta x_i,$$

Figure 1.5

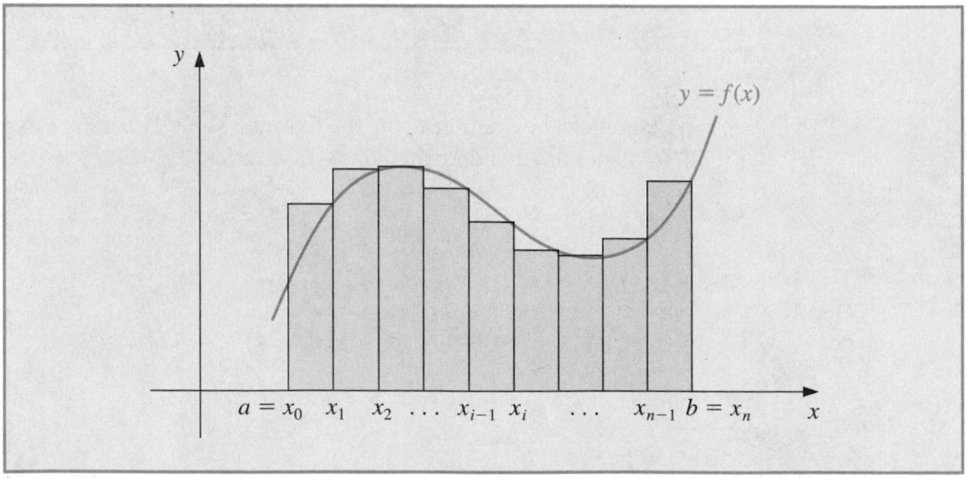

where the numbers $x_0, x_1, \ldots, x_n$ satisfy $a = x_0 < x_1 < \cdots < x_n = b$ and where $\Delta x_i = x_i - x_{i-1}$, for each $i = 1, 2, \ldots, n$, and z_i is arbitrarily chosen in the interval $[x_{i-1}, x_i]$.

A continuous function f on $[a, b]$ is Riemann integrable on the interval. This permits us to choose, for computational convenience, the points x_i to be equally spaced in $[a, b]$ and for each $i = 1, 2, \ldots, n$, to choose $z_i = x_i$. In this case

$$\int_a^b f(x)\,dx = \lim_{n\to\infty} \frac{b-a}{n} \sum_{i=1}^{n} f(x_i),$$

where the numbers shown in Figure 1.5 as x_i are $x_i = a + (i(b-a)/n)$.

Two more basic results are needed in our study of numerical methods. The first is a generalization of the usual Mean Value Theorem for Integrals.

Mean Value Theorem for Integrals

If $f \in C[a, b]$, g is integrable on $[a, b]$ and $g(x)$ does not change sign on $[a, b]$, then there exists a number c in (a, b) with

$$\int_a^b f(x)g(x)\,dx = f(c) \int_a^b g(x)\,dx.$$

The next theorem presented is the Intermediate Value Theorem. Although its statement is not difficult, the proof is beyond the scope of the usual calculus course.

Intermediate Value Theorem

If $f \in C[a, b]$ and K is any number between $f(a)$ and $f(b)$, then there exists a number c in (a, b) for which $f(c) = K$. (Figure 1.6 shows one of the three possibilities for this function and interval.)

Figure 1.6

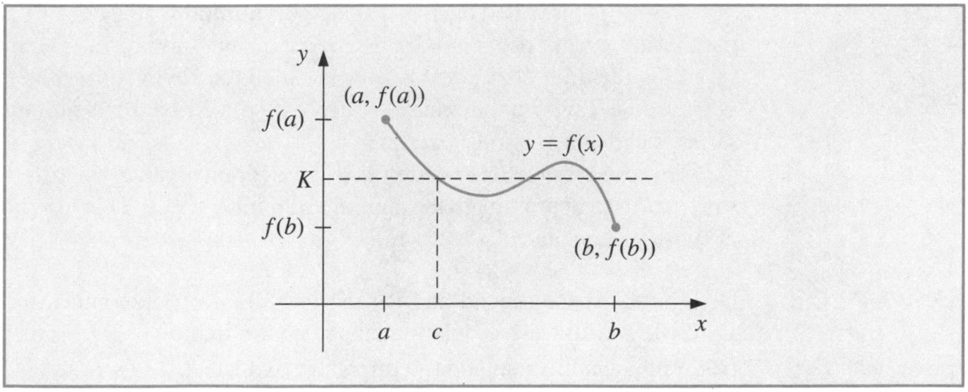

EXAMPLE 2 To show that $x^5 - 2x^3 + 3x^2 - 1 = 0$ has a solution in the interval $[0, 1]$, consider $f(x) = x^5 - 2x^3 + 3x^2 - 1$. Since

$$f(0) = -1 < 0 < 1 = f(1)$$

and f is continuous, the Intermediate Value Theorem implies a number x exists, with $0 < x < 1$, for which $x^5 - 2x^3 + 3x^2 - 1 = 0$. ◆ ◆ ◆

As seen in Example 2, the Intermediate Value Theorem is used to help determine when solutions to certain problems exist. It does not, however, give an efficient means for finding these solutions. This topic is considered in Chapter 2.

The final theorem in this review from calculus describes the development of the Taylor polynomials. The importance of the Taylor polynomials to the study of numerical analysis cannot be overemphasized, and the following result is used repeatedly.

Taylor's Theorem

Suppose $f \in C^n[a, b]$ and $f^{(n+1)}$ exists on $[a, b]$. Let x_0 be a number in $[a, b]$. For every x in $[a, b]$, there exists a number $\xi(x)$ between x_0 and x with

$$f(x) = P_n(x) + R_n(x),$$

where

$$\begin{aligned} P_n(x) &= f(x_0) + f'(x_0)(x - x_0) + \frac{f''(x_0)}{2!}(x - x_0)^2 + \cdots + \frac{f^{(n)}(x_0)}{n!}(x - x_0)^n \\ &= \sum_{k=0}^{n} \frac{f^{(k)}(x_0)}{k!}(x - x_0)^k \end{aligned}$$

and

$$R_n(x) = \frac{f^{(n+1)}(\xi(x))}{(n + 1)!}(x - x_0)^{n+1}.$$

Here $P_n(x)$ is called the **nth Taylor polynomial** for f about x_0, and $R_n(x)$ is called the **truncation error** (or *remainder term*) associated with $P_n(x)$. The infinite series obtained by taking the limit of $P_n(x)$ as $n \to \infty$ is called the *Taylor series* for f about x_0. In the case $x_0 = 0$, the Taylor polynomial is often called a **Maclaurin polynomial**, and the Taylor series is called a *Maclaurin series*.

The term *truncation error* refers to the error involved in using a truncated (that is, finite) summation to approximate the sum of an infinite series. This terminology is reintroduced in subsequent chapters.

EXAMPLE 3 Determine (a) the second and (b) the third Taylor polynomials for $f(x) = \cos x$ about $x_0 = 0$, and use these polynomials to approximate $\cos(0.01)$. (c) Use the third Taylor polynomial and its remainder term to approximate $\int_0^{0.1} \cos x\, dx$.

Since $f \in C^\infty(\mathbb{R})$, Taylor's Theorem can be applied for any $n > 0$. Also,

$$f'(x) = -\sin x,\ f''(x) = -\cos x,\ f'''(x) = \sin x, \quad \text{and} \quad f^{(4)}(x) = \cos x,$$

so

$$f(0) = 1,\ f'(0) = 0,\ f''(0) = -1, \quad \text{and} \quad f'''(0) = 0.$$

a. For $n = 2$ and $x_0 = 0$, we have

$$\cos x = 1 - \frac{1}{2}x^2 + \frac{1}{6}x^3 \sin \xi(x),$$

where $\xi(x)$ is a number between 0 and x. (See Figure 1.7.)

Figure 1.7

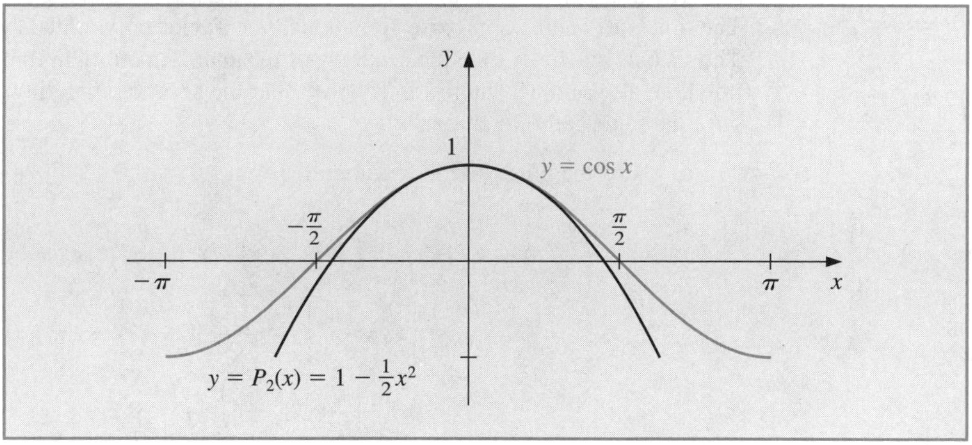

With $x = 0.01$, the Taylor polynomial and remainder term are

$$\begin{aligned}\cos 0.01 &= 1 - \frac{1}{2}(0.01)^2 + \frac{1}{6}(0.01)^3 \sin \xi(x) \\ &= 0.99995 + (0.1\overline{6}) \times 10^{-6} \sin \xi(x),\end{aligned}$$

where $0 < \xi(x) < 0.01$. (The bar over the 6 in $0.1\overline{6}$ is used to indicate that this digit repeats indefinitely.) Since $|\sin \xi(x)| < 1$ for all x, we have

$$|\cos 0.01 - 0.99995| \leq 0.1\overline{6} \times 10^{-6},$$

so the approximation 0.99995 matches at least the first five digits of $\cos 0.01$. Using standard tables we find that $\cos 0.01 = 0.99995000042$, so the approximation actually gives agreement through the first nine digits.

The error bound is much larger than the actual error. This is due in part to the poor bound we used for $|\sin \xi(x)|$. It can be shown that for all values of x, we have $|\sin x| \leq |x|$.

Since $0 < \xi < 0.01$, we could have used the fact that $|\sin \xi(x)| \leq 0.01$ in the error formula, producing the bound $0.1\overline{6} \times 10^{-8}$.

b. Since $f'''(0) = 0$, the third Taylor polynomial and remainder term about $x_0 = 0$ are

$$\cos x = 1 - \frac{1}{2}x^2 + \frac{1}{24}x^4 \cos \tilde{\xi}(x),$$

where $0 < \tilde{\xi}(x) < 0.01$. The approximating polynomial remains the same, and the approximation is still 0.99995, but we now have a much better accuracy assurance. Since $|\cos \tilde{\xi}(x)| \leq 1$ for all x, we have

$$\left| \frac{1}{24}x^4 \cos \tilde{\xi}(x) \right| \leq \frac{1}{24}(0.01)^4(1) \approx 4.2 \times 10^{-10}.$$

The first two parts of the example illustrate the two objectives of numerical analysis. The first is to find an approximation, which the Taylor polynomials in both parts provide. The second is to determine the accuracy of the approximation. In this case the third Taylor polynomial was much more informative than the second, even though both polynomials gave the same approximation.

c. Using the third Taylor polynomial gives

$$\begin{aligned} \int_0^{0.1} \cos x \, dx &= \int_0^{0.1} \left(1 - \frac{1}{2}x^2\right) dx + \frac{1}{24}\int_0^{0.1} x^4 \cos \tilde{\xi}(x) \, dx \\ &= \left[x - \frac{1}{6}x^3\right]_0^{0.1} + \frac{1}{24}\int_0^{0.1} x^4 \cos \tilde{\xi}(x) \, dx \\ &= 0.1 - \frac{1}{6}(0.1)^3 + \frac{1}{24}\int_0^{0.1} x^4 \cos \tilde{\xi}(x) \, dx. \end{aligned}$$

Therefore,

$$\int_0^{0.1} \cos x \, dx \approx 0.1 - \frac{1}{6}(0.1)^3 = 0.0998\overline{3}.$$

A bound for the error in this approximation is determined from the integral of the Taylor remainder term and the fact that $|\cos \tilde{\xi}(x)| \leq 1$ for all x:

$$\frac{1}{24}\left| \int_0^{0.1} x^4 \cos \tilde{\xi}(x) \, dx \right| \leq \frac{1}{24}\int_0^{0.1} x^4 |\cos \tilde{\xi}(x)| \, dx \leq \frac{1}{24}\int_0^{0.1} x^4 \, dx = 8.\overline{3} \times 10^{-8}.$$

Since the true value of this integral is

$$\int_0^{0.1} \cos x \, dx = \sin x \Big]_0^{0.1} = \sin 0.1 \approx 0.099833417,$$

the error for this approximation is within the bound. ◆ ◆ ◆

We could also use Maple in Example 3. Define f by

```
>f:=cos(x);
```

Maple allows us to place multiple statements on a line, and to use a colon to suppress Maple responses. For example; we obtain the third Taylor polynomial with

```
>s3:=taylor(f,x=0,4): p3:=convert(s3, polynom);
```

The first part computes the Taylor series with four terms (degree 3) and remainder expanded about $x_0 = 0$. The second part converts the series s3 to the polynomial p3 by dropping the remainder. To obtain 11 decimal digits of display, we enter

```
>Digits:=11;
```

and evaluate $f(0.01)$, $P_3(0.01)$, and $|f(0.01) - P_3(0.01)|$ with

```
>y1:=evalf(subs(x=0.01,f));
>y2:=evalf(subs(x=0.01,p3));
>error:=abs(y1-y2);
```

This produces $y_1 = f(0.01) = 0.99995000042$, $y_2 = P_3(0.01) = 0.99995000000$, and $|f(0.01) - P_3(0.01)| = .42 \times 10^{-9}$.

To obtain a graph similar to Figure 1.7, enter

```
>plot({f,p3},x=-Pi..Pi);
```

The commands for the integrals are

```
>q1:=int(f,x=0..0.1);
>q2:=int(p3,x=0..0.1);
>error:=abs(q1-q2);
```

which give the values

$$q_1 = \int_0^{0.1} f(x)\,dx = 0.099833416647 \quad \text{and} \quad q_2 = \int_0^{0.1} P_3(x)\,dx = 0.099833333333,$$

with error 0.83314×10^{-7}.

Parts (a) and (b) of the example show how two techniques can produce the same approximation but have differing accuracy assurances. Remember that determining approximations is only part of our objective. The equally important other part is to determine at least a bound for the accuracy of the approximation.

EXERCISE SET 1.2

1. Show that the following equations have at least one solution in the given intervals.
   a. $x\cos x - 2x^2 + 3x - 1 = 0$, $[0.2, 0.3]$ and $[1.2, 1.3]$
   b. $(x-2)^2 - \ln x = 0$, $[1, 2]$ and $[e, 4]$
   c. $2x\cos(2x) - (x-2)^2 = 0$, $[2, 3]$ and $[3, 4]$
   d. $x - (\ln x)^x = 0$, $[4, 5]$

2. Find intervals containing solutions to the following equations.
   a. $x - 3^{-x} = 0$
   b. $4x^2 - e^x = 0$
   c. $x^3 - 2x^2 - 4x + 3 = 0$
   d. $x^3 + 4.001x^2 + 4.002x + 1.101 = 0$

3. Show that the first derivatives of the following functions are zero at least once in the given intervals.
   a. $f(x) = 1 - e^x + (e-1)\sin((\pi/2)x)$, $[0, 1]$
   b. $f(x) = (x-1)\tan x + x\sin \pi x$, $[0, 1]$
   c. $f(x) = x\sin \pi x - (x-2)\ln x$, $[1, 2]$
   d. $f(x) = (x-2)\sin x \ln(x+2)$, $[-1, 3]$

4. Find $\max_{a\le x\le b} |f(x)|$ for the following functions and intervals.
   a. $f(x) = (2 - e^x + 2x)/3$, $[0, 1]$
   b. $f(x) = (4x-3)/(x^2-2x)$, $[0.5, 1]$
   c. $f(x) = 2x\cos(2x) - (x-2)^2$, $[2, 4]$
   d. $f(x) = 1 + e^{-\cos(x-1)}$, $[1, 2]$

5. Let $f(x) = x^3$.
   a. Find the second Taylor polynomial $P_2(x)$ about $x_0 = 0$.
   b. Find $R_2(0.5)$ and the actual error when using $P_2(0.5)$ to approximate $f(0.5)$.
   c. Repeat part (a) with $x_0 = 1$.
   d. Repeat part (b) for the polynomial found in part (c).

6. Let $f(x) = \sqrt{x+1}$.
   a. Find the third Taylor polynomial $P_3(x)$ about $x_0 = 0$.
   b. Use $P_3(x)$ to approximate $\sqrt{0.5}$, $\sqrt{0.75}$, $\sqrt{1.25}$, and $\sqrt{1.5}$.
   c. Determine the actual error of the approximations in part (b).

7. Find the second Taylor polynomial $P_2(x)$ for the function $f(x) = e^x\cos x$ about $x_0 = 0$.
   a. Use $P_2(0.5)$ to approximate $f(0.5)$. Find an upper bound for error $|f(0.5) - P_2(0.5)|$ using the error formula, and compare it to the actual error.
   b. Find a bound for the error $|f(x) - P_2(x)|$ in using $P_2(x)$ to approximate $f(x)$ on the interval $[0, 1]$.
   c. Approximate $\int_0^1 f(x)\,dx$ using $\int_0^1 P_2(x)\,dx$.
   d. Find an upper bound for the error in (c) using $\int_0^1 |R_2(x)\,dx|$, and compare the bound to the actual error.

8. Find the third Taylor polynomial $P_3(x)$ for the function $f(x) = (x-1)\ln x$ about $x_0 = 1$.

a. Use $P_3(0.5)$ to approximate $f(0.5)$. Find an upper bound for error $|f(0.5) - P_3(0.5)|$ using the error formula, and compare it to the actual error.

b. Find a bound for the error $|f(x) - P_3(x)|$ in using $P_3(x)$ to approximate $f(x)$ on the interval $[0.5, 1.5]$.

c. Approximate $\int_{0.5}^{1.5} f(x)\,dx$ using $\int_{0.5}^{1.5} P_3(x)\,dx$.

d. Find an upper bound for the error in (c) using $\int_{0.5}^{1.5} |R_3(x)\,dx|$, and compare the bound to the actual error.

9. Use the error term of a Taylor polynomial to estimate the error involved in using $\sin x \approx x$ to approximate $\sin 1^\circ$.

10. Use a Taylor polynomial about $\pi/4$ to approximate $\cos 42^\circ$ to an accuracy of 10^{-6}.

11. Let $f(x) = e^{x/2}\sin(x/3)$. Use Maple to determine the following.

a. The third Maclaurin polynomial $P_3(x)$.

b. $f^{(4)}(x)$ and a bound for the error $|f(x) - P_3(x)|$ on $[0, 1]$.

12. Let $f(x) = \ln(x^2 + 2)$. Use Maple to determine the following.

a. The Taylor polynomial $P_3(x)$ for f expanded about $x_0 = 1$.

b. The maximum error $|f(x) - P_3(x)|$ for $0 \le x \le 1$.

c. The Maclaurin polynomial $\tilde{P}_3(x)$ for f.

d. The maximum error $|f(x) - \tilde{P}_3(x)|$ for $0 \le x \le 1$.

e. Does $P_3(0)$ approximate $f(0)$ better than $\tilde{P}_3(1)$ approximates $f(1)$?

13. The polynomial $P_2(x) = 1 - \frac{1}{2}x^2$ is to be used to approximate $f(x) = \cos x$ in $[-\frac{1}{2}, \frac{1}{2}]$. Find a bound for the maximum error.

14. The nth Taylor polynomial for a function f at x_0 is sometimes referred to as the polynomial of degree at most n that "best" approximates f near x_0.

a. Explain why this description is accurate.

b. Find the quadratic polynomial that best approximates a function f near $x_0 = 1$ if the tangent line at $x_0 = 1$ has equation $y = 4x - 1$, and if $f''(1) = 6$.

15. The *error function* defined by

$$\operatorname{erf}(x) = \frac{2}{\sqrt{\pi}} \int_0^x e^{-t^2}\,dt$$

gives the probability that any one of a series of trials will lie within x units of the mean, assuming that the trials have a normal distribution with mean 0 and standard deviation $\sqrt{2}/2$. This integral cannot be evaluated in terms of elementary functions, so an approximating technique must be used.

a. Integrate the Maclaurin series for e^{-t^2} to show that

$$\operatorname{erf}(x) = \frac{2}{\sqrt{\pi}} \sum_{k=0}^{\infty} \frac{(-1)^k x^{2k+1}}{(2k+1)k!}.$$

b. The error function can also be expressed in the form

$$\text{erf}(x) = \frac{2}{\sqrt{\pi}} e^{-x^2} \sum_{k=0}^{\infty} \frac{2^k x^{2k+1}}{1 \cdot 3 \cdot 5 \cdots (2k+1)}.$$

Verify that the two series agree for $k = 1, 2, 3$, and 4. [*Hint:* Use the Maclaurin series for e^{-x^2}.]

c. Use the series in part (a) to approximate erf(1) to within 10^{-7}.

d. Use the same number of terms used in part (c) to approximate erf(1) with the series in part (b).

e. Explain why difficulties occur using the series in part (b) to approximate erf(x).

1.3 Round-off Error and Computer Arithmetic

The arithmetic performed by a calculator or computer is different from the arithmetic that we use in our algebra and calculus courses. From your past experience you might expect that we always have as true statements such things as $2+2 = 4, 4 \cdot 4 = 16$, and $(\sqrt{3})^2 = 3$. In standard computational arithmetic we will have the first two but not always the third. To understand why this is true we must explore the world of finite-digit arithmetic.

In our traditional mathematical world we permit numbers with an infinite number of digits. The arithmetic we use in this world defines $\sqrt{3}$ as that unique positive number that when multiplied by itself produces the integer 3. In the computational world, however, each representable number has only a fixed, finite number of digits. This means, for example, that only rational numbers—and not even all these—can be represented exactly. Since $\sqrt{3}$ is not rational, it is given an approximate representation within the machine, a representation whose square will not be precisely 3, although it will likely be sufficiently close to 3 to be acceptable in most situations. In most cases, then, this machine representation and arithmetic is satisfactory and passes without notice or concern, but at times problems arise because of this descrepancy.

Round-off error is produced when a calculator or computer is used to perform real-number calculations. It occurs because the arithmetic performed in a machine involves numbers with only a finite number of digits, with the result that calculations are performed with only approximate representations of the actual numbers. In a typical computer, only a relatively small subset of the real number system is used for the representation of all the real numbers. This subset contains only rational numbers, both positive and negative, and stores the fractional part, together with an exponential part.

In 1985, the IEEE (Institute for Electrical and Electronic Engineers) published a report called *Binary Floating Point Arithmetic Standard 754–1985*. Formats were specified for single, double, and extended precisions, and these standards are generally followed by all microcomputer manufacturers using floating-point hardware. For example, the numerical coprocessor for IBM-compatible microcomputers implements a 64-bit (binary digit) representation for a real number, called a *long real*. The first bit is a sign indicator, denoted s.

This is followed by an 11-bit exponent, c, called the **characteristic**, and a 52-bit binary fraction, f, called the **mantissa**. The base for the exponent is 2.

Since 52 binary digits correspond to between 16 and 17 decimal digits, we can assume that this number has at least 16 decimal digits of precision for the floating-point system. The exponent of 11 binary digits gives a range of 0 to 2047. However, using only positive integers for the exponent does not permit an adequate representation of numbers with small magnitude. To ensure that numbers with small magnitude are equally representable, 1023 is subtracted from the characteristic so that the range of the exponent is actually from -1023 to 1024. To save storage and provide a unique representation for each floating-point number, a normalization is imposed. Using this system gives a floating-point number of the form

$$(-1)^s * 2^{c-1023} * (1+f).$$

Consider for example, the machine number

0 10000000011 1011100100010000000000000000000000000000000000000000.

The leftmost bit is zero, which indicates that the number is positive. The next 11 bits, 10000000011, giving the characteristic, are equivalent to the decimal number

$$c = 1 \cdot 2^{10} + 0 \cdot 2^9 + \cdots + 0 \cdot 2^2 + 1 \cdot 2^1 + 1 \cdot 2^0 = 1024 + 2 + 1 = 1027.$$

The exponential part of the number is, therefore, $2^{1027-1023} = 2^4$. The final 52 bits specify that the mantissa is

$$f = 1 \cdot \left(\frac{1}{2}\right)^1 + 1 \cdot \left(\frac{1}{2}\right)^3 + 1 \cdot \left(\frac{1}{2}\right)^4 + 1 \cdot \left(\frac{1}{2}\right)^5 + 1 \cdot \left(\frac{1}{2}\right)^8 + 1 \cdot \left(\frac{1}{2}\right)^{12}.$$

As a consequence, this machine number precisely represents the decimal number

$$\begin{aligned}
&(-1)^s * 2^{c-1023} * (1+f) \\
&\quad = (-1)^0 \cdot 2^{1027-1023} \left(1 + \left(\frac{1}{2} + \frac{1}{8} + \frac{1}{16} + \frac{1}{32} + \frac{1}{256} + \frac{1}{4096}\right)\right) \\
&\quad = 27.56640625.
\end{aligned}$$

However, the next smallest machine number is

0 10000000011 1011100100001111111111111111111111111111111111111111

and the next largest machine number is

0 10000000011 1011100100010000000000000000000000000000000000000001.

This means that our original machine number represents not only 27.56640625, but also half of the real numbers that are between 27.56640625 and its two nearest machine-number neighbors. To be precise, it represents any real number in the interval

$$[27.566406249999999888977697537484345957636833190917968 75,$$
$$27.566406250000000111022302462515654042363166809082031 25).$$

The smallest normalized positive number that can be represented has all zeros except for the rightmost bit of 1 and is equivalent to

$$2^{-1023} \cdot (1 + 2^{-52}) \approx 10^{-308}$$

and the largest has a leading 0 followed by all 1s and is equivalent to

$$2^{1024} \cdot (2 - 2^{-52}) \approx 10^{308}.$$

Numbers occurring in calculations that have a magnitude less than $2^{-1023} \cdot (1 + 2^{-52})$ result in **underflow** and are generally set to zero. Numbers greater than $2^{1024} \cdot (2 - 2^{-52})$ result in **overflow** and typically cause the computations to halt.

The use of binary digits tends to complicate the computational problems that occur when a finite collection of machine numbers is used to represent all the real numbers. To examine these problems, we now assume, for simplicity, that machine numbers are represented in the normalized *decimal* form

$$\pm 0.d_1d_2 \ldots d_k \times 10^n, \quad 1 \le d_1 \le 9, \quad 0 \le d_i \le 9$$

for each $i = 2, \ldots, k$. Numbers of this form are called *k-digit decimal machine numbers*.

Any positive real number within numerical range of the machine can be normalized to achieve the form

$$y = 0.d_1d_2 \ldots d_kd_{k+1}d_{k+2} \ldots \times 10^n.$$

The **floating-point form** of y, denoted by $fl(y)$, is obtained by terminating the mantissa of y at k decimal digits. There are two ways of performing the termination. One method, called **chopping**, is to simply chop off the digits $d_{k+1}d_{k+2} \ldots$ to obtain

$$fl(y) = 0.d_1d_2 \ldots d_k \times 10^n.$$

The other method, called **rounding**, is to add $5 \times 10^{n-(k+1)}$ to y and then chop the result to obtain $fl(y)$. This means that when rounding we add one to d_k to obtain $fl(y)$ whenever $d_{k+1} \ge 5$, that is, we round up. When $d_{k+1} < 5$, we simply chop off all but the first k digits, so we round down.

EXAMPLE 1 The irrational number π has an infinite decimal expansion of the form $\pi = 3.14159265\ldots$. Written in normalized decimal form, we have

$$\pi = 0.314159265 \ldots \times 10^1.$$

The five-digit floating-point form of π using chopping is

$$fl(\pi) = 0.31415 \times 10^1 = 3.1415.$$

Since the sixth digit of the decimal expansion of π is a 9, the five-digit floating-point form of π using rounding is

$$fl(\pi) = (0.31415 + 0.00001) \times 10^1 = 0.31416 \times 10^1 = 3.1416. \qquad \blacklozenge \ \blacklozenge \ \blacklozenge$$

The error that results from replacing a number with its floating-point form is called **round-off error** (regardless of whether the rounding or chopping method is used). There are two common methods for measuring approximation errors.

The approximation p^* to p has **absolute error** $|p - p^*|$ and **relative error** $|p - p^*|/|p|$, provided that $p \neq 0$.

EXAMPLE 2

a. If $p = 0.3000 \times 10^1$ and $p^* = 0.3100 \times 10^1$, the absolute error is 0.1 and the relative error is $0.333\overline{3} \times 10^{-1}$.

b. If $p = 0.3000 \times 10^{-3}$ and $p^* = 0.3100 \times 10^{-3}$, the absolute error is 0.1×10^{-4}, but the relative error is again $0.333\overline{3} \times 10^{-1}$.

c. If $p = 0.3000 \times 10^4$ and $p^* = 0.3100 \times 10^4$, the absolute error is 0.1×10^3, but the relative error is still $0.333\overline{3} \times 10^{-1}$.

This example shows that the same relative error can occur for widely varying absolute errors. As a measure of accuracy, the absolute error can be misleading and the relative error is more meaningful, since the relative error takes into consideration the size of the true value. ◆ ◆ ◆

The arithmetic operations of addition, subtraction, multiplication, and division performed by a computer on floating-point numbers also introduce error. These arithmetic operations involve manipulating binary digits by various shifting and logical operations, but the actual mechanics of the arithmetic are not pertinent to our discussion. To illustrate the problems that can occur, we simulate this *finite-digit arithmetic* by first performing, at each stage in a calculation, the appropriate operation using exact arithmetic on the floating-point representations of the numbers. We then convert the result to decimal machine-number representation. The most common round-off error producing arithmetic operation involves the subtraction of nearly equal numbers.

EXAMPLE 3

Suppose we use four-digit decimal chopping arithmetic to simulate the problem of performing the computer operation $\pi - \frac{22}{7}$. The floating-point representations of these numbers are

$$fl(\pi) = 0.3141 \times 10^1 \qquad \text{and} \qquad fl\left(\frac{22}{7}\right) = 0.3142 \times 10^1.$$

Performing the exact arithmetic on the floating-point numbers gives

$$fl(\pi) - fl\left(\frac{22}{7}\right) = -0.0001 \times 10^1,$$

which converts to the floating-point approximation of this calculation:

$$p^* = fl\left(fl(\pi) - fl\left(\frac{22}{7}\right)\right) = -0.1000 \times 10^{-2}.$$

Although the relative errors using the floating-point representations for π and $\frac{22}{7}$ are small,

$$\left|\frac{\pi - fl(\pi)}{\pi}\right| \le 0.0002 \quad \text{and} \quad \left|\frac{\frac{22}{7} - fl\left(\frac{22}{7}\right)}{\frac{22}{7}}\right| \le 0.0003,$$

the relative error produced by subtracting the nearly equal numbers π and $\frac{22}{7}$ is about 700 times as large:

$$\left|\frac{\left(\pi - \frac{22}{7}\right) - p^*}{\left(\pi - \frac{22}{7}\right)}\right| \approx 0.2092.$$

◆ ◆ ◆

Rounding arithmetic is easily implemented in Maple. The command

```
>Digits:=t;
```

causes all arithmetic to be rounded to t digits. For example, $fl(fl(x) + fl(y))$ is performed using t-digit rounding arithmetic by

```
>evalf(evalf(x)+evalf(y));
```

Implementing t-digit chopping arithmetic in Maple is more difficult and requires a sequence of steps or a procedure. Exercise 12 explores this problem.

EXERCISE SET 1.3

1. Compute the absolute error and relative error in approximations of p by p^*.
   a. $p = \pi,\ p^* = \frac{22}{7}$
   b. $p = \pi,\ p^* = 3.1416$
   c. $p = e,\ p^* = 2.718$
   d. $p = \sqrt{2},\ p^* = 1.414$
   e. $p = e^{10},\ p^* = 22000$
   f. $p = 10^{\pi},\ p^* = 1400$
   g. $p = 8!,\ p^* = 39900$
   h. $p = 9!,\ p^* = \sqrt{18\pi}\,(9/e)^9$
2. Perform the following computations (i) exactly, (ii) using three-digit chopping arithmetic, and (iii) using three-digit rounding arithmetic. (iv) Compute the relative errors in parts (ii) and (iii).
   a. $\frac{4}{5} + \frac{1}{3}$
   b. $\frac{4}{5} \cdot \frac{1}{3}$
   c. $\left(\frac{1}{3} - \frac{3}{11}\right) + \frac{3}{20}$
   d. $\left(\frac{1}{3} + \frac{3}{11}\right) - \frac{3}{20}$

3. Use three-digit rounding arithmetic to perform the following calculations. Compute the absolute error and relative error with the exact value determined to at least five digits.

   a. $133 + 0.921$ b. $133 - 0.499$
   c. $(121 - 0.327) - 119$ d. $(121 - 119) - 0.327$
   e. $\dfrac{\frac{13}{14} - \frac{6}{7}}{2e - 5.4}$ f. $-10\pi + 6e - \dfrac{3}{62}$
   g. $\left(\dfrac{2}{9}\right) \cdot \left(\dfrac{9}{7}\right)$ h. $\dfrac{\pi - \frac{22}{7}}{\frac{1}{17}}$

4. Repeat Exercise 3 using three-digit chopping arithmetic.
5. Repeat Exercise 3 using four-digit rounding arithmetic.
6. Repeat Exercise 3 using four-digit chopping arithmetic.
7. The first three nonzero terms of the Maclaurin series for the arctan x are $x - \frac{1}{3}x^3 + \frac{1}{5}x^5$. Compute the absolute error and relative error in the following approximations of π using the polynomial in place of the arctan x:

   a. $4\left[\arctan\left(\dfrac{1}{2}\right) + \arctan\left(\dfrac{1}{3}\right)\right]$ b. $16 \arctan\left(\dfrac{1}{5}\right) - 4 \arctan\left(\dfrac{1}{239}\right)$

8. The two-by-two linear system

$$ax + by = e,$$
$$cx + dy = f,$$

where a, b, c, d, e, f are given, can be solved for x and y as follows:

$$\text{set } m = \frac{c}{a}, \quad \text{provided } a \neq 0;$$
$$d_1 = d - mb;$$
$$f_1 = f - me;$$
$$y = \frac{f_1}{d_1};$$
$$x = \frac{(e - by)}{a}.$$

Solve the following linear systems using four-digit rounding arithmetic.

   a. $1.130x - 6.990y = 14.20$
      $8.110x + 12.20y = -0.1370$

   b. $1.013x - 6.099y = 14.22$
      $-18.11x + 112.2y = -0.1376$

9. Suppose the points (x_0, y_0) and (x_1, y_1) are on a straight line with $y_1 \neq y_0$. Two formulas are available to find the x-intercept of the line:

$$x = \frac{x_0 y_1 - x_1 y_0}{y_1 - y_0} \quad \text{and} \quad x = x_0 - \frac{(x_1 - x_0)y_0}{y_1 - y_0}.$$

a. Show that both formulas are algebraically correct.

b. Use the data $(x_0, y_0) = (1.31, 3.24)$ and $(x_1, y_1) = (1.93, 4.76)$ and three-digit rounding arithmetic to compute the x-intercept both ways. Which method is better, and why?

10. The Taylor polynomial of degree n for $f(x) = e^x$ is $\sum_{i=0}^{n} x^i/i!$. Use the Taylor polynomial of degree nine and three-digit chopping arithmetic to find an approximation to e^{-5} by each of the following methods.

a. $e^{-5} \approx \sum_{i=0}^{9} \frac{(-5)^i}{i!} = \sum_{i=0}^{9} \frac{(-1)^i 5^i}{i!}$ **b.** $e^{-5} = \frac{1}{e^5} \approx \frac{1}{\sum_{i=0}^{9} 5^i/i!}$

An approximate value of e^{-5} correct to three digits is 6.74×10^{-3}. Which formula, (a) or (b), gives the most accuracy, and why?

11. A rectangular parallelepiped has sides 3 cm, 4 cm, and 5 cm, measured to the nearest centimeter.

a. What are the best upper and lower bounds for the volume of this parallelepiped?

b. What are the best upper and lower bounds for the surface area?

12. The following Maple procedure chops a floating-point number x to t digits.

```
chop:=proc(x,t);
 if x=0 then 0
 else
 e:=trunc(evalf(log10(abs(x))));
 if e>0 then e:=e+1 fi;
 x2:=evalf(trunc(x*10^(t-e))*10^(e-t));
 fi
end;
```

Verify that the procedure works for the following values.

a. $x = 124.031, t = 5$ **b.** $x = 124.036, t = 5$

c. $x = -0.00653, t = 2$ **d.** $x = -0.00656, t = 2$

1.4 Errors in Scientific Computation

In the previous section we saw how computational devices represent and manipulate numbers using finite-digit arithmetic. We now examine how the problems with this arithmetic can compound and look at ways to arrange arithmetic calculations to reduce this inaccuracy.

The loss of accuracy due to round-off error can often be avoided by a careful sequencing of operations or a reformulation of the problem. This is most easily described by considering a common computational problem.

EXAMPLE 1 The quadratic formula states that the roots of $ax^2 + bx + c = 0$, when $a \neq 0$, are

$$x_1 = \frac{-b + \sqrt{b^2 - 4ac}}{2a} \quad \text{and} \quad x_2 = \frac{-b - \sqrt{b^2 - 4ac}}{2a}.$$

Consider this formula applied, using four-digit rounding arithmetic, to the equation $x^2 + 62.10x + 1 = 0$, whose roots are approximately $x_1 = -0.01610723$ and $x_2 = -62.08390$. In this equation, b^2 is much larger than $4ac$, so the numerator in the calculation for x_1 involves the *subtraction* of nearly equal numbers. Since

$$\sqrt{b^2 - 4ac} = \sqrt{(62.10)^2 - (4.000)(1.000)(1.000)} = \sqrt{3856 - 4.000} = 62.06,$$

we have

$$fl(x_1) = \frac{-62.10 + 62.06}{2.000} = \frac{-0.04000}{2.000} = -0.02000,$$

a poor approximation to $x_1 = -0.01611$ with the large relative error

$$\frac{|-0.01611 + 0.02000|}{|-0.01611|} = 2.4 \times 10^{-1}.$$

On the other hand, the calculation for x_2 involves the *addition* of the nearly equal numbers $-b$ and $-\sqrt{b^2 - 4ac}$. This presents no problem since

$$fl(x_2) = \frac{-62.10 - 62.06}{2.000} = \frac{-124.2}{2.000} = -62.10$$

has the small relative error

$$\frac{|-62.08 + 62.10|}{|-62.08|} = 3.2 \times 10^{-4}.$$

To obtain a more accurate four-digit rounding approximation for x_1, we can change the form of the quadratic formula by *rationalizing the numerator*:

$$x_1 = \left(\frac{-b + \sqrt{b^2 - 4ac}}{2a}\right)\left(\frac{-b - \sqrt{b^2 - 4ac}}{-b - \sqrt{b^2 - 4ac}}\right) = \frac{b^2 - (b^2 - 4ac)}{2a(-b - \sqrt{b^2 - 4ac}\,)},$$

which simplifies to

$$x_1 = \frac{-2c}{b + \sqrt{b^2 - 4ac}}.$$

Using this form of the equation gives

$$fl(x_1) = \frac{-2.000}{62.10 + 62.06} = \frac{-2.000}{124.2} = -0.01610,$$

which has the small relative error 6.2×10^{-4}. ◆ ◆ ◆

The rationalization technique in Example 1 can also be applied to give an alternative formula for x_2:

$$x_2 = \frac{-2c}{b - \sqrt{b^2 - 4ac}}.$$

This is the form to use if b is negative. In Example 1, however, the use of this formula results in the subtraction of nearly equal numbers, which produces the result

$$fl(x_2) = \frac{-2.000}{62.10 - 62.06} = \frac{-2.000}{0.04000} = -50.00,$$

with the large relative error 1.9×10^{-1}.

EXAMPLE 2 Evaluate $f(x) = x^3 - 6.1x^2 + 3.2x + 1.5$ at $x = 4.71$ using three-digit arithmetic.

Table 1.1 gives the intermediate results in the calculations. Carefully verify these results to be sure that your notion of finite-digit arithmetic is correct. Note that the three-digit chopping values simply retain the leading three digits, with no rounding involved, and differ significantly from the three-digit rounding values.

$$\begin{aligned}
\text{Exact:}\quad f(4.71) &= 104.487111 - 135.32301 + 15.072 + 1.5 \\
&= -14.263899; \\
\text{Three-digit (chopping):}\quad f(4.71) &= ((104. - 134.) + 15.0) + 1.5 = -13.5; \\
\text{Three-digit (rounding):}\quad f(4.71) &= ((105. - 135.) + 15.1) + 1.5 = -13.4.
\end{aligned}$$

The relative errors for the three-digit methods are

$$\left| \frac{-14.263899 + 15.0}{-14.263899} \right| \approx 0.05 \quad \text{for chopping}$$

and

$$\left| \frac{-14.263899 + 15.1}{-14.263899} \right| \approx 0.06 \quad \text{for rounding.}$$

Table 1.1

	x	x^2	x^3	$6.1x^2$	$3.2x$
Exact	4.71	22.1841	104.487111	135.32301	15.072
Three-digit (chopping)	4.71	22.1	104.	134.	15.0
Three-digit (rounding)	4.71	22.2	105.	135.	15.1

As an alternative approach, $f(x)$ can be written in a **nested** manner as

$$f(x) = x^3 - 6.1x^2 + 3.2x + 1.5 = ((x - 6.1)x + 3.2)x + 1.5.$$

This gives

$$\text{Three-digit (chopping):} \quad f(4.71) = ((4.71 - 6.1)4.71 + 3.2)4.71 + 1.5 = -14.2$$

and a three-digit rounding answer of -14.3. The new relative errors are

$$\text{Three-digit (chopping):} \quad \left| \frac{-14.263899 + 14.2}{-14.263899} \right| \approx 0.0045;$$

$$\text{Three-digit (rounding):} \quad \left| \frac{-14.263899 + 14.3}{-14.263899} \right| \approx 0.0025.$$

Nesting has reduced the relative error for the chopping approximation to less than one-tenth that obtained initially. For the rounding approximation the improvement has been even more dramatic: the error has been reduced by more than 95%. Nested multiplication should be performed whenever a polynomial is evaluated since it minimizes the number of error producing computations. ◆ ◆ ◆

We are interested in choosing methods that will produce dependably accurate results. One criterion we will impose whenever possible is that small changes in the initial data produce correspondingly small changes in the final results. A method that satisfies this property is called **stable**; otherwise it is **unstable**. Some methods are stable only for certain choices of initial data. These methods are called *conditionally stable*. We attempt to characterize stability properties whenever possible.

To further consider the subject of round-off error growth and its connection to stability, suppose an error with magnitude $E_0 > 0$ is introduced at some stage in the calculations and that the magnitude of the error after n subsequent operations is E_n. There are two distinct cases that often arise in practice. If a constant C exists independent of n, with $E_n \approx CnE_0$, the growth of error is **linear.** If a constant $C > 1$ exists independent of n, with $E_n \approx C^n E_0$, the growth of error is **exponential**. (It would be unlikely to have $E_n \approx C^n E_0$, with $C < 1$, since this implies that the error tends to zero.)

Linear growth of error is usually unavoidable and, when C and E_0 are small, the results are generally acceptable. Methods having exponential growth of error should be avoided, since the term C^n becomes large for even relatively small values of n and E_0. As a consequence, a method that exhibits linear error growth is stable, while one exhibiting exponential error growth is unstable. (See Figure 1.8.)

Since iterative techniques involving sequences are often used, the section concludes with a brief discussion of some terminology used to describe the rate at which convergence occurs when employing a numerical technique. In general, we would like to choose techniques that converge as rapidly as possible. The following definition is used to compare the convergence rates of various methods.

Suppose $\{\beta_n\}_{n=1}^{\infty}$ is a sequence known to converge to zero. If $\{\alpha_n\}_{n=1}^{\infty}$ converges to a number α and K is a positive constant with

$$|\alpha_n - \alpha| \le K|\beta_n| \quad \text{for large } n,$$

then $\{\alpha_n\}$ **converges to** α **with rate of convergence** $O(\beta_n)$ (read "big oh of β_n"). This is indicated by writing $\alpha_n = \alpha + O(\beta_n)$ or $\alpha_n \to \alpha$ with rate of convergence $O(\beta_n)$.

Figure 1.8

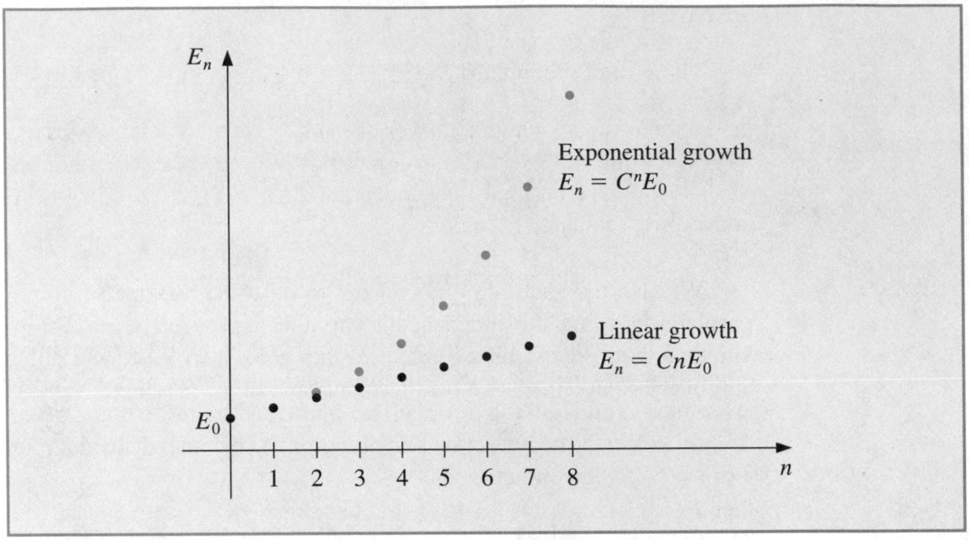

Although we can compare $\{\alpha_n\}$ with an arbitrary sequence $\{\beta_n\}$, in nearly every situation we have

$$\beta_n = \frac{1}{n^p}$$

for some number $p > 0$. We are generally interested in the largest value of p so that $\alpha_n = \alpha + O(1/n^p)$.

EXAMPLE 3 Suppose that the sequences $\{\alpha_n\}$ and $\{\hat{\alpha}_n\}$ are described by $\alpha_n = (n+1)/n^2$ and $\hat{\alpha}_n = (n+3)/n^3$. Although both $\lim_{n\to\infty} \alpha_n = 0$ and $\lim_{n\to\infty} \hat{\alpha}_n = 0$, the sequence $\{\hat{\alpha}_n\}$ converges to this limit much faster than does $\{\alpha_n\}$. This can be seen from the five-digit rounding entries for the sequences shown in Table 1.2.

Table 1.2

n	1	2	3	4	5	6	7
α_n	2.00000	0.75000	0.44444	0.31250	0.24000	0.19444	0.16327
$\hat{\alpha}_n$	4.00000	0.62500	0.22222	0.10938	0.064000	0.041667	0.029155

Since

$$|\alpha_n - 0| = \frac{n+1}{n^2} \le \frac{n+n}{n^2} = 2 \cdot \frac{1}{n}$$

and

$$|\hat{\alpha}_n - 0| = \frac{n+3}{n^3} \le \frac{n+3n}{n^3} = 4 \cdot \frac{1}{n^2},$$

we have

$$\alpha_n = 0 + O\left(\frac{1}{n}\right) \quad \text{and} \quad \hat{\alpha}_n = 0 + O\left(\frac{1}{n^2}\right).$$

This result implies that the convergence of the sequence $\{\alpha_n\}$ is similar to the convergence of $\{1/n\}$ to zero but that the sequence $\{\hat{\alpha}_n\}$ converges in a manner similar to the faster-converging sequence $\{1/n^2\}$. ◆ ◆ ◆

We also use the "big oh" concept to describe the rate of convergence of functions, particularly when the independent variable approaches zero. Suppose G is a function known to converge to zero as h approaches zero. If F is a function that converges to L as h approaches zero and K is a positive constant with

$$|F(h) - L| \le K\,G(h), \qquad \text{as } h \to 0,$$

then $F(h)$ **converges to L with rate of convergence** $O(G(h))$, written $F(h) = L + O(G(h))$. The functions we use for comparison generally have the form $G(h) = h^p$, where $p > 0$. We are generally interested in the largest value of p for which $F(h) = L + O(G(h))$.

EXERCISE SET 1.4

1. (i) Use four-digit rounding arithmetic and the formulas of Example 1 to find the most accurate approximations to the roots of the following quadratic equations. (ii) Compute the absolute errors and relative errors for these approximations.

a. $\frac{1}{3}x^2 - \frac{123}{4}x + \frac{1}{6} = 0$ **b.** $\frac{1}{3}x^2 + \frac{123}{4}x - \frac{1}{6} = 0$

c. $1.002x^2 - 11.01x + 0.01265 = 0$ **d.** $1.002x^2 + 11.01x + 0.01265 = 0$

2. Repeat Exercise 1 using four-digit chopping arithmetic.

3. Let $f(x) = 1.013x^5 - 5.262x^3 - 0.01732x^2 + 0.8389x - 1.912$.

a. Evaluate $f(2.279)$ by first calculating $(2.279)^2$, $(2.279)^3$, $(2.279)^4$, and $(2.279)^5$ using four-digit rounding arithmetic.

b. Evaluate $f(2.279)$ using the formula

$$f(x) = (((1.013x^2 - 5.262)x - 0.01732)x + 0.8389)x - 1.912$$

and four-digit rounding arithmetic.

c. Compute the absolute and relative errors in parts (a) and (b).

4. Repeat Exercise 3 using four-digit chopping arithmetic.

5. The fifth Maclaurin polynomials for e^{2x} and e^{-2x} are

$$P_5(x) = \left(\left(\left(\left(\frac{4}{15}x + \frac{2}{3}\right)x + \frac{4}{3}\right)x + 2\right)x + 2\right)x + 1$$

and

$$\hat{P}_5(x) = \left(\left(\left(\left(-\frac{4}{15}x + \frac{2}{3}\right)x - \frac{4}{3}\right)x + 2\right)x - 2\right)x + 1$$

a. Approximate $e^{-0.98}$ using $\hat{P}_5(0.49)$ and four-digit rounding arithmetic.
b. Compute the absolute and relative error for the approximation in part (a).
c. Approximate $e^{-0.98}$ using $1/P_5(0.49)$ and four-digit rounding arithmetic.
d. Compute the absolute and relative errors for the approximation in part (c).

6. **a.** Show that the polynomial nesting technique described in Example 2 can also be applied to the evaluation of

$$f(x) = 1.01e^{4x} - 4.62e^{3x} - 3.11e^{2x} + 12.2e^{x} - 1.99.$$

b. Use three-digit rounding arithmetic, the assumption that $e^{1.53} = 4.62$, and the fact that $e^{n(1.53)} = (e^{1.53})^n$ to evaluate $f(1.53)$ as given in part (a).
c. Redo the calculation in part (b) by first nesting the calculations.
d. Compare the approximations in parts (b) and (c) to the true three-digit result $f(1.53) = -7.61$.

7. Use three-digit chopping arithmetic to compute the sum $\sum_{i=1}^{10} 1/i^2$ first by $\frac{1}{1} + \frac{1}{4} + \cdots + \frac{1}{100}$ and then by $\frac{1}{100} + \frac{1}{81} + \cdots + \frac{1}{1}$. Which method is more accurate, and why?

8. The Maclaurin series for the arctangent function converges for $-1 < x \le 1$ and is given by

$$\arctan x = \lim_{n\to\infty} P_n(x) = \lim_{n\to\infty} \sum_{i=1}^{n} (-1)^{i+1} \frac{x^{2i-1}}{(2i-1)}.$$

a. Use the fact that $\tan \pi/4 = 1$ to determine the number of terms of the series that need to be summed to ensure that $|4P_n(1) - \pi| < 10^{-3}$.
b. The C programming language requires the value of π to be within 10^{-10}. How many terms of the series would we need to sum to obtain this degree of accuracy?

9. The number e is defined by $e = \sum_{n=0}^{\infty} 1/n!$, where $n! = n(n-1)\cdots 2\cdot 1$, for $n \neq 0$ and $0! = 1$. (i) Use four-digit chopping arithmetic to compute the following approximations to e. (ii) Compute absolute and relative errors for these approximations.

a. $\displaystyle\sum_{n=0}^{5} \frac{1}{n!}$ **b.** $\displaystyle\sum_{j=0}^{5} \frac{1}{(5-j)!}$

c. $\displaystyle\sum_{n=0}^{10} \frac{1}{n!}$ **d.** $\displaystyle\sum_{j=0}^{10} \frac{1}{(10-j)!}$

10. Find the rates of convergence of the following sequences as $n \to \infty$.

a. $\displaystyle\lim_{n\to\infty} \sin\left(\frac{1}{n}\right) = 0$ **b.** $\displaystyle\lim_{n\to\infty} \sin\left(\frac{1}{n^2}\right) = 0$

c. $\displaystyle\lim_{n\to\infty} \left(\sin\left(\frac{1}{n}\right)\right)^2 = 0$ **d.** $\displaystyle\lim_{n\to\infty} [\ln(n+1) - \ln(n)] = 0$

11. Find the rates of convergence of the following functions as $h \to 0$.

a. $\lim_{h\to 0} \frac{\sin h - h\cos h}{h} = 0$ **b.** $\lim_{h\to 0} \frac{1-e^h}{h} = -1$

c. $\lim_{h\to 0} \frac{\sin h}{h} = 1$ **d.** $\lim_{h\to 0} \frac{1-\cos h}{h} = 0$

12. **a.** How many multiplications and additions are required to determine a sum of the form

$$\sum_{i=1}^{n}\sum_{j=1}^{i} a_i b_j?$$

b. Modify the sum in part (a) to an equivalent form that reduces the number of computations.

13. The sequence $\{F_n\}$ described by $F_0 = 1, F_1 = 1$, and $F_{n+2} = F_n + F_{n+1}$, if $n \geq 0$, is called a *Fibonacci sequence*. Its terms occur naturally in many botanical species, particularly those with petals or scales arranged in the form of a logarithmic spiral. Consider the sequence $\{x_n\}$, where $x_n = F_{n+1}/F_n$. Assuming that $\lim_{n\to\infty} x_n = x$ exists, show that x is the *golden ratio* $(1+\sqrt{5})/2$.

14. The Fibonacci sequence also satisfies the equation

$$F_n \equiv \tilde{F}_n = \frac{1}{\sqrt{5}}\left[\left(\frac{1+\sqrt{5}}{2}\right)^n - \left(\frac{1-\sqrt{5}}{2}\right)^n\right].$$

a. Write a Maple procedure to calculate F_{100}.

b. Use Maple with the default value of `Digits` followed by `evalf` to calculate $\tilde{F}_{100}$.

c. Why is the result from part (a) more accurate than the result from part (b)?

d. Why is the result from part (b) obtained more rapidly than the result from part (a)?

e. What results when you use the command `simplify` instead of `evalf` to compute $\tilde{F}_{100}$?

15. The harmonic series $1 + \frac{1}{2} + \frac{1}{3} + \frac{1}{4} + \cdots$ diverges, but the sequence $\gamma_n = 1 + \frac{1}{2} + \cdots + \frac{1}{n} - \ln n$ converges, since $\{\gamma_n\}$ is a bounded, nonincreasing sequence. The limit $\gamma \approx 0.5772156649\ldots$ of the sequence $\{\gamma_n\}$ is called *Euler's constant*.

a. Use the default value of `Digits` in Maple to determine the value of n for γ_n to be within 10^{-2} of γ.

b. Use the default value of `Digits` in Maple to determine the value of n for γ_n to be within 10^{-3} of γ.

c. What happens if you use the default value of `Digits` in Maple to determine the value of n for γ_n to be within 10^{-4} of γ?

1.5 Computer Software

Computer software for approximating the numerical solutions to problems is available in many forms. On the disk accompanying this book, we provide programs written in the programming languages C, FORTRAN, and Pascal and worksheets for the computer algebra systems Maple, *Mathematica*, and MATLAB that can be used to solve the problems given in the examples and exercises. These programs will give satisfactory results for most problems that you will likely need to solve, but they are what we call *special-purpose* programs. We use this term to distinguish these programs from those available in the standard mathematical subroutine libraries. The programs in these packages are called *general-purpose* programs.

The programs in general-purpose software packages differ in their intent from the programs provided with this book. General-purpose software packages consider ways to reduce errors due to machine rounding, underflow, and overflow. They also describe the range of input that will lead to results of a certain specified accuracy. Since these are machine-dependent characteristics, general-purpose software packages use parameters that describe the floating-point characteristics of the machine involved in the computations.

There are many forms of general-purpose numerical software available commercially and in the public domain. Most of the early software was written for mainframe computers. A good reference for this software is *Sources and Development of Mathematical Software*, edited by Crowell [Cr]. Now that the desktop computer has become sufficiently powerful, much of the standard numerical software is available for personal computers and workstations. Most of this numerical software is written in FORTRAN, although some packages are written in C and Pascal.

The professional software packages are highly efficient, accurate, and reliable; they are thoroughly tested, and documentation is readily available. Although the packages are portable, it is a good idea to investigate the machine dependence and read the documentation thoroughly. The programs test for almost all special contingencies that might result in error and failures. At the end of each chapter we discuss some of the appropriate general-purpose packages.

Commercially available packages also represent the state of the art in numerical methods. They are often based on the public-domain packages but include methods in libraries for almost every type of problem.

IMSL (International Mathematical Software Library) consists of the FORTRAN 77 libraries MATH, STAT, and SFUN for numerical mathematics, statistics, and special functions, respectively. These libraries contain more than 700 subroutines that solve most common numerical analysis problems, and are available from IMSL, 2500 Park West Tower One, 2500 City West Boulevard, Houston, TX 77042-3020. The packages are delivered in compiled form with extensive documentation, and there is an example program for each routine, as well as background reference information. IMSL contains methods for linear systems, eigensystem analysis, interpolation and approximation, integration and differentiation, differential equations, transforms, nonlinear equations, optimization, and basic matrix/vector operations.

The Numerical Algorithms Group (NAG) offers more than 1100 subroutines in a FORTRAN 77 library for more than 90 different computers. Subsets of their library are available for IBM personal computers (the PC50 Library consists of 50 of the most frequently used routines) and workstations (the Workstation Library contains 172 routines). The NAG C Library offers many of the same routines as the FORTRAN Library. The NAG user's manual includes instructions and examples, along with sample output for each of the routines. The NAG library contains routines to perform most standard numerical analysis tasks in a manner similar to those in the IMSL. It also includes some statistical routines and a set of graphic routines. The library is available from Numerical Algorithms Group, Inc., 1400 Opus Place, Suite 200, Downers Grove, IL 60515–5702.

The IMSL and NAG packages are designed for the mathematician, scientist, or engineer who wishes to call high-quality FORTRAN subroutines from within a program. The documentation available with the commercial packages illustrates the typical driver program required to use the library routines.

MATLAB is a matrix laboratory that uses subroutines to perform matrix operations, such as finding the eigenvalues of a matrix entered from the command line or from an external file via function calls. It also solves problems associated with nonlinear systems, numerical integration, cubic splines, curve fitting, optimization, ordinary differential equations, and incorporates graphical tools. This is a powerful self-contained system that is especially useful for instruction in an applied linear algebra course. MATLAB is available from The MathWorks Inc., Cochituate Place, 24 Prime Park Way, Natick, MA 01760.

CHAPTER

2 Solutions of Equations of One Variable

2.1 Introduction

In this chapter we consider one of the most basic problems of numerical approximation, the root-finding problem. This process involves finding a **root**, or solution, of an equation of the form $f(x) = 0$. A root of this equation is also called a **zero** of the function f. This is one of the oldest known approximation problems, yet research continues in this area at the present time.

The problem of finding an approximation to the root of an equation can be traced at least as far back as 1700 B.C. A cuneiform table in the Yale Babylonian Collection dating from that period gives a sexigesimal (base-60) number equivalent to 1.414222 as an approximation to $\sqrt{2}$, a result that is accurate to within 10^{-5}. This approximation can be found by applying a technique given in Section 2.4.

2.2 The Bisection Method

The first and most elementary technique we consider is the **Bisection**, or *Binary-Search*, method. The Bisection method is used to determine, to any specified accuracy that your computer will permit, a solution to $f(x) = 0$ on an interval $[a, b]$, provided that f is continuous on the interval and that $f(a)$ and $f(b)$ are of opposite sign. Although the method will work for the case when more than one root is contained in the interval $[a, b]$, we assume for simplicity of our discussion that the root in this interval is unique.

To begin the Bisection method, set $a_1 = a$ and $b_1 = b$, as shown in Figure 2.1, and let p_1 be the midpoint of the interval $[a, b]$:

$$p_1 = a_1 + \frac{b_1 - a_1}{2}.$$

If $f(p_1) = 0$, then the root p is given by $p = p_1$; if $f(p_1) \neq 0$, then $f(p_1)$ has the same sign as either $f(a_1)$ or $f(b_1)$.

Figure 2.1

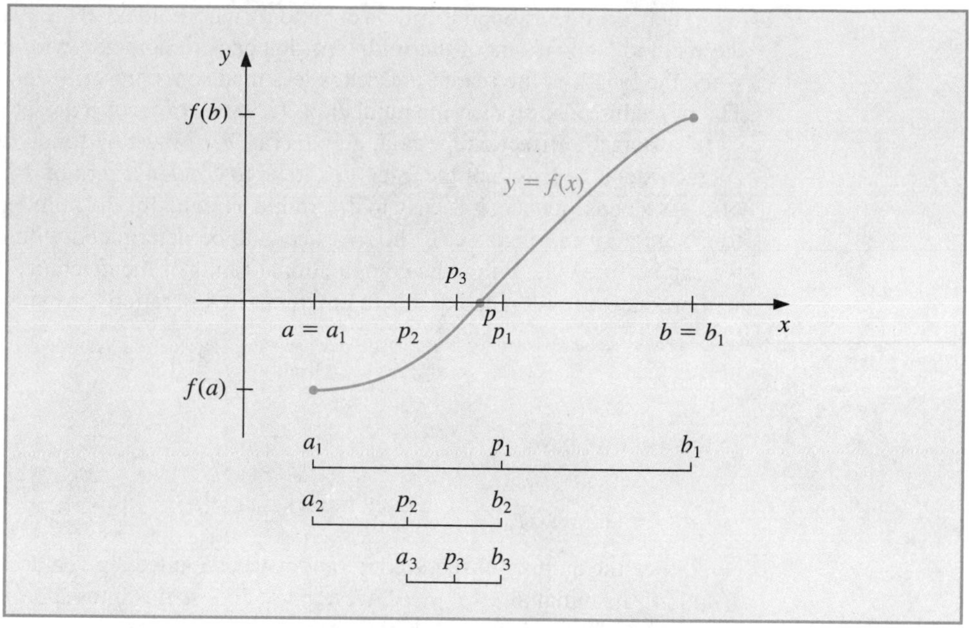

If $f(p_1)$ and $f(a_1)$ have the same sign, then p is in the interval (p_1, b_1), and we set

$$a_2 = p_1 \quad \text{and} \quad b_2 = b_1.$$

If, on the other hand, $f(p_1)$ and $f(a_1)$ have opposite signs, then p is in the interval (a_1, p_1), and we set

$$a_2 = a_1 \quad \text{and} \quad b_2 = p_1.$$

We reapply the process to the interval $[a_2, b_2]$, and continue forming $[a_3, b_3]$, $[a_4, b_4], \ldots$. Each new interval will contain p and have length one half of the length of the preceding interval.

Bisection Method

An interval $[a_{i+1}, b_{i+1}]$ containing an approximation to a root of $f(x) = 0$ is constructed from an interval $[a_i, b_i]$ containing the root by first letting

$$p_i = a_i + \frac{b_i - a_i}{2}.$$

Then set

$$a_{i+1} = a_i \quad \text{and} \quad b_{i+1} = p_i \quad \text{if} \quad f(a_i)f(p_i) < 0,$$

and

$$a_{i+1} = p_i \quad \text{and} \quad b_{i+1} = b_i \quad \text{otherwise.}$$

There are three stopping criteria commonly incorporated in the Bisection method. First, the method stops if one of the midpoints happens to coincide with the root. It also stops when the length of the search interval is less than some prescribed tolerance we call *TOL*. The procedure also stops if the number of iterations exceeds a preset bound N_0.

To start the Bisection method, an interval $[a, b]$ must be found with $f(a) \cdot f(b) < 0$. At each step, the length of the interval known to contain a zero of f is reduced by a factor of 2. As a consequence, it is easy to determine a bound for the number of iterations needed to ensure a given tolerance. If the root needs to be determined within the tolerance *TOL*, we can be sure that, within the computational limits of the machine, the Bisection method will produce a result meeting this requirement in n iterations, provided that

$$n > \log_2 \left(\frac{b-a}{TOL} \right).$$

In this case we have

$$2^n > \frac{b-a}{TOL}, \quad \text{which implies that} \quad |p_n - p| \le \frac{b-a}{2^n} < TOL.$$

Since the number of required iterations to guarantee a given accuracy depends on the length of the initial interval $[a, b]$, we want to choose this interval as small as possible. For example, if $f(x) = 2x^3 - x^2 + x - 1$, we have both

$$f(-4) \cdot f(4) < 0 \quad \text{and} \quad f(0) \cdot f(1) < 0,$$

so the Bisection method could be used on either $[-4, 4]$ or $[0, 1]$. Starting the Bisection method on $[0, 1]$ instead of $[-4, 4]$ reduces by 3 the number of iterations required to achieve a specified accuracy.

EXAMPLE 1 The equation $f(x) = x^3 + 4x^2 - 10 = 0$ has a root in $[1, 2]$ since $f(1) = -5$ and $f(2) = 14$. It is easily seen from a sketch of the graph of f in Figure 2.2 that there is only one root in $[1, 2]$.

Figure 2.2

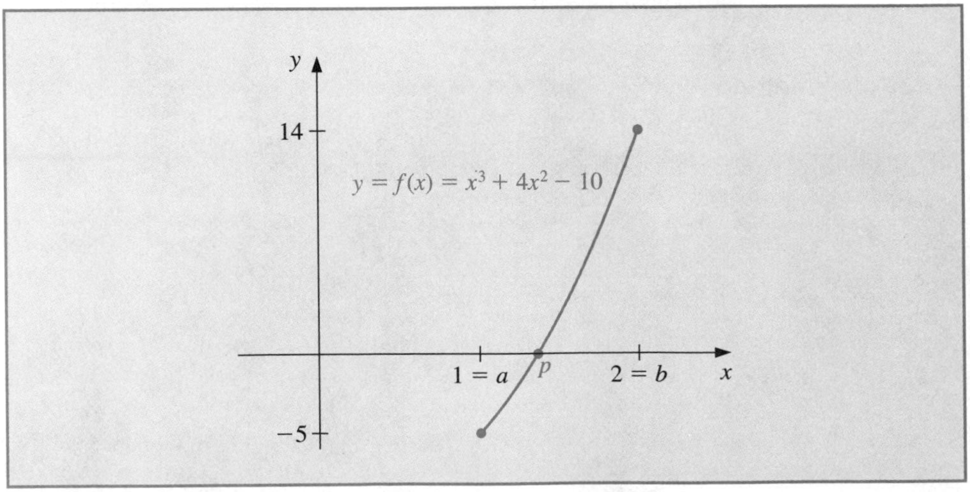

To use Maple to approximate the root, we define the function f by the command

```
>f:=x->x^3+4*x^2-10;
```

The values of a_1 and b_1 are given by

```
>a1:=1; b1:=2;
```

We next compute $f(a_1) = -5$ and $f(b_1) = 14$ by

```
>fa1:=f(a1); fb1:=f(b1);
```

and the midpoint $p_1 = 1.5$ and $f(p_1) = 2.375$ by

```
>p1:=a1+0.5*(b1-a1);
>pf1:=f(p1);
```

Since $f(a_1)$ and $f(p_1)$ have opposite signs, we reject b_1 and let $a_2 = a_1$ and $b_2 = p_1$. This process is continued to find p_2, p_3, and so on.

The program BISECT21, provided with the inputs $a = 1$, $b = 2$, $TOL = 0.0005$, and $N_0 = 20$, gives the values in Table 2.1. The actual root p, to 10 decimal places, is $p = 1.3652300134$, and $|p - p_{11}| < 0.0005$. Since the expected number of iterations is $\log_2((2-1)/0.0005) \approx 10.96$, the bound N_0 was certainly sufficient. ◆ ◆ ◆

Table 2.1

n	a_n	b_n	p_n	$f(p_n)$
1	1.0000000000	2.0000000000	1.5000000000	2.3750000000
2	1.0000000000	1.5000000000	1.2500000000	−1.7968750000
3	1.2500000000	1.5000000000	1.3750000000	0.1621093750
4	1.2500000000	1.3750000000	1.3125000000	−0.8483886719
5	1.3125000000	1.3750000000	1.3437500000	−0.3509826660
6	1.3437500000	1.3750000000	1.3593750000	−0.0964088440
7	1.3593750000	1.3750000000	1.3671875000	0.0323557854
8	1.3593750000	1.3671875000	1.3632812500	−0.0321499705
9	1.3632812500	1.3671875000	1.3652343750	0.0000720248
10	1.3632812500	1.3652343750	1.3642578125	−0.0160466908
11	1.3642578125	1.3652343750	1.3647460938	−0.0079892628

The Bisection method, though conceptually clear, has serious drawbacks. It is slow to converge relative to the other techniques we will discuss, and a good intermediate approximation may be inadvertently discarded. This happened, for example, with p_9 in Example 1. However, the method has the important property that it always converges to a solution and it is easy to determine a bound for the number of iterations needed to ensure a given accuracy. For these reasons, the Bisection method is frequently used as a dependable starting procedure for the more efficient methods presented later in this chapter.

The bound for the number of iterations for the Bisection method assumes that the calculations are performed using infinite-digit arithmetic. When implementing the method on a computer, consideration must be given to the effects of round-off error. For example, the computation of the midpoint of the interval $[a_n, b_n]$ should be found from the equation

$$p_n = a_n + \frac{b_n - a_n}{2}$$

instead of from the algebraically equivalent equation

$$p_n = \frac{a_n + b_n}{2}.$$

The first equation adds a small correction, $(b_n - a_n)/2$, to the known value a_n. When $b_n - a_n$ is near the maximum precision of the machine, this correction might be in error, but the error would not significantly affect the computed value of p_n. However, when $b_n - a_n$ is near the maximum precision of the machine, it is possible for p_n, as defined in the second equation, to return a midpoint that is not even in the interval $[a_n, b_n]$.

A number of tests can be used to see if a root has been found. We would normally use a test of the form

$$|f(p_n)| < \varepsilon,$$

where $\varepsilon > 0$ would be a small number related in some way to the tolerance. However, it is also possible for the value $f(p_n)$ to be small when p_n is not near the root p.

As a final remark, to determine which subinterval of $[a_n, b_n]$ contains a root of f, it is better to make use of **signum** function, which is defined as

$$\operatorname{sgn}(x) = \begin{cases} -1, & \text{if } x < 0, \\ 0, & \text{if } x = 0, \\ 1, & \text{if } x > 0. \end{cases}$$

The test

$$\operatorname{sgn}(f(a_n))\operatorname{sgn}(f(p_n)) < 0 \qquad \text{instead of} \qquad f(a_n)f(p_n) < 0$$

gives the same result but avoids the possibility of overflow or underflow in the multiplication of $f(a_n)$ and $f(p_n)$.

EXERCISE SET 2.2

1. Use the Bisection method to find p_3 for $f(x) = \sqrt{x} - \cos x$ on $[0, 1]$.
2. Let $f(x) = 3(x + 1)(x - \frac{1}{2})(x - 1)$. Use the Bisection method on the following intervals to find p_3 .
   a. $[-2, 1.5]$ **b.** $[-1.25, 2.5]$

3. Use the Bisection method to find solutions accurate to within 10^{-2} for $x^3 - 7x^2 + 14x - 6 = 0$ on each interval.
   a. $[0, 1]$ **b.** $[1, 3.2]$ **c.** $[3.2, 4]$
4. Use the Bisection method to find solutions accurate to within 10^{-2} for $x^4 - 2x^3 - 4x^2 + 4x + 4 = 0$ on each interval.
   a. $[-2, -1]$ **b.** $[0, 2]$
   c. $[2, 3]$ **d.** $[-1, 0]$
5. **a.** Sketch the graphs of $y = x$ and $y = 2\sin x$.
   b. Use the Bisection method to find an approximation to within 10^{-2} to the first positive value of x with $x = 2\sin x$.
6. **a.** Sketch the graphs of $y = x$ and $y = \tan x$.
   b. Use the Bisection method to find an approximation to within 10^{-2} to the first positive value of x with $x = \tan x$.
7. Let $f(x) = (x+2)(x+1)x(x-1)^3(x-2)$. To which zero of f does the Bisection method converge for the following intervals?
   a. $[-3, 2.5]$ **b.** $[-2.5, 3]$
   c. $[-1.75, 1.5]$ **d.** $[-1.5, 1.75]$
8. Let $f(x) = (x+2)(x+1)^2x(x-1)^3(x-2)$. To which zero of f does the Bisection method converge for the following intervals?
   a. $[-1.5, 2.5]$ **b.** $[-0.5, 2.4]$
   c. $[-0.5, 3]$ **d.** $[-3, -0.5]$
9. Use the Bisection method to find an approximation to $\sqrt{3}$ correct to within 10^{-4}. [*Hint:* Consider $f(x) = x^2 - 3$.]
10. Use the Bisection method to find an approximation to $\sqrt[3]{25}$ correct to within 10^{-4}.
11. Find a bound for the number of Bisection method iterations needed to achieve an approximation with accuracy 10^{-3} to the solution of $x^3 + x - 4 = 0$ lying in the interval $[1, 4]$. Find an approximation to the root with this degree of accuracy.
12. Find a bound for the number of Bisection method iterations needed to achieve an approximation with accuracy 10^{-4} to the solution of $x^3 - x - 1 = 0$ lying in the interval $[1, 2]$. Find an approximation to the root with this degree of accuracy.
13. The function defined by $f(x) = \sin \pi x$ has zeros at every integer. Determine an interval $[a, b]$ with $a < 0$ and $b > 2$ for which the Bisection method converges to each value.
    a. 0 **b.** 2 **c.** 1

2.3 The Secant Method

Although the Bisection method always converges, the speed of convergence is usually too slow for general use. Figure 2.3 gives a graphical interpretation of the Bisection method that can be used to discover how improvements on this technique can be derived.

It shows the graph of a continuous function that is negative at a_1 and positive at b_1. The first approximation p_1 to the root p is found by drawing the line joining the points $(a_1, \text{sgn}(f(a_1))) = (a_1, -1)$ and $(b_1, \text{sgn}(f(b_1))) = (b_1, 1)$ and letting p_1 be the point where this line intersects the x-axis. In essence, the line joining $(a_1, -1)$ and $(b_1, 1)$ has been used to approximate the graph of f on the interval $[a_1, b_1]$. Successive approximations apply this same process on subintervals of $[a_1, b_1]$, $[a_2, b_2]$, and so on. Notice that the Bisection method uses no information about the function f except the fact that $f(x)$ is positive and negative at certain values of x.

Figure 2.3

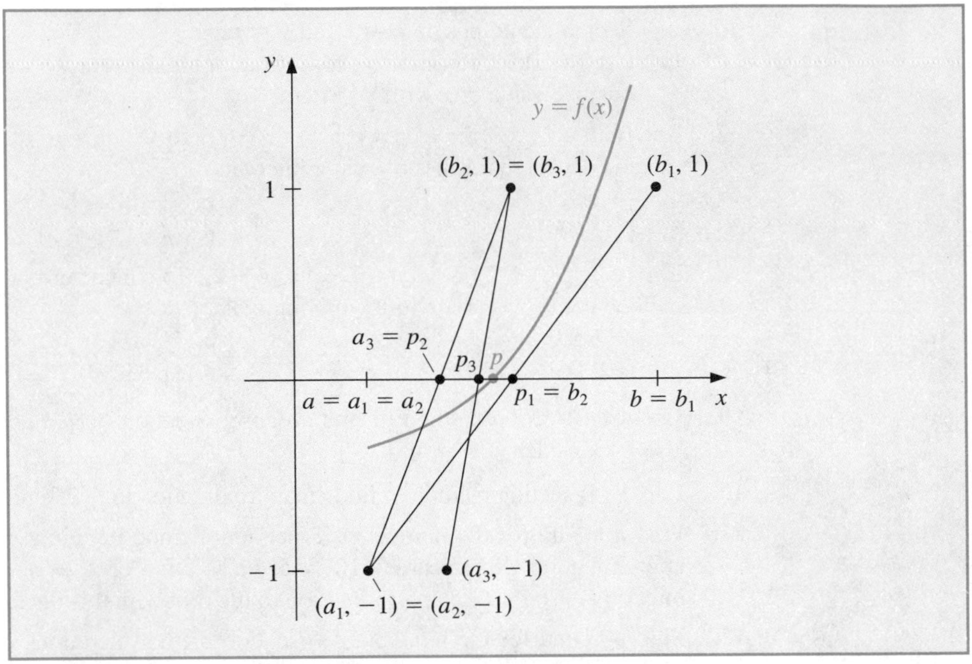

Suppose that in the initial step we know that $|f(a_1)| < |f(b_1)|$. Then we would expect the root p to be closer to a_1 than to b_1. Alternatively, if $|f(b_1)| < |f(a_1)|$, p is likely to be closer to b_1 than to a_1. Instead of choosing the intersection of the line through $(a_1, \text{sgn}(f(a_1))) = (a_1, -1)$ and $(b_1, \text{sgn}(f(b_1))) = (b_1, 1)$ as the approximation to the root p, the *Secant method* chooses the x-intercept of the secant line to the curve, the line through $(a_1, f(a_1))$ and $(b_1, f(b_1))$. This places the approximation closer to the endpoint of the interval for which f has smaller absolute value, as shown in Figure 2.4.

The sequence of approximations generated by the Secant method is started by setting $p_0 = a$ and $p_1 = b$. The equation of the secant line through $(p_0, f(p_0))$ and $(p_1, f(p_1))$ is

$$y = f(p_1) + \frac{f(p_1) - f(p_0)}{p_1 - p_0}(x - p_1).$$

Figure 2.4

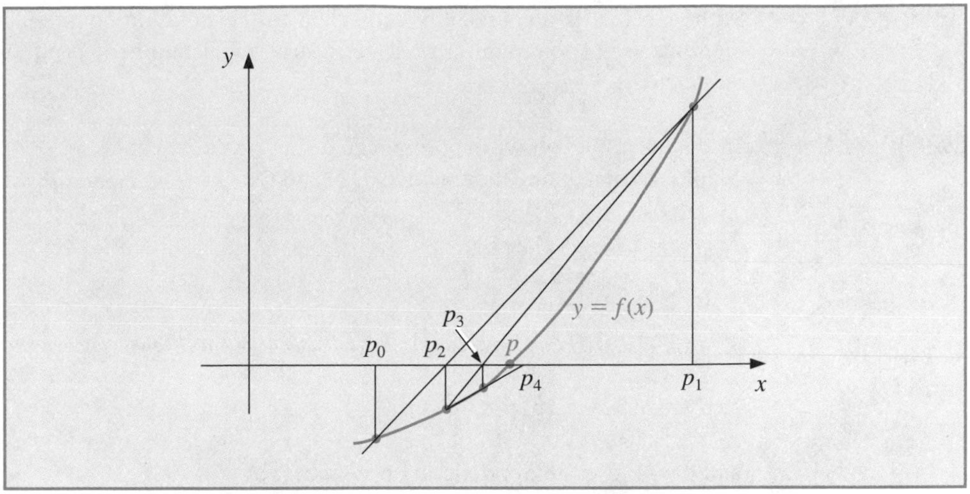

The x-intercept $(p_2, 0)$ of this line satisfies

$$0 = f(p_1) + \frac{f(p_1) - f(p_0)}{p_1 - p_0}(p_2 - p_1)$$

and solving for p_2 gives

$$p_2 = p_1 - \frac{f(p_1)(p_1 - p_0)}{f(p_1) - f(p_0)}.$$

Secant Method

The approximation p_{n+1}, for $n > 1$, to a root of $f(x) = 0$ is computed from the approximations p_n and p_{n-1} using the equation

$$p_{n+1} = p_n - \frac{f(p_n)(p_n - p_{n-1})}{f(p_n) - f(p_{n-1})}.$$

The Secant method does not have the root-bracketing property of the Bisection method. As a consequence, the method does not always converge, but when it does converge, it generally does so much faster than the Bisection method.

We use two stopping conditions in the Secant method. First, we assume that p_i is sufficiently accurate when $|p_i - p_{i-1}|$ is within a given tolerance. Also, a safeguard exit based upon a maximum number of iterations is given in case the method fails to converge as expected.

The iteration equation should not be algebraically simplified to

$$p_{i+1} = p_i - \frac{f(p_i)(p_i - p_{i-1})}{f(p_i) - f(p_{i-1})} = \frac{f(p_{i-1})p_i - f(p_i)p_{i-1}}{f(p_{i-1}) - f(p_i)}.$$

Although this is algebraically equivalent to the iteration equation, it could increase the significance of rounding error if the nearly equal numbers $f(p_{i-1})p_i$ and $f(p_i)p_{i-1}$ are subtracted.

EXAMPLE 1 In this example we will approximate a root of the equation $x^3 + 4x^2 - 10 = 0$. To use Maple we first define the function $f(x)$ and the numbers p_0 and p_1 with the commands

```
>f:=x->x^3+4*x^2-10;
>p0:=1; p1:=2;
```

The values of $f(p_0) = -5$ and $f(p_1) = 14$ are computed by

```
>fp0:=f(p0); fp1:=f(p1);
```

and the first secant approximation, $p_2 = \frac{24}{19}$, by

```
>p2:=p1-fp1*(p1-p0)/(fp1-fp0);
```

The next command forces a floating-point representation for p_2 instead of an exact rational representation.

```
>p2:=evalf(p2);
```

We compute $f(p_2) = -1.602274379$ and continue to compute $p_3 = 1.338827839$ by

```
>fp2:=f(p2);
>p3:=p2-fp2*(p2-p1)/(fp2-fp1);
```

The program SECANT22 with inputs $p_0 = 1$, $p_1 = 2$, $TOL = 0.0005$, and $N_0 = 20$ produces the results in Table 2.2. About half the number of iterations are needed, compared to the Bisection method in Example 1 of Section 2.2. Further, $|p - p_6| = |1.3652300134 - 1.3652300011| < 1.3 \times 10^{-8}$ is much smaller than the tolerance 0.0005. ◆ ◆ ◆

Table 2.2

n	p_n	$f(p_n)$
2	1.2631578947	−1.6022743840
3	1.3388278388	−0.4303647480
4	1.3666163947	0.0229094308
5	1.3652119026	−0.0002990679
6	1.3652300011	−0.0000002032

There are other reasonable choices for generating a sequence of approximations based on the intersection of an approximating line and the x-axis. The **method of False Position** (or *Regula Falsi*) is a hybrid bisection-secant method that constructs approximating lines similar to those of the Secant method but brackets the root in the manner of the Bisection method. As with the Bisection method, the method of False Position requires that an initial interval $[a, b]$ first be found, with $f(a)$ and $f(b)$ of opposite sign. With $a_1 = a$ and $b_1 = b$, the approximation, p_2, is given by

$$p_2 = a_1 - \frac{f(a_1)(b_1 - a_1)}{f(b_1) - f(a_1)}.$$

If $f(p_2)$ and $f(a_1)$ have the same sign, then set $a_2 = p_2$ and $b_2 = b_1$. Alternatively, if $f(p_2)$ and $f(b_1)$ have the same sign, set $a_2 = a_1$ and $b_2 = p_2$. (See Figure 2.5.)

Method of False Position

An interval $[a_{n+1}, b_{n+1}]$, for $n > 1$, containing an approximation to a root of $f(x) = 0$ is found from an interval $[a_n, b_n]$ containing the root by first computing

$$p_{n+1} = a_n - \frac{f(a_n)(b_n - a_n)}{f(b_n) - f(a_n)}.$$

Then set

$$a_{n+1} = a_n \quad \text{and} \quad b_{n+1} = p_{n+1} \quad \text{if} \quad f(a_n)f(p_{n+1}) < 0,$$

and

$$a_{n+1} = p_{n+1} \quad \text{and} \quad b_{n+1} = b_n \quad \text{otherwise.}$$

Figure 2.5

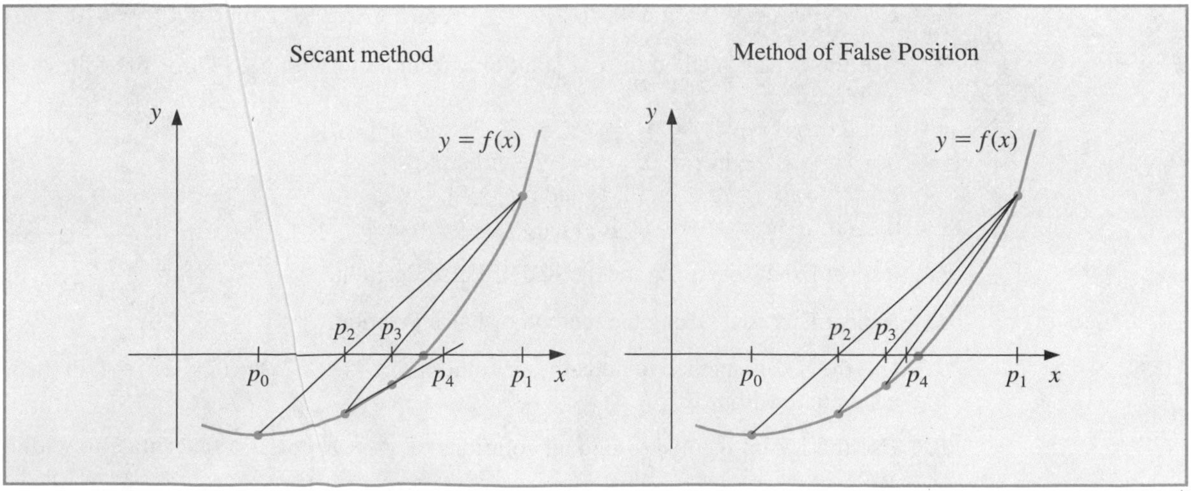

Although the method of False Position may appear superior to the Secant method, it generally converges more slowly, as the results in Table 2.3 indicate for the problem we considered in Example 1. In fact, the method of False Position can converge even more slowly than the Bisection method (as the problem given in Exercise 12 shows), although this is not usually the case. The program FALPOS23 implements the method of False Position.

Table 2.3

n	a_n	b_n	p_{n+1}	$f(p_{n+1})$
1	1.00000000	2.00000000	1.26315789	−1.60227438
2	1.26315789	2.00000000	1.33882784	−0.43036475
3	1.33882784	2.00000000	1.35854634	−0.11000879
4	1.35854634	2.00000000	1.36354744	−0.02776209
5	1.36354744	2.00000000	1.36480703	−0.00698342
6	1.36480703	2.00000000	1.36512372	−0.00175521
7	1.36512372	2.00000000	1.36520330	−0.00044106

EXERCISE SET 2.3

1. Let $f(x) = x^2 - 6$, $p_0 = 3$, and $p_1 = 2$. Find p_3 using each method.
   a. Secant method **b.** method of False Position

2. Let $f(x) = -x^3 - \cos x$, $p_0 = -1$, and $p_1 = 0$. Find p_3 using each method.
   a. Secant method **b.** method of False Position

3. Use the Secant method to find solutions accurate to within 10^{-4} for the following problems.
   a. $x^3 - 2x^2 - 5 = 0$, on $[1, 4]$ **b.** $x^3 + 3x^2 - 1 = 0$, on $[-3, -2]$
   c. $x - \cos x = 0$, on $[0, \pi/2]$ **d.** $x - 0.8 - 0.2 \sin x = 0$, on $[0, \pi/2]$

4. Use the Secant method to find solutions accurate to within 10^{-5} for the following problems.
   a. $2x \cos 2x - (x - 2)^2 = 0$ on $[2, 3]$ and on $[3, 4]$
   b. $(x - 2)^2 - \ln x = 0$ on $[1, 2]$ and on $[e, 4]$
   c. $e^x - 3x^2 = 0$ on $[0, 1]$ and on $[3, 5]$
   d. $\sin x - e^{-x} = 0$ on $[0, 1]$, on $[3, 4]$ and on $[6, 7]$

5. Repeat Exercise 3 using the method of False Position.

6. Repeat Exercise 4 using the method of False Position.

7. Use the Secant method to find all four solutions of $4x \cos(2x) - (x - 2)^2 = 0$ in $[0, 8]$ accurate to within 10^{-5}.

8. Use the Secant method to find all solutions of $x^2 + 10 \cos x = 0$ accurate to within 10^{-5}.

9. Approximate, to within 10^{-4}, the value of x that produces the point on the graph of $y = x^2$ that is closest to $(1, 0)$. [*Hint:* Minimize $[d(x)]^2$, where $d(x)$ represents the distance from (x, x^2) to $(1, 0)$.]

10. Approximate, to within 10^{-4}, the value of x that produces the point on the graph of $y = 1/x$ that is closest to $(2, 1)$.

11. The fourth-degree polynomial

$$f(x) = 230x^4 + 18x^3 + 9x^2 - 221x - 9$$

has two real zeros, one in $[-1, 0]$ and the other in $[0, 1]$. Attempt to approximate these zeros to within 10^{-6} using each method.

a. method of False Position **b.** Secant method

12. The function $f(x) = \tan \pi x - 6$ has a zero at $(1/\pi) \arctan 6 \approx 0.447431543$. Let $p_0 = 0$ and $p_1 = 0.48$ and use 10 iterations of each of the following methods to approximate this root. Which method is most successful and why?

a. Bisection method

b. method of False Position

c. Secant method

13. Use Maple to determine how many iterations of the Secant method with $p_0 = \frac{1}{2}$ and $p_1 = \pi/4$ are needed to find a root of $f(x) = \cos x - x$ to within 10^{-100}.

14. The sum of two numbers is 20. If each number is added to its square root, the product of the two sums is 155.55. Determine the two numbers to within 10^{-4}.

15. A trough of length L has a cross section in the shape of a semicircle with radius r. (See the accompanying figure.) When filled with water to within a distance h of the top, the volume, V, of water is

$$V = L\left[0.5\pi r^2 - r^2 \arcsin\left(\frac{h}{r}\right) - h(r^2 - h^2)^{1/2}\right]$$

Suppose $L = 10$ ft, $r = 1$ ft, and $V = 12.4$ ft^3. Find the depth of water in the trough to within 0.01 ft.

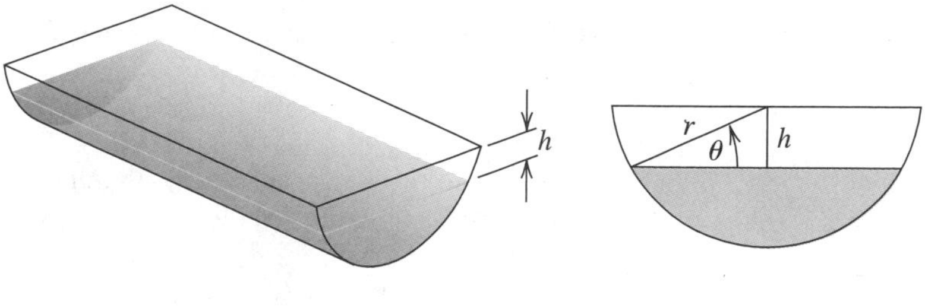

16. A particle starts at rest on a smooth inclined plane whose angle θ is changing at a constant rate

$$\frac{d\theta}{dt} = \omega < 0.$$

At the end of t seconds, the position of the object is given by

$$x(t) = \frac{g}{2\omega^2}\left(\frac{e^{\omega t} - e^{-\omega t}}{2} - \sin \omega t\right).$$

Suppose the particle has moved 1.7 ft in 1 s. Find, to within 10^{-5}, the rate ω at which θ changes. Assume that $g = -32.17$ ft/s^2.

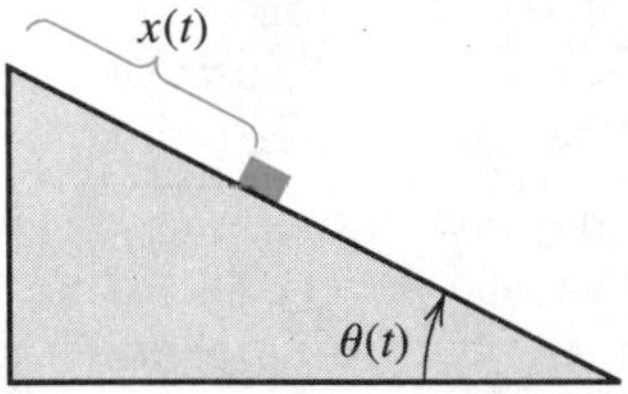

2.4 Newton's Method

The Bisection and Secant methods both have geometric representations that use the zero of an approximating line to the graph of a function f to approximate the solution to $f(x) = 0$. The increase in accuracy of the Secant method over the Bisection method is a consequence of the fact that the secant line to the curve better approximates the graph of f than does the line used to generate the approximations in the Bisection method.

The line that *best* approximates the graph of the function at a point on its graph is the tangent line to the graph at that point. Using this line instead of the secant line produces **Newton's method** (also called the *Newton–Raphson method*), the technique we consider in this section.

Suppose that p_0 is an initial approximation to the root p of the equation $f(x) = 0$ and that f' exists in an interval containing all the approximations to p. The slope of the tangent line to the graph of f at the point $(p_0, f(p_0))$ is $f'(p_0)$, so the equation of this tangent line is

$$y - f(p_0) = f'(p_0)(x - p_0).$$

Since this line crosses the x-axis when the y-coordinate of the point on the line is zero, the next approximation, p_1, to p satisfies

$$0 - f(p_0) = f'(p_0)(p_1 - p_0),$$

which implies that

$$p_1 = p_0 - \frac{f(p_0)}{f'(p_0)},$$

provided that $f'(p_0) \neq 0$. Subsequent approximations are found for p in a similar manner, as shown in Figure 2.6.

Figure 2.6

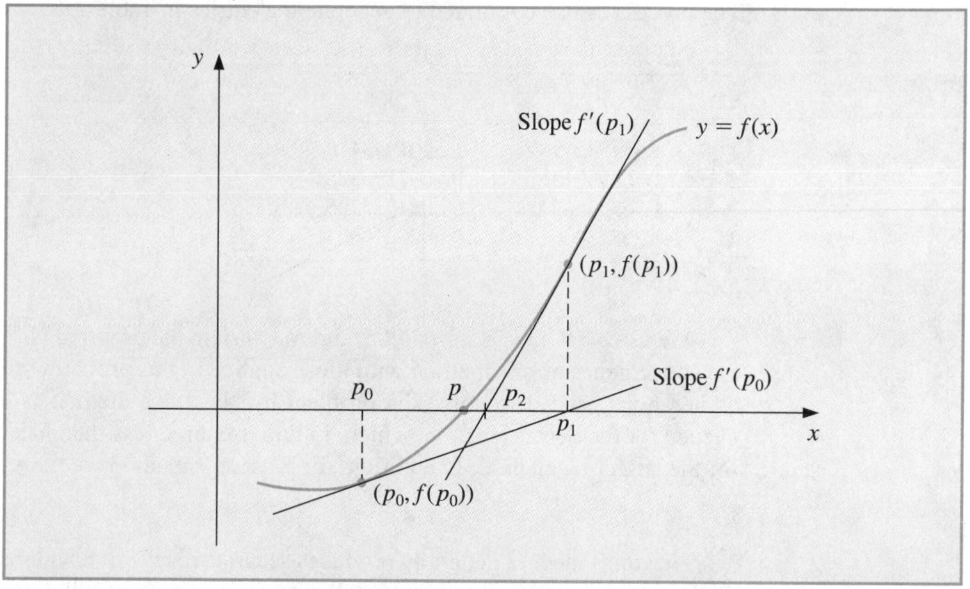

Newton's Method

The approximation p_{n+1} to a root of $f(x) = 0$ is computed from the approximation p_n using the equation

$$p_{n+1} = p_n - \frac{f(p_n)}{f'(p_n)}.$$

EXAMPLE 1 In this example we use Newton's method to approximate the root of the equation $x^3 + 4x^2 - 10 = 0$. Maple is used to find the first iteration of Newton's method with $p_0 = 1$. We define $f(x)$ and compute $f'(x)$ by

```
>f:=x->x^3+4*x^2-10;
>fp:=x->D(f)(x);
>p0:=1;
```

The first iteration of Newton's method gives $p_1 = \frac{16}{11}$, which is obtained with

```
>p1:=p0-f(p0)/fp(p0);
```

A decimal representation of 1.454545455 for p_1 is given by

```
>p1:=evalf(p1);
```

The process can be continued to generate the entries in Table 2.4.

Table 2.4

n	p_n	$f(p_n)$
1	1.4545454545	1.5401953418
2	1.3689004011	0.0607196886
3	1.3652366002	0.0001087706
4	1.3652300134	0.0000000004

We use $p_0 = 1$, $TOL = 0.0005$, and $N_0 = 20$ in the program NEWTON24 to compare the convergence of this method with those applied to this problem previously. The number of iterations needed to solve the problem by Newton's method is less than the number needed for the Secant method, which, in turn, required less than half the iterations needed for the Bisection method. In addition, for Newton's method we have $|p - p_4| \approx 10^{-10}$. ◆ ◆ ◆

Newton's method generally produces accurate results in few iterations. With the aid of Taylor polynomials we can see why this is true. Suppose p is the solution to $f(x) = 0$ and that f'' exists on an interval containing both p and the approximation p_n. Expanding f in the first Taylor polynomial at p_n and evaluating at $x = p$ gives

$$0 = f(p) = f(p_n) + f'(p_n)(p - p_n) + \frac{f''(\xi)}{2}(p - p_n)^2,$$

where ξ lies between p_n and p. Consequently, if $f'(p_n) \neq 0$, we have

$$p - p_n + \frac{f(p_n)}{f'(p_n)} = -\frac{f''(\xi)}{2f'(p_n)}(p - p_n)^2.$$

Since

$$p_{n+1} = p_n - \frac{f(p_n)}{f'(p_n)},$$

this implies that

$$p - p_{n+1} = -\frac{f''(\xi)}{2f'(p_n)}(p - p_n)^2.$$

If the second derivative of f on an interval about p is bounded by M and if p_n is within this interval, then

$$|p - p_{n+1}| \leq \frac{M}{2|f'(p_n)|}|p - p_n|^2.$$

The important feature of this inequality is that the error $|p - p_{n+1}|$ of the $(n + 1)$st approximation is bounded by approximately the square of the error of the nth approximation, $|p - p_n|$. This result implies that Newton's method has the tendency to approximately double the number of digits of accuracy with each successive approximation. Newton's method is not, however, infallible, as the example in Exercise 10 shows.

EXAMPLE 2 Find an approximation to the solution of the equation $x = 3^{-x}$ that is accurate to within 10^{-8}.

A solution of this equation corresponds to a solution of

$$0 = f(x) = x - 3^{-x}.$$

Since f is continuous with $f(0) = -1$ and $f(1) = \frac{2}{3}$, a solution of the equation lies in the interval (0, 1). We have chosen the initial approximation to be the midpoint of this interval, $p_0 = 0.5$. Succeeding approximations are generated by applying the formula

$$p_{n+1} = p_n - \frac{f(p_n)}{f'(p_n)} = p_n - \frac{p_n - 3^{-p_n}}{1 + 3^{-p_n} \ln 3}.$$

These approximations are listed in Table 2.5, together with differences between successive approximations. The difference in successive approximations leads to the correct conclusion that p_3 is accurate at least to the places listed. ◆ ◆ ◆

Table 2.5

n	p_n	$\lvert p_n - p_{n-1}\rvert$
0	0.500000000	
1	0.547329757	0.047329757
2	0.547808574	0.000478817
3	0.547808622	0.000000048

The success of Newton's method is predicated on the assumption that the derivative of f is nonzero at the approximations to the root p. If f' is continuous, this means that the technique will be satisfactory provided that $f'(p) \neq 0$ and that a sufficiently accurate initial approximation is used. The condition $f'(p) \neq 0$ is not trivial; it is true precisely when p is a *simple root*, which means that a function q exists with the property that, for $x \neq p$,

$$f(x) = (x - p)q(x), \qquad \text{where} \quad \lim_{x \to p} q(x) \neq 0.$$

When the root is not simple, Newton's method may converge, but not with the speed we have seen in our previous examples.

EXAMPLE 3 The root $p = 0$ of the equation $f(x) = e^x - x - 1 = 0$ is not simple, since both $f(0) = e^0 - 0 - 1 = 0$ and $f'(0) = e^0 - 1 = 0$. The terms generated by Newton's method with $p_0 = 0$ are shown in Table 2.6 and converge slowly to zero. The graph of f is shown in Figure 2.7. ◆ ◆ ◆

Table 2.6

n	p_n	n	p_n
0	1.0	9	2.7750×10^{-3}
1	0.58198	10	1.3881×10^{-3}
2	0.31906	11	6.9411×10^{-4}
3	0.16800	12	3.4703×10^{-4}
4	0.08635	13	1.7416×10^{-4}
5	0.04380	14	8.8041×10^{-5}
6	0.02206		
7	0.01107		
8	0.005545		

Figure 2.7

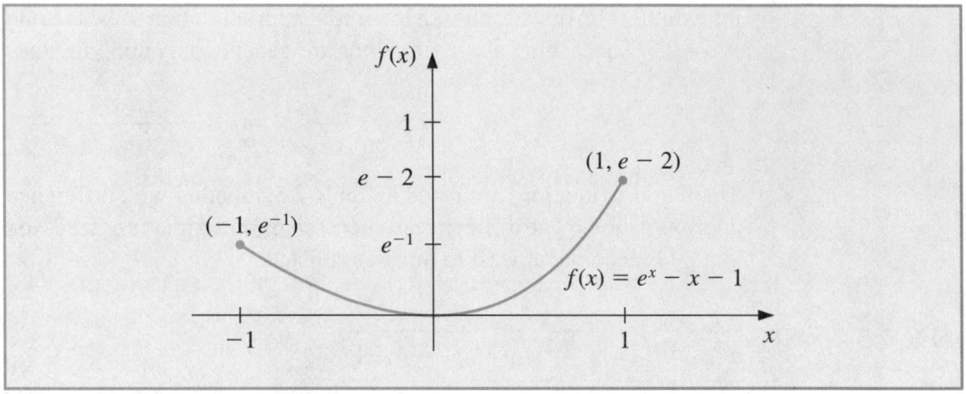

EXERCISE SET 2.4

1. Let $f(x) = x^2 - 6$ and $p_0 = 1$. Use Newton's method to find p_2.
2. Let $f(x) = -x^3 - \cos x$ and $p_0 = -1$. Use Newton's method to find p_2. Could $p_0 = 0$ be used for this problem?
3. Use Newton's method to find solutions accurate to within 10^{-4} for the following problems.
   a. $x^3 - 2x^2 - 5 = 0$, on $[1, 4]$
   b. $x^3 + 3x^2 - 1 = 0$, on $[-3, -2]$
   c. $x - \cos x = 0$, on $[0, \pi/2]$
   d. $x - 0.8 - 0.2\sin x = 0$, on $[0, \pi/2]$
4. Use Newton's method to find solutions accurate to within 10^{-5} for the following problems.
   a. $2x\cos 2x - (x-2)^2 = 0$, on $[2, 3]$ and $[3, 4]$
   b. $(x-2)^2 - \ln x = 0$, on $[1, 2]$ and $[e, 4]$
   c. $e^x - 3x^2 = 0$, on $[0, 1]$ and $[3, 5]$
   d. $\sin x - e^{-x} = 0$, on $[0, 1]$, $[3, 4]$, and $[6, 7]$

5. Use Newton's method to find all four solutions of $4x\cos(2x) - (x-2)^2 = 0$ in $[0, 8]$ accurate to within 10^{-5}.

6. Use Newton's method to find all solutions of $x^2 + 10\cos x = 0$ accurate to within 10^{-5}.

7. Use Newton's method to approximate the solutions of the following equations to within 10^{-5} in the given intervals. In these problems the convergence will be slower than normal since the roots are not simple roots.

a. $x^2 - 2xe^{-x} + e^{-2x} = 0$, on $[0, 1]$

b. $\cos(x + \sqrt{2}) + x(x/\sqrt{2} + 2)/\sqrt{2} = 0$, on $[-2, -1]$

c. $x^3 - 3x^2(2^{-x}) + 3x(4^{-x}) + 8^{-x} = 0$, on $[0, 1]$

d. $e^{6x} + 3(\ln 2)^2 e^{2x} - (\ln 8)e^{4x} - (\ln 2)^3$, on $[-1, 0]$

8. The numerical method defined by

$$p_n = p_{n-1} - \frac{f(p_{n-1})f'(p_{n-1})}{[f'(p_{n-1})]^2 - f(p_{n-1})f''(p_{n-1})},$$

for $n = 1, 2, \ldots$, can be used instead of Newton's method for equations having multiple roots. Repeat Exercise 7 using this method.

9. In Exercise 12 of Section 2.3 we found that for $f(x) = \tan \pi x - 6$, the Bisection method on $[0, 0.48]$ converges more quickly than the method of False Position with $p_0 = 0$ and $p_1 = 0.48$. Also, the Secant method with these values of p_0 and p_1 does not give convergence. Apply Newton's method to this problem with (a) $p_0 = 0$, and (b) $p_0 = 0.48$. (c) Explain the reason for any discrepancies.

10. Use Newton's method to determine the first positive solution to the equation $\tan x = x$, and explain why this problem can give difficulties.

11. Use Newton's method to solve the equation

$$0 = \frac{1}{2} + \frac{1}{4}x^2 - x\sin x - \frac{1}{2}\cos 2x, \quad \text{with } p_0 = \frac{\pi}{2}.$$

Iterate using Newton's method until an accuracy of 10^{-5} is obtained. Explain why the result seems unusual for Newton's method. Also, solve the equation with $p_0 = 5\pi$ and $p_0 = 10\pi$.

12. Use Maple to determine how many iterations of Newton's method with $p_0 = \pi/4$ are needed to find a root of $f(x) = \cos x - x$ to within 10^{-100}.

13. Player A will shut out (win by a score of 21–0) player B in a game of racquetball with probability

$$P = \frac{1+p}{2}\left(\frac{p}{1-p+p^2}\right)^{21},$$

where p denotes the probability A will win any specific rally (independent of the server). (See [K,J], p. 267.) Determine, to within 10^{-3}, the minimal value of p that will ensure that A will shut out B in at least half the matches they play.

14. The function described by $f(x) = \ln(x^2 + 1) - e^{0.4x} \cos \pi x$ has an infinite number of zeros.

a. Determine, within 10^{-6}, the only negative zero.

b. Determine, within 10^{-6}, the four smallest positive zeros.

c. Determine a reasonable initial approximation to find the nth smallest positive zero of f. [*Hint:* Sketch an approximate graph of f.]

d. Use part (c) to determine, within 10^{-6}, the 25th smallest positive zero of f.

15. The accumulated value of a savings account based on regular periodic payments can be determined from the *annuity due equation,*

$$A = \frac{P}{i}[(1+i)^n - 1].$$

In this equation A is the amount in the account, P is the amount regularly deposited, and i is the rate of interest per period for the n deposit periods. An engineer would like to have a savings account valued at \$750,000 upon retirement in 20 years and can afford to put \$1500 per month toward this goal. What is the minimal interest rate at which this amount can be invested, assuming that the interest is compounded monthly?

16. Problems involving the amount of money required to pay off a mortgage over a fixed period of time involve the formula

$$A = \frac{P}{i}[1 - (1+i)^{-n}],$$

known as an *ordinary annuity equation.* In this equation A is the amount of the mortgage, P is the amount of each payment, and i is the interest rate per period for the n payment periods. Suppose that a 30-year home mortgage in the amount of \$135,000 is needed and that the borrower can afford house payments of at most \$1000 per month. What is the maximal interest rate the borrower can afford to pay?

17. A drug administered to a patient produces a concentration in the blood stream given by $c(t) = Ate^{-t/3}$ milligrams per milliliters t hours after A units have been injected. The maximum safe concentration is 1 mg/ml.

a. What amount should be injected to reach this maximum safe concentration and when does this maximum occur?

b. An additional amount of this drug is to be administered to the patient after the concentration falls to 0.25 mg/ml. Determine, to the nearest minute, when this second injection should be given.

c. Assuming that the concentration from consecutive injections is additive and that 75% of the amount originally injected is administered in the second injection, when is it time for the third injection?

18. Let $f(x) = 3^{3x+1} - 7 \cdot 5^{2x}$.

a. Use the Maple commands `solve` and `fsolve` to try to find all roots of f.

b. Plot $f(x)$ to find initial approximations to roots of f.

c. Use Newton's method to find roots of f to within 10^{-16}.

d. Find the exact solutions of $f(x) = 0$ algebraically.

2.5 Error Analysis and Accelerating Convergence

In the previous section we found that Newton's method generally converges very rapidly if a sufficiently accurate initial approximation has been found. This rapid speed of convergence is due to the fact that Newton's method produces *quadratically* convergent approximations.

A method that produces a sequence $\{p_n\}$ of approximations that converge to a number p converges **linearly** if, for large values of n, a constant $0 < M \leq 1$ exists with

$$|p - p_{n+1}| \leq M|p - p_n|.$$

The sequence converges **quadratically** if, for large values of n, a constant $0 < M$ exists with

$$|p - p_{n+1}| \leq M|p - p_n|^2 \qquad \text{for each} \quad n = 0, 1, \ldots.$$

The following example illustrates the advantage of quadratic over linear convergence.

EXAMPLE 1 Suppose that $\{p_n\}$ converges linearly to $p = 0$, $\{\hat{p}_n\}$ converges quadratically to $p = 0$, and the constant $M = 0.5$ is the same in each case. Then

$$|p_1| \leq M|p_0| \leq (0.5) \cdot |p_0| \qquad \text{and} \qquad |\hat{p}_1| \leq M|\hat{p}_0|^2 \leq (0.5) \cdot |\hat{p}_0|^2.$$

Similarly,

$$|p_2| \leq M|p_1| \leq 0.5(0.5) \cdot |p_0| = (0.5)^2|p_0|$$

and

$$|\hat{p}_2| \leq M|\hat{p}_1|^2 \leq 0.5(0.5|\hat{p}_0|^2)^2 = (0.5)^3|\hat{p}_0|^4.$$

Continuing,

$$|p_3| \leq M|p_2| \leq 0.5((0.5)^2|p_0|) = (0.5)^3|p_0|$$

and

$$|\hat{p}_3| \leq M|\hat{p}_2|^2 \leq 0.5((0.5)^3|\hat{p}_0|^4)^2 = (0.5)^7|\hat{p}_0|^8.$$

In general,

$$|p_n| \leq 0.5^n|p_0|, \qquad \text{whereas} \qquad |\hat{p}_n| \leq (0.5)^{2^n-1}|\hat{p}_0|^{2^n}$$

for each $n = 1, 2, \ldots$. Table 2.7 illustrates the relative speed of convergence of these error bounds to zero, assuming that $|p_0| = |\hat{p}_0| = 1$.

The quadratically convergent sequence is within 10^{-38} of zero by the seventh term. At least 126 terms are needed to ensure this accuracy for the linearly convergent sequence. If

Table 2.7

n	Linear Convergence Sequence Bound $(0.5)^n$	Quadratic Convergence Sequence Bound $(0.5)^{2^n-1}$
1	5.0000×10^{-1}	5.0000×10^{-1}
2	2.5000×10^{-1}	1.2500×10^{-1}
3	1.2500×10^{-1}	7.8125×10^{-3}
4	6.2500×10^{-2}	3.0518×10^{-5}
5	3.1250×10^{-2}	4.6566×10^{-10}
6	1.5625×10^{-2}	1.0842×10^{-19}
7	7.8125×10^{-3}	5.8775×10^{-39}

$|\hat{p}_0| < 1$, the bound on the sequence $\{\hat{p}_n\}$ will decrease even more rapidly. No significant change will occur, however, if $|p_0| < 1$. ◆ ◆ ◆

Quadratically convergent sequences generally coverge much more quickly than those that converge only linearly, but many techniques that generate convergent sequences do so linearly. **Aitken's Δ^2 method** is a technique that can be used to accelerate the convergence of a sequence that is linearly convergent, regardless of its origin or application.

Suppose $\{p_n\}_{n=0}^{\infty}$ is a linearly convergent sequence with limit p. To motivate the construction of a sequence $\{\hat{p}_n\}$ that converges more rapidly to p than does $\{p_n\}$, let us first assume that the signs of $p_n - p$, $p_{n+1} - p$, and $p_{n+2} - p$ agree and that n is sufficiently large that

$$\frac{p_{n+1} - p}{p_n - p} \approx \frac{p_{n+2} - p}{p_{n+1} - p}.$$

Then

$$(p_{n+1} - p)^2 \approx (p_{n+2} - p)(p_n - p),$$

so

$$p_{n+1}^2 - 2p_{n+1}p + p^2 \approx p_{n+2}p_n - (p_n + p_{n+2})p + p^2$$

and

$$(p_{n+2} + p_n - 2p_{n+1})p \approx p_{n+2}p_n - p_{n+1}^2.$$

Solving for p gives

$$p \approx \frac{p_{n+2}p_n - p_{n+1}^2}{p_{n+2} - 2p_{n+1} + p_n}.$$

Adding and subtracting the terms p_n^2 and $2p_np_{n+1}$ in the numerator and grouping terms appropriately gives

$$p \approx \frac{p_n^2 + p_n p_{n+2} + 2p_n p_{n+1} - 2p_n p_{n+1} - p_n^2 - p_{n+1}^2}{p_{n+2} - 2p_{n+1} + p_n}$$
$$= \frac{(p_n^2 + p_n p_{n+2} - 2p_n p_{n+1}) - (p_n^2 - 2p_n p_{n+1} + p_{n+1}^2)}{p_{n+2} - 2p_{n+1} + p_n}$$
$$= p_n - \frac{(p_{n+1} - p_n)^2}{p_{n+2} - 2p_{n+1} + p_n}.$$

Aitken's Δ^2 method uses the sequence $\{\hat{p}_n\}_{n=0}^{\infty}$ defined by this approximation to p.

Aitken's Δ^2 Method

$$\hat{p}_n = p_n - \frac{(p_{n+1} - p_n)^2}{p_{n+2} - 2p_{n+1} + p_n}.$$

EXAMPLE 2 The sequence $\{p_n\}_{n=1}^{\infty}$, where $p_n = \cos(1/n)$, converges linearly to $p = 1$. The first few terms of the sequences $\{p_n\}_{n=1}^{\infty}$ and $\{\hat{p}_n\}_{n=1}^{\infty}$ are given in Table 2.8. It certainly appears that $\{\hat{p}_n\}_{n=1}^{\infty}$ converges more rapidly to $p = 1$ than does $\{p_n\}_{n=1}^{\infty}$. ◆ ◆ ◆

Table 2.8

n	p_n	$\hat{p}_n$
1	0.54030	0.96178
2	0.87758	0.98213
3	0.94496	0.98979
4	0.96891	0.99342
5	0.98007	0.99541
6	0.98614	
7	0.98981	

For a given sequence $\{p_n\}_{n=0}^{\infty}$, the **forward difference**, Δp_n, is defined as

$$\Delta p_n = p_{n+1} - p_n, \quad \text{for } n \geq 0.$$

Higher powers, $\Delta^k p_n$, are defined recursively by

$$\Delta^k p_n = \Delta(\Delta^{k-1} p_n), \quad \text{for } k \geq 2.$$

Because of the definition, we have

$$\Delta^2 p_n = \Delta(p_{n+1} - p_n) = \Delta p_{n+1} - \Delta p_n = (p_{n+2} - p_{n+1}) - (p_{n+1} - p_n),$$

so

$$\Delta^2 p_n = p_{n+2} - 2p_{n+1} + p_n.$$

Thus, the formula for $\hat{p}_n$ given in Aitken's Δ^2 method can be written as

$$\hat{p}_n = p_n - \frac{(\Delta p_n)^2}{\Delta^2 p_n}, \quad \text{for all } n \geq 0.$$

The sequence $\{\hat{p}_n\}_{n=1}^{\infty}$ converges to p more rapidly than does the original sequence $\{p_n\}_{n=0}^{\infty}$ in the following sense:

Aitken's Δ^2 Convergence

If $\{p_n\}$ is a sequence that converges linearly to the limit p and $(p_n - p)(p_{n+1} - p) > 0$ for large values of n, then

$$\lim_{n\to\infty} \frac{\hat{p}_n - p}{p_n - p} = 0.$$

We will find occasion to apply this acceleration technique at various times in our study of approximation methods.

EXERCISE SET 2.5

1. The following sequences are linearly convergent. Generate the first five terms of the sequence $\{\hat{p}_n\}$ using Aitken's Δ^2 method.
   a. $p_0 = 0.5, \quad p_n = (2 - e^{p_{n-1}} + p_{n-1}^2)/3, \quad \text{for } n \geq 1$
   b. $p_0 = 0.75, \quad p_n = (e^{p_{n-1}}/3)^{1/2}, \quad \text{for } n \geq 1$
   c. $p_0 = 0.5, \quad p_n = 3^{-p_{n-1}}, \quad \text{for } n \geq 1$
   d. $p_0 = 0.5, \quad p_n = \cos p_{n-1}, \quad \text{for } n \geq 1$
2. Newton's method does not converge quadratically for the following problems. Accelerate the convergence using the Aitken's Δ^2 method. Iterate until $|\hat{p}_n - \hat{p}_{n-1}| < 10^{-4}$.
   a. $x^2 - 2xe^{-x} + e^{-2x} = 0, \quad [0, 1]$
   b. $\cos(x + \sqrt{2}) + x(x/2 + \sqrt{2}) = 0, \quad [-2, -1]$
   c. $x^3 - 3x^2(2^{-x}) + 3x(4^{-x}) - 8^{-x} = 0, \quad [0, 1]$
   d. $e^{6x} + 3(\ln 2)^2 e^{2x} - (\ln 8)e^{4x} - (\ln 2)^3 = 0, \quad [-1, 0]$
3. Consider the function $f(x) = e^{6x} + 3(\ln 2)^2 e^{2x} - (\ln 8)e^{4x} - (\ln 2)^3$. Use Newton's method with $p_0 = 0$ to approximate a zero of f. Generate terms until $|p_{n+1} - p_n| < 0.0002$. Construct the sequence $\{\hat{p}_n\}$. Is the convergence improved?
4. Repeat Exercise 3 with the constants in $f(x)$ replaced by their four-digit approximations, that is, with $f(x) = e^{6x} + 1.441e^{2x} - 2.079e^{4x} - 0.3330$, and compare the solutions to the results in Exercise 3.
5. (i) Show that the following sequences $\{p_n\}$ converge linearly to $p = 0$. (ii) How large must n be before $|p_n - p| \leq 5 \times 10^{-2}$? (iii) Use Aitken's Δ^2 method to generate a sequence $\{\hat{p}_n\}$ until $|\hat{p}_n - p| \leq 5 \times 10^{-2}$.

a. $p_n = \frac{1}{n}$, for $n \geq 1$ **b.** $p_n = \frac{1}{n^2}$, for $n \geq 1$

6. **a.** Show that for any positive integer k, the sequence defined by $p_n = 1/n^k$ converges linearly to $p = 0$.
b. For each pair of integers k and m, determine a number N for which $1/N^k < 10^{-m}$.

7. **a.** Show that the sequence $p_n = 10^{-2^n}$ converges quadratically to zero.
b. Show that the sequence $p_n = 10^{-n^k}$ does not converge to zero quadratically, regardless of the size of the exponent $k > 1$.

8. A sequence $\{p_n\}$ is said to be **superlinearly convergent** to p if a sequence $\{c_n\}$ converging to zero exists with

$$|p_{n+1} - p| \leq c_n |p_n - p|.$$

a. Show that if $\{p_n\}$ is superlinearly convergent to p, then $\{p_n\}$ is linearly convergent to p.
b. Show that $p_n = 1/n^n$ is superlinearly convergent to zero but is not quadratically convergent to zero.

2.6 Müller's Method

There are a number of root-finding problems for which the Secant, False Position, and Newton's methods will not give satisfactory results. They will not give rapid convergence, for example, when the function and its derivative are simultaneously close to zero. In addition, these methods cannot be used to approximate complex roots unless the initial approximation is a complex number with nonzero imaginary part. This often makes them a poor choice for use in approximating the roots of polynomials, which, even with real coefficients, commonly have complex roots occuring in conjugate pairs.

In this section we consider Müller's method, which is a generalization of the Secant method. The Secant method finds the zero of the line passing through points on the graph of the function that corresponds to the two immediately previous approximations, as shown in Figure 2.8(a). Müller's method uses the zero of the parabola through the three immediately previous points on the graph as the new approximation, as shown in part (b) of Figure 2.8.

Suppose that three initial approximations, p_0, p_1, and p_2, are given for a solution of $f(x) = 0$. The derivation of Müller's method for determining the next approximation p_3 begins by considering the quadratic polynomial

$$P(x) = a(x - p_2)^2 + b(x - p_2) + c$$

that passes through $(p_0, f(p_0)), (p_1, f(p_1))$, and $(p_2, f(p_2))$. The constants a, b, and c can be determined from the conditions

$$f(p_0) = a(p_0 - p_2)^2 + b(p_0 - p_2) + c,$$
$$f(p_1) = a(p_1 - p_2)^2 + b(p_1 - p_2) + c,$$

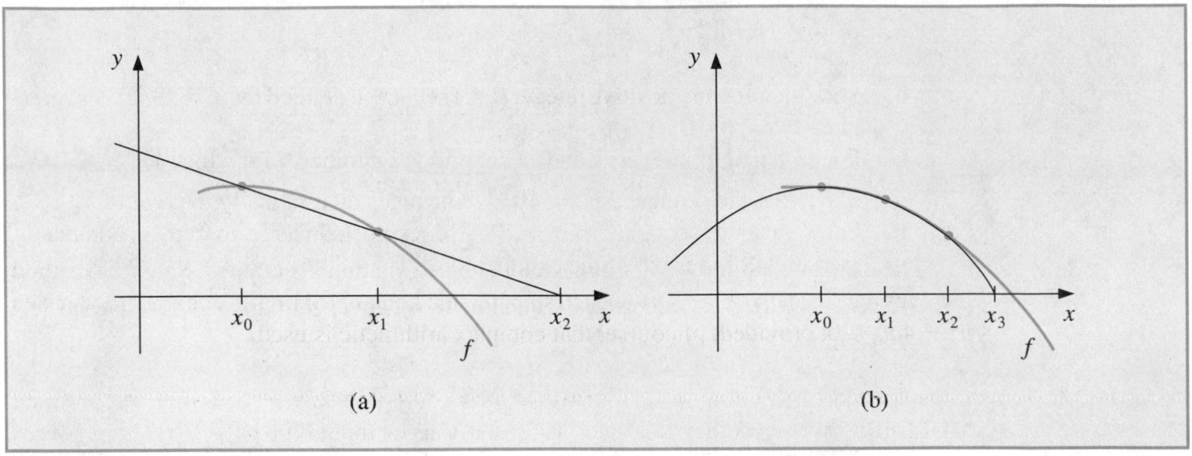

Figure 2.8

and

$$f(p_2) = a \cdot 0^2 + b \cdot 0 + c.$$

To determine p_3, the root of $P(x) = 0$, we apply the quadratic formula to $P(x)$. Because of round-off error problems caused by the subtraction of nearly equal numbers, however, we apply the formula in the manner prescribed in Example 1 of Section 1.4:

$$p_3 - p_2 = \frac{-2c}{b \pm \sqrt{b^2 - 4ac}}.$$

This gives two possibilities for p_3, depending on the sign preceding the radical term. In Müller's method, the sign is chosen to agree with the sign of b. Chosen in this manner, the denominator will be the largest in magnitude, which avoids the possibility of subtracting nearly equal numbers and results in p_3 being selected as the closest root of $P(x) = 0$ to p_2.

Müller's Method

Given initial approximations p_0, p_1, and p_2, generate

$$p_3 = p_2 - \frac{2c}{b + \text{sgn}(b)\sqrt{b^2 - 4ac}},$$

where

$$c = f(p_2),$$

$$b = \frac{(p_0 - p_2)^2[f(p_1) - f(p_2)] - (p_1 - p_2)^2[f(p_0) - f(p_2)]}{(p_0 - p_2)(p_1 - p_2)(p_0 - p_1)},$$

and

$$a = \frac{(p_1 - p_2)[f(p_0) - f(p_2)] - (p_0 - p_2)[f(p_1) - f(p_2)]}{(p_0 - p_2)(p_1 - p_2)(p_0 - p_1)}.$$

Then continue the iteration, with p_1, p_2, and p_3 replacing p_0, p_1, and p_2.

The method continues until a satisfactory approximation is obtained. Since the method involves the radical $\sqrt{b^2 - 4ac}$ at each step, the method approximates complex roots when $b^2 - 4ac < 0$, provided, of course, that complex arithmetic is used.

EXAMPLE 1 Consider the polynomial $f(x) = 16x^4 - 40x^3 + 5x^2 + 20x + 6$. Using the program MULLER25 with accuracy tolerance 10^{-5} and various inputs for p_0, p_1, and p_2 produces the results in Tables 2.9, 2.10, and 2.11. ◆ ◆ ◆

Table 2.9

$p_0 = 0.5$, $p_1 = -0.5$, $p_2 = 0$		
n	p_n	$f(p_n)$
3	$-0.555556 + 0.598352i$	$-29.4007 - 3.89872i$
4	$-0.435450 + 0.102101i$	$1.33223 - 1.19309i$
5	$-0.390631 + 0.141852i$	$0.375057 - 0.670164i$
6	$-0.357699 + 0.169926i$	$-0.146746 - 0.00744629i$
7	$-0.356051 + 0.162856i$	$-0.183868 \times 10^{-2} + 0.539780 \times 10^{-3}i$
8	$-0.356062 + 0.162758i$	$0.286102 \times 10^{-5} + 0.953674 \times 10^{-6}i$

Table 2.10

$p_0 = 0.5$, $p_1 = 1.0$, $p_2 = 1.5$		
n	p_n	$f(p_n)$
3	1.28785	-1.37624
4	1.23746	0.126941
5	1.24160	0.219440×10^{-2}
6	1.24168	0.257492×10^{-4}
7	1.24168	0.257492×10^{-4}

Table 2.11

$p_0 = 2.5$, $p_1 = 2.0$, $p_2 = 2.25$		
n	p_n	$f(p_n)$
3	1.96059	-0.611255
4	1.97056	0.748825×10^{-2}
5	1.97044	-0.295639×10^{-4}
6	1.97044	-0.295639×10^{-4}

To use Maple to generate the first entry in Table 2.9 we define $f(x)$ and the initial approximations with the code

```
>f:=x->16*x^4-40*x^3+5*x^2+20*x+6;
>p0:=0.5; p1:=-0.5; p2:=0.0;
```

We evaluate the polynomial at the initial values

```
>f0:=f(p0); f1:=f(p1); f2:=f(p2);
```

and we compute $c = 6$, $b = 10$, $a = 9$, and $p_3 = -0.5555555558 + 0.5983516452i$ using the Müller's method formulas:

```
>c:=f2;
>b:=((p0-p2)^2*(f1-f2)-(p1-p2)^2*(f0-f2))/((p0-p2)*(p1-p2)*(p0-p1));
>a:=((p1-p2)*(f0-f2)-(p0-p2)*(f1-f2))/((p0-p2)*(p1-p2)*(p0-p1));
>p3:=p2-(2*c)/(b+(b/abs(b))*sqrt(b^2-4*a*c));
```

The value p_3 was generated using complex arithmetic, as is the calculation

```
>f3:=f(p3);
```

which gives $f_3 = -29.40070112 - 3.898724738i$.

The actual values for the roots of the equation are $-0.356062 \pm 0.162758i$, 1.241677, and 1.970446, which demonstrate the accuracy of the approximations from Müller's method.

Example 1 illustrates that Müller's method can approximate the roots of polynomials with a variety of starting values. In fact, the technique generally converges to the root of a polynomial for any initial approximation choice. General-purpose software packages using Müller's method request only one initial approximation per root and, as an option, may even supply this approximation.

Although Müller's method is not quite as efficient as Newton's method, it is generally better than the Secant method. The relative efficiency, however, is not as important as the ease of implementation and the likelihood that a root will be found. Any of these methods will converge quite rapidly once a reasonable initial approximation is determined.

When a sufficiently accurate approximation p^* to a root has been found, $f(x)$ is divided by $x - p^*$ to produce what is called a *deflated* equation. If $f(x)$ is a polynomial of degree n, the deflated polynomial will be of degree $n - 1$, so the computations are simplified. After an approximation to the root of the deflated equation has been determined, either Müller's method or Newton's method can be used in the original function with this root as the initial approximation. This will ensure that the root being approximated is a solution to the true equation, not to the less accurate deflated equation.

EXERCISE SET 2.6

1. Find the approximations to within 10^{-4} to all the real zeros of the following polynomials using Newton's method.
   a. $P(x) = x^3 - 2x^2 - 5$
   b. $P(x) = x^3 + 3x^2 - 1$
   c. $P(x) = x^4 + 2x^2 - x - 3$
   d. $P(x) = x^5 - x^4 + 2x^3 - 3x^2 + x - 4$

2. Find approximations to within 10^{-5} to all the zeros of each of the following polynomials by first finding the real zeros using Newton's method and then reducing to polynomials of lower degree to determine any complex zeros.
   a. $P(x) = x^4 + 5x^3 - 9x^2 - 85x - 136$
   b. $P(x) = x^4 - 2x^3 - 12x^2 + 16x - 40$
   c. $P(x) = x^4 + x^3 + 3x^2 + 2x + 2$
   d. $P(x) = x^5 + 11x^4 - 21x^3 - 10x^2 - 21x - 5$

3. Repeat Exercise 1 using Müller's method.

4. Repeat Exercise 2 using Müller's method.

5. Find, to within 10^{-3}, the zeros and critical points of the following functions. Use this information to sketch the graphs of P.
   a. $P(x) = x^3 - 9x^2 + 12$
   b. $P(x) = x^4 - 2x^3 - 5x^2 + 12x - 5$

6. $P(x) = 10x^3 - 8.3x^2 + 2.295x - 0.21141 = 0$ has a root at $x = 0.29$.
   a. Use Newton's method with $p_0 = 0.28$ to attempt to find this root.
   b. Use Müller's method with $p_0 = 0.275$, $p_1 = 0.28$, and $p_2 = 0.285$ to attempt to find this root.
   c. Explain any discrepancies in parts (a) and (b).

7. Use Maple to find the exact roots of the polynomial $P(x) = x^3 + 4x - 4$.

8. Use Maple to find the exact roots of the polynomial $P(x) = x^3 - 2x - 5$.

9. Use each of the following methods to find a solution accurate to within 10^{-4} for the problem

$$600x^4 - 550x^3 + 200x^2 - 20x - 1 = 0, \quad \text{for } 0.1 \le x \le 1.$$

   a. Bisection method
   b. Newton's method
   c. Secant method
   d. method of False Position
   e. Müller's method

10. Two ladders crisscross an alley of width w. Each ladder reaches from the base of one wall to some point on the opposite wall. The ladders cross at a height H above the pavement. Find W given that the lengths of the ladders are $x_1 = 20$ ft and $x_2 = 30$ ft and that $H = 8$ ft. (See the figure on page 58.)

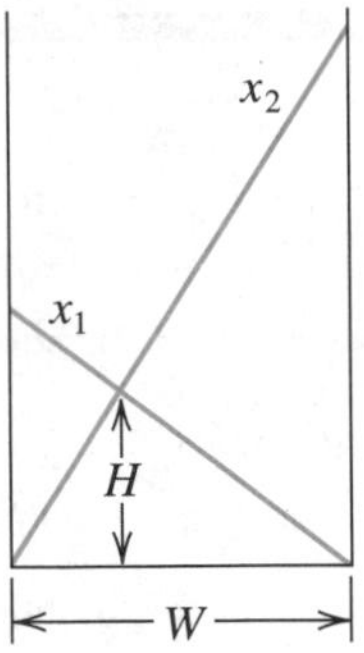

11. A can in the shape of a right circular cylinder is to be constructed to contain 1000 cm^3. The circular top and bottom of the can must have a radius of 0.25 cm more than the radius of the can so that the excess can be used to form a seal with the side. The sheet of material being formed into the side of the can must also be 0.25 cm longer than the circumference of the can so that a seal can be formed. Find, to within 10^{-4}, the minimal amount of material needed to construct the can.

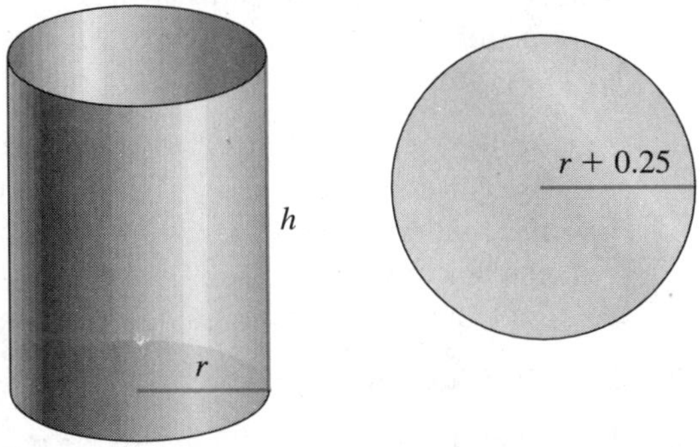

12. In 1224 Leonardo of Pisa, better known as Fibonacci, answered a mathematical challenge of John of Palermo in the presence of Emperor Frederick II. His challenge was to find a root of the equation $x^3 + 2x^2 + 10x = 20$. He first showed that the equation had no rational roots and no Euclidean irrational root—that is, no root in one of the forms $a \pm \sqrt{b}$, $\sqrt{a} \pm \sqrt{b}$, $\sqrt{a \pm \sqrt{b}}$, or $\sqrt{\sqrt{a} \pm \sqrt{b}}$, where a and b are rational numbers. He then approximated the only real root, probably using an algebraic technique of Omar Khayyam involving the intersection of a circle and a parabola. His answer was given in the base-60 number system as

$$1 + 22\left(\frac{1}{60}\right) + 7\left(\frac{1}{60}\right)^2 + 42\left(\frac{1}{60}\right)^3 + 33\left(\frac{1}{60}\right)^4 + 4\left(\frac{1}{60}\right)^5 + 40\left(\frac{1}{60}\right)^6.$$

How accurate was his approximation?

2.7 Survey of Methods and Software

In this chapter we have considered the problem of solving the equation $f(x) = 0$, where f is a given continuous function. All the methods begin with initial approximations and generate a sequence that converges to a root of the equation if the method is successful. If $[a, b]$ is an interval on which $f(a)$ and $f(b)$ are of opposite sign, then the Bisection method and the method of False Position will converge. However, the convergence of these methods may be slow. Faster convergence is generally obtained using the Secant method or the Newton-Raphson method. Good initial approximations are required for these methods, two for the Secant method and one for the Newton-Raphson method, so the Bisection or the False Position method can be used as starter methods for the Secant or Newton-Raphson method.

Müller's method will give rapid convergence without a particularly good initial approximation. It is not quite as efficient as Newton's method but is better than the Secant method. It has the added advantage of being able to approximate complex roots.

Other effective methods are available for determining the roots of polynomials. If this topic is of particular interest, we recommend that consideration be given to Laguerre's method, which can approximate complex roots, the Jenkins-Traub method, and Brent's method, which is based on the Bisection and False-position methods. ISML and NAG supply subroutines based on these methods.

Within MATLAB, the function `ROOTS` is used to compute all the roots, both real and complex, of a polynomial. For an arbitrary function, `FZERO` computes a root near a specified initial approximation to within a specified tolerance.

Maple has the procedure `fsolve` to find roots of equations. For example,

```
>fsolve(x^2 - x - 1);
```

returns the numbers $-.6180339887$ and 1.618033989. You can also specify a particular variable and interval to search. For example,

```
>fsolve(x^2 - x - 1,x,1..2);
```

returns only the number 1.618033989. `fsolve` uses a variety of specialized techniques that depend on the particular form of the equation or system of equations.

Notice that in spite of the diversity of methods, the professionally written packages are based primarily on the methods and principles discussed in this chapter. You should be able to use these packages by reading the manuals accompanying the packages to better understand the parameters and the specifications of the results that are obtained.

There are three books that we consider to be classics on the solution of nonlinear equations, those by Traub [Tr], Ostrowski [Os], and Householder [Ho]. In addition, the book by Brent [Bre] served as the basis for many of the currently used root-finding methods.

CHAPTER 3

Interpolation and Polynomial Approximation

3.1 Introduction

Engineers and scientists commonly assume that relationships between variables in a physical problem can be approximately reproduced from data given by the problem. The ultimate goal might be to determine the values at intermediate points, to approximate the integral or derivative of the underlying function, or to simply give a smooth or continuous representation of the variables in the problem.

Interpolation refers to determining a function that exactly represents a collection of data. The most elementary type of interpolation consists of fitting a polynomial to a collection of data points. Polynomials have derivatives and integrals that are themselves polynomials, so they are a natural choice for approximating derivatives and integrals. We will see in this chapter that approximating polynomials are easily constructed; the following result implies that there are polynomials that are arbitrarily close to any continuous function.

Weierstrass Approximation Theorem

Suppose that f is defined and continuous on $[a, b]$. For each $\varepsilon > 0$, there exists a polynomial $P(x)$ defined on $[a, b]$, with the property that

$$|f(x) - P(x)| < \varepsilon, \qquad \text{for all } x \in [a, b].$$

The Taylor polynomials were introduced in Section 1.2, where they were described as one of the fundamental building blocks of numerical analysis. Given this prominence, you might assume that polynomial interpolation makes heavy use of these functions. However, this is not the case. The Taylor polynomials agree as closely as possible with a given function at a specific point, but they concentrate their accuracy only near that point. A good interpolation polynomial needs to provide a relatively accurate approximation over an entire interval, and Taylor polynomials do not do that. For example, suppose we calculate the first six Taylor polynomials about $x_0 = 0$ for $f(x) = e^x$. Since the derivatives of f are all e^x, which evaluated at $x_0 = 0$ gives 1, the Taylor polynomials are

$$P_0(x) = 1, \quad P_1(x) = 1 + x, \quad P_2(x) = 1 + x + \frac{x^2}{2}, \quad P_3(x) = 1 + x + \frac{x^2}{2} + \frac{x^3}{6},$$
$$P_4(x) = 1 + x + \frac{x^2}{2} + \frac{x^3}{6} + \frac{x^4}{24}, \quad \text{and} \quad P_5(x) = 1 + x + \frac{x^2}{2} + \frac{x^3}{6} + \frac{x^4}{24} + \frac{x^5}{120}.$$

Table 3.1 lists the values of the Taylor polynomial for various values of x. Notice that even for the higher-degree polynomials, the error becomes progressively worse as we move away from zero. This can also be seen from the graphs of the polynomials in Figure 3.1.

Table 3.1

x	$P_0(x)$	$P_1(x)$	$P_2(x)$	$P_3(x)$	$P_4(x)$	$P_5(x)$	e^x
−2.0	1.00000	−1.00000	1.00000	−0.33333	0.33333	0.06667	0.13534
−1.5	1.00000	−0.50000	0.62500	0.06250	0.27344	0.21016	0.22313
−1.0	1.00000	0.00000	0.50000	0.33333	0.37500	0.36667	0.36788
−0.5	1.00000	0.50000	0.62500	0.60417	0.60677	0.60651	0.60653
0.0	1.00000	1.00000	1.00000	1.00000	1.00000	1.00000	1.00000
0.5	1.00000	1.50000	1.62500	1.64583	1.64844	1.64870	1.64872
1.0	1.00000	2.00000	2.50000	2.66667	2.70833	2.71667	2.71828
1.5	1.00000	2.50000	3.62500	4.18750	4.39844	4.46172	4.48169
2.0	1.00000	3.00000	5.00000	6.33333	7.00000	7.26667	7.38906

Figure 3.1

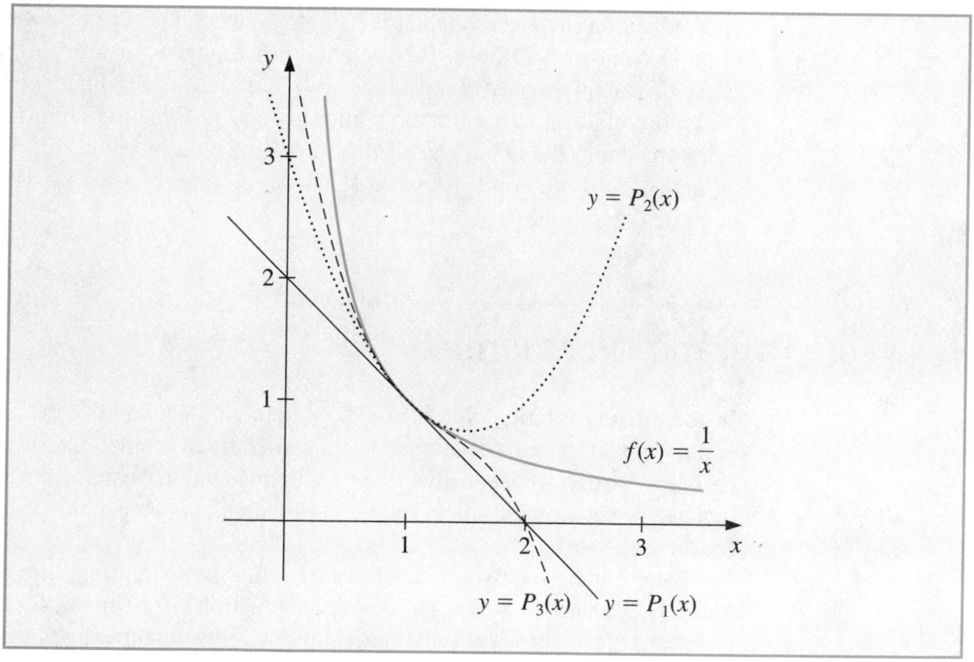

Although better approximations are obtained for this problem if higher-degree Taylor polynomials are used, this situation is not always true. Consider, as an extreme example, using Taylor polynomials of various degrees for $f(x) = 1/x$ expanded about $x_0 = 1$ to approximate $f(3) = \frac{1}{3}$. Since

$$f(x) = x^{-1}, f'(x) = -x^{-2}, f''(x) = (-1)^2 2 \cdot x^{-3},$$

and, in general,

$$f^{(n)}(x) = (-1)^n n! x^{-n-1},$$

the Taylor polynomials for $n \geq 0$ are

$$P_n(x) = \sum_{k=0}^{n} \frac{f^{(k)}(1)}{k!}(x-1)^k = \sum_{k=0}^{n}(-1)^k(x-1)^k.$$

To approximate $f(3) = \frac{1}{3}$ by $P_n(3)$ for increasing values of n, we obtain the values in Table 3.2, rather a dramatic failure.

Table 3.2

n	0	1	2	3	4	5	6	7
$P_n(3)$	1	-1	3	-5	11	-21	43	-85

Since the Taylor polynomials have the property that all the information used in the approximation is concentrated at the single point x_0, the type of difficulty that occurs here is quite common and limits Taylor polynomial approximation to the situation in which approximations are needed only at points close to x_0. For ordinary computational purposes it is more efficient to use methods that include information at various points, which we will consider in the remainder of this chapter. The primary use of Taylor polynomials in numerical analysis is *not* for approximation purposes; instead it is for the derivation of numerical techniques.

3.2 Lagrange Polynomials

In the previous section we discussed the general unsuitability of Taylor polynomials for approximation. These polynomials are useful only over small intervals for functions whose derivatives exist and are easily evaluated. In this section we find approximating polynomials that can be determined simply by specifying certain points on the plane through which they must pass.

Determining a polynomial of degree 1 that passes through the distinct points (x_0, y_0) and (x_1, y_1) is the same as approximating a function f for which $f(x_0) = y_0$ and $f(x_1) = y_1$ by means of a first-degree polynomial interpolating, or agreeing with, the values of f at the given points. We first define the functions

$$L_0(x) = \frac{x - x_1}{x_0 - x_1}, \quad L_1(x) = \frac{x - x_0}{x_1 - x_0},$$

and then define

$$P(x) = L_0(x)f(x_0) + L_1(x)f(x_1).$$

Since

$$L_0(x_0) = 1, \quad L_0(x_1) = 0, \quad L_1(x_0) = 0, \quad \text{and} \quad L_1(x_1) = 1,$$

we have

$$P(x_0) = 1 \cdot f(x_0) + 0 \cdot f(x_1) = f(x_0) = y_0$$

and

$$P(x_1) = 0 \cdot f(x_0) + 1 \cdot f(x_1) = f(x_1) = y_1.$$

So, P is the unique linear function passing through (x_0, y_0) and (x_1, y_1). (See Figure 3.2.)

Figure 3.2

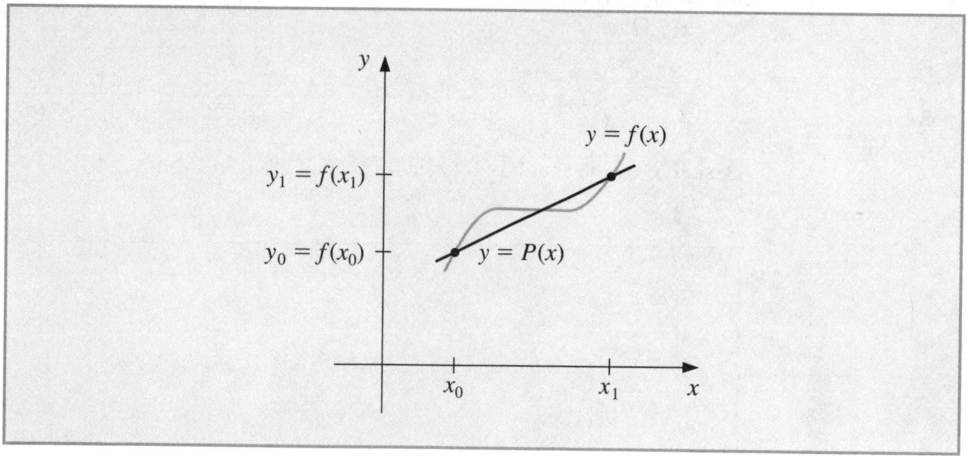

To generalize the concept of linear interpolation to higher-degree polynomials, consider the construction of a polynomial of degree at most n that passes through the $n + 1$ points

$$(x_0, f(x_0)), \ (x_1, f(x_1)), \ldots, (x_n, f(x_n)).$$

(See Figure 3.3.)

Figure 3.3

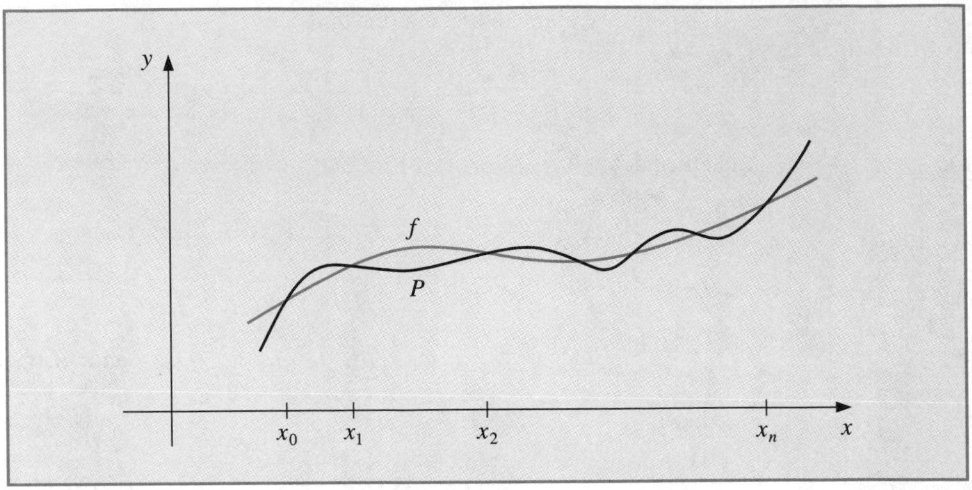

In this case, we need to construct, for each $k = 0, 1, \ldots, n$, a function $L_{n,k}(x)$ with the property that $L_{n,k}(x_i) = 0$ when $i \neq k$ and $L_{n,k}(x_k) = 1$. To satisfy $L_{n,k}(x_i) = 0$ for each $i \neq k$ requires that the numerator of $L_{n,k}(x)$ contains the term

$$(x - x_0)(x - x_1)\cdots(x - x_{k-1})(x - x_{k+1})\cdots(x - x_n)$$

To satisfy $L_{n,k}(x_k) = 1$, the denominator of $L_{n,k}(x)$ must be equal to this term evaluated at $x = x_k$. Thus,

$$L_{n,k}(x) = \frac{(x - x_0)\cdots(x - x_{k-1})(x - x_{k+1})\cdots(x - x_n)}{(x_k - x_0)\cdots(x_k - x_{k-1})(x_k - x_{k+1})\cdots(x_k - x_n)}.$$

A sketch of the graph of a typical $L_{n,k}$ is shown in Figure 3.4.

Figure 3.4

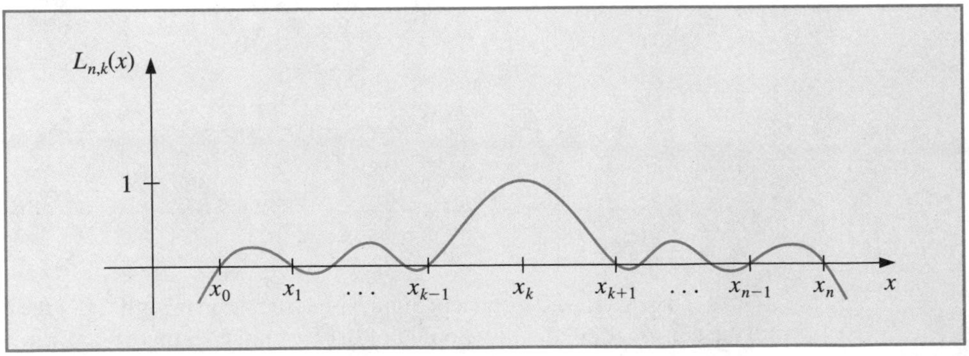

The interpolating polynomial is easily described now that the form of $L_{n,k}(x)$ is known. This polynomial is called the nth Lagrange interpolating polynomial.

nth Lagrange Interpolating Polynomial

$$P_n(x) = f(x_0)L_{n,0}(x) + \cdots + f(x_n)L_{n,n}(x) = \sum_{k=0}^{n} f(x_k)L_{n,k}(x),$$

where

$$L_{n,k}(x) = \frac{(x - x_0)(x - x_1)\cdots(x - x_{k-1})(x - x_{k+1})\cdots(x - x_n)}{(x_k - x_0)(x_k - x_1)\cdots(x_k - x_{k-1})(x_k - x_{k+1})\cdots(x_k - x_n)}$$

for each $k = 0, 1, \ldots, n$.

If $x_0, x_1, \ldots, x_n$ are $(n+1)$ distinct numbers and f is a function whose values are given at these numbers, then $P_n(x)$ is the unique polynomial of degree at most n that agrees with $f(x)$ at $x_0, x_1, \ldots, x_n$. The notation for describing the Lagrange interpolating polynomial $P_n(x)$ is rather complicated. To reduce this somewhat, we will write $L_{n,k}(x)$ simply as $L_k(x)$ when there is no confusion as to its degree.

EXAMPLE 1 Using the numbers, or *nodes*, $x_0 = 2, x_1 = 2.5$, and $x_2 = 4$ to find the second interpolating polynomial for $f(x) = 1/x$ requires that we first determine the coefficient polynomials L_0, L_1, and L_2. In nested form they are

$$L_0(x) = \frac{(x-2.5)(x-4)}{(2-2.5)(2-4)} = (x - 6.5)x + 10,$$

$$L_1(x) = \frac{(x-2)(x-4)}{(2.5-2)(2.5-4)} = \frac{(-4x+24)x - 32}{3},$$

and

$$L_2(x) = \frac{(x-2)(x-2.5)}{(4-2)(4-2.5)} = \frac{(x-4.5)x+5}{3}.$$

Since $f(x_0) = f(2) = 0.5$, $f(x_1) = f(2.5) = 0.4$, and $f(x_2) = f(4) = 0.25$, we have

$$\begin{aligned} P_2(x) &= \sum_{k=0}^{2} f(x_k)L_k(x) \\ &= 0.5((x-6.5)x+10) + 0.4\frac{(-4x+24)x-32}{3} + 0.25\frac{(x-4.5)x+5}{3} \\ &= (0.05x - 0.425)x + 1.15. \end{aligned}$$

An approximation to $f(3) = \frac{1}{3}$ is

$$f(3) \approx P(3) = 0.325.$$

Compare this to Table 3.2, where no Taylor polynomial expanded about $x_0 = 1$ can be used to reasonably approximate $f(3) = \frac{1}{3}$. (See Figure 3.5.) ◆ ◆ ◆

Figure 3.5

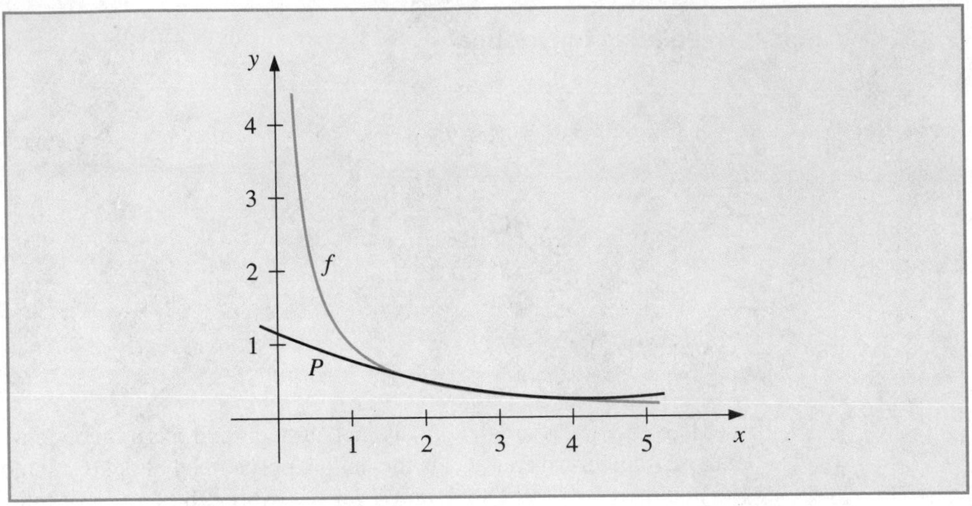

Maple constructs an interpolating polynomial with the command

```
>interp(X,Y,x);
```

where X is the list $[x_0, \ldots, x_n]$, Y is the list $[f(x_0), \ldots, f(x_n)]$, and x is the variable to be used. In this example we can generate the interpolating polynomial $p = .05000000000x^2 - .4250000000x + 1.150000000$ with the command

```
>p:=interp([2,2.5,4],[0.5,0.4,0.25],x);
```

To evaluate $p(x)$ as an approximation to $f(3) = \frac{1}{3}$, enter

```
>subs(x=3,p);
```

which gives .3250000000.

The Lagrange polynomials have remainder terms that are reminiscent of those for the Taylor polynomials. The nth Taylor polynomial about x_0 concentrates all the known information at x_0 and has an error term of the form

$$\frac{f^{(n+1)}(\xi(x))}{(n+1)!}(x - x_0)^{n+1},$$

where $\xi(x)$ is between x and x_0. The nth Lagrange polynomial uses information at the distinct numbers $x_0, x_1, \ldots, x_n$. In place of $(x - x_0)^{n+1}$, its error formula uses a product of the $n + 1$ terms $(x - x_0), (x - x_1), \ldots, (x - x_n)$, and the number $\xi(x)$ can lie anywhere in the interval that contains the points $x_0, x_1, \ldots, x_n$, and x. Otherwise it has the same form as the error formula for the Taylor polynomials.

Lagrange Polynomial Error Formula

$$f(x) = P_n(x) + \frac{f^{(n+1)}(\xi(x))}{(n+1)!}(x - x_0)(x - x_1)\cdots(x - x_n),$$

for some number $\xi(x)$ between $x_0, x_1, \ldots, x_n$ and x.

This error formula is an important theoretical result, because Lagrange polynomials are used extensively for deriving numerical differentiation and integration methods. Error bounds for these techniques are obtained from the Lagrange error formula. The specific use of this error formula, however, is restricted to those functions whose derivatives have known bounds. The next example illustrates interpolation techniques for a situation in which the Lagrange error formula cannot be used. This example shows that we should look for a more efficient way to obtain approximations via interpolation.

EXAMPLE 2 Table 3.3 lists values of a function at various points. The approximations to $f(1.5)$ obtained by various Lagrange polynomials will be compared.

Table 3.3

x	$f(x)$
1.0	0.7651977
1.3	0.6200860
1.6	0.4554022
1.9	0.2818186
2.2	0.1103623

Since 1.5 is between 1.3 and 1.6, the most appropriate linear polynomial uses $x_0 = 1.3$ and $x_1 = 1.6$. The value of the interpolating polynomial at 1.5 is

$$P_1(1.5) = \frac{(1.5 - 1.6)}{(1.3 - 1.6)}(0.6200860) + \frac{(1.5 - 1.3)}{(1.6 - 1.3)}(0.4554022) = 0.5102968.$$

Two polynomials of degree 2 could reasonably be used, one by letting $x_0 = 1.3, x_1 = 1.6$, and $x_2 = 1.9$, which gives

$$\begin{aligned} P_2(1.5) = {} & \frac{(1.5 - 1.6)(1.5 - 1.9)}{(1.3 - 1.6)(1.3 - 1.9)}(0.6200860) + \frac{(1.5 - 1.3)(1.5 - 1.9)}{(1.6 - 1.3)(1.6 - 1.9)}(0.4554022) \\ & + \frac{(1.5 - 1.3)(1.5 - 1.6)}{(1.9 - 1.3)(1.9 - 1.6)}(0.2818186) \\ = {} & 0.5112857, \end{aligned}$$

and the other by letting $x_0 = 1.0, x_1 = 1.3$, and $x_2 = 1.6$, in which case

$$\hat{P}_2(1.5) = 0.5124715.$$

In the third-degree case there are also two reasonable choices for the polynomial. One uses $x_0 = 1.3, x_1 = 1.6, x_2 = 1.9$, and $x_3 = 2.2$, which gives $P_3(1.5) = 0.5118302$. The other is obtained by letting $x_0 = 1.0, x_1 = 1.3, x_2 = 1.6$, and $x_3 = 1.9$, giving $\hat{P}_3(1.5) = 0.5118127$.

The fourth Lagrange polynomial uses all the entries in the table. With $x_0 = 1.0, x_1 = 1.3, x_2 = 1.6, x_3 = 1.9$, and $x_4 = 2.2$, the approximation is $P_4(1.5) = 0.5118200$.

Since $P_3(1.5), \hat{P}_3(1.5)$, and $P_4(1.5)$ all agree to within 2×10^{-5}, we expect $P_4(1.5)$ to be the most accurate approximation and to be correct to within 2×10^{-5}. The actual value of $f(1.5)$ is known to be 0.5118277, so the true accuracies of the approximations are as follows:

$$|P_1(1.5) - f(1.5)| \approx 1.53 \times 10^{-3}, \qquad |P_2(1.5) - f(1.5)| \approx 5.42 \times 10^{-4},$$
$$|\hat{P}_2(1.5) - f(1.5)| \approx 6.44 \times 10^{-4}, \qquad |P_3(1.5) - f(1.5)| \approx 2.5 \times 10^{-6},$$
$$|\hat{P}_3(1.5) - f(1.5)| \approx 1.50 \times 10^{-5}, \qquad |P_4(1.5) - f(1.5)| \approx 7.7 \times 10^{-6}.$$

Although $P_3(1.5)$ is the most accurate approximation, if we had no knowledge of the actual value of $f(1.5)$, we would accept $P_4(1.5)$ as the best approximation, since it includes the most data about the function. The Lagrange error term cannot be applied here, since no knowledge of the fourth derivative of f is available. Unfortunately, this is generally the case. ◆ ◆ ◆

A practical difficulty with Lagrange interpolation is that since the error term is difficult to apply, the degree of the polynomial needed for the desired accuracy is generally not known until the computations are determined. The usual practice is to compute the results given from various polynomials until appropriate agreement is obtained, as was done in the previous example. However, the work done in calculating the approximation by the second polynomial does not lessen the work needed to calculate the third approximation; nor is the fourth approximation easier to obtain once the third approximation is known, and so on. To derive these approximating polynomials in a manner that uses the previous calculations to advantage, we need some new notation.

Let f be a function defined at $x_0, x_1, x_2, \ldots, x_n$ and suppose that $m_1, m_2, \ldots, m_k$ are k distinct integers with $0 \le m_i \le n$ for each i. The Lagrange polynomial that agrees with $f(x)$ at the k points $x_{m_1}, x_{m_2}, \ldots, x_{m_k}$ is denoted $P_{m_1,m_2,\ldots,m_k}(x)$.

EXAMPLE 3 Let $x_0 = 1, x_1 = 2, x_2 = 3, x_3 = 4, x_4 = 6$, and $f(x) = x^3$. Then $P_{1,2,4}(x)$ is the polynomial that agrees with $f(x)$ at $x_1 = 2, x_2 = 3$, and $x_4 = 6$, and

$$P_{1,2,4}(x) = \frac{(x-3)(x-6)}{(2-3)(2-6)}8 + \frac{(x-2)(x-6)}{(3-2)(3-6)}27 + \frac{(x-2)(x-3)}{(6-2)(6-3)}216.$$ ◆ ◆ ◆

The next result describes a method for recursively generating Lagrange polynomial approximations.

Recursively Generated Lagrange Polynomials

Let f be defined at $x_0, x_1, \ldots, x_k$ and x_j, x_i be two numbers in this set. If

$$P(x) = \frac{(x - x_j)P_{0,1,\ldots,j-1,j+1,\ldots,k}(x) - (x - x_i)P_{0,1,\ldots,i-1,i+1,\ldots,k}(x)}{(x_i - x_j)},$$

then $P(x)$ is the kth Lagrange polynomial that interpolates f at the $k + 1$ points $x_0, x_1, \ldots, x_k$.

To see why this recursive formula is true, first let $Q \equiv P_{0,1,\ldots,i-1,i+1,\ldots,k}$ and $\hat{Q} \equiv P_{0,1,\ldots,j-1,j+1,\ldots,k}$. Since $Q(x)$ and $\hat{Q}(x)$ are polynomials of degree at most $k - 1$,

$$P(x) = \frac{(x - x_j)\hat{Q}(x) - (x - x_i)Q(x)}{(x_i - x_j)}$$

must be of degree at most k. If $0 \le r \le k$ with $r \ne i$ and $r \ne j$, then $Q(x_r) = \hat{Q}(x_r) = f(x_r)$, so

$$P(x_r) = \frac{(x_r - x_j)\hat{Q}(x_r) - (x_r - x_i)Q(x_r)}{x_i - x_j} = \frac{(x_i - x_j)}{(x_i - x_j)}f(x_r) = f(x_r).$$

Moreover,

$$P(x_i) = \frac{(x_i - x_j)\hat{Q}(x_i) - (x_i - x_i)Q(x_i)}{x_i - x_j} = \frac{(x_i - x_j)}{(x_i - x_j)}f(x_i) = f(x_i),$$

and similarly, $P(x_j) = f(x_j)$. But there is only one polynomial of degree at most k that agrees with $f(x)$ at $x_0, x_1, \ldots, x_k$, and this polynomial by definition is $P_{0,1,\ldots,k}(x)$. Hence,

$$P_{0,1,\ldots,k}(x) = P(x) = \frac{(x - x_j)P_{0,1,\ldots,j-1,j+1,\ldots,k}(x) - (x - x_i)P_{0,1,\ldots,i-1,i+1,\ldots,k}(x)}{(x_i - x_j)}.$$

This result implies that the approximations from the interpolating polynomials can be generated recursively in the manner shown in Table 3.4. The row-by-row generation is performed to move across the rows as rapidly as possible, since these entries are given by successively higher-degree interpolating polynomials.

Table 3.4

x_0	$P_0 = Q_{0,0}$				
x_1	$P_1 = Q_{1,0}$	$P_{0,1} = Q_{1,1}$			
x_2	$P_2 = Q_{2,0}$	$P_{1,2} = Q_{2,1}$	$P_{0,1,2} = Q_{2,2}$		
x_3	$P_3 = Q_{3,0}$	$P_{2,3} = Q_{3,1}$	$P_{1,2,3} = Q_{3,2}$	$P_{0,1,2,3} = Q_{3,3}$	
x_4	$P_4 = Q_{4,0}$	$P_{3,4} = Q_{4,1}$	$P_{2,3,4} = Q_{4,2}$	$P_{1,2,3,4} = Q_{4,3}$	$P_{0,1,2,3,4} = Q_{4,4}$

This procedure is called **Neville's method** and is implemented with the program NEVLLE31.

The P notation used in Table 3.4 is cumbersome because of the number of subscripts used to represent the entries. Note, however, that as an array is being constructed, only two subscripts are needed. Proceeding down the table corresponds to using consecutive points x_i with larger i, and proceeding to the right corresponds to increasing the degree of the interpolating polynomial. Since the points appear consecutively in each entry, we need to describe only a starting point and the number of additional points used in constructing the approximation. To avoid the cumbersome subscripts we let $Q_{i,j}(x)$, for $0 \le j \le i$, denote the jth interpolating polynomial on the $j+1$ numbers $x_{i-j}, x_{i-j+1}, \ldots, x_{i-1}, x_i$, that is,

$$Q_{i,j} = P_{i-j,i-j+1,\ldots,i-1,i}.$$

Using this notation for Neville's method provides the Q notation in Table 3.4.

EXAMPLE 4 In Example 2, values of various interpolating polynomials at $x = 1.5$ were obtained using the data shown in the first two columns of Table 3.5. Suppose that we want to use Neville's method to calculate the approximation to $f(1.5)$. If $x_0 = 1.0$, $x_1 = 1.3$, $x_2 = 1.6$, $x_3 = 1.9$, and $x_4 = 2.2$, then $f(1.0) = Q_{0,0}$, $f(1.3) = Q_{1,0}$, $f(1.6) = Q_{2,0}$, $f(1.9) = Q_{3,0}$, and $f(2.2) = Q_{4,0}$; so these are the five polynomials of degree zero (constants) that approximate $f(1.5)$. Calculating the approximation $Q_{1,1}(1.5)$ gives

$$\begin{aligned} Q_{1,1}(1.5) &= \frac{(1.5-1.0)Q_{1,0} - (1.5-1.3)Q_{0,0}}{(1.3-1.0)} \\ &= \frac{0.5(0.6200860) - 0.2(0.7651977)}{0.3} = 0.5233449. \end{aligned}$$

Similarly,

$$Q_{2,1}(1.5) = \frac{(1.5-1.3)(0.4554022) - (1.5-1.6)(0.6200860)}{(1.6-1.3)} = 0.5102968,$$

$$Q_{3,1}(1.5) = 0.5132634, \quad \text{and} \quad Q_{4,1}(1.5) = 0.5104270.$$

The best linear approximation is expected to be $Q_{2,1}$, since 1.5 is between $x_1 = 1.3$ and $x_2 = 1.6$.

In a similar manner, the approximations using quadratic polynomials are given by

$$Q_{2,2}(1.5) = \frac{(1.5-1.0)(0.5102968) - (1.5-1.6)(0.5233449)}{(1.6-1.0)} = 0.5124715,$$

$$Q_{3,2}(1.5) = 0.5112857, \quad \text{and} \quad Q_{4,2}(1.5) = 0.5137361.$$

The higher-degree approximations are generated in a similar manner and are shown in Table 3.5. ◆ ◆ ◆

Table 3.5

1.0	0.7651977				
1.3	0.6200860	0.5233449			
1.6	0.4554022	0.5102968	0.5124715		
1.9	0.2818186	0.5132634	0.5112857	0.5118127	
2.2	0.1103623	0.5104270	0.5137361	0.5118302	0.5118200

If the latest approximation, $Q_{4,4}$, is not as accurate as desired, another node, x_5, can be selected and another row can be added to the table:

$$x_5 \quad Q_{5,0} \quad Q_{5,1} \quad Q_{5,2} \quad Q_{5,3} \quad Q_{5,4} \quad Q_{5,5}.$$

Then $Q_{4,4}$, $Q_{5,4}$, and $Q_{5,5}$ can be compared to determine further accuracy.

The function in Example 4 is the Bessel function of the first kind of order zero, whose value at 2.5 is -0.0483838, and a new row of approximations to $f(1.5)$ is

$$\begin{array}{cccc} 2.5 & -0.0483838 & 0.4807699 & 0.5301984 \\ & 0.5119070 & 0.5118430 & 0.5118277. \end{array}$$

The final new entry is correct to seven decimal places.

EXAMPLE 5 Table 3.6 lists the values of $f(x) = \ln x$ accurate to the places given.

Table 3.6

i	x_i	$\ln x_i$
0	2.0	0.6931
1	2.2	0.7885
2	2.3	0.8329

We use Neville's method to approximate $f(2.1) = \ln 2.1$. Completing the table gives the entries in Table 3.7.

Table 3.7

i	x_i	$x - x_i$	Q_{i0}	Q_{i1}	Q_{i2}
0	2.0	0.1	0.6931		
1	2.2	−0.1	0.7885	0.7410	
2	2.3	−0.2	0.8329	0.7441	0.7420

Thus, $P_2(2.1) = Q_{22} = 0.7420$. Since $f(2.1) = \ln 2.1 = 0.7419$ to four decimal places, the absolute error is

$$|f(2.1) - P_2(2.1)| = |0.7419 - 0.7420| = 10^{-4}.$$

However, $f'(x) = 1/x$, $f''(x) = -1/x^2$, and $f'''(x) = 2/x^3$, so the Lagrange error formula gives an error bound of

$$\begin{aligned} |f(2.1) - P_2(2.1)| &= \left| \frac{f'''(\xi)}{3!}(x - x_0)(x - x_1)(x - x_2) \right| \\ &= \left| \frac{1}{3\xi^3}(0.1)(-0.1)(-0.2) \right| \le 8.\overline{3} \times 10^{-5}. \end{aligned}$$

Notice that the actual error, 10^{-4}, exceeds the error bound, $8.\overline{3} \times 10^{-5}$. This apparent contradiction is a consequence of finite-digit computations. We used four-digit approximations, and the Lagrange error formula assumes infinite-digit arithmetic. This is what caused our actual errors to exceed the theoretical error estimate. ◆ ◆ ◆

EXERCISE SET 3.2

1. For the given functions $f(x)$, let $x_0 = 0$, $x_1 = 0.6$, and $x_2 = 0.9$. Construct the Lagrange interpolating polynomials of degree (i) at most 1 and (ii) at most 2 to approximate $f(0.45)$, and find the actual error.
   a. $f(x) = \cos x$ **b.** $f(x) = \sqrt{1+x}$
   c. $f(x) = \ln(x+1)$ **d.** $f(x) = \tan x$

2. Use the Lagrange polynomial error formula to find an error bound for the approximations in Exercise 1.

3. Use appropriate Lagrange interpolating polynomials of degrees 1, 2, and 3 to approximate each of the following:
   a. $f(8.4)$ if $f(8.1) = 16.94410$, $f(8.3) = 17.56492$, $f(8.6) = 18.50515$, $f(8.7) = 18.82091$
   b. $f\left(-\frac{1}{3}\right)$ if $f(-0.75) = -0.07181250$, $f(-0.5) = -0.02475000$, $f(-0.25) = 0.33493750$, $f(0) = 1.10100000$
   c. $f(0.25)$ if $f(0.1) = 0.62049958$, $f(0.2) = -0.28398668$, $f(0.3) = 0.00660095$, $f(0.4) = 0.24842440$
   d. $f(0.9)$ if $f(0.6) = -0.17694460$, $f(0.7) = 0.01375227$, $f(0.8) = 0.22363362$, $f(1.0) = 0.65809197$

4. Use Neville's method to obtain the approximations for Exercise 3.

5. Use Neville's method to approximate $\sqrt{3}$ with the function $f(x) = 3^x$ and the values $x_0 = -2$, $x_1 = -1$, $x_2 = 0$, $x_3 = 1$, and $x_4 = 2$.

6. Use Neville's method to approximate $\sqrt{3}$ with the function $f(x) = \sqrt{x}$ and the values $x_0 = 0$, $x_1 = 1$, $x_2 = 2$, $x_3 = 4$, and $x_4 = 5$. Compare the accuracy with that of Exercise 5.

7. The data for Exercise 3 were generated using the following functions. Use the error formula to find a bound for the error and compare the bound to the actual error for the cases $n = 1$ and $n = 2$.
   a. $f(x) = x \ln x$
   b. $f(x) = x^3 + 4.001x^2 + 4.002x + 1.101$
   c. $f(x) = x\cos x - 2x^2 + 3x - 1$
   d. $f(x) = \sin(e^x - 2)$

8. Use the Lagrange interpolating polynomial of degree 3 or less and four-digit chopping arithmetic to approximate $\cos 0.750$ using the following values. Find an error bound for the approximation.

$$\cos 0.698 = 0.7661 \quad \cos 0.733 = 0.7432$$
$$\cos 0.768 = 0.7193 \quad \cos 0.803 = 0.6946$$

The actual value of cos 0.750 is 0.7317 (to four decimal places). Explain the discrepancy between the actual error and the error bound.

9. Use the following values and four-digit rounding arithmetic to construct a third Lagrange polynomial approximation to $f(1.09)$. The function being approximated is $f(x) = \log_{10}(\tan x)$. Use this knowledge to find a bound for the error in the approximation.

$$f(1.00) = 0.1924 \quad f(1.05) = 0.2414 \quad f(1.10) = 0.2933 \quad f(1.15) = 0.3492$$

10. Repeat Exercise 9 using Maple and ten-digit rounding arithmetic.

11. Let $P_3(x)$ be the interpolating polynomial for the data $(0, 0)$, $(0.5, y)$, $(1, 3)$, and $(2, 2)$. Find y if the coefficient of x^3 in $P_3(x)$ is 6.

12. Neville's method is used to approximate $f(0.5)$, giving the following table.

$x_0 = 0$	$P_0 = 0$		
$x_1 = 0.4$	$P_1 = 2.8$	$P_{01} = 3.5$	
$x_2 = 0.7$	P_2	P_{12}	$P_{012} = \frac{27}{7}$

Determine $P_2 = f(0.7)$.

13. Suppose you need to construct eight-decimal-place tables for the common, or base-10, logarithm function from $x = 1$ to $x = 10$ in such a way that linear interpolation is accurate to within 10^{-6}. Determine a bound for the step size for this table. What choice of step size would you make to ensure that $x = 10$ is included in the table?

14. Suppose $x_j = j$ for $j = 0, 1, 2, 3$ and it is known that

$$P_{0,1}(x) = 2x + 1, \quad P_{0,2}(x) = x + 1, \quad \text{and} \quad P_{1,2,3}(2.5) = 3.$$

Find $P_{0,1,2,3}(2.5)$.

15. Neville's method is used to approximate $f(0)$ using $f(-2)$, $f(-1)$, $f(1)$, and $f(2)$. Suppose $f(-1)$ was overstated by 2 and $f(1)$ was understated by 3. Determine the error in the original calculation of the value of the interpolating polynomial to approximate $f(0)$.

16. The following table lists the population of the United States from 1940 to 1990.

Year	1940	1950	1960	1970	1980	1990
Population (in thousands)	132,165	151,326	179,323	203,302	226,542	249,633

Find the Lagrange polynomial of degree 5 fitting this data, and use this polynomial to estimate the population in the years 1930, 1965, and 2000. The population in 1930 was approximately 123,203,000. How accurate do you think your 1965 and 2000 figures are?

17. In Exercise 15 of Section 1.2 a Maclaurin series was integrated to approximate erf(1), where erf(x) is the normal distribution error function defined by

$$\operatorname{erf}(x) = \frac{2}{\sqrt{\pi}} \int_0^x e^{-t^2}\, dt.$$

a. Use the Maclaurin series to construct a table for erf(x) that is accurate to within 10^{-4} for erf(x_i), where $x_i = 0.2i$, for $i = 0, 1, \ldots, 5$.

b. Use both linear interpolation and quadratic interpolation to obtain an approximation to erf($\frac{1}{3}$). Which approach seems more feasible?

3.3 Divided Differences

Iterated interpolation was used in the previous section to generate successively higher degree polynomial approximations at a specific point. Divided-difference methods introduced in this section are used to successively generate the polynomials themselves.

We first need to introduce the divided-difference notation, which should remind you of the Aitken's Δ^2 notation used in Section 2.5. The **zeroth divided difference** of the function f with respect to x_i, $f[x_i]$, is simply the value of f at x_i:

$$f[x_i] = f(x_i).$$

The remaining divided differences are defined inductively. The **first divided difference** of f with respect to x_i and x_{i+1} is denoted $f[x_i, x_{i+1}]$ and is defined as

$$f[x_i, x_{i+1}] = \frac{f[x_{i+1}] - f[x_i]}{x_{i+1} - x_i}.$$

After the $(k-1)$st divided differences,

$$f[x_i, x_{i+1}, x_{i+2}, \ldots, x_{i+k-1}] \quad \text{and} \quad f[x_{i+1}, x_{i+2}, \ldots, x_{i+k-1}, x_{i+k}],$$

have been determined, the ***k*th divided difference** relative to $x_i, x_{i+1}, x_{i+2}, \ldots, x_{i+k}$ is

$$f[x_i, x_{i+1}, \ldots, x_{i+k-1}, x_{i+k}] = \frac{f[x_{i+1}, x_{i+2}, \ldots, x_{i+k}] - f[x_i, x_{i+1}, \ldots, x_{i+k-1}]}{x_{i+k} - x_i}.$$

With this notation, it can be shown that

$$\begin{aligned} P_n(x) = f[x_0] &+ f[x_0, x_1](x - x_0) \\ &+ f[x_0, x_1, x_2](x - x_0)(x - x_1) + \cdots \\ &+ f[x_0, x_1, \ldots, x_n](x - x_0)(x - x_1)\cdots(x - x_{n-1}). \end{aligned}$$

In simplified form we have the following.

Newton's Interpolatory Divided-Difference Formula

$$P(x) = f[x_0] + \sum_{k=1}^{n} f[x_0, x_1, \ldots, x_k](x - x_0) \cdots (x - x_{k-1}).$$

The generation of the divided differences is outlined in Table 3.8. Two fourth and one fifth difference could also be determined from these data, but they have not been recorded in the table.

Table 3.8

x	$f(x)$	First Divided Differences	Second Divided Differences	Third Divided Differences
x_0	$f[x_0]$			
		$f[x_0, x_1] = \frac{f[x_1]-f[x_0]}{x_1-x_0}$		
x_1	$f[x_1]$		$f[x_0, x_1, x_2] = \frac{f[x_1,x_2]-f[x_0,x_1]}{x_2-x_0}$	
		$f[x_1, x_2] = \frac{f[x_2]-f[x_1]}{x_2-x_1}$		$f[x_0, x_1, x_2, x_3] = \frac{f[x_1,x_2,x_3]-f[x_0,x_1,x_2]}{x_3-x_0}$
x_2	$f[x_2]$		$f[x_1, x_2, x_3] = \frac{f[x_2,x_3]-f[x_1,x_2]}{x_3-x_1}$	
		$f[x_2, x_3] = \frac{f[x_3]-f[x_2]}{x_3-x_2}$		$f[x_1, x_2, x_3, x_4] = \frac{f[x_2,x_3,x_4]-f[x_1,x_2,x_3]}{x_4-x_1}$
x_3	$f[x_3]$		$f[x_2, x_3, x_4] = \frac{f[x_3,x_4]-f[x_2,x_3]}{x_4-x_2}$	
		$f[x_3, x_4] = \frac{f[x_4]-f[x_3]}{x_4-x_3}$		$f[x_2, x_3, x_4, x_5] = \frac{f[x_3,x_4,x_5]-f[x_2,x_3,x_4]}{x_5-x_2}$
x_4	$f[x_4]$		$f[x_3, x_4, x_5] = \frac{f[x_4,x_5]-f[x_3,x_4]}{x_5-x_3}$	
		$f[x_4, x_5] = \frac{f[x_5]-f[x_4]}{x_5-x_4}$		
x_5	$f[x_5]$			

Divided-difference tables are easily constructed by hand calculation. Alternatively, the program DIVDIF32 computes the interpolating polynomial for f at $x_0, x_1, \ldots, x_n$. The form of the output can be modified to produce all the divided differences, as is done in the following example.

EXAMPLE 1 In the previous section we approximated the value at 1.5 given the data shown in the second and third columns of Table 3.9. The remaining entries of this table contain the divided differences computed using the program DIVDIF32.

The coefficients of the Newton forward divided-difference form of the interpolatory polynomial are along the upper diagonal in the table. The polynomial is

$$\begin{aligned} P_4(x) = {} & 0.7651977 - 0.4837057(x - 1.0) - 0.1087339(x - 1.0)(x - 1.3) \\ & + 0.0658784(x - 1.0)(x - 1.3)(x - 1.6) \\ & + 0.0018251(x - 1.0)(x - 1.3)(x - 1.6)(x - 1.9). \end{aligned}$$

Table 3.9

i	x_i	$f[x_i]$	$f[x_{i-1},x_i]$	$f[x_{i-2},x_{i-1},x_i]$	$f[x_{i-3},\dots,x_i]$	$f[x_{i-4},\dots,x_i]$
0	1.0	0.7651977				
			−0.4837057			
1	1.3	0.6200860		−0.1087339		
			−0.5489460		0.0658784	
2	1.6	0.4554022		−0.0494433		0.0018251
			−0.5786120		0.0680685	
3	1.9	0.2818186		0.0118183		
			−0.5715210			
4	2.2	0.1103623				

It is easily verified that $P_4(1.5) = 0.5118200$, which agrees with the result determined in Example 4 of Section 3.2. ◆ ◆ ◆

Newton's interpolatory divided-difference formula has a simpler form when $x_0, x_1, \dots, x_n$ are arranged consecutively with equal spacing. In this case, we introduce the notation $h = x_{i+1} - x_i$ for each $i = 0, 1, \dots, n-1$ and let $x = x_0 + sh$. Then the difference $x - x_i$ can be written as $x - x_i = (s-i)h$, and the divided-difference formula becomes

$$\begin{aligned} P_n(x) = P_n(x_0 + sh) &= f[x_0] + shf[x_0, x_1] + s(s-1)h^2 f[x_0, x_1, x_2] + \cdots \\ &\quad + s(s-1)\cdots(s-n+1)h^n f[x_0, x_1, \dots, x_n] \\ &= f[x_0] + \sum_{k=1}^{n} s(s-1)\cdots(s-k+1)h^k f[x_0, x_1, \dots, x_k]. \end{aligned}$$

Using a generalization of the binomial-coefficient notation,

$$\binom{s}{k} = \frac{s(s-1)\cdots(s-k+1)}{k!},$$

where s need not be an integer, we can express $P_n(x)$ compactly as follows.

Newton Forward Divided-Difference Formula

$$P_n(x) = P_n(x_0 + sh) = f[x_0] + \sum_{k=1}^{n} \binom{s}{k} k!\, h^k f[x_0, x_1, \dots, x_k].$$

Another form is constructed by making use of the forward difference notation introduced in Section 2.5. With this notation

$$\begin{aligned} f[x_0, x_1] &= \frac{f(x_1) - f(x_0)}{x_1 - x_0} = \frac{1}{h}\Delta f(x_0), \\ f[x_0, x_1, x_2] &= \frac{1}{2h}\left[\frac{\Delta f(x_1) - \Delta f(x_0)}{h}\right] = \frac{1}{2h^2}\Delta^2 f(x_0), \end{aligned}$$

and, in general,

$$f[x_0, x_1, \ldots, x_k] = \frac{1}{k!h^k}\Delta^k f(x_0).$$

This gives the following.

Newton Forward-Difference Formula

$$P_n(x) = f[x_0] + \sum_{k=1}^{n} \binom{s}{k} \Delta^k f(x_0).$$

If the interpolating nodes are written from last to first as $x_n, x_{n-1}, \ldots, x_0$, we can write the interpolatory formula as

$$\begin{aligned} P_n(x) = {} & f[x_n] + f[x_{n-1}, x_n](x - x_n) + f[x_{n-2}, x_{n-1}, x_n](x - x_n)(x - x_{n-1}) \\ & + \cdots + f[x_0, \ldots, x_n](x - x_n)(x - x_{n-1}) \cdots (x - x_1). \end{aligned}$$

If the nodes are equally spaced with $x = x_n + sh$ and $x = x_i + (s + n - i)h$, then

$$\begin{aligned} P_n(x) &= P_n(x_n + sh) \\ &= f[x_n] + shf[x_{n-1}, x_n] + s(s+1)h^2 f[x_{n-2}, x_{n-1}, x_n] + \cdots \\ &\quad + s(s+1)\cdots(s+n-1)h^n f[x_0, x_1, \ldots, x_n]. \end{aligned}$$

This form is called the *Newton backward divided-difference formula*. It is used to derive a formula known as the *Newton backward-difference formula*. To discuss this formula, we first need to introduce some notation.

Given the sequence $\{p_n\}_{n=0}^{\infty}$, the **backward difference** ∇p_n (read "nabla p_n") is defined by

$$\nabla p_n \equiv p_n - p_{n-1}, \quad \text{for } n \geq 1$$

and higher powers are defined recursively by

$$\nabla^k p_n = \nabla \left(\nabla^{k-1} p_n\right), \quad \text{for } k \geq 2.$$

This implies that

$$f[x_{n-1}, x_n] = \frac{1}{h}\nabla f(x_n), \quad f[x_{n-2}, x_{n-1}, x_n] = \frac{1}{2h^2}\nabla^2 f(x_n),$$

and, in general,

$$f[x_{n-k}, \ldots, x_{n-1}, x_n] = \frac{1}{k!h^k}\nabla^k f(x_n).$$

Consequently,

$$P_n(x) = f[x_n] + s\nabla f(x_n) + \frac{s(s+1)}{2}\nabla^2 f(x_n) + \cdots + \frac{s(s+1)\cdots(s+n-1)}{n!}\nabla^n f(x_n).$$

Extending the binomial-coefficient notation to include all real values of s, we let

$$\binom{-s}{k} = \frac{-s(-s-1)\cdots(-s-k+1)}{k!} = (-1)^k \frac{s(s+1)\cdots(s+k-1)}{k!}$$

and

$$P_n(x) = f(x_n) + (-1)^1\binom{-s}{1}\nabla f(x_n) + (-1)^2\binom{-s}{2}\nabla^2 f(x_n) + \cdots$$
$$+ (-1)^n\binom{-s}{n}\nabla^n f(x_n),$$

which gives the following result.

Newton Backward-Difference Formula

$$P_n(x) = f[x_n] + \sum_{k=1}^{n}(-1)^k\binom{-s}{k}\nabla^k f(x_n).$$

EXAMPLE 2 Consider the table of data given in Example 1 and reproduced in the first two columns of Table 3.10.

There is only one interpolating polynomial of degree at most 4 using these five data points, but we will organize the data points to obtain the best interpolation approximations of degrees 1, 2, and 3. This will give us a sense of the accuracy of the fourth-degree approximation for the given value of x.

If an approximation to $f(1.1)$ is required, the reasonable choice for $x_0, x_1, \ldots, x_4$ would be $x_0 = 1.0$, $x_1 = 1.3$, $x_2 = 1.6$, $x_3 = 1.9$, and $x_4 = 2.2$, since this choice makes the greatest possible use of the data points closest to $x = 1.1$ and also makes use of the fourth divided difference. These values imply that $h = 0.3$ and $s = \frac{1}{3}$, so the Newton forward divided-difference formula is used with the divided differences that are underlined in Table 3.10.

$$P_4(1.1) = P_4\left(1.0 + \frac{1}{3}(0.3)\right)$$
$$= 0.7651997 + \frac{1}{3}(0.3)(-0.4837057) + \frac{1}{3}\left(-\frac{2}{3}\right)(0.3)^2(-0.1087339)$$
$$+ \frac{1}{3}\left(-\frac{2}{3}\right)\left(-\frac{5}{3}\right)(0.3)^3(0.0658784)$$

$$+\frac{1}{3}\left(-\frac{2}{3}\right)\left(-\frac{5}{3}\right)\left(-\frac{8}{3}\right)(0.3)^4(0.0018251)$$
$$= 0.7196480.$$

Table 3.10

		First Divided Differences	Second Divided Differences	Third Divided Differences	Fourth Divided Differences
1.0	0.7651977				
		−0.4837057			
1.3	0.6200860		−0.1087339		
		−0.5489460		0.0658784	
1.6	0.4554022		−0.0494433		0.0018251
		−0.5786120		0.0680685	
1.9	0.2818186		0.0118183		
		−0.5715210			
2.2	0.1103623				

To approximate a value when x is close to the end of the tabulated values, say, $x = 2.0$, we would again like to make maximum use of the data points closest to x. To do so requires using the Newton backward divided-difference formula with $x_4 = 2.2$, $x_3 = 1.9$, $x_2 = 1.6$, $x_1 = 1.3$, $x_0 = 1.0$, $s = -\frac{2}{3}$ and the divided differences in Table 3.10 that are underlined with a dashed line:

$$P_4(2.0) = P_4\left(2.2 - \frac{2}{3}(0.3)\right)$$
$$= 0.1103623 - \frac{2}{3}(0.3)(-0.5715210) - \frac{2}{3}\left(\frac{1}{3}\right)(0.3)^2(0.0118183)$$
$$-\frac{2}{3}\left(\frac{1}{3}\right)\left(\frac{4}{3}\right)(0.3)^3(0.0680685) - \frac{2}{3}\left(\frac{1}{3}\right)\left(\frac{4}{3}\right)\left(\frac{7}{3}\right)(0.3)^4(0.0018251)$$
$$= 0.2238754.$$

◆ ◆ ◆

The Newton formulas are not appropriate for approximating $f(x)$ when x lies near the center of the table, since employing either the backward or forward method in such a way that the highest-order difference is involved will not allow x_0 to be close to x. A number of divided-difference formulas are available for this situation, each of which has situations when it can be used to maximum advantage. These methods are known as *centered-difference formulas*. There are a number of such methods, but we do not discuss any of these techniques. They can be found in many classical numerical analysis books, including the book by Hildebrand [Hi] that is listed in the bibliography.

EXERCISE SET 3.3

1. Use Newton's interpolatory divided-difference formula to construct interpolating polynomials of degrees 1, 2, and 3 for the following data. Approximate the specified value using each of the polynomials.
   a. $f(8.4)$ if $f(8.1) = 16.94410$, $f(8.3) = 17.56492$, $f(8.6) = 18.50515$, $f(8.7) = 18.82091$
   b. $f(0.9)$ if $f(0.6) = -0.17694460$, $f(0.7) = 0.01375227$, $f(0.8) = 0.22363362$, $f(1.0) = 0.65809197$

2. Use Newton's forward-difference formula to construct interpolating polynomials of degrees 1, 2, and 3 for the following data. Approximate the specified value using each of the polynomials.
   a. $f\left(-\frac{1}{3}\right)$ if $f(-0.75) = -0.07181250$, $f(-0.5) = -0.02475000$, $f(-0.25) = 0.33493750$, $f(0) = 1.10100000$
   b. $f(0.25)$ if $f(0.1) = -0.62049958$, $f(0.2) = -0.28398668$, $f(0.3) = 0.00660095$, $f(0.4) = 0.24842440$

3. Use Newton's backward-difference formula to construct interpolating polynomials of degrees 1, 2, and 3 for the following data. Approximate the specified value using each of the polynomials.
   a. $f\left(-\frac{1}{3}\right)$ if $f(-0.75) = -0.07181250$, $f(-0.5) = -0.02475000$, $f(-0.25) = 0.33493750$, $f(0) = 1.10100000$
   b. $f(0.25)$ if $f(0.1) = -0.62049958$, $f(0.2) = -0.28398668$, $f(0.3) = 0.00660095$, $f(0.4) = 0.24842440$

4. a. Construct the fourth interpolating polynomial for the unequally spaced points given in the following table:

x	$f(x)$
0.0	−6.00000
0.1	−5.89483
0.3	−5.65014
0.6	−5.17788
1.0	−4.28172

   b. Suppose $f(1.1) = -3.99583$ is added to the table. Construct the fifth interpolating polynomial.

5. a. Use the following data and the Newton forward divided-difference formula to approximate $f(0.05)$.

x	0.0	0.2	0.4	0.6	0.8
$f(x)$	1.00000	1.22140	1.49182	1.82212	2.22554

   b. Use the Newton backward divided-difference formula to approximate $f(0.65)$.

6. The following population table was given in Exercise 16 of Section 3.2.

Year	1940	1950	1960	1970	1980	1990
Population (in thousands)	132,165	151,326	179,323	203,302	226,542	249,633

Use an appropriate divided difference method to approximate each value.

a. The population in the year 1930.

b. The population in the year 2000.

7. Show that the polynomial interpolating the following data has degree 3.

x	-2	-1	0	1	2	3
$f(x)$	1	4	11	16	13	-4

8. a. Show that the Newton forward divided-difference polynomials

$$P(x) = 3 - 2(x+1) + 0(x+1)(x) + (x+1)(x)(x-1)$$

and

$$Q(x) = -1 + 4(x+2) - 3(x+2)(x+1) + (x+2)(x+1)(x)$$

both interpolate the data

x	-2	-1	0	1	2
$f(x)$	-1	3	1	-1	3

b. Why does part (a) not violate the uniqueness property of interpolating polynomials?

9. A fourth-degree polynomial $P(x)$ satisfies $\Delta^4 P(0) = 24$, $\Delta^3 P(0) = 6$, and $\Delta^2 P(0) = 0$, where $\Delta P(x) = P(x+1) - P(x)$. Compute $\Delta^2 P(10)$.

10. The following data are given for a polynomial $P(x)$ of unknown degree.

x	0	1	2
$P(x)$	2	-1	4

Determine the coefficient of x^2 in $P(x)$ if all third-order forward differences are 1.

11. The Newton forward divided-difference formula is used to approximate $f(0.3)$ given the following data.

x	0.0	0.2	0.4	0.6
$f(x)$	15.0	21.0	30.0	51.0

Suppose it is discovered that $f(0.4)$ was understated by 10 and $f(0.6)$ was overstated by 5. By what amount should the approximation to $f(0.3)$ be changed?

12. For a function f, the Newton's interpolatory divided-difference formula gives the interpolating polynomial

$$P_3(x) = 1 + 4x + 4x(x - 0.25) + \frac{16}{3}x(x - 0.25)(x - 0.5)$$

on the nodes $x_0 = 0$, $x_1 = 0.25$, $x_2 = 0.5$ and $x_3 = 0.75$. Find $f(0.75)$.

13. For a function f, the forward divided differences are given by

$x_0 = 0$	$f[x_0]$		
		$f[x_0, x_1]$	
$x_1 = 0.4$	$f[x_1]$		$f[x_0, x_1, x_2] = \frac{50}{7}$
		$f[x_1, x_2] = 10$	
$x_2 = 0.7$	$f[x_2] = 6$		

Determine the missing entries in the table.

3.4 Hermite Interpolation

The Lagrange polynomials agree with a function f at specified points. The values of f are often determined from observation, and in some situations it is possible to determine the derivative of f as well. This is likely to be the case, for example, if the independent variable is time and the function describes the position of an object. The derivative of the function in this case is the velocity, which might be available.

In this section we will consider Hermite interpolation, which determines a polynomial that agrees with the function and its first derivative at specified points. If $n+1$ data points and values of the function, $(x_0, f(x_0)), (x_1, f(x_1)), \ldots, (x_n, f(x_n))$, are given, then the Lagrange polynomial agreeing with f at these points will generally have degree n. If we require, in addition, that the derivative of the Hermite polynomial agree with the derivative of f at $x_0, x_1, \ldots, x_n$, then the additional $n + 1$ conditions raise the expected degree of the Hermite polynomial to $2n + 1$.

Hermite Polynomial

Suppose that $f \in C^1[a, b]$ and that $x_0, \ldots, x_n$ in $[a, b]$ are distinct. The unique polynomial of least degree agreeing with f and f' at $x_0, \ldots, x_n$ is the polynomial of degree at most $2n + 1$ given by

$$H_{2n+1}(x) = \sum_{j=0}^{n} f(x_j)H_{n,j}(x) + \sum_{j=0}^{n} f'(x_j)\hat{H}_{n,j}(x),$$

where

$$H_{n,j}(x) = [1 - 2(x - x_j)L'_{n,j}(x_j)]L^2_{n,j}(x)$$

and

$$\hat{H}_{n,j}(x) = (x - x_j)L^2_{n,j}(x).$$

Here, $L_{n,j}$ denotes the jth Lagrange coefficient polynomial of degree n.

The error term for the Hermite polynomial is similar to that of the Lagrange polynomial, with the only modifications being those needed to accommodate the increased amount of data used in the Hermite polynomial.

Hermite Polynomial Error Formula

If $f \in C^{(2n+2)}[a, b]$, then

$$f(x) = H_{2n+1}(x) + \frac{f^{(2n+2)}(\xi(x))}{(2n+2)!}(x - x_0)^2 \cdots (x - x_n)^2$$

for some $\xi(x)$ with $a < \xi(x) < b$.

Although the Hermite formula provides a complete description of the Hermite polynomials, the need to determine and evaluate the Lagrange polynomials and their derivatives makes the procedure tedious even for small values of n. An alternative method for generating Hermite approximations is based on the connection between the nth divided difference and the nth derivative of f.

Divided-Difference Relationship to the Derivative

If $f \in C^n[a, b]$ and $x_0, x_1, \ldots, x_n$ are distinct in $[a, b]$, then a number ξ in (a, b) exists with

$$f[x_0, x_1, \ldots, x_n] = \frac{f^{(n)}(\xi)}{n!}.$$

To use this result to generate the Hermite polynomial, first suppose the distinct numbers $x_0, x_1, \ldots, x_n$ are given together with the values of f and f' at these numbers. Define a new sequence $z_0, z_1, \ldots, z_{2n+1}$ by

$$z_{2k} = z_{2k+1} = x_k, \quad \text{for each } k = 0, 1, \ldots, n.$$

Now construct the divided-difference table using the variables $z_0, z_1, \ldots, z_{2n+1}$.

Since $z_{2k} = z_{2k+1} = x_k$ for each k, we cannot define $f[z_{2k}, z_{2k+1}]$ by the basic divided-difference relation:

$$f[z_{2k}, z_{2k+1}] = \frac{f[z_{2k+1}] - f[z_{2k}]}{z_{2k+1} - z_{2k}}.$$

But, for each k we have $f[x_k, x_{k+1}] = f'(\xi_k)$ for some number ξ_k in (x_k, x_{k+1}), and $\lim_{x_{k+1}\to x_k} f[x_k, x_{k+1}] = f'(x_k)$. So a reasonable substitution in this situation is $f[z_{2k}, z_{2k+1}] = f'(x_k)$, and we use the entries

$$f'(x_0), f'(x_1), \ldots, f'(x_n)$$

in place of the first divided differences

$$f[z_0, z_1], f[z_2, z_3], \ldots, f[z_{2n}, z_{2n+1}].$$

The remaining divided differences are produced as usual, and the appropriate divided differences are employed in Newton's interpolatory divided-difference formula.

Divided-Difference Form of the Hermite Polynomial

If $f \in C^1[a, b]$ and $x_0, x_1, \ldots, x_n$ are distinct in $[a, b]$, then

$$H_{2n+1}(x) = f[z_0] + \sum_{k=1}^{2n+1} f[z_0, z_1, \ldots, z_k](x - z_0), \ldots, (x - z_{k-1}),$$

where $z_{2k} = z_{2k+1} = x_k$ and $f[z_{2k}, z_{2k+1}] = f'(x_k)$ for each $k = 0, 1, \ldots, n$.

Table 3.11 shows the entries that are used for the first three divided-difference columns when determining the Hermite polynomial $H_5(x)$ for x_0, x_1, and x_2. The remaining entries are generated in the usual divided-difference manner.

Table 3.11

z	$f(z)$	First Divided Differences	Second Divided Differences
$z_0 = x_0$	$f[z_0] = f(x_0)$		
		$f[z_0, z_1] = f'(x_0)$	
$z_1 = x_0$	$f[z_1] = f(x_0)$		$f[z_0, z_1, z_2] = \frac{f[z_1,z_2]-f[z_0,z_1]}{z_2-z_0}$
		$f[z_1, z_2] = \frac{f[z_2]-f[z_1]}{z_2-z_1}$	
$z_2 = x_1$	$f[z_2] = f(x_1)$		$f[z_1, z_2, z_3] = \frac{f[z_2,z_3]-f[z_1,z_2]}{z_3-z_1}$
		$f[z_2, z_3] = f'(x_1)$	
$z_3 = x_1$	$f[z_3] = f(x_1)$		$f[z_2, z_3, z_4] = \frac{f[z_3,z_4]-f[z_2,z_3]}{z_4-z_2}$
		$f[z_3, z_4] = \frac{f[z_4]-f[z_3]}{z_4-z_3}$	
$z_4 = x_2$	$f[z_4] = f(x_2)$		$f[z_3, z_4, z_5] = \frac{f[z_4,z_5]-f[z_3,z_4]}{z_5-z_3}$
		$f[z_4, z_5] = f'(x_2)$	
$z_5 = x_2$	$f[z_5] = f(x_2)$		

EXAMPLE 1 The entries in Table 3.12 use the data in the examples we have previously considered, together with known values of the derivative. The underlined entries are the given data; the remainder are generated by the standard divided-difference method. They produce the following approximation to $f(1.5)$.

$$\begin{aligned} H_5(1.5) &= 0.6200860 - 0.5220232(1.5-1.3) - 0.0897427(1.5-1.3)^2 \\ &\quad + 0.0663657(1.5-1.3)^2(1.5-1.6) + 0.0026663(1.5-1.3)^2(1.5-1.6)^2 \\ &\quad - 0.0027738(1.5-1.3)^2(1.5-1.6)^2(1.5-1.9) \\ &= 0.5118277. \end{aligned}$$

◆ ◆ ◆

Table 3.12

1.3	0.6200860					
		−0.5220232				
1.3	0.6200860		−0.0897427			
		−0.5489460		0.0663657		
1.6	0.4554022		−0.0698330		0.0026663	
		−0.5698959		0.0679655		−0.0027738
1.6	0.4554022		−0.0290537		0.0010020	
		−0.5786120		0.0685667		
1.9	0.2818186		−0.0084837			
		−0.5811571				
1.9	0.2818186					

The program HERMIT33 generates the coefficients for the Hermite polynomials using this modified Newton interpolatory divided-difference formula. The structure of the program is slightly different from the discussion, to take advantage of efficiency of computation.

EXERCISE SET 3.4

1. Use Hermite interpolation to construct an approximating polynomial for the following data.

a.

x	$f(x)$	$f'(x)$
8.3	17.56492	3.116256
8.6	18.50515	3.151762

b.

x	$f(x)$	$f'(x)$
0.8	0.22363362	2.1691753
1.0	0.65809197	2.0466965

c.

x	$f(x)$	$f'(x)$
−0.5	−0.0247500	0.7510000
−0.25	0.3349375	2.1890000
0	1.1010000	4.0020000

d.

x	$f(x)$	$f'(x)$
0.1	−0.62049958	3.58502082
0.2	−0.28398668	3.14033271
0.3	0.00660095	2.66668043
0.4	0.24842440	2.16529366

2. The data in Exercise 1 were generated using the following functions. For the given value of x, use the polynomials constructed in Exercise 1 to approximate $f(x)$, and calculate the actual error.

a. $f(x) = x \ln x$; approximate $f(8.4)$.

b. $f(x) = \sin(e^x - 2)$; approximate $f(0.9)$.

c. $f(x) = x^3 + 4.001x^2 + 4.002x + 1.101$; approximate $f(-\frac{1}{3})$.

d. $f(x) = x \cos x - 2x^2 + 3x - 1$; approximate $f(0.25)$.

3. a. Use the following values and five-digit rounding arithmetic to construct the Hermite interpolating polynomial to approximate $\sin 0.34$.

x	$\sin x$	$D_x \sin x = \cos x$
0.30	0.29552	0.95534
0.32	0.31457	0.94924
0.35	0.34290	0.93937

b. Determine an error bound for the approximation in part (a) and compare to the actual error.

c. Add $\sin 0.33 = 0.32404$ and $\cos 0.33 = 0.94604$ to the data and redo the calculations.

4. Let $f(x) = 3xe^x - e^{2x}$.

a. Approximate $f(1.03)$ by the Hermite interpolating polynomial of degree at most 3 using $x_0 = 1$ and $x_1 = 1.05$. Compare the actual error to the error bound.

b. Repeat (a) with the Hermite interpolating polynomial of degree at most 5, using $x_0 = 1, x_1 = 1.05$, and $x_2 = 1.07$.

5. Use the error formula and Maple to find a bound for the errors in the approximations of $f(x)$ in parts (a) and (c) of Exercise 2.

6. The following table lists data for the function described by $f(x) = e^{0.1x^2}$. Approximate $f(1.25)$ by using $H_5(1.25)$ and $H_3(1.25)$, where H_5 uses the nodes $x_0 = 1, x_1 = 2$, and $x_2 = 3$ and H_3 uses the nodes $\bar{x}_0 = 1$ and $\bar{x}_1 = 1.5$. Find error bounds for these approximations.

x	$f(x) = e^{0.1x^2}$	$f'(x) = 0.2xe^{0.1x^2}$
$x_0 = \bar{x}_0 = 1$	1.105170918	0.2210341836
$\bar{x}_1 = 1.5$	1.252322716	0.3756968148
$x_1 = 2$	1.491824698	0.5967298792
$x_2 = 3$	2.459603111	1.475761867

7. A car traveling along a straight road is clocked at a number of points. The data from the observations are given in the following table, where the time is in seconds, the distance is in feet, and the speed is in feet per second.

Time	0	3	5	8	13
Distance	0	225	383	623	993
Speed	75	77	80	74	72

a. Use a Hermite polynomial to predict the position of the car and its speed when $t = 10$ s.

b. Use the derivative of the Hermite polynomial to determine whether the car ever exceeds a 55-mi/h speed limit on the road. If so, what is the first time the car exceeds this speed?

c. What is the predicted maximum speed for the car?

8. Let $z_0 = x_0$, $z_1 = x_0$, $z_2 = x_1$, and $z_3 = x_1$. Form the following divided-difference table.

$z_0 = x_0$	$f[z_0] = f(x_0)$			
		$f[z_0, z_1] = f'(x_0)$		
$z_1 = x_0$	$f[z_1] = f(x_0)$		$f[z_0, z_1, z_2]$	
		$f[z_1, z_2]$		$f[z_0, z_1, z_2, z_3]$
$z_2 = x_1$	$f[z_2] = f(x_1)$		$f[z_1, z_2, z_3]$	
		$f[z_2, z_3] = f'(x_1)$		
$z_3 = x_1$	$f[z_3] = f(x_1)$			

Show if

$$P(x) = f[z_0] + f[z_0, z_1](x - x_0) + f[z_0, z_1, z_2](x - x_0)^2 + f[z_0, z_1, z_2, z_3](x - x_0)^2(x - x_1),$$

then

$$P(x_0) = f(x_0),\ P(x_1) = f(x_1),\ P'(x_0) = f'(x_0),\ \text{and } P'(x_1) = f'(x_1),$$

which implies that $P(x) \equiv H_3(x)$.

3.5 Spline Interpolation

The previous sections use polynomials to approximate arbitrary functions. However, we have seen that relatively high-degree polynomials are needed for accurate approximation and that these have some serious disadvantages. They all have an oscillatory nature, and a fluctuation over a small portion of the interval can induce large fluctuations over the entire range.

An alternative approach is to divide the interval into a collection of subintervals and construct a different approximating polynomial on each subinterval. This is called **piecewise polynomial approximation**.

The simplest piecewise polynomial approximation consists of joining a set of data points $(x_0, f(x_0)), (x_1, f(x_1)), \ldots, (x_n, f(x_n))$ by a series of straight lines, such as those shown in Figure 3.6.

A disadvantage of linear approximation is that the approximation is generally not differentiable at the endpoints of the subintervals, so the interpolating function is not

Figure 3.6

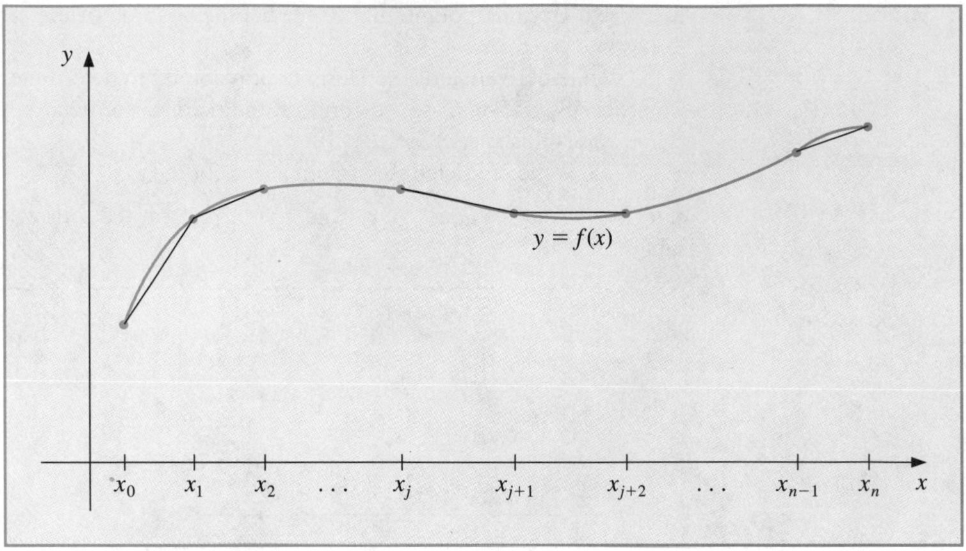

"smooth" at these points. It is often clear from physical conditions that smoothness is required, and the approximating function must be continuously differentiable.

One remedy for this problem is to use a piecewise polynomial of Hermite type. For example, if the values of f and f' are known at each of the points $x_0 < x_1 < \cdots < x_n$, a cubic Hermite polynomial can be used on each of the subintervals $[x_0, x_1], [x_1, x_2], \ldots, [x_{n-1}, x_n]$ to obtain an approximating function that has a continuous derivative on the interval $[x_0, x_n]$. To determine the appropriate cubic Hermite polynomial on a given interval, we simply compute the function $H_3(x)$ for that interval.

Figure 3.7

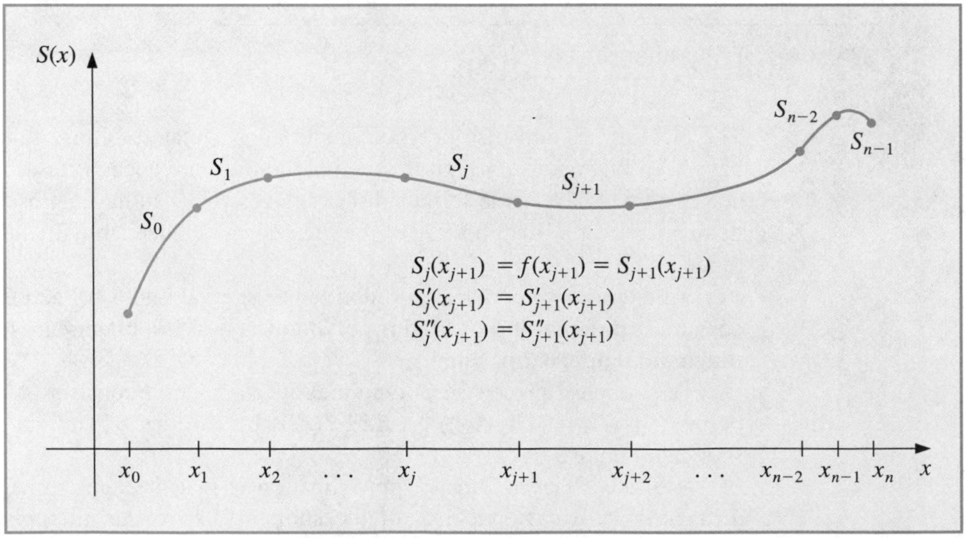

The Hermite polynomials are commonly used in application problems to study the motion of particles in space. The difficulty with using Hermite piecewise polynomials for general interpolation problems concerns the need to know the derivative of the function being approximated. The remainder of this section considers approximation using piecewise polynomials that require no derivative information, except perhaps at the endpoints of the interval on which the function is being approximated.

The most common piecewise polynomial approximation uses cubic polynomials between pairs of nodes and is called **cubic spline** interpolation. A general cubic polynomial involves four constants; so there is sufficient flexibility in the cubic spline procedure to ensure that the interpolant has two continuous derivatives on the interval. The derivatives of the cubic spline do not, in general, however, agree with the derivatives of the function, even at the nodes. (See Figure 3.7.)

Cubic Spline Interpolation

Given a function f defined on $[a, b]$ and a set of nodes, $a = x_0 < x_1 < \cdots < x_n = b$, a cubic spline interpolant, S, for f is a function that satisfies the following conditions:

(a) For each $j = 0, 1, \ldots, n-1$, $S(x)$ is a cubic polynomial, denoted by $S_j(x)$, on the subinterval $[x_j, x_{j+1})$.
(b) $S_j(x_j) = f(x_j)$ for each $j = 0, 1, \ldots, n$.
(c) $S_{j+1}(x_{j+1}) = S_j(x_{j+1})$ for each $j = 0, 1, \ldots, n-2$.
(d) $S'_{j+1}(x_{j+1}) = S'_j(x_{j+1})$ for each $j = 0, 1, \ldots, n-2$.
(e) $S''_{j+1}(x_{j+1}) = S''_j(x_{j+1})$ for each $j = 0, 1, \ldots, n-2$.
(f) One of the following sets of boundary conditions is satisfied:
 (i) $S''(x_0) = S''(x_n) = 0$ (natural or free boundary);
 (ii) $S'(x_0) = f'(x_0)$ and $S'(x_n) = f'(x_n)$ (clamped boundary).

Although cubic splines are defined with other boundary conditions, the conditions given in (f) are used most frequently in practice. When the natural boundary conditions are used, the spline assumes the shape that a long flexible rod would take if forced to go through the points $\{(x_0, f(x_0)), (x_1, f(x_1)), \ldots, (x_n, f(x_n))\}$. This spline continues linearly when $x \le x_0$ and when $x \ge x_n$.

In general, clamped boundary conditions lead to more accurate approximations since they include more information about the function. However, for this type of boundary condition, we need values of the derivative at the endpoints or an accurate approximation to those values.

To construct the cubic spline interpolant for a given function f, the conditions in the definition are applied to the cubic polynomials

$$S_j(x) = a_j + b_j(x - x_j) + c_j(x - x_j)^2 + d_j(x - x_j)^3$$

for each $j = 0, 1, \ldots, n-1$.

Since

$$S_j(x_j) = a_j = f(x_j),$$

condition (c) can be applied to obtain

$$a_{j+1} = S_{j+1}(x_{j+1}) = S_j(x_{j+1}) = a_j + b_j(x_{j+1} - x_j) + c_j(x_{j+1} - x_j)^2 + d_j(x_{j+1} - x_j)^3$$

for each $j = 0, 1, \ldots, n-2$.

Since the term $x_{j+1} - x_j$ is used repeatedly in this development, it is convenient to introduce the simpler notation

$$h_j = x_{j+1} - x_j,$$

for each $j = 0, 1, \ldots, n-1$. If we also define $a_n = f(x_n)$, then the equation

$$a_{j+1} = a_j + b_j h_j + c_j h_j^2 + d_j h_j^3 \tag{3.1}$$

holds for each $j = 0, 1, \ldots, n-1$.

In a similar manner, define $b_n = S'(x_n)$ and observe that

$$S_j'(x) = b_j + 2c_j(x - x_j) + 3d_j(x - x_j)^2 \tag{3.2}$$

implies that $S_j'(x_j) = b_j$ for each $j = 0, 1, \ldots, n-1$. Applying condition (d) gives

$$b_{j+1} = b_j + 2c_j h_j + 3d_j h_j^2, \tag{3.3}$$

for each $j = 0, 1, \ldots, n-1$.

Another relation between the coefficients of S_j is obtained by defining $c_n = S''(x_n)/2$ and applying condition (e). In this case,

$$c_{j+1} = c_j + 3d_j h_j, \tag{3.4}$$

for each $j = 0, 1, \ldots, n-1$.

Solving for d_j in Eq. (3.4) and substituting this value into Eqs. (3.1) and (3.3) gives the new equations

$$a_{j+1} = a_j + b_j h_j + \frac{h_j^2}{3}(2c_j + c_{j+1}) \tag{3.5}$$

and

$$b_{j+1} = b_j + h_j(c_j + c_{j+1}) \tag{3.6}$$

for each $j = 0, 1, \ldots, n-1$.

The final relationship involving the coefficients is obtained by solving the appropriate equation in the form of Eq. (3.5) for b_j,

$$b_j = \frac{1}{h_j}(a_{j+1} - a_j) - \frac{h_j}{3}(2c_j + c_{j+1}), \tag{3.7}$$

and then, with a reduction of the index, for b_{j-1}, which gives

$$b_{j-1} = \frac{1}{h_{j-1}}(a_j - a_{j-1}) - \frac{h_{j-1}}{3}(2c_{j-1} + c_j).$$

Substituting these values into the equation derived from Eq. (3.6), when the index is reduced by 1, gives the linear system of equations

$$h_{j-1}c_{j-1} + 2(h_{j-1} + h_j)c_j + h_jc_{j+1} = \frac{3}{h_j}(a_{j+1} - a_j) - \frac{3}{h_{j-1}}(a_j - a_{j-1}) \tag{3.8}$$

for each $j = 1, 2, \ldots, n-1$. This system involves only $\{c_j\}_{j=0}^n$ as unknowns since the values of $\{h_j\}_{j=0}^{n-1}$ and $\{a_j\}_{j=0}^n$ are given by the spacing of the nodes $\{x_j\}_{j=0}^n$ and the values $\{f(x_j)\}_{j=0}^n$.

Once the values of $\{c_j\}_{j=0}^n$ are determined, it is a simple matter to find the remainder of the constants $\{b_j\}_{j=0}^{n-1}$ from Eq. (3.7) and $\{d_j\}_{j=0}^{n-1}$ from Eq. (3.4) and to construct the cubic polynomials $\{S_j(x)\}_{j=0}^{n-1}$. In the case of the clamped spline, we also need equations involving the $\{c_j\}$ that ensure that $S'(x_0) = f'(x_0)$ and $S'(x_n) = f'(x_n)$. In Eq. (3.2) we have $S_j'(x)$ in terms of b_j, c_j, and d_j. Since we now know b_j and d_j in terms of c_j, we can use this equation to show that the appropriate equations are

$$2h_0c_0 + h_0c_1 = \frac{3}{h_0}(a_1 - a_0) - 3f'(x_0) \tag{3.9}$$

and

$$h_{n-1}c_{n-1} + 2h_{n-1}c_n = 3f'(x_n) - \frac{3}{h_{n-1}}(a_n - a_{n-1}). \tag{3.10}$$

The solution to the cubic spline problem with the natural boundary conditions $S''(x_0) = S''(x_n) = 0$ can be obtained by applying the program NCUBSP34. The program CCUBSP35 determines the cubic spline with the clamped boundary conditions $S'(x_0) = f'(x_0)$ and $S'(x_n) = f'(x_n)$.

EXAMPLE 1 Determine the clamped cubic spline for $f(x) = x \sin 4x$ using the nodes $x_0 = 0$, $x_1 = 0.25$, $x_2 = 0.4$, and $x_3 = 0.6$.

We first define the function by

```
>f:=y->y*sin(4*y);
```

We use Maple to do the calculations. We first define the nodes and step sizes as

```
>x[0]:=0; x[1]:=0.25; x[2]:=0.4; x[3]:=0.6;
```

and

```
>h[0]:=x[1]-x[0]; h[1]:=x[2]-x[1]; h[2]:=x[3]-x[2];
```

The function is evaluated at the nodes with the commands

```
>a[0]:=f(x[0]); a[1]:=f(x[1]); a[2]:=f(x[2]); a[3]:=f(x[3]);
```

Eq. (3.8) with $j = 1$ and $j = 2$ gives us the following:

```
>eq1:=h[0]*c[0]+2*(h[0]+h[1])*c[1]+h[1]*c[2]
 =2*(a[2]-a[1])/h[1]-3*(a[1]-a[0])/h[0];
>eq2:=h[1]*c[1]+2*(h[1]+h[2])*c[2]+h[2]*c[3]
 =3*(a[3]-a[2])/h[2]-3*(a[2]-a[1]/h[1];
```

Since we are constructing a clamped spline, we need the derivative $f'(x) \equiv fp(x)$ at the endpoints 0 and 0.6, so we also define

```
>fp:=y->D(f)(y);
```

and the fact that $f'(x_0) = s'(x_0)$ and $f'(x_3) = s'(x_3)$ to obtain two equations

```
>eq0:=2*h[0]*c[0]+h[0]*c[1]=3*(a[1]-a[0])/h[0]-3*fp(x[0]);
>eq3:=h[2]*c[2]+2*h[2]*c[3]=3*fp(x[3])-3*(a[3]-a[2])/h[2];
```

The Maple function `solve` is used to obtain values for $c_0 = c[0]$, $c_1 = c[1]$, $c_2 = c[2]$, and $c_3 = c[3]$.

```
>g:=solve({eq0,eq1,eq2,eq3},{c[0],c[1],c[2],c[3]});
```

The values produced by this Maple command are then assigned using

```
>c[0]:=4.649673229; c[1]:=.7983053588; c[2]:=-3.574944315;
 c[3]:=-6.623958383;
```

Then we use equations (3.7) and (3.4) to obtain

```
>b[0]:=evalf((a[1]-a[0])/h[0]-h[0]*(2*c[0]+c[1])/3);
>b[1]:=(a[2]-a[1])/h[1]-h[1]*(2*c[1]+c[2])/3;
>b[2]:=(a[3]-a[2])/h[2]-h[2]*(2*c[2]+c[3])/3;
>d[0]:=(c[1]-c[0])/(3*h[0]);
>d[1]:=(c[2]-c[1])/(3*h[1]);
>d[2]:=(c[3]-c[2])/(3*h[2]);
```

The three pieces of the spline are now defined by

```
>s1:=y->a[0]+b[0]*(y-x[0])+c[0]*(y-x[0])^2+d[0]*(y-x[0])^3;
>s2:=y->a[1]+b[1]*(y-x[1])+c[1]*(y-x[1])^2+d[1]*(y-x[1])^3;
>s3:=y->a[2]+b[2]*(y-x[2])+c[2]*(y-x[2])^2+d[2]*(y-x[2])^3;
```

Table 3.13

j	x_j	a_j	b_j	c_j	d_j
0	0	0	0.82×10^{-8}	4.650	−5.135
1	0.25	0.210	1.362	0.798	−9.718
2	0.4	0.400	0.945	−3.575	−5.082
3	0.6	0.405	—	−6.624	—

The values of the coefficients to three decimal places are listed in Table 3.13.

◆ ◆ ◆

EXAMPLE 2 Figure 3.8 shows a ruddy duck in flight.

Figure 3.8

We have chosen points along the top profile of the duck through which we want an approximating curve to pass. Table 3.14 lists the coordinates of 21 data points relative to the superimposed coordinate system shown in Figure 3.9 on page 94.

Table 3.14

x	0.9	1.3	1.9	2.1	2.6	3.0	3.9	4.4	4.7	5.0	6.0
$f(x)$	1.3	1.5	1.85	2.1	2.6	2.7	2.4	2.15	2.05	2.1	2.25
x	7.0	8.0	9.2	10.5	11.3	11.6	12.0	12.6	13.0	13.3	
$f(x)$	2.3	2.25	1.95	1.4	0.9	0.7	0.6	0.5	0.4	0.25	

Notice that more points are used when the curve is changing rapidly than when it is changing more slowly.

Figure 3.9

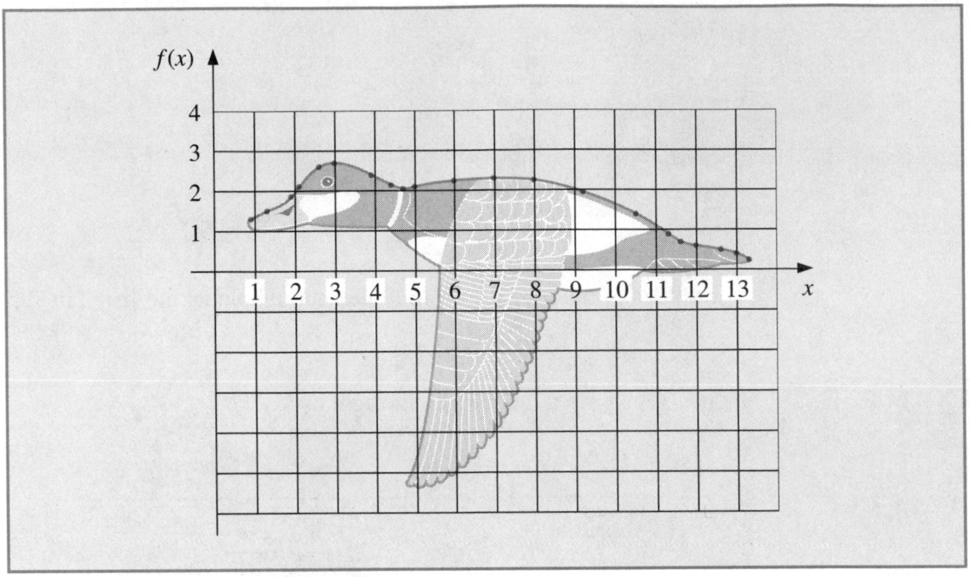

Using NCUBSP34 to generate the natural cubic spline for this data produces the coefficients shown in Table 3.15. This spline curve is nearly identical to the profile, as shown in Figure 3.10.

Table 3.15

j	x_j	a_j	b_j	c_j	d_j
0	0.9	1.3	5.40	0.00	−0.25
1	1.3	1.5	0.42	−0.30	0.95
2	1.9	1.85	1.09	1.41	−2.96
3	2.1	2.1	1.29	−0.37	−0.45
4	2.6	2.6	0.59	−1.04	0.45
5	3.0	2.7	−0.02	−0.50	0.17
6	3.9	2.4	−0.50	−0.03	0.08
7	4.4	2.15	−0.48	0.08	1.31
8	4.7	2.05	−0.07	1.27	−1.58
9	5.0	2.1	0.26	−0.16	0.04
10	6.0	2.25	0.08	−0.03	0.00
11	7.0	2.3	0.01	−0.04	−0.02
12	8.0	2.25	−0.14	−0.11	0.02
13	9.2	1.95	−0.34	−0.05	−0.01
14	10.5	1.4	−0.53	−0.10	−0.02
15	11.3	0.9	−0.73	−0.15	1.21
16	11.6	0.7	−0.49	0.94	−0.84
17	12.0	0.6	−0.14	−0.06	0.03
18	12.6	0.5	−0.18	0.00	−0.43
19	13.0	0.4	−0.39	−0.52	0.49
20	13.3	0.25			

Figure 3.10

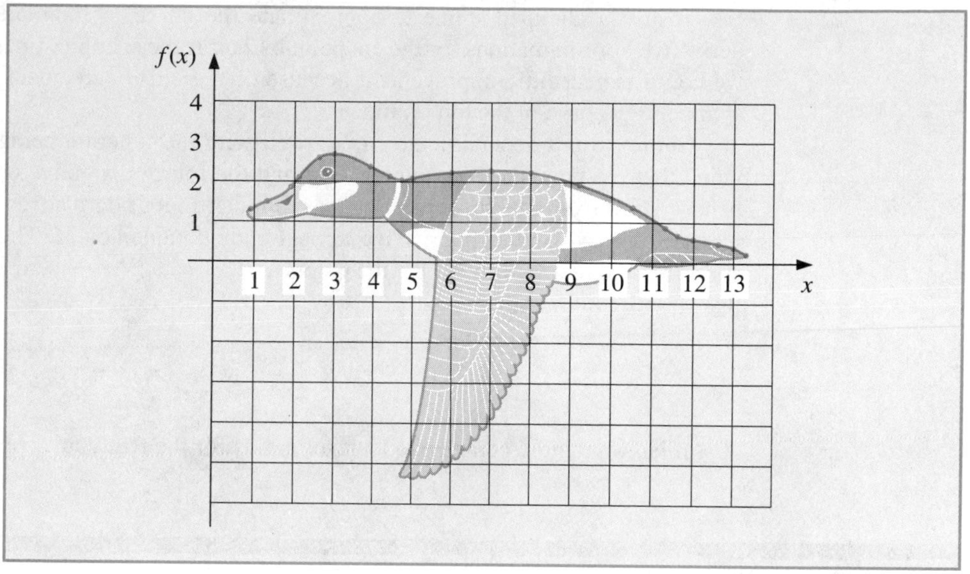

For comparison purposes, Figure 3.11 gives an illustration of the curve generated using a Lagrange interpolating polynomial to fit this same data. This produces a very strange illustration of the back of a duck, in flight or otherwise. The interpolating polynomial in this case is of degree 20 and oscillates wildly. ◆ ◆ ◆

Figure 3.11

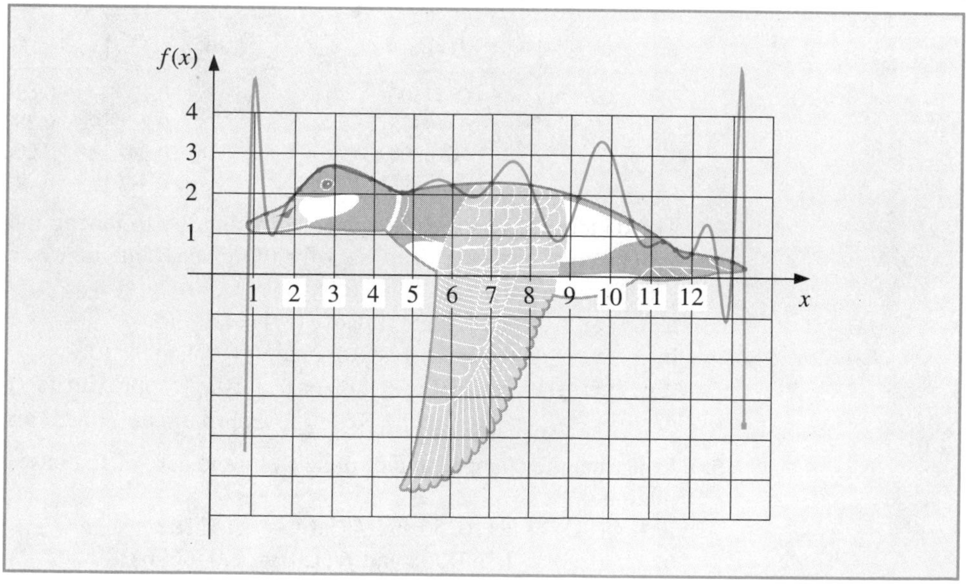

To use a clamped spline to approximate the curve in Example 2, we would need derivative approximations for the endpoints. Even if these approximations were available, we could expect little improvement because of the close agreement of the natural cubic spline to the curve of the top profile.

Cubic splines generally agree quite well with the function being approximated, provided that the points are not too far apart and the fourth derivative of the function is well behaved. For example, suppose that f has four continuous derivatives on $[a, b]$ and that the fourth derivative on this interval has a magnitude bounded by M. Then the clamped cubic spline $S(x)$ agreeing with $f(x)$ at the points $a = x_0 < x_1 < \cdots < x_n = b$ has the property that for all x in $[a, b]$,

$$|f(x) - S(x)| \leq \frac{5M}{384} \max_{0 \leq j \leq n-1} (x_{j+1} - x_j)^4. \tag{3.11}$$

A similar—but more complicated—result holds for the free cubic splines.

EXERCISE SET 3.5

1. Determine the free cubic spline S that interpolates the data $f(0) = 0$, $f(1) = 1$, and $f(2) = 2$.
2. Determine the clamped cubic spline s that interpolates the data $f(0) = 0$, $f(1) = 1$, $f(2) = 2$ and satisfies $s'(0) = s'(2) = 1$.
3. Construct the free cubic spline for the following data.

a.

x	$f(x)$
8.3	17.56492
8.6	18.50515

b.

x	$f(x)$
0.8	0.22363362
1.0	0.65809197

c.

x	$f(x)$
−0.5	−0.0247500
−0.25	0.3349375
0	1.1010000

d.

x	$f(x)$
0.1	−0.62049958
0.2	−0.28398668
0.3	0.00660095
0.4	0.24842440

4. The data in Exercise 3 were generated using the following functions. Use the cubic splines constructed in Exercise 3 for the given value of x to approximate $f(x)$ and $f'(x)$, and calculate the actual error.
   - **a.** $f(x) = x \ln x$; approximate $f(8.4)$ and $f'(8.4)$.
   - **b.** $f(x) = \sin(e^x - 2)$; approximate $f(0.9)$ and $f'(0.9)$.
   - **c.** $f(x) = x^3 + 4.001x^2 + 4.002x + 1.101$; approximate $f(-\frac{1}{3})$ and $f'(-\frac{1}{3})$.
   - **d.** $f(x) = x \cos x - 2x^2 + 3x - 1$; approximate $f(0.25)$ and $f'(0.25)$.
5. Construct the clamped cubic spline using the data of Exercise 3 and the given information.
   - **a.** $f'(8.3) = 3.116256$ and $f'(8.6) = 3.151762$
   - **b.** $f'(0.8) = 2.1691753$ and $f'(1.0) = 2.0466965$

c. $f'(-0.5) = 0.7510000$ and $f'(0) = 4.0020000$
d. $f'(0.1) = 3.58502082$ and $f'(0.4) = 2.16529366$

6. Repeat Exercise 4 using the cubic splines constructed in Exercise 5.

7. **a.** Construct a free cubic spline to approximate $f(x) = \cos \pi x$ by using the values given by $f(x)$ at $x = 0, 0.25, 0.5, 0.75$, and 1.0.
b. Integrate the spline over $[0, 1]$, and compare the result to $\int_0^1 \cos \pi x \, dx = 0$.
c. Use the derivatives of the spline to approximate $f'(0.5)$ and $f''(0.5)$, and compare these approximations to the actual values.

8. **a.** Construct a free cubic spline to approximate $f(x) = e^{-x}$ by using the values given by $f(x)$ at $x = 0, 0.25, 0.75$, and 1.0.
b. Integrate the spline over $[0, 1]$, and compare the result to $\int_0^1 e^{-x} \, dx = 1 - 1/e$.
c. Use the derivatives of the spline to approximate $f'(0.5)$ and $f''(0.5)$, and compare the approximations to the actual values.

9. Repeat Exercise 7, constructing instead the clamped cubic spline with $f'(0) = f'(1) = 0$.

10. Repeat Exercise 8, constructing instead the clamped cubic spline with $f'(0) = -1, f'(1) = -e^{-1}$.

11. A natural cubic spline S on $[0, 2]$ is defined by

$$S(x) = \begin{cases} S_0(x) = 1 + 2x - x^3, & \text{if } 0 \le x < 1, \\ S_1(x) = a + b(x-1) + c(x-1)^2 + d(x-1)^3, & \text{if } 1 \le x \le 2. \end{cases}$$

Find a, b, c, and d.

12. A clamped cubic spline s for a function f is defined on $[1, 3]$ by

$$s(x) = \begin{cases} s_0(x) = 3(x-1) + 2(x-1)^2 - (x-1)^3, & \text{if } 1 \le x < 2, \\ s_1(x) = a + b(x-2) + c(x-2)^2 + d(x-2)^3, & \text{if } 2 \le x \le 3. \end{cases}$$

Given $f'(1) = f'(3)$, find a, b, c, and d.

13. A natural cubic spline S is defined by

$$S(x) = \begin{cases} S_0(x) = 1 + B(x-1) - D(x-1)^3, & \text{if } 1 \le x < 2, \\ S_1(x) = 1 + b(x-2) - \frac{3}{4}(x-2)^2 + d(x-2)^3, & \text{if } 2 \le x \le 3. \end{cases}$$

If S interpolates the data $(1, 1)$, $(2, 1)$, and $(3, 0)$, find B, D, b, and d.

14. A clamped cubic spline s for a function f is defined by

$$s(x) = \begin{cases} s_0(x) = 1 + Bx + 2x^2 - 2x^3, & \text{if } 0 \le x < 1, \\ s_1(x) = 1 + b(x-1) - 4(x-1)^2 + 7(x-1)^3, & \text{if } 1 \le x \le 2. \end{cases}$$

Find $f'(0)$ and $f'(2)$.

15. Suppose that $f(x)$ is a polynomial of degree 3. Show that $f(x)$ is its own clamped cubic spline but that it cannot be its own free cubic spline.

16. Suppose the data $\{x_i, f(x_i)\}_{i=1}^{n}$ lie on a straight line. What can be said about the free and clamped cubic splines for the function f? [*Hint:* Take a cue from the results of Exercises 1 and 2.]

17. The data in the following table give the population of the United States for the years 1940 to 1990 and were considered in Exercise 16 of Section 3.2.

Year	1940	1950	1960	1970	1980	1990
Population (in thousands)	132,165	151,326	179,323	203,302	226,542	249,633

Find a free cubic spline agreeing with these data, and use the spline to predict the population in the years 1930, 1965, and 2000. Compare your approximations with those previously obtained. If you had to make a choice, which interpolation procedure would you choose?

18. A car traveling along a straight road is clocked at a number of points. The data from the observations are given in the following table, where the time is in seconds, the distance is in feet, and the speed is in feet per second.

Time	0	3	5	8	13
Distance	0	225	383	623	993
Speed	75	77	80	74	72

a. Use a clamped cubic spline to predict the position of the car and its speed when $t = 10$ s.

b. Use the derivative of the spline to determine whether the car ever exceeds a 55-mi/h speed limit on the road; if so, what is the first time the car exceeds this speed?

c. What is the predicted maximum speed for the car?

19. The 1995 Kentucky Derby was won by a horse named Thunder Gulch in a time of 2:01$\frac{1}{5}$ (2 min 1$\frac{1}{5}$ s) for the 1$\frac{1}{4}$-mi race. Times at the quarter-mile, half-mile, and mile poles were 22$\frac{2}{5}$, 45$\frac{4}{5}$, and 1:35$\frac{3}{5}$.

a. Use these values together with the starting time to construct a free cubic spline for Thunder Gulch's race.

b. Use the spline to predict the time at the three-quarter-mile pole, and compare this to the actual time of 1:10$\frac{1}{5}$.

c. Use the spline to approximate Thunder Gulch's starting speed and speed at the finish line.

20. It is suspected that the high amounts of tannin in mature oak leaves inhibit the growth of the winter moth (*Operophtera bromata L., Geometridae*) larvae that extensively damage these trees in certain years. The following table lists the average weight of two samples of larvae at times in the first 28 days after birth. The first sample was reared on young oak leaves, whereas the second sample was reared on mature leaves from the same tree.

a. Use a free cubic spline to approximate the average weight curve for each sample.
b. Find an approximate maximum average weight for each sample by determining the maximum of the spline.

Day	0	6	10	13	17	20	28
Sample 1 average weight (mg)	6.67	17.33	42.67	37.33	30.10	29.31	28.74
Sample 2 average weight (mg)	6.67	16.11	18.89	15.00	10.56	9.44	8.89

3.6 Parametric Curves

None of the techniques we have developed can be used to generate curves of the form shown in Figure 3.12, since this curve cannot be expressed as a function of one coordinate variable in terms of the other. In this section we will see how to represent general curves by using a parameter to express both the x- and y-coordinate variables. This technique can be extended to represent general curves and surfaces in space.

Figure 3.12

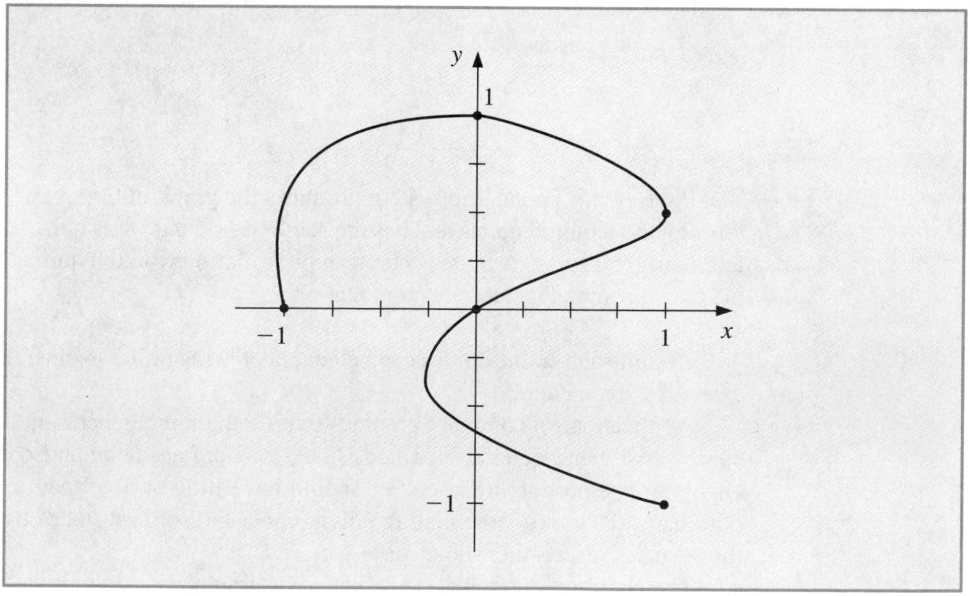

A straightforward parametric technique for determining a polynomial or piecewise polynomial to connect the points $(x_0, y_0), (x_1, y_1), \ldots, (x_n, y_n)$ is to use a parameter t on an interval $[t_0, t_n]$, with $t_0 < t_1 < \cdots < t_n$, and construct approximation functions with

$$x_i = x(t_i) \quad \text{and} \quad y_i = y(t_i) \quad \text{for each } i = 0, 1, \ldots, n.$$

The following example demonstrates the technique when both approximating functions are Lagrange interpolating polynomials.

EXAMPLE 1 Construct a pair of Lagrange polynomials to approximate the curve shown in Figure 3.12, using the data points shown on the curve.

There is flexibility in choosing the parameter, and we will choose the points $\{t_i\}$ equally spaced in [0, 1]. In this case, we have the data in Table 3.16.

Table 3.16

i	0	1	2	3	4
t_i	0	0.25	0.5	0.75	1
x_i	−1	0	1	0	1
y_i	0	1	0.5	0	−1

This produces the Lagrange polynomials

$$x(t) = \left(\left(\left(64t - \frac{352}{3}\right)t + 60\right)t - \frac{14}{3}\right)t - 1$$

and

$$y(t) = \left(\left(\left(-\frac{64}{3}t + 48\right)t - \frac{116}{3}t\right) + 11\right)t.$$

Plotting this parametric system produces the graph in Figure 3.13. Although it passes through the required points and has the same basic shape, it is quite a crude approximation to the original curve. A more accurate approximation would require additional nodes, with the accompanying increase in computation. ◆ ◆ ◆

Hermite and spline curves can be generated in a similar manner, but these also require extensive computation.

Applications in computer graphics require the rapid generation of smooth curves that can be easily and quickly modified. Also, for both aesthetic and computational reasons, changing one portion of the curves should have little or no effect on other portions. This eliminates the use of interpolating polynomials and splines, since changing one portion of these curves affects the whole curve.

The choice of curve for use in computer graphics is generally a form of the piecewise cubic Hermite polynomial. Each portion of a cubic Hermite polynomial is completely determined by specifying its endpoints and the derivatives at these endpoints. As a consequence, one portion of the curve can be changed while leaving most of the curve the same. Only the adjacent portions need to be modified if we want to ensure smoothness at the endpoints.

Figure 3.13

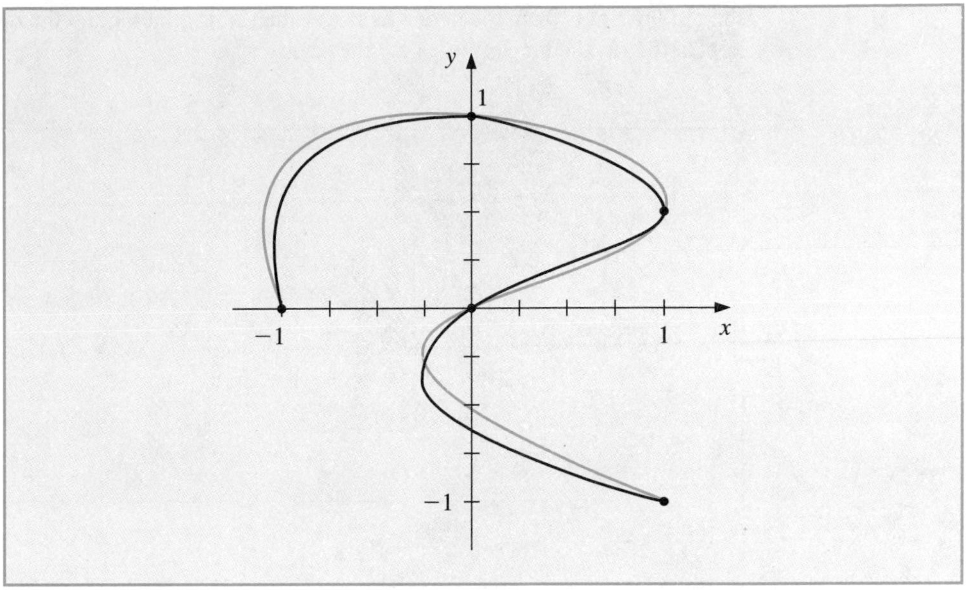

The computations can be performed quickly, and the curve can be modified a section at a time.

The problem with Hermite interpolation is the need to specify the derivatives at the endpoints of each section of the curve. Suppose the curve has $n + 1$ data points $(x_0, y_0), \ldots, (x_n, y_n)$, and we wish to parameterize the cubic to allow complex features. Then if $(x_i, y_i) = (x(t_i), y(t_i))$ for each $i = 0, 1, \ldots, n$, we must specify $x'(t_i)$ and $y'(t_i)$. This is not as difficult as it would first appear, however, since each portion of the curve is generated independently. Essentially, then, we can simplify the process to one of determining a pair of cubic Hermite polynomials in the parameter t, where $t_0 = 0$, $t_1 = 1$, given the endpoint data $(x(0), y(0))$, and $(x(1), y(1))$ and the derivatives dy/dx (at $t = 0$) and dy/dx (at $t = 1$).

Notice that we are specifying only six conditions, and each cubic polynomial has four parameters, for a total of eight. This provides considerable flexibility in choosing the pair of cubic Hermite polynomials to satisfy these conditions, because the natural form for determining $x(t)$ and $y(t)$ requires that we specify $x'(0)$, $x'(1)$, $y'(0)$, and $y'(1)$. The explicit Hermite curve in x and y requires specifying only the quotients

$$\frac{dy}{dx}(\text{at } t = 0) = \frac{y'(0)}{x'(0)} \quad \text{and} \quad \frac{dy}{dx}(\text{at } t = 1) = \frac{y'(1)}{x'(1)}.$$

By multiplying $x'(0)$ and $y'(0)$ by a common scaling factor, the tangent line to the curve at $(x(0), y(0))$ remains the same, but the shape of the curve varies. The larger the scaling factor, the closer the curve comes to approximating the tangent line near $(x(0), y(0))$. A similar situation exists at the other endpoint $(x(1), y(1))$.

To further simplify the process, the derivative at an endpoint is specified graphically by describing a second point, called a *guidepoint*, on the desired tangent line. The farther the guidepoint is from the node, the larger the scaling factor and the more closely the curve approximates the tangent line near the node.

Figure 3.14

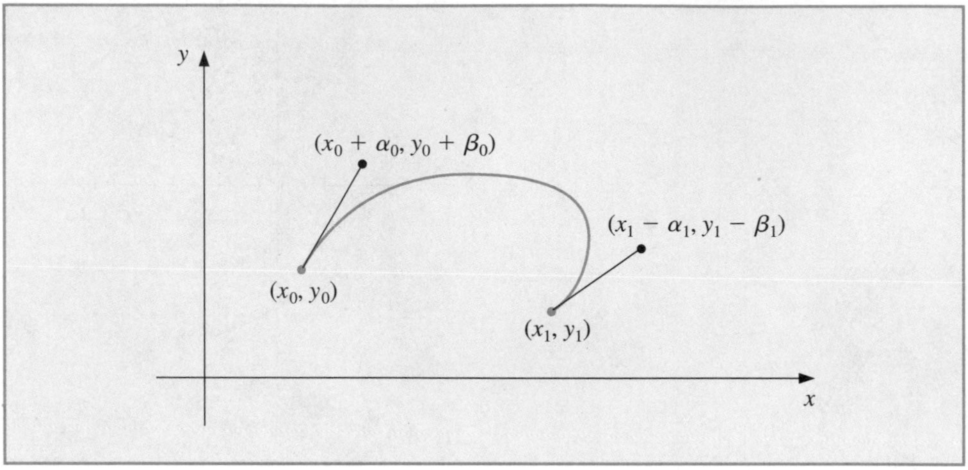

In Figure 3.14, the nodes occur at (x_0, y_0) and (x_1, y_1), the guidepoint for (x_0, y_0) is $(x_0 + \alpha_0, y_0 + \beta_0)$, and the guidepoint for (x_1, y_1) is $(x_1 - \alpha_1, y_1 - \beta_1)$. The cubic Hermite polynomial $x(t)$ on $[0, 1]$ must satisfy

$$x(0) = x_0, \qquad x(1) = x_1, \qquad x'(0) = \alpha_0, \quad \text{and} \quad x'(1) = \alpha_1.$$

The unique cubic polynomial satisfying these conditions is

$$x(t) = [2(x_0 - x_1) + (\alpha_0 + \alpha_1)]t^3 + [3(x_1 - x_0) - (\alpha_1 + 2\alpha_0)]t^2 + \alpha_0 t + x_0.$$

In a similar manner, the unique cubic polynomial for y satisfying $y(0) = y_0$, $y(1) = y_1$, $y'(0) = \beta_0$, and $y'(1) = \beta_1$ is

$$y(t) = [2(y_0 - y_1) + (\beta_0 + \beta_1)]t^3 + [3(y_1 - y_0) - (\beta_1 + 2\beta_0)]t^2 + \beta_0 t + y_0.$$

EXAMPLE 2 The graphs in Figure 3.15 show some possibilities that occur when the nodes are $(0, 0)$ and $(1, 0)$, and the slopes at these nodes are 1 and -1, respectively. The specification of the slope at the endpoints requires only that $\alpha_0 = \beta_0$ and $\alpha_1 = -\beta_1$, since the ratios $\alpha_0/\beta_0 = 1$ and $\alpha_1/\beta_1 = -1$ give the slopes at the left and right endpoints, respectively.

◆ ◆ ◆

The standard procedure for determining curves in an interactive graphics mode is to first use an input device, such as a mouse or trackball, to set the nodes and guidepoints to generate a first approximation to the curve. These can be set manually, but most graphics systems permit you to use your input device to draw the curve on the screen freehand and will select appropriate nodes and guidepoints for your freehand curve.

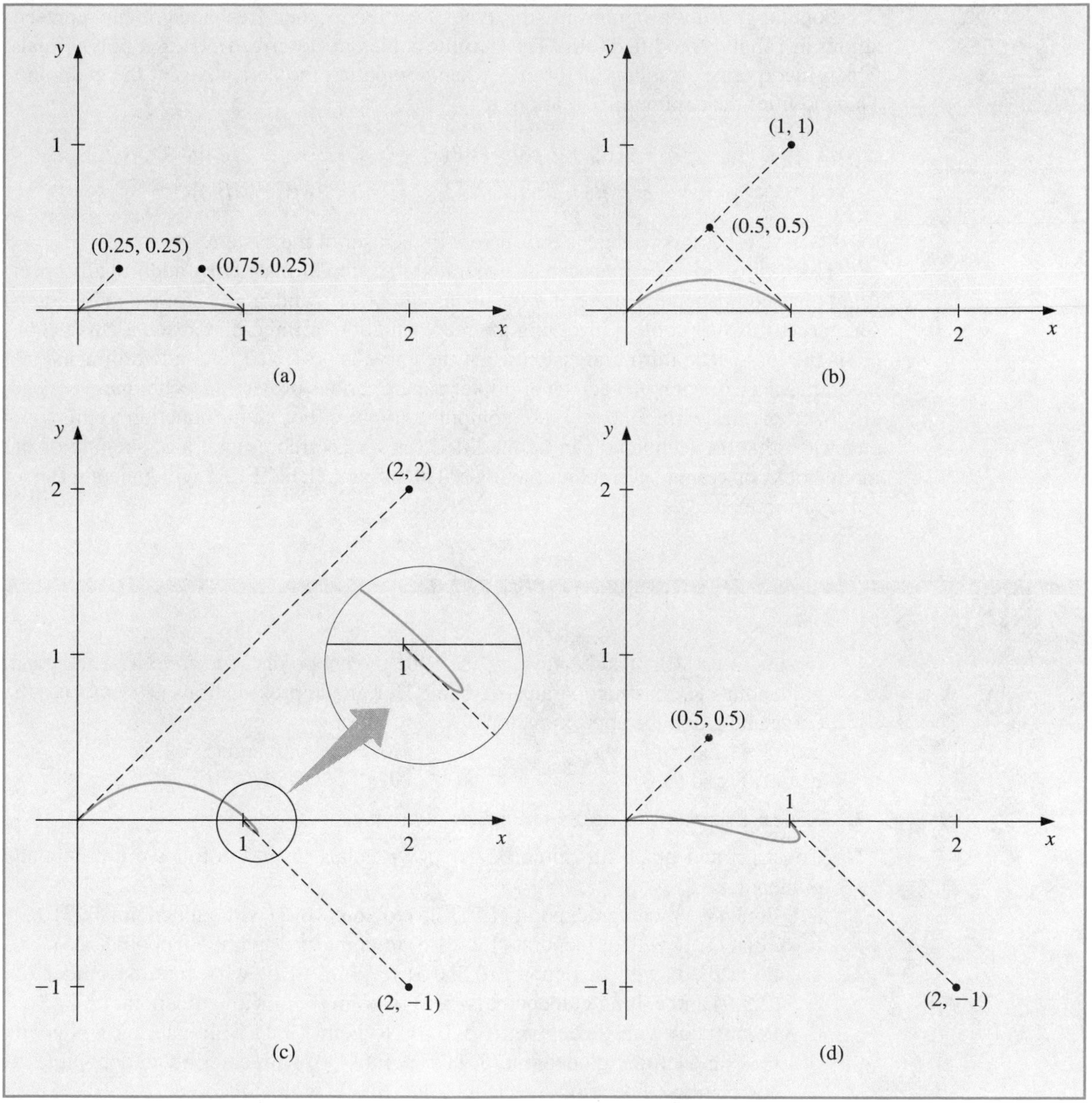

Figure 3.15

The nodes and guidepoints can then be manipulated into a position that produces an aesthetically satisfying curve. Since the computation is minimal, the curve is determined so quickly that the resulting change can be seen almost immediately. Moreover, all the data needed to compute the curves are imbedded in coordinates of the nodes and guidepoints, so no analytical knowledge is required of the user of the system.

Popular graphics programs use this type of system for their freehand graphic representations in a slightly modified form. The Hermite cubics are described as Bézier polynomials, which incorporate a scaling factor of 3 when computing the derivatives at the endpoints. This modifies the parametric equations to

$$x(t) = [2(x_0 - x_1) + 3(\alpha_0 + \alpha_1)]t^3 + [3(x_1 - x_0) - 3(\alpha_1 + 2\alpha_0)]t^2 + 3\alpha_0 t + x_0,$$
$$y(t) = [2(y_0 - y_1) + 3(\beta_0 + \beta_1)]t^3 + [3(y_1 - y_0) - 3(\beta_1 + 2\beta_0)]t^2 + 3\beta_0 t + y_0,$$

for $0 \le t \le 1$, but this change is transparent to the user of the system.

Three-dimensional curves can be generated in a similar manner by additionally specifying third components z_0 and z_1 for the nodes and $z_0 + \gamma_0$ and $z_1 - \gamma_1$ for the guidepoints. The more difficult problem involving the representation of three-dimensional curves concerns the loss of the third dimension when the curve is projected onto a two-dimensional medium such as a computer screen or printer paper. Various projection techniques are used, but this topic lies within the realm of computer graphics. For an introduction to this topic and ways that the technique can be modified for surface representations, see one of the many books on computer graphics methods. The program BEZIER36 will generate Bézier curves from input data.

EXERCISE SET 3.6

1. Let $(x_0, y_0) = (0, 0)$ and $(x_1, y_1) = (5, 2)$ be the endpoints of a curve. Use the given guidepoints to construct parametric cubic Hermite approximations $(x(t), y(t))$ to the curve and graph the approximations.
   a. $(1, 1)$ and $(6, 1)$
   b. $(0.5, 0.5)$ and $(5.5, 1.5)$
   c. $(1, 1)$ and $(6, 3)$
   d. $(2, 2)$ and $(7, 0)$
2. Repeat Exercise 1 using cubic Bézier polynomials.
3. Construct and graph the cubic Bézier polynomials given the following points and guidepoints.
   a. Point $(1, 1)$ with guidepoint $(1.5, 1.25)$ to point $(6, 2)$ with guidepoint $(7, 3)$
   b. Point $(1, 1)$ with guidepoint $(1.25, 1.5)$ to point $(6, 2)$ with guidepoint $(5, 3)$
   c. Point $(0, 0)$ with guidepoint $(0.5, 0.5)$ to point $(4, 6)$ with entering guidepoint $(3.5, 7)$ and exiting guidepoint $(4.5, 5)$ to point $(6, 1)$ with guidepoint $(7, 2)$
   d. Point $(0, 0)$ with guidepoint $(0.5, 0.25)$ to point $(2, 1)$ with entering guidepoint $(3, 1)$ and exiting guidepoint $(3, 1)$ to point $(4, 0)$ with entering guidepoint $(5, 1)$ and exiting guidepoint $(3, -1)$ to point $(6, -1)$ with guidepoint $(6.5, -0.25)$
4. Use the data in the following table to approximate the letter $\mathcal{N}$.

i	x_i	y_i	α_i	β_i	α_i'	β_i'
0	3	6	3.3	6.5		
1	2	2	2.8	3.0	2.5	2.5
2	6	6	5.8	5.0	5.0	5.8
3	5	2	5.5	2.2	4.5	2.5
4	6.5	3			6.4	2.8

3.7 Survey of Methods and Software

In this chapter we have considered approximating a function using polynomials and piecewise polynomials. The function can be specified by a given defining equation or by providing points in the plane through which the graph of the function passes. A set of nodes $x_0, x_1, \ldots, x_n$, is given in each case, and more information, such as the value of various derivatives, may also be required. We need to find an approximating function that satisfies the conditions specified by these data.

The interpolating polynomial $P(x)$ is the polynomial of least degree that satisfies, for a function f,

$$P(x_i) = f(x_i) \qquad \text{for each } i = 0, 1, \ldots, n.$$

Although there is a unique interpolating polynomial, it can take many different forms. The Lagrange form is most often used for interpolating tables when n is small and for deriving formulas for approximating derivatives and integrals. Neville's method is used for evaluating several interpolating polynomials at the same value of x. Newton's forms of the polynomial are more appropriate for computation and are also used extensively for deriving formulas for solving differential equations. However, polynomial interpolation has the inherent weaknesses of oscillation, particularly if the number of nodes is large. In this case there are other methods that can be better applied.

The Hermite polynomials interpolate a function and its derivative at the nodes. They can be very accurate but require more information about the function being approximated. When you have a large number of nodes, the Hermite polynomials also exhibit oscillation weaknesses.

The most commonly used form of interpolation is piecewise polynomial interpolation. If function and derivative values are available, piecewise cubic Hermite interpolation is recommended. This is the preferred method for interpolating values of a function that is the solution to a differential equation. When only the function values are available, free cubic spline interpolation can be used. This spline forces the second derivative of the spline to be zero at the endpoints. The free boundary conditions will generally give less accurate results than the clamped conditions near the ends of the interval $[x_0, x_n]$ unless the function f happens to nearly satisfy $f''(x_0) = f''(x_n) = 0$. An alternative to the free boundary condition that does not require knowledge of the derivative of f is the *not-a-knot* condition. (See de Boor [De], pp. 55–56.) This condition requires that $S'''(x)$ be continuous at x_1 and at x_{n-1}.

There are other methods of interpolation that are commonly used. Trigonometric interpolation is used with large amounts of data when the function has a periodic nature. In particular, the fast Fourier transform discussed in Chapter 8 is employed. Interpolation by rational functions is also used. If the data are suspected to be inaccurate, smoothing techniques can be applied, and some form of least squares fit of data is recommended. Polynomials, trigonometric functions, rational functions, and splines can be used in least squares fitting of data. We consider these topics in Chapter 8.

Interpolation routines included in the IMSL and the NAG Library are based on the book *A Practical Guide to Splines* by de Boor [De] and use interpolation by cubic splines.

The libraries contain subroutines for spline interpolation with user supplied end conditions, periodic end conditions, and the not-a-knot condition. There are also cubic splines to minimize oscillations or to preserve concavity. Methods for two-dimensional interpolation by bicubic splines are also included.

The MATLAB function POLYFIT can be used to find an interpolating function of degree at most n that passes through $n + 1$ specified points. Cubic splines can be produced with the function SPLINE.

The natural cubic spline can be constructed with Maple. First enter

```
>readlib(spline);
```

to make the package available. The command

```
>spline(X,Y,x,3);
```

constructs the natural cubic spline interpolating `X=[x[0],...,x[n]]` and `Y=[y[0],...,y[n]]`, where `x` is the variable and 3 refers to the degree of the cubic spline. Linear and quadratic splines can also be created.

General references to the methods in this chapter are the books by Powell [Po] and by Davis [Da]. The seminal paper on splines is due to Schoenberg [Scho]. Important books on splines are by Schultz [Schu] and de Boor [De].

CHAPTER 4

Numerical Integration and Differentiation

4.1 Introduction

Many techniques are described in calculus courses for the exact evaluation of integrals, but these are given more for illustration of algebraic manipulation and simplification than for use in the evaluation of integrals that occur in real-life problems. Exact techniques fail to solve many problems that arise in the physical world; for these we need approximation methods of the type we consider in this chapter. The basic techniques are discussed in Section 4.2, and refinements and special applications of these procedures are given in the next six sections. The final section in the chapter considers approximating the derivatives of functions.

You might wonder why there is so much more emphasis on approximating integrals than on approximating derivatives. This is because determining the derivative of a function is a constructive process that leads to straightforward rules for evaluation. Although the definition of the integral is also constructive, the principal tool for evaluating a definite integral is the Fundamental Theorem of Calculus. To apply this theorem we must determine the antiderivative of the function we wish to evaluate. This is not generally a constructive process, and it leads to the need for accurate approximation procedures.

In this chapter we will also discover one of the more interesting facts in the study of numerical methods. The approximation of integrals—a task that is frequently needed—can usually be accomplished very accurately and often with little effort. The accurate approximation of derivatives—which is needed far less frequently—is a more difficult problem. We think that there is something satisfying about a subject that provides good approximation methods for problems that need them but is less successful for problems that don't.

4.2 Basic Quadrature Rules

The basic procedure for approximating the definite integral of a function f on the interval $[a, b]$ is to determine an interpolating polynomial that approximates f and then integrate this polynomial. In this section we determine approximations that arise when some basic polynomials are used for the approximation and determine error bounds for these approximations.

The approximations we consider use interpolating polynomials at equally spaced points in the interval $[a, b]$. The first of these is the Midpoint rule, which uses the midpoint of $[a, b]$, $(a+b)/2$, as its only interpolation point. The Midpoint rule approximation is easy to generate geometrically, as shown in Figure 4.1, but to establish the pattern for the higher-order methods and to determine an error formula for the technique, we will use a basic tool for these derivations, the Newton interpolatory divided-difference formula we discussed in Section 3.3.

Figure 4.1

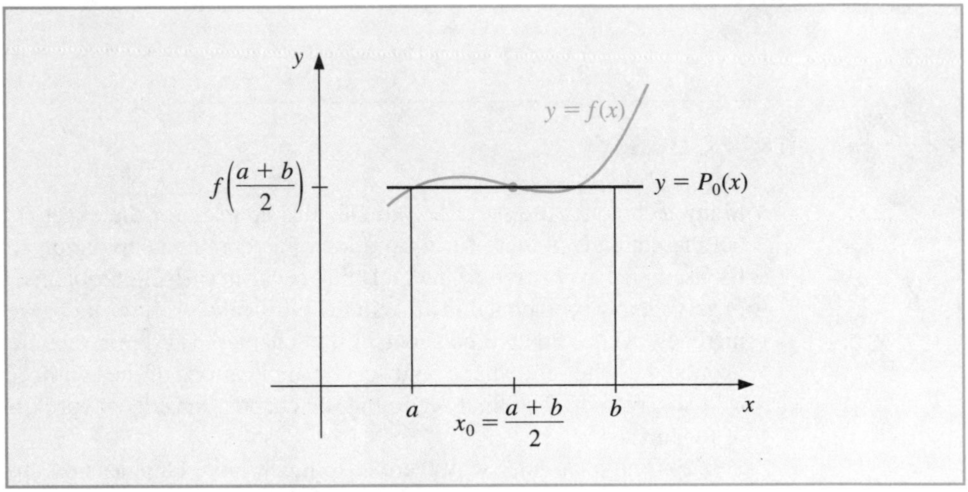

The Newton interpolatory divided-difference formula states that the interpolating polynomial for the function f using the nodes $x_0, x_1, \ldots, x_n$ can be expressed in the form

$$P_{0,1,\ldots,n}(x) = f[x_0] + f[x_0, x_1](x - x_0) + f[x_0, x_1, x_2](x - x_0)(x - x_1) + \cdots + f[x_0, x_1, \ldots, x_n](x - x_0)(x - x_1)\cdots(x - x_{n-1}).$$

Since this is equivalent to the nth Lagrange polynomial, the error formula has the form

$$f(x) - P_{0,1,\ldots,n}(x) = \frac{f^{(n+1)}(\xi(x))}{(n+1)!}(x - x_0)(x - x_1)\cdots(x - x_n),$$

where $\xi(x)$ lies between $x, x_0, x_1, \ldots, x_n$.

To derive the Midpoint rule we could use the constant interpolating polynomial with $x_0 = (a+b)/2$ to produce

$$\int_a^b f(x)\,dx \approx \int_a^b f[x_0]\,dx = f[x_0](b-a) = f\left(\frac{a+b}{2}\right)(b-a),$$

but we could also use a linear interpolating polynomial with this value of x_0 and an arbitrary value of x_1. This is due to the fact that the integral of the second term in the

Newton interpolatory divided-difference formula is zero for our choice of x_0, independent of the value of x_1, and as such does not contribute to the approximation:

$$\begin{aligned}\int_a^b f[x_0,x_1](x-x_0)\,dx &= \frac{f[x_0,x_1]}{2}(x-x_0)^2\Big]_a^b\\ &= \frac{f[x_0,x_1]}{2}\left(x-\frac{a+b}{2}\right)^2\Big]_a^b\\ &= \frac{f[x_0,x_1]}{2}\left[\left(b-\frac{a+b}{2}\right)^2-\left(a-\frac{a+b}{2}\right)^2\right]\\ &= \frac{f[x_0,x_1]}{2}\left[\left(\frac{b-a}{2}\right)^2-\left(\frac{a-b}{2}\right)^2\right]=0.\end{aligned}$$

In general, the higher the degree of the approximation, the higher the order of the error term, so we will integrate the error for $P_{0,1}(x)$ instead of $P_0(x)$ to determine an error formula for the Midpoint rule.

Suppose that the arbitrary x_1 was chosen to be the same value as x_0. (In fact, this is the only value that we *cannot* have for x_1, but we will ignore this problem for the moment.) Then the integral of the error formula for the interpolating polynomial $P_0(x)$ has the form

$$\int_a^b \frac{(x-x_0)(x-x_1)}{2}f''(\xi(x))\,dx = \int_a^b \frac{(x-x_0)^2}{2}f''(\xi(x))\,dx.$$

Since the term $(x-x_0)^2$ does not change sign on the interval (a,b), the Mean Value Theorem for Integrals implies that a number ξ in (a,b) exists with

$$\begin{aligned}\int_a^b \frac{(x-x_0)^2}{2}f''(\xi(x))\,dx &= f''(\xi)\int_a^b \frac{(x-x_0)^2}{2}\,dx = \frac{f''(\xi)}{6}(x-x_0)^3\Big]_a^b\\ &= \frac{f''(\xi)}{6}\left[\left(b-\frac{b+a}{2}\right)^3-\left(a-\frac{b+a}{2}\right)^3\right]\\ &= \frac{f''(\xi)}{6}\frac{(b-a)^3}{4} = \frac{f''(\xi)}{24}(b-a)^3.\end{aligned}$$

As a consequence, the Midpoint rule with its error formula has the following form:

Midpoint Rule

$$\int_a^b f(x)\,dx = (b-a)f\left(\frac{a+b}{2}\right)+\frac{f''(\xi)}{24}(b-a)^3,$$

for some number ξ in (a,b).

The invalid assumption, $x_1 = x_0$, that leads to this result can be avoided by taking x_1 close, but not equal to, x_0 and using limits to show that the error formula is valid.

The Midpoint rule uses a constant interpolating polynomial disguised as a linear interpolating polynomial. The next method we consider uses a true linear interpolating polynomial, one with the distinct nodes $x_0 = a$ and $x_1 = b$. This approximation is also easy to generate geometrically, as is shown in Figure 4.2, and is aptly called the Trapezoidal rule. If we integrate the linear interpolating polynomial with $x_0 = a$ and $x_1 = b$ we also produce this formula:

$$\begin{aligned}\int_a^b \{f[x_0] + f[x_0, x_1](x - x_0)\}\, dx &= \left[f[a]x + f[a,b]\frac{(x-a)^2}{2}\right]_a^b \\ &= f(a)(b-a) + \frac{f(b)-f(a)}{b-a}\left[\frac{(b-a)^2}{2} - \frac{(a-a)^2}{2}\right] \\ &= (b-a)\frac{f(a)+f(b)}{2}.\end{aligned}$$

Figure 4.2

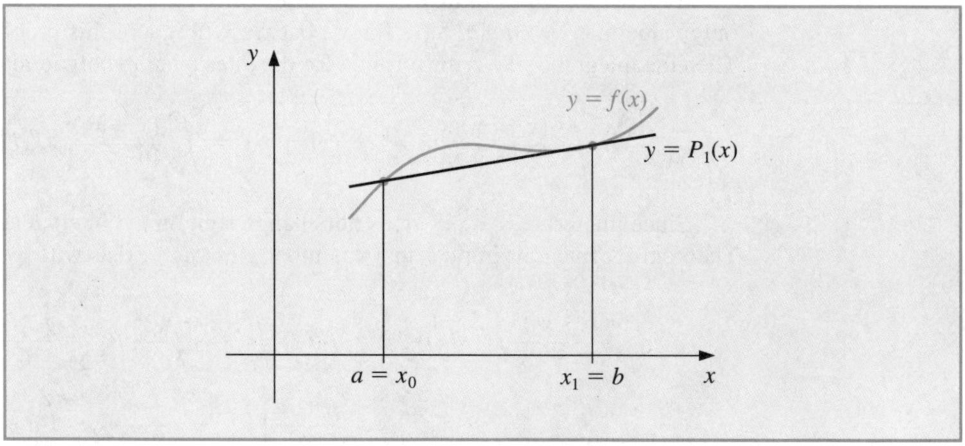

The error for the Trapezoidal rule follows from integrating the error term for $P_{0,1}(x)$ when $x_0 = a$ and $x_1 = b$. Since $(x - x_0)(x - x_1) = (x - a)(x - b)$ does not change sign in the interval (a, b), we can again apply the Mean Value Theorem for Integrals. In this case it implies that a number ξ in (a, b) exists with

$$\begin{aligned}\int_a^b \frac{(x-a)(x-b)}{2} f''(\xi(x))\, dx &= \frac{f''(\xi)}{2}\int_a^b (x-a)[(x-a)+(a-b)]\, dx \\ &= \frac{f''(\xi)}{2}\left[\frac{(x-a)^3}{3} + \frac{(x-a)^2}{2}(a-b)\right]_a^b \\ &= \frac{f''(\xi)}{2}\left[\frac{(b-a)^3}{3} + \frac{(b-a)^2}{2}(a-b)\right].\end{aligned}$$

Simplifying this equation gives the Trapezoidal rule with its error formula.

Trapezoidal Rule

$$\int_a^b f(x)\,dx = (b-a)\frac{f(a)+f(b)}{2} - \frac{f''(\xi)}{12}(b-a)^3,$$

for some ξ in (a, b).

We cannot improve on the order of this error formula, as we did in the case of the Midpoint rule, because the integral of the next higher term in the Newton interpolatory divided-difference formula is

$$\int_a^b f[x_0, x_1, x_2](x-x_0)(x-x_1)\,dx = f[x_0, x_1, x_2]\int_a^b (x-a)(x-b)\,dx.$$

Since $(x-a)(x-b) < 0$ for all x in (a, b), this term will not be zero unless $f[x_0, x_1, x_2] = 0$. As a consequence, the error formulas for the Midpoint and the Trapezoidal rules are both of order 3, even though they are derived from interpolation formulas with error formulas of order 1 and 2, respectively.

Next in line is an integration formula based on approximating the function f by a quadratic polynomial that agrees with f at the equally spaced points $x_0 = a$, $x_1 = (a+b)/2$, and $x_2 = b$. This formula is not easy to generate geometrically, although the approximation is illustrated in Figure 4.3.

Figure 4.3

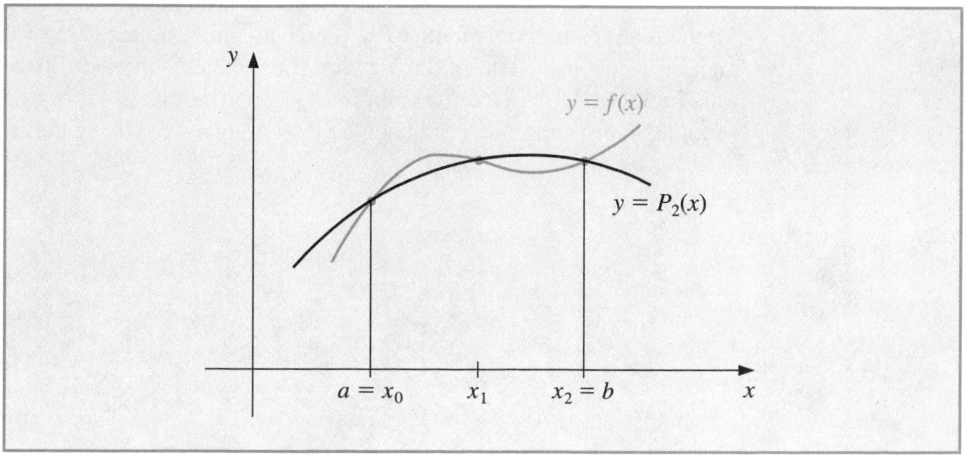

To derive the formula we integrate $P_{0,1,2}(x)$.

$$
\begin{aligned}
\int_a^b & P_{0,1,2}(x)\,dx \\
&= \int_a^b \left\{ f(a) + f\left[a, \frac{a+b}{2}\right](x-a) + f\left[a, \frac{a+b}{2}, b\right](x-a)\left(x - \frac{a+b}{2}\right)\right\} dx \\
&= \left[f(a)x + f\left[a, \frac{a+b}{2}\right]\frac{(x-a)^2}{2}\right]_a^b \\
&\quad + f\left[a, \frac{a+b}{2}, b\right]\int_a^b (x-a)\left[(x-a) + \left(a - \frac{a+b}{2}\right)\right] dx \\
&= f(a)(b-a) + \frac{f(\frac{a+b}{2}) - f(a)}{\frac{a+b}{2} - a}\,\frac{(b-a)^2}{2} \\
&\quad + \frac{f[\frac{a+b}{2}, b] - f[a, \frac{a+b}{2}]}{b-a}\left[\frac{(x-a)^3}{3} + \frac{(x-a)^2}{2}\left(\frac{a-b}{2}\right)\right]_a^b \\
&= (b-a)\left[f(a) + f\left(\frac{a+b}{2}\right) - f(a)\right] \\
&\quad + \left(\frac{1}{b-a}\right)\left[\frac{f(b) - f(\frac{a+b}{2})}{\frac{b-a}{2}} - \frac{f(\frac{a+b}{2}) - f(a)}{\frac{b-a}{2}}\right]\left[\frac{(b-a)^3}{3} - \frac{(b-a)^3}{4}\right] \\
&= (b-a)f\left(\frac{a+b}{2}\right) + \frac{2}{(b-a)^2}\left[f(b) - 2f\left(\frac{a+b}{2}\right) + f(a)\right]\frac{(b-a)^3}{12}.
\end{aligned}
$$

Simplifying this equation gives the approximation method known as Simpson's rule:

$$\int_a^b f(x)\,dx \approx \frac{(b-a)}{6}\left[f(a) + 4f\left(\frac{a+b}{2}\right) + f(b)\right].$$

An order 4 error formula for Simpson's rule can be derived by using the error formula for the interpolating polynomial $P_{0,1,2}(x)$, but, similar to the case of the Midpoint rule, the integral of the next term in the Newton interpolatory divided-difference formula is zero. This implies that the error formula for $P_{0,1,2,3}(x)$ can be used to produce an order 5 error formula. When simplified, Simpson's rule with this error formula is as follows:

Simpson's Rule

$$\int_a^b f(x)\,dx = \frac{(b-a)}{6}\left[f(a) + 4f\left(\frac{a+b}{2}\right) + f(b)\right] - \frac{f^{(4)}(\xi)}{2880}(b-a)^5,$$

for some ξ in (a, b).

This higher-order error term makes Simpson's rule significantly superior to the Midpoint and Trapezoidal rules in almost all situations if the width of the interval is small. This is illustrated in the following example.

EXAMPLE 1 Tables 4.1 and 4.2 show results of the Midpoint, Trapezoidal, and Simpson's rules applied to a variety of functions integrated over the intervals $[1, 1.2]$ and $[0, 2]$. All the methods give useful results for the function on $[1, 1.2]$, but only Simpson's rule gives reasonable accuracy on $[0, 2]$. ◆ ◆ ◆

Table 4.1 Integrals on the Interval $[1, 1.2]$

$f(x)$	x^2	x^4	$1/(x+1)$	$\sqrt{1+x^2}$	$\sin x$	e^x
Exact value	0.24267	0.29766	0.09531	0.29742	0.17794	0.60184
Midpoint	0.24200	0.29282	0.09524	0.29732	0.17824	0.60083
Trapezoidal	0.24400	0.30736	0.09545	0.29626	0.17735	0.60384
Simpson's	0.24267	0.29767	0.09531	0.29742	0.17794	0.60184

Table 4.2 Integrals on the Interval $[0, 2]$

$f(x)$	x^2	x^4	$1/(x+1)$	$\sqrt{1+x^2}$	$\sin x$	e^x
Exact value	2.667	6.400	1.099	2.958	1.416	6.389
Midpoint	2.000	2.000	1.000	2.818	1.682	5.436
Trapezoidal	4.000	16.000	1.333	3.326	0.909	8.389
Simpson's	2.667	6.667	1.111	2.964	1.425	6.421

To demonstrate the error terms for the Midpoint, Trapezoidal, and Simpson's methods, we will find bounds for the errors in approximating $\int_0^2 \sqrt{1+x^2}\,dx$. With $f(x) = (1+x^2)^{1/2}$, we have

$$f'(x) = \frac{x}{\sqrt{1+x^2}}, \quad f''(x) = \frac{1}{(1+x^2)^{3/2}}, \quad \text{and} \quad f'''(x) = \frac{-3x}{(1+x^2)^{5/2}}.$$

Maximum and minimum values for f'' on $[0, 2]$ can occur only when $x = 0$, $x = 2$, or when $f'''(x) = 0$. Since $f'''(x) = 0$ only when $x = 0$, we have

$$\max_{0\le x\le 2} |f''(x)| = \max\{|f''(0)|, |f''(2)|\} = \max\left\{1, \frac{\sqrt{5}}{25}\right\} = 1.$$

To bound the error in the Midpoint method we have

$$\left|\frac{f''(\xi)}{24}(b-a)^3\right| \le \frac{1}{24}(2-0)^3 = \frac{1}{3} = 0.\overline{3}.$$

The actual error is within this bound, since $|2.958 - 2.818| = 0.14$. For the Trapezoidal method we have

$$\left| -\frac{f''(\xi)}{12}(b-a)^3 \right| \leq \frac{1}{12}(2-0)^3 = \frac{2}{3} = 0.\overline{6},$$

and the actual error is $|2.958 - 3.326| = 0.368$. We need more derivatives for Simpson's rule:

$$f^{(4)}(x) = \frac{12x^2 - 3}{(1+x^2)^{7/2}} \quad \text{and} \quad f^{(5)}(x) = \frac{45x - 60x^3}{(1+x^2)^{9/2}}.$$

Since $f^{(5)}(x) = 0$ implies

$$0 = 45x - 60x^3 = 15x(3 - 4x^2),$$

$f^{(4)}(x)$ has critical points $0, \pm\sqrt{3}/2$. Thus,

$$|f^{(4)}(\xi)| \leq \max_{0 \leq x \leq 2} |f^{(4)}(x)| = \max \left\{ |f^{(4)}(0)|, \left| f^{(4)}\left(\frac{\sqrt{3}}{2}\right) \right|, |f^{(4)}(2)| \right\}$$

$$= \max \left\{ |-3|, \frac{768\sqrt{7}}{2401}, \frac{9\sqrt{5}}{125} \right\} = 3.$$

The error for Simpson's rule is bounded by

$$\left| -\frac{f^{(4)}(\xi)}{2880}(b-a)^5 \right| \leq \frac{3}{2880}(2-0)^5 = \frac{96}{2880} = 0.0\overline{3},$$

and the actual error is $|2.958 - 2.964| = 0.006$.

There are higher-order formulas that can be used to improve the accuracy when integrating over large intervals, some of which are considered in the exercises, but a better solution to the problem will be considered in the next section.

EXERCISE SET 4.2

1. Use the Midpoint rule to approximate the following integrals.

   a. $\displaystyle\int_{0.5}^{1} x^4\,dx$ b. $\displaystyle\int_{0}^{0.5} \frac{2}{x-4}\,dx$

   c. $\displaystyle\int_{1}^{1.5} x^2 \ln x\,dx$ d. $\displaystyle\int_{0}^{1} x^2 e^{-x}\,dx$

   e. $\displaystyle\int_{1}^{1.6} \frac{2x}{x^2-4}\,dx$ f. $\displaystyle\int_{1}^{1.6} \frac{2}{x^2-4}\,dx$

g. $\int_0^{\pi/4} x \sin x \, dx$ h. $\int_0^{\pi/4} e^{3x} \sin 2x \, dx$

2. Use the error formula to find a bound for the error in Exercise 1, and compare the bound to the actual error.
3. Repeat Exercise 1 using the Trapezoidal rule.
4. Repeat Exercise 2 using the Trapezoidal rule and the results of Exercise 3.
5. Repeat Exercise 1 using Simpson's rule.
6. Repeat Exercise 2 using Simpson's rule and the results of Exercise 5.

Other quadrature formulas with error terms are given by

(i) $\int_a^b f(x)\,dx = \frac{3h}{8}[f(a)+3f(a+h)+3f(a+2h)+f(b)] - \frac{3h^5}{80} f^{(4)}(\xi)$, where $h = \frac{b-a}{3}$;

(ii) $\int_a^b f(x)\,dx = \frac{3h}{2}[f(a+h)+f(a+2h)] + \frac{3h^3}{4} f''(\xi)$, where $h = \frac{b-a}{3}$.

7. Repeat Exercises 1 and 2 using Formula (i).
8. Repeat Exercises 1 and 2 using Formula (ii).
9. The Trapezoidal rule applied to $\int_0^2 f(x)\,dx$ gives the value 4, and Simpson's rule gives the value 2. What is $f(1)$?
10. The Trapezoidal rule applied to $\int_0^2 f(x)\,dx$ gives the value 5, and the Midpoint rule gives the value 4. What value does Simpson's rule give?
11. Find the constants c_0, c_1, and x_1 so that the quadrature formula

$$\int_0^1 f(x)\,dx = c_0 f(0) + c_1 f(x_1)$$

gives exact results for all polynomials of degree at most 2.

12. Find the constants x_0, x_1, and c_1 so that the quadrature formula

$$\int_0^1 f(x)\,dx = \frac{1}{2} f(x_0) + c_1 f(x_1)$$

gives exact results for all polynomials of degree at most 3.

13. Given the function f at the following values:

x	1.8	2.0	2.2	2.4	2.6
$f(x)$	3.12014	4.42569	6.04241	8.03014	10.46675

a. Approximate $\int_{1.8}^{2.6} f(x)\,dx$ using each of the following.
(i) the Midpoint rule **(ii)** the Trapezoidal rule **(iii)** Simpson's rule

b. Suppose the data have round-off errors given by the following table:

x	1.8	2.0	2.2	2.4	2.6
Error in $f(x)$	2×10^{-6}	-2×10^{-6}	-0.9×10^{-6}	-0.9×10^{-6}	2×10^{-6}

Calculate the errors due to round-off in each of the approximation methods.

4.3 Composite Quadrature Rules

The basic notions underlying numerical integration were derived in the previous section, but the techniques given there are not satisfactory for many problems. We saw an example of this at the end of that section, where the approximations were poor for integrals of functions on the interval $[0, 2]$. To see why this occurs, let us consider Simpson's method, generally the most accurate of these techniques. With its error formula, it is given by

$$\begin{aligned}\int_a^b f(x)\,dx &= \frac{b-a}{6}\left[f(a) + 4f\left(\frac{a+b}{2}\right) + f(b)\right] - \frac{(b-a)^5}{2880}f^{(4)}(\xi)\\ &= \frac{h}{3}[f(a) + 4f(a+h) + f(b)] - \frac{h^5}{90}f^{(4)}(\xi)\end{aligned}$$

where $h = (b-a)/2$ and ξ lies in (a, b). If the fourth derivative of f is bounded on (a, b), the error term in this formula is $O(h^5)$. That is, the error term behaves in a manner similar to h^5 as h approaches zero. As a consequence, we expect the error to be small provided that the fourth derivative of f is well behaved and the interval $[a, b]$ is small. The first assumption we can live with, but the second might be quite unreasonable. There is no reason, in general, to expect that the interval over which the integration is performed is small, and if it is not, the $O(h^5)$ portion in the error term can dominate the calculations.

We circumvent the problem of a large interval of integration by subdividing the interval $[a, b]$ into a collection of intervals that are sufficiently small so that the error over each is kept under control.

EXAMPLE 1 Consider finding an approximation to $\int_0^2 e^x\,dx$. If Simpson's rule is used with $h = 1$,

$$\int_0^2 e^x\,dx \approx \frac{1}{3}(e^0 + 4e^1 + e^2) = 6.4207278.$$

Since the exact answer in this case is $e^2 - e^0 = 6.3890561$, the error of magnitude 0.0316717 is larger than would generally be regarded as acceptable. To apply a piecewise technique to this problem, subdivide $[0, 2]$ into $[0, 1]$ and $[1, 2]$, and use Simpson's rule twice with $h = \frac{1}{2}$, giving

$$\int_0^2 e^x\,dx = \int_0^1 e^x\,dx + \int_1^2 e^x\,dx \approx \frac{1}{6}\left[e^0 + 4e^{0.5} + e^1\right] + \frac{1}{6}\left[e^1 + 4e^{1.5} + e^2\right]$$
$$= \frac{1}{6}\left[e^0 + 4e^{0.5} + 2e^1 + 4e^{1.5} + e^2\right] = 6.3912102.$$

The magnitude of the error now has been reduced by more than 90% to 0.0021541. If we subdivide each of the intervals [0, 1] and [1, 2] and use Simpson's rule with $h = \frac{1}{4}$, we get

$$\begin{aligned}\int_0^2 e^x\,dx &= \int_0^{0.5} e^x\,dx + \int_{0.5}^1 e^x\,dx + \int_1^{1.5} e^x\,dx + \int_{1.5}^2 e^x\,dx\\ &\approx \frac{1}{12}\left[e^0 + 4e^{0.25} + e^{0.5}\right] + \frac{1}{12}\left[e^{0.5} + 4e^{0.75} + e^1\right]\\ &\quad + \frac{1}{12}\left[e^1 + 4e^{1.25} + e^{1.5}\right] + \frac{1}{12}\left[e^{1.5} + 4e^{1.75} + e^2\right]\\ &= \frac{1}{12}\left[e^0 + 4e^{0.25} + 2e^{0.5} + 4e^{0.75} + 2e^1 + 4e^{1.25} + 2e^{1.5} + 4e^{1.75} + e^2\right]\\ &= 6.3891937.\end{aligned}$$

The magnitude of the error for this approximation is 0.0001376, only 0.4% of the original error. ◆ ◆ ◆

The generalization of the procedure considered in this example is called the Composite Simpson's rule and is described as follows.

Choose an even integer, subdivide the interval $[a, b]$ into n subintervals, and use Simpson's rule on each consecutive pair of subintervals. Let $h = (b - a)/n$ and $a = x_0 < x_1 < \cdots < x_n = b$, where $x_j = x_0 + jh$ for each $j = 0, 1, \ldots, n$. Then

$$\begin{aligned}\int_a^b f(x)\,dx &= \sum_{j=1}^{n/2} \int_{x_{2j-2}}^{x_{2j}} f(x)\,dx\\ &= \sum_{j=1}^{n/2} \left\{ \frac{h}{3}[f(x_{2j-2}) + 4f(x_{2j-1}) + f(x_{2j})] - \frac{h^5}{90} f^{(4)}(\xi_j) \right\}\end{aligned}$$

for some ξ_j with $x_{2j-2} < \xi_j < x_{2j}$, provided that $f \in C^4[a, b]$. To simplify this formula, first note that for each $j = 1, 2, \ldots, \frac{n}{2} - 1$, we have $f(x_{2j})$ appearing in the term corresponding to the interval $[x_{2j-2}, x_{2j}]$ and also in the term corresponding to the interval $[x_{2j}, x_{2j+2}]$. Combining these terms gives (see Figure 4.4 on page 118)

$$\int_a^b f(x)\,dx = \frac{h}{3}\left[f(x_0) + 2\sum_{j=1}^{(n/2)-1} f(x_{2j}) + 4\sum_{j=1}^{n/2} f(x_{2j-1}) + f(x_n)\right] - \frac{h^5}{90}\sum_{j=1}^{n/2} f^{(4)}(\xi_j).$$

Figure 4.4

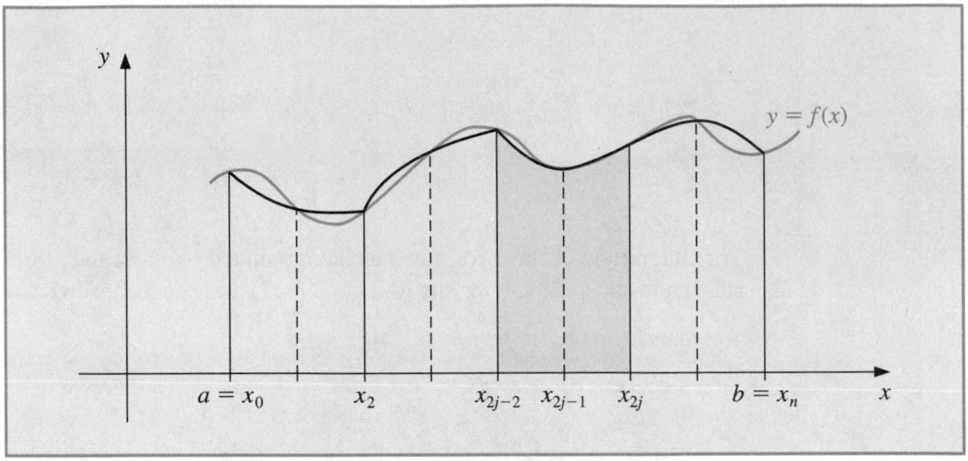

The error associated with the approximation is

$$E(f) = -\frac{h^5}{90}\sum_{j=1}^{n/2} f^{(4)}(\xi_j),$$

where $x_{2j-2} < \xi_j < x_{2j}$ for each $j = 1, 2, \ldots, \frac{n}{2}$. If $f \in C^4[a, b]$, the Extreme Value Theorem implies that $f^{(4)}$ assumes its maximum and minimum in $[a, b]$. Since

$$\min_{x\in[a,b]} f^{(4)}(x) \leq f^{(4)}(\xi_j) \leq \max_{x\in[a,b]} f^{(4)}(x),$$

we have

$$\frac{n}{2}\min_{x\in[a,b]} f^{(4)}(x) \leq \sum_{j=1}^{n/2} f^{(4)}(\xi_j) \leq \frac{n}{2}\max_{x\in[a,b]} f^{(4)}(x),$$

and

$$\min_{x\in[a,b]} f^{(4)}(x) \leq \frac{2}{n}\sum_{j=1}^{n/2} f^{(4)}(\xi_j) \leq \max_{x\in[a,b]} f^{(4)}(x).$$

The term in the middle lies between values of $f^{(4)}$, so the Intermediate Value Theorem implies that a number μ in (a, b) exists with

$$f^{(4)}(\mu) = \frac{2}{n}\sum_{j=1}^{n/2} f^{(4)}(\xi_j).$$

Since $h = (b - a)/n$, the error formula simplifies to

$$E(f) = -\frac{h^5}{90}\sum_{j=1}^{n/2} f^{(4)}(\xi_j) = -\frac{h^4(b-a)}{180} f^{(4)}(\mu).$$

Summarizing these results, we have the following: If $f \in C^4[a,b]$ and n is even, the Composite Simpson's rule for n subintervals of $[a,b]$ can be expressed with error term as follows:

Composite Simpson's Rule

$$\int_a^b f(x)\,dx = \frac{h}{3}\left[f(a) + 2\sum_{j=1}^{(n/2)-1} f(x_{2j}) + 4\sum_{j=1}^{n/2} f(x_{2j-1}) + f(b)\right] - \frac{(b-a)h^4}{180} f^{(4)}(\mu),$$

for some μ in (a,b).

The program CSIMPR41 implements the Composite Simpson's rule on n subintervals. This is the most frequently used general-purpose integral approximation technique.

The subdivision approach can be applied to any of the formulas we saw in the preceding section. Since the Trapezoidal rule (see Figure 4.5) uses only interval endpoint values to determine its approximation, the integer n can be odd or even.

Figure 4.5

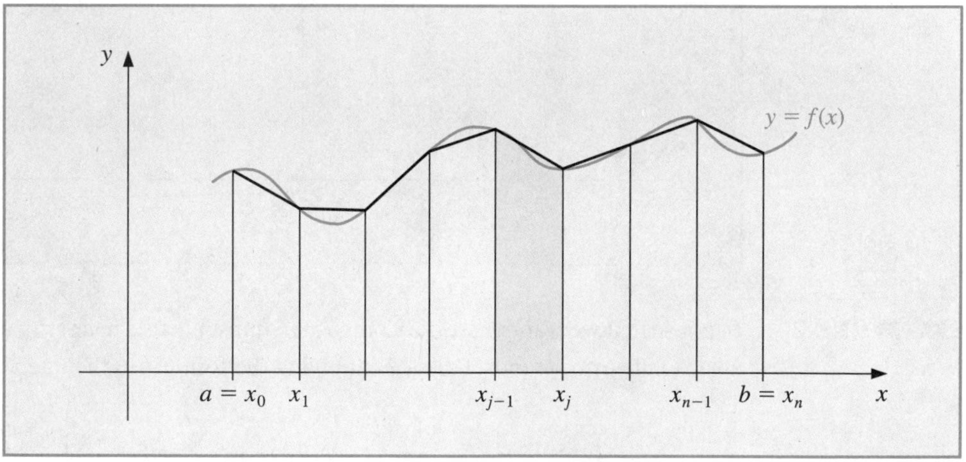

Let $f \in C^2[a,b]$. With $h = (b-a)/n$ and $x_j = a + jh$ for each $j = 0, 1, \ldots, n$, the Composite Trapezoidal rule for n subintervals is as follows.

Composite Trapezoidal Rule

$$\int_a^b f(x)\,dx = \frac{h}{2}\left[f(a) + f(b) + 2\sum_{j=1}^{n-1} f(x_j)\right] - \frac{(b-a)h^2}{12} f''(\mu),$$

for some $\mu \in (a,b)$.

Let $f \in C^2[a,b]$ and n be an even integer. With $h = (b-a)/(n+2)$ and $x_j = a + (j+1)h$ for each $j = -1, 0, \ldots, n+1$, the Composite Midpoint rule for $n+2$ subintervals is as follows (see Figure 4.6).

Composite Midpoint Rule

$$\int_a^b f(x)\,dx = 2h\sum_{j=0}^{n/2} f(x_{2j}) + \frac{(b-a)h^2}{6} f''(\mu),$$

for some $\mu \in (a,b)$.

Figure 4.6

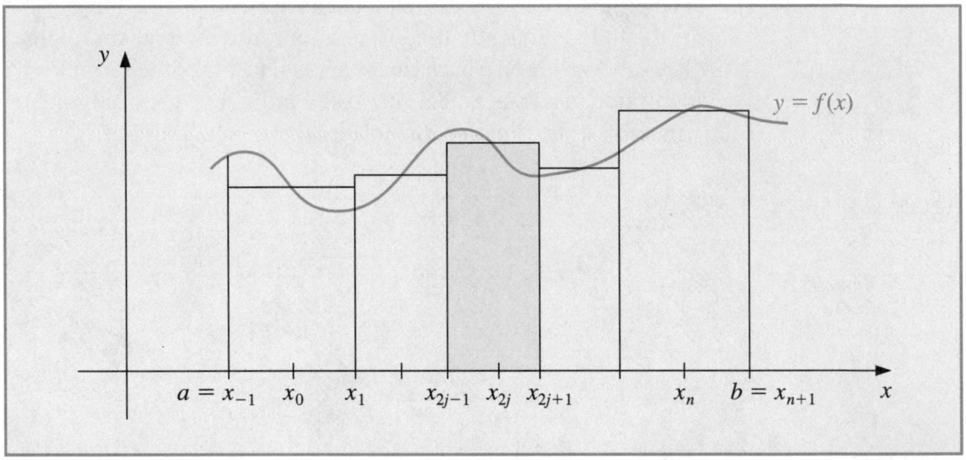

EXAMPLE 2 Suppose that we want to use the Composite Simpson's rule to approximate $\int_0^\pi \sin x\,dx$ with an absolute error at most 0.00002. Applying the formula to the integral $\int_0^\pi \sin x\,dx$ gives

$$\int_0^\pi \sin x\,dx = \frac{h}{3}\left[2\sum_{j=1}^{(n/2)-1} \sin x_{2j} + 4\sum_{j=1}^{n/2} \sin x_{2j-1}\right] - \frac{\pi h^4}{180}\sin\mu.$$

Since the absolute error is to be less than 0.00002, the inequality

$$\left|\frac{\pi h^4}{180}\sin\mu\right| \le \frac{\pi h^4}{180}\cdot 1 = \frac{\pi^5}{180 n^4} \le 0.00002$$

is used to determine n and then h. Completing these calculations gives $n \ge 18$. Verifying this, we find that when $n = 20$ and $h = \pi/20$, Composite Simpson's rule gives

$$\int_0^\pi \sin x\,dx \approx \frac{\pi}{60}\left[2\sum_{j=1}^{9}\sin\left(\frac{j\pi}{10}\right) + 4\sum_{j=1}^{10}\sin\left(\frac{(2j-1)\pi}{20}\right)\right] = 2.000006,$$

compared to the exact value of 2.

To be assured of this degree of accuracy using the Composite Trapezoidal rule requires that

$$\left|\frac{\pi h^2}{12}\sin\mu\right| \le \frac{\pi h^2}{12} = \frac{\pi^3}{12n^2} < 0.00002,$$

which implies that $n \ge 360$.

For comparison purposes, the Composite Midpoint rule with $n = 18$ and $h = \pi/20$ gives

$$\int_0^\pi \sin x\, dx \approx \frac{\pi}{10}\sum_{j=0}^{9}\sin\left(\frac{(2j+1)\pi}{20}\right) = 2.0082484$$

and the Composite Trapezoidal rule with $n = 20$ and $h = \pi/20$ gives

$$\int_0^\pi \sin x\, dx \approx \frac{\pi}{40}\left[2\sum_{j=1}^{19}\sin\left(\frac{j\pi}{20}\right) + \sin 0 + \sin\pi\right]$$
$$= \frac{\pi}{40}\left[2\sum_{j=1}^{19}\sin\left(\frac{j\pi}{20}\right)\right] = 1.9958860.$$

Simpson's rule with $h = \pi/20$ gave an answer well within the required error bound of 0.00002, whereas the Midpoint and Trapezoidal rules with $h = \pi/20$ clearly do not.

◆ ◆ ◆

Maple incorporates all three composite rules. To obtain access to the library where they are defined, enter

```
>with(student);
```

The calls for the methods are `middlesum(f,x=a..b,n)`, `trapezoid(f,x=a..b,n)`, and `simpson(f,x=a..b,n)`. For our example the following commands are used.

```
>f:=sin(x);
```

$$f := \sin(x)$$

```
>evalf(middlesum(f,x=0..Pi,20));
```

2.002057648

```
>evalf(trapezoid(f,x=0..Pi,20));
```

1.995885974

```
>evalf(simpson(f,x=0..Pi,20));
```

2.000006785

We will illustrate the Composite Midpoint Rule using loops. First we define $f(x)$ and a, b, n, and h.

```
>f:=x->sin(x);
>a:=0; b:=Pi; n:=18; h:=(b-a)/(n+2);
```

We need a variable to calculate the running sum, and we initialize it at 0.

```
>Tot:=0;
```

In Maple the counter-controlled loop is defined by

```
for loop control variable from initial-value to terminal value do
   statement;
   statement;
   ⋮
   statement;
od;
```

In the following example the loop control variable is j, which begins at 0 and goes to $n/2 = 9$ in steps of 1. For each value of $j = 0, 1, \ldots, 9$ the loop is traversed and each calculation inside the loop is performed until the word od is encountered. The reserved words involved are `for`, `from`, `to`, `do` and `od`. Note that no ; folllows the do statement.

```
>for j from 0 to n/2 do
> xj:=a+(2*j+1)*h;
> Tot:=evalf(Tot+f(xj))
>od;
```

This produces a series of results culminating in the final summation

$$Tot = \sum_{j=0}^{n/2} f(x_{2j}) = \sum_{j=0}^{9} f(x_{2j}) = 6.392453222.$$

We then multiply by $2h$ to finish the Composite Midpoint Method

```
>Tot:=evalf(2*h*Tot);
```

$$Tot := 2.008248408$$

An important property shared by all the composite rules is stability with respect to round-off error. To demonstrate this, suppose we apply the Composite Simpson's rule with n subintervals to a function on $[a, b]$ and determine the maximum bound for the round-off error. Assume that $f(x_i)$ is approximated by $\tilde{f}(x_i)$ and that

$$f(x_i) = \tilde{f}(x_i) + e_i, \qquad \text{for each } i = 0, 1, \ldots, n,$$

where e_i denotes the round-off error associated with using $\tilde{f}(x_i)$ to approximate $f(x_i)$. Then the accumulated round-off error, $e(h)$, in the Composite Simpson's rule is

$$e(h) = \left| \frac{h}{3} \left[e_0 + 2 \sum_{j=1}^{(n/2)-1} e_{2j} + 4 \sum_{j=1}^{n/2} e_{2j-1} + e_n \right] \right|$$
$$\leq \frac{h}{3} \left[|e_0| + 2 \sum_{j=1}^{(n/2)-1} |e_{2j}| + 4 \sum_{j=1}^{n/2} |e_{2j-1}| + |e_n| \right].$$

If the round-off errors are uniformly bounded by some known tolerance ϵ, then

$$e(h) \leq \frac{h}{3} \left[\epsilon + 2 \left(\frac{n}{2} - 1 \right) \epsilon + 4 \left(\frac{n}{2} \right) \epsilon + \epsilon \right] = \frac{h}{3} 3n\epsilon = nh\epsilon.$$

But $nh = b - a$, so $e(h) \leq (b - a)\epsilon$, a bound independent of h and n. This means that even though we may need to divide an interval into more parts to ensure accuracy, the increased computation that this requires does not increase the round-off error.

EXERCISE SET 4.3

1. Use the Composite Trapezoidal rule with the indicated values of n to approximate the following integrals.

   a. $\int_1^2 x \ln x \, dx, \quad n = 4$

   b. $\int_{-2}^2 x^3 e^x \, dx, \quad n = 4$

   c. $\int_0^2 \frac{2}{x^2 + 4} \, dx, \quad n = 6$

   d. $\int_0^\pi x^2 \cos x \, dx, \quad n = 6$

   e. $\int_0^2 e^{2x} \sin 3x \, dx, \quad n = 8$

   f. $\int_1^3 \frac{x}{x^2 + 4} \, dx, \quad n = 8$

   g. $\int_3^5 \frac{1}{\sqrt{x^2 - 4}} \, dx, \quad n = 8$

   h. $\int_0^{3\pi/8} \tan x \, dx, \quad n = 8$

2. Use the Composite Simpson's rule to approximate the integrals in Exercise 1.

3. Use the Composite Midpoint rule with $n + 2$ subintervals to approximate the integrals in Exercise 1.

4. Approximate $\int_0^2 x^2 e^{-x^2} \, dx$ using $h = 0.25$.

   a. Use the Composite Trapezoidal rule.
   b. Use the Composite Simpson's rule.
   c. Use the Composite Midpoint rule.

5. Determine the values of n and h required to approximate

$$\int_0^2 e^{2x} \sin 3x \, dx$$

to within 10^{-4}.

a. Use the Composite Trapezoidal rule.
b. Use the Composite Simpson's rule.
c. Use the Composite Midpoint rule.

6. Repeat Exercise 5 for the integral $\int_0^{\pi} x^2 \cos x \, dx$.

7. Determine the values of n and h required to approximate

$$\int_0^2 \frac{1}{x+4} \, dx$$

to within 10^{-5} and compute the approximation.

a. Use the Composite Trapezoidal rule.
b. Use the Composite Simpson's rule.
c. Use the Composite Midpoint rule.

8. Repeat Exercise 7 for the integral $\int_1^2 x \ln x \, dx$.

9. Suppose that $f(0.25) = f(0.75) = \alpha$. Find α if the Composite Trapezoidal rule with $n = 2$ gives the value 2 for $\int_0^1 f(x)\,dx$ and with $n = 4$ gives the value 1.75.

10. The Midpoint rule for approximating $\int_{-1}^1 f(x)\,dx$ gives the value 12, the Composite Midpoint rule with $n = 2$ gives 5, and Composite Simpson's rule gives 6. Use the fact that $f(-1) = f(1)$ and $f(-0.5) = f(0.5) - 1$ to determine $f(-1)$, $f(-0.5)$, $f(0)$, $f(0.5)$, and $f(1)$.

11. In multivariable calculus and in statistics courses it is shown that

$$\int_{-\infty}^{\infty} \frac{1}{\sigma\sqrt{2\pi}} e^{-(1/2)(x/\sigma)^2} \, dx = 1$$

for any positive σ. The function

$$f(x) = \frac{1}{\sigma\sqrt{2\pi}} e^{-(1/2)(x/\sigma)^2}$$

is the *normal density function* with *mean* $\mu = 0$ and *standard deviation* σ. The probability that a randomly chosen value described by this distribution lies in $[a, b]$ is given by $\int_a^b f(x)\,dx$. Approximate to within 10^{-5} the probability that a randomly chosen value described by this distribution will lie in

a. $[-2\sigma, 2\sigma]$ **b.** $[-3\sigma, 3\sigma]$

12. Determine to within 10^{-6} the length of the graph of the ellipse with equation $4x^2 + 9y^2 = 36$.

13. A car laps a race track in 84 s. The speed of the car at each 6-s interval is determined using a radar gun and is given from the beginning of the lap, in feet/second, by the entries in the following table:

Time	0	6	12	18	24	30	36	42	48	54	60	66	72	78	84
Speed	124	134	148	156	147	133	121	109	99	85	78	89	104	116	123

How long is the track?

14. A particle of mass m moving through a fluid is subjected to a viscous resistance R, which is a function of the velocity v. The relationship between the resistance R, velocity v, and time t is given by the equation

$$t = \int_{v(t_0)}^{v(t)} \frac{m}{R(u)}\, du.$$

Suppose that $R(v) = -v\sqrt{v}$ for a particular fluid, where R is in newtons and v is in meters/second. If $m = 10$ kg and $v(0) = 10$ m/s, approximate the time required for the particle to slow to $v = 5$ m/s.

15. The equation

$$\int_0^x \frac{1}{\sqrt{2\pi}} e^{-t^2/2}\, dt = 0.45$$

can be solved for x by using Newton's method with

$$f(x) = \int_0^x \frac{1}{\sqrt{2\pi}} e^{-t^2/2}\, dt - 0.45$$

and

$$f'(x) = \frac{1}{\sqrt{2\pi}} e^{-x^2/2}.$$

To evaluate f at the approximation p_k, we need a quadrature formula to approximate

$$\int_0^{p_k} \frac{1}{\sqrt{2\pi}} e^{-t^2/2}\, dt.$$

a. Find a solution to $f(x) = 0$ accurate to within 10^{-5} using Newton's method with $p_0 = 0.5$ and the Composite Simpson's rule.

b. Repeat (a) using the Composite Trapezoidal rule in place of the Composite Simpson's rule.

4.4 Romberg Integration

Extrapolation is used to accelerate the convergence of many approximation techniques. It can be applied whenever it is known that the approximation technique has an error term with a predictable form, one that depends on a parameter, usually the step size h. We will first consider the general form of extrapolation, and then apply the technique to determine integration approximations.

The Trapezoidal rule is one of the simplest of the integration formulas, but it is seldom sufficiently accurate. Romberg Integration uses the Composite Trapezoidal rule to give preliminary approximations, and then applies Richardson extrapolation to obtain improved approximations. Our first step is to describe the extrapolation process.

Suppose that $N(h)$ is a formula involving a step size h that approximates an unknown value M, and that it is known that the error for $N(h)$ has the form

$$M - N(h) = K_1 h + K_2 h^2 + K_3 h^3 + \cdots,$$

or

$$M = N(h) + K_1 h + K_2 h^2 + K_3 h^3 + \cdots. \tag{4.1}$$

for some unspecified collection of constants, $K_1, K_2, K_3 \ldots$. We assume here that $h > 0$ can be arbitrarily chosen and that improved approximations occur as h becomes small. The objective of extrapolation is to improve the formula from one of order $O(h)$ to one of higher order. Do not be misled by the relative simplicity of Eq. (4.1). It may be quite difficult to obtain the approximation $N(h)$, particularly for small values of h.

Since Eq. (4.1) is assumed to hold for all positive h, consider the result when we replace the parameter h by half its value. This gives the formula

$$M = N\left(\frac{h}{2}\right) + K_1 \frac{h}{2} + K_2 \frac{h^2}{4} + K_3 \frac{h^3}{8} + \cdots.$$

Subtracting (4.1) from twice this equation eliminates the term involving K_1 and gives

$$M = \left[N\left(\frac{h}{2}\right) + \left(N\left(\frac{h}{2}\right) - N(h)\right)\right] + K_2\left(\frac{h^2}{2} - h^2\right) + K_3\left(\frac{h^3}{4} - h^3\right) + \cdots.$$

To facilitate the discussion, define $N_1(h) \equiv N(h)$ and

$$N_2(h) = N_1\left(\frac{h}{2}\right) + \left[N_1\left(\frac{h}{2}\right) - N_1(h)\right].$$

Then we have an $O(h^2)$ approximation formula for M:

$$M = N_2(h) - \frac{K_2}{2}h^2 - \frac{3K_3}{4}h^3 - \cdots. \tag{4.2}$$

If we now replace h by $h/2$ in Eq. (4.2) we have

$$M = N_2\left(\frac{h}{2}\right) - \frac{K_2}{8}h^2 - \frac{3K_3}{32}h^3 - \cdots. \tag{4.3}$$

This can be combined with Eq. (4.2) to eliminate the h^2 term. Specifically, subtracting (4.2) from 4 times Eq. (4.3) gives

$$3M = 4N_2\left(\frac{h}{2}\right) - N_2(h) + \frac{3K_3}{4}\left(-\frac{h^3}{2} + h^3\right) + \cdots,$$

which simplifies to the $O(h^3)$ formula for approximating M:

$$M = \left[N_2\left(\frac{h}{2}\right) + \frac{N_2(h/2) - N_2(h)}{3}\right] + \frac{K_3}{8}h^3 + \cdots.$$

Defining

$$N_3(h) \equiv N_2\left(\frac{h}{2}\right) + \frac{N_2(h/2) - N_2(h)}{3}$$

simplifies this to the $O(h^3)$ formula:

$$M = N_3(h) + \frac{K_3}{8}h^3 + \cdots.$$

The process is continued by constructing the $O(h^4)$ approximation

$$N_4(h) = N_3\left(\frac{h}{2}\right) + \frac{N_3(h/2) - N_3(h)}{7},$$

the $O(h^5)$ approximation

$$N_5(h) = N_4\left(\frac{h}{2}\right) + \frac{N_4(h/2) - N_4(h)}{15},$$

and so on. In general, if M can be written in the form

$$M = N(h) + \sum_{j=1}^{m-1} K_j h^j + O(h^m),$$

then for each $j = 2, 3, \ldots, m$, we have an $O(h^j)$ approximation of the form

$$N_j(h) = N_{j-1}\left(\frac{h}{2}\right) + \frac{N_{j-1}(h/2) - N_{j-1}(h)}{2^{j-1} - 1}.$$

These approximations are generated by rows to take advantage of the highest order formulas. The first four rows are shown in Table 4.3.

Table 4.3

$O(h)$	$O(h^2)$	$O(h^3)$	$O(h^4)$
$N_1(h) \equiv N(h)$			
$N_1\left(\frac{h}{2}\right) \equiv N\left(\frac{h}{2}\right)$	$N_2(h)$		
$N_1\left(\frac{h}{4}\right) \equiv N\left(\frac{h}{4}\right)$	$N_2\left(\frac{h}{2}\right)$	$N_3(h)$	
$N_1\left(\frac{h}{8}\right) \equiv N\left(\frac{h}{8}\right)$	$N_2\left(\frac{h}{4}\right)$	$N_3\left(\frac{h}{2}\right)$	$N_4(h)$

We now consider a technique called Romberg integration, which applies extrapolation to approximations generated using the Composite Trapezoidal rule. Recall that the Composite Trapezoidal rule for approximating the integral of a function f on an interval $[a, b]$ using m subintervals is

$$\int_a^b f(x)\,dx = \frac{h}{2}\left[f(a) + f(b) + 2\sum_{j=1}^{m-1} f(x_j)\right] - \frac{(b-a)}{12}h^2 f''(\mu),$$

where $a < \mu < b$, $h = (b-a)/m$ and $x_j = a + jh$ for each $j = 0, 1, \ldots, m$ (see Figure 4.7).

Figure 4.7

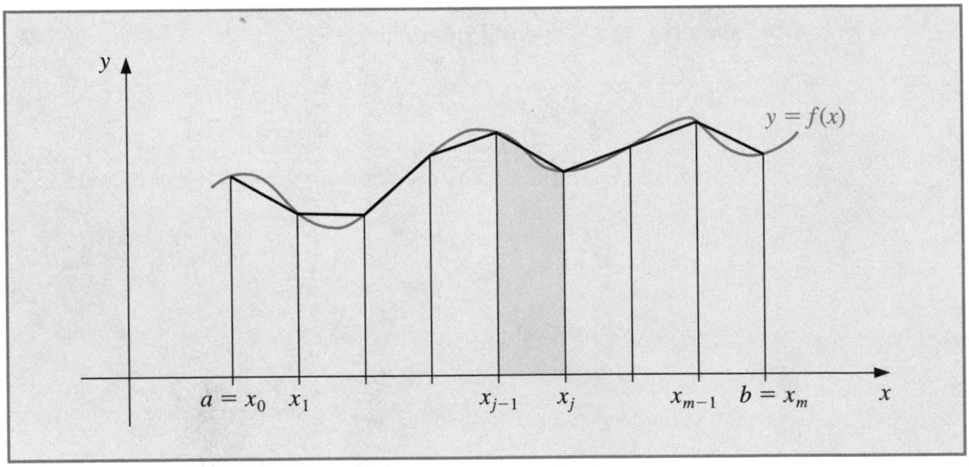

The first step in the Romberg process obtains the Composite Trapezoidal rule approximations with $m_1 = 1, m_2 = 2, m_3 = 4, \ldots,$ and $m_n = 2^{n-1}$, where n is some positive integer. The step size h_k corresponding to m_k is $h_k = (b-a)/m_k = (b-a)/2^{k-1}$. With this notation the Trapezoidal rule becomes

$$\int_a^b f(x)\,dx = \frac{h_k}{2}\left[f(a) + f(b) + 2\left(\sum_{i=1}^{2^{k-1}-1} f(a + ih_k)\right)\right] - \frac{b-a}{12}h_k^2 f''(\mu_k),$$

where μ_k is a number in (a, b). If the notation $R_{k,1}$ is introduced to denote the portion of this equation that is used for the trapezoidal approximation, then (see Figure 4.8):

$$R_{1,1} = \frac{h_1}{2}[f(a) + f(b)] = \frac{(b-a)}{2}[f(a) + f(b)];$$
$$R_{2,1} = \frac{1}{2}\left[R_{1,1} + h_1 f(a + h_2)\right];$$
$$R_{3,1} = \frac{1}{2}\{R_{2,1} + h_2[f(a + h_3) + f(a + 3h_3)]\}.$$

In general, we have the following:

Composite Trapezoidal Approximations

$$R_{k,1} = \frac{1}{2}\left[R_{k-1,1} + h_{k-1}\sum_{i=1}^{2^{k-2}} f(a + (2i-1)h_k)\right]$$

for each $k = 2, 3, \ldots, n$.

Figure 4.8

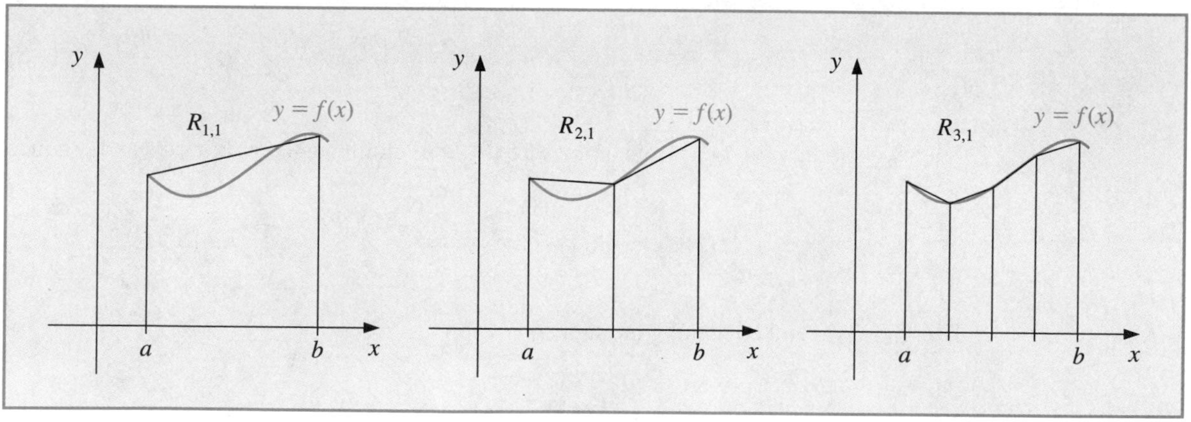

EXAMPLE 1 Using the Composite Trapezoidal rule to perform the first step of the Romberg Integration scheme for approximating $\int_0^\pi \sin x\, dx$ with $n = 6$ leads to

$$R_{1,1} = \frac{\pi}{2}[\sin 0 + \sin \pi] = 0,$$
$$R_{2,1} = \frac{1}{2}\left[R_{1,1} + \pi \sin \frac{\pi}{2}\right] = 1.57079633,$$
$$R_{3,1} = \frac{1}{2}\left[R_{2,1} + \frac{\pi}{2}\left(\sin \frac{\pi}{4} + \sin \frac{3\pi}{4}\right)\right] = 1.89611890,$$

$$R_{4,1} = \frac{1}{2}\left[R_{3,1} + \frac{\pi}{4}\left(\sin\frac{\pi}{8} + \sin\frac{3\pi}{8} + \sin\frac{5\pi}{8} + \sin\frac{7\pi}{8}\right)\right] = 1.97423160,$$

$$R_{5,1} = 1.99357034, \quad \text{and} \quad R_{6,1} = 1.99839336.$$

◆ ◆ ◆

Since the correct value for the integral in Example 1 is 2, it is clear that, although the calculations involved are not difficult, the convergence is slow. To speed the convergence, the Richardson extrapolation will be performed.

We must first have an approximation method with its error term in the form prescribed at the beginning of the section. It can be shown, although not easily, that if $f \in C^\infty[a, b]$, the Composite Trapezoidal rule can be written with an alternative error term of the form

$$\int_a^b f(x)\,dx - R_{k,1} = \sum_{i=1}^{\infty} K_i h_k^{2i} = K_1 h_k^2 + \sum_{i=2}^{\infty} K_i h_k^{2i}, \tag{4.4}$$

where K_i for each i is independent of h_k and depends only on $f^{(2i-1)}(a)$ and $f^{(2i-1)}(b)$. Because the powers of h_k in these equations are all even, the first averaging step produces an $O(h^4)$ approximation, the second, an $O(h^6)$ approximation, and so on. As a consequence, the accuracy acceleration is much faster than in our initial discussion.

The first step is to rewrite Eq. (4.4) with h_k replaced by $h_{k+1} = h_k/2$:

$$\int_a^b f(x)\,dx - R_{k+1,1} = \sum_{i=1}^{\infty} K_i h_{k+1}^{2i} = \sum_{i=1}^{\infty} \frac{K_i h_k^{2i}}{2^{2i}} = \frac{K_1 h_k^2}{4} + \sum_{i=2}^{\infty} \frac{K_i h_k^{2i}}{4^i}. \tag{4.5}$$

Then we subtract Eq. (4.4) from 4 times Eq. (4.5) to eliminate the $K_1 h_k^2$ terms and produce

$$3\int_a^b f(x)\,dx - 4R_{k+1,1} + R_{k,1} = 4\sum_{i=2}^{\infty} \frac{K_i h_k^{2i}}{4^i} - \sum_{i=2}^{\infty} K_i h_k^{2i}.$$

Dividing by 3 and combining the sums gives

$$\int_a^b f(x)\,dx - \frac{4R_{k+1,1} - R_{k,1}}{3} = \sum_{i=2}^{\infty} \frac{K_i}{3}\left(\frac{h_k^{2i}}{4^{i-1}} - h_k^{2i}\right),$$

or

$$\int_a^b f(x)\,dx - \left[R_{k+1,1} + \frac{R_{k+1,1} - R_{k,1}}{3}\right] = \sum_{i=2}^{\infty} \frac{K_i}{3}\left(\frac{1 - 4^{i-1}}{4^{i-1}}\right) h_k^{2i}.$$

Extrapolation can now be applied to this formula to eliminate the $O(h_k^4)$ term and obtain an $O(h_k^6)$ result, and so on. To simplify the notation, we define

$$R_{k,2} = R_{k,1} + \frac{R_{k,1} - R_{k-1,1}}{3}$$

for each $k = 2, 3, \ldots, n$, and apply extrapolation to these values. Continuing this notation, we have, for each $k = 2, 3, 4, \ldots, n$ and $j = 2, \ldots, k$, an $O(h_k^{2j})$ approximation formula defined by

$$R_{k,j} = R_{k,j-1} + \frac{R_{k,j-1} - R_{k-1,j-1}}{4^{j-1} - 1}.$$

The results that are generated from these formulas are shown in Table 4.4.

Table 4.4

$R_{1,1}$					
$R_{2,1}$	$R_{2,2}$				
$R_{3,1}$	$R_{3,2}$	$R_{3,3}$			
$R_{4,1}$	$R_{4,2}$	$R_{4,3}$	$R_{4,4}$		
$\vdots$	$\vdots$	$\vdots$		$\ddots$	
$R_{n,1}$	$R_{n,2}$	$R_{n,3}$	$R_{n,4}$	$\cdots$	$R_{n,n}$

The Romberg technique has the desirable feature that it allows an entire new row in the table to be calculated by doing one application of the Composite Trapezoidal rule. Then it uses a simple averaging on the previously calculated values to obtain the succeeding entries in the row. The method used to construct a table of this type calculates the entries row by row, that is, in the order $R_{1,1}, R_{2,1}, R_{2,2}, R_{3,1}, R_{3,2}, R_{3,3}$, etc. This is the process followed in the program ROMBRG42.

EXAMPLE 2 In Example 1, the values for $R_{1,1}$ through $R_{n,1}$ were obtained by approximating $\int_0^\pi \sin x\, dx$ with $n = 6$. The output from ROMBRG42 produces the Romberg table shown in Table 4.5. Although there are 21 entries in the table, only the 6 in the first column require function evaluations since these are the only entries generated by the integration technique. The other entries are obtained by a simple averaging process. ◆ ◆ ◆

Table 4.5

0					
1.57079633	2.09439511				
1.89611890	2.00455976	1.99857073			
1.97423160	2.00026917	1.99998313	2.00000555		
1.99357034	2.00001659	1.99999975	2.00000001	1.99999999	
1.99839336	2.00000103	2.00000000	2.00000000	2.00000000	2.00000000

ROMBRG32 requires a preset integer n to determine the number of rows to be generated. It is often also useful to prescribe an error tolerance for the approximation and generate R_{nn}, for n within some upper bound, until consecutive diagonal entries $R_{n,n}$ agree to within the tolerance.

EXERCISE SET 4.4

1. Use Romberg integration to compute $R_{3,3}$ for the following integrals.

 a. $\int_1^{1.5} x^2 \ln x \, dx$
 b. $\int_0^1 x^2 e^{-x} \, dx$
 c. $\int_0^{0.35} \frac{2}{x^2 - 4} \, dx$
 d. $\int_0^{\pi/4} x^2 \sin x \, dx$
 e. $\int_0^{\pi/4} e^{3x} \sin 2x \, dx$
 f. $\int_1^{1.6} \frac{2x}{x^2 - 4} \, dx$
 g. $\int_3^{3.5} \frac{x}{\sqrt{x^2 - 4}} \, dx$
 h. $\int_0^{\pi/4} (\cos x)^2 \, dx$

2. Calculate $R_{4,4}$ for the integrals in Exercise 1.

3. Use Romberg integration to approximate the integrals in Exercise 1 to within 10^{-6}. Compute the Romberg table until $|R_{n-1,n-1} - R_{n,n}| < 10^{-6}$, or until $n = 10$. Compare your results to the exact values of the integrals.

4. Apply Romberg integration to the following integrals until $R_{n-1,n-1}$ and $R_{n,n}$ agree to within 10^{-4}.

 a. $\int_0^1 x^{1/3} \, dx$
 b. $\int_0^{0.3} f(x) \, dx$, where

$$f(x) = \begin{cases} x^3 + 1, & 0 \le x \le 0.1, \\ 1.001 + 0.03(x - 0.1) \\ \quad + 0.3(x - 0.1)^2 + 2(x - 0.1)^3, & 0.1 < x \le 0.2, \\ 1.009 + 0.15(x - 0.2) \\ \quad + 0.9(x - 0.2)^2 + 2(x - 0.2)^3, & 0.2 < x \le 0.3. \end{cases}$$

5. Use the following data to approximate $\int_1^5 f(x) \, dx$ as accurately as possible.

x	1	2	3	4	5
$f(x)$	2.4142	2.6734	2.8974	3.0976	3.2804

6. Romberg integration is used to approximate

$$\int_0^1 \frac{x^2}{1 + x^3} \, dx.$$

If $R_{11} = 0.250$ and $R_{22} = 0.2315$, what is R_{21}?

7. Romberg integration is used to approximate

$$\int_2^3 f(x)\,dx.$$

If $f(2) = 0.51342, f(3) = 0.36788, R_{31} = 0.43687$, and $R_{33} = 0.43662$, find $f(2.5)$.

8. Romberg integration for approximating $\int_0^1 f(x)\,dx$ gives $R_{11} = 4$ and $R_{22} = 5$. Find $f(\frac{1}{2})$.

9. Romberg integration for approximating $\int_a^b f(x)\,dx$ gives $R_{11} = 8$, $R_{22} = \frac{16}{3}$, and $R_{33} = \frac{208}{45}$. Find R_{31}.

10. Suppose that $N(h)$ is an approximation to M for every $h > 0$ and that

$$M = N(h) + K_1 h + K_2 h^2 + K_3 h^3 + \cdots$$

for some constants $K_1, K_2, K_3, \ldots$. Use the values $N(h)$, $N\left(\frac{h}{3}\right)$, and $N\left(\frac{h}{9}\right)$ to produce an $O(h^3)$ approximation to M.

11. Suppose that $N(h)$ is an approximation to M for every $h > 0$ and that

$$M = N(h) + K_1 h^2 + K_2 h^4 + K_3 h^6 + \cdots$$

for some constants $K_1, K_2, K_3, \ldots$. Use the values $N(h)$, $N\left(\frac{h}{3}\right)$, and $N\left(\frac{h}{9}\right)$ to produce an $O(h^6)$ approximation to M.

12. We learn in calculus that $e = \lim_{h\to 0}(1 + h)^{1/h}$.

a. Determine approximations to e corresponding to $h = 0.04, 0.02$, and 0.01.

b. Use extrapolation on the approximations, assuming that constants $K_1, K_2, \ldots,$ exist with

$$e = (1 + h)^{1/h} + K_1 h + K_2 h^2 + K_3 h^3 + \cdots$$

to produce an $O(h^3)$ approximation to e, where $h = 0.04$.

c. Do you think that the assumption in part (b) is correct?

13. **a.** Show that

$$\lim_{h\to 0}\left(\frac{2+h}{2-h}\right)^{1/h} = e.$$

b. Compute approximations to e using the formula

$$N(h) = \left(\frac{2+h}{2-h}\right)^{1/h}$$

for $h = 0.04, 0.02$, and 0.01.

c. Assume that $e = N(h) + K_1 h + K_2 h^2 + K_3 h^3 + \cdots$. Use extrapolation, with at least 16 digits of precision, to compute an $O(h^3)$ approximation to e with $h = 0.04$. Do you think the assumption is correct?

d. Show that $N(-h) = N(h)$.

e. Use part (d) to show that $K_1 = K_3 = K_5 = \cdots = 0$ in the formula

$$e = N(h) + K_1 h + K_2 h^2 + K_3 h^3 + K_4 h^4 + K_5^5 + \cdots,$$

so that the formula reduces to

$$e = N(h) + K_2 h^2 + K_4 h^4 + K_6 h^6 + \cdots.$$

f. Use the results of part (e) and extrapolation to compute an $O(h^6)$ approximation to e with $h = 0.04$.

14. Suppose the following extrapolation table has been constructed to approximate the number M with $M = N_1(h) + K_1 h^2 + K_2 h^4 + K_3 h^6$:

$N_1(h)$		
$N_1\left(\frac{h}{2}\right)$	$N_2(h)$	
$N_1\left(\frac{h}{4}\right)$	$N_2\left(\frac{h}{2}\right)$	$N_3(h)$

a. Show that the linear interpolating polynomial $P_{0,1}(h)$ through $(h^2, N_1(h))$ and $(h^2/4, N_1(h/2))$ satisfies $P_{0,1}(0) = N_2(h)$. Similarly, show that $P_{1,2}(0) = N_2(h/2)$.

b. Show that the linear interpolating polynomial $P_{0,2}(h)$ through $(h^4, N_2(h))$ and $(h^4/16, N_2(h/2))$ satisfies $P_{0,2}(0) = N_3(h)$.

4.5 Gaussian Quadrature

The formulas in Section 4.2 for approximating the integral of a function were derived by integrating successively higher-order interpolating polynomials. The error term in the interpolating polynomial of degree n involves the $(n + 1)$st derivative of the function being approximated. Since every polynomial of degree less than or equal to n has zero for its $(n + 1)$st derivative, applying a formula of this type to such polynomials gives an exact result.

All the formulas in Section 4.2 use values of the function at equally spaced points. This is convenient when the formulas are combined to form the composite rules we considered in Section 4.3, but this restriction can significantly decrease the accuracy of the approximation. Consider, for example, the Trapezoidal rule applied to determine the integrals of the functions shown in Figure 4.9.

The Trapezoidal rule approximates the integral of the function by integrating the linear function that joins the endpoints of the graph of the function. But this is not likely the best

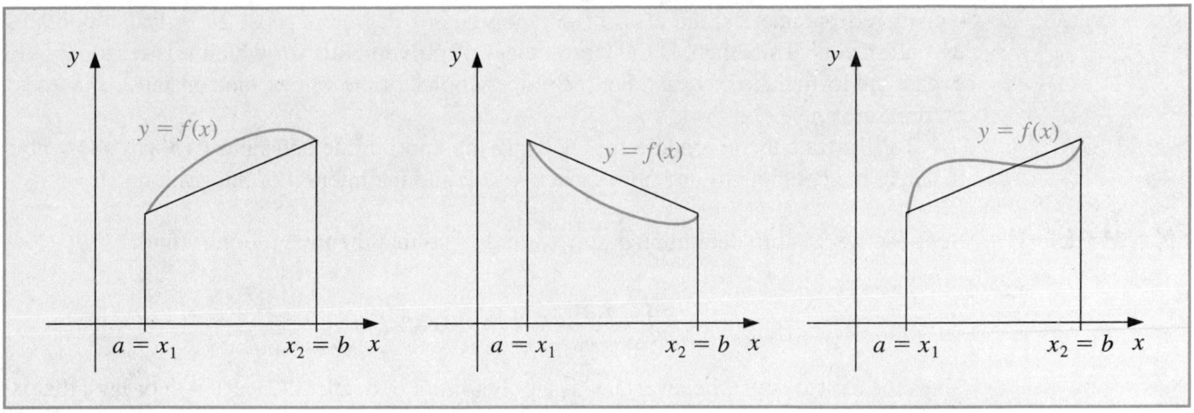

Figure 4.9

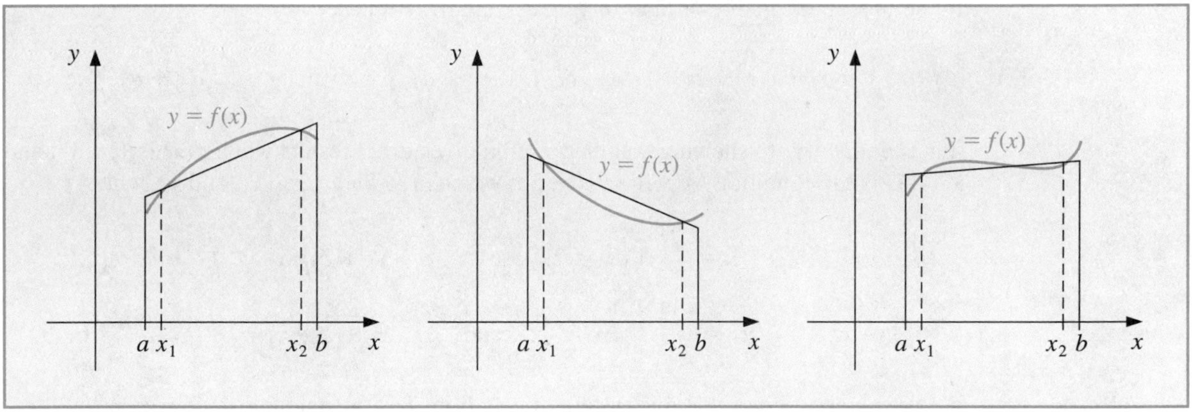

Figure 4.10

line for approximating the integral. Lines such at those shown in Figure 4.10 would give better approximations in most cases.

Gaussian quadrature chooses the points for evaluation in an optimal, rather than equally spaced, manner. The nodes $x_1, x_2, \ldots, x_n$ in the interval $[a, b]$ and coefficients $c_1, c_2, \ldots, c_n$ are chosen to minimize the expected error obtained in the approximation

$$\int_a^b f(x)\,dx \approx \sum_{i=1}^{n} c_i f(x_i).$$

To measure this accuracy, we assume that the best choice of these values is that which produces the exact result for the largest class of polynomials.

The coefficients $c_1, c_2, \ldots, c_n$ in the approximation formula are arbitrary, and the nodes $x_1, x_2, \ldots, x_n$ are restricted only by the fact that they must lie in $[a, b]$, the interval of integration. This gives $2n$ parameters to choose. If the coefficients of a polynomial are considered parameters, the class of polynomials of degree at most $2n-1$ also contains $2n$ parameters. This, then, is the largest class of polynomials for which it is reasonable to expect the formula to be exact. For the proper choice of the values and constants, exactness on this set can be obtained.

To illustrate the procedure for choosing the appropriate constants, we will show how to select the coefficients and nodes when $n=2$ and the interval of integration is $[-1,1]$.

EXAMPLE 1 Suppose we want to determine c_1, c_2, x_1, and x_2 so that the integration formula

$$\int_{-1}^{1} f(x)\,dx \approx c_1 f(x_1) + c_2 f(x_2)$$

gives the exact result whenever $f(x)$ is a polynomial of degree $2(2)-1=3$ or less, that is, when

$$f(x) = a_0 + a_1 x + a_2 x^2 + a_3 x^3$$

for some collection of constants, a_0, a_1, a_2, and a_3. Because

$$\int (a_0 + a_1 x + a_2 x^2 + a_3 x^3)\,dx = a_0 \int 1\,dx + a_1 \int x\,dx + a_2 \int x^2\,dx + a_3 \int x^3\,dx,$$

this is equivalent to showing that the formula gives exact results when $f(x)$ is $1, x, x^2$, and x^3; this is the condition we will satisfy. So, we need to find c_1, c_2, x_1, and x_2, with

$$c_1 \cdot 1 + c_2 \cdot 1 = \int_{-1}^{1} 1\,dx = 2, \qquad c_1 \cdot x_1 + c_2 \cdot x_2 = \int_{-1}^{1} x\,dx = 0,$$

$$c_1 \cdot x_1^2 + c_2 \cdot x_2^2 = \int_{-1}^{1} x^2\,dx = \frac{2}{3}, \quad \text{and} \quad c_1 \cdot x_1^3 + c_2 \cdot x_2^3 = \int_{-1}^{1} x^3\,dx = 0.$$

A little algebra shows that this system of equations has the unique solution

$$c_1 = 1, \qquad c_2 = 1, \qquad x_1 = -\frac{\sqrt{3}}{3}, \quad \text{and} \quad x_2 = \frac{\sqrt{3}}{3}.$$

This result produces the following integral approximation formula:

$$\int_{-1}^{1} f(x)\,dx \approx f\left(\frac{-\sqrt{3}}{3}\right) + f\left(\frac{\sqrt{3}}{3}\right),$$

which gives the exact result for every polynomial of degree 3 or less. ◆ ◆ ◆

The technique in Example 1 can be used to determine the nodes and coefficients for formulas that give exact results for higher-degree polynomials, but an alternative method obtains them more easily. In Section 8.3 we will consider various collections of orthogonal polynomials, functions that have the property that a particular definite integral of the product of any two of them is zero. The set that is relevant to our problem is the set of *Legendre polynomials*, a collection $\{P_0(x), P_1(x), \ldots, P_n(x), \ldots\}$ that has the following properties:

1. For each n, $P_n(x)$ is a polynomial of degree n.
2. $\int_{-1}^{1} P_i(x)P_j(x)\,dx = 0$ whenever $i \neq j$.

The first few Legendre polynomials are

$$P_0(x) = 1, \qquad P_1(x) = x, \qquad P_2(x) = x^2 - \frac{1}{3},$$

$$P_3(x) = x^3 - \frac{3}{5}x, \qquad \text{and} \qquad P_4(x) = x^4 - \frac{6}{7}x^2 + \frac{3}{35}.$$

The roots of these polynomials are distinct, lie in the interval $(-1, 1)$, have a symmetry with respect to the origin, and, most importantly, are the correct choice for determining the nodes that solve our problem.

The nodes $x_1, x_2, \ldots, x_n$ needed to produce an integral approximation formula that will give exact results for any polynomial of degree $2n - 1$ or less are the roots of the nth-degree Legendre polynomial. In addition, once the roots are known, the appropriate coefficients for the function evaluations at these nodes can be found from the fact that for each $i = 1, 2, \ldots, n$, we have

$$c_i = \int_{-1}^{1} \prod_{\substack{j=1 \\ j\neq i}}^{n} \frac{(x - x_j)}{(x_i - x_j)}\,dx.$$

However, both the roots of the Legendre polynomials and the coefficients are extensively tabulated, so it is not necessary to perform these evaluations. A small sample is given in Table 4.6, and listings for higher-degree polynomials are given, for example, in the book by Stroud and Secrest [StS].

Table 4.6

n	Roots $r_{n,i}$	Coefficients $c_{n,i}$
2	0.5773502692	1.0000000000
	−0.5773502692	1.0000000000
3	0.7745966692	0.5555555556
	0.0000000000	0.8888888889
	−0.7745966692	0.5555555556
4	0.8611363116	0.3478548451
	0.3399810436	0.6521451549
	−0.3399810436	0.6521451549
	−0.8611363116	0.3478548451
5	0.9061798459	0.2369268850
	0.5384693101	0.4786286705
	0.0000000000	0.5688888889
	−0.5384693101	0.4786286705
	−0.9061798459	0.2369268850

This completes the solution to the approximation problem for definite integrals of functions on the interval $[-1, 1]$. But this solution is sufficient for any closed interval since the simple linear relation

$$t = \frac{2x - a - b}{b - a}$$

transforms the interval $[a, b]$ into $[-1, 1]$. Then the Legendre polynomials can be used to approximate

$$\int_a^b f(x)\,dx = \int_{-1}^{1} f\left(\frac{(b-a)t + b + a}{2}\right)\frac{(b-a)}{2}\,dt.$$

Using the roots $r_{n,1}, r_{n,2}, \ldots, r_{n,n}$ and the coefficients $c_{n,1}, c_{n,2}, \ldots, c_{n,n}$ given in Table 4.6 produces the following approximation formula, which gives the exact result for a polynomial of degree $2n + 1$ or less.

Gaussian Quadrature

$$\int_a^b f(x)\,dx = \frac{b-a}{2}\sum_{j=1}^{n} c_{n,j} f\left(\frac{(b-a)r_{n,j} + b + a}{2}\right).$$

EXAMPLE 2 Consider the problem of finding approximations to $\int_1^{1.5} e^{-x^2}\,dx$, whose value to seven decimal places is 0.1093643.

Gaussian quadrature applied to this problem requires that the integral be transformed into one whose interval of integration is $[-1, 1]$:

$$\int_1^{1.5} e^{-x^2}\,dx = \frac{1}{4}\int_{-1}^{1} e^{-(t+5)^2/16}\,dt.$$

The values in Table 4.6 give the following Gaussian quadrature approximations.

$n = 2$:

$$\int_1^{1.5} e^{-x^2}\,dx \approx \frac{1}{4}\left[e^{-(5+0.5773502692)^2/16} + e^{-(5-0.5773502692)^2/16}\right] = 0.1094003,$$

$n = 3$:

$$\int_1^{1.5} e^{-x^2}\,dx \approx \frac{1}{4}\Big[(0.5555555556)e^{-(5+0.7745966692)^2/16} + (0.8888888889)e^{-(5)^2/16}$$
$$+ (0.5555555556)e^{-(5-0.7745966692)^2/16}\Big] = 0.1093642.$$

Using Gaussian quadrature with $n = 3$ required 3 function evaluations and produces an approximation that is accurate to within 10^{-7}. The same number of function evaluations is needed if Simpson's rule is applied to the original integral using $h = (1.5 - 1)/2 = 0.25$.

This application of Simpson's rule gives the approximation

$$\int_1^{1.5} e^{-x^2}\,dx \approx \frac{0.25}{3}\left(e^{-1} + 4e^{-(1.25)^2} + e^{-(1.5)^2}\right) = 0.1093104,$$

a result that is accurate only to within 5×10^{-5}. ◆ ◆ ◆

For small problems, Composite Simpson's rule may be acceptable to avoid the computational complexity of Gaussian quadrature, but for problems requiring expensive function evaluations, the Gaussian procedure should certainly be considered. Gaussian quadrature is particularly important for approximating multiple integrals since the number of function evaluations increases as a power of the number of integrals being evaluated. This topic is considered in Section 4.7.

EXERCISE SET 4.5

1. Approximate the following integrals using Gaussian quadrature with $n = 2$ and compare your results to the exact values of the integrals.

a. $\displaystyle\int_1^{1.5} x^2 \ln x\,dx$ **b.** $\displaystyle\int_0^{1} x^2 e^{-x}\,dx$

c. $\displaystyle\int_0^{0.35} \frac{2}{x^2-4}\,dx$ **d.** $\displaystyle\int_0^{\pi/4} x^2 \sin x\,dx$

e. $\displaystyle\int_0^{\pi/4} e^{3x} \sin 2x\,dx$ **f.** $\displaystyle\int_1^{1.6} \frac{2x}{x^2-4}\,dx$

g. $\displaystyle\int_3^{3.5} \frac{x}{\sqrt{x^2-4}}\,dx$ **h.** $\displaystyle\int_0^{\pi/4} (\cos x)^2\,dx$

2. Repeat Exercise 1 with $n = 3$.

3. Repeat Exercise 1 with $n = 4$.

4. Repeat Exercise 1 with $n = 5$.

5. Determine constants a, b, c, and d that will produce a quadrature formula

$$\int_{-1}^{1} f(x)\,dx = af(-1) + bf(1) + cf'(-1) + df'(1)$$

that gives exact results for polynomials of degree 3 or less.

6. Determine constants a, b, c, and d that will produce a quadrature formula

$$\int_{-1}^{1} f(x)\,dx = af(-1) + bf(0) + cf(1) + df'(-1) + ef'(1)$$

that gives exact results for polynomials of degree 4 or less.

7. Verify the entries for the values of $n = 2$ and 3 in Table 4.6 by finding the roots of the respective Legendre polynomials and use the equations preceding this table to find the coefficients associated with the values.

4.6 Adaptive Quadrature

In Section 4.3 we used composite methods of approximation to break an integral over a large interval into integrals over smaller subintervals. The approach used in that section involved equally-sized subintervals, which permitted us to combine the individual approximations into a convenient form. Although this is satisfactory for most problems, it leads to increased computation when the function being integrated varies widely on some, but not all, parts of the interval of integration. In this case, techniques that adapt to the differing accuracy needs of the interval are superior.

In this section we consider an Adaptive quadrature method and see how it can be used not only to reduce approximation error, but also to predict an error estimate for the approximation that does not rely on knowledge of higher derivatives of the function.

Suppose that we want to approximate $\int_a^b f(x)\,dx$ to within a specified tolerance $\epsilon > 0$. We first apply Simpson's rule with step size $h = (b-a)/2$, which gives (see Figure 4.11)

$$\int_a^b f(x)\,dx = S(a,b) - \frac{(b-a)^5}{2880} f^{(4)}(\xi) = S(a,b) - \frac{h^5}{90} f^{(4)}(\xi), \qquad \textbf{(4.6)}$$

for some ξ in (a,b), where

$$S(a,b) = \frac{h}{3}[f(a) + 4f(a+h) + f(b)].$$

Figure 4.11

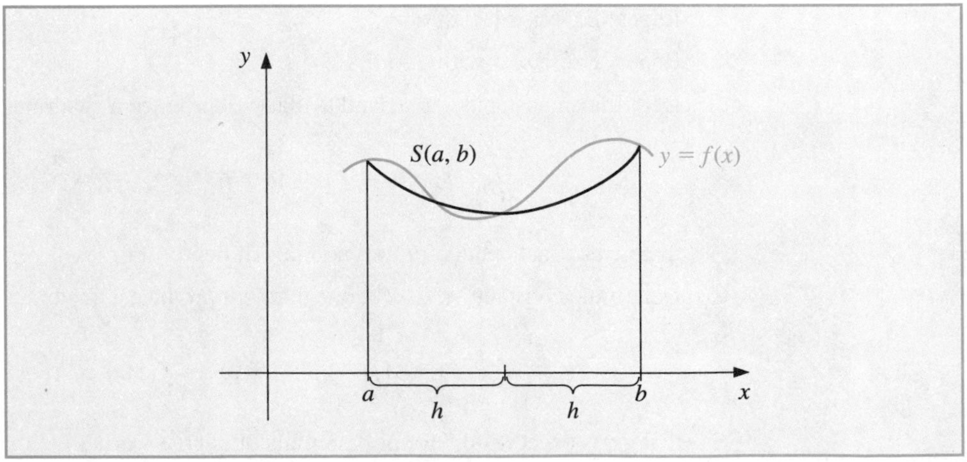

Next we determine an estimate for the accuracy of our approximation, in particular, one that does not require determining $f^{(4)}(\xi)$. To do this, we apply the Composite Simpson's rule to the problem with $n = 4$ and step size $(b-a)/4 = h/2$. Thus

$$\int_a^b f(x)\,dx = \frac{h}{6}\left[f(a) + 4f\left(a+\frac{h}{2}\right) + 2f(a+h) + 4f\left(a+\frac{3h}{2}\right) + f(b)\right] - \left(\frac{h}{2}\right)^4 \frac{(b-a)}{180} f^{(4)}(\tilde{\xi}), \tag{4.7}$$

for some $\tilde{\xi}$ in (a, b). To simplify notation, let

$$S\left(a, \frac{a+b}{2}\right) = \frac{h}{6}\left[f(a) + 4f\left(a+\frac{h}{2}\right) + f(a+h)\right]$$

and

$$S\left(\frac{a+b}{2}, b\right) = \frac{h}{6}\left[f(a+h) + 4f\left(a+\frac{3h}{2}\right) + f(b)\right].$$

Then Eq. (4.7) can be rewritten (see Figure 4.12) as

$$\int_a^b f(x)\,dx = S\left(a, \frac{a+b}{2}\right) + S\left(\frac{a+b}{2}, b\right) - \frac{1}{16}\left(\frac{h^5}{90}\right) f^{(4)}(\tilde{\xi}). \tag{4.8}$$

Figure 4.12

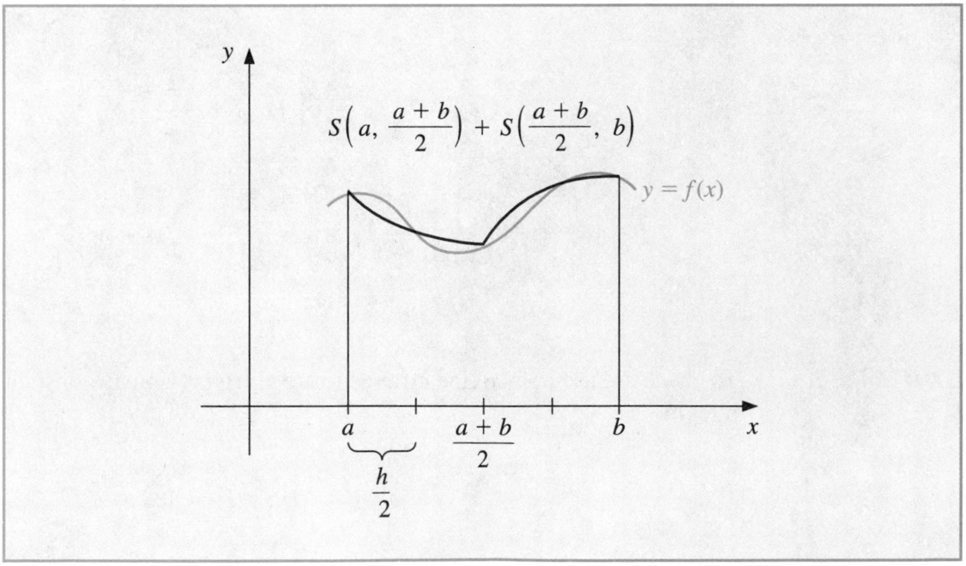

The error estimate is derived by assuming that $f^{(4)}(\xi) \approx f^{(4)}(\tilde{\xi})$, and the success of the technique depends on the accuracy of this assumption. If it is accurate, then equating the integrals in Eqs. (4.6) and (4.8) implies that

$$S\left(a, \frac{a+b}{2}\right) + S\left(\frac{a+b}{2}, b\right) - \frac{1}{16}\left(\frac{h^5}{90}\right) f^{(4)}(\xi) \approx S(a,b) - \frac{h^5}{90} f^{(4)}(\xi),$$

so

$$\frac{h^5}{90} f^{(4)}(\xi) \approx \frac{16}{15}\left[S(a,b) - S\left(a, \frac{a+b}{2}\right) - S\left(\frac{a+b}{2}, b\right)\right].$$

Using this estimate in Eq. (4.8) produces the error estimation

$$\left|\int_a^b f(x)\,dx - S\left(a, \frac{a+b}{2}\right) - S\left(\frac{a+b}{2}, b\right)\right|$$
$$\approx \frac{1}{15}\left|S(a,b) - S\left(a, \frac{a+b}{2}\right) - S\left(\frac{a+b}{2}, b\right)\right|.$$

This implies that $S(a, (a+b)/2) + S((a+b)/2, b)$ approximates $\int_a^b f(x)\,dx$ 15 times better than it agrees with the known value $S(a,b)$. As a consequence, $S(a, (a+b)/2) + S((a+b)/2, b)$ approximates $\int_a^b f(x)\,dx$ to within ϵ provided that the two approximations $S(a, (a+b)/2) + S((a+b)/2, b)$ and $S(a,b)$ differ by less than 15ϵ.

Adaptive Quadrature Error Estimate

If

$$\left|S(a,b) - S\left(a, \frac{a+b}{2}\right) - S\left(\frac{a+b}{2}, b\right)\right| < 15\epsilon,$$

then

$$\left|\int_a^b f(x)\,dx - S\left(a, \frac{a+b}{2}\right) - S\left(\frac{a+b}{2}, b\right)\right| < \epsilon.$$

EXAMPLE 1 To check the accuracy of the error estimate given previously consider its application to the integral

$$\int_0^{\pi/2} \sin x\,dx = 1.$$

In this case,

$$S\left(0, \frac{\pi}{2}\right) = \frac{(\pi/4)}{3}\left[\sin 0 + 4\sin\frac{\pi}{4} + \sin\frac{\pi}{2}\right] = \frac{\pi}{12}(2\sqrt{2} + 1) = 1.002279878$$

and

$$S\left(0,\frac{\pi}{4}\right)+S\left(\frac{\pi}{4},\frac{\pi}{2}\right)=\frac{(\pi/8)}{3}\left[\sin 0+4\sin\frac{\pi}{8}+2\sin\frac{\pi}{4}+4\sin\frac{3\pi}{8}+\sin\frac{\pi}{2}\right]$$
$$=1.000134585.$$

So,

$$\frac{1}{15}\left|S\left(0,\frac{\pi}{2}\right)-S\left(0,\frac{\pi}{4}\right)-S\left(\frac{\pi}{4},\frac{\pi}{2}\right)\right|=0.000143020.$$

This closely approximates the actual error,

$$\left|\int_0^{\pi/2}\sin x\,dx-1.000134585\right|=0.000134585,$$

even though $D_x^4 \sin x = \sin x$ varies significantly in the interval $(0, \pi/2)$. ◆ ◆ ◆

If an error tolerance $\epsilon > 0$ is specified and

$$\left|s(a,b)-s\left(a,\frac{a+b}{2}\right)-s\left(\frac{a+b}{2},b\right)\right|<15\epsilon,$$

then we assume that $s\left(a,\frac{a+b}{2}\right)+s\left(\frac{a+b}{2},b\right)$ is within ϵ of the value of $\int_a^b f(x)\,dx$.

When the approximation is not sufficiently accurate, the error-estimation procedure can be applied individually to the subintervals $[a,(a+b)/2]$ and $[(a+b)/2,b]$ to determine if the approximation to the integral on each subinterval is within a tolerance of $\epsilon/2$. If so, the sum of the approximations agrees with $\int_a^b f(x)\,dx$ to within the tolerance ϵ.

If the approximation on one of the subintervals fails to be within the tolerance $\epsilon/2$, that subinterval is itself subdivided, and each of its subintervals is analyzed to determine if the integral approximation on that subinterval is accurate to within $\epsilon/4$.

The halving procedure is continued until each portion is within the required tolerance. Although problems can be constructed for which this tolerance will never be met, the technique is successful for most problems because each subdivision increases the accuracy of the approximation by a factor of approximately 15 while requiring an increased accuracy factor of only 2.

The program ADAPQR43 applies the adaptive quadrature procedure for Simpson's rule. Some technical difficulties require the implementation of the method to differ slightly from the preceding discussion. The tolerance between successive approximations has been conservatively set at 10ϵ rather than the derived value of 15ϵ to compensate for possible error in the assumption $f^{(4)}(\xi) \approx f^{(4)}(\tilde{\xi})$. In problems when $f^{(4)}$ is known to be widely varying, it would be reasonable to lower this bound further.

EXAMPLE 2 The graph of the function $f(x) = (100/x^2)\sin(10/x)$ for x in $[1,3]$ is shown in Figure 4.13. The program ADAPQR43 with tolerance 10^{-4} to approximate $\int_1^3 f(x)\,dx$ produces -1.426014, a result that is accurate to within 1.1×10^{-5}. The approximation required that

Simpson's rule with $n = 4$ be performed on the 23 subintervals whose endpoints are shown on the horizontal axis in Figure 4.13. The total number of functional evaluations required for this approximation is 93.

Figure 4.13

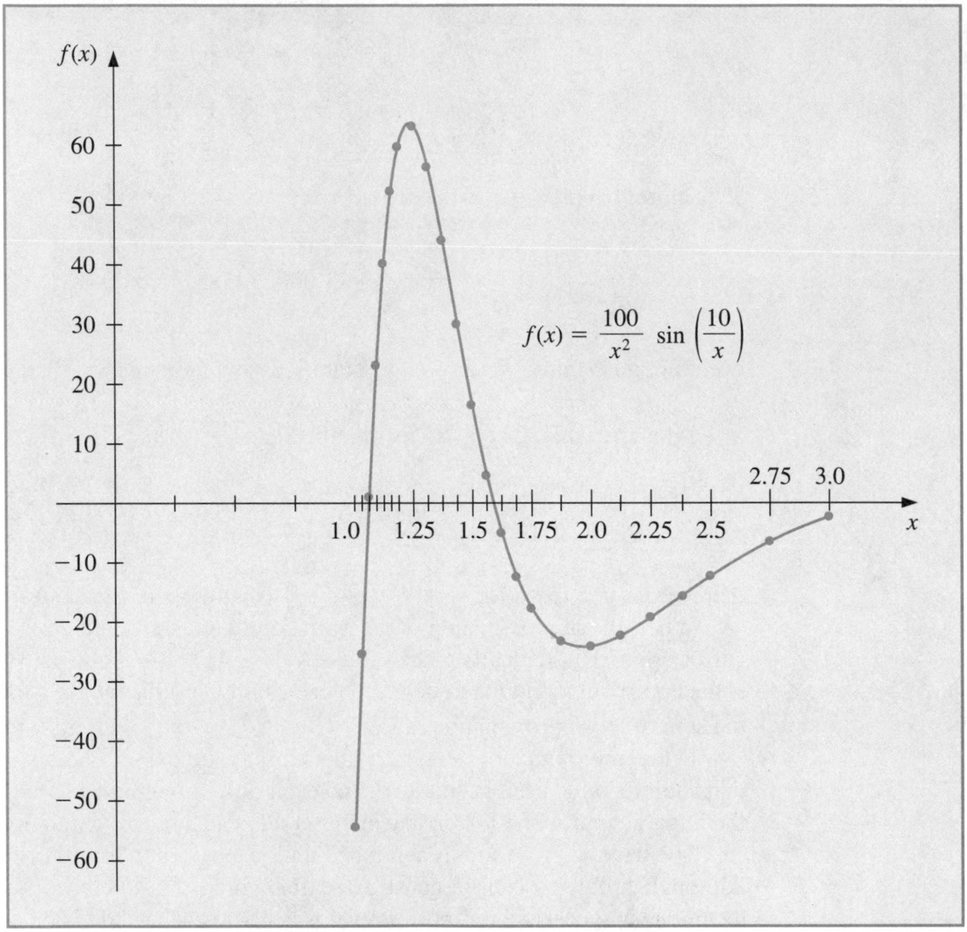

The largest value of h for which the standard Composite Simpson's rule gives 10^{-4} accuracy is $h = \frac{1}{88}$. This application requires 177 function evaluations, nearly twice as many as the adaptive technique. ◆ ◆ ◆

EXERCISE SET 4.6

1. Compute the three Simpson's rule approximations $S(a, b)$, $S(a, (a + b)/2)$, and $S((a + b)/2, b)$ for the following integrals, and verify the estimate given in the approximation formula.

a. $\int_1^{1.5} x^2 \ln x \, dx$

b. $\int_0^1 x^2 e^{-x} \, dx$

c. $\int_0^{0.35} \frac{2}{x^2 - 4} \, dx$

d. $\int_0^{\pi/4} x^2 \sin x \, dx$

e. $\int_0^{\pi/4} e^{3x} \sin 2x \, dx$

f. $\int_1^{1.6} \frac{2x}{x^2 - 4} \, dx$

g. $\int_3^{3.5} \frac{x}{\sqrt{x^2 - 4}} \, dx$

h. $\int_0^{\pi/4} (\cos x)^2 \, dx$

2. Use Adaptive quadrature to find approximations to within 10^{-3} for the integrals in Exercise 1. Do not use a computer program to generate these results.

3. Use Adaptive quadrature to approximate the following integrals to within 10^{-5}.

a. $\int_1^3 e^{2x} \sin 3x \, dx$

b. $\int_1^3 e^{3x} \sin 2x \, dx$

c. $\int_0^5 \left[2x \cos(2x) - (x - 2)^2\right] dx$

d. $\int_0^5 \left[4x \cos(2x) - (x - 2)^2\right] dx$

4. Use Simpson's Composite rule with $n = 4, 6, 8, \ldots$, until successive approximations to the following integrals agree to within 10^{-6}. Determine the number of nodes required. Use Adaptive quadrature to approximate the integral to within 10^{-6} and count the number of nodes. Did Adaptive quadrature produce any improvement?

a. $\int_0^{\pi} x \cos x^2 \, dx$

b. $\int_0^{\pi} x \sin x^2 \, dx$

c. $\int_0^{\pi} x^2 \cos x \, dx$

d. $\int_0^{\pi} x^2 \sin x \, dx$

5. Sketch the graphs of $\sin(1/x)$ and $\cos(1/x)$ on $[0.1, 2]$. Use Adaptive quadrature to approximate the integrals

$$\int_{0.1}^2 \sin \frac{1}{x} \, dx \quad \text{and} \quad \int_{0.1}^2 \cos \frac{1}{x} \, dx$$

to within 10^{-3}.

6. Let $T(a, b)$ and $T(a, \frac{a+b}{2}) + T(\frac{a+b}{2}, b)$ be the single and double applications of the Trapezoidal rule to $\int_a^b f(x) \, dx$. Determine a relationship between

$$\left| T(a, b) - \left(T\left(a, \frac{a+b}{2}\right) + T\left(\frac{a+b}{2}, b\right) \right) \right|$$

and

$$\left| \int_a^b f(x) \, dx - \left(T\left(a, \frac{a+b}{2}\right) + T\left(\frac{a+b}{2}, b\right) \right) \right|.$$

7. The differential equation

$$mu''(t) + ku(t) = F_0 \cos \omega t$$

describes a spring-mass system with mass m, spring constant k, and no applied damping. The term $F_0 \cos \omega t$ describes a periodic external force applied to the system. The solution to the equation when the system is initially at rest ($u'(0) = u(0) = 0$) is

$$u(t) = \frac{2F_0}{m(\omega_0^2 - \omega^2)} \sin \frac{(\omega_0 - \omega)}{2} t \sin \frac{(\omega_0 + \omega)}{2} t, \quad \text{where} \quad \omega_0 = \sqrt{\frac{k}{m}} \neq \omega.$$

Sketch the graph of u when $m = 1, k = 9, F_0 = 1, \omega = 2$, and $t \in [0, 2\pi]$. Approximate $\int_0^{2\pi} u(t)\, dt$ to within 10^{-4}.

8. If the term $cu'(t)$ is added to the left side of the motion equation in Exercise 7, the resulting differential equation describes a spring-mass system that is damped with damping constant c. The solution to this equation when the solution is initially at rest is

$$u(t) = c_1 e^{r_1 t} + c_2 e^{r_2 t} + \frac{F_0}{\sqrt{m^2(\omega_0^2 - \omega^2)^2 + c^2\omega^2}} \cos(\omega t - \delta),$$

where

$$\delta = \arctan\left(\frac{c\omega}{m(\omega_0^2 - \omega^2)}\right), \qquad r_1 = \frac{-c + \sqrt{c^2 - 4\omega_0^2 m^2}}{2m},$$

and

$$r_2 = \frac{-c - \sqrt{c^2 - 4\omega_0^2 m^2}}{2m}.$$

Sketch the graph of u when $m = 1, k = 9, F_0 = 1, c = 10, \omega = 2$ and $t \in [0, 2\pi]$. Approximate $\int_0^{2\pi} u(t)\, dt$ to within 10^{-4}.

9. The study of light diffraction at a rectangular aperture involves the Fresnel integrals

$$c(t) = \int_0^t \cos \frac{\pi}{2} w^2\, dw \quad \text{and} \quad s(t) = \int_0^t \sin \frac{\pi}{2} w^2\, dw.$$

Construct a table of values for $c(t)$ and $s(t)$ that is accurate to within 10^{-4} for values of $t = 0.1, 0.2, \ldots, 1.0$.

4.7 Multiple Integrals

The techniques discussed in the previous sections can be modified in a straightforward manner for use in the approximation of multiple integrals. Let us first consider the double integral

$$\iint_R f(x, y)\, dA,$$

where R is a rectangular region in the plane:

$$R = \{(x, y) \mid a \leq x \leq b, \quad c \leq y \leq d\}$$

for some constants a, b, c, and d (see Figure 4.14). We will employ the Composite Simpson's rule to illustrate the approximation technique, although any other approximation formula could be used without major modifications.

Figure 4.14

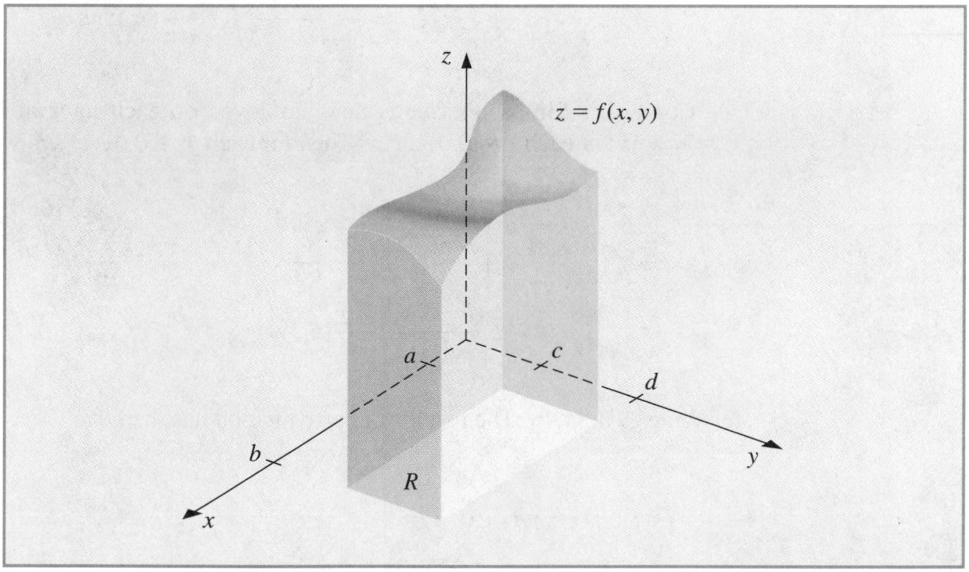

Suppose that even integers n and m are chosen to determine the step sizes $h = (b-a)/n$ and $k = (d - c)/m$. We first write the double integral as an iterated integral,

$$\iint_R f(x, y)\, dA = \int_a^b \left(\int_c^d f(x, y)\, dy \right) dx,$$

and use the Composite Simpson's rule to approximate

$$\int_c^d f(x, y)\, dy,$$

treating x as a constant. Let $y_j = c + jk$ for each $j = 0, 1, \ldots, m$. Then

$$\int_c^d f(x,y)\,dy = \frac{k}{3}\left[f(x,y_0) + 2\sum_{j=1}^{(m/2)-1} f(x,y_{2j}) + 4\sum_{j=1}^{m/2} f(x,y_{2j-1}) + f(x,y_m)\right] - \frac{(d-c)k^4}{180}\frac{\partial^4 f}{\partial y^4}(x,\mu)$$

for some μ in (c, d). Thus,

$$\int_a^b\int_c^d f(x,y)\,dy\,dx = \frac{k}{3}\int_a^b f(x,y_0)\,dx + \frac{2k}{3}\sum_{j=1}^{(m/2)-1}\int_a^b f(x,y_{2j})\,dx + \frac{4k}{3}\sum_{j=1}^{m/2}\int_a^b f(x,y_{2j-1})\,dx + \frac{k}{3}\int_a^b f(x,y_m)\,dx - \frac{(d-c)k^4}{180}\int_a^b \frac{\partial^4 f}{\partial y^4}(x,\mu)\,dx.$$

Composite Simpson's rule is now employed on each integral in this equation. Let $x_i = a + ih$ for each $i = 0, 1, \ldots, n$. Then for each $j = 0, 1, \ldots, m$, we have

$$\int_a^b f(x,y_j)\,dx = \frac{h}{3}\left[f(x_0,y_j) + 2\sum_{i=1}^{(n/2)-1} f(x_{2i},y_j) + 4\sum_{i=1}^{n/2} f(x_{2i-1},y_j) + f(x_n,y_j)\right] - \frac{(b-a)h^4}{180}\frac{\partial^4 f}{\partial x^4}(\xi_j,y_j)$$

for some ξ_j in (a, b). The resulting approximation has the form

$$\begin{aligned}\int_a^b\int_c^d f(x,y)\,dy\,dx &\approx \frac{hk}{9}\Bigg\{\left[f(x_0,y_0) + 2\sum_{i=1}^{(n/2)-1} f(x_{2i},y_0) + 4\sum_{i=1}^{n/2} f(x_{2i-1},y_0) + f(x_n,y_0)\right]\\ &+ 2\left[\sum_{j=1}^{(m/2)-1} f(x_0,y_{2j}) + 2\sum_{j=1}^{(m/2)-1}\sum_{i=1}^{(n/2)-1} f(x_{2i},y_{2j})\right.\\ &\left. + 4\sum_{j=1}^{(m/2)-1}\sum_{i=1}^{n/2} f(x_{2i-1},y_{2j}) + \sum_{j=1}^{(m/2)-1} f(x_n,y_{2j})\right]\\ &+ 4\left[\sum_{j=1}^{m/2} f(x_0,y_{2j-1}) + 2\sum_{j=1}^{m/2}\sum_{i=1}^{(n/2)-1} f(x_{2i},y_{2j-1})\right.\end{aligned}$$

$$+ 4\sum_{j=1}^{m/2}\sum_{i=1}^{n/2} f(x_{2i-1}, y_{2j-1}) + \sum_{j=1}^{m/2} f(x_n, y_{2j-1})\Bigg]$$

$$+ \left[f(x_0, y_m) + 2\sum_{i=1}^{(n/2)-1} f(x_{2i}, y_m) + 4\sum_{i=1}^{n/2} f(x_{2i-1}, y_m) + f(x_n, y_m)\right]\Bigg\}.$$

The error term, E, is given by

$$E = \frac{-k(b-a)h^4}{540}\left[\frac{\partial^4 f}{\partial x^4}(\xi_0, y_0) + 2\sum_{j=1}^{(m/2)-1}\frac{\partial^4 f}{\partial x^4}(\xi_{2j}, y_{2j}) + 4\sum_{j=1}^{m/2}\frac{\partial^4 f}{\partial x^4}(\xi_{2j-1}, y_{2j-1})\right.$$

$$\left. + \frac{\partial^4 f}{\partial x^4}(\xi_m, y_m)\right] - \frac{(d-c)k^4}{180}\int_a^b \frac{\partial^4 f}{\partial y^4}(x, \mu)\, dx.$$

If $\partial^4 f/\partial x^4$ and $\partial^4 f/\partial y^4$ are continuous, the Intermediate Value Theorem and Mean Value Theorem for Integrals can be used to show that the error formula can be simplified to

$$E = \frac{-(d-c)(b-a)}{180}\left[h^4\frac{\partial^4 f}{\partial x^4}(\bar{\eta}, \bar{\mu}) + k^4\frac{\partial^4 f}{\partial y^4}(\hat{\eta}, \hat{\mu})\right]$$

for some $(\bar{\eta}, \bar{\mu})$ and $(\hat{\eta}, \hat{\mu})$ in R.

EXAMPLE 1 The Composite Simpson's rule applied to approximate

$$\int_{1.4}^{2.0}\int_{1.0}^{1.5} \ln(x + 2y)\, dy\, dx$$

with $n = 4$ and $m = 2$ uses the step sizes $h = 0.15$ and $k = 0.25$. The region of integration R is shown in Figure 4.15 together with the nodes (x_i, y_j) for $i = 0, 1, 2, 3, 4$ and $j = 0, 1, 2$, and the coefficients $w_{i,j}$ of $f(x_i, y_i) = \ln(x_i + 2y_i)$ in the sum.

Figure 4.15

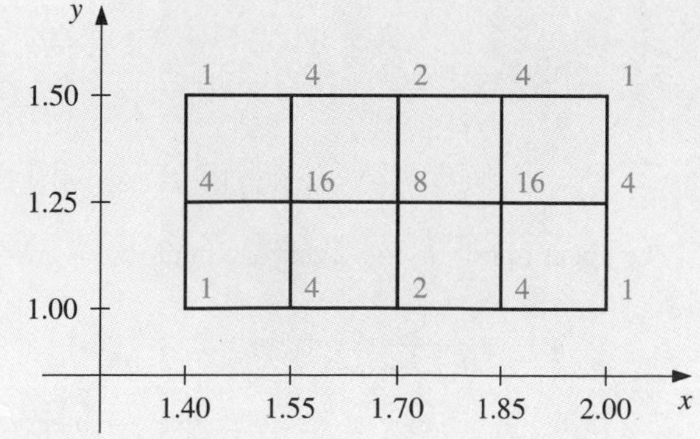

The approximation is

$$\int_{1.4}^{2.0}\int_{1.0}^{1.5} \ln(x+2y)\,dy\,dx \approx \frac{(0.15)(0.25)}{9}\sum_{i=0}^{4}\sum_{j=0}^{2} w_{i,j}\ln(x_i+2y_j)$$
$$= 0.4295524387.$$

Since

$$\frac{\partial^4 f}{\partial x^4}(x,y) = \frac{-6}{(x+2y)^4} \quad \text{and} \quad \frac{\partial^4 f}{\partial y^4}(x,y) = \frac{-96}{(x^2+2y)^4},$$

the error is bounded by

$$|E| \le \frac{(0.5)(0.6)}{180}\left[(0.15)^4 \max_{(x,y) \text{ in } R}\frac{6}{(x+2y)^4} + (0.25)^4 \max_{(x,y) \text{ in } R}\frac{96}{(x+2y)^4}\right] \le 4.72 \times 10^{-6}.$$

The actual value of the integral to 10 decimal places is

$$\int_{1.4}^{2.0}\int_{1.0}^{1.5} \ln(x+2y)\,dy\,dx = 0.4295545265,$$

so the approximation is accurate to within 2.1×10^{-6}. ◆ ◆ ◆

The same techniques can be applied for the approximation of triple integrals, as well as higher integrals for functions of more than three variables. The number of functional evaluations required for the approximation is the product of the number required when the method is applied to each variable.

To reduce the number of functional evaluations, more efficient methods such as Gaussian quadrature, Romberg integration, or Adaptive quadrature can be incorporated in place of Simpson's formula. The following example illustrates the use of Gaussian quadrature for the integral considered in Example 1.

EXAMPLE 2 Consider the double integral given in Example 1. Before employing a Gaussian quadrature technique to approximate this integral, we must transform the region of integration

$$R = \{(x,y) \mid 1.4 \le x \le 2.0, \quad 1.0 \le y \le 1.5\}$$

into

$$\hat{R} = \{(u,v) \mid -1 \le u \le 1, \quad -1 \le v \le 1\}.$$

The linear transformations that accomplish this are

$$u = \frac{1}{2.0-1.4}(2x - 1.4 - 2.0), \quad \text{and} \quad v = \frac{1}{1.5-1.0}(2y - 1.0 - 1.5).$$

Employing this change of variables gives an integral on which Gaussian quadrature can be applied:

$$\int_{1.4}^{2.0}\int_{1.0}^{1.5} \ln(x+2y)\,dy\,dx = 0.075\int_{-1}^{1}\int_{-1}^{1} \ln(0.3u+0.5v+4.2)\,dv\,du.$$

The Gaussian quadrature formula for $n = 3$ in both u and v requires that we use the nodes

$$u_1 = v_1 = r_{3,2} = 0, \quad u_0 = v_0 = r_{3,1} = -0.7745966692,$$
$$\text{and} \quad u_2 = v_2 = r_{3,3} = 0.7745966692.$$

The associated weights are found in Table 4.3 (Section 4.5) to be $c_{3,2} = 0.8\overline{8}$ and $c_{3,1} = c_{3,3} = 0.5\overline{5}$, so

$$\begin{aligned}\int_{1.4}^{2.0}\int_{1.0}^{1.5} \ln(x+2y)\,dy\,dx &\approx 0.075\sum_{i=1}^{3}\sum_{j=1}^{3} c_{3,i}c_{3,j}\ln(0.3r_{3,i}+0.5r_{3,j}+4.2)\\ &= 0.4295545313.\end{aligned}$$

Even though this result requires only 9 functional evaluations compared to 15 for the Composite Simpson's rule considered in Example 1, the result is accurate to within 4.8×10^{-9}, compared to an accuracy of only 2×10^{-6} for Simpson's rule. ◆ ◆ ◆

The use of approximation methods for double integrals is not limited to integrals with rectangular regions of integration. The techniques previously discussed can be modified to approximate double integrals with variable inner limits—that is, integrals of the form

$$\int_a^b\int_{c(x)}^{d(x)} f(x,y)\,dy\,dx.$$

For this type of integral we begin as before by applying Simpson's Composite rule to integrate with respect to both variables. The step size for the variable x is $h = (b-a)/2$, but the step size $k(x)$ for y varies with x (see Figure 4.16 on page 152):

$$k(x) = \frac{d(x)-c(x)}{2}.$$

Consequently,

$$\begin{aligned}\int_a^b\int_{c(x)}^{d(x)} f(x,y)\,dy\,dx &\approx \int_a^b \frac{k(x)}{3}[f(x,c(x)) + 4f(x,c(x)+k(x)) + f(x,d(x))]\,dx\\
&\approx \frac{h}{3}\bigg\{\frac{k(a)}{3}[f(a,c(a)) + 4f(a,c(a)+k(a)) + f(a,d(a))]\\
&\quad + \frac{4k(a+h)}{3}[f(a+h,c(a+h)) + 4f(a+h,c(a+h)\\
&\quad + k(a+h)) + f(a+h,d(a+h))]\\
&\quad + \frac{k(b)}{3}[f(b,c(b)) + 4f(b,c(b)+k(b)) + f(b,d(b))]\bigg\}.\end{aligned}$$

Figure 4.16

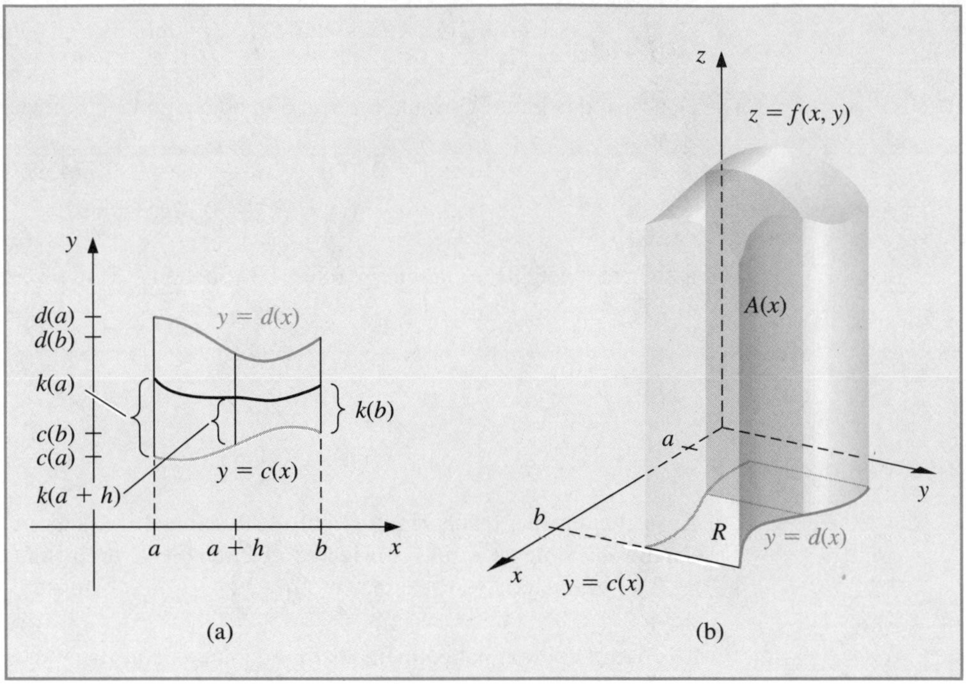

The program DINTGL44 applies the Composite Simpson's rule to a double integral in this form and is also appropriate, of course, when $c(x) \equiv c$ and $d(x) \equiv d$.

To apply Gaussian quadrature to the double integral first requires transforming, for each x in $[a, b]$, the interval $[c(x), d(x)]$ to $[-1, 1]$ and then applying Gaussian quadrature. This results in the formula

$$\int_a^b \int_{c(x)}^{d(x)} f(x, y)\, dy\, dx$$

$$\approx \int_a^b \frac{d(x) - c(x)}{2} \sum_{j=1}^{n} c_{n,j} f\left(x, \frac{(d(x) - c(x))r_{n,j} + d(x) + c(x)}{2}\right) dx,$$

where, as before, the roots $r_{n,j}$ and coefficients $c_{n,j}$ come from Table 4.6. Now the interval $[a, b]$ is transformed to $[-1, 1]$, and Gaussian quadrature is applied to approximate the integral on the right side of this equation. The program DGQINT45 uses this technique.

EXAMPLE 3 Applying Simpson's double integral program DINTGL44 with $n = m = 10$ to

$$\int_{0.1}^{0.5} \int_{x^3}^{x^2} e^{y/x}\, dy\, dx$$

requires 121 evaluations of the function $f(x, y) = e^{y/x}$ and produces the approximation 0.0333054, accurate to nearly 7 decimal places, to the volume of the solid shown in Figure

4.17. Applying the Gaussian quadrature program DGQINT45 with $n = m = 5$ requires only 25 function evaluations and gives the approximation, 0.3330556611, which is accurate to 11 decimal places. ◆ ◆ ◆

Figure 4.17

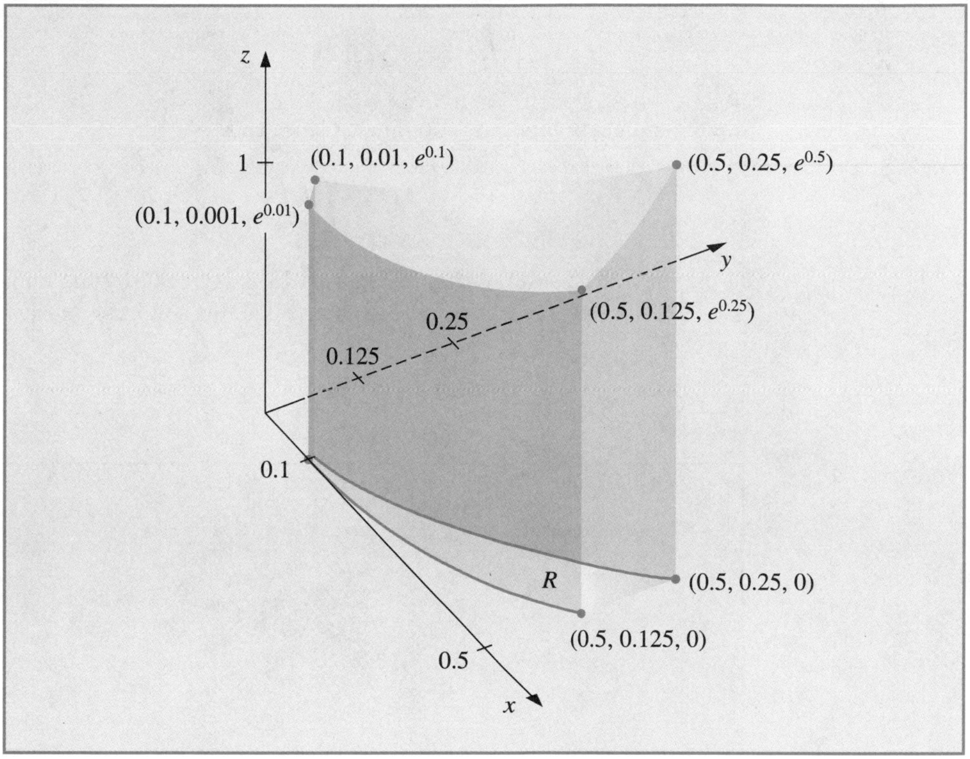

Triple integrals of the form

$$\int_a^b \int_{c(x)}^{d(x)} \int_{\alpha(x,y)}^{\beta(x,y)} f(x, y, z)\, dz\, dy\, dx$$

are approximated in a similar manner. Because of the number of calculations involved, Gaussian quadrature is the method of choice. The program TINTGL46 implements this procedure.

The following example requires the evaluation of four triple integrals.

EXAMPLE 4 The center of mass of a solid region D with density function σ occurs at

$$(\bar{x}, \bar{y}, \bar{z}) = \left(\frac{M_{yz}}{M}, \frac{M_{xz}}{M}, \frac{M_{xy}}{M} \right),$$

where

$$M_{yz} = \iiint_D x\sigma(x,y,z)\,dV, \qquad M_{xz} = \iiint_D y\sigma(x,y,z)\,dV,$$

and

$$M_{xy} = \iiint_D z\sigma(x,y,z)\,dV$$

are the moments about the coordinate planes and

$$M = \iiint_D \sigma(x,y,z)\,dV$$

is the mass. The solid shown in Figure 4.18 is bounded by the upper nappe of the cone $z^2 = x^2 + y^2$ and the plane $z = 2$. Suppose that this solid has density function given by

$$\sigma(x,y,z) = \sqrt{x^2+y^2}.$$

Figure 4.18

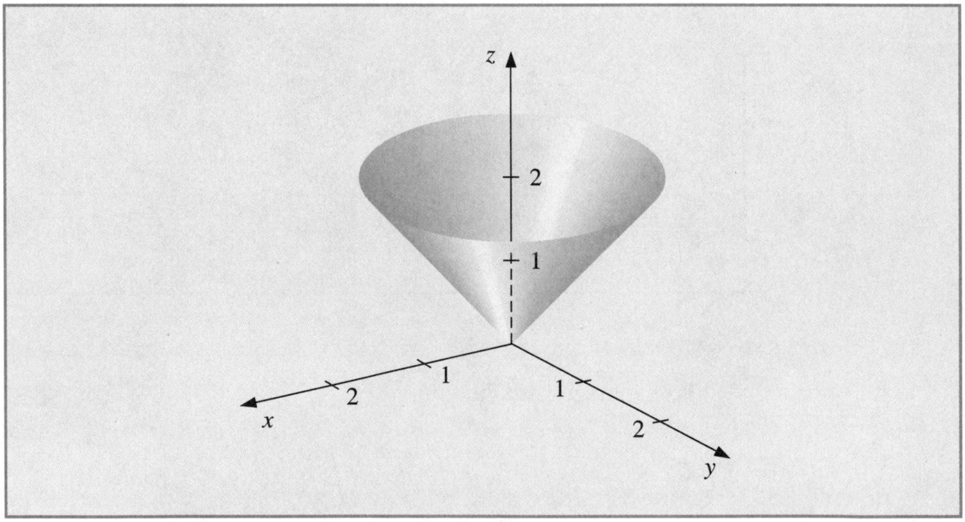

Applying TINTGL46 with $n = m = p = 5$ requires 125 function evaluations per integral and gives the following approximations:

$$M = \int_{-2}^{2}\int_{-\sqrt{4-x^2}}^{\sqrt{4-x^2}}\int_{\sqrt{x^2+y^2}}^{2} \sqrt{x^2+y^2}\,dz\,dy\,dx$$
$$= 4\int_{0}^{2}\int_{0}^{\sqrt{4-x^2}}\int_{\sqrt{x^2+y^2}}^{2} \sqrt{x^2+y^2}\,dz\,dy\,dx \approx 8.37504476,$$

$$M_{yz} = \int_{-2}^{2} \int_{-\sqrt{4-x^2}}^{\sqrt{4-x^2}} \int_{\sqrt{x^2+y^2}}^{2} x\sqrt{x^2+y^2}\, dz\, dy\, dx \approx -5.55111512 \times 10^{-17},$$

$$M_{xz} = \int_{-2}^{2} \int_{-\sqrt{4-x^2}}^{\sqrt{4-x^2}} \int_{\sqrt{x^2+y^2}}^{2} y\sqrt{x^2+y^2}\, dz\, dy\, dx \approx -8.01513675 \times 10^{-17},$$

$$M_{xy} = \int_{-2}^{2} \int_{-\sqrt{4-x^2}}^{\sqrt{4-x^2}} \int_{\sqrt{x^2+y^2}}^{2} z\sqrt{x^2+y^2}\, dz\, dy\, dx \approx 13.40038156.$$

This implies that the approximate location of the center of mass is $\sqrt{(\bar{x}, \bar{y}, \bar{z})} = (0, 0, 1.60003701)$. These integrals are quite easy to evaluate directly. If you do this, you will find that the center of mass occurs at $(0, 0, 1.6)$. ◆ ◆ ◆

EXERCISE SET 4.7

1. Use Composite Simpson's rule for double integrals with $n = m = 4$ to approximate the following double integrals. Compare the results to the exact answer.

   a. $\displaystyle\int_{2.1}^{2.5} \int_{1.2}^{1.4} xy^2\, dy\, dx$ b. $\displaystyle\int_{0}^{0.5} \int_{0}^{0.5} e^{y-x}\, dy\, dx$

   c. $\displaystyle\int_{2}^{2.2} \int_{x}^{2x} (x^2 + y^3)\, dy\, dx$ d. $\displaystyle\int_{1}^{1.5} \int_{0}^{x} (x^2 + \sqrt{y})\, dy\, dx$

2. Find the smallest values for $n = m$ so that Composite Simpson's rule for double integrals can be used to approximate the integrals in Exercise 1 to within 10^{-6} of the actual value.

3. Use Composite Simpson's rule for double integrals with $n = 4, m = 8$ and $n = 8, m = 4$ and $n = m = 6$ to approximate the following double integrals. Compare the results to the exact answer.

   a. $\displaystyle\int_{0}^{\pi/4} \int_{\sin x}^{\cos x} (2y \sin x + \cos^2 x)\, dy\, dx$ b. $\displaystyle\int_{1}^{e} \int_{1}^{x} \ln xy\, dy\, dx$

   c. $\displaystyle\int_{0}^{1} \int_{x}^{2x} (x^2 + y^3)\, dy\, dx$ d. $\displaystyle\int_{0}^{1} \int_{x}^{2x} (y^2 + x^3)\, dy\, dx$

   e. $\displaystyle\int_{0}^{\pi} \int_{0}^{x} \cos x\, dy\, dx$ f. $\displaystyle\int_{0}^{\pi} \int_{0}^{x} \cos y\, dy\, dx$

   g. $\displaystyle\int_{0}^{\pi/4} \int_{0}^{\sin x} \frac{1}{\sqrt{1-y^2}}\, dy\, dx$ h. $\displaystyle\int_{-\pi}^{3\pi/2} \int_{0}^{2\pi} (y \sin x + x \cos y)\, dy\, dx$

4. Find the smallest values for $n = m$ so that Composite Simpson's rule for double integrals can be used to approximate the integrals in Exercise 3 to within 10^{-6} of the actual value.

5. Use Gaussian quadrature for double integrals with $n = m = 2$ to approximate the integrals in Exercise 1 and compare the results to those obtained in Exercise 1.

6. Find the smallest values of $n = m$ so that Gaussian quadrature for double integrals may be used to approximate the integrals in Exercise 1 to within 10^{-6}. Do not continue beyond $n = m = 5$. Compare the number of functional evaluations required to the number required in Exercise 2.

7. Use Gaussian quadrature for double integrals with $n = m = 3$; $n = 3, m = 4$; $n = 4, m = 3$ and $n = m = 4$ to approximate the integrals in Exercise 3.

8. Use Gaussian quadrature for double integrals with $n = m = 5$ to approximate the integrals in Exercise 3. Compare the number of functional evaluations required to the number required in Exercise 4.

9. Use Gaussian quadrature for triple integrals with $n = m = p = 2$ to approximate the following triple integrals, and compare the results to the exact answer.

a. $\int_0^1 \int_1^2 \int_0^{0.5} e^{x+y+z}\, dz\, dy\, dx$ **b.** $\int_0^1 \int_x^1 \int_0^y y^2 z\, dz\, dy\, dx$

c. $\int_0^1 \int_{x^2}^x \int_{x-y}^{x+y} y\, dz\, dy\, dx$ **d.** $\int_0^1 \int_{x^2}^x \int_{x-y}^{x+y} z\, dz\, dy\, dx$

e. $\int_0^\pi \int_0^x \int_0^{xy} \frac{1}{y} \sin \frac{z}{y}\, dz\, dy\, dx$ **f.** $\int_0^1 \int_0^1 \int_{-xy}^{xy} e^{x^2+y^2}\, dz\, dy\, dx$

10. Repeat Exercise 9 using $n = m = p = 3$.

11. Use Composite Simpson's rule for double integrals with $n = m = 14$ and Gaussian quadrature for double integrals with $n = m = 4$ to approximate

$$\iint_R e^{-(x+y)}\, dA$$

for the region R in the plane bounded by the curves $y = x^2$ and $y = \sqrt{x}$.

12. Use Composite Simpson's rule for double integrals to approximate

$$\iint_R \sqrt{xy + y^2}\, dA,$$

where R is the region in the plane bounded by the lines $x + y = 6$, $3y - x = 2$, and $3x - y = 2$. First partition R into two regions, R_1 and R_2, on which Composite Simpson's rule for double integrals can be applied. Use $n = m = 6$ on both R_1 and R_2.

13. The area of the surface described by $z = f(x, y)$ for (x, y) in R is given by

$$\iint_R \sqrt{[f_x(x, y)]^2 + [f_y(x, y)]^2 + 1}\, dA.$$

Find an approximation to the area of the surface on the hemisphere $x^2 + y^2 + z^2 = 9$, $z \geq 0$ that lies above the region $R = \{(x, y) \mid 0 \leq x \leq 1, 0 \leq y \leq 1\}$, using each program.

a. DINTGL44 with $n = m = 8$

b. DGQINT45 with $n = m = 4$

14. A plane lamina is defined to be a thin sheet of continuously distributed mass. If σ is a function describing the density of a lamina having the shape of a region R in the xy-plane, then the center of the mass of the lamina $(\bar{x}, \bar{y})$ is defined by

$$\bar{x} = \frac{\iint_R x\sigma(x, y)\, dA}{\iint_R \sigma(x, y)\, dA}, \qquad \bar{y} = \frac{\iint_R y\sigma(x, y)\, dA}{\iint_R \sigma(x, y)\, dA}.$$

Find the center of mass of the lamina described by $R = \{(x, y) \mid 0 \leq x \leq 1, 0 \leq y \leq \sqrt{1 - x^2}\}$ with the density function $\sigma(x, y) = e^{-(x^2+y^2)}$ using each program.

a. DINTGL44 with $n = m = 14$

b. DGQINT45 with $n = m = 5$

15. Use Gaussian quadrature for triple integrals with $n = m = p = 4$ to approximate

$$\iiint_S xy \sin(yz)\, dV,$$

where S is the solid bounded by the coordinate planes and the planes $x = \pi$, $y = \pi/2$, $z = \pi/3$. Compare this approximation to the exact result.

16. Use Gaussian quadrature for triple integrals with $n = m = p = 5$ to approximate

$$\iiint_S \sqrt{xyz}\, dV,$$

when S is the region in the first octant bounded by the cylinder $x^2 + y^2 = 4$, the sphere $x^2 + y^2 + z^2 = 4$, and the plane $x + y + z = 8$. How many functional evaluations are required for the approximation?

4.8 Improper Integrals

Improper integrals result when the notion of integration is extended either to an interval of integration on which the function is unbounded or to an interval with one or more infinite endpoints. In either circumstance the normal rules of integral approximation must be modified.

We will first handle the situation when the integrand is unbounded at the left endpoint of the interval of integration, as shown in Figure 4.19 on page 158. We will then show that, by a suitable manipulation, the other improper integrals can be reduced to problems of this form.

Figure 4.19

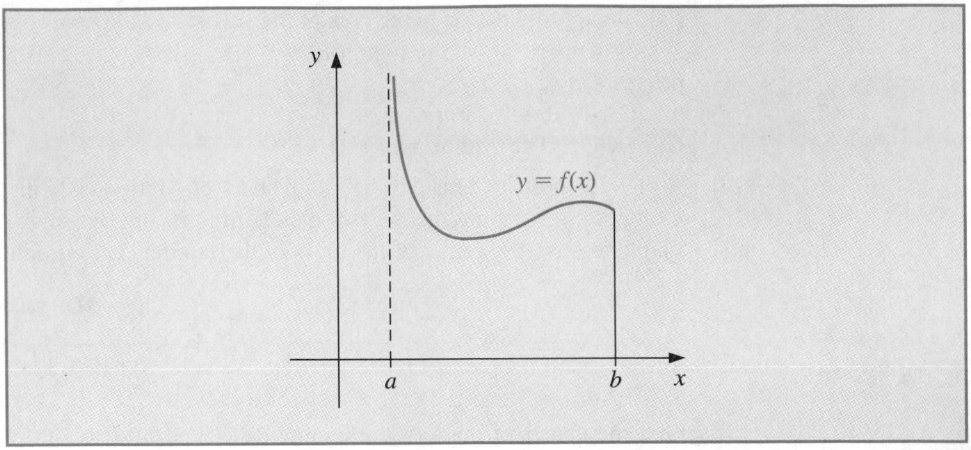

The improper integral with a singularity at the left endpoint,

$$\int_a^b \frac{1}{(x-a)^p}\,dx,$$

converges if and only if $0 < p < 1$, and in this case,

$$\int_a^b \frac{1}{(x-a)^p}\,dx = \frac{(b-a)^{1-p}}{1-p}.$$

If f is a function that can be written in the form

$$f(x) = \frac{g(x)}{(x-a)^p},$$

where $0 < p < 1$ and g is continuous on $[a, b]$, then the improper integral

$$\int_a^b f(x)\,dx$$

also exists. We will approximate this integral using Composite Simpson's rule. If $g \in C^5[a, b]$, we can construct the fourth Taylor polynomial, $P_4(x)$, for g about a:

$$P_4(x) = g(a) + g'(a)(x-a) + \frac{g''(a)}{2!}(x-a)^2 + \frac{g'''(a)}{3!}(x-a)^3 + \frac{g^{(4)}(a)}{4!}(x-a)^4,$$

and write

$$\int_a^b f(x)\,dx = \int_a^b \frac{g(x) - P_4(x)}{(x-a)^p}\,dx + \int_a^b \frac{P_4(x)}{(x-a)^p}\,dx.$$

We can exactly determine the value of

$$\int_a^b \frac{P_4(x)}{(x-a)^p}\,dx = \sum_{k=0}^{4}\int_a^b \frac{g^{(k)}(a)}{k!}(x-a)^{k-p}\,dx = \sum_{k=0}^{4}\frac{g^{(k)}(a)}{k!(k+1-p)}(b-a)^{k+1-p}. \quad \textbf{(4.9)}$$

This is generally the dominant portion of the approximation, especially when the Taylor polynomial $P_4(x)$ agrees closely with the function g throughout the interval $[a, b]$.

To approximate the integral of f, we need to add this value to the approximation of

$$\int_a^b \frac{g(x)-P_4(x)}{(x-a)^p}\,dx.$$

To determine this, we first define

$$G(x) = \begin{cases} \dfrac{g(x)-P_4(x)}{(x-a)^p}, & \text{if } a < x \le b, \\ 0, & \text{if } x = a. \end{cases}$$

Since $0 < p < 1$ and $P_4^{(k)}(a)$ agrees with $g^{(k)}(a)$ for each $k = 0, 1, 2, 3, 4$, we have $G \in C^4[a, b]$. This implies that Composite Simpson's rule can be applied to approximate the integral of G on $[a, b]$. Adding this approximation to the value from Eq. (4.9) gives an approximation to the improper integral of f on $[a, b]$, within the accuracy of the Composite Simpson's rule approximation.

EXAMPLE 1 We will use the Composite Simpson's rule with $h = 0.25$ to approximate the value of the improper integral

$$\int_0^1 \frac{e^x}{\sqrt{x}}\,dx.$$

Since the fourth Taylor polynomial for e^x about $x = 0$ is

$$P_4(x) = 1 + x + \frac{x^2}{2} + \frac{x^3}{6} + \frac{x^4}{24},$$

we have

$$\begin{aligned} \int_0^1 \frac{P_4(x)}{\sqrt{x}}\,dx &= \int_0^1 \left(x^{-1/2} + x^{1/2} + \frac{1}{2}x^{3/2} + \frac{1}{6}x^{5/2} + \frac{1}{24}x^{7/2}\right) dx \\ &= \lim_{M\to 0^+}\left[2x^{1/2} + \frac{2}{3}x^{3/2} + \frac{1}{5}x^{5/2} + \frac{1}{21}x^{7/2} + \frac{1}{108}x^{9/2}\right]_M^1 \\ &= 2 + \frac{2}{3} + \frac{1}{5} + \frac{1}{21} + \frac{1}{108} \approx 2.9235450. \end{aligned}$$

Table 4.7 lists the values needed for the Composite Simpson's rule of

$$G(x) = \begin{cases} \dfrac{e^x - P_4(x)}{\sqrt{x}}, & \text{when } 0 < x \le 1, \\ 0, & \text{when } x = 0. \end{cases}$$

Table 4.7

x	$G(x)$
0.00	0
0.25	0.0000170
0.50	0.0004013
0.75	0.0026026
1.00	0.0099485

Applying the Composite Simpson's rule using these data gives

$$\int_0^1 G(x)\,dx \approx \frac{0.25}{3}[0 + 4(0.0000170) + 2(0.0004013) + 4(0.0026026) + 0.0099485]$$
$$= 0.0017691.$$

Hence,

$$\int_0^1 \frac{e^x}{\sqrt{x}}\,dx \approx 2.9235450 + 0.0017691 = 2.9253141.$$

This result is accurate within the accuracy of the Composite Simpson's rule approximation for the function G. Since $|G^{(4)}(x)| < 1$ on $[0, 1]$, the error is bounded by

$$\frac{1-0}{180}(0.25)^4(1) = 0.0000217.$$

◆ ◆ ◆

To approximate the improper integral with a singularity at the right endpoint, we can apply the technique we used previously but expand in terms of the right endpoint b instead of the left endpoint a. Alternatively, we can make the substitution $z = -x$, $dz = -dx$ to change the improper integral into one of the form

$$\int_a^b f(x)\,dx = \int_{-b}^{-a} f(-z)\,dz,$$

which has its singularity at the left endpoint. (See Figure 4.20.)

An improper integral with a singularity at c, where $a < c < b$, is treated as the sum of improper integrals with endpoint singularities since

$$\int_a^b f(x)\,dx = \int_a^c f(x)\,dx + \int_c^b f(x)\,dx.$$

Figure 4.20

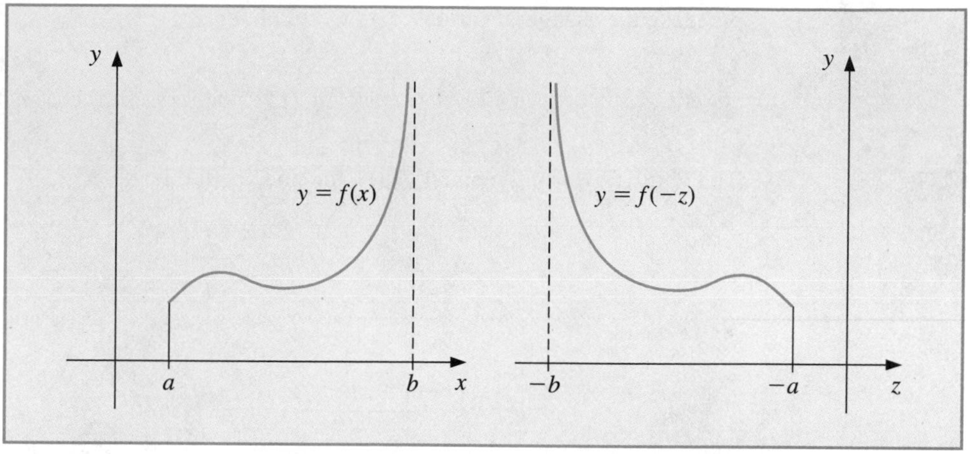

The other type of improper integral involves infinite limits of integration. A basic convergent integral of this type has the form

$$\int_a^\infty \frac{1}{x^p}\,dx,$$

for $p > 1$, which is converted to an integral with left-endpoint singularity by making the integration substitution

$$t = x^{-1}, \quad dt = -x^{-2}\,dx, \quad \text{so} \quad dx = -x^2\,dt = -t^{-2}\,dt.$$

Then

$$\int_a^\infty \frac{1}{x^p}\,dx = \int_{1/a}^0 -\frac{t^p}{t^2}\,dt = \int_0^{1/a} \frac{1}{t^{2-p}}\,dt.$$

In a similar manner, the variable change $t = x^{-1}$ converts the improper integral $\int_a^\infty f(x)\,dx$ into one that has a left-endpoint singularity at zero:

$$\int_a^\infty f(x)\,dx = \int_0^{1/a} t^{-2} f\left(\frac{1}{t}\right)\,dt.$$

It can now be approximated using a quadrature formula of the type described earlier.

EXAMPLE 2 To approximate the value of the improper integral

$$I = \int_1^\infty x^{-3/2} \sin\frac{1}{x}\,dx$$

we make the change of variable $t = x^{-1}$ to obtain

$$I = \int_0^1 t^{-1/2} \sin t \, dt.$$

The fourth Taylor polynomial, $P_4(t)$, for $\sin t$ about 0 is

$$P_4(t) = t - \frac{1}{6}t^3,$$

so we have

$$\begin{aligned} I &= \int_0^1 \frac{\sin t - t + \frac{1}{6}t^3}{t^{1/2}} \, dt + \int_0^1 t^{1/2} - \frac{1}{6}t^{5/2} \, dt \\ &= \int_0^1 \frac{\sin t - t + \frac{1}{6}t^3}{t^{1/2}} \, dt + \left[\frac{2}{3}t^{3/2} - \frac{1}{21}t^{7/2} \right]\Bigg|_0^1 \\ &= \int_0^1 \frac{\sin t - t + \frac{1}{6}t^3}{t^{1/2}} \, dt + 0.61904761. \end{aligned}$$

The Composite Simpson's rule with $n = 16$ for the remaining integral is 0.0014890097. This gives a final approximation of

$$I = 0.0014890097 + 0.61904761 = 0.62053661,$$

which is accurate to within 4.0×10^{-8}. ◆ ◆ ◆

EXERCISE SET 4.8

1. Use Composite Simpson's rule and the given values of n to approximate the following improper integrals.

   a. $\displaystyle\int_0^1 x^{-1/4} \sin x \, dx$ with $n = 4$

   b. $\displaystyle\int_0^1 \frac{e^{2x}}{\sqrt[5]{x^2}} \, dx$ with $n = 6$

   c. $\displaystyle\int_1^2 \frac{\ln x}{(x-1)^{1/5}} \, dx$ with $n = 8$

   d. $\displaystyle\int_0^1 \frac{\cos 2x}{x^{1/3}} \, dx$ with $n = 6$

2. Use the Composite Simpson's rule and the given values of n to approximate the following improper integrals.

   a. $\displaystyle\int_0^1 \frac{e^{-x}}{\sqrt{1-x}} \, dx$ with $n = 6$

   b. $\displaystyle\int_0^2 \frac{xe^x}{\sqrt[3]{(x-1)^2}} \, dx$ with $n = 8$

3. Use the transformation $t = x^{-1}$ and then the Composite Simpson's rule and the given values of n to approximate the following improper integrals.

a. $\displaystyle\int_1^{\infty} \frac{1}{x^2+9}\,dx$ with $n = 4$ b. $\displaystyle\int_1^{\infty} \frac{1}{1+x^4}\,dx$ with $n = 4$

c. $\displaystyle\int_1^{\infty} \frac{\cos x}{x^3}\,dx$ with $n = 6$ d. $\displaystyle\int_1^{\infty} x^{-4}\sin x\,dx$ with $n = 6$

4. The improper integral $\int_0^{\infty} f(x)\,dx$ cannot be converted into an integral with finite limits using the substitution $t = 1/x$ because the limit at zero becomes infinite. The problem is resolved by first writing $\int_0^{\infty} f(x)\,dx = \int_0^1 f(x)\,dx + \int_1^{\infty} f(x)\,dx$. Apply this technique to approximate the following improper integrals to within 10^{-6}.

a. $\displaystyle\int_0^{\infty} \frac{1}{1+x^4}\,dx$ b. $\displaystyle\int_0^{\infty} \frac{1}{(1+x^2)^3}\,dx$

5. The improper integral $\int_{-\infty}^{\infty} f(x)\,dx$ can be written as $\int_{-\infty}^{0} f(x)\,dx + \int_0^{\infty} f(x)\,dx$ where $\int_{-\infty}^{0} f(x)\,dx = \int_0^{\infty} f(-x)\,dx$. Use the techniques of Exercises 3 and 4 to approximate the following integrals.

a. $\displaystyle\int_{-\infty}^{\infty} \frac{1}{1+x^2}\,dx$ with $n = 4$ b. $\displaystyle\int_{-\infty}^{\infty} \frac{x}{(x^2+1)^4}\,dx$ with $n = 4$

c. $\displaystyle\int_{-\infty}^{\infty} \frac{1}{1+x^4}\,dx$ with $n = 6$ d. $\displaystyle\int_{-\infty}^{\infty} \frac{1}{(1+x^2)^3}\,dx$ with $n = 6$

6. Suppose a body of mass m is traveling vertically upward starting at the surface of the earth. If all resistance except gravity is neglected, the escape velocity v is given by

$$v^2 = 2gR\int_1^{\infty} z^{-2}\,dz, \quad \text{where } z = \frac{x}{R},$$

$R = 3960$ mi is the radius of the earth, and $g = 0.00609$ mi/s^2 is the force of gravity at the earth's surface. Approximate the escape velocity v.

4.9 Numerical Differentiation

The derivative of the function f at x_0 is defined as

$$f'(x_0) = \lim_{h\to 0} \frac{f(x_0+h) - f(x_0)}{h}.$$

This formula gives an obvious way to generate an approximation to $f'(x_0)$; simply compute

$$\frac{f(x_0+h) - f(x_0)}{h}$$

for small values of h. Although this may be obvious, it is not very successful, due to our old nemesis, round-off error. But, it is certainly the place to start.

To approximate $f'(x_0)$, suppose first that $x_0 \in (a, b)$, where $f \in C^2[a, b]$, and that $x_1 = x_0 + h$ for some $h \neq 0$ that is sufficiently small to ensure that $x_1 \in [a, b]$. We construct the first Lagrange polynomial, $P_{0,1}$, for f determined by x_0 and x_1 with its error term

$$\begin{aligned} f(x) &= P_{0,1}(x) + \frac{(x - x_0)(x - x_1)}{2!} f''(\xi(x)) \\ &= \frac{f(x_0)(x - x_0 - h)}{-h} + \frac{f(x_0 + h)(x - x_0)}{h} + \frac{(x - x_0)(x - x_0 - h)}{2} f''(\xi(x)) \end{aligned}$$

for some $\xi(x)$ in $[a, b]$. Differentiating this equation gives

$$\begin{aligned} f'(x) &= \frac{f(x_0 + h) - f(x_0)}{h} + D_x \left[\frac{(x - x_0)(x - x_0 - h)}{2} f''(\xi(x)) \right] \\ &= \frac{f(x_0 + h) - f(x_0)}{h} + \frac{2(x - x_0) - h}{2} f''(\xi(x)) \\ &\quad + \frac{(x - x_0)(x - x_0 - h)}{2} D_x(f''(\xi(x))), \end{aligned}$$

so

$$f'(x) \approx \frac{f(x_0 + h) - f(x_0)}{h}.$$

There are two terms for the error in this approximation. The first term involves $f''(\xi(x))$, which can be bounded if we have a bound for the second derivative of f. The second part of the truncation error involves $D_x f''(\xi(x)) = f'''(\xi(x)) \cdot \xi'(x)$, which generally cannot be estimated because it contains the unknown term $\xi'(x)$. However, when x is x_0, the coefficient of $D_x f''(\xi(x))$ is zero. In this case the formula simplifies to the following:

Two-Point Formula

$$f'(x_0) = \frac{f(x_0 + h) - f(x_0)}{h} - \frac{h}{2} f''(\xi),$$

where ξ lies between x_0 and $x_0 + h$.

For small values of h, the difference quotient $[f(x_0 + h) - f(x_0)]/h$ can be used to approximate $f'(x_0)$ with an error bounded by $M|h|/2$, if M is a bound on $|f''(x)|$ for $x \in [a, b]$. This is a two-point formula known as the **forward-difference formula** if $h > 0$ (see Figure 4.21) and the **backward-difference formula** if $h < 0$.

EXAMPLE 1 Let $f(x) = \ln x$ and $x_0 = 1.8$. The forward-difference formula,

$$\frac{f(1.8 + h) - f(1.8)}{h},$$

Figure 4.21

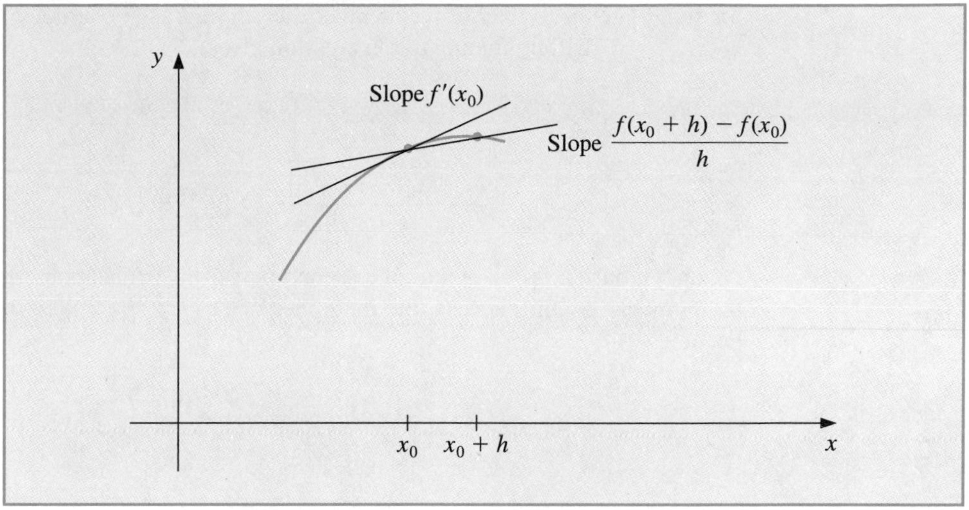

can be used to approximate $f'(1.8)$ with error

$$\frac{|hf''(\xi)|}{2} = \frac{|h|}{2\xi^2} \le \frac{|h|}{2(1.8)^2}, \qquad \text{where } 1.8 < \xi < 1.8 + h.$$

The results in Table 4.8 are produced when $h = 0.1, 0.01$, and 0.001.

Table 4.8

h	$f(1.8 + h)$	$\frac{f(1.8 + h) - f(1.8)}{h}$	$\frac{\lvert h \rvert}{2(1.8)^2}$
0.1	0.64185389	0.5406722	0.0154321
0.01	0.59332685	0.5540180	0.0015432
0.001	0.58834207	0.5554013	0.0001543

Since $f'(x) = 1/x$, the exact value of $f'(1.8)$ is $0.\bar{5}$, and the error bounds are quite close to the true approximation error. ◆ ◆ ◆

To obtain general derivative approximation formulas, suppose that $x_0, x_1, \ldots, x_n$ are $(n + 1)$ distinct numbers in some interval I and that $f \in C^{n+1}(I)$. Then

$$f(x) = \sum_{j=0}^{n} f(x_j)L_j(x) + \frac{(x - x_0)\cdots(x - x_n)}{(n + 1)!} f^{(n+1)}(\xi(x))$$

for some $\xi(x)$ in I, where $L_j(x)$ denotes the jth Lagrange coefficient polynomial for f at $x_0, x_1, \ldots, x_n$. Differentiating this expression gives

$$f'(x) = \sum_{j=0}^{n} f(x_j)L_j'(x) + D_x\left[\frac{(x-x_0)\cdots(x-x_n)}{(n+1)!}\right] f^{(n+1)}(\xi(x)) + \frac{(x-x_0)\cdots(x-x_n)}{(n+1)!} D_x[f^{(n+1)}(\xi(x))].$$

Again we have a problem with the second part of the truncation error unless x is one of the numbers x_k. In this case, the multiplier of $D_x[f^{(n+1)}(\xi(x))]$ is zero, and the formula becomes

$$f'(x_k) = \sum_{j=0}^{n} f(x_j)L_j'(x_k) + \frac{f^{(n+1)}(\xi(x_k))}{(n+1)!} \prod_{\substack{j=0 \\ j\neq k}}^{n} (x_k - x_j).$$

Applying this technique using the second Lagrange polynomial at x_0, $x_1 = x_0 + h$, and $x_2 = x_0 + 2h$ produces the following formula.

Three-Point Endpoint Formula

$$f'(x_0) = \frac{1}{2h}[-3f(x_0) + 4f(x_0+h) - f(x_0+2h)] + \frac{h^2}{3}f^{(3)}(\xi),$$

where ξ lies between x_0 and $x_0 + 2h$.

This formula is most useful when approximating the derivative at the endpoint of an interval. This situation occurs, for example, when approximations are needed for the derivatives used for the clamped cubic splines discussed in Section 3.5. Left endpoint approximations are found using $h > 0$, and right endpoint approximations using $h < 0$.

When approximating the derivative of a function at an interior point of an interval, it is better to use the formula that is produced from the second Lagrange polynomial at $x_0 - h$, x_0, and $x_0 + h$.

Three-Point Midpoint Formula

$$f'(x_0) = \frac{1}{2h}[f(x_0+h) - f(x_0-h)] - \frac{h^2}{6}f^{(3)}(\xi),$$

where ξ lies between $x_0 - h$ and $x_0 + h$.

The error in the Midpoint formula is approximately half the error in the Endpoint formula and f needs to be evaluated at only two points, whereas in the Endpoint formula three

Figure 4.22

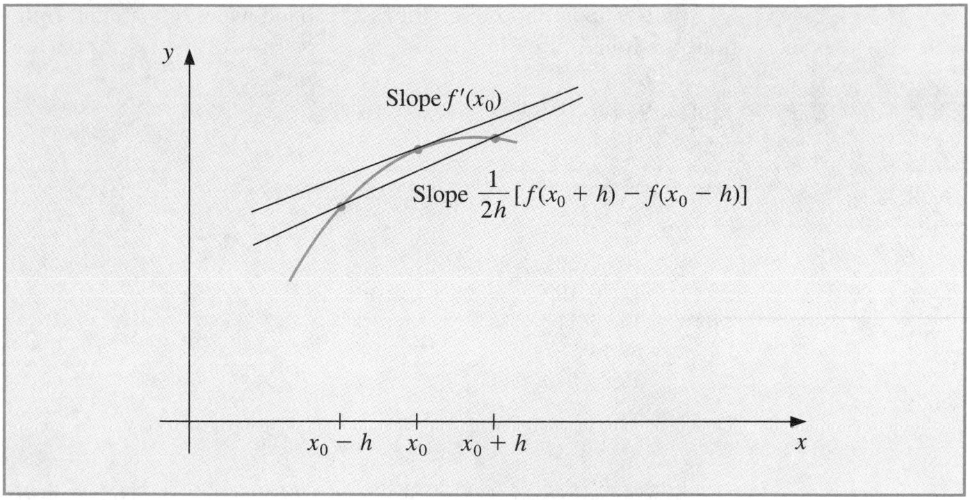

evaluations are required. Figure 4.22 gives an illustration of the approximation produced from the Midpoint formula.

These methods are called *three-point formulas* (even though the third point, $f(x_0)$, does not appear in the Midpoint formula). Similarly, there are methods known as *five-point formulas* that involve evaluating the function at two additional points, whose error term is $O(h^4)$. These formulas are generated by differentiating fourth Lagrange polynomials that pass through the evaluation points. The most useful is the interior-point formula:

Five-Point Midpoint Formula

$$f'(x_0) = \frac{1}{12h}[f(x_0-2h) - 8f(x_0-h) + 8f(x_0+h) - f(x_0+2h)] + \frac{h^4}{30}f^{(5)}(\xi),$$

where ξ lies between $x_0 - 2h$ and $x_0 + 2h$.

There is another five-point formula that is useful, particularly with regard to the clamped cubic spline interpolation of Section 3.5.

Five-Point Endpoint Formula

$$f'(x_0) = \frac{1}{12h}[-25f(x_0) + 48f(x_0+h) - 36f(x_0+2h) + 16f(x_0+3h) - 3f(x_0+4h)] + \frac{h^4}{5}f^{(5)}(\xi),$$

where ξ lies between x_0 and $x_0 + 4h$.

Left-endpoint approximations are found using $h > 0$, and right-endpoint approximations are found using $h < 0$.

EXAMPLE 2 Table 4.9 gives values for $f(x) = xe^x$.

Table 4.9

x	$f(x)$
1.8	10.889365
1.9	12.703199
2.0	14.778112
2.1	17.148957
2.2	19.855030

Since $f'(x) = (x + 1)e^x$, we have $f'(2.0) = 22.167168$. Approximating $f'(2.0)$ using the various three- and five-point formulas produces the following results.

Three-Point formulas

Endpoint with $h = 0.1$: $\frac{1}{0.2}[-3f(2.0) + 4f(2.1) - f(2.2)] = 22.032310.$

Endpoint with $h = -0.1$: $\frac{1}{-0.2}[-3f(2.0) + 4f(1.9) - f(1.8)] = 22.054525.$

Midpoint with $h = 0.1$: $\frac{1}{0.2}[f(2.1) - f(1.9)] = 22.228790.$

Midpoint with $h = 0.2$: $\frac{1}{0.4}[f(2.2) - f(1.8)] = 22.414163.$

Five-Point formula

Midpoint with $h = 0.1$ (the only five-point formula applicable):

$$\frac{1}{1.2}[f(1.8) - 8f(1.9) + 8f(2.1) - f(2.2)] = 22.166999.$$

The errors are approximately 1.35×10^{-1}, 1.13×10^{-1}, -6.16×10^{-2}, -2.47×10^{-1}, and 1.69×10^{-4}, respectively. The five-point formula is clearly the superior result. Note also that the error from the midpoint formula with $h = 0.1$ is approximately half of the magnitude of the error produced using the endpoint formula with either $h = 0.1$ or $h = -0.1$. ◆ ◆ ◆

Round-off error is particularly important in numerical differentiation. In Section 4.3 we found that by reducing the step size in the Composite Simpson's rule we could reduce the truncation error, and, even though the amount of calculation increased, total round-off error in the method was not affected. In a numerical differentiation technique, the truncation error will also decrease if the step size is reduced, but only at the expense of increased round-off error. To see why this occurs, let us examine more closely the Three-Point Midpoint formula:

$$f'(x_0) = \frac{1}{2h}[f(x_0 + h) - f(x_0 - h)] - \frac{h^2}{6}f^{(3)}(\xi).$$

Suppose that, in evaluating $f(x_0+h)$ and $f(x_0-h)$, we encounter round-off errors $e(x_0+h)$ and $e(x_0-h)$; that is, our computed values $\tilde{f}(x_0+h)$ and $\tilde{f}(x_0-h)$ are related to the true values $f(x_0+h)$ and $f(x_0-h)$ by the formulas

$$f(x_0+h) = \tilde{f}(x_0+h) + e(x_0+h) \quad \text{and} \quad f(x_0-h) = \tilde{f}(x_0-h) + e(x_0-h).$$

In this case, the total error in the approximation,

$$f'(x_0) - \frac{\tilde{f}(x_0+h) - \tilde{f}(x_0-h)}{2h} = \frac{e(x_0+h) - e(x_0-h)}{2h} - \frac{h^2}{6} f^{(3)}(\xi),$$

is due in part to round-off and in part to truncating. If we assume that the round-off errors $e(x_0 \pm h)$ are bounded by some number $\epsilon > 0$ and that the third derivative of f is bounded by a number $M > 0$, then

$$\left| f'(x_0) - \frac{\tilde{f}(x_0+h) - \tilde{f}(x_0+h)}{2h} \right| \le \frac{\epsilon}{h} + \frac{h^2}{6} M.$$

To reduce the truncation portion of the error, $h^2M/6$, we must reduce h. But as h is reduced, the round-off portion of the error, ϵ/h, grows. In practice, then, it is seldom advantageous to let h be too small, since the round-off error will dominate the calculations.

EXAMPLE 3 Consider using the values in Table 4.10 to approximate $f'(0.900)$ for $f(x) = \sin x$. The true value is $\cos(0.900) = 0.62161$. Using the formula

$$f'(0.900) \approx \frac{f(0.900+h) - f(0.900-h)}{2h}$$

with different values of h gives the approximations in Table 4.11.

It appears that an optimal choice for h lies between 0.005 and 0.05. The minimal value of the error term,

$$e(h) = \frac{\epsilon}{h} + \frac{h^2}{6} M,$$

occurs when $0 = e'(h) = -\frac{\epsilon}{h^2} + \frac{h}{3}M$, that is, when $h = \sqrt[3]{3\epsilon/M}$.

Table 4.10

x	$\sin x$	x	$\sin x$
0.800	0.71736	0.901	0.78395
0.850	0.75128	0.902	0.78457
0.880	0.77074	0.905	0.78643
0.890	0.77707	0.910	0.78950
0.895	0.78021	0.920	0.79560
0.898	0.78208	0.950	0.81342
0.899	0.78270	1.000	0.84147

Table 4.11

h	Approximation to $f'(0.900)$	Error
0.001	0.62500	0.00339
0.002	0.62250	0.00089
0.005	0.62200	0.00039
0.010	0.62150	−0.00011
0.020	0.62150	−0.00011
0.050	0.62140	−0.00021
0.100	0.62055	−0.00106

Note that

$$M = \max_{x\in[0.800,1.00]} |f'''(x)| = \max_{x\in[0.800,1.00]} |\cos x| \approx 0.69671.$$

Also, if the values of f are given to five decimal places, it is reasonable to assume that $\epsilon = 0.000005$. Therefore, the optimal choice of h is approximately

$$h = \sqrt[3]{\frac{3(0.000005)}{0.69671}} \approx 0.028,$$

which is consistent with the results in Table 4.11. In practice, though, we cannot compute an optimal h to use in approximating the derivative, since we have no knowledge of the third derivative of the function. ◆ ◆ ◆

We have considered only the round-off error problems that are presented by the Three-Point Midpoint formula, but similar difficulties occur with all the differentiation formulas. The reason for the problems can be traced to the need to divide by a power of h. As we found in Section 1.4 (see, in particular, Example 1), division by small numbers tends to exaggerate round-off error, and this operation should be avoided if possible. In the case of numerical differentiation, it is impossible to avoid the problem entirely, although the higher-order methods reduce the difficulty.

Keep in mind that as an approximation method, numerical differentiation is unstable, since the small values of h needed to reduce truncation error cause the round-off error to grow. This is the first class of unstable methods we have encountered, and these techniques would be avoided if it were possible. However, it is not, since in addition to being used for computational purposes, the formulas we have derived are needed for approximating the solutions of ordinary and partial-differential equations.

Methods for approximating higher derivatives of functions can be derived as was done for approximating the first derivative or by using an averaging technique that is similar to that used for extrapolation. These techniques, of course, suffer from the same stability weaknesses as the approximation methods for first derivatives, but they are needed for approximating the solution to boundary value problems in differential equations. The only one we will need is a three-point midpoint formula, which has the following form.

Three-Point Midpoint Formula for Approximating f''

$$f''(x_0) = \frac{1}{h^2}[f(x_0 - h) - 2f(x_0) + f(x_0 + h)] - \frac{h^2}{12}f^{(4)}(\xi),$$

where ξ lies between $x_0 - h$ and $x_0 + h$.

EXERCISE SET 4.9

1. Use the forward-differencc formulas and backward-difference formulas to determine approximations that will complete the following tables.

a.

x	$f(x)$	$f'(x)$
0.5	0.4794	
0.6	0.5646	
0.7	0.6442	

b.

x	$f(x)$	$f'(x)$
0.0	0.00000	
0.2	0.74140	
0.4	1.3718	

2. The data in Exercise 1 were taken from the following functions. Compute the actual errors in Exercise 1, and find error bounds using the error formulas.
   a. $f(x) = \sin x$ **b.** $f(x) = e^x - 2x^2 + 3x - 1$

3. Use the most appropriate three-point formula to determine approximations that will complete the following tables.

a.

x	$f(x)$	$f'(x)$
1.1	9.025013	
1.2	11.02318	
1.3	13.46374	
1.4	16.44465	

b.

x	$f(x)$	$f'(x)$
8.1	16.94410	
8.3	17.56492	
8.5	18.19056	
8.7	18.82091	

c.

x	$f(x)$	$f'(x)$
2.9	−4.827866	
3.0	−4.240058	
3.1	−3.496909	
3.2	−2.596792	

d.

x	$f(x)$	$f'(x)$
2.0	3.6887983	
2.1	3.6905701	
2.2	3.6688192	
2.3	3.6245909	

4. The data in Exercise 3 were taken from the following functions. Compute the actual errors in Exercise 3 and find error bounds using the error formulas.
   a. $f(x) = e^{2x}$ **b.** $f(x) = x \ln x$
   c. $f(x) = x \cos x - x^2 \sin x$ **d.** $f(x) = 2(\ln x)^2 + 3 \sin x$

5. Use the most accurate formula possible to determine approximations that will complete the following tables.

a.

x	$f(x)$	$f'(x)$
2.1	−1.709847	
2.2	−1.373823	
2.3	−1.119214	
2.4	−0.9160143	
2.5	−0.7470223	
2.6	−0.6015966	

b.

x	$f(x)$	$f'(x)$
−3.0	9.367879	
−2.8	8.233241	
−2.6	7.180350	
−2.4	6.209329	
−2.2	5.320305	
−2.0	4.513417	

6. The data in Exercise 5 were taken from the given functions. Compute the actual errors in Exercise 5 and find error bounds using the error formulas and Maple.

a. $f(x) = \tan x$ **b.** $f(x) = e^{x/3} + x^2$

7. Let $f(x) = \cos \pi x$. Use the Three-point Midpoint formula for f'' and the values of $f(x)$ at $x = 0.25$, 0.5, and 0.75 to approximate $f''(0.5)$. Compare this result to the exact value and to the approximation found in Exercise 7 of Section 3.5. Explain why this method is particularly accurate for this problem.

8. Let $f(x) = 3xe^x - \cos x$. Use the following data and the Three-point Midpoint formula for f'' to approximate $f''(1.3)$ with $h = 0.1$ and $h = 0.01$, and compare your results to $f''(1.3)$.

x	1.20	1.29	1.30	1.31	1.40
$f(x)$	11.59006	13.78176	14.04276	14.30741	16.86187

9. Use the following data and the knowledge that the first five derivatives of f were bounded on $[1, 5]$ by 2, 3, 6, 12 and 23, respectively, to approximate $f'(3)$ as accurately as possible. Find a bound for the error.

x	1	2	3	4	5
$f(x)$	2.4142	2.6734	2.8974	3.0976	3.2804

10. Repeat Exercise 9, assuming instead that the third derivative of f is bounded on $[1, 5]$ by 4.

11. Analyze the round-off errors for the formula

$$f'(x_0) = \frac{f(x_0 + h) - f(x_0)}{h} - \frac{h}{2} f''(\xi_0).$$

Find an optimal $h > 0$ in terms of a bound M for f'' on $(x_0, x_0 + h)$.

12. All calculus students know that the derivative of a function f at x can be defined as

$$f'(x) = \lim_{h \to 0} \frac{f(x + h) - f(x)}{h}.$$

Choose your favorite function f, nonzero number x, and computer or calculator.

Generate approximations $f_n'(x)$ to $f'(x)$ by

$$f_n'(x) = \frac{f(x + 10^{-n}) - f(x)}{10^{-n}}$$

for $n = 1, 2, \ldots, 20$ and describe what happens.

13. Consider the function

$$e(h) = \frac{\varepsilon}{h} + \frac{h^2}{6}M,$$

where M is a bound for the third derivative of a function. Show that $e(h)$ has a minimum at $\sqrt[3]{3\varepsilon/M}$.

14. The forward-difference formula can be expressed as

$$f'(x_0) = \frac{1}{h}[f(x_0 + h) - f(x_0)] - \frac{h}{2}f''(x_0) - \frac{h^2}{6}f'''(x_0) + O(h^3).$$

Use extrapolation on this formula to derive an $O(h^3)$ formula for $f'(x_0)$.

15. In Exercise 7 of Section 3.4, data were given describing a car traveling on a straight road. That problem asked to predict the position and speed of the car when $t = 10$ s. Use the following times and positions to predict the speed at each time listed.

Time	0	3	5	8	10	13
Distance	0	225	383	623	742	993

16. In a circuit with impressed voltage $\mathcal{E}(t)$ and inductance L, Kirchhoff's first law gives the relationship

$$\mathcal{E} = L\frac{di}{dt} + Ri,$$

where R is the resistance in the circuit and i is the current. Suppose we measure the current for several values of t and obtain:

t	1.00	1.01	1.02	1.03	1.0
i	3.10	3.12	3.14	3.18	3.24

where t is measured in seconds, i is in amperes, the inductance, L, is a constant 0.98 henries, and the resistance is 0.142 ohms. Approximate the voltage $\mathcal{E}$ at the values $t = 1.00$, 1.01, 1.02, 1.03, and 1.04.

4.10 Survey of Methods and Software

In this chapter we considered approximating integrals of functions of one, two, or three variables and approximating the derivatives of a function of a single real variable.

The Midpoint rule, Trapezoidal rule, and Simpson's rule were studied to introduce the techniques and error analysis of quadrature methods. Composite Simpson's rule is easy to use and produces accurate approximations unless the function oscillates in a subinterval of the interval of integration. Adaptive quadrature can be used if the function is suspected of oscillatory behavior. To minimize the number of nodes while maintaining accuracy, we studied Gaussian quadrature. Romberg integration was introduced to take advantage of the easily applied Composite Trapezoidal rule and extrapolation.

Most software for integrating a function of a single real variable is based on the adaptive approach or extremely accurate Gaussian formulas. Cautious Romberg integration is an adaptive technique that includes a check to make sure that the integrand is smoothly behaved over subintervals of the interval of integration. This method has been successfully used in software libraries. Multiple integrals are generally approximated by extending good adaptive methods to higher dimensions. Gaussian-type quadrature is also recommended to decrease the number of function evaluations.

The main routines in both the IMSL and NAG Libraries are based on QUADPACK: A Subroutine Package for Automatic Integration by Piessens, de Doncker-Kapenga, Uberhuber, and Kahaner published by Springer-Verlag in 1983 [PDUK]. The routines are also available as public domain software. The main technique is an adaptive integration scheme based on the 21-point Gaussian-Kronrod rule using the 10-point Gaussian rule for error estimation. The Gaussian rule uses the 10 points $x_1, \ldots, x_{10}$ and weights $w_1, \ldots, w_{10}$ to give the quadrature formula $\sum_{i=1}^{10} w_i f(x_i)$ to approximate $\int_a^b f(x)\,dx$. The additional points x_{11}, $\ldots, x_{21}$ and the new weights $v_1, \ldots, v_{21}$ are then used in the Kronrod formula, $\sum_{i=1}^{21} v_i f(x_i)$. The results of the two formulas are compared to eliminate error. The advantage in using $x_1, \ldots, x_{10}$ in each formula is that f needs to be evaluated at only 21 points. If independent 10- and 21-point Gaussian rules were used, 31 function evaluations would be needed. This procedure also permits endpoint singularities in the integrand. Other subroutines allow user specified singularities and infinite intervals of integration. Methods are also available for multiple integrals.

The Maple function call

```
>int(f,x=a..b);
```

computes the definite integral $\int_a^b f(x)\,dx$. The numerical method applies singularity handling routines and then uses Clenshaw-Curtis quadrature, which is described in [CC]. If this fails, an adaptive Newton-Cotes formula is applied. The method attempts to achieve a relative error tolerance $0.5 \times 10^{(1-\texttt{Digits})}$, where `Digits` is a variable in Maple that specifies the number of digits of rounding Maple uses for numerical calculation. The default value for `Digits` is 10, but it can be changed to any positive integer `n` by the command `Digits:=n;`

Although numerical differentiation is unstable, derivative approximation formulas are needed for solving differential equations. The NAG Library includes a subroutine for the numerical differentiation of a function of one real variable, with differentiation to the fourteenth derivative being possible. An IMSL function uses an adaptive change in step size for finite differences to approximate a derivative of f at x to within a given tolerance. Both packages allow the differentiation and integration of interpolatory cubic splines.

For further reading on numerical integration we recommend the books by Engels [E] and by Davis and Rabinowitz [DR]. For more information on Gaussian quadrature, see Stroud and Secrest [StS].

CHAPTER 5

Numerical Solution of Initial-Value Problems

5.1 Introduction

Differential equations are used to model problems that involve the change of some variable with respect to another. These problems require the solution to an initial-value problem—that is, the solution to a differential equation that satisfies a given initial condition.

In many real-life situations, the differential equation that models the problem is too complicated to solve exactly, and one of two approaches is taken to approximate the solution. The first approach is to simplify the differential equation to one that can be solved exactly and then use the solution of the simplified equation to approximate the solution to the original equation. The other approach, the one we examine in this chapter, involves finding methods for directly approximating the solution of the original problem. This is the approach commonly taken since more accurate results and realistic error information can be obtained.

The methods we consider in this chapter do not produce a continuous approximation to the solution of the initial-value problem. Rather, approximations are found at certain specified, and often equally spaced, points. Some method of interpolation, commonly a form of Hermite, is used if intermediate values are needed.

The first part of the chapter concerns approximating the solution $y(t)$ to a problem of the form

$$\frac{dy}{dt} = f(t, y), \quad \text{for } a \leq t \leq b,$$

subject to an initial condition

$$y(a) = \alpha.$$

These techniques form the core of the study since more general procedures use these as a base. Later in the chapter we deal with the extension of these methods to a system of first-order differential equations in the form

$$\frac{dy_1}{dt} = f_1(t, y_1, y_2, \ldots, y_n),$$
$$\frac{dy_2}{dt} = f_2(t, y_1, y_2, \ldots, y_n),$$
$$\vdots$$
$$\frac{dy_n}{dt} = f_n(t, y_1, y_2, \ldots, y_n),$$

for $a \le t \le b$, subject to the initial conditions

$$y_1(a) = \alpha_1, \quad y_2(a) = \alpha_2, \quad \ldots, \quad y_n(a) = \alpha_n.$$

We also examine the relationship of a system of this type to the general nth-order initial-value problem of the form

$$y^{(n)} = f\left(t, y, y', y'', \ldots, y^{(n-1)}\right)$$

for $a \le t \le b$, subject to the initial conditions

$$y(a) = \alpha_0, \quad y'(a) = \alpha_1, \quad \ldots, \quad y^{(n-1)}(a) = \alpha_{n-1}.$$

Before describing the methods for approximating the solution to our basic problem, we consider some situations when we know the solution will exist. In fact, since we will not be solving the given problem, only an approximation to the problem, we need to know when problems that are close to the given problem have solutions that accurately approximate the solution to the given problem. This property of an initial-value problem is called **well-posed**, and these are the problems for which numerical methods are appropriate. The following result shows that the class of well-posed problems is quite broad.

Well-Posed Condition

Suppose that f and f_y, its first partial derivative with respect to y, are continuous for t in $[a, b]$ and for all y. Then the initial-value problem

$$y' = f(t, y), \quad \text{for } a \le t \le b, \quad \text{with } y(a) = \alpha,$$

has a unique solution $y(t)$ for $a \le t \le b$, and the problem is well-posed.

EXAMPLE 1 Consider the initial-value problem

$$y' = 1 + t\sin(ty), \quad \text{for } 0 \le t \le 2, \quad \text{with } y(0) = 0.$$

Since the functions

$$f(t, y) = 1 + t\sin(ty) \qquad \text{and} \qquad f_y(t, y) = t^2\cos(ty)$$

are both continuous for $0 \le t \le 2$ and for all y, a unique solution exists to this well-posed initial-value problem. ◆ ◆ ◆

If you have taken a course in differential equations, you might attempt to determine the solution to the problem in Example 1 by using one of the techniques you learned in that course.

Maple can be used to solve many initial-value problems. Consider the problem

$$\frac{dy}{dt} = y - t^2 + 1, \quad \text{for } 0 \le t \le 2, \quad \text{with } y(0) = 0.5.$$

To define the differential equation, enter

```
>deq:=D(y)(t)=y(t)-t*t+1;
```

and the initial condition

```
>init:=y(0)=0.5;
```

The names `deq` and `init` are chosen by the user. The command to solve the initial-value problems is

```
>deqsol:=dsolve({deq,init},y(t));
```

which gives the response

$$deqsol = y(t) = t^2 + 2t + 1 - 0.5e^t$$

To use the solution to obtain $y(1.5)$, we enter

```
>q:=rhs(deqsol);
>evalf(subs(t=1.5,q));
```

The function `rhs` is used to assign the solution of the initial-value problem to the function q, which we then evaluate at $t = 1.5$ to obtain $y(1.5) = 4.009155465$. The function `dsolve` can fail if an explicit solution to the initial value problem cannot be found. For example, the command

```
>deqsol2:=dsolve({D(y)(t)=1+t*sin(t*y(t)),y(0)=0},y(t));
```

does not succeed, because Maple cannot find an explicit solution. In this case a numerical method must be used.

5.2 Taylor Methods

Many of the numerical methods we saw in the first four chapters have an underlying derivation from Taylor's Theorem. The approximation of the solution to initial-value problems is no exception. In this case, the function we need to expand in a Taylor polynomial is the (unknown) solution to the problem, $y(t)$. In its most elementary form this leads to **Euler's Method**. Although Euler's method is seldom used in practice, the simplicity of its derivation illustrates the technique used for more advanced procedures, without the cumbersome algebra that accompanies these constructions.

The object of Euler's method is to find an approximation to the solution of a problem of the form

$$\frac{dy}{dt} = f(t, y), \quad \text{for } a \le t \le b, \quad \text{with } y(a) = \alpha$$

at the equally spaced **mesh points** $\{t_0, t_1, t_2, \ldots, t_N\}$ (see Figure 5.1), where

$$t_i = a + ih, \quad \text{for each } i = 0, 1, \ldots N.$$

The common distance between the points, $h = (b - a)/N$, is called the **step size**. Approximations at other values of t in $[a, b]$ can then be found using interpolation.

Figure 5.1

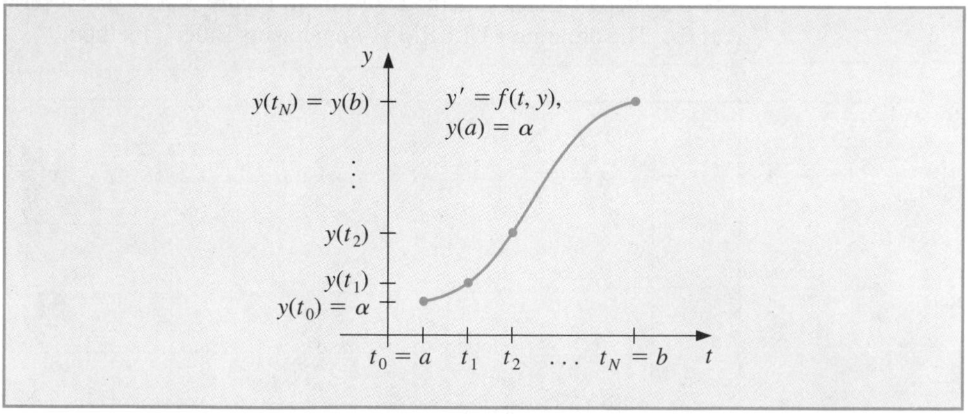

Suppose that $y(t)$, the solution to the problem, has two continuous derivatives on $[a, b]$, so that for each $i = 0, 1, 2, \ldots, N - 1$, Taylor's Theorem implies that

$$y(t_{i+1}) = y(t_i) + (t_{i+1} - t_i)y'(t_i) + \frac{(t_{i+1} - t_i)^2}{2}y''(\xi_i),$$

for some ξ_i in (t_i, t_{i+1}). Letting $h = t_{i+1} - t_i$, we have

$$y(t_{i+1}) = y(t_i) + hy'(t_i) + \frac{h^2}{2}y''(\xi_i),$$

and, since $y(t)$ satisfies the differential equation $y'(t) = f(t, y(t))$,

$$y(t_{i+1}) = y(t_i) + hf(t_i, y(t_i)) + \frac{h^2}{2}y''(\xi_i).$$

Euler's method constructs the approximation w_i to $y(t_i)$ for each $i = 1, 2, \ldots, N$ by deleting the error term in this equation. This produces a *difference equation* that approximates the differential equation. The term **local error** refers to the error at the given step if it is assumed that the previous results are all exact. The true, or accumulated, error of the method is called **global error**.

Euler's Method

$$w_0 = \alpha,$$
$$w_{i+1} = w_i + hf(t_i, w_i),$$

where $i = 0, 1, \ldots, N-1$, with local error $\frac{1}{2}y''(\xi_i)h^2$ for some ξ_i in (t_i, t_{i+1}).

To interpret Euler's method geometrically, note that when w_i is a close approximation to $y(t_i)$, the assumption that the problem is well-posed implies that

$$f(t_i, w_i) \approx y'(t_i) = f(t_i, y(t_i)).$$

The first step of Euler's method appears in Figure 5.2(a), and a series of steps appears in part (b). The program EULERM51 implements Euler's method.

Figure 5.2

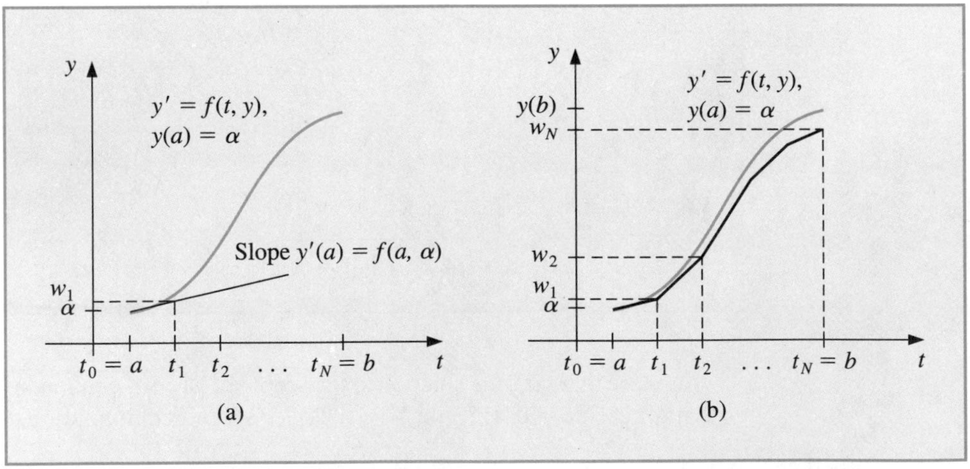

EXAMPLE 1 Suppose that Euler's method is used to approximate the solution to the initial-value problem

$$y' = y - t^2 + 1, \quad \text{for } 0 \le t \le 2, \quad \text{with } y(0) = 0.5,$$

assuming that $N = 10$. Then $h = 0.2$ and $t_i = 0.2i$.

Since $f(t, y) = y - t^2 + 1$ and $w_0 = y(0) = 0.5$, we have

$$w_{i+1} = w_i + h(w_i - t_i^2 + 1) = w_i + 0.2[w_i - 0.04i^2 + 1] = 1.2w_i - 0.008i^2 + 0.2$$

for $i = 0, 1, \ldots, 9$.

The exact solution is $y(t) = (t + 1)^2 - 0.5e^t$. Table 5.1 shows the comparison between the approximate values at t_i and the actual values. ◆ ◆ ◆

Table 5.1

t_i	$y_i = y(t_i)$	w_i	$\lvert y_i - w_i \rvert$
0.0	0.5000000	0.5000000	0.0000000
0.2	0.8292986	0.8000000	0.0292986
0.4	1.2140877	1.1520000	0.0620877
0.6	1.6489406	1.5504000	0.0985406
0.8	2.1272295	1.9884800	0.1387495
1.0	2.6408591	2.4581760	0.1826831
1.2	3.1799415	2.9498112	0.2301303
1.4	3.7324000	3.4517734	0.2806266
1.6	4.2834838	3.9501281	0.3333557
1.8	4.8151763	4.4281538	0.3870225
2.0	5.3054720	4.8657845	0.4396874

Euler's method is available in Maple. To access the procedure we enter the commands

```
>with(share);
>readshare(ODE,plots);
```

We define the differential equation of Example 1 with

```
>eq:=(t,y)->y-t^2+1;
```

Euler's method is called with the command

```
>eul:=firsteuler(eq,[0,0.5],0.2,10);
```

which responds with

$$eul := \text{array}(0..10, [$$
$$(0) = [0, .5]$$
$$(1) = [.2, .80]$$
$$(2) = [.4, 1.152]$$
$$(3) = [.6, 1.5504]$$

$(4) = [.8, 1.98848]$
$(5) = [1.0, 2.458176]$
$(6) = [1.2, 2.9498112]$
$(7) = [1.4, 3.45177344]$
$(8) = [1.6, 3.950128128]$
$(9) = [1.8, 4.428153754]$
$(10) = [2.0, 4.865784505]$
])

In this call, $[0, 0.5]$ refers to the initial condition $y(0) = 0.5$, and we use $h = 0.2$ to perform 10 steps of Euler's method. The procedure `firsteuler` returns an array *eul* representing i, t_i, and w_i. To access w_8 and name it $w8$, for example, we use

```
>w8:=eul[8][2];
```

Since Euler's method is derived from a Taylor polynomial whose error term involves the square of the step size h, the local error at each step is proportional to h^2. However, the total error, or global error, accumulates these local errors, so it generally grows at a much faster rate. The following result describes the global error of Euler's method and is typical of the type of results for initial-value approximation methods.

Euler's Method Error Bound

Let $y(t)$ denote the unique solution to the initial-value problem

$$y' = f(t, y), \quad \text{for } a \le t \le b, \quad \text{with } y(a) = \alpha,$$

and $w_0, w_1, \ldots, w_N$ be the approximations generated by Euler's method for some positive integer N. Suppose that f is continuous for all t in $[a, b]$ and all y in $(-\infty, \infty)$, and constants L and M exist with

$$\left|\frac{\partial f}{\partial t}(t, y(t))\right| \le L \quad \text{and} \quad |y''(t)| \le M.$$

Then, for each $i = 0, 1, 2, \ldots, N$,

$$|y(t_i) - w_i| \le \frac{hM}{2L}[e^{L(t_i - a)} - 1].$$

The important feature of this result is that the global error depends linearly on h, which is a reduction from quadratically on h for the local error. The reduction of one power of h from local to global error is typical of initial-value techniques. Even though we have a reduction in order from local to global errors, the formula shows that the error tends to zero with h.

EXAMPLE 2 Returning to the initial-value problem

$$y' = y - t^2 + 1, \quad \text{for } 0 \le t \le 2, \quad \text{with } y(0) = 0.5,$$

considered in Example 1, we see that with $f(t, y) = y - t^2 + 1$, we have $f_y(t, y) = 1$ for all y and $L = 1$. For this problem we know that the exact solution is $y(t) = (t+1)^2 - \frac{1}{2}e^t$, so $y''(t) = 2 - 0.5e^t$, and

$$|y''(t)| \le 0.5e^2 - 2 \quad \text{for } 0 \le t \le 2.$$

Using the inequality in the error bound for Euler's method with $h = 0.2$, $L = 1$, and $M = 0.5e^2 - 2$ gives

$$|y_i - w_i| \le 0.1(0.5e^2 - 2)(e^{t_i} - 1).$$

Table 5.2 lists the actual error found in Example 1, together with this error bound.

◆ ◆ ◆

Table 5.2

t_i	0.2	0.4	0.6	0.8	1.0	1.2	1.4	1.6	1.8	2.0
Actual Error	0.02930	0.06209	0.09854	0.13875	0.18268	0.23013	0.28063	0.33336	0.38702	0.43969
Error Bound	0.03752	0.08334	0.13931	0.20767	0.29117	0.39315	0.51771	0.66985	0.85568	1.08264

Since Euler's method was derived using Taylor's Theorem with $n = 2$, the first attempt to find methods for improving the accuracy of difference methods is to extend this technique of derivation to larger values of n. Suppose the solution $y(t)$ to the initial-value problem

$$y' = f(t, y), \quad \text{for } a \le t \le b, \quad \text{with } y(a) = \alpha,$$

has $n + 1$ continuous derivatives. If we expand the solution $y(t)$ in terms of its nth Taylor polynomial about t_i, we obtain

$$y(t_{i+1}) = y(t_i) + hy'(t_i) + \frac{h^2}{2}y''(t_i) + \cdots + \frac{h^n}{n!}y^{(n)}(t_i) + \frac{h^{n+1}}{(n+1)!}y^{(n+1)}(\xi_i)$$

for some ξ_i in (t_i, t_{i+1}).

Successive differentiation of the solution $y(t)$ gives

$$y'(t) = f(t, y(t)), \quad y''(t) = f'(t, y(t)), \quad \text{and, generally,} \quad y^{(k)}(t) = f^{(k-1)}(t, y(t)),$$

and substituting these results into the Taylor expansion gives

$$\begin{aligned} y(t_{i+1}) = y(t_i) + hf(t_i, y(t_i)) + \frac{h^2}{2}f'(t_i, y(t_i)) + \cdots \\ + \frac{h^n}{n!}f^{(n-1)}(t_i, y(t_i)) + \frac{h^{n+1}}{(n+1)!}f^{(n)}(\xi_i, y(\xi_i)). \end{aligned}$$

The difference-equation method corresponding to this equation is obtained by deleting the remainder term involving ξ_i.

Taylor Method of Order n

$$w_0 = \alpha,$$
$$w_{i+1} = w_i + hT^{(n)}(t_i, w_i)$$

for each $i = 0, 1, \ldots, N-1$, where

$$T^{(n)}(t_i, w_i) = f(t_i, w_i) + \frac{h}{2}f'(t_i, w_i) + \cdots + \frac{h^{n-1}}{n!}f^{(n-1)}(t_i, w_i).$$

The local error is $\frac{1}{(n+1)!}y^{(n+1)}(\xi_i)h^{n+1}$ for some ξ_i in (t_i, t_{i+1}).

The formula for $T^{(n)}$ is easily expressed, but it is difficult to use, because it requires the derivatives of f with respect to t. Since f is described as a multivariable function of both t and y, the chain rule implies that the total derivative of f with respect to t, which we denoted $f'(t, y(t))$, is obtained by

$$\begin{aligned} f'(t, y(t)) &= \frac{\partial f}{\partial t}(t, y(t)) \cdot \frac{dt}{dt} + \frac{\partial f}{\partial y}(t, y(t))\frac{dy(t)}{dt}, \\ &= \frac{\partial f}{\partial t}(t, y(t)) + \frac{\partial f}{\partial y}(t, y(t))y'(t) \end{aligned}$$

or, since $y'(t) = f(t, y(t))$,

$$f'(t, y(t)) = \frac{\partial f}{\partial t}(t, y(t)) + f(t, y(t))\frac{\partial f}{\partial y}(t, y(t)).$$

Subsequent derivatives increase in complication. For example, $f''(t, y(t))$ involves the partial derivatives of all the terms on the right side of this equation with respect to both t and y.

EXAMPLE 3 Suppose that we want to apply Taylor's method of orders 2 and 4 to the initial-value problem

$$y' = y - t^2 + 1, \quad \text{for } 0 \le t \le 2, \quad \text{with } y(0) = 0.5,$$

which was studied in Examples 1 and 2. We must find the first three derivatives of $f(t, y(t)) = y(t) - t^2 + 1$ with respect to the variable t:

$$\begin{aligned} f'(t, y(t)) &= \frac{d}{dt}(y - t^2 + 1) = y' - 2t = y - t^2 + 1 - 2t \\ f''(t, y(t)) &= \frac{d}{dt}(y - t^2 + 1 - 2t) = y' - 2t - 2 \\ &= y - t^2 + 1 - 2t - 2 = y - t^2 - 2t - 1 \end{aligned}$$

and

$$f'''(t, y(t)) = \frac{d}{dt}(y - t^2 - 2t - 1) = y' - 2t - 2 = y - t^2 - 2t - 1.$$

So

$$\begin{aligned} T^{(2)}(t_i, w_i) &= f(t_i, w_i) + \frac{h}{2}f'(t_i, w_i) = w_i - t_i^2 + 1 + \frac{h}{2}(w_i - t_i^2 - 2t + 1) \\ &= \left(1 + \frac{h}{2}\right)(w_i - t_i^2 + 1) - ht_i \end{aligned}$$

and

$$\begin{aligned} T^{(4)}(t_i, w_i) &= f(t_i, w_i) + \frac{h}{2}f'(t_i, w_i) + \frac{h^2}{6}f''(t_i, w_i) + \frac{h^3}{24}f'''(t_i, w_i) \\ &= w_i - t_i^2 + 1 + \frac{h}{2}(w_i - t_i^2 - 2t_i + 1) + \frac{h^2}{6}(w_i - t_i^2 - 2t_i - 1) \\ &\quad + \frac{h^3}{24}(w_i - t_i^2 - 2t_i - 1) \\ &= \left(1 + \frac{h}{2} + \frac{h^2}{6} + \frac{h^3}{24}\right)(w_i - t_i^2) - \left(1 + \frac{h}{3} + \frac{h^2}{12}\right)ht_i \\ &\quad + 1 + \frac{h}{2} - \frac{h^2}{6} - \frac{h^3}{24}. \end{aligned}$$

The Taylor methods of orders 2 and 4 are, consequently,

$$\begin{aligned} w_0 &= 0.5, \\ w_{i+1} &= w_i + h\left[\left(1 + \frac{h}{2}\right)(w_i - t_i^2 + 1) - ht_i\right] \end{aligned}$$

and

$$\begin{aligned} w_0 &= 0.5, \\ w_{i+1} &= w_i + h\left[\left(1 + \frac{h}{2} + \frac{h^2}{6} + \frac{h^3}{24}\right)(w_i - t_i^2) \right. \\ &\quad \left. - \left(1 + \frac{h}{3} + \frac{h^2}{12}\right)ht_i + 1 + \frac{h}{2} - \frac{h^2}{6} - \frac{h^3}{24}\right]. \end{aligned}$$

If $h = 0.2$, then $N = 10$ and $t_i = 0.2i$ for each $i = 1, 2, \ldots, 10$, so the second-order method becomes

$$\begin{aligned} w_0 &= 0.5, \\ w_{i+1} &= w_i + 0.2\left[\left(1 + \frac{0.2}{2}\right)(w_i - 0.04i^2 + 1) - 0.04i\right] \\ &= 1.22w_i - 0.0088i^2 - 0.008i + 0.22, \end{aligned}$$

and the fourth-order method becomes

$$\begin{aligned} w_{i+1} &= w_i + 0.2\left[\left(1 + \frac{0.2}{2} + \frac{0.04}{6} + \frac{0.008}{24}\right)(w_i - 0.04i^2)\right. \\ &\quad \left. - \left(1 + \frac{0.2}{3} + \frac{0.04}{12}\right)(0.04i) + 1 + \frac{0.2}{2} - \frac{0.04}{6} - \frac{0.008}{24}\right] \\ &= 1.2214w_i - 0.008856i^2 - 0.00856i + 0.2186, \end{aligned}$$

for each $i = 0, 1, \ldots, 9$.

Table 5.3 lists the actual values of the solution $y(t) = (t+1)^2 - 0.5e^t$, the results from the Taylor methods of orders 2 and 4, and the actual errors involved with these methods.

Table 5.3

t_i	Exact $y(t_i)$	Taylor Order 2 w_i	Error $\lvert y(t_i) - w_i\rvert$	Taylor Order 4 w_i	Error $\lvert y(t_i) - w_i\rvert$
0.0	0.5000000	0.5000000	0	0.5000000	0
0.2	0.8292986	0.8300000	0.0007014	0.8293000	0.0000014
0.4	1.2140877	1.2158000	0.0017123	1.2140910	0.0000034
0.6	1.6489406	1.6520760	0.0031354	1.6489468	0.0000062
0.8	2.1272295	2.1323327	0.0051032	2.1272396	0.0000101
1.0	2.6408591	2.6486459	0.0077868	2.6408744	0.0000153
1.2	3.1799415	3.1913480	0.0114065	3.1799640	0.0000225
1.4	3.7324000	3.7486446	0.0162446	3.7324321	0.0000321
1.6	4.2834838	4.3061464	0.0226626	4.2835285	0.0000447
1.8	4.8151763	4.8462986	0.0311223	4.8152377	0.0000615
2.0	5.3054720	5.3476843	0.0422123	5.3055554	0.0000834

Suppose we need to determine an approximation to an intermediate point in the table, for example at $t = 1.25$. If we use linear interpolation on the Taylor method of order 4 approximations at $t = 1.2$ and $t = 1.4$, we have

$$y(1.25) \approx \frac{1.25 - 1.4}{1.2 - 1.4} 3.1799640 + \frac{1.25 - 1.2}{1.4 - 1.2} 3.7324321 = 3.3180810.$$

Since the true value is $y(1.25) = 3.3173285$, this approximation has an error of 0.0007525, which is nearly 30 times the average of the approximation errors at 1.2 and 1.4.

To improve the approximation to $y(1.25)$, we can use cubic Hermite interpolation. This requires approximations to $y'(1.2)$ and $y'(1.4)$, as well as approximations to $y(1.2)$ and $y(1.4)$. But the derivative approximations are available from the differential equation, since $y'(t) = f(t, y(t))$. In our example $y'(t) = y(t) - t^2 + 1$, so

$$y'(1.2) = y(1.2) - (1.2)^2 + 1 \approx 3.1799640 - 1.44 + 1 = 2.7399640$$

and

$$y'(1.4) = y(1.4) - (1.4)^2 + 1 \approx 3.7324327 - 1.96 + 1 = 2.7724321.$$

The divided-difference procedure in Section 3.3 gives the information in Table 5.4. The underlined entries come from the data and the other entries use the divided-difference formulas.

Table 5.4

1.2	3.1799640			
		2.7399640		
1.2	3.1799640		0.1118825	
		2.7623405		−0.3071225
1.4	3.7324321		0.0504580	
		2.7724321		
1.4	3.7324321			

The cubic Hermite polynomial is

$$y(t) \approx 3.1799640 + 2.7399640(t-1.2) + 0.1118825(t-1.2)^2 - 0.3071225(t-1.2)^2(t-1.4),$$

so

$$y(1.25) \approx 3.1799640 + 0.1369982 + 0.0002797 + 0.0001152 = 3.3173571,$$

a result that is accurate to within 0.0000286. This is less than twice the average error at 1.2 and 1.4, or about 4% of the error obtained using linear interpolation. ◆ ◆ ◆

Error estimates for the Taylor methods are similar to those for Euler's method. If sufficient differentiability conditions are met, an nth-order Taylor method will have local error $O(h^{n+1})$ and global error $O(h^n)$.

EXERCISE SET 5.2

1. Use Euler's method to approximate the solutions for each of the following initial-value problems.
   a. $y' = te^{3t} - 2y$, for $0 \le t \le 1$, with $y(0) = 0$ and $h = 0.5$
   b. $y' = 1 + (t-y)^2$, for $2 \le t \le 3$, with $y(2) = 1$ and $h = 0.5$
   c. $y' = 1 + \dfrac{y}{t}$, for $1 \le t \le 2$, with $y(1) = 2$ and $h = 0.25$
   d. $y' = \cos 2t + \sin 3t$, for $0 \le t \le 1$, with $y(0) = 1$ and $h = 0.25$
2. The actual solutions to the initial-value problems in Exercise 1 are given here. Compare the actual error at each step to the error bound.
   a. $y(t) = \frac{1}{5}te^{3t} - \frac{1}{25}e^{3t} + \frac{1}{25}e^{-2t}$ **b.** $y(t) = t + (1-t)^{-1}$
   c. $y(t) = t\ln t + 2t$ **d.** $y(t) = \frac{1}{2}\sin 2t - \frac{1}{3}\cos 3t + \frac{4}{3}$
3. Use Euler's method to approximate the solutions for each of the following initial-value problems.

a. $y' = \frac{y}{t} - \left(\frac{y}{t}\right)^2$, for $1 \le t \le 2$, with $y(1) = 1$ and $h = 0.1$

b. $y' = 1 + \frac{y}{t} + \left(\frac{y}{t}\right)^2$, for $1 \le t \le 3$, with $y(1) = 0$ and $h = 0.2$

c. $y' = -(y+1)(y+3)$, for $0 \le t \le 2$, with $y(0) = -2$ and $h = 0.2$

d. $y' = -5y + 5t^2 + 2t$, for $0 \le t \le 1$, with $y(0) = 1/3$ and $h = 0.1$

4. The actual solutions to the initial-value problems in Exercise 3 are given here. Compute the actual error in the approximations of Exercise 3.

a. $y(t) = t(1 + \ln t)^{-1}$

b. $y(t) = t \tan(\ln t)$

c. $y(t) = -3 + 2(1 + e^{-2t})^{-1}$

d. $y(t) = t^2 + \frac{1}{3}e^{-5t}$

5. Repeat Exercise 1 using Taylor's method of order 2.

6. Repeat Exercise 3 using Taylor's method of order 2.

7. Repeat Exercise 3 using Taylor's method of order 4.

8. Given the initial-value problem

$$y' = \frac{2}{t}y + t^2e^t, \quad 1 \le t \le 2, \quad y(1) = 0$$

with exact solution $y(t) = t^2(e^t - e)$:

a. Use Euler's method with $h = 0.1$ to approximate the solution and compare it with the actual values of y.

b. Use the answers generated in part (a) and linear interpolation to approximate the following values of y and compare them to the actual values.

(i) $y(1.04)$ **(ii)** $y(1.55)$ **(iii)** $y(1.97)$

c. Use Taylor's method of order 2 with $h = 0.1$ to approximate the solution and compare it with the actual values of y.

d. Use the answers generated in part (c) and linear interpolation to approximate y at the following values and compare them to the actual values of y.

(i) $y(1.04)$ **(ii)** $y(1.55)$ **(iii)** $y(1.97)$

e. Use Taylor's method of order 4 with $h = 0.1$ to approximate the solution and compare it with the actual values of y.

f. Use the answers generated in part (e) and piecewise cubic Hermite interpolation to approximate y at the following values and compare them to the actual values of y.

(i) $y(1.04)$ **(ii)** $y(1.55)$ **(iii)** $y(1.97)$

9. Given the initial-value problem

$$y' = \frac{1}{t^2} - \frac{y}{t} - y^2, \quad 1 \le t \le 2, \quad y(1) = -1$$

with exact solution $y(t) = -1/t$.

a. Use Euler's method with $h = 0.05$ to approximate the solution and compare it with the actual values of y.

b. Use the answers generated in part (a) and linear interpolation to approximate the following values of y and compare them to the actual values.
(i) $y(1.052)$ **(ii)** $y(1.555)$ **(iii)** $y(1.978)$

c. Use Taylor's method of order 2 with $h = 0.05$ to approximate the solution and compare it with the actual values of y.

d. Use the answers generated in part (c) and linear interpolation to approximate the following values of y and compare them to the actual values.
(i) $y(1.052)$ **(ii)** $y(1.555)$ **(iii)** $y(1.978)$

e. Use Taylor's method of order 4 with $h = 0.05$ to approximate the solution and compare it with the actual values of y.

f. Use the answers generated in part (e) and piecewise cubic Hermite interpolation to approximate the following values of y and compare them to the actual values.
(i) $y(1.052)$ **(ii)** $y(1.555)$ **(iii)** $y(1.978)$

10. In an electrical circuit with impressed voltage $\mathcal{E}$, having resistance R, inductance L, and capacitance C in parallel, the current i satisfies the differential equation

$$\frac{di}{dt} = C\frac{d^2\mathcal{E}}{dt^2} + \frac{1}{R}\frac{d\mathcal{E}}{dt} + \frac{1}{L}\mathcal{E}.$$

Suppose $i(0) = 0$, $C = 0.3$ farads, $R = 1.4$ ohms, $L = 1.7$ henries, and the voltage is given by

$$\mathcal{E}(t) = e^{-0.06\pi t}\sin(2t - \pi).$$

Use Euler's method to find the current i for the values $t = 0.1j$, $j = 0, 1, \ldots, 100$.

11. In a book entitled *Looking at History Through Mathematics*, Rashevsky [Ra] considers a model for a problem involving the production of nonconformists in society. Suppose that a society has a population of $x(t)$ individuals at time t, in years, and that all nonconformists who mate with other nonconformists have offspring who are also nonconformists, while a fixed proportion r of all other offspring are also nonconformist. If the birth and death rates for all individuals are assumed to be the constants b and d, respectively, and if conformists and nonconformists mate at random, the problem can be expressed by the differential equations

$$\frac{dx(t)}{dt} = (b - d)x(t) \quad \text{and} \quad \frac{dx_n(t)}{dt} = (b - d)x_n(t) + rb(x(t) - x_n(t)),$$

where $x_n(t)$ denotes the number of nonconformists in the population at time t.

a. If the variable $p(t) = x_n(t)/x(t)$ is introduced to represent the proportion of nonconformists in the society at time t, show that these equations can be combined and simplified to the single differential equation

$$\frac{dp(t)}{dt} = rb(1 - p(t)).$$

b. Assuming that $p(0) = 0.01$, $b = 0.02$, $d = 0.015$, and $r = 0.1$, use Euler's method to approximate the solution $p(t)$ from $t = 0$ to $t = 50$ when the step size is $h = 1$ year.

c. Solve the differential equation for $p(t)$ exactly, and compare your result in part (b) when $t = 50$ with the exact value at that time.

12. A projectile of mass $m = 0.11$ kg shot vertically upward with initial velocity $v(0) = 8$ m/s is slowed due to the force of gravity $F_g = mg$ and due to air resistance $F_r = -kv|v|$, where $g = -9.8$ m/s^2 and $k = 0.002$ kg/m. The differential equation for the velocity v is given by

$$mv' = mg - kv|v|.$$

a. Find the velocity after $0.1, 0.2, \ldots, 1.0$ s.

b. To the nearest tenth of a second, determine when the projectile reaches its maximum height and begins falling.

5.3 Runge-Kutta Methods

In the last section we saw how Taylor methods of arbitrary high order can be generated. The application of one of the higher-order methods to a specific problem is anything but easy, however, due to the need to determine and evaluate high-order derivatives with respect to t on the right side of the differential equation. In this section we consider Runge-Kutta methods, which modify the Taylor methods so that the high-order error bounds are preserved, but the need to determine and evaluate the high-order partial derivatives is eliminated. The strategy behind these techniques involves approximating a Taylor method with a method that is easier to evaluate. This approximation increases the error, but the increase does not exceed the order of the error that was already present in the Taylor method. As a consequence, the new error does not significantly influence the calculations.

The Runge-Kutta techniques make use of the Taylor expansion of f, the function on the right side of the differential equation. Since f is a function of two variables, t and y, we must first consider the generalization of Taylor's Theorem to functions of this type. This generalization appears more complicated than the original form, but this is only because of all the partial derivatives of the function f.

Taylor's Theorem for Two Variables

If f and all its partial derivatives of order less than or equal to n are continuous on $D = \{(t, y) | a \le t \le b,\ c \le y \le d\}$ and (t, y) and $(t + h, y + k)$ both belong to D, then

$$f(t+h, y+k) \approx f(t,y) + \left[h\frac{\partial f}{\partial t}(t,y) + k\frac{\partial f}{\partial y}(t,y)\right] + \left[\frac{h^2}{2}\frac{\partial^2 f}{\partial t^2}(t,y) + hk\frac{\partial^2 f}{\partial t\, \partial y}(t,y) + \frac{k^2}{2}\frac{\partial^2 f}{\partial y^2}(t,y)\right] + \cdots + \frac{1}{n!}\sum_{j=0}^{n}\binom{n}{j} h^{n-j}k^j \frac{\partial^n f}{\partial t^{n-j}\partial y^j}(t,y).$$

The error term in this approximation is similar to that given in Taylor's Theorem, with the added complications that arise because of the incorporation of all the partial derivatives of order $n+1$.

To illustrate the use of this formula in developing the Runge-Kutta methods, let us consider the Runge-Kutta method of order 2. We saw in the previous section that the Taylor method of order 2 comes from

$$\begin{aligned} y(t_{i+1}) &= y(t_i) + hy'(t_i) + \frac{h^2}{2}y''(t_i) + \frac{h^3}{3!}y'''(\xi) \\ &= y(t_i) + hf(t_i, y(t_i)) + \frac{h^2}{2}f'(t_i, y(t_i)) + \frac{h^3}{3!}y'''(\xi), \end{aligned}$$

or, since

$$f'(t_i, y(t_i)) = \frac{\partial f}{\partial t}(t_i, y(t_i)) + \frac{\partial f}{\partial y}(t_i, y(t_i))y'(t_i)$$

and $y'(t_i) = f(t_i, y(t_i))$, we have

$$y(t_{i+1}) = y(t_i) + h\left\{f(t_i, y(t_i)) + \frac{h}{2}\frac{\partial f}{\partial t}(t_i, y(t_i)) + \frac{h}{2}\frac{\partial f}{\partial y}(t_i, y(t_i))f(t_i, y(t_i))\right\} + \frac{h^3}{3!}y'''(\xi).$$

Taylor's Theorem of two variables permits us to replace the term in the large braces with a multiple of a function evaluation of f of the form $a_1 f(t_i + \alpha, y(t_i) + \beta)$. If we expand this term using the theorem with $n = 1$, we have

$$\begin{aligned} a_1 f(t_i + \alpha, y(t_i) + \beta) &\approx a_1\left[f(t_i, y(t_i)) + \alpha\frac{\partial f}{\partial t}(t_i, y(t_i)) + \beta\frac{\partial f}{\partial y}(t_i, y(t_i))\right] \\ &= a_1 f(t_i, y(t_i)) + a_1\alpha\frac{\partial f}{\partial t}(t_i, y(t_i)) + a_1\beta\frac{\partial f}{\partial y}(t_i, y(t_i)). \end{aligned}$$

Equating this expression with the terms enclosed in the large braces in the preceding equation implies that a, α, and β should be chosen so that

$$1 = a_1, \quad \frac{h}{2} = a_1\alpha, \quad \text{and} \quad \frac{h}{2}f(t_i, y(t_i)) = a_1\beta;$$

that is,

$$a_1 = 1, \quad \alpha = \frac{h}{2}, \quad \text{and} \quad \beta = \frac{h}{2} f(t_i, y(t_i)).$$

The error introduced by replacing the term in the Taylor method with its approximation has the same order as the error term for the method, so the Runge-Kutta method produced in this way, called the **Midpoint method**, is also a second-order method. As a consequence, the local error of the method is proportional to the cube of the step size, and the global error is proportional to the square of the step size.

Midpoint Method

$$w_0 = \alpha$$

$$w_{i+1} = w_i + h\left[f\left(t_i + \frac{h}{2}, w_i + \frac{h}{2} f(t_i, w_i)\right)\right],$$

where $i = 0, 1, \ldots, N-1$, with local error $O(h^3)$.

Using $a_1 f(t + \alpha, y + \beta)$ to replace the term in the Taylor method is the easiest choice, but it is not the only one. If we instead use a term of the form

$$a_1 f(t, y) + a_2 f(t + \alpha, y + \beta f(t, y)),$$

the extra parameter in this formula provides an infinite number of second-order Runge-Kutta formulas. When $a_1 = a_2 = \frac{1}{2}$ and $\alpha = \beta = h$, we have the **Modified Euler method**.

Modified Euler Method

$$w_0 = \alpha$$

$$w_{i+1} = w_i + \frac{h}{2}[f(t_i, w_i) + f(t_{i+1}, w_i + hf(t_i, w_i))]$$

where $i = 0, 1, \ldots, N-1$, with local error $O(h^3)$.

When $a_1 = \frac{1}{4}, a_2 = \frac{3}{4}$, and $\alpha = \beta = \frac{2}{3}h$, we have **Heun's method.**

Heun's Method

$$w_0 = \alpha$$

$$w_{i+1} = w_i + \frac{h}{4}\left[f(t_i, w_i) + 3f\left(t_i + \frac{2}{3}h, w_i + \frac{2}{3}hf(t_i, w_i)\right)\right]$$

where $i = 0, 1, \ldots, N-1$, with local error $O(h^3)$.

EXAMPLE 1 Suppose we apply the Runge-Kutta methods of order 2 to our usual example,

$$y' = y - t^2 + 1, \quad \text{for } 0 \le t \le 2, \quad \text{with } y(0) = 0.5,$$

where $N = 10, h = 0.2, t_i = 0.2i$, and $w_0 = 0.5$ in each case. The difference equations produced from the various formulas are

$$\begin{aligned}
\text{Midpoint method:} \quad & w_{i+1} = 1.22w_i - 0.0088i^2 - 0.008i + 0.218;\\
\text{Modified Euler method:} \quad & w_{i+1} = 1.22w_i - 0.0088i^2 - 0.008i + 0.216;\\
\text{Heun's method:} \quad & w_{i+1} = 1.22w_i - 0.0088i^2 - 0.008i + 0.217\overline{3};
\end{aligned}$$

for each $i = 0, 1, \ldots, 9$. Table 5.5 lists the results of these calculations. ◆ ◆ ◆

Table 5.5

t_i	$y(t_i)$	Midpoint Method	Error	Modified Euler Method	Error	Heun's Method	Error
0.0	0.5000000	0.5000000	0	0.5000000	0	0.5000000	0
0.2	0.8292986	0.8280000	0.0012986	0.8260000	0.0032986	0.8273333	0.0019653
0.4	1.2140877	1.2113600	0.0027277	1.2069200	0.0071677	1.2098800	0.0042077
0.6	1.6489406	1.6446592	0.0042814	1.6372424	0.0116982	1.6421869	0.0067537
0.8	2.1272295	2.1212842	0.0059453	2.1102357	0.0169938	2.1176014	0.0096281
1.0	2.6408591	2.6331668	0.0076923	2.6176876	0.0231715	2.6280070	0.0128521
1.2	3.1799415	3.1704634	0.0094781	3.1495789	0.0303627	3.1635019	0.0164396
1.4	3.7324000	3.7211654	0.0112346	3.6936862	0.0387138	3.7120057	0.0203944
1.6	4.2834838	4.2706218	0.0128620	4.2350972	0.0483866	4.2587802	0.0247035
1.8	4.8151763	4.8009586	0.0142177	4.7556185	0.0595577	4.7858452	0.0293310
2.0	5.3054720	5.2903695	0.0151025	5.2330546	0.0724173	5.2712645	0.0342074

Higher-order Taylor formulas can be converted into Runge-Kutta techniques in a similar way, but the algebra becomes extremely tedious. The most common Runge-Kutta method is of order 4 and is obtained by expanding an expression that involves only four function evaluations. Deriving this expression requires solving a system of equations involving 12 unknowns. Once the algebra has been performed, the method has the following simple representation.

Runge-Kutta Method of Order 4

$$\begin{aligned}
w_0 &= \alpha,\\
k_1 &= hf(t_i, w_i),\\
k_2 &= hf\left(t_i + \frac{h}{2}, w_i + \frac{1}{2}k_1\right),
\end{aligned}$$

$$k_3 = hf\left(t_i + \frac{h}{2}, w_i + \frac{1}{2}k_2\right),$$
$$k_4 = hf(t_{i+1}, w_i + k_3),$$
$$w_{i+1} = w_i + \frac{1}{6}(k_1 + 2k_2 + 2k_3 + k_4),$$

where $i = 0, 1, \ldots, N-1$, with local error $O(h^5)$.

The program RKO4M52 implements this method.

EXAMPLE 2 The Runge-Kutta method of order 4 applied to the initial-value problem

$$y' = y - t^2 + 1, \quad \text{for } 0 \le t \le 2, \quad \text{with } y(0) = 0.5,$$

with $h = 0.2, N = 10$, and $t_i = 0.2i$, gives the results and errors listed in Table 5.6. ◆ ◆ ◆

Table 5.6

t_i	Exact $y_i = y(t_i)$	Runge-Kutta Order 4 w_i	Error $\lvert y_i - w_i\rvert$
0.0	0.5000000	0.5000000	0
0.2	0.8292986	0.8292933	0.0000053
0.4	1.2140877	1.2140762	0.0000114
0.6	1.6489406	1.6489220	0.0000186
0.8	2.1272295	2.1272027	0.0000269
1.0	2.6408591	2.6408227	0.0000364
1.2	3.1799415	3.1798942	0.0000474
1.4	3.7324000	3.7323401	0.0000599
1.6	4.2834838	4.2834095	0.0000743
1.8	4.8151763	4.8150857	0.0000906
2.0	5.3054720	5.3053630	0.0001089

We generate the first entry in Table 5.6 using Maple. First, we define the function $f(t, y)$ with the command

```
>f:=(t,y)->y-t^2+1;
```

The values of a, b, N, h, and $y(0)$ are defined by

```
>a:=0; b:=2; N:=10; h:=(b-a)/N; alpha:=0.5;
```

and we initialize w_0 and t_0 with

```
>w0:=alpha; t0:=a;
```

We compute $k_1 = 0.3$, $k_2 = 0.328$, $k_3 = 0.3308$, and $k_4 = 0.35816$ with

```
>k1:=h*f(t0,w0);

>k2:=h*f(t0+h/2,w0+k1/2);

>k3:=h*f(t0+h/2,w0+k2/2);

>k4:=h*f(t0+h,w0+k3);
```

The approximation $w_1 = 0.8292933334$ at $t_1 = 0.2$ is obtained from

```
>w1:=w0+(k1+2*k2+2*k3+k4)/6;
```

The remaining entries in Table 5.6 are generated in a similar fashion.

The main computational effort in applying the Runge-Kutta methods involves the function evaluations of f. In the second-order methods, the local error is $O(h^3)$ and the cost is two functional evalutions per step. The Runge-Kutta method of order 4 requires four evaluations per step and the local error is $O(h^5)$. The relationship between the number of evaluations per step, and the order of the local error is shown in Table 5.7. Because of the relative decrease in the order for n greater than 4, the methods of order less than 5 with smaller step size are used in preference to the higher-order methods using a larger step size.

Table 5.7

Evaluations per step	2	3	4	$5 \le n \le 7$	$8 \le n \le 9$	$10 \le n$
Best possible local error	$O(h^3)$	$O(h^4)$	$O(h^5)$	$O(h^n)$	$O(h^{n-1})$	$O(h^{n-2})$

One way to compare the lower-order Runge-Kutta methods is described as follows: The Runge-Kutta method of order 4 requires four evaluations per step, so to be superior to Euler's method it should give more accurate answers than when Euler's method uses one-fourth the Runge-Kutta mesh size (by *mesh size* we mean the difference between consecutive mesh points). Similarly, if the Runge-Kutta method of order 4 is to be superior to the second-order Runge-Kutta methods, it should give more accuracy with step size h than a second-order method with step size $\frac{1}{2}h$. An illustration of the superiority of the Runge-Kutta Fourth-Order method by this measure is shown in the following example.

EXAMPLE 3 For the problem

$$y' = y - t^2 + 1, \quad \text{for } 0 \le t \le 2, \quad \text{with } y(0) = 0.5,$$

Euler's method with $h = 0.025$, the Modified Euler's method with $h = 0.05$, and the Runge-Kutta method of order 4 with $h = 0.1$ are compared at the common mesh points of the three methods, 0.1, 0.2, 0.3, 0.4, and 0.5. Each of these techniques requires 20 functional evaluations to approximate $y(0.5)$. (See Table 5.8.) In this example, the fourth-order method is clearly superior, as it is in most situations. ◆ ◆ ◆

Table 5.8

t_i	Exact	Euler $h = 0.025$	Modified Euler $h = 0.05$	Runge-Kutta Order 4 $h = 0.1$
0.0	0.5000000	0.5000000	0.5000000	0.5000000
0.1	0.6574145	0.6554982	0.6573085	0.6574144
0.2	0.8292986	0.8253385	0.8290778	0.8292983
0.3	1.0150706	1.0089334	1.0147254	1.0150701
0.4	1.2140877	1.2056345	1.2136079	1.2140869
0.5	1.4256394	1.4147264	1.4250141	1.4256384

The Modified Euler method and the Runge-Kutta Fourth-Order method are also available in Maple. To implement the procedure we first enter the commands

```
>with(share);
```

```
>readshare(ODE,plots);
```

Define the function $f(t, y)$ by

```
>eq:=(t,y)->y-t*t+1;
```

The Runge-Kutta values given in Table 5.8 are produced with

```
>eqrk:=rungekutta(eq,[0,0.5],0.1,5);
```

We can access a specific entry from the output with a command of the form

```
>eqrk[3][2];
```

The second coordinate of the item labeled (3) is $w_3 = 1.015070059$.

We can use the Modified Euler method in a similar manner with the command `impeuler`.

EXERCISE SET 5.3

1. Use the Midpoint method to approximate the solutions to each of the following initial-value problems, and compare the results to the actual values.
   a. $y' = te^{3t} - 2y$, for $0 \le t \le 1$, with $y(0) = 0$ and $h = 0.5$; actual solution $y(t) = \frac{1}{5}te^{3t} - \frac{1}{25}e^{3t} + \frac{1}{25}e^{-2t}$.
   b. $y' = 1 + (t - y)^2$, for $2 \le t \le 3$, with $y(2) = 1$ and $h = 0.5$; actual solution $y(t) = t + 1/(1 - t)$.

c. $y' = 1 + \dfrac{y}{t}$, for $1 \leq t \leq 2$, with $y(1) = 2$ and $h = 0.25$; actual solution $y(t) = t \ln t + 2t$.

d. $y' = \cos 2t + \sin 3t$, for $0 \leq t \leq 1$, with $y(0) = 1$ and $h = 0.25$; actual solution $y(t) = \frac{1}{2}\sin 2t - \frac{1}{3}\cos 3t + \frac{4}{3}$.

2. Repeat Exercise 1 using Heun's method.

3. Repeat Exercise 1 using the Modified Euler method.

4. Use the Modified Euler method to approximate the solutions to each of the following initial-value problems, and compare the results to the actual values.

a. $y' = \dfrac{y}{t} - \left(\dfrac{y}{t}\right)^2$, for $1 \leq t \leq 2$, with $y(1) = 1$ and $h = 0.1$; actual solution $y(t) = t/(1 + \ln t)$.

b. $y' = 1 + \dfrac{y}{t} + \left(\dfrac{y}{t}\right)^2$, for $1 \leq t \leq 3$, with $y(1) = 0$ and $h = 0.2$; actual solution $y(t) = t\tan(\ln t)$.

c. $y' = -(y + 1)(y + 3)$, for $0 \leq t \leq 2$, with $y(0) = -2$ and $h = 0.2$; actual solution $y(t) = -3 + 2(1 + e^{-2t})^{-1}$.

d. $y' = -5y + 5t^2 + 2t$, for $0 \leq t \leq 1$, with $y(0) = \frac{1}{3}$ and $h = 0.1$; actual solution $y(t) = t^2 + \frac{1}{3}e^{-5t}$.

5. Use the results of Exercise 4 and linear interpolation to approximate values of $y(t)$, and compare the results to the actual values.

a. $y(1.25)$ and $y(1.93)$
b. $y(2.1)$ and $y(2.75)$
c. $y(1.3)$ and $y(1.93)$
d. $y(0.54)$ and $y(0.94)$

6. Repeat Exercise 4 using Heun's method.

7. Repeat Exercise 5 using the results of Exercise 6.

8. Repeat Exercise 4 using the Midpoint method.

9. Repeat Exercise 5 using the results of Exercise 8.

10. Repeat Exercise 1 using the Runge-Kutta method of order 4.

11. Repeat Exercise 4 using the Runge-Kutta method of order 4.

12. Use the results of Exercise 11 and Cubic Hermite interpolation to approximate values of $y(t)$ and compare the approximations to the actual values.

a. $y(1.25)$ and $y(1.93)$
b. $y(2.1)$ and $y(2.75)$
c. $y(1.3)$ and $y(1.93)$
d. $y(0.54)$ and $y(0.94)$

13. Show that the Midpoint method, the Modified Euler method, and Heun's method give the same approximations to the initial-value problem

$$y' = -y + t + 1, \quad 0 \leq t \leq 1, \quad y(0) = 1,$$

for any choice of h. Why is this true?

14. Water flows from an inverted conical tank with circular orifice at the rate

$$\frac{dx}{dt} = -0.6\pi r^2\sqrt{-2g}\frac{\sqrt{x}}{A(x)},$$

where r is the radius of the orifice, x is the height of the liquid level from the vertex of the cone, and $A(x)$ is the area of the cross section of the tank x units above the orifice. Suppose $r = 0.1$ ft, $g = -32.17$ ft/s^2, and the tank has an initial water level of 8 ft and initial volume of $512(\pi/3)$ ft^3.

a. Compute the water level after 10 min with $h = 20$ s.

b. Determine, to within 1 min, when the tank will be empty.

15. The irreversible chemical reaction in which two molecules of solid potassium dichromate ($K_2Cr_2O_7$), two molecules of water (H_2O), and three atoms of solid sulfur (S) combine to yield three molecules of the gas sulfur dioxide (SO_2), four molecules of solid potassium hydroxide (KOH), and two molecules of solid chromic oxide (Cr_2O_3) can be represented symbolically by the stoichiometric equation

$$2K_2Cr_2O_7 + 2H_2O + 3S \longrightarrow 4KOH + 2Cr_2O_3 + 3SO_2.$$

If n_1 molecules of $K_2Cr_2O_7$, n_2 molecules of H_2O, and n_3 molecules of S are originally available, the following differential equation describes the amount $x(t)$ of KOH after time t:

$$\frac{dx}{dt} = k\left(n_1 - \frac{x}{2}\right)^2\left(n_2 - \frac{x}{2}\right)^2\left(n_3 - \frac{3x}{4}\right)^3,$$

where k is the velocity constant of the reaction. If $k = 6.22 \times 10^{-19}$, $n_1 = n_2 = 2 \times 10^3$, and $n_3 = 3 \times 10^3$, how many units of potassium hydroxide will have been formed after 0.2 s?

5.4 Predictor-Corrector Methods

The Taylor and Runge-Kutta methods are examples of **one-step methods** for approximating the solution to initial-value problems. These methods use w_i in the approximation w_{i+1} to $y(t_{i+1})$ but do not involve any of the prior approximations $w_0, w_1, \ldots, w_{i-1}$. Generally some functional evaluations of f are required at intermediate points, but these are discarded as soon as w_{i+1} is obtained.

Since $|y(t_j) - w_j|$ decreases in accuracy as j increases, better approximation methods can be derived if, when approximating $y(t_{i+1})$, we include in the method some of the approximations prior to w_i. Methods developed using this philosophy are called **multistep methods**. In brief, one-step methods consider what occurred at only one previous step; multistep methods consider what happened at more than one previous step.

To derive a multistep method, first note that the solution to the initial-value problem

$$\frac{dy}{dt} = f(t, y), \quad a \le t \le b,$$

subject to an initial condition

$$y(a) = \alpha,$$

if integrated over the interval $[t_i, t_{i+1}]$, has the property that

$$y(t_{i+1}) - y(t_i) = \int_{t_i}^{t_{i+1}} y'(t)\,dt = \int_{t_i}^{t_{i+1}} f(t, y(t))\,dt.$$

Consequently,

$$y(t_{i+1}) = y(t_i) + \int_{t_i}^{t_{i+1}} f(t, y(t))\,dt.$$

Since we cannot integrate $f(t, y(t))$ without knowing $y(t)$, which is the solution to the problem, we instead integrate an interpolating polynomial, $P(t)$, determined by some of the previously obtained data points $(t_0, w_0), (t_1, w_1), \ldots, (t_i, w_i)$. When we assume, in addition, that $y(t_i) \approx w_i$, we have

$$y(t_{i+1}) \approx w_i + \int_{t_i}^{t_{i+1}} P(t)\,dt.$$

If w_{m+1} is the first approximation generated by the multistep method, then we need to supply starting values $w_0, w_1, \ldots, w_m$ for the method. These starting values are generated using a Runge-Kutta One-Step method with the same error characteristics as the multistep method.

There are two distinct classes of multistep methods. In an **explicit method,** w_{i+1} does not depend on the function evaluation $f(t_{i+1}, w_{i+1})$. A method that does depend in part on $f(t_{i+1}, w_{i+1})$ is **implicit**.

Some of the explicit multistep methods, together with their required starting values and local error terms, are given next.

Adams-Bashforth Two-Step Explicit Method

$$w_0 = \alpha,\ w_1 = \alpha_1,$$
$$w_{i+1} = w_i + \frac{h}{2}[3f(t_i, w_i) - f(t_{i-1}, w_{i-1})],$$

where $i = 1, 2, \ldots, N-1$, with local error $\frac{5}{12}y'''(\mu_i)h^3$ for some μ_i in (t_{i-1}, t_{i+1}).

Adams-Bashforth Three-Step Explicit Method

$$w_0 = \alpha, \ w_1 = \alpha_1, \ w_2 = \alpha_2,$$
$$w_{i+1} = w_i + \frac{h}{12}[23f(t_i, w_i) - 16f(t_{i-1}, w_{i-1}) + 5f(t_{i-2}, w_{i-2})]$$

where $i = 2, 3, \ldots, N-1$, with local error $\frac{3}{8}y^{(4)}(\mu_i)h^4$ for some μ_i in (t_{i-2}, t_{i+1}).

Adams-Bashforth Four-Step Explicit Method

$$w_0 = \alpha, \ w_1 = \alpha_1, \ w_2 = \alpha_2, \ w_3 = \alpha_3,$$
$$w_{i+1} = w_i + \frac{h}{24}[55f(t_i, w_i) - 59f(t_{i-1}, w_{i-1}) + 37f(t_{i-2}, w_{i-2}) - 9f(t_{i-3}, w_{i-3})]$$

where $i = 3, 4, \ldots, N-1$, with local error $\frac{251}{720}y^{(5)}(\mu_i)h^5$ for some μ_i in (t_{i-3}, t_{i+1}).

Adams-Bashforth Five-Step Explicit Method

$$w_0 = \alpha, \ w_1 = \alpha_1, \ w_2 = \alpha_2, \ w_3 = \alpha_3, \ w_4 = \alpha_4$$
$$w_{i+1} = w_i + \frac{h}{720}[1901f(t_i, w_i) - 2774f(t_{i-1}, w_{i-1}) + 2616f(t_{i-2}, w_{i-2}) - 1274f(t_{i-3}, w_{i-3}) + 251f(t_{i-4}, w_{i-4})]$$

where $i = 4, 5, \ldots, N-1$, with local error $\frac{95}{288}y^{(6)}(\mu_i)h^6$ for some μ_i in (t_{i-4}, t_{i+1}).

Implicit methods use $(t_{i+1}, f(t_{i+1}, y(t_{i+1})))$ as an additional interpolation node in the approximation of the integral

$$\int_{t_i}^{t_{i+1}} f(t, y(t))\, dt.$$

Some of the more common implicit methods are listed next. Notice that the local error of an $(m-1)$-step implicit method is $O(h^{m+1})$, the same as that of an m-step explicit method. They both use m function evaluations, however, since the implicit methods use $f(t_{i+1}, w_{i+1})$, but the explicit methods do not.

Adams-Moulton Two-Step Implicit Method

$$w_0 = \alpha, \ w_1 = \alpha_1$$
$$w_{i+1} = w_i + \frac{h}{12}[5f(t_{i+1}, w_{i+1}) + 8f(t_i, w_i) - f(t_{i-1}, w_{i-1})]$$

where $i = 1, 2, \ldots, N-1$, with local error $-\frac{1}{24}y^{(4)}(\mu_i)h^4$ for some μ_i in (t_{i-1}, t_{i+1}).

Adams-Moulton Three-Step Implicit Method

$$w_0 = \alpha,\ w_1 = \alpha_1,\ w_2 = \alpha_2,$$

$$w_{i+1} = w_i + \frac{h}{24}[9f(t_{i+1}, w_{i+1}) + 19f(t_i, w_i) - 5f(t_{i-1}, w_{i-1}) + f(t_{i-2}, w_{i-2})],$$

where $i = 2, 3, \ldots, N-1$, with local error $-\frac{19}{720}y^{(5)}(\mu_i)h^5$ for some μ_i in (t_{i-2}, t_{i+1}).

Adams-Moulton Four-Step Implicit Method

$$w_0 = \alpha,\ w_1 = \alpha_1,\ w_2 = \alpha_2,\ w_3 = \alpha_3,$$

$$w_{i+1} = w_i + \frac{h}{720}[251f(t_{i+1}, w_{i+1}) + 646f(t_i, w_i) - 246f(t_{i-1}, w_{i-1}) + 106f(t_{i-2}, w_{i-2}) - 19f(t_{i-3}, w_{i-3})]$$

where $i = 3, 4, \ldots, N-1$, with local error $-\frac{3}{160}y^{(6)}(\mu_i)h^6$ for some μ_i in (t_{i-3}, t_{i+1}).

It is interesting to compare an m-step Adams-Bashforth explicit method to an $(m-1)$-step Adams-Moulton implicit method. Both require m evaluations of f per step, and both have the terms $y^{(m+1)}(\mu_i)h^{m+1}$ in their local errors. In general, the coefficients of the terms involving f in the approximation and those in the local error are smaller for the implicit methods than for the explicit methods. This leads to smaller round-off errors for the implicit methods.

EXAMPLE 1 Consider the initial-value problem

$$y' = y - t^2 + 1, \quad \text{for } 0 \le t \le 2, \quad \text{with } y(0) = 0.5,$$

and the approximations given by the Adams-Bashforth Four-Step method and the Adams-Moulton Three-Step method, both using $h = 0.2$. The explicit Adams-Bashforth method has the difference equation

$$w_{i+1} = w_i + \frac{h}{24}[55f(t_i, w_i) - 59f(t_{i-1}, w_{i-1}) + 37f(t_{i-2}, w_{i-2}) - 9f(t_{i-3}, w_{i-3})],$$

for $i = 3, 4, \ldots, 9$. When simplified using $f(t, y) = y - t^2 + 1, h = 0.2$, and $t_i = 0.2i$, it becomes

$$w_{i+1} = \frac{1}{24}[35w_i - 11.8w_{i-1} + 7.4w_{i-2} - 1.8w_{i-3} - 0.192i^2 - 0.192i + 4.736].$$

The implicit Adams-Moulton method has the difference equation

$$w_{i+1} = w_i + \frac{h}{24}[9f(t_{i+1}, w_{i+1}) + 19f(t_i, w_i) - 5f(t_{i-1}, w_{i-1})] + f(t_{i-2}, w_{i-2})],$$

for $i = 2, 3, \ldots, 9$. This reduces to

$$w_{i+1} = \frac{1}{24}[1.8w_{i+1} + 27.8w_i - w_{i-1} + 0.2w_{i-2} - 0.192i^2 - 0.192i + 4.736].$$

To use this method explicitly, we solve for w_{i+1}, which gives

$$w_{i+1} = \frac{1}{22.2}[27.8w_i - w_{i-1} + 0.2w_{i-2} - 0.192i^2 - 0.192i + 4.736]$$

for $i = 2, 3, \ldots, 9$. The results in Table 5.9 were obtained using the exact values from $y(t) = (t+1)^2 - 0.5e^t$ for α, α_1, α_2, and α_3 in the explicit Adams-Bashforth case and for α, α_1, and α_2 in the implicit Adams-Moulton case. ◆ ◆ ◆

Table 5.9

t_i	$y_i = y(t_i)$	Adams Bashforth w_i	Error	Adams Moulton w_i	Error
0.0	0.5000000	0.5000000	0	0.5000000	0
0.2	0.8292986	0.8292986	0.0000000	0.8292986	0.0000000
0.4	1.2140877	1.2140877	0.0000000	1.2140877	0.0000000
0.6	1.6489406	1.6489406	0.0000000	1.6489341	0.0000065
0.8	2.1272295	2.1273124	0.0000828	2.1272136	0.0000160
1.0	2.6408591	2.6410810	0.0002219	2.6408298	0.0000293
1.2	3.1799415	3.1803480	0.0004065	3.1798937	0.0000478
1.4	3.7324000	3.7330601	0.0006601	3.7323270	0.0000731
1.6	4.2834838	4.2844931	0.0010093	4.2833767	0.0001071
1.8	4.8151763	4.8166575	0.0014812	4.8150236	0.0001527
2.0	5.3054720	5.3075838	0.0021119	5.3052587	0.0002132

In Example 1, the implicit Adams-Moulton method gave considerably better results than the explicit Adams-Bashforth method of the same order. Although this is generally the case, the implicit methods have the inherent weakness of first having to convert the method algebraically to an explicit representation for w_{i+1}. That this procedure can become difficult, if not impossible, can be seen by considering the elementary initial-value problem

$$y' = e^y, \quad \text{for } 0 \le t \le 0.25, \quad \text{with } y(0) = 1.$$

Since $f(t, y) = e^y$, the Adams-Moulton Three-Step method has

$$w_{i+1} = w_i + \frac{h}{24}[9e^{w_{i+1}} + 19e^{w_i} - 5e^{w_{i-1}} + e^{w_{i-2}}]$$

as its difference equation, and this equation cannot be solved explicitly for w_{i+1}. We could use Newton's method or the Secant method to approximate w_{i+1}, but this complicates the procedure considerably.

In practice, implicit multistep methods are not used alone. Rather, they are used to improve approximations obtained by explicit methods. The combination of an explicit and implicit technique is called a **predictor-corrector method**. The explicit method predicts an approximation, and the implicit method corrects this prediction.

Consider the following fourth-order method for solving an initial-value problem. The first step is to calculate the starting values w_0, w_1, w_2, and w_3 for the Adams-Bashforth Four-Step method. To do this, we use a fourth-order one-step method, the Runge-Kutta method of order 4. The next step is to calculate an approximation, $w_4^{(0)}$, to $y(t_4)$ using the explicit Adams-Bashforth Four-Step method as predictor:

$$w_4^{(0)} = w_3 + \frac{h}{24}[55f(t_3, w_3) - 59f(t_2, w_2) + 37f(t_1, w_1) - 9f(t_0, w_0)].$$

This approximation is improved by use of the implicit Adams-Moulton Three-Step method as corrector:

$$w_4^{(1)} = w_3 + \frac{h}{24}\left[9f\left(t_4, w_4^{(0)}\right) + 19f(t_3, w_3) - 5f(t_2, w_2) + f(t_1, w_1)\right].$$

The value $w_4 \equiv w_4^{(1)}$ is then used as the approximation to $y(t_4)$, and the technique of using the Adams-Bashforth method as a predictor and the Adams-Moulton method as a corrector is repeated to find $w_5^{(0)}$ and $w_5^{(1)}$, the initial and final approximations to $y(t_5)$, and so on.

The program PRCORM53 is based on the Adams-Bashforth Four-Step method as predictor and one iteration of the Adams-Moulton Three-Step method as corrector, with the starting values obtained from the Runge-Kutta method of order 4.

EXAMPLE 2 Table 5.10 lists the results obtained by using the program PRCORM53 for the initial-value problem

$$y' = y - t^2 + 1, \quad \text{for } 0 \le t \le 2, \quad \text{with } y(0) = 0.5,$$

with $N = 10$. The results here are more accurate than those in Example 1, which used only the corrector (that is, the Adams-Moulton method). This would not generally be the case. ◆ ◆ ◆

Table 5.10

t_i	$y_i = y(t_i)$	w_i	Error $\lvert y_i - w_i \rvert$
0.0	0.5000000	0.5000000	0
0.2	0.8292986	0.8292933	0.0000053
0.4	1.2140877	1.2140762	0.0000114
0.6	1.6489406	1.6489220	0.0000186
0.8	2.1272295	2.1272056	0.0000239
1.0	2.6408591	2.6408286	0.0000305
1.2	3.1799415	3.1799026	0.0000389
1.4	3.7324000	3.7323505	0.0000495
1.6	4.2834838	4.2834208	0.0000630
1.8	4.8151763	4.8150964	0.0000799
2.0	5.3054720	5.3053707	0.0001013

Other multistep methods can be derived using integration of interpolating polynomials over intervals of the form $[t_j, t_{i+1}]$ for $j \leq i-1$, where some of the data points are omitted. Milne's method is an explicit technique that results when a Newton Backward-Difference interpolating polynomial is integrated over $[t_{i-3}, t_{i+1}]$.

Milne's Method

$$w_{i+1} = w_{i-3} + \frac{4h}{3}[2f(t_i, w_i) - f(t_{i-1}, w_{i-1}) + 2f(t_{i-2}, w_{i-2})],$$

where $i = 3, 4, \ldots, N-1$, with local error $\frac{14}{45}h^5 y^{(5)}(\mu_i)$ for some μ_i in (t_{i-3}, t_{i+1}).

This method is used as a predictor for an implicit method called Simpson's method. Its name comes from the fact that it can be derived using Simpson's rule for approximating integrals.

Simpson's Method

$$w_{i+1} = w_{i-1} + \frac{h}{3}[f(t_{i+1}, w_{i+1}) + 4f(t_i, w_i) + f(t_{i-1}, w_{i-1})],$$

where $i = 1, 2, \ldots, N-1$, with local error $-\frac{1}{90}h^5 y^{(5)}(\mu_i)$ for some μ_i in (t_{i-1}, t_{i+1}).

Although the local error involved with a predictor-corrector method of the Milne-Simpson type is generally smaller than that of the Adams-Bashforth-Moulton method, the technique has limited use because of round-off error problems, which do not occur with the Adams procedure.

EXERCISE SET 5.4

1. Use all the Adams-Bashforth methods to approximate the solutions to the following initial-value problems. In each case use exact starting values and compare the results to the actual values.
   a. $y' = te^{3t} - 2y$, for $0 \leq t \leq 1$, with $y(0) = 0$ and $h = 0.2$; actual solution $y(t) = \frac{1}{5}te^{3t} - \frac{1}{25}e^{3t} + \frac{1}{25}e^{-2t}$.
   b. $y' = 1 + (t-y)^2$, for $2 \leq t \leq 3$, with $y(2) = 1$ and $h = 0.2$; actual solution $y(t) = t + 1/(1-t)$.
   c. $y' = 1 + \frac{y}{t}$, for $1 \leq t \leq 2$, with $y(1) = 2$ and $h = 0.2$; actual solution $y(t) = t\ln t + 2t$.
   d. $y' = \cos 2t + \sin 3t$, for $0 \leq t \leq 1$ with $y(0) = 1$ and $h = 0.2$; actual solution $y(t) = \frac{1}{2}\sin 2t - \frac{1}{3}\cos 3t + \frac{4}{3}$.

2. Use all the Adams-Moulton methods to approximate the solutions to the Exercises 1(a), 1(c), and 1(d). In each case use exact starting values and explicitly solve for w_{i+1}. Compare the results to the actual values.

3. Use each of the Adams-Bashforth methods to approximate the solutions to the following initial-value problems. In each case use starting values obtained from the Runge-Kutta method of order 4. Compare the results to the actual values.
   a. $y' = \frac{y}{t} - \left(\frac{y}{t}\right)^2$, for $1 \le t \le 2$, with $y(1) = 1$ and $h = 0.1$; actual solution $y(t) = t/(1 + \ln t)$.
   b. $y' = 1 + \frac{y}{t} + \left(\frac{y}{t}\right)^2$, for $1 \le t \le 3$, with $y(1) = 0$ and $h = 0.2$; actual solution $y(t) = t\tan(\ln t)$.
   c. $y' = -(y + 1)(y + 3)$, for $0 \le t \le 2$, with $y(0) = -2$ and $h = 0.1$; actual solution $y(t) = -3 + 2/(1 + e^{-2t})$.
   d. $y' = -5y + 5t^2 + 2t$, for $0 \le t \le 1$, with $y(0) = 1/3$ and $h = 0.1$; actual solution $y(t) = t^2 + \frac{1}{3}e^{-5t}$.

4. Use the predictor-corrector method based on the Adams-Bashforth Four-Step method and the Adams-Moulton Three-Step method to approximate the solutions to the initial-value problems in Exercise 1.

5. Use the predictor-corrector method based on the Adams-Bashforth Four-Step method and the Adams-Moulton Three-Step method to approximate the solutions to the initial-value problem in Exercise 3.

6. The initial-value problem

$$y' = e^y, \quad \text{for } 0 \le t \le 0.20, \quad \text{with } y(0) = 1$$

has solution

$$y(t) = 1 - \ln(1 - et).$$

Applying the Adams-Moulton Three-Step method to this problem is equivalent to finding the fixed point w_{i+1} of

$$g(w) = w_i + \frac{h}{24}[9e^w + 19e^{w_i} - 5e^{w_{i-1}} + e^{w_{i-2}}].$$

   a. With $h = 0.01$, obtain w_{i+1} by functional iteration for $i = 2, \ldots, 19$ using exact starting values w_0, w_1, and w_2. At each step use w_i to initially approximate w_{i+1}.
   b. Will Newton's method speed the convergence over functional iteration?

7. Use the Milne-Simpson Predictor-Corrector method to approximate the solutions to the initial-value problems in Exercise 3.

8. Use the Milne-Simpson Predictor-Corrector method to approximate the solution to

$$y' = -5y, \quad \text{for } 0 \le t \le 2, \quad \text{with } y(0) = e,$$

with $h = 0.1$. Repeat the procedure with $h = 0.05$. Are the answers consistent with the local error?

5.5 Extrapolation Methods

Extrapolation was used in Section 4.5 for the approximation of definite integrals, where we found that by correctly averaging relatively inaccurate trapezoidal approximations we could produce new approximations that are exceedingly accurate. In this section we will apply extrapolation to increase the accuracy of approximations to the solution of initial-value problems. As we have previously seen, the original approximations must have an error expansion of a specific form for the procedure to be successful.

To apply extrapolation to solve initial-value problems, we use a technique based on the Midpoint method:

$$w_{i+1} = w_{i-1} + 2hf(t_i, w_i), \quad \text{for } i \geq 1. \tag{5.1}$$

This technique requires two starting values, since both w_0 and w_1 are needed before the first midpoint approximation, w_2, can be determined. As usual, we use the initial condition for $w_0 = y(a) = \alpha$. To determine the second starting value, w_1, we apply Euler's method. Subsequent approximations are obtained from Eq. (5.1). After a series of approximations of this type are generated ending at a value t, an endpoint correction is performed that involves the final two midpoint approximations. This produces an approximation $w(t, h)$ to $y(t)$ that has the form

$$y(t) = w(t, h) + \sum_{k=1}^{\infty} \delta_k h^{2k}, \tag{5.2}$$

where the δ_k are constants related to the derivatives of the solution $y(t)$. The important point is that the δ_k do not depend on the step size h.

To illustrate the extrapolation technique for solving

$$y'(t) = f(t, y), \quad \text{for } a \leq t \leq b, \quad \text{with } y(a) = \alpha,$$

let us assume that we have a fixed step size h and that we wish to approximate $y(a + h)$.

As the first step we let $h_0 = h/2$ and use Euler's method with $w_0 = \alpha$ to approximate $y(a + h_0) = y(a + h/2)$ as

$$w_1 = w_0 + h_0 f(a, w_0).$$

We then apply the Midpoint method with $t_{i-1} = a$ and $t_i = a + h_0 = a + h/2$ to produce a first approximation to $y(a + h) = y(a + 2h_0)$,

$$w_2 = w_0 + 2h_0 f(a + h_0, w_1).$$

The endpoint correction is applied to obtain the final approximation to $y(a + h)$ for the step size h_0. This results in an $O(h_0^2)$ approximation to $y(t_1)$ given by

$$y_{1,1} = \frac{1}{2}[w_2 + w_1 + h_0 f(a + 2h_0, w_2)].$$

We save the approximation $y_{1,1}$ and discard the intermediate results, w_1 and w_2.

To obtain the next approximation, $y_{2,1}$, to $y(t_1)$, we let $h_1 = h/4$ and use Euler's method with $w_0 = \alpha$ to obtain an approximation w_1 to $y(a + h_1) = y(a + h/4)$; that is,

$$w_1 = w_0 + h_1 f(a, w_0).$$

Next we produce an approximation w_2 to $y(a + 2h_1) = y(a + h/2)$ and w_3 to $y(a + 3h_1) = y(a + 3h/4)$ given by

$$w_2 = w_0 + 2h_1 f(a + h_1, w_1) \quad \text{and} \quad w_3 = w_1 + 2h_1 f(a + 2h_1, w_2).$$

Then we produce the approximation w_4 to $y(a + 4h_1) = y(t_1)$ given by

$$w_4 = w_2 + 2h_1 f(a + 3h_1, w_3).$$

The endpoint correction is now applied to w_3 and w_4 to produce the improved $O(h_1^2)$ approximation to $y(t_1)$,

$$y_{2,1} = \frac{1}{2}[w_4 + w_3 + h_1 f(a + 4h_1, w_4)].$$

Because of the form of the error given in Eq. (5.2), the two approximations to $y(a + h)$ have the property that

$$y(a + h) = y_{1,1} + \delta_1 \left(\frac{h}{2}\right)^2 + \delta_2 \left(\frac{h}{2}\right)^4 + \cdots = y_{1,1} + \delta_1 \frac{h^2}{4} + \delta_2 \frac{h^4}{16} + \cdots$$

and

$$y(a + h) = y_{2,1} + \delta_1 \left(\frac{h}{4}\right)^2 + \delta_2 \left(\frac{h}{4}\right)^4 + \cdots = y_{2,1} + \delta_1 \frac{h^2}{16} + \delta_2 \frac{h^4}{256} + \cdots.$$

We can eliminate the $O(h^2)$ portion of this truncation error by averaging these two formulas appropriately. Specifically, if we subtract the first from 4 times the second and divide the result by 3, we have

$$y(a + h) = y_{2,1} + \frac{1}{3}(y_{2,1} - y_{1,1}) - \delta_2 \frac{h^4}{64} + \cdots.$$

So the approximation

$$y_{2,2} = y_{2,1} + \frac{1}{3}(y_{2,1} - y_{1,1})$$

has error of order $O(h^4)$.

Continuing in this manner, we next let $h_2 = h/6$ and apply Euler's method once, followed by the Midpoint method five times. Then we use the endpoint correction to determine the h^2 approximation, $y_{3,1}$, to $y(a + h)$. This approximation can be averaged with $y_{2,1}$ to produce a second $O(h^4)$ approximation that we denote $y_{3,2}$. Then $y_{3,2}$ and $y_{2,2}$ are averaged to eliminate the $O(h^4)$ error terms and produce an approximation with error of order $O(h^6)$. Higher-order formulas are generated by continuing the process.

The only significant difference between the extrapolation performed here and that used for Romberg integration in Section 4.4 results from the way the subdivisions are chosen. In Romberg integration there is a convenient formula for representing the Composite Trapezoidal rule approximations that uses consecutive divisions of the step size by the integers 1, 2, 4, 8, 16, 32, 64,.... This procedure permits the averaging process to proceed in an easily followed manner. We do not have a means for easily producing refined approximations for initial-value problems, so the divisions for the extrapolation technique are chosen to minimize the number of required function evaluations. The averaging procedure arising from this choice of subdivision is not as elementary, but, other than that, the process is the same as that used for Romberg integration.

The progam EXTRAP54 uses the extrapolation technique with the sequence of integers $q_0 = 2, q_1 = 4, q_2 = 6, q_3 = 8, q_4 = 12, q_5 = 16, q_6 = 24, q_7 = 32$. A basic step size, h, is selected, and the method progresses by using $h_j = h/q_j$, for each $j = 0, \dots, 7$, to approximate $y(t + h)$. The error is controlled by requiring that the approximations $y_{1,1}, y_{2,2}, \dots$ be computed until $|y_{i,i} - y_{i-1,i-1}|$ is less than a given tolerance. If the tolerance is not achieved by $i = 8$, then h is reduced, and the process is reapplied. Minimum and maximum values of h, $hmin$, and $hmax$, respectively, are specified in the program to ensure control over the method.

If $y_{i,i}$ is found to be acceptable, then w_1 is set to $y_{i,i}$, and computations begin again to determine w_2, which will approximate $y(t_2) = y(a + 2h)$. The process is repeated until the approximation w_N to $y(b)$ is determined.

EXAMPLE 1 Consider the initial-value problem

$$y' = y - t^2 + 1, \quad \text{for } 0 \le t \le 2, \quad \text{with } y(0) = 0.5,$$

which has solution $y(t) = (t + 1)^2 - 0.5e^t$. The program EXTRAP54 applied to this problem with $h = 0.25$, $TOL = 10^{-10}$, $hmax = 0.25$, and $hmin = 0.01$ gives the values in Table 5.11. ◆ ◆ ◆

Table 5.11

$y_{1,1} = 0.9187011719$				
$y_{2,1} = 0.9200379848$	$y_{2,2} = 0.9204835892$			
$y_{3,1} = 0.9202873689$	$y_{3,2} = 0.9204868761$	$y_{3,3} = 0.9204872870$		
$y_{4,1} = 0.9203747896$	$y_{4,2} = 0.9204871876$	$y_{4,3} = 0.9204872914$	$y_{4,4} = 0.9204872917$	
$y_{5,1} = 0.9204372763$	$y_{5,2} = 0.9204872656$	$y_{5,3} = 0.9204872916$	$y_{5,4} = 0.9204872917$	$y_{5,5} = 0.9204872917$

We will also compute the first three entries in Table 5.11 using Maple. Define $f(t, y)$ with the command

```
>f:=(t,y)->y-t^2+1;
```

The variables a, b, α, and h are defined by

```
>a:=0; b:=2; alpha:=0.5; h:=0.25;
```

and $t_0 = t0$ and $w_0 = w0$ are initialized by

```
>t0:=1; w0:=alpha;
```

We use Euler's method with $h_0 = h0 = h/2 = 0.125$ to obtain $w_1 = w1$ as 0.6875 with

```
>h0:=h/2;

>w1:=w0+h0*f(t0,w0);
```

We then use the Midpoint method to obtain $w_2 = w2$ as 0.91796875 with

```
>t:=t0+h0;

>w2:=w0+2*h0*f(t,w1);

>t:=t0+2*h0;
```

The endpoint correction gives $y_{1,1} = y11 = 0.9187011719$ with

```
>y11:=(w2+w1+h0*f(t,w2))/2;
```

We then proceed to the next iteration, defining

```
>h1:=h/4;
```

and using Euler's method

```
>w1:=w0+h1*f(t0,w0);

>t:=t0+h1;
```

to give $w_1 = 0.59375$. We then use the Midpoint method three times to obtain

```
>w2:=w0+2*h1*f(t,w1); t:=t0+2*h1;

>w3:=w1+2*h1*f(t,w2); t:=t0+3*h1;

>w4:=w2+2*h1*f(t,w3); t:=t0+4*h1;
```

which produces $w_2 = 0.6987304688$, $w_3 = 0.8041381836$, and $w_4 = 0.9198532106$.

We have $w2 \approx y(0.125)$, $w3 \approx y(0.1875)$, and $w4 \approx y(0.25)$, and we use the endpoint correction to obtain $y_{2,1} = y21 = 0.920037985$ from

```
>y21:=(w4+w3+h1*f(t,w4))/2;
```

Extrapolation gives the entry $y_{2,2} = y22 = 0.9204835891$ with the command

```
>y22:=y21+(h/4)^2*(y21-y11)/((h/2)^2-(h/4)^2);
```

The computations stopped with $w_1 = y_{5,5}$ because $|y_{5,5} - y_{4,4}| \leq 10^{-10}$ and $y_{5,5}$ is accepted as the approximation to $y(t_1) = y(0.25)$. The complete set of approximations accurate to the places listed is given in Table 5.12.

Table 5.12

t_i	$y_i = y(t_i)$	w_i	h_i	k
0.25	0.9204872917	0.9204872917	0.25	5
0.50	1.4256393646	1.4256393646	0.25	5
0.75	2.0039999917	2.0039999917	0.25	5
1.00	2.6408590858	2.6408590858	0.25	5
1.25	3.3173285213	3.3173285212	0.25	4
1.50	4.0091554648	4.0091554648	0.25	3
1.75	4.6851986620	4.6851986619	0.25	3
2.00	5.3054719505	5.3054719505	0.25	3

EXERCISE SET 5.5

1. The initial-value problem

$$y' = \sqrt{2 - y^2}e^t, \quad \text{for } 0 \leq t \leq 0.8, \quad \text{with } y(0) = 0$$

has actual solution $y(t) = \sqrt{2}\sin(e^t - 1)$. Use extrapolation with $h = 0.1$ to find an approximation for $y(0.1)$ to within a tolerance of $TOL = 10^{-5}$. Compare the approximation to the actual value.

2. The initial-value problem

$$y' = -y + 1 - \frac{y}{t}, \quad \text{for } 1 \leq t \leq 2, \quad \text{with } y(1) = 1$$

has actual solution $y(t) = 1 + (e^{1-t} - 1)/t$. Use extrapolation with $h = 0.2$ to find an approximation for $y(1.2)$ to within a tolerance of $TOL = 0.00005$. Compare the approximation to the actual value.

3. Use the extrapolation program EXTRAP54 with $TOL = 10^{-4}$ to approximate the solutions to the following initial-value problems:

a. $y' = \left(\frac{y}{t}\right)^2 + \left(\frac{y}{t}\right)$, for $1 \leq t \leq 1.2$, with $y(1) = 1$, $hmax = 0.05$, and $hmin = 0.02$.

b. $y' = \sin t + e^{-t}$, for $0 \leq t \leq 1$, with $y(0) = 0$, $hmax = 0.25$, and $hmin = 0.02$.

c. $y' = (y^2 + y)t^{-1}$, for $1 \leq t \leq 3$, with $y(1) = -2$, $hmax = 0.5$, and $hmin = 0.02$.

d. $y' = -ty + 4ty^{-1}$, for $0 \leq t \leq 1$, with $y(0) = 1$, $hmax = 0.25$, and $hmin = 0.02$.

4. Use the extrapolation program EXTRAP54 with tolerance $TOL = 10^{-6}$, $hmax = 0.5$, and $hmin = 0.05$ to approximate the solutions to the following initial-value problems. Compare the results to the actual values.

a. $y' = \frac{y}{t} - \frac{y^2}{t^2}$, for $1 \le t \le 4$, with $y(1) = 1$; actual solution $y(t) = t/(1 + \ln t)$.

b. $y' = 1 + \frac{y}{t} + \left(\frac{y}{t}\right)^2$, for $1 \le t \le 3$, with $y(1) = 0$; actual solution $y(t) = t\tan(\ln t)$.

c. $y' = -(y + 1)(y + 3)$, for $0 \le t \le 3$, with $y(0) = -2$; actual solution $y(t) = -3 + 2(1 + e^{-2t})^{-1}$.

d. $y' = (t + 2t^3)y^3 - ty$, for $0 \le t \le 2$, with $y(0) = \frac{1}{3}$; actual solution $y(t) = (3 + 2t^2 + 6e^{t^2})^{-1/2}$.

5. Let $P(t)$ be the number of individuals in a population at time t, measured in years. If the average birth rate b is constant and the average death rate d is proportional to the size of the population (due to overcrowding), then the growth rate of the population is given by the *logistic equation*

$$\frac{dP(t)}{dt} = bP(t) - k[P(t)]^2$$

where $d = kP(t)$. Suppose $P(0) = 50{,}976$, $b = 2.9 \times 10^{-2}$, and $k = 1.4 \times 10^{-7}$. Find the population after 5 years.

5.6 Adaptive Techniques

The appropriate use of varying step size was seen in Section 4.6 to produce integral approximating methods that are efficient in the amount of computation required. This might not be sufficient to favor these methods due to the increased complication of applying them, but they have another important feature. The step-size selection procedure produces an estimate of the local error that does not require the approximation of the higher derivatives of the function. These methods are called *adaptive* because they adapt the number and position of the nodes used in the approximation to keep the local error within a specified bound.

There is a close connection between the problem of approximating the value of a definite integral and that of approximating the solution to an initial-value problem. It is not surprising, then, that there are adaptive methods for approximating the solutions to initial-value problems, and that these methods are not only efficient but incorporate the control of error.

An ideal difference-equation method,

$$w_{i+1} = w_i + h_i\phi(t_i, w_i, h_i), \quad i = 0, 1, \ldots, N - 1,$$

for approximating the solution, $y(t)$, to the initial-value problem

$$y' = f(t, y), \quad \text{for } a \le t \le b, \quad \text{with } y(a) = \alpha$$

would have the property that given a tolerance $\epsilon > 0$, the minimal number of mesh points would be used to ensure that the global error, $|y(t_i) - w_i|$, would not exceed ϵ for any $i = 0, 1, \ldots, N$. Having a minimal number of mesh points and also controlling the global error of a difference method is, not surprisingly, inconsistent with the points being equally spaced in the interval. In this section we examine techniques used to control the error of a difference-equation method in an efficient manner by the appropriate choice of mesh points.

Although we cannot generally determine the global error of a method, we saw in Section 5.2 that there is often a close connection between the local error and the global error. If a method has local error of order $n + 1$, then the global error of the method is of order n. By using methods of differing order we can predict the local error and, using this prediction, choose a step size that will keep the global error in check.

To illustrate the technique, suppose that we have two approximation techniques. The first is an nth-order method obtained from an nth-order Taylor method of the form

$$y(t_{i+1}) = y(t_i) + h\phi(t_i, y(t_i), h) + O(h^{n+1}).$$

This method produces approximations

$$\begin{aligned} w_0 &= \alpha \\ w_{i+1} &= w_i + h\phi(t_i, w_i, h) \quad \text{for} \quad i > 0, \end{aligned}$$

which satisfy, for some K and all relevant h and i,

$$|y(t_i) - w_i| < Kh^n.$$

In general, the method is generated by applying a Runge-Kutta modification to the Taylor method, but the specific derivation is unimportant.

The second method is similar but of higher order. For example, let us suppose it comes from an $(n + 1)$st-order Taylor method of the form

$$y(t_{i+1}) = y(t_i) + h\tilde{\phi}(t_i, y(t_i), h) + O(h^{n+2}),$$

producing approximations

$$\begin{aligned} \tilde{w}_0 &= \alpha \\ \tilde{w}_{i+1} &= \tilde{w}_i + h\tilde{\phi}(t_i, \tilde{w}_i, h) \quad \text{for} \quad i > 0, \end{aligned}$$

which satisfy, for some $\tilde{K}$ and all relevant h and i,

$$|y(t_i) - \tilde{w}_i| < \tilde{K}h^{n+1}.$$

We assume now that at the point t_i we have

$$w_i = \tilde{w}_i = z(t_i),$$

where $z(t)$ is the solution to the differential equation that does not satisfy the original initial condition but instead satisfies the condition $z(t_i) = w_i$. The typical difference between $y(t)$ and $z(t)$ is shown in Figure 5.3.

Figure 5.3

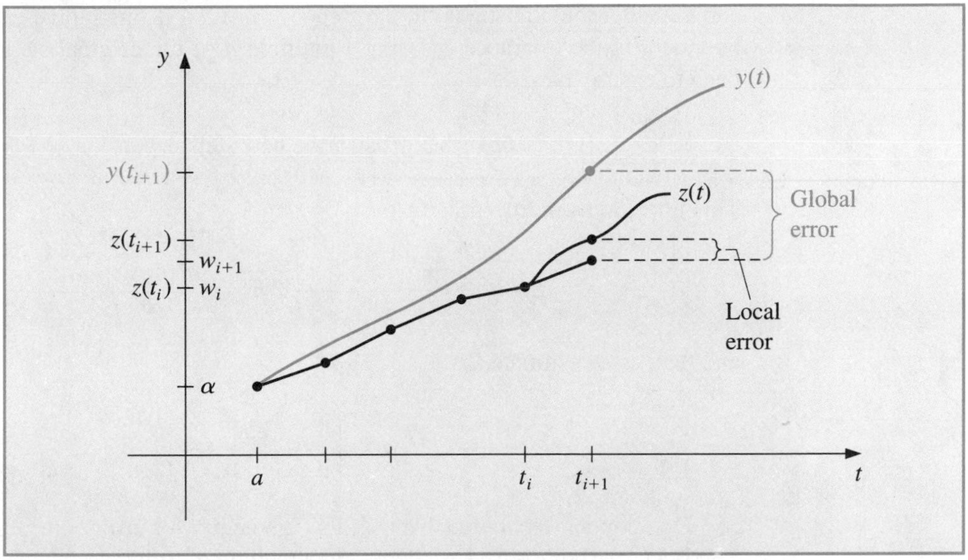

Applying the two methods to the differential equation with the fixed step size h produces two approximations, w_{i+1} and $\tilde{w}_{i+1}$, whose differences from $y(t_i + h)$ represent global errors but whose differences from $z(t_i + h)$ represent local errors.

Now consider

$$z(t_i + h) - w_{i+1} = (\tilde{w}_{i+1} - w_{i+1}) + (z(t_i + h) - \tilde{w}_{i+1}).$$

The term on the left side of this equation is $O(h^{n+1})$, the local error of the method, but the second term on the right side is $O(h^{n+2})$. This implies that the dominant portion on the right side comes from the first term; that is,

$$z(t_i + h) - w_{i+1} \approx \tilde{w}_{i+1} - w_{i+1}.$$

Since $z(t_{i+1}) - w_{i+1}$ is $O(h^{n+1})$, we assume that a constant K exists with

$$Kh^{n+1} = |z(t_i + h) - w_{i+1}| \approx |\tilde{w}_{i+1} - w_{i+1}|.$$

Then K can be approximated by

$$K \approx \frac{|\tilde{w}_{i+1} - w_{i+1}|}{h^{n+1}}. \tag{5.3}$$

Let us now return to the global error associated with the problem we really want to solve,

$$y' = f(t, y), \quad \text{for } a \le t \le b, \quad \text{with } y(a) = \alpha,$$

and consider the adjustment to the step size needed if this global error is expected to be bounded by the tolerance ϵ. Using a multiple q of the original step size implies that we need to ensure that

$$|y(t_i + qh) - w_{i+1}(\text{using the new step size } qh)| < K(qh)^n < \epsilon.$$

This implies, from Eq. (5.3), that

$$Kq^n h^n \approx \frac{|\tilde{w}_{i+1} - w_{i+1}|}{h^{n+1}} q^n h^n = \frac{q^n |\tilde{w}_{i+1} - w_{i+1}|}{h} < \epsilon$$

and that, as an estimate,

$$q < \left[\frac{\epsilon h}{|\tilde{w}_{i+1} - w_{i+1}|} \right]^{1/n}.$$

One popular technique that uses this inequality for error control is the **Runge-Kutta-Fehlberg method**. It uses a Runge-Kutta method of order 5,

$$\tilde{w}_{i+1} = w_i + \frac{16}{135}k_1 + \frac{6656}{12825}k_3 + \frac{28561}{56430}k_4 - \frac{9}{50}k_5 + \frac{2}{55}k_6,$$

to estimate the local error in a Runge-Kutta method of order 4,

$$w_{i+1} = w_i + \frac{25}{216}k_1 + \frac{1408}{2565}k_3 + \frac{2197}{4104}k_4 - \frac{1}{5}k_5,$$

where

$$\begin{aligned}
k_1 &= hf(t_i, w_i),\\
k_2 &= hf\left(t_i + \frac{h}{4}, w_i + \frac{1}{4}k_1\right),\\
k_3 &= hf\left(t_i + \frac{3h}{8}, w_i + \frac{3}{32}k_1 + \frac{9}{32}k_2\right),\\
k_4 &= hf\left(t_i + \frac{12h}{13}, w_i + \frac{1932}{2197}k_1 - \frac{7200}{2197}k_2 + \frac{7296}{2197}k_3\right),\\
k_5 &= hf\left(t_i + h, w_i + \frac{439}{216}k_1 - 8k_2 + \frac{3680}{513}k_3 - \frac{845}{4104}k_4\right),\\
k_6 &= hf\left(t_i + \frac{h}{2}, w_i - \frac{8}{27}k_1 + 2k_2 - \frac{3544}{2565}k_3 + \frac{1859}{4104}k_4 - \frac{11}{40}k_5\right).
\end{aligned}$$

An advantage of this method is that only six evaluations of f are required per step, whereas arbitrary Runge-Kutta methods of order 4 and 5 used together would require (see Table 5.7 on page 195) at least four evaluations of f for the fourth-order method and an additional six for the fifth-order method.

In the theory of error control, an initial value of h at the ith step was used to find the first values of w_{i+1} and $\tilde{w}_{i+1}$, which led to the determination of q for that step. Then the calculations were repeated with the step size h replaced by qh. This procedure requires twice the number of functional evaluations per step as without error control. In practice, the value of q to be used is chosen somewhat differently in order to make the increased functional-evaluation cost worthwhile. The value of q determined at the ith step is used for two purposes:

1. To reject the initial choice of h at the ith step and repeat the calculations using qh if the error is not within the required bound, and
2. To predict an appropriate initial choice of h for the $(i + 1)$st step.

Because of the penalty in terms of functional evaluations that must be paid if many steps are repeated, q tends to be chosen conservatively. In fact, for the Runge-Kutta-Fehlberg method with $n = 4$, the usual choice is

$$q = \left(\frac{\epsilon h}{2|\tilde{w}_{i+1} - w_{i+1}|}\right)^{1/4} \approx 0.84\left(\frac{\epsilon h}{|\tilde{w}_{i+1} - w_{i+1}|}\right)^{1/4}.$$

The program RKFVSM55, which implements the Runge-Kutta-Fehlberg method, incorporates a technique to eliminate large modifications in step size. This is done to avoid spending too much time with very small step sizes in regions with irregularities in the derivatives of y and to avoid large step sizes, which may result in skipping sensitive regions nearby. In some instances the step-size-increase procedure is omitted completely and the step-size-decrease procedure is modified to be incorporated only when needed to bring the error under control.

EXAMPLE 1 The Runge-Kutta-Fehlberg method will be used to approximate the solution to the initial-value problem

$$y' = y - t^2 + 1, \quad \text{for } 0 \leq t \leq 2, \quad \text{with } y(0) = 0.5,$$

which has solution $y(t) = (t + 1)^2 - 0.5e^t$. The input consists of tolerance, $TOL = 10^{-5}$, a maximum step size, $hmax = 0.25$, and a minimum step size, $hmin = 0.01$. The results from the program RKFVSM55 are shown in Table 5.13 on the following page.

◆ ◆ ◆

An implementation of the Runge-Kutta-Fehlberg method is available in Maple using the `dsolve` command with the numeric option. Consider the initial value problem of Example 1. The command

```
>g:=dsolve({D(y)(t)=y(t)-t*t+1,y(0)=0.5},y(t),numeric);
```

Table 5.13

t_i	$y_i = y(t_i)$	w_i	h_i	R_i	$\lvert y_i - w_i\rvert$
0.0000000	0.5000000	0.5000000	0	0	0.0000000
0.2500000	0.9204873	0.9204886	0.2500000	0.0000062	0.0000013
0.4865522	1.3964884	1.3964910	0.2365522	0.0000045	0.0000026
0.7293332	1.9537446	1.9537488	0.2427810	0.0000043	0.0000042
0.9793332	2.5864198	2.5864260	0.2500000	0.0000038	0.0000062
1.2293332	3.2604520	3.2604605	0.2500000	0.0000024	0.0000085
1.4793332	3.9520844	3.9520955	0.2500000	0.0000007	0.0000111
1.7293332	4.6308127	4.6308268	0.2500000	0.0000015	0.0000141
1.9793332	5.2574687	5.2574861	0.2500000	0.0000043	0.0000173
2.0000000	5.3054720	5.3054896	0.0206668	0.0000000	0.0000177

uses the Runge-Kutta-Fehlberg method. We can approximate $y(2)$ in our example by using

```
>g(2.0);
```

which gives

$$[t = 2.0, y(t) = 5.305471958400194]$$

The Runge-Kutta-Fehlberg method is popular for error control because at each step it provides, at little additional cost, *two* approximations that can be compared and related to the local error. Predictor-corrector techniques always generate two approximations at each step, so they are natural candidates for error-control adaptation.

To demonstrate the procedure, we construct a variable-step-size predictor-corrector method using the explicit Adams-Bashforth Four-Step method as predictor and the implicit Adams-Moulton Three-Step method as corrector.

The Adams-Bashforth Four-Step method comes from the equation

$$\begin{aligned} y(t_{i+1}) = y(t_i) + \frac{h}{24}[55f(t_i, y(t_i)) - 59f(t_{i-1}y(t_{i-1})) \\ + 37f(t_{i-2}, y(t_{i-2})) - 9f(t_{i-3}, y(t_{i-3}))] + \frac{251}{720}y^{(5)}(\hat{\mu}_i)h^5 \end{aligned}$$

for some μ_i in (t_{i-3}, t_{i+1}). Suppose we assume that the approximations $w_0, w_1, \ldots, w_i$ are all exact and, as in the case of the one-step methods, we let z represent the solution to the differential equation satisfying the initial condition $z(t_i) = w_i$. Then

$$z(t_{i+1}) - w_{i+1}^{(0)} = \frac{251}{720}z^{(5)}(\hat{\mu}_i)h^5 \quad \text{for some } \hat{\mu}_i \text{ in } (t_{i-3}, t_i). \tag{5.4}$$

A similar analysis of the Adams-Moulton Three-Step method leads to the local error

$$z(t_{i+1}) - w_{i+1} = -\frac{19}{720} z^{(5)}(\tilde{\mu}_i) h^5 \quad \text{for some } \tilde{\mu}_i \text{ in } (t_{i-2}, t_{i+1}). \tag{5.5}$$

To proceed further, we must make the assumption that for small values of h,

$$z^{(5)}(\hat{\mu}_i) \approx z^{(5)}(\tilde{\mu}_i),$$

and the effectiveness of the error-control technique depends directly on this assumption.

If we subtract Eq. (5.5) from Eq. (5.4), we have

$$w_{i+1} - w_{i+1}^{(0)} = \frac{h^5}{720}\left[251 z^{(5)}(\hat{\mu}_i) + 19 z^{(5)}(\tilde{\mu}_i)\right] \approx \frac{3}{8} h^5 z^{(5)}(\tilde{\mu}_i),$$

so

$$z^{(5)}(\tilde{\mu}_i) \approx \frac{8}{3h^5}\left(w_{i+1} - w_{i+1}^{(0)}\right).$$

Using this result to eliminate the term involving $h^5 z^{(5)}(\tilde{\mu}_i)$ from Eq. (5.5) gives the approximation to the error

$$|z(t_{i+1}) - w_{i+1}| \approx \frac{19h^5}{720} \cdot \frac{8}{3h^5}\left|w_{i+1} - w_{i+1}^{(0)}\right| = \frac{19\left|w_{i+1} - w_{i+1}^{(0)}\right|}{270}.$$

This expression was derived under the assumption that $w_0, w_1, \ldots, w_i$ are all exact, which means that this is an approximation to the local error. As in the case of the one-step methods, the global error is of order one degree less, so for the function y that is the solution to the original initial-value problem,

$$y' = f(t, y), \quad \text{for } a \le t \le b, \quad \text{with } y(a) = \alpha,$$

the global error can be estimated by

$$|y(t_{i+1}) - w_{i+1}| \approx \frac{|z(t_{i+1}) - w_{i+1}|}{h} \approx \frac{19\left|w_{i+1} - w_{i+1}^{(0)}\right|}{270h}.$$

Suppose we now reconsider the situation with a new step size qh generating new approximations $\hat{w}_{i+1}^{(0)}$ and $\hat{w}_{i+1}$. To control the global error to within ϵ, we want to choose q so that

$$\frac{|z(t_i + qh) - \hat{w}_{i+1} \text{ (using the step size } qh)|}{qh} < \epsilon.$$

But from Eq. (5.5),

$$\frac{|z(t_i + qh) - \hat{w}_{i+1}(\text{using } qh)|}{qh} = \frac{19}{720}\left|z^{(5)}(\tilde{\mu}_i)\right| q^4 h^4 \approx \frac{19}{720}\left[\frac{8}{3h^5}\left|w_{i+1} - w_{i+1}^{(0)}\right|\right] q^4 h^4,$$

so we need to choose q with

$$\frac{19}{720}\left[\frac{8}{3h^5}\left|w_{i+1}-w_{i+1}^{(0)}\right|\right]q^4h^4 = \frac{19}{270}\frac{\left|w_{i+1}-w_{i+1}^{(0)}\right|}{h}q^4 < \epsilon.$$

Consequently,

$$q < \left(\frac{270}{19}\frac{h\epsilon}{\left|w_{i+1}-w_{i+1}^{(0)}\right|}\right)^{1/4} \approx 2\left(\frac{h\epsilon}{\left|w_{i+1}-w_{i+1}^{(0)}\right|}\right)^{1/4}.$$

A number of approximation assumptions have been made in this development, so in actual practice q is chosen conservatively, usually as

$$q = 1.5\left(\frac{h\epsilon}{\left|w_{i+1}-w_{i+1}^{(0)}\right|}\right)^{1/4}.$$

A change in step size for a multistep method is more costly in terms of functional evaluations than for a one-step method, since new equally spaced starting values must be computed. As a consequence, it is also common practice to ignore the step-size change whenever the global error is between $\epsilon/10$ and ϵ; that is, when

$$\frac{\epsilon}{10} < |y(t_{i+1}) - w_{i+1}| \approx \frac{19\left|w_{i+1}-w_{i+1}^{(0)}\right|}{270h} < \epsilon.$$

Table 5.14

t_i	$y_i = y(t_i)$	w_i	h_i	σ_i	$\lvert y_i - w_i\rvert$
0.1257017	0.7002323	0.7002318	0.1257017	4.051×10^{-6}	0.0000005
0.2514033	0.9230960	0.9230949	0.1257017	4.051×10^{-6}	0.0000011
0.3771050	1.1673894	1.1673877	0.1257017	4.051×10^{-6}	0.0000017
0.5028066	1.4317502	1.4317480	0.1257017	4.051×10^{-6}	0.0000022
0.6285083	1.7146334	1.7146306	0.1257017	4.610×10^{-6}	0.0000028
0.7542100	2.0142869	2.0142834	0.1257017	5.210×10^{-6}	0.0000035
0.8799116	2.3287244	2.3287200	0.1257017	5.913×10^{-6}	0.0000043
1.0056133	2.6556930	2.6556877	0.1257017	6.706×10^{-6}	0.0000054
1.1313149	2.9926385	2.9926319	0.1257017	7.604×10^{-6}	0.0000066
1.2570166	3.3366642	3.3366562	0.1257017	8.622×10^{-6}	0.0000080
1.3827183	3.6844857	3.6844761	0.1257017	9.777×10^{-6}	0.0000097
1.4857283	3.9697541	3.9697433	0.1030100	7.029×10^{-6}	0.0000108
1.5887383	4.2527830	4.2527711	0.1030100	7.029×10^{-6}	0.0000120
1.6917483	4.5310269	4.5310137	0.1030100	7.029×10^{-6}	0.0000133
1.7947583	4.8016639	4.8016488	0.1030100	7.029×10^{-6}	0.0000151
1.8977683	5.0615660	5.0615488	0.1030100	7.760×10^{-6}	0.0000172
1.9233262	5.1239941	5.1239764	0.0255579	3.918×10^{-8}	0.0000177
1.9488841	5.1854932	5.1854751	0.0255579	3.918×10^{-8}	0.0000181
1.9744421	5.2460056	5.2459870	0.0255579	3.918×10^{-8}	0.0000186
2.0000000	5.3054720	5.3054529	0.0255579	3.918×10^{-8}	0.0000191

In addition, q is generally given an upper bound to ensure that a single unusually accurate approximation does not result in too large a step size. The program VPRCOR56 incorporates this safeguard with an upper bound of four times the previous step size.

It should be emphasized that since the multistep methods require equal step sizes for the starting values, any change in step size necessitates recalculating new starting values at that point. In the program VPRCOR56, this is done by incorporating RKO4M53, the Runge-Kutta method of order 4, as a subroutine.

EXAMPLE 2 Table 5.14 lists the results obtained using the program VPRCOR56 to find approximations to the solution of the initial-value problem

$$y' = y - t^2 + 1, \quad \text{for } 0 \leq t \leq 2, \quad \text{with } y(0) = 0.5,$$

which has solution $y(t) = (t+1)^2 - 0.5e^t$. Included in the input is the tolerance, $TOL = 10^{-5}$, maximum step size, $hmax = 0.25$, and minimum step size, $hmin = 0.01$. ◆ ◆ ◆

EXERCISE SET 5.6

1. The initial-value problem

$$y' = \sqrt{2 - y^2 e^t}, \quad \text{for } 0 \leq t \leq 0.8, \quad \text{with } y(0) = 0$$

has actual solution $y(t) = \sqrt{2}\sin(e^t - 1)$.

   a. Use the Runge-Kutta-Fehlberg method with tolerance $TOL = 10^{-4}$ to find w_1. Compare the approximate solution to the actual solution.

   b. Use the Adams Variable-Step-Size Predictor-Corrector method with tolerance $TOL = 10^{-4}$ and starting values from the Runge-Kutta method of order 4 to find w_4. Compare the approximate solution to the actual solution.

2. The initial-value problem

$$y' = -y + 1 - \frac{y}{t}, \quad \text{for } 1 \leq t \leq 2, \quad \text{with } y(1) = 1$$

has actual solution $y(t) = 1 + (e^{1-t} - 1)t^{-1}$.

   a. Use the Runge-Kutta-Fehlberg method with tolerance $TOL = 10^{-3}$ to find w_1 and w_2. Compare the approximate solutions to the actual solution.

   b. Use the Adams Variable-Step-Size Predictor-Corrector method with tolerance $TOL = 0.002$ and starting values from the Runge-Kutta method of order 4 to find w_4 and w_5. Compare the approximate solutions to the actual solution.

3. Use the Runge-Kutta-Fehlberg method with tolerance $TOL = 10^{-4}$ to approximate the solution to the following initial-value problems.

   a. $y' = \left(\frac{y}{t}\right)^2 + \frac{y}{t}$, for $1 \leq t \leq 1.2$, with $y(1) = 1$, $hmax = 0.05$, and $hmin = 0.02$.

   b. $y' = \sin t + e^{-t}$, for $0 \leq t \leq 1$, with $y(0) = 0$, $hmax = 0.25$, and $hmin = 0.02$.

   c. $y' = (y^2 + y)t^{-1}$, for $1 \leq t \leq 3$, with $y(1) = -2$, $hmax = 0.5$, and $hmin = 0.02$.

   d. $y' = -ty + 4ty^{-1}$, for $0 \leq t \leq 1$, with $y(0) = 1$, $hmax = 0.2$, and $hmin = 0.01$.

4. Use the Runge-Kutta-Fehlberg method with tolerance $TOL = 10^{-6}$, $hmax = 0.5$, and $hmin = 0.05$ to approximate the solutions to the following initial-value problems. Compare the results to the actual values.

   a. $y' = \dfrac{y}{t} - \dfrac{y^2}{t^2}$, for $1 \le t \le 4$, with $y(1) = 1$; actual solution $y(t) = t/(1 + \ln t)$.

   b. $y' = 1 + \dfrac{y}{t} + \left(\dfrac{y}{t}\right)^2$, for $1 \le t \le 3$, with $y(1) = 0$; actual solution $y(t) = t\tan(\ln t)$.

   c. $y' = -(y + 1)(y + 3)$, for $0 \le t \le 3$, with $y(0) = -2$; actual solution $y(t) = -3 + 2(1 + e^{-2t})^{-1}$.

   d. $y' = (t + 2t^3)y^3 - ty$, for $0 \le t \le 2$, with $y(0) = \frac{1}{3}$; actual solution $y(t) = (3 + 2t^2 + 6e^{t^2})^{-1/2}$.

5. Use the Adams Variable-Step-Size Predictor-Corrector method with $TOL = 10^{-4}$ to approximate the solutions to the initial-value problems in Exercise 3.

6. Use the Adams Variable-Step-Size Predictor-Corrector method with tolerance $TOL = 10^{-6}$, $hmax = 0.5$, and $hmin = 0.02$ to approximate the solutions to the initial-value problems in Exercise 4.

7. An electrical circuit consists of a capacitor of constant capacitance $C = 1.1$ farads in series with a resistor of constant resistance $R_0 = 2.1$ ohms. A voltage $\mathcal{E}(t) = 110 \sin t$ is applied at time $t = 0$. When the resistor heats up, the resistance becomes a function of the current i,

$$R(t) = R_0 + ki, \quad \text{where } k = 0.9,$$

   and the differential equation for i becomes

$$\left(1 + \frac{2k}{R_0}i\right)\frac{di}{dt} + \frac{1}{R_0 C}i = \frac{1}{R_0 C}\frac{d\mathcal{E}}{dt}.$$

   Find the current i after 2 s, assuming $i(0) = 0$.

5.7 Methods for Systems of Equations

The most common application of numerical methods for approximating the solution of initial-value problems concerns not a single problem, but a linked system of differential equations. Why, then, have we spent the majority of this chapter considering the solution of a single equation? The answer is simple: to approximate the solution of a system of initial-value problems, we successively apply the techniques that we used to solve a single problem. As is so often the case in mathematics, the key to the methods for systems can be found by examining the easier problem and then modifying it to treat the more complicated situation.

An ***m*th-order system** of first-order initial-value problems has the form

$$\begin{aligned}\frac{du_1}{dt} &= f_1(t, u_1, u_2, \dots, u_m),\\ \frac{du_2}{dt} &= f_2(t, u_1, u_2, \dots, u_m),\\ &\vdots\\ \frac{du_m}{dt} &= f_m(t, u_1, u_2, \dots, u_m),\end{aligned}$$

for $a \le t \le b$, with the initial conditions

$$u_1(a) = \alpha_1, \quad u_2(a) = \alpha_2, \dots, u_m(a) = \alpha_m.$$

The object is to find m functions $u_1, u_2, \dots, u_m$ that satisfy the system of differential equations together with all the initial conditions.

Methods to solve systems of first-order differential equations are generalizations of the methods for a single first-order equation presented earlier in this chapter. For example, the classical Runge-Kutta method of order 4 given by

$$\begin{aligned}w_0 &= \alpha,\\ k_1 &= hf(t_i, w_i),\\ k_2 &= hf\left(t_i + \frac{h}{2}, w_i + \frac{1}{2}k_1\right),\\ k_3 &= hf\left(t_i + \frac{h}{2}, w_i + \frac{1}{2}k_2\right),\\ k_4 &= hf(t_{i+1}, w_i + k_3),\end{aligned}$$

and

$$w_{i+1} = w_i + \frac{1}{6}[k_1 + 2k_2 + 2k_3 + k_4],$$

for each $i = 0, 1, \dots, N-1$, is used to solve the first-order initial-value problem

$$y' = f(t, y), \quad \text{for } a \le t \le b, \quad \text{with } y(a) = \alpha.$$

It is generalized as follows.

Let an integer $N > 0$ be chosen and set $h = (b - a)/N$. Partition the interval $[a, b]$ into N subintervals with the mesh points

$$t_j = a + jh \quad \text{for each } j = 0, 1, \dots, N.$$

Use the notation w_{ij} for each $j = 0, 1, \dots, N$ and $i = 1, 2, \dots, m$ to denote an approximation to $u_i(t_j)$; that is, w_{ij} will approximate the ith solution $u_i(t)$ of the system at the jth mesh point t_j. For the initial conditions, set

$$w_{1,0} = \alpha_1, \ w_{2,0} = \alpha_2, \ \ldots, \ w_{m,0} = \alpha_m.$$

Figure 5.4 gives an illustration of this notation.

Figure 5.4

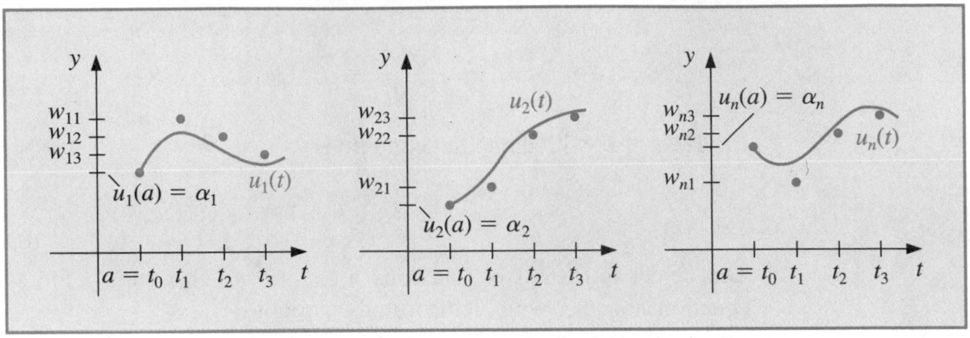

Suppose that the values $w_{1,j}, w_{2,j}, \ldots, w_{m,j}$ have been computed. We obtain $w_{1,j+1}$, $w_{2,j+1}, \ldots, w_{m,j+1}$ by first calculating, for each $i = 1, 2, \ldots, m$,

$$k_{1,i} = hf_i(t_j, w_{1,j}, w_{2,j}, \ldots, w_{m,j}),$$

and then finding, for each i,

$$k_{2,i} = hf_i\left(t_j + \frac{h}{2}, w_{1,j} + \frac{1}{2}k_{1,1}, w_{2,j} + \frac{1}{2}k_{1,2}, \ldots, w_{m,j} + \frac{1}{2}k_{1,m}\right).$$

We next determine all the terms

$$k_{3,i} = hf_i\left(t_j + \frac{h}{2}, w_{1,j} + \frac{1}{2}k_{2,1}, w_{2,j} + \frac{1}{2}k_{2,2}, \ldots, w_{m,j} + \frac{1}{2}k_{2,m}\right)$$

and, finally, calculate all the terms

$$k_{4,i} = hf_i(t_j + h, w_{1,j} + k_{3,1}, w_{2,j} + k_{3,2}, \ldots, w_{m,j} + k_{3,m}).$$

Combining these values gives

$$w_{i,j+1} = w_{i,j} + \frac{1}{6}[k_{1,i} + 2k_{2,i} + 2k_{3,i} + k_{4,i}]$$

for each $i = 1, 2, \ldots m$.

Note that all the values $k_{1,1}, k_{1,2}, \ldots, k_{1,m}$ must be computed before any of the terms of the form $k_{2,i}$ can be determined. In general, each $k_{l,1}, k_{l,2}, \ldots, k_{l,m}$ must be computed before any of the expressions $k_{l+1,i}$. The program RKO4SY57 implements the Runge-Kutta Fourth-Order method for systems.

EXAMPLE 1 Kirchhoff's Law states that the sum of all instantaneous voltage changes around a closed electrical circuit is zero. This implies that the current $I(t)$ in a closed circuit containing a resistance of R ohms, a capacitance of C farads, an inductance of L henrys, and a voltage source of $E(t)$ volts must satisfy the equation

$$LI'(t) + RI(t) + \frac{1}{C}\int I(t)\,dt = E(t).$$

The currents $I_1(t)$ and $I_2(t)$ in the left and right loops, respectively, of the circuit shown in Figure 5.5 are the solutions to the system of equations

$$2I_1(t) + 6[I_1(t) - I_2(t)] + 2I_1'(t) = 12,$$
$$\frac{1}{0.5}\int I_2(t)\,dt + 4I_2(t) + 6[I_2(t) - I_1(t)] = 0.$$

Figure 5.5

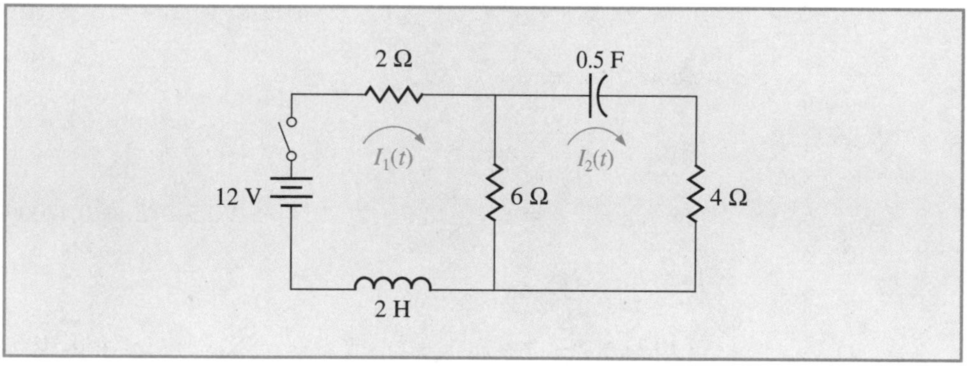

If we assume that the switch in the circuit is closed at time $t = 0$, then $I_1(0) = I_2(0) = 0$. Differentiating the second equation, solving for $I_1'(t)$ in the first equation, and substituting this into the second equation gives the following system of initial-value problems:

$$I_1' = f_1(t, I_1, I_2) = -4I_1 + 3I_2 + 6, \; I_1(0) = 0,$$
$$I_2' = f_2(t, I_1, I_2) = 0.6I_1' - 0.2I_2 = -2.4I_1 + 1.6I_2 + 3.6, \; I_2(0) = 0.$$

The exact solution to this system can be shown to be

$$I_1(t) = -3.375e^{-2t} + 1.875e^{-0.4t} + 1.5,$$
$$I_2(t) = -2.25e^{-2t} + 2.25e^{-0.4t}.$$

Suppose we apply the Runge-Kutta method of order 4 to this system with $h = 0.1$. Since $w_{1,0} = I_1(0) = 0$ and $w_{2,0} = I_2(0) = 0$,

$$k_{1,1} = hf_1(t_0, w_{1,0}, w_{2,0}) = 0.1\, f_1(0, 0, 0) = 0.1[-4(0) + 3(0) + 6] = 0.6,$$
$$k_{1,2} = hf_2(t_0, w_{1,0}, w_{2,0}) = 0.1\, f_2(0, 0, 0) = 0.1[-2.4(0) + 1.6(0) + 3.6] = 0.36,$$

$$k_{2,1} = hf_1\left(t_0 + \frac{1}{2}h, w_{1,0} + \frac{1}{2}k_{1,1}, w_{2,0} + \frac{1}{2}k_{1,2}\right) = 0.1\ f_1(0.05, 0.3, 0.18)$$
$$= 0.1[-4(0.3) + 3(0.18) + 6] = 0.534,$$
$$k_{2,2} = hf_2\left(t_0 + \frac{1}{2}h, w_{1,0} + \frac{1}{2}k_{1,1}, w_{2,0} + \frac{1}{2}k_{1,2}\right) = 0.1\ f_2(0.05, 0.3, 0.18)$$
$$= 0.1[-2.4(0.3) + 1.6(0.18) + 3.6] = 0.3168.$$

Generating the remaining entries in a similar manner produces

$$k_{3,1} = (0.1)f_1(0.05, 0.267, 0.1584) = 0.54072,$$
$$k_{3,2} = (0.1)f_2(0.05, 0.267, 0.1584) = 0.321264,$$
$$k_{4,1} = (0.1)f_1(0.1, 0.54072, 0.321264) = 0.4800912,$$

and

$$k_{4,2} = (0.1)f_2(0.1, 0.54072, 0.321264) = 0.28162944.$$

As a consequence,

$$I_1(0.1) \approx w_{1,1} = w_{1,0} + \frac{1}{6}[k_{1,1} + 2k_{2,1} + 2k_{3,1} + k_{4,1}]$$
$$= 0 + \frac{1}{6}[0.6 + 2(0.534) + 2(0.54072) + 0.4800912] = 0.5382552$$

and

$$I_2(0.1) \approx w_{2,1} = w_{2,0} + \frac{1}{6}[k_{1,2} + 2k_{2,2} + 2k_{3,2} + k_{4,2}] = 0.3196263.$$

The remaining entries in Table 5.15 are generated in a similar manner. ◆ ◆ ◆

Table 5.15

t_j	$w_{1,j}$	$w_{2,j}$	$\lvert I_1(t_j) - w_{1,j}\rvert$	$\lvert I_2(t_j) - w_{2,j}\rvert$
0.0	0	0	0	0
0.1	0.5382550	0.3196263	0.8285×10^{-5}	0.5803×10^{-5}
0.2	0.9684983	0.5687817	0.1514×10^{-4}	0.9596×10^{-5}
0.3	1.310717	0.7607328	0.1907×10^{-4}	0.1216×10^{-4}
0.4	1.581263	0.9063208	0.2098×10^{-4}	0.1311×10^{-4}
0.5	1.793505	1.014402	0.2193×10^{-4}	0.1240×10^{-4}

Maple's command `dsolve` can be used to solve systems of first-order differential equations. The system in Example 1 is defined with

```
>sys2:=D(u1)(t)=-4*u1(t)+3*u2(t)+6,
 D(u2)(t)=-2.4*u1(t)+1.6*u2(t)+3.6;
```

and the initial conditions with

```
>init2:=u1(0)=0,u2(0)=0;
```

The system is solved with the command

```
>sol2:=dsolve({sys2,init2},{u1(t),u2(t)});
```

to obtain

$$\begin{aligned} sol2 := \{u2(t) &= 2.25e^{-0.4t} - 2.25e^{-2t}, \\ u1(t) &= 1.5 - 3.375e^{-2t} + 1.875e^{-0.4t}\} \end{aligned}$$

To access the solution use

```
>r1:=rhs(sol2[2]);
```

$$r1 := 1.5 - 3.375e^{-2t} + 1.875e^{-0.4t}$$

and

```
>r2:=rhs(sol2[1]);
```

which gives a similar response.

To evaluate $u_1(0.5)$ and $u_2(0.5)$, use

```
>evalf(subs(t=0.5,r1));evalf(subs(t=0.5,r2);
```

to get $u_1(0.5) = 1.793527048$ and $u_2(0.5) = 1.014415451$.

The Runge-Kutta Fourth-Order method for systems is available in Maple. To apply the method to the system in Example 1, first define the functions f_1 and f_2 by

```
>f1:=(t,u1,u2) -> -4*u1+3*u2+6;
```

```
>f2:=(t,u1,u2) -> -2.4*u1+1.6*u2+3.6;
```

We need to access the differential equation solving procedure, so we enter

```
>with(share);
```

```
>readshare(ODE,plots);
```

The procedure is called with

```
>rhp:=rungekutta([f1,f2],[0,0,0],0.1,5);
```

The second parameter refers to the initial conditions $[a, \alpha_1, \alpha_2]$, and the third and fourth refer to $h = 0.1$ and $n = 5$. Maple responds with an array that gives the approximation at t_i, for $i = 0, 1, \ldots, 5$.
To access the entry with subscript 3, for example, enter

```
>rhp[3][2]; rhp[3][3];
```

which gives the results $u_1(0.3) \approx 1.310719039$ and $u_2(0.3) \approx 0.767331320$.

Many important physical problems—for example, electrical circuits and vibrating systems—involve initial-value problems whose equations have order higher than 1. New techniques are not required for solving these problems, since by relabeling the variables we can reduce a higher-order differential equation into a system of first-order differential equations and then apply one of the methods we have already discussed.

A general mth-order initial-value problem has the form

$$y^{(m)}(t) = f\left(t, y, y', \ldots, y^{(m-1)}\right),$$

for $a \leq t \leq b$, with initial conditions

$$y(a) = \alpha_1, y'(a) = \alpha_2, \ldots, y^{(m-1)}(a) = \alpha_m.$$

To convert this into a system of first-order differential equations, define

$$u_1(t) = y(t), u_2(t) = y'(t), \ldots, u_m(t) = y^{(m-1)}(t).$$

Using this notation, we obtain the first-order system

$$\begin{aligned}
\frac{du_1}{dt} &= \frac{dy}{dt} = u_2, \\
\frac{du_2}{dt} &= \frac{dy'}{dt} = u_3, \\
&\vdots \\
\frac{du_{m-1}}{dt} &= \frac{dy^{(m-2)}}{dt} = u_m,
\end{aligned}$$

and

$$\frac{du_m}{dt} = \frac{dy^{(m-1)}}{dt} = y^{(m)} = f\left(t, y, y', \ldots, y^{(m-1)}\right) = f(t, u_1, u_2, \ldots, u_m),$$

with initial conditions

$$u_1(a) = y(a) = \alpha_1, \; u_2(a) = y'(a) = \alpha_2, \ldots, u_m(a) = y^{(m-1)}(a) = \alpha_m.$$

EXAMPLE 2 Consider the second-order initial-value problem

$$y'' - 2y' + 2y = e^{2t} \sin t, \quad \text{for } 0 \leq t \leq 1, \quad \text{with } y(0) = -0.4 \text{ and } y'(0) = -0.6.$$

With $u_1(t) = y(t)$ and $u_2(t) = y'(t)$, this equation is transformed into the system

$$u_1'(t) = u_2(t),$$
$$u_2'(t) = e^{2t} \sin t - 2u_1(t) + 2u_2(t),$$

with initial conditions

$$u_1(0) = -0.4, \ u_2(0) = -0.6.$$

The set of values $w_{1,j}$ and $w_{2,j}$ for $j = 0, 1, \ldots, 10$, obtained using the Runge-Kutta Fourth-Order method, are presented in Table 5.15 and compared to the actual values of $u_1(t) = 0.2e^{2t}(\sin t - 2\cos t)$ and $u_2(t) = u_1'(t) = 0.2e^{2t}(4\sin t - 3\cos t)$. ◆ ◆ ◆

Table 5.16

t_j	$y(t_j) = u_1(t_j)$	$w_{1,j}$	$y'(t_j) = u_2(t_j)$	$w_{2,j}$	$\|y(t_j) - w_{1,j}\|$	$\|y'(t_j) - w_{2,j}\|$
0.0	−0.40000000	−0.40000000	−0.60000000	−0.60000000	0	0
0.1	−0.46173297	−0.46173334	−0.63163105	−0.63163124	3.7×10^{-7}	7.75×10^{-7}
0.2	−0.52555905	−0.52555988	−0.64014866	−0.64014895	8.3×10^{-7}	1.01×10^{-6}
0.3	−0.58860005	−0.58860144	−0.61366361	−0.61366381	1.39×10^{-6}	8.34×10^{-7}
0.4	−0.64661028	−0.64661231	−0.53658220	−0.53658203	2.03×10^{-6}	1.79×10^{-7}
0.5	−0.69356395	−0.69356666	−0.38873906	−0.38873810	2.71×10^{-6}	5.96×10^{-7}
0.6	−0.72114849	−0.72115190	−0.14438322	−0.14438087	3.41×10^{-6}	7.75×10^{-7}
0.7	−0.71814890	−0.71815295	0.22899243	0.22899702	4.05×10^{-6}	2.03×10^{-6}
0.8	−0.66970677	−0.66971133	0.77198383	0.77199180	4.56×10^{-6}	5.30×10^{-6}
0.9	−0.55643814	−0.55644290	1.53476862	0.15347815	4.76×10^{-6}	9.54×10^{-6}
1.0	−0.35339436	−0.35339886	2.57874662	0.25787663	4.50×10^{-6}	1.34×10^{-5}

We can use `dsolve` from Maple on those higher-order equations that can be solved exactly. The nth derivative $y^{(n)}(t)$ is specified in Maple by `(D@@n)(y)(t)`. To define the differential equation of Example 2, use

```
>def2:=(D@@2)(y)(t)-2*D(y)(t)+2*y(t)=exp(2*t)*sin(t);
```

and to specify the initial conditions use

```
>init2:=y(0)=-0.4, D(y)(0)=-0.6;
```

The solution is obtained by the command

```
>sol2:=dsolve({def2,init2},y(t));
```

to obtain

$$\text{sol2} := y(t) = \frac{1}{10}e^{2t}\sin t \sin 2t - \frac{1}{5}e^{2t}\sin t \cos 2t - \frac{3}{5}e^{2t}\cos t + \frac{1}{5}e^{2t}(\cos t)^3 + \frac{2}{5}e^{2t}(\cos t)^2 \sin t.$$

We isolate the solution in function form using

```
>g:=rhs(sol2);
```

To obtain $y(1.0) = g(1.0) = -0.3533943558$, enter

```
>evalf(subs(t=1.0,g));
```

Other one-step approximation methods can be extended to systems. If the Runge-Kutta-Fehlberg method is extended, then each component of the numerical solution w_{1j}, $w_{2j}, \ldots, w_{mj}$ must be examined for accuracy. If any of the components fail to be sufficiently accurate, the entire numerical solution must be recomputed.

The multistep methods and predictor-corrector techniques are also extended easily to systems. Again, if error control is used, each component must be accurate. The extension of the extrapolation technique to systems can also be done, but the notation becomes quite involved.

EXERCISE SET 5.7

1. Use the Runge-Kutta method for systems to approximate the solutions of the following systems of first-order differential equations and compare the results to the actual solutions.
   a. $u_1' = 3u_1 + 2u_2 - (2t^2+1)e^{2t}$, for $0 \le t \le 1$ with $u_1(0) = 1$;
   $u_2' = 4u_1 + u_2 + (t^2 + 2t - 4)e^{2t}$, for $0 \le t \le 1$ with $u_2(0) = 1$;
   $h = 0.2$; actual solutions $u_1(t) = \frac{1}{3}e^{5t} - \frac{1}{3}e^{-t} + e^{2t}$ and $u_2(t) = \frac{1}{3}e^{5t} + \frac{2}{3}e^{-t} + t^2e^{2t}$.
   b. $u_1' = -4u_1 - 2u_2 + \cos t + 4\sin t$, for $0 \le t \le 2$ with $u_1(0) = 0$;
   $u_2' = 3u_1 + u_2 - 3\sin t$, for $0 \le t \le 2$ with $u_2(0) = -1$;
   $h = 0.1$; actual solutions $u_1(t) = 2e^{-t} - 2e^{-2t} + \sin t$ and $u_2(t) = -3e^{-t} + 2e^{-2t}$.
   c. $u_1' = u_2$, for $0 \le t \le 2$ with $u_1(0) = 1$;
   $u_2' = -u_1 - 2e^t + 1$, for $0 \le t \le 2$ with $u_2(0) = 0$;
   $u_3' = -u_1 - e^t + 1$, for $0 \le t \le 2$ with $u_3(0) = 1$;
   $h = 0.5$; actual solutions $u_1(t) = \cos t + \sin t - e^t + 1$, $u_2(t) = -\sin t + \cos t - e^t$, and $u_3(t) = -\sin t + \cos t$.
   d. $u_1' = u_2 - u_3 + t$, for $0 \le t \le 1$ with $u_1(0) = 1$;
   $u_2' = 3t^2$, for $0 \le t \le 1$ with $u_2(0) = 1$;
   $u_3' = u_2 + e^{-t}$, for $0 \le t \le 1$ with $u_3(0) = -1$;
   $h = 0.1$; actual solutions $u_1(t) = -0.05t^5 + 0.25t^4 + t + 2 - e^{-t}$, $u_2(t) = t^3 + 1$, and $u_3(t) = 0.25t^4 + t - e^{-t}$.

2. Use the Runge-Kutta method for systems to approximate the solutions of the following higher-order differential equations and compare the results to the actual solutions.
   a. $y'' - 2y' + y = te^t - t$, for $0 \le t \le 1$ with $y(0) = y'(0) = 0$ and $h = 0.1$; actual solution $y(t) = \frac{1}{6}t^3e^t - te^t + 2e^t - t - 2$.

b. $t^2y'' - 2ty' + 2y = t^3 \ln t$, for $1 \le t \le 2$ with $y(1) = 1$, $y'(1) = 0$, and $h = 0.1$; actual solution $y(t) = \frac{7}{4}t + \frac{1}{2}t^3 \ln t - \frac{3}{4}t^3$.

c. $y''' + 2y'' - y' - 2y = e^t$, for $0 \le t \le 3$ with $y(0) = 1$, $y'(0) = 2$, $y''(0) = 0$, and $h = 0.2$; actual solution $y(t) = \frac{43}{36}e^t + \frac{1}{4}e^{-t} - \frac{4}{9}e^{-2t} + \frac{1}{6}te^t$.

d. $t^3y''' - t^2y'' + 3ty' - 4y = 5t^3 \ln t + 9t^3$, for $1 \le t \le 2$ with $y(1) = 0$, $y'(1) = 1$, $y''(1) = 3$, and $h = 0.1$; actual solution $y(t) = -t^2 + t\cos(\ln t) + t\sin(\ln t) + t^3 \ln t$.

3. Change the Adams Fourth-Order Predictor-Corrector method to obtain approximate solutions to systems of first-order equations.

4. Repeat Exercise 1 using the method developed in Exercise 3.

5. The study of mathematical models for predicting the population dynamics of competing species has its origin in independent works published in the early part of this century by A. J. Lotka and V. Volterra. Consider the problem of predicting the population of two species, one of which is a predator, whose population at time t is $x_2(t)$, feeding on the other, which is the prey, whose population is $x_1(t)$. We will assume that the prey always has an adequate food supply and that its birth rate at any time is proportional to the number of prey alive at that time; that is, birth rate (prey) is $k_1x_1(t)$. The death rate of the prey depends on both the number of prey and predators alive at that time. For simplicity, we assume death rate (prey) $= k_2x_1(t)x_2(t)$. The birth rate of the predator, on the other hand, depends on its food supply, $x_1(t)$, as well as on the number of predators available for reproduction purposes. For this reason, we assume that the birth rate (predator) is $k_3x_1(t)x_2(t)$. The death rate of the predator will be taken as simply proportional to the number of predators alive at the time; that is, death rate (predator) $= k_4x_2(t)$.

Since $x_1'(t)$ and $x_2'(t)$ represent the change in the prey and predator populations, respectively, with respect to time, the problem is expressed by the system of nonlinear differential equations

$$x_1'(t) = k_1x_1(t) - k_2x_1(t)x_2(t) \quad \text{and} \quad x_2'(t) = k_3x_1(t)x_2(t) - k_4x_2(t).$$

Solve this system for $0 \le t \le 4$, assuming that the initial population of the prey is 1000 and of the predators is 500 and that the constants are $k_1 = 3$, $k_2 = 0.002$, $k_3 = 0.0006$, and $k_4 = 0.5$. Is there a stable solution to this population model? If so, for what values x_1 and x_2 is the solution stable?

6. In Exercise 5 we considered the problem of predicting the population in a predator-prey model. Another problem of this type is concerned with two species competing for the same food supply. If the numbers of species alive at time t are denoted by $x_1(t)$ and $x_2(t)$, it is often assumed that, although the birth rate of each of the species is simply proportional to the number of species alive at that time, the death rate of each species depends on the population of both species. We will assume that the population of a particular pair of species is described by the equations

$$\frac{dx_1(t)}{dt} = x_1(t)[4 - 0.0003x_1(t) - 0.0004x_2(t)]$$

and

$$\frac{dx_2(t)}{dt} = x_2(t)[2 - 0.0002x_1(t) - 0.0001x_2(t)].$$

If it is known that the initial population of each species is 10,000, find the solution to this system for $0 \leq t \leq 4$. Is there a stable solution to this population model? If so, for what values of x_1 and x_2 is the solution stable?

5.8 Stiff Differential Equations

All the methods for approximating the solution to initial-value problems have error terms that involve a higher derivative of the solution of the equation. If the derivative can be reasonably bounded, then the method will have a predictable error bound that can be used to estimate the accuracy of the approximation. Even if the derivative grows as the steps increase, the error can be kept in relative control, provided that the solution also grows in magnitude. Problems frequently arise, however, where the magnitude of the derivative increases, but the solution does not. In this situation, the error can grow so large that it dominates the calculations. Initial-value problems for which this is likely to occur are called **stiff equations** and are quite common, particularly in the study of vibrations, chemical reactions, and electrical circuits. Stiff systems derive their name from the motion of spring and mass systems that have large spring constants.

Stiff differential equations are characterized as those whose exact solution has a term of the form e^{-ct}, where c is a large positive constant. This is usually only a part of the solution, called the *transient* solution, the more important portion of the solution is called the *steady-state* solution. A transient portion of a stiff equation will rapidly decay to zero as t increases, but since the nth derivative of this term has magnitude $c^n e^{-ct}$, the derivative does not decay as quickly. In fact, since the derivative in the error term is evaluated not at t, but at a number between zero and t, the derivative terms may increase as t increases—and very rapidly indeed. Fortunately, stiff equations can generally be predicted from the physical problem from which the equation is derived, and with care the error can be kept under control. The manner in which this is done is considered in this section.

EXAMPLE 1 The system of initial-value problems

$$u_1' = 9u_1 + 24u_2 + 5\cos t - \frac{1}{3}\sin t, \quad \text{with } u_1(0) = \frac{4}{3}$$
$$u_2' = -24u_1 - 51u_2 - 9\cos t + \frac{1}{3}\sin t, \quad \text{with } u_2(0) = \frac{2}{3}$$

has the unique solution

$$u_1(t) = 2e^{-3t} - e^{-39t} + \frac{1}{3}\cos t,$$

$$u_2(t) = -e^{-3t} + 2e^{-39t} - \frac{1}{3}\cos t.$$

The transient term e^{-39t} in the solution causes this system to be stiff. Applying the Runge-Kutta Fourth-Order method for systems gives results listed in Table 5.17. Accurate approximations occur when $h = 0.05$. Increasing the step-size to $h = 0.1$, however, leads to the disastrous results shown in the table. ◆ ◆ ◆

Table 5.17

t	$u_1(t)$	$w_1(t)$ $h = 0.05$	$w_1(t)$ $h = 0.1$	$u_2(t)$	$w_2(t)$ $h = 0.05$	$w_2(t)$ $h = 0.1$
0.1	1.793061	1.712219	−2.645169	−1.032001	−0.8703152	7.844527
0.2	1.423901	1.414070	−18.45158	−0.8746809	−0.8550148	38.87631
0.3	1.131575	1.130523	−87.47221	−0.7249984	−0.7228910	176.4828
0.4	0.9094086	0.9092763	−934.0722	−0.6082141	−0.6079475	789.3540
0.5	0.7387877	0.7387506	−1760.016	−0.5156575	−0.5155810	3520.00
0.6	0.6057094	0.6056833	−7848.550	−0.4404108	−0.4403558	15697.84
0.7	0.4998603	0.4998361	−34989.63	−0.3774038	−0.3773540	69979.87
0.8	0.4136714	0.4136490	−155979.4	−0.3229535	−0.3229078	311959.5
0.9	0.3416143	0.3415939	−695332.0	−0.2744088	−0.2743673	1390664.
1.0	0.2796748	0.2796568	−3099671.	−0.2298877	−0.2298511	6199352.

Although stiffness is usually associated with systems of differential equations, the approximation characteristics of a particular numerical method applied to a stiff system can be predicted by examining the error produced when the method is applied to a simple *test equation*,

$$y' = \lambda y, \quad \text{with } y(0) = \alpha,$$

where λ is a negative real number. The solution to this equation contains the transient solution $e^{\lambda t}$ and the steady-state solution is zero, so the approximation characteristics of a method are easy to determine. (A more complete discussion of the round-off error associated with stiff systems requires examining the test equation when λ is a complex number with negative imaginary part.)

Suppose that we apply Euler's method to the test equation. Letting $h = (b - a)/N$ and $t_j = jh$, for $j = 0, 1, 2, \ldots, N$, implies that

$$w_0 = \alpha$$

and

$$w_{j+1} = w_j + h(\lambda w_j) = (1 + h\lambda)w_j,$$

so

$$w_{j+1} = (1 + h\lambda)^{j+1} w_0 = (1 + h\lambda)^{j+1}\alpha, \quad \text{for } j = 0, 1, \ldots, N-1. \tag{5.6}$$

Since the exact solution is $y(t) = \alpha e^{\lambda t}$, the absolute error is

$$|y(t_j) - w_j| = |e^{jh\lambda} - (1 + h\lambda)^j|\,|\alpha| = |(e^{h\lambda})^j - (1 + h\lambda)^j|\,|\alpha|,$$

and the accuracy is determined by how well the term $1 + h\lambda$ approximates $e^{h\lambda}$. When $\lambda < 0$, the exact solution, $(e^{h\lambda})^j$, decays to zero as j increases, but by Eq. (5.6), the approximation will have this property only if $|1 + h\lambda| < 1$. This effectively restricts the step size h for Euler's method to satisfy $|1 + h\lambda| < 1$, which implies that $h < 2/|\lambda|$.

Suppose now that a round-off error δ_0 is introduced in the initial condition for Euler's method,

$$w_0 = \alpha + \delta_0.$$

At the jth step the round-off error is

$$\delta_j = (1 + h\lambda)^j \delta_0.$$

Since $\lambda < 0$, the condition for the control of the growth of round-off error is the same as the condition for controlling the absolute error: $h < 2/|\lambda|$.

The situation is similar for other one-step methods. In general, a function Q exists with the property that the difference method, when applied to the test equation, gives

$$w_{j+1} = Q(h\lambda) w_j.$$

The accuracy of the method depends upon how well $Q(h\lambda)$ approximates $e^{h\lambda}$, and the error will grow without bound if $|Q(h\lambda)| > 1$.

The problem is more complicated in the case of multistep methods, due to the interplay of previous approximations at each step. The problem tends to be particularly acute in the case of the explicit techniques, which carries over to predictor-corrector methods. In practice, the techniques used for stiff systems are implicit multistep methods. Generally, w_{i+1} is obtained by iteratively solving a nonlinear equation or nonlinear system, often by Newton's method. To illustrate the procedure, consider the following implicit technique.

Implicit Trapezoidal Method

$$w_0 = \alpha$$

$$w_{j+1} = w_j + \frac{h}{2}[f(t_{j+1}, w_{j+1}) + f(t_j, w_j)]$$

where $j = 0, 1, \ldots, N-1$.

To determine w_1 using this technique, we apply Newton's method to find the root of the equation

$$0 = F(w) = w - w_0 - \frac{h}{2}[f(t_0, w_0) + f(t_1, w)] = w - \alpha - \frac{h}{2}[f(a, \alpha) + f(t_1, w)].$$

To approximate this solution, select $w_1^{(0)}$ (usually as w_0) and generate $w_1^{(k)}$ by applying Newton's method to obtain

$$\begin{aligned} w_1^{(k)} &= w_1^{(k-1)} - \frac{F(w_1^{(k-1)})}{F'(w_1^{(k-1)})} \\ &= w_1^{(k-1)} - \frac{w_1^{(k-1)} - \alpha - \frac{h}{2}[f(a, \alpha) + f(t_1, w_1^{(k-1)})]}{1 - \frac{h}{2} f_y(t_1, w_1^{(k-1)})} \end{aligned}$$

until $|w_1^{(k)} - w_1^{(k-1)}|$ is sufficiently small. Normally only three or four iterations are required.

Once a satisfactory approximation for w_1 has been determined, the method is repeated to find w_2 and so on. This is the procedure incorporated in TRAPNT58, which implements this technique.

The Secant method can be used as an alternative to Newton's method, but then two distinct initial approximations to w_{j+1} are required. To determine these, the usual practice is to let $w_{j+1}^{(0)} = w_j$ and obtain $w_{j+1}^{(1)}$ from some explicit multistep method. When a system of stiff equations is involved, a generalization is required for either Newton's or the Secant method. These topics are considered in Chapter 10.

EXAMPLE 2 The stiff initial-value problem

$$y' = 5e^{5t}(y - t)^2 + 1, \quad \text{for } 0 \le t \le 1, \quad \text{with } y(0) = -1$$

has solution $y(t) = t - e^{-5t}$. To show the effects of stiffness, the Trapezoidal method and the Runge-Kutta fourth-order method are applied both with $N = 4$, giving $h = 0.25$, and with $N = 5$, giving $h = 0.20$. The Trapezoidal method performs well in both cases, using $M = 10$ and $TOL = 10^{-6}$, as does Runge-Kutta with $h = 0.2$, as shown in Table 5.18. However, for $h = 0.25$ the Runge-Kutta method gives inaccurate results, as shown in Table 5.19 on the following page. ◆ ◆ ◆

Table 5.18

	Runge-Kutta Method		**Trapezoidal Method**	
	$h = 0.2$		$h = 0.2$	
t_i	w_i	$\lvert y(t_i) - w_i \rvert$	w_i	$\lvert y(t_i) - w_i \rvert$
0.0	−1.0000000	0	−1.0000000	0
0.2	−0.1488521	1.9027×10^{-2}	−0.1414969	2.6383×10^{-2}
0.4	0.2684884	3.8237×10^{-3}	0.2748614	1.0197×10^{-2}
0.6	0.5519927	1.7798×10^{-3}	0.5539828	3.7700×10^{-3}
0.8	0.7822857	6.0131×10^{-4}	0.7830720	1.3876×10^{-3}
1.0	0.9934905	2.2845×10^{-4}	0.9937726	5.1050×10^{-4}

Table 5.19

	Runge-Kutta Method		Trapezoidal Method	
	$h = 0.25$		$h = 0.25$	
t_i	w_i	$\lvert y(t_i) - w_i\rvert$	w_i	$\lvert y(t_i) - w_i\rvert$
0.0	−1.0000000	0	−1.0000000	0
0.25	0.4014315	4.37936×10^{-1}	0.0054557	4.1961×10^{-2}
0.5	3.4374753	3.01956	0.4267572	8.8422×10^{-3}
0.75	1.44639×10^{23}	1.44639×10^{23}	0.7291528	2.6706×10^{-3}
1.0	Overflow		0.9940199	7.5790×10^{-4}

EXERCISE SET 5.8

1. Solve the following stiff initial-value problems using Euler's method and compare the results with the actual solution.
   a. $y' = -9y$, for $0 \le t \le 1$, with $y(0) = e$ and $h = 0.1$; actual solution $y(t) = e^{1-9t}$.
   b. $y' = -20(y - t^2) + 2t$, for $0 \le t \le 1$, with $y(0) = \frac{1}{3}$ and $h = 0.1$; actual solution $y(t) = t^2 + \frac{1}{3}e^{-20t}$.
   c. $y' = -20y + 20\sin t + \cos t$, for $0 \le t \le 2$, with $y(0) = 1$ and $h = 0.25$; actual solution $y(t) = \sin t + e^{-20t}$.
   d. $y' = \dfrac{50}{y} - 50y$, for $0 \le t \le 1$, with $y(0) = \sqrt{2}$ and $h = 0.1$; actual solution $y(t) = (1 + e^{-100t})^{1/2}$.
2. Repeat Exercise 1 using the Runge-Kutta Fourth-Order method.
3. Repeat Exercise 1 using the Adams Fourth-Order Predictor-Corrector method.
4. Repeat Exercise 1 using the Trapezoidal method with a tolerance of 10^{-5}.
5. The Backward Euler One-Step method is defined by

$$w_{i+1} = w_i + hf(t_{i+1}, w_{i+1}) \quad \text{for } i = 0, 1, \ldots, N - 1.$$

   Repeat Exercise 1 using the Backward Euler method incorporating Newton's method to solve for w_{i+1}.
6. In Exercise 11 of Section 5.2, the differential equation

$$\frac{dp(t)}{dt} = rb(1 - p(t))$$

   was obtained as a model for studying the proportion $p(t)$ of nonconformists in a society whose birth rate was b and where r represented the rate at which offspring would become nonconformists when at least one of their parents was a conformist. That exercise required that an approximation for $p(t)$ be found by using Euler's method for integral values of t when given $p(0) = 0.01, b = 0.02$, and $r = 0.1$, and then the approximation for $p(50)$ be compared with the actual value. Use the Trapezoidal method to obtain another approximation for $p(50)$, again assuming that $h = 1$ year.

5.9 Survey of Methods and Software

In this chapter we have considered methods to approximate the solutions to initial-value problems for ordinary differential equations. We began with a discussion of the most elementary numerical technique, Euler's method. This procedure was not sufficiently accurate to be of use in applications, but it illustrated the general behavior of the more powerful techniques, without the accompanying algebraic difficulties. The Taylor methods were then considered as generalizations of Euler's method. They were found to be accurate but cumbersome because of the need to determine extensive partial derivatives of the defining function of the differential equation. The Runge-Kutta formulas simplified the Taylor methods, while not significantly increasing the error. To this point we had considered only one-step methods, techniques that use data only at the most recently computed point.

Multistep methods were discussed in Section 5.4, where explicit methods of Adams-Bashforth type and implicit methods of Adams-Moulton type were considered. These culminate in predictor-corrector methods, which use an explicit method, such as an Adams-Bashforth, to predict the solution and then apply a corresponding implicit method, such as an Adams-Moulton, to correct the approximation.

Section 5.7 illustrated how these techniques can be used to solve higher order initial value problems and systems of initial value problems.

These one- and multistep methods serve as an introduction to numerical methods for ordinary differential equations since the more accurate adaptive methods are based on these relatively uncomplicated techniques. In particular, we saw in Section 5.6 that the Runge-Kutta-Fehlberg method is a one-step procedure that seeks to select mesh spacing to keep the local error of the approximation under control. The Variable Step-Size Predictor-Corrector method also presented in Section 5.6 is based on the four-step Adams-Bashforth method and three-step Adams-Moulton method. It also changes the step size to keep the local error within a given tolerance. The Extrapolation method discussed in Section 5.5 is based on a modification of the Midpoint method and incorporates extrapolation to maintain a desired accuracy of approximation.

The final topic in the chapter concerned the difficulty that is inherent in the approximation of the solution to a stiff equation, a differential equation whose exact solution contains a portion of the form $e^{-\lambda t}$, where λ is a positive constant. Special caution must be taken with problems of this type, or the results can be overwhelmed by round-off error.

Methods of the Runge-Kutta-Fehlberg type are generally sufficient for nonstiff problems, where moderate accuracy is required. The extrapolation procedures are recommended for nonstiff problems, where high accuracy is required. Finally, extensions of the Implicit Trapezoidal method to variable-order and variable step-size implicit Adams-type methods are used for stiff initial-value problems.

The ISML Library includes three subroutines for approximating the solutions of initial-value problems. One is a variable step-size subroutine similar to the Runge-Kutta-Fehlberg method but based on fifth- and sixth-order formulas. The second is an extrapolation method based on rational functions approximations. The third subroutine is designed for stiff systems and uses implicit multistep methods of order up to 12. The NAG Library contains a Runge-Kutta type formula with a variable step size. A variable order, variable-step-size backward-difference method for stiff systems is also available.

There are many books specializing in the numerical solution of initial-value problems. Two classics are by Henrici [He] and Gear [G]. Two books by Hairer, Nörsett, and Warner provide comprehensive discussions on nonstiff [HNW1] and stiff [HNW2] problems.

CHAPTER

6 Direct Methods for Solving Linear Systems

6.1 Introduction

Systems of equations are used to represent physical problems that involve the interaction of various properties. The variables in the system represent the properties being studied, and the equations describe the interaction between the variables. The system is easiest to study when the equations are all linear. Often the number of equations is the same as the number of variables, for only in this case is it likely that a unique solution will exist.

Although not all physical problems can be reasonably represented using a linear system with the same number of equations as unknowns, the solutions to many problems either have this form or can be approximated by such a system. In fact, this is quite often the only approach that can give quantitative information about a physical problem.

In this chapter we consider direct methods for approximating the solution of a system of n linear equations in n unknowns. A direct method is one that gives the exact solution to the system, if it is assumed that all calculations can be performed without round-off error effects. This assumption is idealized. We will need to consider quite carefully the role of finite-digit arithmetic error in the approximation to the solution to the system and how to arrange the calculations to minimize its effect.

6.2 Gaussian Elimination

If you have studied linear algebra or matrix theory, you probably have been introduced to Gaussian elimination, the most elementary method for systematically determining the solution of a system of linear equations. Variables are eliminated from the equations until one equation involves only one variable, a second equation involves only that variable and one other, a third has only these two and one additional, and so on. The solution is found by solving for the variable in the single equation, using this to reduce the second equation to one that now contains a single variable, and so on, until values for all the variables are found.

Three operations are permitted on a system of equations (E_n).

Operations on Systems of Equations

1. Equation E_i can be multiplied by any nonzero constant λ, with the resulting equation used in place of E_i. This operation is denoted $(\lambda E_i) \rightarrow (E_i)$.
2. Equation E_j can be multiplied by any constant λ, and added to equation E_i, with the resulting equation used in place of E_i. This operation is denoted $(E_i + \lambda E_j) \rightarrow (E_i)$.
3. Equations E_i and E_j can be transposed in order. This operation is denoted $(E_i) \leftrightarrow (E_j)$.

By a sequence of the operations just given, a linear system can be transformed to a more easily solved linear system with the same solutions. The sequence of operations is illustrated in the next example.

EXAMPLE 1 The four equations

$$\begin{aligned}
E_1: &\quad x_1 + x_2 + 3x_4 = 4, \\
E_2: &\quad 2x_1 + x_2 - x_3 + x_4 = 1, \\
E_3: &\quad 3x_1 - x_2 - x_3 + 2x_4 = -3, \\
E_4: &\quad -x_1 + 2x_2 + 3x_3 - x_4 = 4,
\end{aligned}$$

will be solved for x_1, x_2, x_3, and x_4. First use equation E_1 to eliminate the unknown x_1 from E_2, E_3, and E_4 by performing $(E_2 - 2E_1) \rightarrow (E_2)$, $(E_3 - 3E_1) \rightarrow (E_3)$, and $(E_4 + E_1) \rightarrow (E_4)$. The resulting system is

$$\begin{aligned}
E_1: &\quad x_1 + x_2 + 3x_4 = 4, \\
E_2: &\quad -x_2 - x_3 - 5x_4 = -7, \\
E_3: &\quad -4x_2 - x_3 - 7x_4 = -15, \\
E_4: &\quad 3x_2 + 3x_3 + 2x_4 = 8,
\end{aligned}$$

where, for simplicity, the new equations are again labeled E_1, E_2, E_3, and E_4.

In the new system, E_2 is used to eliminate x_2 from E_3 and E_4 by $(E_3 - 4E_2) \rightarrow (E_3)$ and $(E_4 + 3E_2) \rightarrow (E_4)$, resulting in

$$\begin{aligned}
E_1: &\quad x_1 + x_2 + 3x_4 = 4, \\
E_2: &\quad -x_2 - x_3 - 5x_4 = -7, \\
E_3: &\quad 3x_3 + 13x_4 = 13, \\
E_4: &\quad -13x_4 = -13.
\end{aligned}$$

The system of equations is now in *triangular* (or *reduced*) form and can be solved for the unknowns by a backward-substitution process. Noting that E_4 implies $x_4 = 1$, we can solve E_3 for x_3:

$$x_3 = \frac{1}{3}(13 - 13x_4) = \frac{1}{3}(13 - 13) = 0.$$

Continuing, E_2 gives

$$x_2 = -(-7 + 5x_4 + x_3) = -(-7 + 5 + 0) = 2,$$

and E_1 gives

$$x_1 = 4 - 3x_4 - x_2 = 4 - 3 - 2 = -1.$$

The solution is, therefore, $x_1 = -1$, $x_2 = 2$, $x_3 = 0$, and $x_4 = 1$. It is easy to verify that these values solve the original system of equations. ◆ ◆ ◆

When performing the calculations of Example 1, we did not need to write out the full equations at each step or to carry the variables x_1, x_2, x_3, and x_4 through the calculations, since they always remained in the same column. The only variation from system to system occurred in the coefficients of the unknowns and in the values on the right side of the equations. For this reason, a linear system is often replaced by a **matrix**, a rectangular array of elements in which not only is the value of an element important, but also its position in the array. The matrix contains all the information about the system that is necessary to determine its solution, but in a compact form.

The notation for an $n \times m$ (n by m) matrix will be a capital letter, such as A, for the matrix and lowercase letters with double subscripts, such as a_{ij}, to refer to the entry at the intersection of the ith row and jth column; that is,

$$A = (a_{ij}) = \begin{bmatrix} a_{11} & a_{12} & \cdots & a_{1m} \\ a_{21} & a_{22} & \cdots & a_{2m} \\ \vdots & \vdots & & \vdots \\ a_{n1} & a_{n2} & \cdots & a_{nm} \end{bmatrix}.$$

EXAMPLE 2 The matrix

$$A = \begin{bmatrix} 2 & -1 & 7 \\ 3 & 1 & 0 \end{bmatrix}$$

is a 2×3 matrix with $a_{11} = 2$, $a_{12} = -1$, $a_{13} = 7$, $a_{21} = 3$, $a_{22} = 1$, and $a_{23} = 0$. ◆ ◆ ◆

The $1 \times n$ matrix $A = [a_{11}\ a_{12}\ \cdots a_{1n}]$ is called an ***n*-dimensional row vector**, and an $n \times 1$ matrix

$$A = \begin{bmatrix} a_{11} \\ a_{21} \\ \vdots \\ a_{n1} \end{bmatrix}$$

is called an ***n*-dimensional column vector**. Usually the unnecessary subscript is omitted for vectors and a boldface lowercase letter is used for notation. So,

$$\mathbf{x} = \begin{bmatrix} x_1 \\ x_2 \\ \vdots \\ x_n \end{bmatrix}$$

denotes a column vector, and $\mathbf{y} = [y_1\ y_2\ \cdots\ y_n]$ denotes a row vector.

An $n \times (n+1)$ matrix can be used to represent the linear system

$$\begin{aligned} a_{11}x_1 + a_{12}x_2 + \cdots + a_{1n}x_n &= b_1, \\ a_{21}x_1 + a_{22}x_2 + \cdots + a_{2n}x_n &= b_2, \\ &\vdots \\ a_{n1}x_1 + a_{n2}x_2 + \cdots + a_{nn}x_n &= b_n, \end{aligned}$$

by first constructing

$$A = (a_{ij}) = \begin{bmatrix} a_{11} & a_{12} & \cdots & a_{1n} \\ a_{21} & a_{22} & \cdots & a_{2n} \\ \vdots & \vdots & & \vdots \\ a_{n1} & a_{n2} & \cdots & a_{nn} \end{bmatrix} \quad \text{and} \quad \mathbf{b} = \begin{bmatrix} b_1 \\ b_2 \\ \vdots \\ b_n \end{bmatrix}$$

and then combining these matrices to form the *augmented matrix*:

$$[A, \mathbf{b}] = \left[\begin{array}{cccc:c} a_{11} & a_{12} & \cdots & a_{1n} & b_1 \\ a_{21} & a_{22} & \cdots & a_{2n} & b_2 \\ \vdots & \vdots & & \vdots & \vdots \\ a_{n1} & a_{n2} & \cdots & a_{nn} & b_n \end{array}\right],$$

where the vertical dotted line is used to separate the coefficients of the unknowns from the values on the right-hand side of the equations.

Repeating the operations involved in Example 1 with the matrix notation results in first considering the augmented matrix:

$$\left[\begin{array}{rrrr:r} 1 & 1 & 0 & 3 & 4 \\ 2 & 1 & -1 & 1 & 1 \\ 3 & -1 & -1 & 2 & -3 \\ -1 & 2 & 3 & -1 & 4 \end{array}\right].$$

Performing the operations as described in that example produces the matrices

$$\begin{bmatrix} 1 & 1 & 0 & 3 & \vdots & 4 \\ 0 & -1 & -1 & -5 & \vdots & -7 \\ 0 & -4 & -1 & -7 & \vdots & -15 \\ 0 & 3 & 3 & 2 & \vdots & 8 \end{bmatrix} \quad \text{and} \quad \begin{bmatrix} 1 & 1 & 0 & 3 & \vdots & 4 \\ 0 & -1 & -1 & -5 & \vdots & -7 \\ 0 & 0 & 3 & 13 & \vdots & 13 \\ 0 & 0 & 0 & -13 & \vdots & -13 \end{bmatrix}.$$

The latter matrix can now be transformed into its corresponding linear system and solutions for x_1, x_2, x_3, and x_4 obtained. The procedure involved in this process is called **Gaussian Elimination with Backward Substitution**.

The general Gaussian elimination procedure applied to the linear system

$$\begin{aligned} E_1:&\quad a_{11}x_1 + a_{12}x_2 + \cdots + a_{1n}x_n = b_1, \\ E_2:&\quad a_{21}x_1 + a_{22}x_2 + \cdots + a_{2n}x_n = b_2, \\ &\quad\vdots \\ E_n:&\quad a_{n1}x_1 + a_{n2}x_2 + \cdots + a_{nn}x_n = b_n, \end{aligned}$$

is handled in a similar manner. First form the augmented matrix $\tilde{A}$:

$$\tilde{A} = [A, \mathbf{b}] = \begin{bmatrix} a_{11} & a_{12} & \cdots & a_{1n} & \vdots & a_{1,n+1} \\ a_{21} & a_{22} & \cdots & a_{2n} & \vdots & a_{2,n+1} \\ \vdots & \vdots & & \vdots & \vdots & \vdots \\ a_{n1} & a_{n2} & \cdots & a_{nn} & \vdots & a_{n,n+1} \end{bmatrix},$$

where A denotes the matrix formed by the coefficients and the entries in the $(n+1)$st column are the values of $\mathbf{b}$; that is, $a_{i,n+1} = b_i$ for each $i = 1, 2, \ldots, n$.

Suppose the **pivot element** $a_{11} \neq 0$. To simplify the discussion we introduce the *multiplier* $m_{k1} = a_{k1}/a_{11}$ and perform the operations corresponding to $(E_k - m_{k1}E_1) \longrightarrow (E_k)$ for each $k = 2, 3, \ldots, n$ to eliminate (that is, change to zero) the coefficient of x_1 in each of these rows:

$$\begin{bmatrix} a_{11} & a_{12} & \cdots & a_{1n} & \vdots & b_1 \\ a_{21} & a_{22} & \cdots & a_{2n} & \vdots & b_2 \\ \vdots & \vdots & & \vdots & \vdots & \vdots \\ a_{n1} & a_{n2} & \cdots & a_{nn} & \vdots & b_n \end{bmatrix} \quad \begin{matrix} E_2 - m_{21}E_1 \longrightarrow E_2 \\ E_3 - m_{31}E_1 \longrightarrow E_3 \\ \vdots \\ E_n - m_{n1}E_1 \longrightarrow E_n \end{matrix} \quad \begin{bmatrix} a_{11} & a_{12} & \cdots & a_{1n} & \vdots & b_1 \\ 0 & a_{22} & \cdots & a_{2n} & \vdots & b_2 \\ \vdots & \vdots & & \vdots & \vdots & \vdots \\ 0 & a_{n2} & \cdots & a_{nn} & \vdots & b_n \end{bmatrix}.$$

Although the entries in rows $2, 3, \ldots, n$ are expected to change, for ease of notation, we again denote the entry in the ith row and the jth column by a_{ij}.

If the pivot element $a_{22} \neq 0$, we form the multipliers $m_{k2} = a_{k2}/a_{22}$ and perform the operations $(E_k - m_{k2}E_2) \longrightarrow E_k$ for each $k = 3, \ldots, n$ obtaining

$$\begin{bmatrix} a_{11} & a_{12} & \cdots & a_{1n} & \vdots & b_1 \\ 0 & a_{22} & \cdots & a_{2n} & \vdots & b_2 \\ \vdots & \vdots & & \vdots & \vdots & \vdots \\ 0 & a_{n2} & \cdots & a_{nn} & \vdots & b_n \end{bmatrix} \quad \begin{matrix} E_3 - m_{32}E_2 \rightarrow E_3 \\ \vdots \\ E_n - m_{n2}E_2 \rightarrow E_n \end{matrix} \quad \begin{bmatrix} a_{11} & a_{12} & \cdots & a_{1n} & \vdots & b_1 \\ 0 & a_{22} & \cdots & a_{2n} & \vdots & b_2 \\ \vdots & \vdots & & \vdots & \vdots & \vdots \\ 0 & 0 & \cdots & a_{nn} & \vdots & b_n \end{bmatrix}$$

We then follow this sequential procedure for the rows $i = 3 \ldots, n-1$. Define the multiplier $m_{ki} = a_{ki}/a_{ii}$ and perform the operation

$$(E_k - m_{ki}E_i) \rightarrow (E_k)$$

for each $k = i+1, i+2, \ldots, n$, provided the pivot element a_{ii} is nonzero. This eliminates x_i in each row below the ith for all values of $i = 1, 2, \ldots, n-1$. The resulting matrix has the form

$$\tilde{\tilde{A}} = \begin{bmatrix} a_{11} & a_{12} & \cdots & a_{1n} & \vdots & a_{1,n+1} \\ 0 & a_{22} & \cdots & a_{2n} & \vdots & a_{2,n+1} \\ \vdots & \ddots & \ddots & \vdots & \vdots & \vdots \\ 0 & \cdots & 0 & a_{nn} & \vdots & a_{n,n+1} \end{bmatrix},$$

where, except in the first row, the values of a_{ij} are not expected to agree with those in the original matrix $\tilde{A}$. The matrix $\tilde{\tilde{A}}$ represents a linear system with the same solution set as the original system. Since the new linear system is triangular,

$$\begin{aligned} a_{11}x_1 + a_{12}x_2 + \cdots + a_{1n}x_n &= a_{1,n+1}, \\ a_{22}x_2 + \cdots + a_{2n}x_n &= a_{2,n+1}, \\ \ddots \qquad \vdots \quad &\quad \vdots \\ a_{nn}x_n &= a_{n,n+1}, \end{aligned}$$

backward substitution can be performed. Solving the nth equation for x_n gives

$$x_n = \frac{a_{n,n+1}}{a_{nn}}.$$

Then solving the $(n-1)$st equation for x_{n-1} and using x_n yields

$$x_{n-1} = \frac{a_{n-1,n+1} - a_{n-1,n}x_n}{a_{n-1,n-1}}.$$

Continuing this process, we obtain

$$x_i = \frac{a_{i,n+1} - a_{i,n}x_n - a_{i,n-1}x_{n-1} - \cdots - a_{i,i+1}x_{i+1}}{a_{ii}} = \frac{a_{i,n+1} - \sum_{j=i+1}^{n} a_{ij}x_j}{a_{ii}}$$

for each $i = n-1, n-2, \ldots, 2, 1$.

The procedure will not work if at the ith step the pivot element a_{ii} is zero, for then either the multipliers $m_{ki} = a_{ki}/a_{ii}$ are not defined (this occurs if $a_{ii} = 0$ for some $i < n$) or the backward substitution cannot be performed (if $a_{nn} = 0$). This does not necessarily mean that the system has no solution, but rather that the technique for finding the solution must be altered. An illustration is given in the following example.

EXAMPLE 3 Consider the linear system

$$\begin{aligned}
E_1:&\quad x_1 - x_2 + 2x_3 - x_4 = -8,\\
E_2:&\quad 2x_1 - 2x_2 + 3x_3 - 3x_4 = -20,\\
E_3:&\quad x_1 + x_2 + x_3 = -2,\\
E_4:&\quad x_1 - x_2 + 4x_3 + 3x_4 = 4.
\end{aligned}$$

The augmented matrix is

$$\left[\begin{array}{rrrr:r} 1 & -1 & 2 & -1 & -8 \\ 2 & -2 & 3 & -3 & -20 \\ 1 & 1 & 1 & 0 & -2 \\ 1 & -1 & 4 & 3 & 4 \end{array}\right].$$

Performing the operations

$$(E_2 - 2E_1) \rightarrow (E_2),\ (E_3 - E_1) \rightarrow (E_3),\ \text{and}\ (E_4 - E_1) \rightarrow (E_4),$$

we have

$$\left[\begin{array}{rrrr:r} 1 & -1 & 2 & -1 & -8 \\ 0 & 0 & -1 & -1 & -4 \\ 0 & 2 & -1 & 1 & 6 \\ 0 & 0 & 2 & 4 & 12 \end{array}\right].$$

Since the element a_{22} in this matrix is zero, the procedure cannot continue in its present form. But the operation $(E_i) \leftrightarrow (E_p)$ is permitted, so a search is made of the elements a_{32} and a_{42} for the first nonzero element. Since $a_{32} \neq 0$, the operation $(E_2) \leftrightarrow (E_3)$ is performed to obtain a new matrix:

$$\left[\begin{array}{rrrr:r} 1 & -1 & 2 & -1 & -8 \\ 0 & 2 & -1 & 1 & 6 \\ 0 & 0 & -1 & -1 & -4 \\ 0 & 0 & 2 & 4 & 12 \end{array}\right].$$

The variable x_2 is already eliminated from E_3 and E_4, so the computations continue with the operation $(E_4 + 2E_3) \longrightarrow (E_4)$, giving

$$\left[\begin{array}{rrrr:r} 1 & -1 & 2 & -1 & -8 \\ 0 & 2 & -1 & 1 & 6 \\ 0 & 0 & -1 & -1 & -4 \\ 0 & 0 & 0 & 2 & 4 \end{array}\right].$$

Finally, the backward substitution is applied:

$$x_4 = \frac{4}{2} = 2, \quad x_3 = \frac{[-4 - (-1)x_4]}{-1} = 2,$$

$$x_2 = \frac{[6 - x_4 - (-1)x_3]}{2} = 3, \quad x_1 = \frac{[-8 - (-1)x_4 - 2x_3 - (-1)x_2]}{1} = -7.$$

◆ ◆ ◆

To define matrices and perform Gaussian elimination using Maple, you must first access the linear algebra library using the command

```
>with(linalg);
```

To define the matrix $\tilde{A}^{(1)}$ of Example 3, which we will call AA, use the command

```
>AA:=matrix(4,5,[1,-1,2,-1,-8,2,-2,3,-3,-20,1,1,1,0,-2,1,-1,4,3,4]);
```

The first two parameters, 4 and 5, give the number of rows and columns, respectively, and the last parameter is a list of the entries of $\tilde{A}^{(1)} \equiv AA$. The function `addrow(AA,i,j,m)` performs the operation $(E_j + mE_i) \longrightarrow (E_j)$ and the function `swaprow(AA,i,j)` performs the operation $(E_i) \leftrightarrow (E_j)$. So, the sequence of operations

```
>AA:=addrow(AA,1,2,-2);
>AA:=addrow(AA,1,3,-1);
>AA:=addrow(AA,1,4,-1);
>AA:=swaprow(AA,2,3);
>AA:=addrow(AA,3,4,2);
```

gives the final reduction, which is again called AA. Alternatively, the single command `AA:=gausselim(AA);` returns the reduced matrix. The final operation,

```
>x:=backsub(AA);
```

produces the solution $x := [-7, 3, 2, 2]$.

Example 3 illustrates what is done if one of the pivot elements is zero. If the ith pivot element is zero, the ith column of the matrix is searched from the ith row downward for the first nonzero entry, and a row interchange is performed to obtain the new matrix. Then the procedure continues as before. If no nonzero entry is found the procedure stops, and the linear system does not have a unique solution; it might have no solution or an infinite number of solutions. The program GAUSEL61 implements Gaussian Elimination with Backward Substitution and incorporates pivoting when required.

The computations in the program are performed using only one $n \times (n+1)$ array for storage by replacing at each step the previous value of a_{ij} by the new one. In addition, the multipliers are stored in the locations of a_{ki} known to have zero values—that is, when $i < n$ and $k = i+1, i+2, \ldots, n$. Thus, the original matrix A is overwritten by the multipliers below the main diagonal and by the nonzero entries of the final reduced matrix on and above the main diagonal. These values can be used to solve other linear systems involving the original matrix A, as we will see in Section 6.5.

Both the amount of time required to complete the calculations and the subsequent round-off error depend on the number of floating-point arithmetic operations needed to solve a routine problem. In general, the amount of time required to perform a multiplication or division on a computer is approximately the same and is considerably greater than that required to perform an addition or subtraction. Even though the actual differences in execution time depend on the particular computing system being used, the count of the additions/subtractions are kept separate from the count of the multiplications/divisions because of the time differential. The total number of arithmetic operations depends on the size n, as follows:

$$\text{Multiplications/divisions:} \quad \frac{n^3}{3} + n^2 - \frac{n}{3}.$$

$$\text{Additions/subtractions:} \quad \frac{n^3}{3} + \frac{n^2}{2} - \frac{5n}{6}.$$

For large n, the total number of multiplications and divisions is approximately $n^3/3$, that is, $O(n^3)$, as is the total number of additions and subtractions. Thus, the amount of computation and the time required increases with n in proportion to n^3, as shown in Table 6.1.

Table 6.1

n	Multiplications/Divisions	Additions/Subtractions
3	17	11
10	430	375
50	44,150	42,875
100	343,300	338,250

EXERCISE SET 6.2

1. Obtain a solution by graphical methods of the following linear systems, if possible.

   a. $x_1 + 2x_2 = 3,$
   $x_1 - x_2 = 0.$

   b. $x_1 + 2x_2 = 0,$
   $x_1 - x_2 = 0.$

c.
$$\begin{aligned} x_1 + 2x_2 &= 3, \\ 2x_1 + 4x_2 &= 6. \end{aligned}$$

d.
$$\begin{aligned} x_1 + 2x_2 &= 3, \\ -2x_1 - 4x_2 &= 6. \end{aligned}$$

e.
$$\begin{aligned} x_1 + 2x_2 &= 0, \\ 2x_1 + 4x_2 &= 0. \end{aligned}$$

f.
$$\begin{aligned} 2x_1 + x_2 &= -1, \\ x_1 + x_2 &= 2, \\ x_1 - 3x_2 &= 5. \end{aligned}$$

g.
$$\begin{aligned} 2x_1 + x_2 &= -1, \\ 4x_1 + 2x_2 &= -2, \\ x_1 - 3x_2 &= 5. \end{aligned}$$

h.
$$\begin{aligned} 2x_1 + x_2 + x_3 &= 1, \\ 2x_1 + 4x_2 - x_3 &= -1. \end{aligned}$$

2. Use Gaussian elimination and two-digit rounding arithmetic to solve the following linear systems. Do not reorder the equations. (The exact solution to each system is $x_1 = 1, x_2 = -1, x_3 = 3$.)

a.
$$\begin{aligned} 4x_1 - x_2 + x_3 &= 8, \\ 2x_1 + 5x_2 + 2x_3 &= 3, \\ x_1 + 2x_2 + 4x_3 &= 11. \end{aligned}$$

b.
$$\begin{aligned} 4x_1 + x_2 + 2x_3 &= 9, \\ 2x_1 + 4x_2 - x_3 &= -5, \\ x_1 + x_2 - 3x_3 &= -9. \end{aligned}$$

3. Use Gaussian elimination to solve the following linear systems, if possible, and determine whether row interchanges are necessary:

a.
$$\begin{aligned} x_1 - x_2 + 3x_3 &= 2, \\ 3x_1 - 3x_2 + x_3 &= -1, \\ x_1 + x_2 &= 3. \end{aligned}$$

b.
$$\begin{aligned} 2x_1 - 1.5x_2 + 3x_3 &= 1, \\ -x_1 + 2x_3 &= 3, \\ 4x_1 - 4.5x_2 + 5x_3 &= 1. \end{aligned}$$

c.
$$\begin{aligned} 2x_1 &= 3, \\ x_1 + 1.5x_2 &= 4.5, \\ -3x_2 + 0.5x_3 &= -6.6, \\ 2x_1 - 2x_2 + x_3 + x_4 &= 0.8. \end{aligned}$$

d.
$$\begin{aligned} x_1 - \tfrac{1}{2}x_2 + x_3 &= 4, \\ 2x_1 - x_2 - x_3 + x_4 &= 5, \\ x_1 + x_2 &= 2, \\ x_1 - \tfrac{1}{2}x_2 + x_3 + x_4 &= 5. \end{aligned}$$

e.
$$\begin{aligned} x_1 + x_2 + x_4 &= 2, \\ 2x_1 + x_2 - x_3 + x_4 &= 1, \\ 4x_1 - x_2 - 2x_3 + 2x_4 &= 0, \\ 3x_1 - x_2 - x_3 + 2x_4 &= -3. \end{aligned}$$

f.
$$\begin{aligned} x_1 + x_2 + x_4 &= 2, \\ 2x_1 + x_2 - x_3 + x_4 &= 1, \\ -x_1 + 2x_2 + 3x_3 - x_4 &= 4, \\ 3x_1 - x_2 - x_3 + 2x_4 &= -3. \end{aligned}$$

4. Use Maple with `Digits` set to 7 and Gaussian elimination to solve the following linear systems.

a.
$$\begin{aligned} \tfrac{1}{4}x_1 + \tfrac{1}{5}x_2 + \tfrac{1}{6}x_3 &= 9, \\ \tfrac{1}{3}x_1 + \tfrac{1}{4}x_2 + \tfrac{1}{5}x_3 &= 8, \\ \tfrac{1}{2}x_1 + x_2 + 2x_3 &= 8. \end{aligned}$$

b.
$$\begin{aligned} 3.333x_1 + 15920x_2 - 10.333x_3 &= 15913, \\ 2.222x_1 + 16.71x_2 + 9.612x_3 &= 28.544, \\ 1.5611x_1 + 5.1791x_2 + 1.6852x_3 &= 8.4254. \end{aligned}$$

c.
$$\begin{aligned}
x_1 + \tfrac{1}{2}x_2 + \tfrac{1}{3}x_3 + \tfrac{1}{4}x_4 &= \tfrac{1}{6},\\
\tfrac{1}{2}x_1 + \tfrac{1}{3}x_2 + \tfrac{1}{4}x_3 + \tfrac{1}{5}x_4 &= \tfrac{1}{7},\\
\tfrac{1}{3}x_1 + \tfrac{1}{4}x_2 + \tfrac{1}{5}x_3 + \tfrac{1}{6}x_4 &= \tfrac{1}{8},\\
\tfrac{1}{4}x_1 + \tfrac{1}{5}x_2 + \tfrac{1}{6}x_3 + \tfrac{1}{7}x_4 &= \tfrac{1}{9}.
\end{aligned}$$

d.
$$\begin{aligned}
2x_1 + x_2 - x_3 + x_4 - 3x_5 &= 7,\\
x_1 + 2x_3 - x_4 + x_5 &= 2,\\
-2x_2 - x_3 + x_4 - x_5 &= -5,\\
3x_1 + x_2 - 4x_3 + 5x_5 &= 6,\\
x_1 - x_2 - x_3 - x_4 + x_5 &= 3.
\end{aligned}$$

5. Given the linear system

$$\begin{aligned}
2x_1 - 6\alpha x_2 &= 3,\\
3\alpha x_1 - x_2 &= \tfrac{3}{2}.
\end{aligned}$$

a. Find value(s) of α for which the system has no solutions.
b. Find value(s) of α for which the system has an infinite number of solutions.
c. Assuming a unique solution exists for a given α, find the solution.

6. Given the linear system

$$\begin{aligned}
x_1 - x_2 + \alpha x_3 &= -2,\\
-x_1 + 2x_2 - \alpha x_3 &= 3,\\
\alpha x_1 + x_2 + x_3 &= 2.
\end{aligned}$$

a. Find value(s) of α for which the system has no solutions.
b. Find value(s) of α for which the system has an infinite number of solutions.
c. Assuming a unique solution exists for a given α, find the solution.

7. Suppose that in a biological system there are n species of animals and m sources of food. Let x_j represent the population of the jth species for each $j = 1, \ldots, n$; b_i represent the available daily supply of the ith food; and a_{ij} represent the amount of the ith food consumed on the average by a member of the jth species. The linear system

$$\begin{aligned}
a_{11}x_1 + a_{12}x_2 + \cdots + a_{1n}x_n &= b_1,\\
a_{21}x_1 + a_{22}x_2 + \cdots + a_{2n}x_n &= b_2,\\
\vdots \qquad \vdots \qquad\quad\;\; \vdots \quad &\quad \vdots\\
a_{m1}x_1 + a_{m2}x_2 + \cdots + a_{mn}x_n &= b_m
\end{aligned}$$

represents an equilibrium where there is a daily supply of food to precisely meet the average daily consumption of each species.

a. Let

$$A = (a_{ij}) = \begin{bmatrix} 1 & 2 & 0 & 3 \\ 1 & 0 & 2 & 2 \\ 0 & 0 & 1 & 1 \end{bmatrix},$$

$\mathbf{x} = (x_j) = [1000, 500, 350, 400]$, and $\mathbf{b} = (b_i) = [3500, 2700, 900]$. Is there sufficient food to satisfy the average daily consumption?

b. What is the maximum number of animals of each species that could be individually added to the system with the supply of food still meeting the consumption?

c. If species 1 became extinct, how much of an individual increase of each of the remaining species could be supported?

d. If species 2 became extinct, how much of an individual increase of each of the remaining species could be supported?

8. A Fredholm integral equation of the second kind is an equation of the form

$$u(x) = f(x) + \int_a^b K(x,t)u(t)\,dt,$$

where a and b and the functions f and K are given. To approximate the function u on the interval $[a, b]$, a partition $x_0 = a < x_1 < \cdots < x_{m-1} < x_m = b$ is selected and the equations

$$u(x_i) = f(x_i) + \int_a^b K(x_i,t)u(t)\,dt, \quad \text{for each } i = 0, \ldots, m,$$

are solved for $u(x_0), u(x_1), \ldots, u(x_m)$. The integrals are approximated using quadrature formulas based on the nodes $x_0, \ldots, x_m$. In our problem, $a = 0$, $b = 1$, $f(x) = x^2$, and $K(x,t) = e^{|x-t|}$.

a. Show that the linear system

$$u(0) = f(0) + \tfrac{1}{2}[K(0,0)u(0) + K(0,1)u(1)],$$
$$u(1) = f(1) + \tfrac{1}{2}[K(1,0)u(0) + K(1,1)u(1)]$$

must be solved when the Trapezoidal rule is used.

b. Set up and solve the linear system that results when the Composite Trapezoidal rule is used with $n = 4$.

c. Repeat part (b) using the Composite Simpson's rule.

6.3 Pivoting Strategies

If all the calculations could be done using exact arithmetic, we could nearly end the chapter with the previous section. We now know how many calculations are needed to perform Gaussian elimination on a system, and from this we should be able to determine whether our computational device can solve our problem in reasonable time. In a practical situation, however, we do not have exact arithmetic, and the large number of arithmetic computations, on the order of $O(n^3)$, makes the consideration of computational round-off error necessary. In this section we will see how the calculations in Gaussian elimination can be arranged to reduce the effect of this error.

In deriving the Gaussian elimination method, we found that a row interchange is needed when one of the pivot elements, a_{ii}, is zero. This row interchange has the form $(E_i) \leftrightarrow (E_p)$, where p is the smallest integer greater than i with $a_{pi} \neq 0$. To reduce the round-off error associated with finite-digit arithmetic, it is often necessary to perform row interchanges even when the pivot elements are not zero.

If a_{ii} is small in magnitude compared to a_{ki}, the magnitude of the multiplier

$$m_{ki} = \frac{a_{ki}}{a_{ii}}$$

will be much larger than 1. A round-off error introduced in the computation of one of the terms a_{il} is multiplied by m_{ki} when computing a_{kl}, compounding the original error. Also, when performing the backward substitution for

$$x_i = \frac{a_{i,n+1} - \sum_{j=i+1}^{n} a_{ij}}{a_{ii}}$$

with a small value of a_{ii}, any round-off error in the numerator is dramatically increased when dividing by a_{ii}. An illustration of this difficulty is given in the following example.

EXAMPLE 1 The linear system

$$\begin{aligned} E_1&: \quad 0.003000x_1 + 59.14x_2 = 59.17, \\ E_2&: \quad 5.291x_1 - 6.130x_2 = 46.78 \end{aligned}$$

has the solution $x_1 = 10.00$ and $x_2 = 1.000$. Suppose Gaussian elimination is performed on this system using four-digit arithmetic with rounding.

The first pivot element, $a_{11} = 0.003000$, is small, and its associated multiplier,

$$m_{21} = \frac{5.291}{0.003000} = 1763.\bar{6},$$

rounds to the large number 1764. Performing $(E_2 - m_{21}E_1) \rightarrow (E_2)$ and the appropriate rounding gives

$$\begin{aligned} 0.003000x_1 + 59.14x_2 &= 59.17 \\ -104300x_2 &\approx -104400 \end{aligned}$$

instead of the precise values

$$\begin{aligned} 0.003000x_1 + 59.14x_2 &= 59.17 \\ -104309.37\bar{6}x_2 &= -104309.37\bar{6}. \end{aligned}$$

The disparity in the magnitudes of $m_{21}a_{13}$ and a_{23} has introduced round-off error, but the error has not yet been propagated. Backward substitution yields

$$x_2 \approx 1.001,$$

which is a close approximation to the actual value, $x_2 = 1.000$. However, because of the small pivot $a_{11} = 0.003000$,

$$x_1 \approx \frac{59.17 - (59.14)(1.001)}{0.003000} = -10.00$$

contains the small error of 0.001 multiplied by $59.14/0.003000 \approx 20000$. This ruins the approximation to the actual value $x_1 = 10.00$. (See Figure 6.1.) ◆ ◆ ◆

Figure 6.1

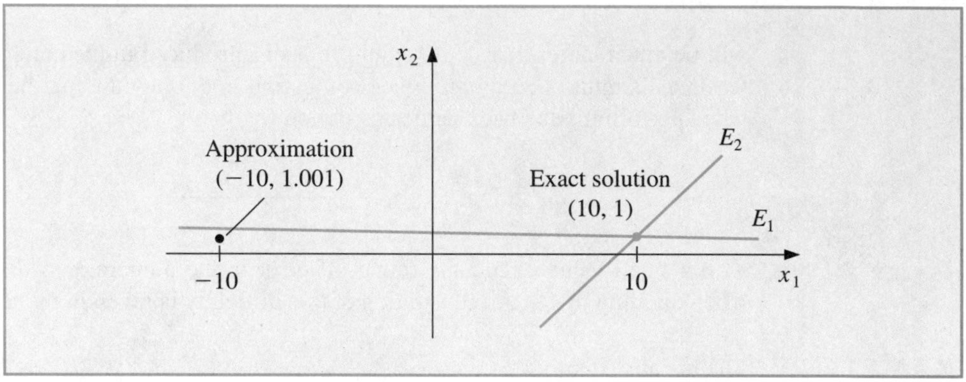

Example 1 shows how difficulties arise when the pivot element a_{ii} is small relative to the entries a_{kj} for $i \le k \le n$ and $i \le j \le n$. To avoid this problem, pivoting is performed by selecting a larger element a_{pq} for the pivot and interchanging the ith and pth rows, followed by the interchange of the ith and qth columns, if necessary.

The simplest strategy is to select, at the ith step, the element in the same column that is below the diagonal and has the largest absolute value; that is, to determine the smallest $p \ge i$ such that

$$|a_{pi}| = \max_{i \le k \le n} |a_{ki}|$$

and perform $(E_i) \leftrightarrow (E_p)$. In this case no interchange of columns is used.

EXAMPLE 2 Reconsider the system

$$\begin{aligned} E_1:&\quad 0.003000x_1 + 59.14x_2 = 59.17,\\ E_2:&\quad 5.291x_1 - 6.130x_2 = 46.78. \end{aligned}$$

The pivoting procedure just described results in first finding

$$\max\{|a_{11}|, |a_{21}|\} = \max\{|0.003000|, |5.291|\} = |5.291| = |a_{21}|.$$

The operation $(E_2) \leftrightarrow (E_1)$ is then performed to give the system

$$\begin{aligned} E_1&: \quad 5.291x_1 - 6.130x_2 = 46.78, \\ E_2&: \quad 0.003000x_1 + 59.14x_2 = 59.17. \end{aligned}$$

The multiplier for this system is

$$m_{21} = \frac{a_{21}}{a_{11}} = 0.0005670,$$

and the operation $(E_2 - m_{21}E_1) \rightarrow (E_2)$ reduces the system to

$$\begin{aligned} 5.291x_1 - 6.130x_2 &= 46.78, \\ 59.14x_2 &\approx 59.14. \end{aligned}$$

The four-digit answers resulting from the backward substitution are the correct values, $x_1 = 10.00$ and $x_2 = 1.000$. ◆ ◆ ◆

The technique just described is called **partial pivoting**, or *maximal column pivoting*, and is implemented in the program GAUMPP62.

Although partial pivoting is sufficient for many linear systems, situations do arise when it is inadequate. For example, the linear system

$$\begin{aligned} E_1&: \quad 30.00x_1 + 591400x_2 = 591700, \\ E_2&: \quad 5.291x_1 - \quad 6.130x_2 = 46.78 \end{aligned}$$

is the same as that in Examples 1 and 2 except that all entries in the first equation have been multiplied by 10^4. Partial pivoting with four-digit arithmetic leads to the same results as obtained in Example 1 since no row interchange would be performed. A technique known as **scaled partial pivoting** is needed for this system. The first step in this procedure is to define a scale factor s_k for each row:

$$s_k = \max_{1 \le j \le n} |a_{kj}|.$$

The appropriate row interchange to place zeros in the first column is determined by choosing the first integer p with

$$\frac{|a_{p1}|}{s_p} = \max_{1 \le k \le n} \frac{|a_{k1}|}{s_k}$$

and performing $(E_1) \leftrightarrow (E_p)$. The effect of scaling is to ensure that the largest element in each row has a *relative* magnitude of 1 before the comparison for row interchange is performed.

In a similar manner, before eliminating variable x_i using the operations

$$E_k - m_{ki}E_i \longrightarrow E_k \quad \text{for } k = i + 1, \ldots, n$$

we select the smallest integer $p \geq i$ with

$$\frac{|a_{pi}|}{s_p} = \max_{i \leq k \leq n} \frac{|a_{ki}|}{s_k}$$

and perform the row interchange $E_i \leftrightarrow E_p$ if $i \neq p$. We must note that the scale factors $s_1, \ldots, s_n$ are computed only once at the start of the procedure and must also be interchanged when row interchanges are performed.

In the program GAUSPP63 the scaling is done only for comparison purposes, so the division by scaling factors produces no round-off error in the system.

EXAMPLE 3 Applying scaled partial pivoting to the system in Example 1 gives

$$s_1 = \max\{|30.00|, |591400|\} = 591400 \quad \text{and} \quad s_2 = \max\{|5.291|, |-6.130|\} = 6.130.$$

Consequently,

$$\frac{|a_{11}|}{s_1} = \frac{30.00}{591400} = 0.5073 \times 10^{-4} \quad \text{and} \quad \frac{|a_{21}|}{s_2} = \frac{5.291}{6.130} = 0.8631$$

and the interchange $(E_1) \leftrightarrow (E_2)$ is made. Applying Gaussian elimination to the new system produces the correct results: $x_1 = 10.00$ and $x_2 = 1.000$. ◆ ◆ ◆

EXAMPLE 4 Use scaled partial pivoting to solve the linear system using three-digit rounding arithmetic.

$$\begin{aligned}
2.11x_1 - 4.21x_2 + 0.921x_3 &= 2.01,\\
4.01x_1 + 10.2x_2 - 1.12x_3 &= -3.09,\\
1.09x_1 + 0.987x_2 + 0.832x_3 &= 4.21.
\end{aligned}$$

To obtain three-digit rounding arithmetic, enter

```
>Digits:=3;
```

We have $s_1 = 4.21$, $s_2 = 10.2$, and $s_3 = 1.09$. So

$$\frac{|a_{11}|}{s_1} = \frac{2.11}{4.21} = 0.501, \quad \frac{|a_{21}|}{s_1} = \frac{4.01}{10.2} = 0.393, \quad \text{and} \quad \frac{|a_{31}|}{s_3} = \frac{1.09}{1.09} = 1.$$

The augmented matrix AA is defined by

```
>AA:=matrix(3,4,[2.11,-4.21,0.921,2.01,4.01,10.2,-1.12,-3.09,1.09,
0.987,0.832,4.21]);
```

which gives

$$AA := \begin{bmatrix} 2.11 & -4.21 & .921 & 2.01 \\ 4.01 & 10.2 & -1.12 & -3.09 \\ 1.09 & .987 & .832 & 4.21 \end{bmatrix}.$$

Since $|a_{31}|/s_3$ is largest, we perform $(E_1) \leftrightarrow (E_3)$ using

```
>AA:=swaprow(AA,1,3);
```

to obtain

$$AA := \begin{bmatrix} 1.09 & .987 & .832 & 4.21 \\ 4.01 & 10.2 & -1.12 & -3.09 \\ 2.11 & -4.21 & .921 & 2.01 \end{bmatrix}.$$

We compute the multipliers $m_{21} = 3.68$ and $m_{31} = 1.94$ using

```
>m21:=4.01/1.09;
```

and

```
>m31:=2.11/1.09;
```

We perform the first two eliminations using

```
>AA:=addrow(AA,1,2,-m21);
```

and

```
>AA:=addrow(AA,1,3,-m31);
```

to obtain

$$AA := \begin{bmatrix} 1.09 & .987 & .832 & 4.21 \\ 0 & 6.57 & -4.18 & -18.6 \\ 0 & -6.12 & -.689 & -6.16 \end{bmatrix}.$$

Since

$$\frac{|a_{22}|}{s_2} = \frac{6.57}{10.2} = 0.644 < \frac{|a_{32}|}{s_3} = \frac{6.12}{4.21} = 1.45,$$

we perform

```
>AA:=swaprow(AA,2,3);
```

giving

$$AA := \begin{bmatrix} 1.09 & .987 & .832 & 4.21 \\ 0 & -6.12 & -.689 & -6.16 \\ 0 & 6.57 & -4.18 & -18.6 \end{bmatrix}.$$

The multiplier $m_{32} = -1.07$ is computed by

```
>m32:=6.57/(-6.12);
```

The elimination step

```
>AA:=addrow(AA,2,3,-m32);
```

gives

$$AA := \begin{bmatrix} 1.09 & .987 & .832 & 4.21 \\ 0 & -6.12 & -.689 & -6.16 \\ 0 & .02 & -4.92 & -25.2 \end{bmatrix}.$$

We cannot use `backsub` because of the entry .02 in the (3, 2) position. This entry is nonzero due to rounding, but we can remedy this minor problem using the command

```
>AA[3,2]:=0;
```

which replaces the entry .02 with a 0. To see this enter

```
>evalm(AA);
```

which displays the matrix AA. Finally,

```
>x:=backsub(AA);
```

gives the solution

$$x := [-.431 \quad .430 \quad 5.12].$$

◆ ◆ ◆

The scaled partial pivoting procedure adds a total of

$$\frac{3}{2}n(n-1) \quad \text{comparisons}$$

and

$$\frac{n(n+1)}{2} - 1 \quad \text{divisions}$$

to the Gaussian elimination procedure. The time required to perform a comparison is slightly more than that of an addition/subtraction. Since the total time to perform the basic Gaussian elimination procedure is $O(n^3/3)$ multiplications/divisions and $O(n^3/3)$ additions/subtractions, scaled partial pivoting does not add significantly to the computational time required to solve a system for large values of n.

If a system has computation difficulties that scaled partial pivoting cannot resolve, *maximal* (also called *total* or *full*) *pivoting* can be used. Maximal pivoting at the ith step searches all the entries a_{kj}, for $k = i, i+1, \ldots, n$ and $j = i, i+1, \ldots, n$, to find the entry with the largest magnitude. Both row and column interchanges are performed to bring this entry to the pivot position. The additional time required to incorporate maximal pivoting into Gaussian elimination is

$$\frac{n(n-1)(2n+5)}{6} \quad \text{comparisons.}$$

This approximately doubles the amount of addition/subtraction time over ordinary Gaussian elimination.

EXERCISE SET 6.3

1. Use standard Gaussian elimination to find the row interchanges that are required to solve the following linear systems.

   a. $$\begin{aligned} x_1 - 5x_2 + x_3 &= 7 \\ 10x_1 + 20x_3 &= 6 \\ 5x_1 - x_3 &= 4 \end{aligned}$$

   b. $$\begin{aligned} x_1 + x_2 - x_3 &= 1 \\ x_1 + x_2 + 4x_3 &= 2 \\ 2x_1 - x_2 + 2x_3 &= 3 \end{aligned}$$

   c. $$\begin{aligned} 2x_1 - 3x_2 + 2x_3 &= 5 \\ -4x_1 + 2x_2 - 6x_3 &= 14 \\ 2x_1 + 2x_2 + 4x_3 &= 8 \end{aligned}$$

   d. $$\begin{aligned} x_2 + x_3 &= 6 \\ x_1 - 2x_2 - x_3 &= 4 \\ x_1 - x_2 + x_3 &= 5 \end{aligned}$$

2. Repeat Exercise 1 using Gaussian elimination with partial pivoting.
3. Repeat Exercise 1 using Gaussian elimination with scaled partial pivoting.
4. Repeat Exercise 1 using Gaussian elimination with complete pivoting.
5. Use Gaussian elimination and three-digit chopping arithmetic to solve the following linear systems, and compare the approximations to the actual solution.

   a. $$\begin{aligned} 0.03x_1 + 58.9x_2 &= 59.2 \\ 5.31x_1 - 6.10x_2 &= 47.0 \end{aligned}$$
   Actual solution $x_1 = 10, x_2 = 1$.

   b. $$\begin{aligned} 58.9x_1 + 0.03x_2 &= 59.2 \\ -6.10x_1 + 5.31x_2 &= 47.0 \end{aligned}$$
   Actual solution $x_1 = 1, x_2 = 10$.

   c. $$\begin{aligned} 3.03x_1 - 12.1x_2 + 14x_3 &= -119 \\ -3.03x_1 + 12.1x_2 - 7x_3 &= 120 \\ 6.11x_1 - 14.2x_2 + 21x_3 &= -139 \end{aligned}$$
   Actual solution $x_1 = 0$, $x_2 = 10$, $x_3 = \frac{1}{7}$.

   d. $$\begin{aligned} 3.3330x_1 + 15920x_2 + 10.333x_3 &= 7953 \\ 2.2220x_1 + 16.710x_2 + 9.6120x_3 &= 0.965 \\ -1.5611x_1 + 5.1792x_2 - 1.6855x_3 &= 2.714 \end{aligned}$$
   Actual solution $x_1 = 1, x_2 = 0.5, x_3 = -1$.

   e. $$\begin{aligned} 1.19x_1 + 2.11x_2 - 100x_3 + x_4 &= 1.12 \\ 14.2x_1 - 0.122x_2 + 12.2x_3 - x_4 &= 3.44 \\ 100x_2 - 99.9x_3 + x_4 &= 2.15 \\ 15.3x_1 + 0.110x_2 - 13.1x_3 - x_4 &= 4.16 \end{aligned}$$

Actual solution $x_1 = 0.17682530$, $x_2 = 0.01269269$, $x_3 = -0.02065405$, $x_4 = -1.18260870$.

f.
$$\begin{aligned}
\pi x_1 - e x_2 + \sqrt{2} x_3 - \sqrt{3} x_4 &= \sqrt{11}\\
\pi^2 x_1 + e x_2 - e^2 x_3 + \tfrac{3}{7} x_4 &= 0\\
\sqrt{5} x_1 - \sqrt{6} x_2 + x_3 - \sqrt{2} x_4 &= \pi\\
\pi^3 x_1 + e^2 x_2 - \sqrt{7} x_3 + \tfrac{1}{9} x_4 &= \sqrt{2}
\end{aligned}$$

Actual solution $x_1 = 0.78839378$, $x_2 = -3.12541367$, $x_3 = 0.16759660$, $x_4 = 4.55700252$.

6. Repeat Exercise 5 using three-digit rounding arithmetic.
7. Repeat Exercise 5 using Gaussian elimination with partial pivoting.
8. Repeat Exercise 5 using Gaussian elimination with scaled partial pivoting.
9. Suppose that

$$\begin{aligned}
2x_1 + x_2 + 3x_3 &= 1\\
4x_1 + 6x_2 + 8x_3 &= 5\\
6x_1 + \alpha x_2 + 10x_3 &= 5
\end{aligned}$$

with $|\alpha| < 10$. For which of the following values of α will there be no row interchange required when solving this system using scaled partial pivoting?

a. $\alpha = 6$ **b.** $\alpha = 9$ **c.** $\alpha = -3$

6.4 Linear Algebra and Matrix Inversion

Early in this chapter we illustrated the convenience of matrix notation for the study of linear systems of equations, but there is a wealth of additional material in linear algebra that finds application in the study of approximation techniques. In this section we introduce some basic notation and results that are needed for both theory and application. All the topics discussed here should be familiar to anyone who has studied matrix theory at the undergraduate level. This section could be omitted, but it is advisable to read the section to see the results from linear algebra that will be frequently called upon for service.

Two matrices A and B are **equal** if both are of the same size, say, $n \times m$, and if $a_{ij} = b_{ij}$ for each $i = 1, 2, \ldots, n$ and $j = 1, 2, \ldots, m$.

This defintion means, for example, that

$$\begin{bmatrix} 2 & -1 & 7 \\ 3 & 1 & 0 \end{bmatrix} \neq \begin{bmatrix} 2 & 3 \\ -1 & 1 \\ 7 & 0 \end{bmatrix}$$

since they differ in dimension.

If A and B are $n \times m$ matrices and λ is a real number, then the **sum** of A and B, denoted $A + B$, is the $n \times m$ matrix whose entries are $a_{ij} + b_{ij}$, and the **scalar product** of λ and A, denoted λA, is the $n \times m$ matrix whose entries are λa_{ij}.

If A is an $n \times m$ matrix and B is an $m \times p$ matrix, the **matrix product** of A and B, denoted AB, is an $n \times p$ matrix C whose entries c_{ij} are given by

$$c_{ij} = \sum_{k=1}^{m} a_{ik}b_{kj} = a_{i1}b_{1j} + a_{i2}b_{2j} + \cdots + a_{im}b_{mj},$$

for each $i = 1, 2, \ldots n$ and $j = 1, 2, \ldots, p$.

The computation of c_{ij} can be viewed as the multiplication of the entries of the ith row of A with corresponding entries in the jth column of B, followed by a summation; that is,

$$[a_{i1}, a_{i2}, \ldots, a_{im}] \begin{bmatrix} b_{1j} \\ b_{2j} \\ \vdots \\ b_{mj} \end{bmatrix} = [c_{ij}],$$

where

$$c_{ij} = a_{i1}b_{1j} + a_{i2}b_{2j} + \cdots + a_{im}b_{mj} = \sum_{k=1}^{m} a_{ik}b_{kj}.$$

This explains why the number of columns of A must equal the number of rows of B for the product AB to be defined.

EXAMPLE 1 Let

$$A = \begin{bmatrix} 2 & 1 & -1 \\ 3 & 1 & 2 \\ 0 & -2 & -3 \end{bmatrix}, \quad B = \begin{bmatrix} 3 & 2 \\ -1 & 1 \\ 6 & 4 \end{bmatrix},$$

$$C = \begin{bmatrix} 2 & 1 & 0 \\ -1 & 3 & 2 \end{bmatrix}, \quad \text{and} \quad D = \begin{bmatrix} 1 & -1 & 1 \\ 2 & -1 & 2 \\ 3 & 0 & 3 \end{bmatrix}.$$

Then,

$$AD = \begin{bmatrix} 1 & -3 & 1 \\ 11 & -4 & 11 \\ -13 & 2 & -13 \end{bmatrix} \neq \begin{bmatrix} -1 & -2 & -6 \\ 1 & -3 & -10 \\ 6 & -3 & -12 \end{bmatrix} = DA.$$

Further,

$$BC = \begin{bmatrix} 4 & 9 & 4 \\ -3 & 2 & 2 \\ 8 & 18 & 8 \end{bmatrix} \quad \text{and} \quad CB = \begin{bmatrix} 5 & 5 \\ 6 & 9 \end{bmatrix}$$

are not even the same size. Finally,

$$AB = \begin{bmatrix} -1 & 1 \\ 20 & 15 \\ -16 & -14 \end{bmatrix}$$

but BA cannot be computed. ◆ ◆ ◆

A **square** matrix has the same number of rows as columns. A **diagonal** matrix is a square matrix $D = (d_{ij})$ with $d_{ij} = 0$ whenever $i \neq j$. The **identity** matrix of order n, $I_n = (\delta_{ij})$, is a diagonal matrix with entries

$$\delta_{ij} = \begin{cases} 1, & \text{if } i = j, \\ 0, & \text{if } i \neq j. \end{cases}$$

When the size of I_n is clear, this matrix is generally written simply as I. For example, the identity matrix of order three is

$$I = \begin{bmatrix} 1 & 0 & 0 \\ 0 & 1 & 0 \\ 0 & 0 & 1 \end{bmatrix}.$$

If A is any $n \times n$ matrix, then $AI = IA = A$.

An $n \times n$ **upper-triangular** matrix $U = (u_{ij})$ has, for each $j = 1, 2, \ldots, n$, the entries

$$u_{ij} = 0, \quad \text{for each } i = j+1, j+2, \ldots, n;$$

and a **lower-triangular** matrix $L = (l_{ij})$ has, for each $j = 1, 2, \ldots, n$, the entries

$$l_{ij} = 0, \quad \text{for each } i = 1, 2, \ldots, j-1.$$

(A diagonal matrix is both upper and lower triangular.)

In Example 1 we found that, in general, $AB \neq BA$, even when both products are defined. However, the other arithmetic properties associated with multiplication do hold. For example, when A, B, and C are matrices of the appropriate size and λ is a scalar, we have

$$A(BC) = (AB)C, \quad A(B+C) = AB + AC, \quad \text{and} \quad \lambda(AB) = (\lambda A)B = A(\lambda B).$$

An $n \times n$ matrix A is **nonsingular**, or *invertible*, if an $n \times n$ matrix A^{-1} exists with $AA^{-1} = A^{-1}A = I$. The matrix A^{-1} is the **inverse** of A. A matrix without an inverse is **singular**, or *noninvertible*.

EXAMPLE 2 Let

$$A = \begin{bmatrix} 1 & 2 & -1 \\ 2 & 1 & 0 \\ -1 & 1 & 2 \end{bmatrix} \quad \text{and} \quad B = \frac{1}{9}\begin{bmatrix} -2 & 5 & -1 \\ 4 & -1 & 2 \\ -3 & 3 & 3 \end{bmatrix}.$$

Since

$$AB = \begin{bmatrix} 1 & 2 & -1 \\ 2 & 1 & 0 \\ -1 & 1 & 2 \end{bmatrix} \cdot \frac{1}{9}\begin{bmatrix} -2 & 5 & -1 \\ 4 & -1 & 2 \\ -3 & 3 & 3 \end{bmatrix} = \begin{bmatrix} 1 & 0 & 0 \\ 0 & 1 & 0 \\ 0 & 0 & 1 \end{bmatrix}$$

and

$$BA = \frac{1}{9}\begin{bmatrix} -2 & 5 & -1 \\ 4 & -1 & 2 \\ -3 & 3 & 3 \end{bmatrix} \cdot \begin{bmatrix} 1 & 2 & -1 \\ 2 & 1 & 0 \\ -1 & 1 & 2 \end{bmatrix} = \begin{bmatrix} 1 & 0 & 0 \\ 0 & 1 & 0 \\ 0 & 0 & 1 \end{bmatrix},$$

A and B are nonsingular with $B = A^{-1}$ and $A = B^{-1}$. ◆ ◆ ◆

The reason for introducing this matrix operation at this time is that the linear system

$$\begin{aligned} a_{11}x_1 + a_{12}x_2 + \cdots + a_{1n}x_n &= b_1, \\ a_{21}x_1 + a_{22}x_2 + \cdots + a_{2n}x_n &= b_2, \\ \vdots \qquad\qquad & \quad \vdots \\ a_{n1}x_1 + a_{n2}x_2 + \cdots + a_{nn}x_n &= b_n, \end{aligned}$$

can be viewed as the matrix equation $A\mathbf{x} = \mathbf{b}$, where

$$A = \begin{bmatrix} a_{11} & a_{12} & \cdots & a_{1n} \\ a_{21} & a_{22} & \cdots & a_{2n} \\ \vdots & \vdots & & \vdots \\ a_{n1} & a_{n2} & \cdots & a_{nn} \end{bmatrix}, \quad \mathbf{x} = \begin{bmatrix} x_1 \\ x_2 \\ \vdots \\ x_n \end{bmatrix}, \quad \text{and} \quad \mathbf{b} = \begin{bmatrix} b_1 \\ b_2 \\ \vdots \\ b_n \end{bmatrix}.$$

If A is a nonsingular matrix, then the solution $\mathbf{x}$ to the linear system $A\mathbf{x} = \mathbf{b}$ is given by $\mathbf{x} = A^{-1}(A\mathbf{x}) = A^{-1}\mathbf{b}$. In general, however, it is more difficult to determine A^{-1} than it is to solve the system $A\mathbf{x} = \mathbf{b}$, because the number of operations involved in determining A^{-1} is larger. Even so, it is useful from a conceptual standpoint to describe a method for determining the inverse of a matrix.

EXAMPLE 3 To determine the inverse of the matrix

$$A = \begin{bmatrix} 1 & 2 & -1 \\ 2 & 1 & 0 \\ -1 & 1 & 2 \end{bmatrix},$$

let us first consider the product AB, where B is an arbitrary 3×3 matrix.

$$\begin{aligned} AB &= \begin{bmatrix} 1 & 2 & -1 \\ 2 & 1 & 0 \\ -1 & 1 & 2 \end{bmatrix} \begin{bmatrix} b_{11} & b_{12} & b_{13} \\ b_{21} & b_{22} & b_{23} \\ b_{31} & b_{32} & b_{33} \end{bmatrix} \\ &= \begin{bmatrix} b_{11} + 2b_{21} - b_{31} & b_{12} + 2b_{22} - b_{32} & b_{13} + 2b_{23} - b_{33} \\ 2b_{11} + b_{21} & 2b_{12} + b_{22} & 2b_{13} + b_{23} \\ -b_{11} + b_{21} + 2b_{31} & -b_{12} + b_{22} + 2b_{32} & -b_{13} + b_{23} + 2b_{33} \end{bmatrix}. \end{aligned}$$

If $B = A^{-1}$, then $AB = I$, so we must have

$$\begin{aligned} b_{11} + 2b_{21} - b_{31} &= 1, & b_{12} + 2b_{22} - b_{32} &= 0, & b_{13} + 2b_{23} - b_{33} &= 0 \\ 2b_{11} + b_{21} &= 0, & 2b_{12} + b_{22} &= 1, & 2b_{13} + b_{23} &= 0 \\ -b_{11} + b_{21} + 2b_{31} &= 0, & -b_{12} + b_{22} + 2b_{32} &= 0, & -b_{13} + b_{23} + 2b_{33} &= 1 \end{aligned}$$

Notice that the coefficients in each of the systems of equations are the same; the only change in the systems occurs on the right side of the equations. As a consequence, the computations can be performed on the larger augmented matrix, formed by combining the matrices for each of the systems

$$\left[\begin{array}{rrr:rrr} 1 & 2 & -1 & 1 & 0 & 0 \\ 2 & 1 & 0 & 0 & 1 & 0 \\ -1 & 1 & 2 & 0 & 0 & 1 \end{array}\right].$$

First, performing $(E_2 - 2E_1) \to (E_1)$ and $(E_3 + E_1) \to (E_3)$ gives

$$\left[\begin{array}{rrr:rrr} 1 & 2 & -1 & 1 & 0 & 0 \\ 0 & -3 & 2 & -2 & 1 & 0 \\ 0 & 3 & 1 & 1 & 0 & 1 \end{array}\right].$$

Next, performing $(E_3 + E_2) \to (E_3)$ produces

$$\left[\begin{array}{rrr:rrr} 1 & 2 & -1 & 1 & 0 & 0 \\ 0 & -3 & 2 & -2 & 1 & 0 \\ 0 & 0 & 3 & -1 & 1 & 1 \end{array}\right].$$

Backward substitution is performed on each of the three augmented matrices,

$$\left[\begin{array}{rrr:r} 1 & 2 & -1 & 1 \\ 0 & -3 & 2 & -2 \\ 0 & 0 & 3 & -1 \end{array}\right], \quad \left[\begin{array}{rrr:r} 1 & 2 & -1 & 0 \\ 0 & -3 & 2 & 1 \\ 0 & 0 & 3 & 1 \end{array}\right], \quad \left[\begin{array}{rrr:r} 1 & 2 & -1 & 0 \\ 0 & -3 & 2 & 0 \\ 0 & 0 & 3 & 1 \end{array}\right],$$

to eventually give

$$\begin{aligned} b_{11} &= -\tfrac{2}{9}, & b_{12} &= \tfrac{5}{9}, & & b_{13} = -\tfrac{1}{9}, \\ b_{21} &= \tfrac{4}{9}, & b_{22} &= -\tfrac{1}{9}, & \text{and} \quad & b_{23} = \tfrac{2}{9}, \\ b_{31} &= -\tfrac{1}{3}, & b_{32} &= \tfrac{1}{3}, & & b_{33} = \tfrac{1}{3}. \end{aligned}$$

These are the entries of A^{-1}:

$$A^{-1} = \begin{bmatrix} -\frac{2}{9} & \frac{5}{9} & -\frac{1}{9} \\ \frac{4}{9} & -\frac{1}{9} & \frac{2}{9} \\ -\frac{1}{3} & \frac{1}{3} & \frac{1}{3} \end{bmatrix} = \frac{1}{9}\begin{bmatrix} -2 & 5 & -1 \\ 4 & -1 & 2 \\ -3 & 3 & 3 \end{bmatrix}.$$

◆ ◆ ◆

The **transpose** of an $n \times m$ matrix $A = (a_{ij})$ is the $m \times n$ matrix $A^t = (a_{ji})$. A square matrix A is **symmetric** if $A = A^t$.

EXAMPLE 4 The matrices

$$A = \begin{bmatrix} 7 & 2 & 0 \\ 3 & 5 & -1 \\ 0 & 5 & -6 \end{bmatrix}, \quad B = \begin{bmatrix} 2 & 4 & 7 \\ 3 & -5 & -1 \end{bmatrix}, \quad C = \begin{bmatrix} 6 & 4 & -3 \\ 4 & -2 & 0 \\ -3 & 0 & 1 \end{bmatrix}$$

have transposes

$$A^t = \begin{bmatrix} 7 & 3 & 0 \\ 2 & 5 & 5 \\ 0 & -1 & -6 \end{bmatrix}, \quad B^t = \begin{bmatrix} 2 & 3 \\ 4 & -5 \\ 7 & -1 \end{bmatrix}, \quad C^t = \begin{bmatrix} 6 & 4 & -3 \\ 4 & -2 & 0 \\ -3 & 0 & 1 \end{bmatrix}.$$

The matrix C is symmetric, since $C^t = C$. The matrices A and B are not symmetric.

◆ ◆ ◆

The following operations involving the transpose of a matrix hold whenever the operation is possible.

Transpose Facts

(a) $(A^t)^t = A$.
(b) $(A + B)^t = A^t + B^t$.
(c) $(AB)^t = B^t A^t$.
(d) If A^{-1} exists, $(A^{-1})^t = (A^t)^{-1}$.

The determinant of a square matrix is a number that can be useful in determining the existence and uniqueness of solutions to linear systems. We denote the determinant of a matrix A by $\det A$, but it is also common to use the notation $|A|$.

Determinant of a Matrix

(a) If $A = [a]$ is a 1×1 matrix, then $\det A = a$.
(b) If A is an $n \times n$ matrix, the **minor** M_{ij} is the determinant of the $(n-1) \times (n-1)$ submatrix of A obtained by deleting the ith row and jth column of the matrix A.

Then the determinant of A is given either by

$$\det A = \sum_{j=1}^{n} (-1)^{i+j} a_{ij} M_{ij} \quad \text{for any } i = 1, 2, \ldots, n,$$

or by

$$\det A = \sum_{i=1}^{n} (-1)^{i+j} a_{ij} M_{ij} \quad \text{for any } j = 1, 2, \ldots, n.$$

It can be shown that, if $n > 1$, to calculate the determinant of a general $n \times n$ matrix requires $O(n!)$ multiplications/divisions and additions/subtractions. Even for relatively small values of n, the number of calculations becomes unwieldy.

Although it appears that there are $2n$ different definitions of det A, depending on which row or column is chosen, all definitions give the same numerical result. The flexibility in the definition is used in the following example. It is most convenient to compute det A across the row or down the column with the most zeros.

EXAMPLE 5 Let

$$A = \begin{bmatrix} 2 & -1 & 3 & 0 \\ 4 & -2 & 7 & 0 \\ -3 & -4 & 1 & 5 \\ 6 & -6 & 8 & 0 \end{bmatrix}.$$

To compute det A, it is easiest to expand about the fourth column:

$$\begin{aligned} \det A &= -a_{14}M_{14} + a_{24}M_{24} - a_{34}M_{34} + a_{44}M_{44} = -5M_{34} \\ &= -5 \det \begin{bmatrix} 2 & -1 & 3 \\ 4 & -2 & 7 \\ 6 & -6 & 8 \end{bmatrix} \\ &= -5\left\{2 \det \begin{bmatrix} -2 & 7 \\ -6 & 8 \end{bmatrix} - (-1) \det \begin{bmatrix} 4 & 7 \\ 6 & 8 \end{bmatrix} + 3 \det \begin{bmatrix} 4 & -2 \\ 6 & -6 \end{bmatrix}\right\} \\ &= -5\{(2(-16+42) - (-1)(32-42) + 3(-24+12)\} = -30. \end{aligned}$$

◆ ◆ ◆

The following properties of determinants are useful in relating linear systems and Gaussian elimination to determinants.

Determinant Facts

Suppose A is an $n \times n$ matrix:

(a) If any row or column of A has only zero entries, then $\det A = 0$.
(b) If $\tilde{A}$ is obtained from A by the operation $(E_i) \leftrightarrow (E_k)$, with $i \neq k$, then $\det \tilde{A} = -\det A$.
(c) If A has two rows or two columns the same, then $\det A = 0$.
(d) If $\tilde{A}$ is obtained from A by the operation $(\lambda E_i) \rightarrow (E_i)$, then $\det \tilde{A} = \lambda \det A$.
(e) If $\tilde{A}$ is obtained from A by the operation $(E_i + \lambda E_k) \rightarrow (E_i)$ with $i \neq k$, then $\det \tilde{A} = \det A$.
(f) If B is also an $n \times n$ matrix, then $\det AB = \det A \det B$.
(g) $\det A^t = \det A$.

(h) If A^{-1} exists, then $\det A^{-1} = \dfrac{1}{\det A}$.

(i) If A is an upper triangular, lower triangular, or diagonal matrix, then $\det A = \prod_{i=1}^{n} a_{ii}$.

EXAMPLE 6 We will compute the determinant of the matrix

$$A = \begin{bmatrix} 2 & 1 & -1 & 1 \\ 1 & 1 & 0 & 3 \\ -1 & 2 & 3 & -1 \\ 3 & -1 & -1 & 2 \end{bmatrix}$$

using Determinant Facts (b), (d), and (e) and Maple. Matrix A is defined by

```
>A:=matrix(4,4,[2,1,-1,1,1,1,0,3,-1,2,3,-1,3,-1,-1,2]);
```

The sequence of operations in Table 6.2 produces the matrix

$$A8 = \begin{bmatrix} 1 & \frac{1}{2} & -\frac{1}{2} & \frac{1}{2} \\ 0 & 1 & 1 & 5 \\ 0 & 0 & 3 & 13 \\ 0 & 0 & 0 & -13 \end{bmatrix}.$$

By fact (i), $\det A8 = (1)(1)(3)(-13) = -39$, so $\det A = -\det A8 = -39$. ◆ ◆ ◆

Table 6.2

Operation	Maple	Effect
$\frac{1}{2}E_1 \to E_1$	`A1:= mulrow(A,1,0.5)`	$\det A1 = \frac{1}{2}\det A$
$E_2 - E_1 \to E_2$	`A2:= addrow(A1,1,2,-1)`	$\det A2 = \det A1 = \frac{1}{2}\det A$
$E_3 + E_1 \to E_3$	`A3:= addrow(A2,1,3,1)`	$\det A3 = \det A2 = \frac{1}{2}\det A$
$E_4 - 3E_1 \to E_4$	`A4:= addrow(A3,1,4,-3)`	$\det A4 = \det A3 = \frac{1}{2}\det A$
$2E_2 \to E_2$	`A5:= mulrow(A4,2,2)`	$\det A5 = 2\det A4 = \det A$
$E_3 - \frac{5}{2}E_2 \to E_3$	`A6:= addrow(A5,2,3,-2.5)`	$\det A6 = \det A5 = \det A$
$E_4 + \frac{5}{2}E_2 \to E_4$	`A7:= addrow(A6,2,4,2.5)`	$\det A7 = \det A6 = \det A$
$E_3 \leftrightarrow E_4$	`A8:= swaprow(A7,3,4)`	$\det A8 = -\det A7 = -\det A$

The key result relating nonsingularity, Gaussian elimination, linear systems, and determinants is that the following statements are equivalent.

Equivalent Statements about an $n \times n$ Matrix A

(a) The equation $A\mathbf{x} = \mathbf{0}$ has the unique solution $\mathbf{x} = \mathbf{0}$.
(b) The system $A\mathbf{x} = \mathbf{b}$ has a unique solution for any n-dimensional column vector $\mathbf{b}$.
(c) The matrix A is nonsingular; that is, A^{-1} exists.
(d) $\det A \neq 0$.
(e) Gaussian elimination with row interchanges can be performed on the system $A\mathbf{x} = \mathbf{b}$ for any n-dimensional column vector $\mathbf{b}$.

Maple can be used to perform the arithmetic operations on matrices. Matrix addition is done with `matadd(A,B)` or `evalm(A+B)`. Scalar multiplication is defined by `scalarmul(A,C)` or `evalm(C*A)`. Matrix multiplication is done using `multiply(A,B)` or `evalm(A&*B)`. The matrix operation of transposition is achieved with `transpose(A)`, matrix inversion with `inverse(A)`, and the determinant with `det(A)`.

EXERCISE SET 6.4

1. Compute the following matrix products.

a. $\begin{bmatrix} 1 & 0 & 0 \\ -1 & 1 & 0 \\ 2 & 3 & 1 \end{bmatrix} \cdot \begin{bmatrix} 1 & 0 & 0 \\ 2 & 2 & 0 \\ 1 & -1 & 1 \end{bmatrix}$ **b.** $\begin{bmatrix} 1 & 0 & 0 \\ 2 & 1 & 0 \\ -2 & -1 & 1 \end{bmatrix} \cdot \begin{bmatrix} 1 & -1 & 2 \\ 0 & 1 & 3 \\ 0 & 0 & 2 \end{bmatrix}$

c. $\begin{bmatrix} 1 & 0 & 0 \\ 0 & 1 & 0 \\ 0 & -2 & 1 \end{bmatrix} \cdot \begin{bmatrix} 1 & 0 & 0 \\ 2 & 1 & 0 \\ -3 & 0 & 1 \end{bmatrix}$ **d.** $\begin{bmatrix} 2 & -1 & 4 \\ 0 & -1 & 2 \\ 0 & 0 & 3 \end{bmatrix} \cdot \begin{bmatrix} 3 & -3 & 4 \\ 0 & 1 & 1 \\ 0 & 0 & 2 \end{bmatrix}$

2. Determine which of the following matrices are nonsingular and compute the inverse of these matrices.

a. $\begin{bmatrix} 4 & 2 & 6 \\ 3 & 0 & 7 \\ -2 & -1 & -3 \end{bmatrix}$ **b.** $\begin{bmatrix} 1 & 2 & 0 \\ 2 & 1 & -1 \\ 3 & 1 & 1 \end{bmatrix}$

c. $\begin{bmatrix} 4 & 0 & 0 \\ 0 & 0 & 0 \\ 0 & 0 & 3 \end{bmatrix}$ **d.** $\begin{bmatrix} 1 & 1 & -1 & 1 \\ 1 & 2 & -4 & -2 \\ 2 & 1 & 1 & 5 \\ -1 & 0 & -2 & -4 \end{bmatrix}$

e. $\begin{bmatrix} 4 & 0 & 0 & 0 \\ 6 & 7 & 0 & 0 \\ 9 & 11 & 1 & 0 \\ 5 & 4 & 1 & 1 \end{bmatrix}$ **f.** $\begin{bmatrix} 2 & 0 & 1 & 2 \\ 1 & 1 & 0 & 2 \\ 2 & -1 & 3 & 1 \\ 3 & -1 & 4 & 3 \end{bmatrix}$

3. Compute the determinants of the matrices in Exercise 2 and the determinants of the inverse matrices of those that are nonsingular.

4. Consider the four 3×3 linear systems having the same coefficient matrix:

$$\begin{aligned} 2x_1 - 3x_2 + x_3 &= 2, \\ x_1 + x_2 - x_3 &= -1, \\ -x_1 + x_2 - 3x_3 &= 0, \end{aligned} \qquad \begin{aligned} 2x_1 - 3x_2 + x_3 &= 6, \\ x_1 + x_2 - x_3 &= 4, \\ -x_1 + x_2 - 3x_3 &= 5, \end{aligned}$$

$$\begin{aligned} 2x_1 - 3x_2 + x_3 &= 0, \\ x_1 + x_2 - x_3 &= 1, \\ -x_1 + x_2 - 3x_3 &= -3, \end{aligned} \qquad \begin{aligned} 2x_1 - 3x_2 + x_3 &= -1, \\ x_1 + x_2 - x_3 &= 0, \\ -x_1 + x_2 - 3x_3 &= 0. \end{aligned}$$

a. Solve the linear systems by applying Gaussian elimination to the augmented matrix

$$\left[\begin{array}{rrr|rrrr} 2 & -3 & 1 & 2 & 6 & 0 & -1 \\ 1 & 1 & -1 & -1 & 4 & 1 & 0 \\ -1 & 1 & -3 & 0 & 5 & -3 & 0 \end{array}\right].$$

b. Solve the linear systems by finding and multiplying by the inverse of

$$A = \begin{bmatrix} 2 & -3 & 1 \\ 1 & 1 & -1 \\ -1 & 1 & -3 \end{bmatrix}.$$

c. Which method requires more operations?

5. Show that the following statements are true or provide counterexamples to show they are not.

a. The product of two symmetric matrices is symmetric.

b. The inverse of a nonsingular symmetric matrix is a nonsingular symmetric matrix.

c. If A and B are $n \times n$ matrices, then $(AB)^t = A^t B^t$.

6. **a.** Show that the product of two $n \times n$ lower triangular matrices is lower triangular.

b. Show that the product of two $n \times n$ upper triangular matrices is upper triangular.

c. Show that the inverse of a nonsingular $n \times n$ lower triangular matrix is lower triangular.

7. The solution by **Cramer's rule** to the linear system

$$\begin{aligned} a_{11}x_1 + a_{12}x_2 + a_{13}x_3 &= b_1, \\ a_{21}x_1 + a_{22}x_2 + a_{23}x_3 &= b_2, \\ a_{31}x_1 + a_{32}x_2 + a_{33}x_3 &= b_3 \end{aligned}$$

has

$$x_1 = \frac{1}{D} \det \begin{bmatrix} b_1 & a_{12} & a_{13} \\ b_2 & a_{22} & a_{23} \\ b_3 & a_{32} & a_{33} \end{bmatrix} \equiv \frac{D_1}{D},$$

$$x_2 = \frac{1}{D} \det \begin{bmatrix} a_{11} & b_1 & a_{13} \\ a_{21} & b_2 & a_{23} \\ a_{31} & b_3 & a_{33} \end{bmatrix} \equiv \frac{D_2}{D},$$

and

$$x_3 = \frac{1}{D} \det \begin{bmatrix} a_{11} & a_{12} & b_1 \\ a_{21} & a_{22} & b_2 \\ a_{31} & a_{32} & b_3 \end{bmatrix} \equiv \frac{D_3}{D},$$

where

$$D = \det \begin{bmatrix} a_{11} & a_{12} & a_{13} \\ a_{21} & a_{22} & a_{23} \\ a_{31} & a_{32} & a_{33} \end{bmatrix}.$$

a. Use Cramer's rule to find the solution to the linear system

$$\begin{aligned} 2x_1 + 3x_2 - x_3 &= 4, \\ x_1 - 2x_2 + x_3 &= 6, \\ x_1 - 12x_2 + 5x_3 &= 10. \end{aligned}$$

b. Show that the linear system

$$\begin{aligned} 2x_1 + 3x_2 - x_3 &= 4, \\ x_1 - 2x_2 + x_3 &= 6, \\ -x_1 - 12x_2 + 5x_3 &= 9 \end{aligned}$$

does not have a solution. Compute D_1, D_2, and D_3.

c. Show that the linear system

$$\begin{aligned} 2x_1 + 3x_2 - x_3 &= 4, \\ x_1 - 2x_2 + x_3 &= 6, \\ -x_1 - 12x_2 + 5x_3 &= 10 \end{aligned}$$

has an infinite number of solutions. Compute D_1, D_2, and D_3.

d. Suppose that a 3×3 linear system with $D = 0$ has solutions. Explain why we must also have $D_1 = D_2 = D_3 = 0$.

8. In a paper entitled "Population Waves," Bernadelli [Ber] hypothesizes a type of simplified beetle, which has a natural life span of 3 years. The female of this species has a survival rate of $\frac{1}{2}$ in the first year of life, has a survival rate of $\frac{1}{3}$ from the second to third years, and gives birth to an average of six new females before expiring at the end of the third year. A matrix can be used to show the contribution an individual female beetle makes, in a probabilistic sense, to the female population of the species by letting a_{ij} in the matrix $A = (a_{ij})$ denote the contribution that a single female beetle of age j will make to the next year's female population of age i; that is,

$$A = \begin{bmatrix} 0 & 0 & 6 \\ \frac{1}{2} & 0 & 0 \\ 0 & \frac{1}{3} & 0 \end{bmatrix}.$$

a. The contribution that a female beetle makes to the population 2 years hence is determined from the entries of A^2, of 3 years hence from A^3, and so on. Construct A^2 and A^3, and try to make a general statement about the contribution of a female beetle to the population in n years' time for any positive integral value of n.

b. Use your conclusions from part (a) to describe what will occur in future years to a population of these beetles that initially consists of 6000 female beetles in each of the three age groups.

c. Construct A^{-1} and describe its significance regarding the population of this species.

9. The study of food chains is an important topic in the determination of the spread and accumulation of environmental pollutants in living matter. Suppose that a food chain has three links. The first link consists of vegetation of types $v_1, v_2, \ldots, v_n$, which provide all the food requirements for herbivores of species $h_1, h_2, \ldots, h_m$ in the second link. The third link consists of carnivorous animals $c_1, c_2, \ldots, c_k$, which depend entirely on the herbivores in the second link for their food supply. The coordinate a_{ij} of the matrix

$$A = \begin{bmatrix} a_{11} & a_{12} & \cdots & a_{1m} \\ a_{21} & a_{22} & \cdots & a_{2m} \\ \vdots & \vdots & & \vdots \\ a_{n1} & a_{n2} & \cdots & a_{nm} \end{bmatrix}$$

represents the total number of plants of type v_i eaten by the herbivores in the species h_j, whereas b_{ij} in

$$B = \begin{bmatrix} b_{11} & b_{12} & \cdots & b_{1k} \\ b_{21} & b_{22} & \cdots & b_{2k} \\ \vdots & \vdots & & \vdots \\ b_{m1} & b_{m2} & \cdots & b_{mk} \end{bmatrix}$$

describes the number of herbivores in species h_i that are devoured by the animals of type c_j.

a. Show that the number of plants of type v_i that eventually end up in the animals of species c_j is given by the entry in the ith row and jth column of the matrix AB.

b. What physical significance is associated with the matrices A^{-1}, B^{-1}, and $(AB)^{-1} = B^{-1}A^{-1}$?

10. In Section 3.6 we found that the parametric form $(x(t), y(t))$ of the cubic Hermite polynomials through $(x(0), y(0)) = (x_0, y_0)$ and $(x(1), y(1)) = (x_1, y_1)$ with guide-points $(x_0 + \alpha_0, y_0 + \beta_0)$ and $(x_1 - \alpha_1, y_1 - \beta_1)$, respectively, is given by

$$x(t) = [2(x_0 - x_1) + (\alpha_0 + \alpha_1)]t^3 + [3(x_1 - x_0) - \alpha_1 - 2\alpha_0]t^2 + \alpha_0 t + x_0$$

and

$$y(t) = [2(y_0 - y_1) + (\beta_0 + \beta_1)]t^3 + [3(y_1 - y_0) - \beta_1 - 2\beta_0]t^2 + \beta_0 t + y_0.$$

The Bézier cubic polynomials have the form

$$\hat{x}(t) = [2(x_0 - x_1) + 3(\alpha_0 + \alpha_1)]t^3 + [3(x_1 - x_0) - 3(\alpha_1 + 2\alpha_0)]t^2 + 3\alpha_0 t + x_0$$

and

$$\hat{y}(t) = [2(y_0 - y_1) + 3(\beta_0 + \beta_1)]t^3 + [3(y_1 - y_0) - 3(\beta_1 + 2\beta_0)]t^2 + 3\beta_0 t + y_0.$$

a. Show that the matrix

$$A = \begin{bmatrix} 7 & 4 & 4 & 0 \\ -6 & -3 & -6 & 0 \\ 0 & 0 & 3 & 0 \\ 0 & 0 & 0 & 1 \end{bmatrix}$$

maps the Hermite polynomial coefficients onto the Bézier polynomial coefficients.

b. Determine a matrix B that maps the Bézier polynomial coefficients onto the Hermite polynomial coefficients.

11. Consider the 2×2 linear system $(A + iB)(\mathbf{x} + i\mathbf{y}) = \mathbf{c} + i\mathbf{d}$ with complex entries in component form:

$$\begin{aligned} (a_{11} + ib_{11})(x_1 + iy_1) + (a_{12} + ib_{12})(x_2 + iy_2) &= c_1 + id_1, \\ (a_{21} + ib_{21})(x_1 + iy_1) + (a_{22} + ib_{22})(x_2 + iy_2) &= c_2 + id_2. \end{aligned}$$

a. Use the properties of complex numbers to convert this system to the equivalent 4×4 real linear system

$$\begin{aligned} \text{Real part:} \quad & A\mathbf{x} - B\mathbf{y} = \mathbf{c}, \\ \text{Imaginary part:} \quad & B\mathbf{x} + A\mathbf{y} = \mathbf{d}. \end{aligned}$$

b. Solve the linear system

$$\begin{aligned} (1 - 2i)(x_1 + iy_1) + (3 + 2i)(x_2 + iy_2) &= 5 + 2i, \\ (2 + i)(x_1 + iy_1) + (4 + 3i)(x_2 + iy_2) &= 4 - i. \end{aligned}$$

6.5 Matrix Factorization

Gaussian elimination is the principal tool for the direct solution of linear systems of equations, so it should come as no surprise to learn that it appears in other guises. In this section we will see that the steps used to solve a system of the form $A\mathbf{x} = \mathbf{b}$ by Gaussian

Elimination can be used to factor the matrix A into a product of matrices that are easier to manipulate. The factorization is particularly useful when it has the form $A = LU$, where L is lower triangular and U is upper triangular. Although not all matrices have this type of representation, many do that occur frequently in the application of numerical techniques.

In Section 6.2 we found that Gaussian elimination applied to an arbitrary nonsingular system requires $O(n^3)$ operations to determine $\mathbf{x}$. If A has been factored into the triangular form $A = LU$, then we can solve for $\mathbf{x}$ more easily by using a two-step process. First we let $\mathbf{y} = U\mathbf{x}$ and solve the system $L\mathbf{y} = \mathbf{b}$ for $\mathbf{y}$. Since L is lower triangular, determining $\mathbf{y}$ from this equation requires only $O(n^2)$ operations. Once $\mathbf{y}$ is known, the upper triangular system $U\mathbf{x} = \mathbf{y}$ requires only an additional $O(n^2)$ operations to determine the solution $\mathbf{x}$. This means that the total number of operations needed to solve the system $A\mathbf{x} = \mathbf{b}$ is reduced from $O(n^3)$ to $O(n^2)$. In systems greater than 100 by 100, this can reduce the amount of calculation by more than 99%. Not surprisingly, the reductions from the factorization do not come free; determining the specific matrices L and U requires $O(n^3)$ operations. But once the factorization is determined, systems involving the matrix A can be solved in this simplified manner for any number of vectors $\mathbf{b}$.

EXAMPLE 1 The linear system

$$\begin{aligned} x_1 + x_2 \qquad\quad + 3x_4 &= 4, \\ 2x_1 + x_2 - x_3 + x_4 &= 1, \\ 3x_1 - x_2 - x_3 + 2x_4 &= -3, \\ -x_1 + 2x_2 + 3x_3 - x_4 &= 4 \end{aligned}$$

was considered in Section 6.2. The sequence of operations $(E_2 - 2E_1) \to (E_2)$, $(E_3 - 3E_1) \to (E_3)$, $(E_4 - (-1)E_1) \to (E_4)$, $(E_3 - 4E_2) \to (E_3)$, $(E_4 - (-3)E_2) \to (E_4)$ converts the system to one that has the upper triangular form

$$\begin{aligned} x_1 + x_2 \qquad\quad + 3x_4 &= 4, \\ -x_2 - x_3 - 5x_4 &= -7, \\ 3x_3 + 13x_4 &= 13, \\ -13x_4 &= -13. \end{aligned}$$

Let U be the upper triangular matrix with these coefficients as its entries and L be the lower triangular matrix with 1s along the diagonal and the multipliers m_{kj} as entries below the diagonal. Then we have the factorization

$$A = \begin{bmatrix} 1 & 1 & 0 & 3 \\ 2 & 1 & -1 & 1 \\ 3 & -1 & -1 & 2 \\ -1 & 2 & 3 & -1 \end{bmatrix} = \begin{bmatrix} 1 & 0 & 0 & 0 \\ 2 & 1 & 0 & 0 \\ 3 & 4 & 1 & 0 \\ -1 & -3 & 0 & 1 \end{bmatrix} \begin{bmatrix} 1 & 1 & 0 & 3 \\ 0 & -1 & -1 & -5 \\ 0 & 0 & 3 & 13 \\ 0 & 0 & 0 & -13 \end{bmatrix} = LU.$$

This factorization permits us to easily solve any system involving the matrix A. For example, to solve

$$\begin{bmatrix} 1 & 0 & 0 & 0 \\ 2 & 1 & 0 & 0 \\ 3 & 4 & 1 & 0 \\ -1 & -3 & 0 & 1 \end{bmatrix} \begin{bmatrix} 1 & 1 & 0 & 3 \\ 0 & -1 & -1 & -5 \\ 0 & 0 & 3 & 13 \\ 0 & 0 & 0 & -13 \end{bmatrix} \begin{bmatrix} x_1 \\ x_2 \\ x_3 \\ x_4 \end{bmatrix} = \begin{bmatrix} 8 \\ 7 \\ 14 \\ -7 \end{bmatrix},$$

we first introduce the substitution $\mathbf{y} = U\mathbf{x}$. Then $L\mathbf{y} = \mathbf{b}$; that is,

$$\begin{bmatrix} 1 & 0 & 0 & 0 \\ 2 & 1 & 0 & 0 \\ 3 & 4 & 1 & 0 \\ -1 & -3 & 0 & 1 \end{bmatrix} \begin{bmatrix} y_1 \\ y_2 \\ y_3 \\ y_4 \end{bmatrix} = \begin{bmatrix} 8 \\ 7 \\ 14 \\ -7 \end{bmatrix}.$$

This system is solved for $\mathbf{y}$ by a simple forward substitution process:

$$\begin{aligned} y_1 &= 8, \\ 2y_1 + y_2 &= 7, \quad \text{so} \quad y_2 = 7 - 2y_1 = -9 \\ 3y_1 + 4y_2 + y_3 &= 14, \quad \text{so} \quad y_3 = 14 - 3y_1 - 4y_2 = 26 \\ -y_1 - 3y_2 + y_4 &= -7, \quad \text{so} \quad y_4 = -7 + y_1 + 3y_2 = -26. \end{aligned}$$

We then solve $U\mathbf{x} = \mathbf{y}$ for $\mathbf{x}$, the solution of the original system; that is,

$$\begin{bmatrix} 1 & 1 & 0 & 3 \\ 0 & -1 & -1 & -5 \\ 0 & 0 & 3 & 13 \\ 0 & 0 & 0 & -13 \end{bmatrix} \begin{bmatrix} x_1 \\ x_2 \\ x_3 \\ x_4 \end{bmatrix} = \begin{bmatrix} 8 \\ -9 \\ 26 \\ -26 \end{bmatrix}.$$

Using backward substitution we obtain $x_4 = 2$, $x_3 = 0$, $x_2 = -1$, $x_1 = 3$. ◆ ◆ ◆

In the factorization of $A = LU$, we generate the first column of L and the first row of U using the equations

$$l_{11}u_{11} = a_{11}$$

and, for each $j = 2, 3, \ldots, n$,

$$l_{j1} = \frac{a_{j1}}{u_{11}} \quad \text{and} \quad u_{1j} = \frac{a_{1j}}{l_{11}}.$$

For each $i = 2, 3, \ldots, n-1$, we select the diagonal entries u_{ii} and l_{ii} and generate the remaining entries in the ith column of L and the ith row of U. The required equations are

$$l_{ii}u_{ii} = a_{ii} - \sum_{k=1}^{i-1} l_{ik}u_{ki};$$

and, for each $j = i + 1, \ldots, n$,

$$l_{ji} = \frac{1}{u_{ii}}\left[a_{ji} - \sum_{k=1}^{i-1} l_{jk}u_{ki}\right] \quad \text{and} \quad u_{ij} = \frac{1}{l_{ii}}\left[a_{ij} - \sum_{k=1}^{i-1} l_{ik}u_{kj}\right].$$

Finally, l_{nn} and u_{nn} are selected to satisfy

$$l_{nn}u_{nn} = a_{nn} - \sum_{k=1}^{n-1} l_{nk}u_{kn}.$$

A general procedure for factoring matrices into a product of triangular matrices is performed in the program LUFACT64. Although new matrices L and U are constructed, the values generated replace the corresponding entries of A that are no longer needed. Thus, the new matrix has entries $a_{ij} = l_{ij}$ for each $i = 2, 3, \ldots, n$ and $j = 1, 2, \ldots, i-1$ and $a_{ij} = u_{ij}$ for each $i = 1, 2, \ldots, n$ and $j = i+1, i+2, \ldots, n$.

The factorization is particularly useful when a number of linear systems involving A must be solved, since the bulk of the operations need to be performed only once. To solve $LU\mathbf{x} = \mathbf{b}$, we first solve $L\mathbf{y} = \mathbf{b}$ for $\mathbf{y}$. Since L is lower triangular, we have

$$y_1 = \frac{b_1}{l_{11}}$$

and

$$y_i = \frac{1}{l_{ii}}\left[b_i - \sum_{j=1}^{i-1} l_{ij}y_j\right], \qquad \text{for each } i = 2, 3, \ldots, n.$$

Once $\mathbf{y}$ is calculated by this forward substitution process, the upper triangular system $U\mathbf{x} = \mathbf{y}$ is solved by backward substitution using the equations

$$x_n = \frac{y_n}{u_{nn}} \quad \text{and} \quad x_i = \frac{1}{u_{ii}}\left[y_i - \sum_{j=i+1}^{n} u_{ij}x_j\right].$$

In the previous discussion we assumed that A is such that a linear system of the form $A\mathbf{x} = \mathbf{b}$ can be solved using Gaussian elimination that does not require row interchanges. From a practical standpoint, this factorization is useful only when row interchanges are not required to control the round-off error resulting from the use of finite-digit arithmetic. Although many systems we encounter when using approximation methods are of this type, factorization modifications must be made when row interchanges are required. We begin the discussion with the introduction of a class of matrices that are used to rearrange, or permute, rows of a given matrix.

An $n \times n$ **permutation matrix** P is a matrix with precisely one entry whose value is 1 in each column and each row and all of whose other entries are 0.

EXAMPLE 2 The matrix

$$P = \begin{bmatrix} 1 & 0 & 0 \\ 0 & 0 & 1 \\ 0 & 1 & 0 \end{bmatrix}$$

is a 3×3 permutation matrix. For any 3×3 matrix A, multiplying on the left by P has the effect of interchanging the second and third rows of A:

$$PA = \begin{bmatrix} 1 & 0 & 0 \\ 0 & 0 & 1 \\ 0 & 1 & 0 \end{bmatrix} \begin{bmatrix} a_{11} & a_{12} & a_{13} \\ a_{21} & a_{22} & a_{23} \\ a_{31} & a_{32} & a_{33} \end{bmatrix} = \begin{bmatrix} a_{11} & a_{12} & a_{13} \\ a_{31} & a_{32} & a_{33} \\ a_{21} & a_{22} & a_{23} \end{bmatrix}.$$

Similarly, multiplying on the right by P interchanges the second and third columns of A.

◆ ◆ ◆

There are two useful properties of permutation matrices that relate to Gaussian elimination. The first of these is illustrated in the previous example and states that if $k_1, \ldots, k_n$ is a permutation of the integers $1, \ldots, n$ and the permutation matrix $P = (p_{ij})$ is defined by

$$p_{ij} = \begin{cases} 1, & \text{if } j = k_i \\ 0, & \text{otherwise,} \end{cases}$$

then

$$PA = \begin{bmatrix} a_{k_1,1} & a_{k_1,2} & \cdots & a_{k_1,n} \\ a_{k_2,1} & a_{k_2,2} & \cdots & a_{k_2,n} \\ \vdots & \vdots & & \vdots \\ a_{k_n,1} & a_{k_n,2} & \cdots & a_{k_n,n} \end{bmatrix}.$$

The second is that if P is a permutation matrix, then P^{-1} exists and $P^{-1} = P^t$.

EXAMPLE 3 Since $a_{11} = 0$, the matrix

$$A = \begin{bmatrix} 0 & 1 & -1 & 1 \\ 1 & 1 & -1 & 2 \\ -1 & -1 & 1 & 0 \\ 1 & 2 & 0 & 2 \end{bmatrix}$$

does not have an LU factorization. However, using the row interchange $(E_1) \leftrightarrow (E_2)$, followed by $(E_3 + E_1) \rightarrow E_3$ and $(E_4 - E_1) \rightarrow E_4$, produces

$$\begin{bmatrix} 1 & 1 & -1 & 2 \\ 0 & 1 & -1 & 1 \\ 0 & 0 & 0 & 2 \\ 0 & 1 & 1 & 0 \end{bmatrix}.$$

Then the row interchange $(E_3) \leftrightarrow (E_4)$, followed by $(E_3 - E_2) \rightarrow E_3$, gives the matrix

$$U = \begin{bmatrix} 1 & 1 & -1 & 2 \\ 0 & 1 & -1 & 1 \\ 0 & 0 & 2 & -1 \\ 0 & 0 & 0 & 2 \end{bmatrix}.$$

The permutation matrix associated with the row interchanges $(E_1) \leftrightarrow (E_2)$ and $(E_3) \leftrightarrow (E_4)$ is

$$P = \begin{bmatrix} 0 & 1 & 0 & 0 \\ 1 & 0 & 0 & 0 \\ 0 & 0 & 0 & 1 \\ 0 & 0 & 1 & 0 \end{bmatrix}.$$

Gaussian elimination can be performed on PA without row interchanges to give the LU factorization

$$PA = \begin{bmatrix} 1 & 0 & 0 & 0 \\ 0 & 1 & 0 & 0 \\ 1 & 1 & 1 & 0 \\ -1 & 0 & 0 & 1 \end{bmatrix} \begin{bmatrix} 1 & 1 & -1 & -2 \\ 0 & 1 & -1 & 1 \\ 0 & 0 & 2 & -1 \\ 0 & 0 & 0 & 2 \end{bmatrix} = LU.$$

So

$$A = P^{-1}(LU) = P^t(LU) = (P^tL)U = \begin{bmatrix} 0 & 1 & 0 & 0 \\ 1 & 0 & 0 & 0 \\ -1 & 0 & 0 & 1 \\ 1 & 1 & 1 & 0 \end{bmatrix} \begin{bmatrix} 1 & 1 & -1 & 2 \\ 0 & 1 & -1 & 1 \\ 0 & 0 & 2 & -1 \\ 0 & 0 & 0 & 2 \end{bmatrix}.$$

◆ ◆ ◆

Maple has the command `LUdecomp` to compute a factorization of the form $A = PLU$ of the matrix A. If the matrix A has been created, the function call

```
>U:=LUdecomp(A,P='G',L='H');
```

returns the upper triangular matrix U as the value of the function and returns the lower triangular matrix L in H and the permutation matrix P in G.

EXERCISE SET 6.5

1. Solve the following linear systems.

   a. $$\begin{bmatrix} 1 & 0 & 0 \\ 2 & 1 & 0 \\ -1 & 0 & 1 \end{bmatrix} \begin{bmatrix} 2 & 3 & -1 \\ 0 & -2 & 1 \\ 0 & 0 & 3 \end{bmatrix} \begin{bmatrix} x_1 \\ x_2 \\ x_3 \end{bmatrix} = \begin{bmatrix} 2 \\ -1 \\ 1 \end{bmatrix}$$

b. $\begin{bmatrix} 2 & 0 & 0 \\ -1 & 1 & 0 \\ 3 & 2 & -1 \end{bmatrix} \begin{bmatrix} 1 & 1 & 1 \\ 0 & 1 & 2 \\ 0 & 0 & 1 \end{bmatrix} \begin{bmatrix} x_1 \\ x_2 \\ x_3 \end{bmatrix} = \begin{bmatrix} -1 \\ 3 \\ 0 \end{bmatrix}$

2. Factor the following matrices into the LU decomposition with $l_{ii} = 1$ for all i.

a. $\begin{bmatrix} 2 & -1 & 1 \\ 3 & 3 & 9 \\ 3 & 3 & 5 \end{bmatrix}$

b. $\begin{bmatrix} 1.012 & -2.132 & 3.104 \\ -2.132 & 4.096 & -7.013 \\ 3.104 & -7.013 & 0.014 \end{bmatrix}$

c. $\begin{bmatrix} 2 & 0 & 0 & 0 \\ 1 & 1.5 & 0 & 0 \\ 0 & -3 & 0.5 & 0 \\ 2 & -2 & 1 & 1 \end{bmatrix}$

d. $\begin{bmatrix} 2.1756 & 4.0231 & -2.1732 & 5.1967 \\ -4.0231 & 6.0000 & 0 & 1.1973 \\ -1.0000 & -5.2107 & 1.1111 & 0 \\ 6.0235 & 7.0000 & 0 & -4.1561 \end{bmatrix}$

3. Obtain factorizations of the form $A = P^tLU$ for the following matrices.

a. $A = \begin{bmatrix} 0 & 2 & 3 \\ 1 & 1 & -1 \\ 0 & -1 & 1 \end{bmatrix}$

b. $A = \begin{bmatrix} 1 & 2 & -1 \\ 1 & 2 & 3 \\ 2 & -1 & 4 \end{bmatrix}$

c. $A = \begin{bmatrix} 1 & -2 & 3 & 0 \\ 3 & -6 & 9 & 3 \\ 2 & 1 & 4 & 1 \\ 1 & -2 & 2 & -2 \end{bmatrix}$

d. $A = \begin{bmatrix} 1 & -2 & 3 & 0 \\ 1 & -2 & 3 & 1 \\ 1 & -2 & 2 & -2 \\ 2 & 1 & 3 & -1 \end{bmatrix}$

4. Suppose $A = P^tLU$, where P is a permutation matrix, L is a lower-triangular matrix with 1s on the diagonal, and U is an upper-triangular matrix.

a. Count the number of operations needed to compute P^tLU for a given matrix A.

b. Show that if P contains k row interchanges, then

$$\det P = \det P^t = (-1)^k.$$

c. Use $\det A = \det P^t \det L \det U = (-1)^k \det U$ to count the number of operations for determining $\det A$ by factoring.

d. Compute $\det A$ and count the number of operations when

$$A = \begin{bmatrix} 0 & 2 & 1 & 4 & -1 & 3 \\ 1 & 2 & -1 & 3 & 4 & 0 \\ 0 & 1 & 1 & -1 & 2 & -1 \\ 2 & 3 & -4 & 2 & 0 & 5 \\ 1 & 1 & 1 & 3 & 0 & 2 \\ -1 & -1 & 2 & -1 & 2 & 0 \end{bmatrix}.$$

6.6 Techniques for Special Matrices

Although this chapter has been concerned primarily with the effective application of Gaussian elimination for finding the solution to a linear system of equations, many of the results have wider application. It might be said that Gaussian elimination is the hub about which the chapter revolves, but the wheel itself is of equal interest and has application in many forms in the study of numerical methods. In this section we consider some matrices that are of special types, forms that will be used in other chapters of the book.

The $n \times n$ matrix A is said to be **strictly diagonally dominant** when

$$|a_{ii}| > \sum_{\substack{j=1,\\ j\neq i}}^{n} |a_{ij}|$$

holds for each $i = 1, 2, \ldots, n$.

EXAMPLE 1 Consider the matrices

$$A = \begin{bmatrix} 7 & 2 & 0 \\ 3 & 5 & -1 \\ 0 & 5 & -6 \end{bmatrix} \quad \text{and} \quad B = \begin{bmatrix} 6 & 4 & -3 \\ 4 & -2 & 0 \\ -3 & 0 & 1 \end{bmatrix}.$$

The nonsymmetric matrix A is strictly diagonally dominant, since $|7| > |2| + |0|$, $|5| > |3| + |-1|$, and $|-6| > |0| + |5|$. The symmetric matrix B is not strictly diagonally dominant, because, for example, in the first row the absolute value of the diagonal element is $|6| < |4| + |-3| = 7$. It is interesting to note that A^t is not strictly diagonally dominant, nor, of course, is $B^t = B$. ◆ ◆ ◆

Strictly Diagonally Dominant Matrices

A strictly diagonally dominant matrix A is nonsingular. Moreover, in this case, Gaussian elimination can be performed on any linear system of the form $A\mathbf{x} = \mathbf{b}$ to obtain its unique solution without row or column interchanges, and the computations are stable with respect to the growth of round-off error.

A matrix A is **positive definite** if it is symmetric and if $\mathbf{x}^t A\mathbf{x} > 0$ for every n-dimensional column vector $\mathbf{x} \neq \mathbf{0}$. Using the definition to determine whether a matrix is positive definite can be difficult. Fortunately, there are more easily verified criteria for identifying members that are and are not of this important class.

Positive Definite Matrix Properties

If A is an $n \times n$ positive definite matrix, then

(a) A is nonsingular;
(b) $a_{ii} > 0$ for each $i = 1, 2, \ldots, n$;
(c) $\max_{1 \le k, j \le n} |a_{kj}| \le \max_{1 \le i \le n} |a_{ii}|$;
(d) $(a_{ij})^2 < a_{ii} a_{jj}$ for each $i \neq j$.

Not all authors require symmetry of a positive definite matrix. For example, Golub and Van Loan [GV], a standard reference in matrix methods, requires only that $\mathbf{x}^t A \mathbf{x} > 0$ for each nonzero vector $\mathbf{x}$. Matrices we call positive definite are called symmetric positive definite in [GV]. Keep this discrepancy in mind if you are using material from other sources.

The next result parallels the strictly diagonally dominant results presented previously.

Positive Definite Matrix Equivalences

The following are equivalent for any $n \times n$ symmetric matrix A:

(a) A is positive definite.
(b) Gaussian elimination without row interchanges can be performed on the linear system $A\mathbf{x} = \mathbf{b}$ with all pivot elements positive. (This ensures that the computations are stable with respect to the growth of round-off error.)
(c) A can be factored in the form LL^t, where L is lower triangular with positive diagonal entries.
(d) A can be factored in the form LDL^t, where L is lower triangular with 1s on its diagonal and D is a diagonal matrix with positive diagonal entries.
(e) For each $i = 1, 2, \ldots, n$, we have

$$\det \begin{bmatrix} a_{11} & a_{12} & \cdots & a_{1i} \\ a_{21} & a_{22} & \cdots & a_{2i} \\ \vdots & \vdots & & \vdots \\ a_{i1} & a_{i2} & \cdots & a_{ii} \end{bmatrix} > 0.$$

Maple also has a useful command to determine the positive definiteness of a matrix. The command

```
>definite(A,positive_def);
```

returns *true* or *false* as an indication. Consistent with our definition, symmetry is required for a *true* result to be produced.

The factorization in part (c) can be obtained by Choleski's factorization method. Set $l_{11} = \sqrt{a_{11}}$ and generate the remainder of the first column of L using the equation

$$l_{j1} = \frac{a_{j1}}{l_{11}} \quad \text{for each } j = 2, 3, \ldots, n.$$

For each $i = 2, 3, \ldots, n-1$, determine the ith column of L by

$$l_{ii} = \left(a_{ii} - \sum_{k=1}^{i-1} l_{ik}^2 \right)^{1/2}$$

and, for each $j = i+1, i+2, \ldots, n$,

$$l_{ji} = \frac{1}{l_{ii}} \left(a_{ji} - \sum_{k=1}^{i-1} l_{jk} l_{ik} \right).$$

Finally,

$$l_{nn} = \left(a_{nn} - \sum_{k=1}^{n-1} l_{nk}^2 \right)^{1/2}.$$

These equations can be derived by writing out the system associated with $A = LL^t$. Choleski's method gives the LL^t factorization and can be implemented using the program CHOLFC65.

The Choleski factorization of A is computed in Maple using the statement

```
>L:=cholesky(A);
```

In a similar manner to the general LU factorization, the factorization $A = LDL^t$ uses the equations $d_1 = a_{11}$ and $l_{j1} = a_{j1}/d_1$ for each $j = 2, 3, \ldots, n$, to generate the first column of L. For each $i = 2, 3, \ldots, n-1$, compute d_i and the ith column of L as follows:

$$d_i = a_{ii} - \sum_{j=1}^{i-1} l_{ij}^2 d_j$$

and

$$l_{ji} = \frac{1}{d_i} \left[a_{ji} - \sum_{k=1}^{i-1} l_{jk} l_{ik} d_k \right]$$

for each $j = i+1, i+2, \ldots, n$. The last entry in D is

$$d_n = a_{nn} - \sum_{j=1}^{n-1} l_{nj}^2 d_j.$$

The LDL^t factorization can be accomplished with the program LDLFCT66.

EXAMPLE 2 The matrix

$$A = \begin{bmatrix} 4 & -1 & 1 \\ -1 & 4.25 & 2.75 \\ 1 & 2.75 & 3.5 \end{bmatrix}$$

is positive definite. The factorization LDL^t of A is

$$A = LDL^t = \begin{bmatrix} 1 & 0 & 0 \\ -0.25 & 1 & 0 \\ 0.25 & 0.75 & 1 \end{bmatrix} \begin{bmatrix} 4 & 0 & 0 \\ 0 & 4 & 0 \\ 0 & 0 & 1 \end{bmatrix} \begin{bmatrix} 1 & -0.25 & 0.25 \\ 0 & 1 & 0.75 \\ 0 & 0 & 1 \end{bmatrix},$$

and Choleski's method produces the factorization

$$A = LL^t = \begin{bmatrix} 2 & 0 & 0 \\ -0.5 & 2 & 0 \\ 0.5 & 1.5 & 1 \end{bmatrix} \begin{bmatrix} 2 & -0.5 & 0.5 \\ 0 & 2 & 1.5 \\ 0 & 0 & 1 \end{bmatrix}.$$

◆ ◆ ◆

We can solve the linear system $A\mathbf{x} = \mathbf{b}$ when A is positive definite by using Choleski's method to first factor A into the form LL^t. Let $\mathbf{y} = L^t\mathbf{x}$. The linear system $L\mathbf{y} = \mathbf{b}$ is solved using forward substitution,

$$y_1 = \frac{b_1}{l_{11}}$$

and, for $i = 2, 3, \ldots, n$,

$$y_i = \frac{1}{l_{ii}}\left[b_i - \sum_{j=1}^{i-1} l_{ij}y_j\right].$$

Then the solution to the original system is obtained by using backward substitution to solve $L^t\mathbf{x} = \mathbf{y}$ with the equations

$$x_n = \frac{y_n}{l_{nn}}$$

and, for $i = n-1, n-2, \ldots, 1$,

$$x_i = \frac{1}{l_{ii}}\left[y_i - \sum_{j=i+1}^{n} l_{ji}x_j\right].$$

If $A\mathbf{x} = \mathbf{b}$ is to be solved and the factorization $A = LDL^t$ is known, then we let $\mathbf{y} = DL^t\mathbf{x}$ and solve the system $L\mathbf{y} = \mathbf{b}$ using forward substitution

$$y_1 = b_1$$

and

$$y_i = b_i - \sum_{j=1}^{i-1} l_{ij}y_j.$$

The system $D\mathbf{z} = \mathbf{y}$ is solved as

$$z_i = \frac{y_i}{d_i}, \qquad \text{for each } i = 1, 2, \ldots, n.$$

Then the system $L^t\mathbf{x} = \mathbf{z}$ is solved by backward substitution

$$x_n = z_n$$

and, for $i = n-1, n-2, \ldots, 1$,

$$x_i = z_i - \sum_{j=i+1}^{n} l_{ji}x_j.$$

Any symmetric matrix A for which Gaussian elimination can be applied without row interchanges can be factored into the form LDL^t. In this general case, L is lower triangular with 1s on its diagonal, and D is the diagonal matrix with the Gaussian elimination pivots on its diagonal. This result is widely applied, since symmetric matrices are common and easily recognized.

The last class of matrices considered are *band matrices*. In many applications the band matrices are also strictly diagonally dominant or positive definite. This combination of properties is very useful.

An $n \times n$ matrix is called a **band matrix** if integers p and q, with $1 < p, q < n$, exist with the property that $a_{ij} = 0$ whenever $i + p \leq j$ or $j + q \leq i$. The **bandwidth** is $w = p + q - 1$.

For example, the matrix

$$A = \begin{bmatrix} 7 & 2 & 0 \\ 3 & 5 & -1 \\ 0 & -5 & -6 \end{bmatrix}$$

is a band matrix with $p = q = 2$ and bandwidth 3.

Band matrices concentrate all their nonzero entries about the diagonal. Two special cases of band matrices that occur often have $p = q = 2$ and $p = q = 4$.

Matrices of bandwidth 3, occurring when $p = q = 2$, are called **tridiagonal**, because they have the form

$$A = \begin{bmatrix} \alpha_1 & \gamma_1 & 0 & \cdots & \cdots & 0 \\ \beta_2 & \alpha_2 & \gamma_2 & \ddots & & \vdots \\ 0 & \beta_3 & \alpha_3 & \ddots & \ddots & 0 \\ \vdots & \ddots & \ddots & \ddots & \ddots & \gamma_{n-1} \\ 0 & \cdots & \cdots & 0 & \beta_n & \alpha_n \end{bmatrix}.$$

Since the entries of tridiagonal matrices are predominantly zero, it is common to avoid the double subscript notation by relabeling the entries as indicated.

Tridiagonal matrices will appear in Chapter 11 in connection with the study of piecewise linear approximations to boundary-value problems. The case of $p = q = 4$ will also be used in that chapter for the solution of boundary-value problems, when the approximating functions assume the form of cubic splines.

The factorization methods can be simplified considerably in the case of band matrices, because a large number of zeros appear in regular patterns. It is particularly interesting to observe the form the Crout (where $u_{ii} = 1$) and Doolittle (where $l_{ii} = 1$) methods assume in this case. To illustrate the situation, suppose a tridiagonal matrix

$$A = \begin{bmatrix} \alpha_1 & \gamma_1 & 0 & \cdots & 0 \\ \beta_2 & \alpha_2 & \gamma_2 & \ddots & \vdots \\ 0 & \beta_3 & \alpha_3 & \ddots & 0 \\ \vdots & \ddots & \ddots & \ddots & \gamma_{n-1} \\ 0 & \cdots & 0 & \beta_n & \alpha_n \end{bmatrix}.$$

can be factored into the triangular matrices L and U in the Crout form

$$L = \begin{bmatrix} l_1 & 0 & \cdots & \cdots & 0 \\ \beta_2 & l_2 & \ddots & & \vdots \\ 0 & \beta_3 & l_3 & \ddots & \vdots \\ \vdots & \ddots & \ddots & \ddots & 0 \\ 0 & \cdots & 0 & \beta_n & l_n \end{bmatrix} \quad \text{and} \quad U = \begin{bmatrix} 1 & u_1 & 0 & \cdots & 0 \\ 0 & 1 & u_2 & \ddots & \vdots \\ \vdots & \ddots & 1 & \ddots & 0 \\ \vdots & & \ddots & \ddots & u_{n-1} \\ 0 & \cdots & \cdots & 0 & 1 \end{bmatrix}.$$

The entries are given by the following equations:

$$l_1 = \alpha_1 \quad \text{and} \quad u_1 = \frac{\gamma_1}{l_1};$$

for each $i = 2, 3, \ldots, n-1$,

$$l_i = \alpha_i - \beta_i u_{i-1} \quad \text{and} \quad u_i = \frac{\gamma_i}{l_i};$$

and

$$l_n = \alpha_n - \beta_n u_{n-1}.$$

The linear system $A\mathbf{x} = LU\mathbf{x} = \mathbf{b}$ is solved using the equations

$$y_1 = \frac{b_1}{l_1}$$

and, for each $i = 2, 3, \ldots, n$,

$$y_i = \frac{1}{l_i}\left[b_i - \beta_i y_{i-1}\right],$$

which determines $\mathbf{y}$ in the linear system $L\mathbf{y} = \mathbf{b}$. The linear system $U\mathbf{x} = \mathbf{y}$ is solved using the equations

$$x_n = y_n$$

and, for each $i = n-1, n-2, \ldots, 1$,

$$x_i = y_i - u_i x_{i+1}.$$

The Crout factorization of a tridiagonal matrix can be performed with the program CRTRLS67.

EXAMPLE 3 To illustrate the procedure for tridiagonal matrices, consider the tridiagonal system of equations

$$\begin{aligned} 2x_1 - x_2 \qquad\qquad &= 1,\\ -x_1 + 2x_2 - x_3 \qquad &= 0,\\ -x_2 + 2x_3 - x_4 &= 0,\\ -x_3 + 2x_4 &= 1, \end{aligned}$$

whose augmented matrix is

$$\left[\begin{array}{rrrr:r} 2 & -1 & 0 & 0 & 1\\ -1 & 2 & -1 & 0 & 0\\ 0 & -1 & 2 & -1 & 0\\ 0 & 0 & -1 & 2 & 1 \end{array}\right].$$

The LU factorization is given by

$$\begin{bmatrix} 2 & -1 & 0 & 0\\ -1 & 2 & -1 & 0\\ 0 & -1 & 2 & -1\\ 0 & 0 & -1 & 2 \end{bmatrix} = \begin{bmatrix} 2 & 0 & 0 & 0\\ -1 & \frac{3}{2} & 0 & 0\\ 0 & -1 & \frac{4}{5} & 0\\ 0 & 0 & -1 & \frac{5}{4} \end{bmatrix} \begin{bmatrix} 1 & -\frac{1}{2} & 0 & 0\\ 0 & 1 & -\frac{2}{3} & 0\\ 0 & 0 & 1 & -\frac{3}{4}\\ 0 & 0 & 0 & 1 \end{bmatrix} = LU.$$

Solving the system $L\mathbf{y} = \mathbf{b}$ gives $\mathbf{y} = (\frac{1}{2}, \frac{1}{3}, \frac{1}{4}, 1)^t$, and the solution of $U\mathbf{x} = \mathbf{y}$ is $\mathbf{x} = (1, 1, 1, 1)^t$. ◆ ◆ ◆

The tridiagonal factorization can be applied whenever $l_i \neq 0$ for each $i = 1, 2, \ldots, n$. Two conditions, either of which ensure that this is true, are that the coefficient matrix of the system is positive definite or that it is strictly diagonally dominant. An additional condition that ensures this method can be applied is as follows.

Nonsingular Tridiagonal Matrices

Suppose that A is tridiagonal with $\beta_i \neq 0$ and $\gamma_i \neq 0$ for each $i = 2, 3, \ldots, n-1$. If $|\alpha_1| > |\gamma_1|$, $|\alpha_n| > |\beta_n|$, and $|\alpha_i| \geq |\beta_i| + |\gamma_i|$ for each $i = 2, 3, \ldots, n-1$, then A is nonsingular, and the values of l_i are nonzero for each $i = 1, 2, \ldots, n$.

EXERCISE SET 6.6

1. Determine which of the following matrices are (i) symmetric, (ii) singular, (iii) strictly diagonally dominant, (iv) positive definite.

a. $\begin{bmatrix} 2 & 1 \\ 1 & 3 \end{bmatrix}$ b. $\begin{bmatrix} -2 & 1 \\ 1 & -3 \end{bmatrix}$

c. $\begin{bmatrix} 2 & 1 & 0 \\ 0 & 3 & 0 \\ 1 & 0 & 4 \end{bmatrix}$ d. $\begin{bmatrix} 2 & 1 & 0 \\ 0 & 3 & 2 \\ 1 & 2 & 4 \end{bmatrix}$

e. $\begin{bmatrix} 4 & 2 & 6 \\ 3 & 0 & 7 \\ -2 & -1 & -3 \end{bmatrix}$ f. $\begin{bmatrix} 2 & -1 & 0 \\ -1 & 4 & 2 \\ 0 & 2 & 2 \end{bmatrix}$

g. $\begin{bmatrix} 4 & 0 & 0 & 0 \\ 6 & 7 & 0 & 0 \\ 9 & 11 & 1 & 0 \\ 5 & 4 & 1 & 1 \end{bmatrix}$ h. $\begin{bmatrix} 2 & 3 & 1 & 2 \\ -2 & 4 & -1 & 5 \\ 3 & 7 & 1.5 & 1 \\ 6 & -9 & 3 & 7 \end{bmatrix}$

2. Find a factorizaton of the form $A = LDL^t$ for the following symmetric matrices:

a. $A = \begin{bmatrix} 2 & -1 & 0 \\ -1 & 2 & -1 \\ 0 & -1 & 2 \end{bmatrix}$ b. $A = \begin{bmatrix} 4 & 1 & 1 & 1 \\ 1 & 3 & -1 & 1 \\ 1 & -1 & 2 & 0 \\ 1 & 1 & 0 & 2 \end{bmatrix}$

c. $A = \begin{bmatrix} 4 & 1 & -1 & 0 \\ 1 & 3 & -1 & 0 \\ -1 & -1 & 5 & 2 \\ 0 & 0 & 2 & 4 \end{bmatrix}$ d. $A = \begin{bmatrix} 6 & 2 & 1 & -1 \\ 2 & 4 & 1 & 0 \\ 1 & 1 & 4 & -1 \\ -1 & 0 & -1 & 3 \end{bmatrix}$

3. Find a factorization of the form $A = LL^t$ for the matrices in Exercise 2.

4. Use the factorization in Exercise 2 to solve the following linear systems.

a.
$$\begin{aligned} 2x_1 - x_2 \phantom{{}+2x_3} &= 3, \\ -x_1 + 2x_2 - x_3 &= -3, \\ -x_2 + 2x_3 &= 1. \end{aligned}$$

b.
$$\begin{aligned} 4x_1 + x_2 + x_3 + x_4 &= 0.65, \\ x_1 + 3x_2 - x_3 + x_4 &= 0.05, \\ x_1 - x_2 + 2x_3 \phantom{{}+x_4} &= 0, \\ x_1 + x_2 \phantom{{}+x_3} + 2x_4 &= 0.5. \end{aligned}$$

c.
$$\begin{aligned} 4x_1 + x_2 - x_3 \phantom{{}+2x_4} &= 7, \\ x_1 + 3x_2 - x_3 \phantom{{}+2x_4} &= 8, \\ -x_1 - x_2 + 5x_3 + 2x_4 &= -4, \\ 2x_3 + 4x_4 &= 6. \end{aligned}$$

d.
$$\begin{aligned} 6x_1 + 2x_2 + x_3 - x_4 &= 0, \\ 2x_1 + 4x_2 + x_3 \phantom{{}-x_4} &= 7, \\ x_1 + x_2 + 4x_3 - x_4 &= -1, \\ -x_1 \phantom{{}+x_2} - x_3 + 3x_4 &= -2. \end{aligned}$$

5. Use Crout factorization for tridiagonal systems to solve the following linear systems.

a.
$$\begin{aligned} x_1 - x_2 \phantom{{}-2x_3} &= 0, \\ -2x_1 + 4x_2 - 2x_3 &= -1, \\ -x_2 + 2x_3 &= 1.5. \end{aligned}$$

b.
$$\begin{aligned} 3x_1 + x_2 \phantom{{}+x_3} &= -1, \\ 2x_1 + 4x_2 + x_3 &= 7, \\ 2x_2 + 5x_3 &= 9. \end{aligned}$$

c.
$$\begin{aligned} 2x_1 - x_2 \quad\quad &= 3, \\ -x_1 + 2x_2 - x_3 &= -3, \\ -x_2 + 2x_3 &= 1. \end{aligned}$$

d.
$$\begin{aligned} 0.5x_1 + 0.25x_2 \quad\quad\quad\quad &= 0.35, \\ 0.35x_1 + 0.8x_2 + 0.4x_3 \quad\quad &= 0.77, \\ 0.25x_2 + x_3 + 0.5x_4 &= -0.5, \\ x_3 - 2x_4 &= -2.25. \end{aligned}$$

6. Let A be the 10×10 tridiagonal matrix given by $a_{ii} = 2, a_{i,i+1} = a_{i,i-1} = -1$, for each $i = 2, \ldots, 9$, and $a_{11} = a_{10,10} = 2, a_{12} = a_{10,9} = -1$. Let $\mathbf{b}$ be the 10-dimensional column vector given by $b_1 = b_{10} = 1$ and $b_i = 0$ for each $i = 2, 3, \ldots, 9$. Solve $A\mathbf{x} = \mathbf{b}$ using the Crout factorization for tridiagonal systems.

7. Suppose that A and B are positive definite $n \times n$ matrices.

a. Is $-A$ positive definite?
b. Is A^t positive definite?
c. Is $A + B$ positive definite?
d. Is A^2 positive definite?
e. Is $A - B$ positive definite?

8. Let

$$A = \begin{bmatrix} 1 & 0 & -1 \\ 0 & 1 & 1 \\ -1 & 1 & \alpha \end{bmatrix}.$$

Find all values of α for which

a. A is singular.
b. A is strictly diagonally dominant.
c. A is symmetric.
d. A is positive definite.

9. Let

$$A = \begin{bmatrix} \alpha & 1 & 0 \\ \beta & 2 & 1 \\ 0 & 1 & 2 \end{bmatrix}.$$

Find all values of α and β for which

a. A is singular.
b. A is strictly diagonally dominant.
c. A is symmetric.
d. A is positive definite.

10. Suppose A and B commute; that is, $AB = BA$. Must A^t and B^t also commute?

11. In a paper by Dorn and Burdick [DB], it is reported that the average wing length that resulted from mating three mutant varieties of fruit flies (*Drosophila melanogaster*) can be expressed in the symmetric matrix form

$$A = \begin{bmatrix} 1.59 & 1.69 & 2.13 \\ 1.69 & 1.31 & 1.72 \\ 2.13 & 1.72 & 1.85 \end{bmatrix},$$

where a_{ij} denotes the average wing length of an offspring resulting from the mating of a male of type i with a female of type j.

a. What physical significance is associated with the symmetry of this matrix?

b. Is this matrix positive definite? If so, prove it; if not, find a nonzero vector $\mathbf{x}$ for which $\mathbf{x}^t A\mathbf{x} \leq 0$.

6.7 Survey of Methods and Software

In this chapter we have looked at direct methods for solving linear systems. A linear system consists of n equations in n unknowns expressed in matrix notation as $A\mathbf{x} = \mathbf{b}$. These techniques use a finite sequence of arithmetic operations to determine the exact solution of the system subject only to round-off error. We found that the linear system $A\mathbf{x} = \mathbf{b}$ has a unique solution if and only if A^{-1} exists, which is equivalent to $\det A \neq 0$. The solution of the linear system is the vector $\mathbf{x} = A^{-1}\mathbf{b}$.

Pivoting techniques were introduced to minimize the effects of round-off error, which can dominate the solution when using direct methods. We studied partial pivoting, scaled partial pivoting, and total pivoting. We recommend the partial or scaled partial pivoting methods for most problems since these decrease the effects of round-off error without adding much extra computation. Total pivoting should be used if round-off error is suspected to be large. In Section 7.6 we will see some procedures for estimating this round-off error.

Gaussian elimination with minor modifications was shown to yield a factorization of the matrix A into LU, where L is lower triangular with 1s on the diagonal and U is upper triangular. This process is called Crout factorization. Not all nonsingular matrices can be factored this way, but a permutation of the rows will always give a factorization of the form $PA = LU$, where P is the permutation matrix used to rearrange the rows of A. The advantage of the factorization is that the work is reduced when solving linear systems $A\mathbf{x} = \mathbf{b}$ with the same coefficient matrix A and different vectors $\mathbf{b}$.

Factorizations take a simpler form when the matrix A is positive definite. For example, the Choleski factorization has the form $A = LL^t$, where L is lower triangular. Positive definite matrices can also be factored in the form $A = LDL^t$, where D is diagonal and L is lower triangular with 1s on the diagonal.

If A is tridiagonal, the LU factorization takes a particularly simple form, with L having 1s on the main diagonal and 0s elsewhere, except on the diagonal immediately below the main diagonal. In addition, U has its only nonzero entries on the main diagonal and one diagonal above.

The direct methods are the methods of choice for most linear systems. For tridiagonal, banded, and positive definite matrices, the special methods are recommended. For the general case, Gaussian elimination or LU factorization methods, which allow pivoting, are recommended. In these cases, the effects of round-off error should be monitored. In Section 7.6 we discuss estimating errors in direct methods.

Large linear systems with primarily zero entries occurring in regular patterns can be solved efficiently using an iterative procedure, such as those discussed in Chapter 7. Systems of this type arise naturally, for example, when finite-difference techniques are used to solve boundary-value problems, a common application in the numerical solution of partial-differential equations.

It can be very difficult to solve a large linear system that has primarily nonzero entries or one where the zero entries are not in a predictable pattern. The matrix associated with the system can be placed in secondary storage in partitioned form and portions read into main memory only as needed for calculation. Methods that require secondary storage can be either iterative or direct, but they generally require techniques from the fields of data structures and graph theory. A study of the problems involved in theory and implementation is beyond the scope of this text.

The software for matrix operations and the direct solution of linear systems implemented in IMSL and NAG is based on LAPACK, a subroutine package in the public domain. There is excellent documentation available with it and from the books written about it. Accompanying LAPACK is a set of lower-level operations called Basic Linear Algebra Subprograms (BLAS). Level 1 of BLAS generally consists of vector operations with input data and operation counts of $O(n)$. Level 2 consists of the matrix-vector operations with input data and operation counts of $O(n^2)$.

The subroutines in LAPACK for solving linear systems first factor the matrix A. The factorization depends on the type of matrix in the following way:

1. General matrix $PA = LU$;
2. Positive definite matrix $A = LL^t$;
3. Symmetric matrix $A = LDL^t$;
4. Tridiagonal matrix $A = LU$ (in banded form).

Linear systems are then solved based on factorization. It is also possible to compute determinants and inverses and to estimate the round-off error involved.

Many of the subroutines in LAPACK can be implemented using MATLAB. A nonsingular matrix A is factored using the command

$$[LUP] = lu(A)$$

into the form $PA = LU$, where P is the permutation matrix defined by performing partial pivoting to solve a linear system involving A. If the nonsingular matrix A and the vector $\mathbf{b}$ have been defined in MATLAB, the command

$$x = A\backslash b$$

solves the linear system by first using the $PA = LU$ factoring command. Then it solves the lower-triangular system $L\mathbf{z} = \mathbf{b}$ for $\mathbf{z}$ using its command

$$z = L\backslash b$$

This is followed by a solution to the upper-triangular system $U\mathbf{x} = \mathbf{z}$ using the command

$$x = U\backslash b$$

Other MATLAB commands include computing the inverse, transpose, and determinant of matrix A by issuing the commands $inv(A)$, A', and $\det(A)$, respectively. The Maple command `linsolve(A,b)` is used to solve the linear system $Ax = b$. If the system has no solution, x is given the value NULL.

Further information on the numerical solution of linear systems and matrices can be found in Golub and Van Loan [GV], Forsythe and Moler [FM], and Stewart [Ste]. The use of direct techniques for solving large sparse systems is discussed in detail in George and Liu [GL] and in Pissanetzky [Pi].

CHAPTER 7

Iterative Methods for Solving Linear Systems

7.1 Introduction

The previous chapter considered the approximation of the solution of a linear system using direct methods, techniques that would produce the exact solution if all the calculations are performed using exact arithmetic. In this chapter we will describe some popular *iterative* techniques. These methods would not return the exact solution even if all the calculations could be performed using exact arithmetic. In many instances, however, they are more effective than the direct methods, since they can require far less computational effort and round-off error is reduced. This is particularly true when the matrix is **sparse**—that is, has a high percentage of zero entries.

Some additional material from linear algebra is needed to describe the convergence of the iterative methods. Principally, we need to have a measure of how close two vectors are to one another, since the object of an iterative method is to determine an approximation that is within a certain tolerance of the exact solution. In Section 7.2, the notion of a norm is used to show how various forms of distance between vectors can be described. We will also see how this concept can be extended to describe the norm of—and, consequently, the distance between—matrices. In Section 7.3, matrix eigenvalues and eigenvectors are described, and we consider the connection between these concepts and the convergence of an iterative method.

Section 7.4 describes the elementary Jacobi and Gauss-Seidel iterative methods. By analyzing the size of the largest eigenvalue of a matrix associated with an iterative method, we can determine conditions that predict the likelihood of convergence of the method. In Section 7.5 we introduce the SOR method. This is a commonly applied iterative technique, since it reduces the approximation errors faster than the Jacobi and Gauss-Seidel methods.

The final two sections in the chapter discuss some of the concerns that should be addressed when applying either an iterative or direct technique for approximating the solution to a linear system.

7.2 Convergence of Vectors

The distance between the real numbers x and y is $|x - y|$. In Chapter 2 we saw that the stopping techniques for the iterative root-finding techniques used this measure to estimate the accuracy of approximate solutions and to determine when an approximation was sufficiently accurate. The iterative methods for solving systems of equations use similar logic, so the first step is to determine a measurement procedure for n-dimensional vectors, the form that is taken by the solution to a system of equations.

Let $\mathbb{R}^n$ denote the set of all n-dimensional column vectors with real number coefficients. It is a space-saving convenience to use the transpose notation presented in Section 6.4 when such a vector is represented in terms of its components. For example, the vector

$$\mathbf{x} = \begin{bmatrix} x_1 \\ x_2 \\ \vdots \\ x_n \end{bmatrix}$$

is generally written $\mathbf{x} = (x_1, x_2, \ldots, x_n)^t$.

Vector Norm on $\mathbb{R}^n$

A vector norm on $\mathbb{R}^n$ is a function, $\|\cdot\|$, from $\mathbb{R}^n$ into $\mathbb{R}$ with the following properties:

(i) $\|\mathbf{x}\| \geq 0$ for all $\mathbf{x} \in \mathbb{R}^n$,
(ii) $\|\mathbf{x}\| = 0$ if and only if $\mathbf{x} = (0, 0, \ldots, 0)^t \equiv \mathbf{0}$,
(iii) $\|\alpha\mathbf{x}\| = |\alpha|\|\mathbf{x}\|$ for all $\alpha \in \mathbb{R}$ and $\mathbf{x} \in \mathbb{R}^n$,
(iv) $\|\mathbf{x} + \mathbf{y}\| \leq \|\mathbf{x}\| + \|\mathbf{y}\|$ for all $\mathbf{x}, \mathbf{y} \in \mathbb{R}^n$.

For our purposes, we need only two specific norms on $\mathbb{R}^n$. (A third is presented in Exercise 2.)

The l_2 and l_∞ norms for the vector $\mathbf{x} = (x_1, x_2, \ldots, x_n)^t$ are defined by

$$\|\mathbf{x}\|_2 = \left\{\sum_{i=1}^{n} x_i^2\right\}^{1/2} \quad \text{and} \quad \|\mathbf{x}\|_\infty = \max_{1 \leq i \leq n} |x_i|.$$

The l_2 norm is called the **Euclidean norm** of the vector $\mathbf{x}$, since it represents the usual notion of distance from the origin in case $\mathbf{x}$ is in $\mathbb{R}^1 \equiv \mathbb{R}$, $\mathbb{R}^2$, or $\mathbb{R}^3$. For example, the l_2 norm of the vector $\mathbf{x} = (x_1, x_2, x_3)^t$ gives the length of the straight line joining the points $(0, 0, 0)$ and (x_1, x_2, x_3); that is, the length of the shortest path between those two points. Figure 7.1 shows the boundary of those vectors in $\mathbb{R}^2$ and $\mathbb{R}^3$ that have l_2 norm less than 1. Figure 7.2 gives a similar illustration for the l_∞ norm.

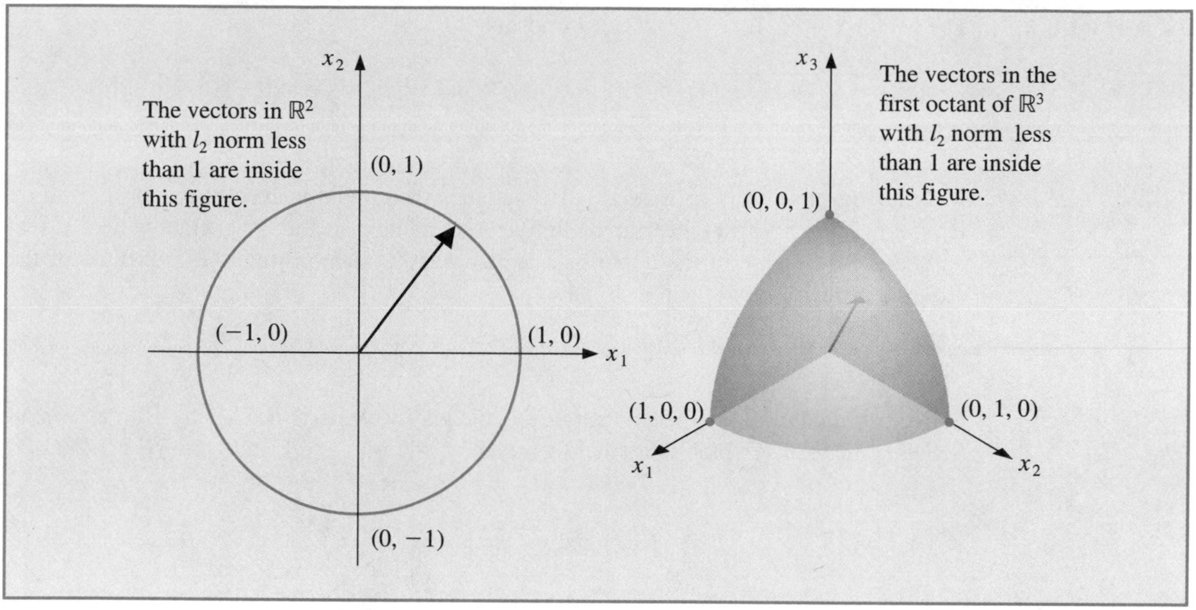

Figure 7.1

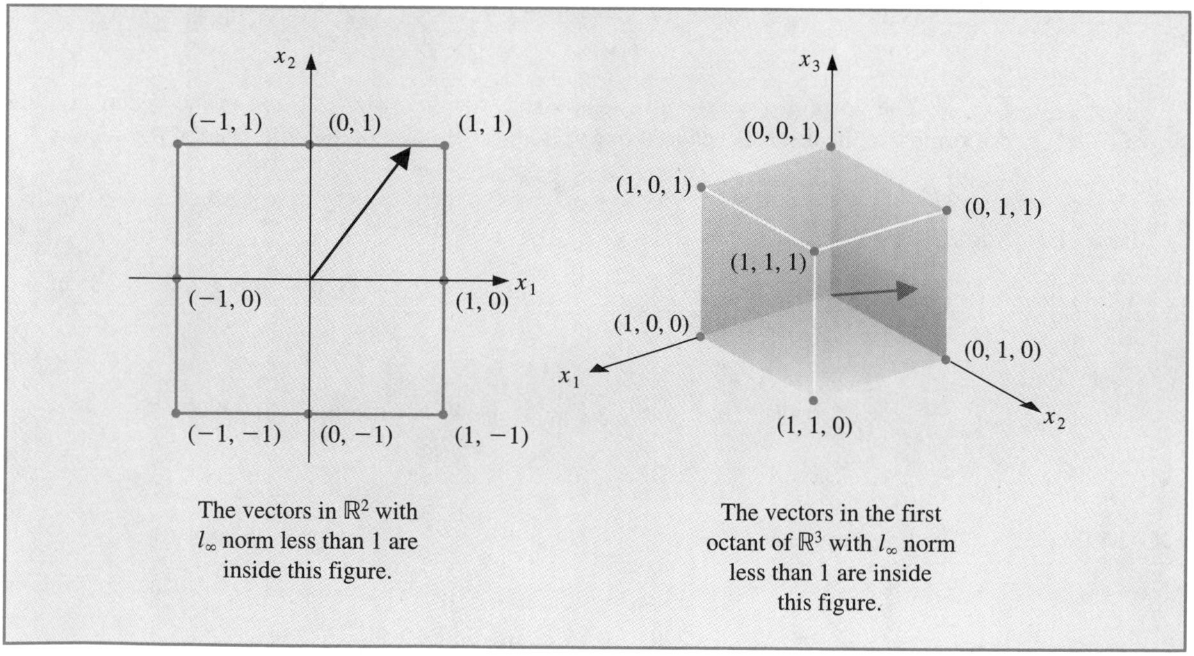

Figure 7.2

EXAMPLE 1 The vector $\mathbf{x} = (-1, 1, -2)^t$ in $\mathbb{R}^3$ has norms

$$\|\mathbf{x}\|_2 = \sqrt{(-1)^2 + (1)^2 + (-2)^2} = \sqrt{6} \quad \text{and} \quad \|\mathbf{x}\|_\infty = \max\{|-1|, |1|, |-2|\} = 2.$$

◆ ◆ ◆

Showing that $\|\mathbf{x}\|_\infty = \max_{1\le i\le n} |x_i|$ satisfies the conditions necessary for a norm on $\mathbb{R}^n$ follows directly from the truth of similar statements concerning absolute values of real numbers. In the case of the l_2 norm, it is also easy to demonstrate the first three of the required properties, but the fourth,

$$\|\mathbf{x} + \mathbf{y}\|_2 \le \|\mathbf{x}\|_2 + \|\mathbf{y}\|_2,$$

is more difficult to show. To demonstrate this inequality we need the Cauchy-Buniakowsky-Schwarz inequality, which states that for any $\mathbf{x} = (x_1, x_2, \ldots, x_n)^t$ and $\mathbf{y} = (y_1, y_2, \ldots, y_n)^t$,

$$\sum_{i=1}^n |x_i y_i| \le \left\{\sum_{i=1}^n x_i^2\right\}^{1/2} \left\{\sum_{i=1}^n y_i^2\right\}^{1/2}.$$

With this it follows that $\|\mathbf{x} + \mathbf{y}\|_2 \le \|\mathbf{x}\|_2 + \|\mathbf{y}\|_2$ since

$$\begin{aligned}\|\mathbf{x} + \mathbf{y}\|_2^2 &= \sum_{i=1}^n x_i^2 + 2\sum_{i=1}^n x_i y_i + \sum_{i=1}^n y_i^2 \le \sum_{i=1}^n x_i^2 + 2\sum_{i=1}^n |x_i y_i| + \sum_{i=1}^n y_i^2 \\ &\le \sum_{i=1}^n x_i^2 + 2\left\{\sum_{i=1}^n x_i^2\right\}^{1/2} \left\{\sum_{i=1}^n y_i^2\right\}^{1/2} + \sum_{i=1}^n y_i^2 = \left(\|\mathbf{x}\|_2 + \|\mathbf{y}\|_2\right)^2.\end{aligned}$$

The norm of a vector gives a measure for the distance between the vector and the origin, so the distance between two vectors is the norm of the difference of the vectors.

Distance Between Two Vectors

If $\mathbf{x} = (x_1, x_2, \ldots, x_n)^t$ and $\mathbf{y} = (y_1, y_2, \ldots, y_n)^t$ are vectors in $\mathbb{R}^n$, the l_2 and l_∞ distances between $\mathbf{x}$ and $\mathbf{y}$ are defined by

$$\|\mathbf{x} - \mathbf{y}\|_2 = \left\{\sum_{i=1}^n (x_i - y_i)^2\right\}^{1/2} \quad \text{and} \quad \|\mathbf{x} - \mathbf{y}\|_\infty = \max_{1\le i\le n} |x_i - y_i|.$$

EXAMPLE 2 The linear system

$$\begin{aligned}3.3330x_1 + 15920x_2 - 10.333x_3 &= 15913, \\ 2.2220x_1 + 16.710x_2 + 9.6120x_3 &= 28.544, \\ 1.5611x_1 + 5.1791x_2 + 1.6852x_3 &= 8.4254\end{aligned}$$

has solution $(x_1, x_2, x_3)^t = (1.0000, 1.0000, 1.0000)^t$. If Gaussian elimination is performed in five-digit rounding arithmetic using partial pivoting, the solution obtained is

$$\tilde{\mathbf{x}} = (\tilde{x}_1, \tilde{x}_2, \tilde{x}_3)^t = (1.2001, 0.99991, 0.92538)^t.$$

Measurements of $\mathbf{x} - \tilde{\mathbf{x}}$ are given by

$$\begin{aligned}\|\mathbf{x} - \tilde{\mathbf{x}}\|_\infty &= \max\{|1.0000 - 1.2001|, |1.0000 - 0.99991|, |1.0000 - 0.92538|\}\\ &= \max\{0.2001, 0.00009, 0.07462\} = 0.2001\end{aligned}$$

and

$$\begin{aligned}\|\mathbf{x} - \tilde{\mathbf{x}}\|_2 &= \left[(1.0000 - 1.2001)^2 + (1.0000 - 0.99991)^2 + (1.0000 - 0.92538)^2\right]^{1/2}\\ &= \left[(0.2001)^2 + (0.00009)^2 + (0.07462)^2\right]^{1/2} = 0.21356.\end{aligned}$$

Although the components $\tilde{x}_2$ and $\tilde{x}_3$ are good approximations to x_2 and x_3, the component $\tilde{x}_1$ is a poor approximation to x_1, and $|x_1 - \tilde{x}_1|$ dominates both norms. ◆ ◆ ◆

The distance concept in $\mathbb{R}^n$ is used to define a limit of a sequence of vectors. A sequence $\{\mathbf{x}^{(k)}\}_{k=1}^{\infty}$ of vectors in $\mathbb{R}^n$ is said to **converge** to $\mathbf{x}$ with respect to the norm $\|\cdot\|$ if, given any $\epsilon > 0$, there exists an integer $N(\epsilon)$ such that

$$\|\mathbf{x}^{(k)} - \mathbf{x}\| < \epsilon \quad \text{for all } k \geq N(\epsilon).$$

EXAMPLE 3 Let $\mathbf{x}^{(k)} \in \mathbb{R}^4$ be defined by

$$\mathbf{x}^{(k)} = \left(x_1^{(k)}, x_2^{(k)}, x_3^{(k)}, x_4^{(k)}\right)^t = \left(1, 2 + \frac{1}{k}, \frac{3}{k^2}, e^{-k}\sin k\right)^t.$$

Since $\lim_{k\to\infty} 1 = 1$, $\lim_{k\to\infty}(2 + 1/k) = 2$, $\lim_{k\to\infty} 3/k^2 = 0$, and $\lim_{k\to\infty} e^{-k}\sin k = 0$, and for any given ϵ an integer $N(\epsilon)$ can be found so that we simultaneously have $|x_1^{(k)} - 1|$, $|x_2^{(k)} - 2|$, $|x_3^{(k)} - 0|$, and $|x_4^{(k)} - 0|$ less than ϵ. This implies that the sequence $\{\mathbf{x}^{(k)}\}$ converges to $(1, 2, 0, 0)^t$ with respect to $\|\cdot\|_\infty$. ◆ ◆ ◆

To show directly that the sequence in Example 3 converges to $(1, 2, 0, 0)^t$ with respect to the l_2 norm is quite complicated. However, suppose that j is an index with the property that

$$\|\mathbf{x}\|_\infty = \max_{i=1,\ldots,n} |x_i| = |x_j|.$$

Then

$$\|\mathbf{x}\|_\infty^2 = |x_j|^2 = x_j^2 \leq \sum_{i=1}^{n} x_i^2 = \|\mathbf{x}\|_2^2$$

and

$$\|\mathbf{x}\|_2 = \sum_{i=1}^{n} x_i^2 \leq \sum_{i=1}^{n} x_j^2 = n x_j^2 = n\|\mathbf{x}\|_\infty^2.$$

These inequalities imply that the sequence of vectors $\{\mathbf{x}^{(k)}\}$ converges to $\mathbf{x}$ in $\mathbb{R}^n$ with respect to $\|\cdot\|_2$ if and only if $\lim_{k\to\infty} x_i^{(k)} = x_i$ for each $i = 1, 2, \ldots, n$, since this is when the sequence converges in the l_∞ norm.

In fact, it can be shown that all norms on $\mathbb{R}^n$ are equivalent with respect to convergence; that is, if $\|\cdot\|$ and $\|\cdot\|'$ are any two norms on $\mathbb{R}^n$ and $\{\mathbf{x}^{(k)}\}_{k=1}^{\infty}$ has the limit $\mathbf{x}$ with respect to $\|\cdot\|$, then $\{\mathbf{x}^{(k)}\}_{k=1}^{\infty}$ has the limit $\mathbf{x}$ with respect to $\|\cdot\|'$. Since a vector sequence converges in the l_∞ norm precisely when each of its component sequences converges, we have the following.

Vector Sequence Convergence

The sequence of vectors $\{\mathbf{x}^{(k)}\}$ converges to $\mathbf{x}$ in $\mathbb{R}^n$ if and only if $\lim_{k\to\infty} x_i^{(k)} = x_i$ for each $i = 1, 2, \ldots, n$.

In the subsequent sections, we will need methods for determining the distance between $n \times n$ matrices. This again requires the use of a norm.

Matrix Norm

A matrix norm on the set of all $n \times n$ matrices is a real-valued function, $\|\cdot\|$, defined on this set, satisfying for all $n \times n$ matrices A and B and all real numbers α:

(i) $\|A\| \geq 0$,
(ii) $\|A\| = 0$, if and only if A is O, the matrix with all zero entries,
(iii) $\|\alpha A\| = |\alpha|\|A\|$,
(iv) $\|A + B\| \leq \|A\| + \|B\|$,
(v) $\|AB\| \leq \|A\|\|B\|$.

A **distance between $n \times n$ matrices** A and B with respect to this matrix norm is $\|A - B\|$. Although matrix norms can be obtained in various ways, the only norms we consider are those that are natural consequences of a vector norm.

Natural Matrix Norm

If $\|\cdot\|$ is a vector norm on $\mathbb{R}^n$, the natural matrix norm on the set of $n \times n$ matrices given by $\|\cdot\|$ is defined by

$$\|A\| = \max_{\|\mathbf{x}\|=1} \|A\mathbf{x}\|.$$

As a consequence, the matrix norms we will consider have the forms

$$\|A\|_\infty = \max_{\|\mathbf{x}\|_\infty=1} \|A\mathbf{x}\|_\infty, \quad \text{(the } l_\infty \text{ norm)}$$

and

$$\|A\|_2 = \max_{\|\mathbf{x}\|_2=1} \|A\mathbf{x}\|_2, \quad \text{(the } l_2 \text{ norm).}$$

When $n = 2$ these norms have the geometric representations shown in Figures 7.3 and 7.4.

Figure 7.3

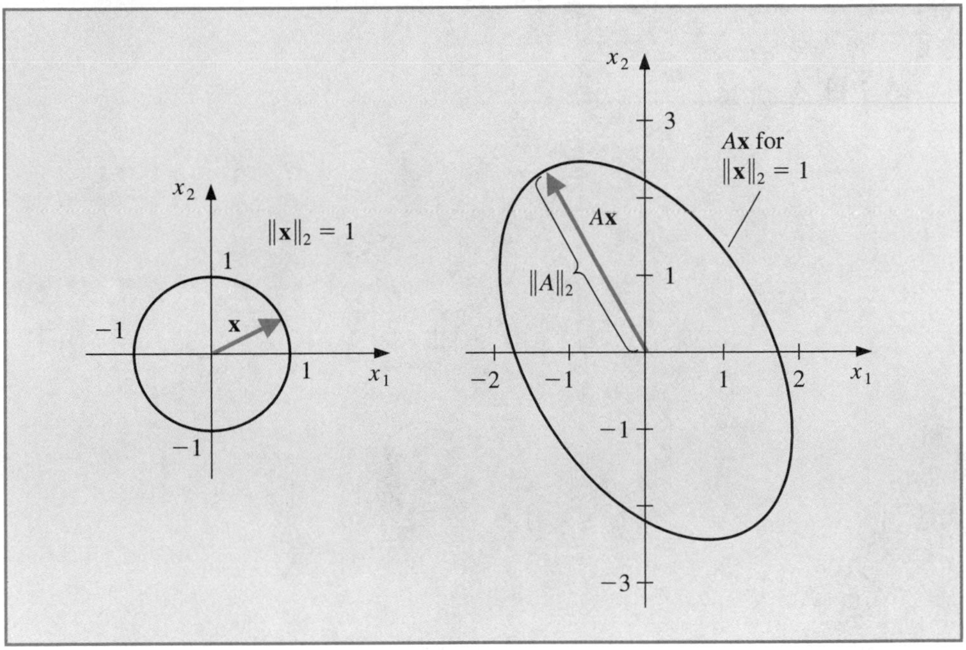

Figure 7.4

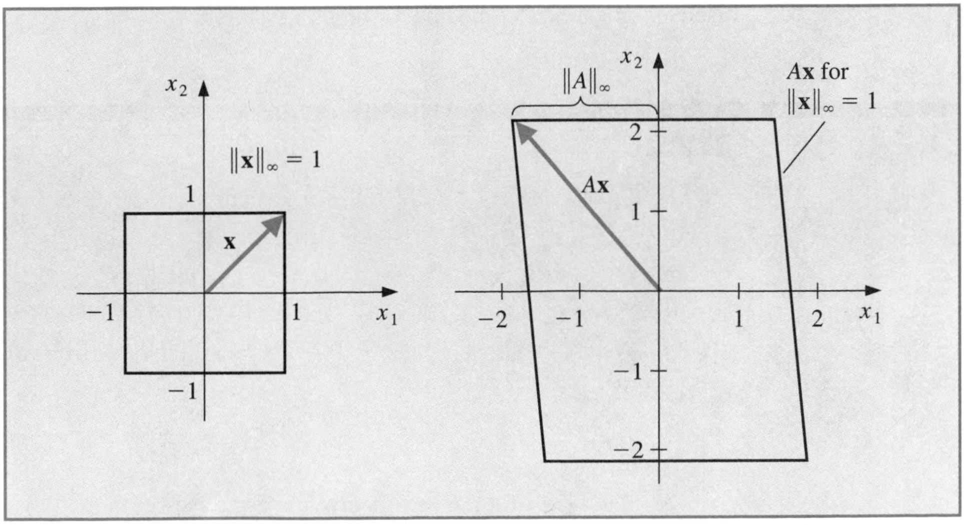

The l_∞ norm of a matrix has a representation with respect to the entries of the matrix that makes it particularly easy to compute.

l_∞ Norm of a Matrix

$$\|A\|_\infty = \max_{1\le i\le n} \sum_{j=1}^{n} |a_{ij}|.$$

EXAMPLE 4 If

$$A = \begin{bmatrix} 1 & 2 & -1 \\ 0 & 3 & -1 \\ 5 & -1 & 1 \end{bmatrix},$$

then

$$\sum_{j=1}^{3} |a_{1j}| = |1| + |2| + |-1| = 4, \qquad \sum_{j=1}^{3} |a_{2j}| = |0| + |3| + |-1| = 4,$$

and

$$\sum_{j=1}^{3} |a_{3j}| = |5| + |-1| + |1| = 7.$$

So

$$\|A\|_\infty = \max\{4, 4, 7\} = 7.$$

◆ ◆ ◆

The l_2 norm of a matrix is not as easily determined, but in the next section we will discover an alternative method for finding this norm.

EXERCISE SET 7.2

1. Find $\|\mathbf{x}\|_\infty$ and $\|\mathbf{x}\|_2$ for the following vectors.
   - **a.** $\mathbf{x} = \left(3, -4, 0, \frac{3}{2}\right)^t$
   - **b.** $\mathbf{x} = (2, 1, -3, 4)^t$
   - **c.** $\mathbf{x} = (\sin k, \cos k, 2^k)^t$ for a fixed positive integer k
   - **d.** $\mathbf{x} = (4/(k+1), 2/k^2, k^2 e^{-k})^t$ for a fixed positive integer k
2. **a.** Verify that $\|\cdot\|_1$ is a norm for $\mathbb{R}^n$ (called the l_1 norm), where

   $$\|\mathbf{x}\|_1 = \sum_{i=1}^{n} |x_i|.$$

   b. Find $\|\mathbf{x}\|_1$ for the vectors given in Exercise 1.

3. Show that the following sequences are convergent, and find their limits.
   a. $\mathbf{x}^{(k)} = (1/k, e^{1-k}, -2/k^2)^t$
   b. $\mathbf{x}^{(k)} = \left(e^{-k}\cos k, k\sin(1/k), 3 + k^{-2}\right)^t$
   c. $\mathbf{x}^{(k)} = \left(ke^{-k^2}, (\cos k)/k, \sqrt{k^2+k} - k\right)^t$
   d. $\mathbf{x}^{(k)} = (e^{1/k}, (k^2+1)/(1-k^2), (1/k^2)(1+3+5+\cdots+(2k-1)))^t$

4. Find $\|\cdot\|_\infty$ for the following matrices.

   a. $\begin{bmatrix} 10 & 15 \\ 0 & 1 \end{bmatrix}$ b. $\begin{bmatrix} 10 & 0 \\ 15 & 1 \end{bmatrix}$

   c. $\begin{bmatrix} 2 & -1 & 0 \\ -1 & 2 & -1 \\ 0 & -1 & 2 \end{bmatrix}$ d. $\begin{bmatrix} 4 & -1 & 7 \\ -1 & 4 & 0 \\ -7 & 0 & 4 \end{bmatrix}$

5. The following linear systems $A\mathbf{x} = \mathbf{b}$ have $\mathbf{x}$ as the actual solution and $\tilde{\mathbf{x}}$ as an approximate solution. Compute $\|\mathbf{x} - \tilde{\mathbf{x}}\|_\infty$ and $\|A\tilde{\mathbf{x}} - \mathbf{b}\|_\infty$.

   a. $$\frac{1}{2}x_1 + \frac{1}{3}x_2 = \frac{1}{63},$$
   $$\frac{1}{3}x_1 + \frac{1}{4}x_2 = \frac{1}{168},$$
   $$\mathbf{x} = \left(\frac{1}{7}, -\frac{1}{6}\right)^t,$$
   $$\tilde{\mathbf{x}} = (0.142, -0.166)^t.$$

   b. $$x_1 + 2x_2 + 3x_3 = 1,$$
   $$2x_1 + 3x_2 + 4x_3 = -1,$$
   $$3x_1 + 4x_2 + 6x_3 = 2,$$
   $$\mathbf{x} = (0, -7, 5)^t,$$
   $$\tilde{\mathbf{x}} = (-0.33, -7.9, 5.8)^t.$$

   c. $$x_1 + 2x_2 + 3x_3 = 1,$$
   $$2x_1 + 3x_2 + 4x_3 = -1,$$
   $$3x_1 + 4x_2 + 6x_3 = 2,$$
   $$\mathbf{x} = (0, -7, 5)^t,$$
   $$\tilde{\mathbf{x}} = (-0.2, -7.5, 5.4)^t.$$

   d. $$0.04x_1 + 0.01x_2 - 0.01x_3 = 0.06,$$
   $$0.2x_1 + 0.5x_2 - 0.2x_3 = 0.3,$$
   $$x_1 + 2x_2 + 4x_3 = 11,$$
   $$\mathbf{x} = (1.827586, 0.6551724, 1.965517)^t,$$
   $$\tilde{\mathbf{x}} = (1.8, 0.64, 1.9)^t.$$

6. The l_1 matrix norm, defined by $\|A\|_1 = \max_{\|\mathbf{x}\|_1=1} \|A\mathbf{x}\|_1$, can be computed using the formula

$$\|A\|_1 = \max_{1\le j\le n} \sum_{i=1}^{n} |a_{ij}|,$$

where the l_1 vector norm is defined in Exercise 2. Find the l_1 norm of the matrices in Exercise 4.

7. Show by example that $\|\cdot\|_{\textcircled{\infty}}$, defined by $\|A\|_{\textcircled{\infty}} = \max_{1\le i,j\le n} |a_{ij}|$, does not define a matrix norm.

8. Show that $\|\cdot\|_{\textcircled{1}}$, defined by

$$\|A\|_{\textcircled{1}} = \sum_{i=1}^{n}\sum_{j=1}^{n} |a_{ij}|,$$

is a matrix norm. Find $\|\cdot\|_{\textcircled{1}}$ for the matrices in Exercise 4.

9. Show that if $\|\cdot\|$ is a vector norm on $\mathbb{R}^n$, then $\|A\| = \max_{\|\mathbf{x}\|=1} \|A\mathbf{x}\|$ is a matrix norm.

10. The following excerpt from *Mathematics Magazine* [Sz] gives a way to prove the Cauchy-Buniakowsky-Schwarz Inequality.

a. Show that when $\mathbf{x} \ne \mathbf{0}$ and $\mathbf{y} \ne \mathbf{0}$, we have

$$\frac{\sum_{i=1}^{n} x_i y_i}{\left(\sum_{i=1}^{n} x_i^2\right)^{1/2}\left(\sum_{i=1}^{n} y_i^2\right)^{1/2}} = 1 - \frac{1}{2}\sum_{i=1}^{n}\left(\frac{x_i}{\left(\sum_{j=1}^{n} x_j^2\right)^{1/2}} - \frac{y_i}{\left(\sum_{j=1}^{n} y_j^2\right)^{1/2}}\right)^2.$$

b. Use the result in part (a) to show that

$$\sum_{i=1}^{n} x_i y_i \le \left(\sum_{i=1}^{n} x_i^2\right)^{1/2}\left(\sum_{i=1}^{n} y_i^2\right)^{1/2}.$$

c. Replace y_i with $-y_i$ whenever $x_i y_i < 0$ in the inequality in (b) and explain why this inequality then implies that

$$\sum_{i=1}^{n} |x_i y_i| \le \left(\sum_{i=1}^{n} x_i^2\right)^{1/2}\left(\sum_{i=1}^{n} y_i^2\right)^{1/2}.$$

7.3 Eigenvalues and Eigenvectors

An $n \times m$ matrix can be considered as a function that uses matrix multiplication to take m-dimensional vectors into n-dimensional vectors. A square matrix A takes the set of n-dimensional vectors into itself. In this case certain nonzero vectors have $\mathbf{x}$ and $A\mathbf{x}$ parallel, which means that a constant λ exists with $A\mathbf{x} = \lambda\mathbf{x}$; that is, $(A - \lambda I)\mathbf{x} = \mathbf{0}$. There is a close connection between these numbers λ and the likelihood that an iterative method will converge. We will consider this connection in this section.

For a square $n \times n$ matrix A, the **characteristic polynomial** of A is defined by

$$p(\lambda) = \det(A - \lambda I).$$

Because of the way the determinant of a matrix is defined, p is an nth-degree polynomial and, consequently, has at most n distinct zeros, some of which may be complex. These zeros of p are called the **eigenvalues** of the matrix A.

If λ is an eigenvalue, then $\det(A - \lambda I) = 0$, and the equivalence result at the end of Section 6.4 implies that $A - \lambda I$ is a singular matrix. As a consequence, the linear system defined by $(A - \lambda I)\mathbf{x} = \mathbf{0}$ has a solution other than the zero vector. If $(A - \lambda I)\mathbf{x} = \mathbf{0}$ and $\mathbf{x} \neq \mathbf{0}$, then $\mathbf{x}$ is called an **eigenvector** of A corresponding to the eigenvalue λ.

If $\mathbf{x}$ is an eigenvector associated with the eigenvalue λ, then $A\mathbf{x} = \lambda\mathbf{x}$, so the matrix A takes the vector $\mathbf{x}$ into a scalar multiple of itself. When λ is a real number and $\lambda > 1$, A has the effect of stretching $\mathbf{x}$ by a factor of λ. When $0 < \lambda < 1$, A shrinks $\mathbf{x}$ by a factor of λ. When $\lambda < 0$, the effects are similar, but the direction is reversed (see Figure 7.5).

Figure 7.5

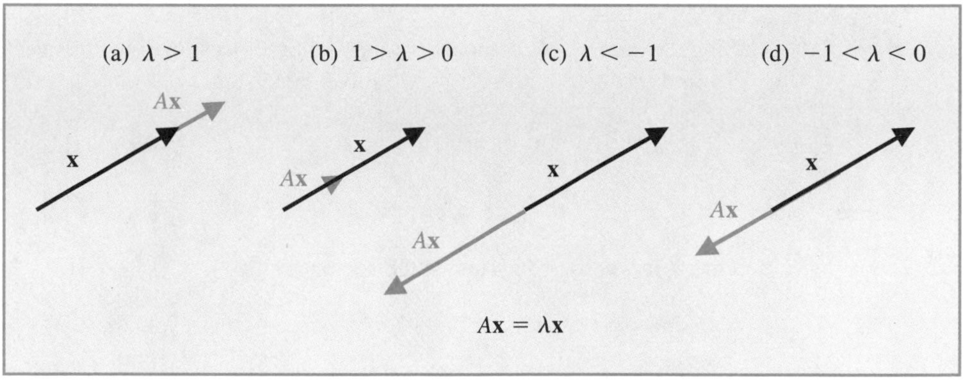

EXAMPLE 1 Let

$$A = \begin{bmatrix} 1 & 0 & 2 \\ 0 & 1 & -1 \\ -1 & 1 & 1 \end{bmatrix}.$$

To compute the eigenvalues of A, consider

$$p(\lambda) = \det(A - \lambda I) = \det \begin{bmatrix} 1-\lambda & 0 & 2 \\ 0 & 1-\lambda & -1 \\ -1 & 1 & 1-\lambda \end{bmatrix} = (1-\lambda)(\lambda^2 - 2\lambda + 4).$$

The eigenvalues of A are the solutions of $p(\lambda) = 0$: $\lambda_1 = 1, \lambda_2 = 1 + \sqrt{3}i$, and $\lambda_3 = 1 - \sqrt{3}i$.

An eigenvector $\mathbf{x}$ of A associated with λ_1 is a solution of the system $(A - \lambda_1 I)\mathbf{x} = \mathbf{0}$:

$$\begin{bmatrix} 0 & 0 & 2 \\ 0 & 0 & -1 \\ -1 & 1 & 0 \end{bmatrix} \cdot \begin{bmatrix} x_1 \\ x_2 \\ x_3 \end{bmatrix} = \begin{bmatrix} 0 \\ 0 \\ 0 \end{bmatrix}.$$

Thus

$$2x_3 = 0, \quad -x_3 = 0, \quad \text{and} \quad -x_1 + x_2 = 0,$$

which implies that

$$x_3 = 0, \quad x_2 = x_1, \quad \text{and} \quad x_1 \text{ is arbitrary.}$$

The choice $x_1 = 1$ produces the eigenvector $(1, 1, 0)^t$ with $\|(1, 1, 0)\|_\infty = 1$, corresponding to the eigenvalue $\lambda_1 = 1$.

The choice $x_1 = \frac{\sqrt{2}}{2}$ produces an eigenvector corresponding to λ_1 with

$$\left\| \left(\frac{\sqrt{2}}{2}, \frac{\sqrt{2}}{2}, 0 \right) \right\|_2 = 1.$$

Since λ_2 and λ_3 are complex numbers, their corresponding eigenvectors are also complex. To find an eigenvector for λ_2, we solve the system

$$\begin{bmatrix} 1-(1+\sqrt{3}i) & 0 & 2 \\ 0 & 1-(1+\sqrt{3}i) & -1 \\ -1 & 1 & 1-(1+\sqrt{3}i) \end{bmatrix} \begin{bmatrix} x_1 \\ x_2 \\ x_3 \end{bmatrix} = \begin{bmatrix} 0 \\ 0 \\ 0 \end{bmatrix}.$$

One solution to this system is the vector

$$\left(-\frac{2\sqrt{3}}{3}i, \frac{\sqrt{3}}{3}i, 1 \right)^t.$$

Similarly, the vector

$$\left(\frac{2\sqrt{3}}{3}i, -\frac{\sqrt{3}}{3}i, 1 \right)^t$$

is an eigenvector corresponding to the eigenvalue $\lambda_3 = 1 - \sqrt{3}i$. ◆ ◆ ◆

Maple provides the function `Eigenvals` to compute eigenvalues and, optionally, eigenvectors of a matrix. For the example we enter the following:

```
>with(linalg);
>A:=matrix(3,3,[1,0,2,0,1,-1,-1,1,1]);
>evalf(Eigenvals(A));
```

$$[1.000000000 + 1.732050807I, \quad 1.000000000 - 1.732050807I, \quad 1.000000000]$$

This computes the eigenvalues

$$\lambda_2 = 1 + \sqrt{3}i, \quad \lambda_3 = 1 - \sqrt{3}i, \quad \lambda_1 = 1.$$

To compute both the eigenvalues and eigenvectors, use

```
>evalf(Eigenvals(A,B));
```

The eigenvalues are computed and displayed as before, and the eigenvectors are indicated by the columns of B. If the eigenvalues are all real, each column of B gives an eigenvector. To see what happens when we have a complex eigenvalue, let us display B using

```
>evalm(B);
```

$$B = \begin{bmatrix} 1.154700538 & .6324555321\ 10^{-10} & .7453559925 \\ -.5773502680 & .1264911064\ 10^{-9} & .7453559926 \\ -.2581988896\ 10^{-19} & 1.000000000 & -.72572776\ 10^{-11} \end{bmatrix}$$

The first two columns correspond to the real and imaginary parts of the eigenvectors corresponding to eigenvalues λ_2 and λ_3. Thus, an eigenvector for λ_2 is

$$\begin{bmatrix} 1.154700538 \\ -.5773502680 \\ -.2581988896\ 10^{-19} \end{bmatrix} + \begin{bmatrix} .6324555321\ 10^{-10} \\ .1264911064\ 10^{-9} \\ 1.000000000 \end{bmatrix} i \approx \begin{bmatrix} 1.154700538 \\ -.5773502680 \\ 0 \end{bmatrix} + \begin{bmatrix} 0 \\ 0 \\ 1 \end{bmatrix} i,$$

or,

$$(1.54700538, -0.5773502680, i)^t = \left(\frac{2\sqrt{3}}{3}, -\frac{\sqrt{3}}{3}, i\right)^t.$$

Since any multiple of an eigenvector is also an eigenvector, multiplying each coordinate by $-i$ gives the previously determined eigenvector

$$\left(-\frac{2\sqrt{3}}{3}i, \frac{\sqrt{3}}{3}i, 1\right)^t.$$

Similarly, we can obtain the eigenvector

$$\left(\frac{2\sqrt{3}}{3}i, -\frac{\sqrt{3}}{3}i, 1\right)^t$$

for the eigenvalue λ_3. Since λ_1 is real, the third column of B is an eigenvector corresponding to λ_1. We can obtain the eigenvector $(1, 1, 0)^t$ from the third column by setting the third coordinate to zero and dividing by the first coordinate.

The notions of eigenvalues and eigenvectors are introduced here for a specific computational convenience, but these concepts arise frequently in the study of physical systems. In fact, they are of sufficient interest that Chapter 9 is devoted to their numerical approximation.

The **spectral radius** $\rho(A)$ of a matrix A is defined by

$$\rho(A) = \max |\lambda|, \quad \text{where } \lambda \text{ is an eigenvalue of } A.$$

(*Note*: For complex $\lambda = \alpha + \beta i$, we have $|\lambda| = (\alpha^2 + \beta^2)^{1/2}$.)

EXAMPLE 2 For the matrix considered in Example 1,

$$\rho(A) = \max\{1, |1 + \sqrt{3}i|, |1 - \sqrt{3}i|\} = \max\{1, 2, 2\} = 2. \quad \blacklozenge \blacklozenge \blacklozenge$$

The spectral radius is closely related to the norm of a matrix.

l_2 Matrix Norm Characterization

If A is an $n \times n$ matrix, then

(a) $\|A\|_2 = [\rho(A^t A)]^{1/2}$;
(b) $\rho(A) \le \|A\|$ for any natural norm.

The first part of this result is the computational method for determining the l_2 norm of matrices that we mentioned at the end of the previous section.

EXAMPLE 3 If

$$A = \begin{bmatrix} 1 & 1 & 0 \\ 1 & 2 & 1 \\ -1 & 1 & 2 \end{bmatrix},$$

then

$$A^t A = \begin{bmatrix} 1 & 1 & -1 \\ 1 & 2 & 1 \\ 0 & 1 & 2 \end{bmatrix} \begin{bmatrix} 1 & 1 & 0 \\ 1 & 2 & 1 \\ -1 & 1 & 2 \end{bmatrix} = \begin{bmatrix} 3 & 2 & -1 \\ 2 & 6 & 4 \\ -1 & 4 & 5 \end{bmatrix}.$$

To calculate $\rho(A^t A)$, we need the eigenvalues of $A^t A$:

$$0 = \det(A^t A - \lambda I) = \det \begin{bmatrix} 3-\lambda & 2 & -1 \\ 2 & 6-\lambda & 4 \\ -1 & 4 & 5-\lambda \end{bmatrix}$$
$$= -\lambda^3 + 14\lambda^2 - 42\lambda = -\lambda(\lambda^2 - 14\lambda + 42).$$

So

$$\lambda = 0 \quad \text{or} \quad \lambda = 7 \pm \sqrt{7},$$

and

$$\|A\|_2 = \sqrt{\rho(A^t A)} = \sqrt{\max\{0, 7 - \sqrt{7}, 7 + \sqrt{7}\}} = \sqrt{7 + \sqrt{7}} \approx 3.106. \quad \blacklozenge \blacklozenge \blacklozenge$$

The operations in Example 3 can also be performed using Maple:

```
>with(linalg);
>A:=matrix(3,3,[1,1,0,1,2,1,-1,1,2]);
>B:=transpose(A);
>C:=multiply(A,B);
>evalf(Eigenvals(C));
```

$$[0.1097678465\ 10^{-8}, 4.354248690, 9.645751311]$$

Since $\|A\|_2 = \sqrt{\rho(A^t A)} = \sqrt{\rho(C)}$, we have

$$\|A\|_2 = \sqrt{9.645751311} = 3.105760987.$$

Maple also permits us to obtain $\|A\|_2 = \sqrt{7 + \sqrt{7}}$ directly with the command

```
>norm(A,2);
```

The l_∞ norm of A can be determined with `norm(A)` or `norm (A,infinity)`. For the matrix we are considering, $\|A\|_\infty = 4$.

In studying iterative matrix techniques, it is of particular importance to know when powers of a matrix become small (that is, when all of the entries approach zero). We call an $n \times n$ matrix A **convergent** if for each $i = 1, 2, \ldots, n$ and $j = 1, 2, \ldots, n$ we have

$$\lim_{k \to \infty} (A^k)_{ij} = 0.$$

EXAMPLE 4 Let

$$A = \begin{bmatrix} \frac{1}{2} & 0 \\ \frac{1}{4} & \frac{1}{2} \end{bmatrix}.$$

Computing powers of A, we obtain:

$$A^2 = \begin{bmatrix} \frac{1}{4} & 0 \\ \frac{1}{4} & \frac{1}{4} \end{bmatrix}, \quad A^3 = \begin{bmatrix} \frac{1}{8} & 0 \\ \frac{3}{16} & \frac{1}{8} \end{bmatrix}, \quad A^4 = \begin{bmatrix} \frac{1}{16} & 0 \\ \frac{1}{8} & \frac{1}{16} \end{bmatrix},$$

and, in general,

$$A^k = \begin{bmatrix} \left(\frac{1}{2}\right)^k & 0 \\ \frac{k}{2^{k+1}} & \left(\frac{1}{2}\right)^k \end{bmatrix}.$$

Since

$$\lim_{k \to \infty} \left(\frac{1}{2}\right)^k = 0 \quad \text{and} \quad \lim_{k \to \infty} \frac{k}{2^{k+1}} = 0,$$

A is a convergent matrix. Note that $\rho(A) = \frac{1}{2}$, since $\frac{1}{2}$ is the only eigenvalue of A.

◆ ◆ ◆

An important connection exists between the spectral radius of a matrix and the convergence of the matrix.

Convergent Matrix Equivalences

(a) A is a convergent matrix.
(b) $\lim_{n\to\infty} \|A^n\| = 0$, for some natural norm.
(c) $\lim_{n\to\infty} \|A^n\| = 0$, for all natural norms.
(d) $\rho(A) < 1$.
(e) $\lim_{n\to\infty} A^n\mathbf{x} = \mathbf{0}$, for every $\mathbf{x}$.

EXERCISE SET 7.3

1. Compute the eigenvalues and associated eigenvectors of the following matrices.

a. $\begin{bmatrix} 2 & -1 \\ -1 & 2 \end{bmatrix}$ **b.** $\begin{bmatrix} 0 & 1 \\ 1 & 1 \end{bmatrix}$

c. $\begin{bmatrix} 0 & \frac{1}{2} \\ \frac{1}{2} & 0 \end{bmatrix}$ **d.** $\begin{bmatrix} 1 & 1 \\ -2 & -2 \end{bmatrix}$

e. $\begin{bmatrix} 2 & 1 & 0 \\ 1 & 2 & 0 \\ 0 & 0 & 3 \end{bmatrix}$ **f.** $\begin{bmatrix} -1 & 2 & 0 \\ 0 & 3 & 4 \\ 0 & 0 & 7 \end{bmatrix}$

g. $\begin{bmatrix} 2 & 1 & 1 \\ 2 & 3 & 2 \\ 1 & 1 & 2 \end{bmatrix}$ **h.** $\begin{bmatrix} 3 & 2 & -1 \\ 1 & -2 & 3 \\ 2 & 0 & 4 \end{bmatrix}$

2. Find the spectral radius for each matrix in Exercise 1.

3. Show that

$$A_1 = \begin{bmatrix} 1 & 0 \\ \frac{1}{4} & \frac{1}{2} \end{bmatrix}$$

is not convergent, but

$$A_2 = \begin{bmatrix} \frac{1}{2} & 0 \\ 16 & \frac{1}{2} \end{bmatrix}$$

is convergent.

4. Which of the matrices in Exercise 1 are convergent?

5. Find the $\|\cdot\|_2$ norms of the matrices in Exercise 1.

6. Show that if λ is an eigenvalue of a matrix A and $\|\cdot\|$ is a vector norm, then an eigenvector $\mathbf{x}$ associated with λ exists with $\|\mathbf{x}\| = 1$.

7. Find matrices A and B for which $\rho(A + B) > \rho(A) + \rho(B)$. (This shows that $\rho(A)$ cannot be a matrix norm.)

8. Show that if A is symmetric, then $\|A\|_2 = \rho(A)$.

9. In Exercise 8 of Section 6.4, we assumed that the contribution a female beetle of a certain type made to the future years' beetle population could be expressed in terms of the matrix

$$A = \begin{bmatrix} 0 & 0 & 6 \\ \frac{1}{2} & 0 & 0 \\ 0 & \frac{1}{3} & 0 \end{bmatrix},$$

where the entry in the ith row and jth column represents the probabilistic contribution of a beetle of age j onto the next year's female population of age i.

a. Does the matrix A have any real eigenvalues? If so, determine them and any associated eigenvectors.

b. If a sample of this species was needed for laboratory test purposes that would have a constant proportion in each age group from year to year, what criteria could be imposed on the initial population to ensure that this requirement would be satisfied?

7.4 The Jacobi and Gauss-Seidel Methods

In this section we describe the elementary Jacobi and Gauss-Seidel iterative methods. These are classic methods that date to the late eighteenth century, but they find current application in problems where the matrix is large and has mostly zero entries in predictable locations. Applications of this type are common, for example, in the study of large integrated circuits and in the numerical solution of boundary-value problems and partial-differential equations.

An iterative technique for solving the $n \times n$ linear system $A\mathbf{x} = \mathbf{b}$ starts with an initial approximation $\mathbf{x}^{(0)}$ to the solution $\mathbf{x}$ and generates a sequence of vectors $\{\mathbf{x}^{(k)}\}_{k=1}^{\infty}$ that converges to $\mathbf{x}$. These iterative techniques involve a process that converts the system $A\mathbf{x} = \mathbf{b}$ into an equivalent system of the form $\mathbf{x} = T\mathbf{x} + \mathbf{c}$ for some $n \times n$ matrix T and vector $\mathbf{c}$.

After the initial vector $\mathbf{x}^{(0)}$ is selected, the sequence of approximate solution vectors is generated by computing

$$\mathbf{x}^{(k)} = T\mathbf{x}^{(k-1)} + \mathbf{c}$$

for each $k = 1, 2, 3, \ldots$.

The following result provides an important connection between the eigenvalues of the matrix T and the expectation that the iterative method will converge.

Convergence and the Spectral Radius

For any $\mathbf{x}^{(0)}$ in $\mathbb{R}^n$, the sequence

$$\mathbf{x}^{(k)} = T\mathbf{x}^{(k-1)} + \mathbf{c}$$

converges to the unique solution of $\mathbf{x} = T\mathbf{x} + \mathbf{c}$ if and only if $\rho(T) < 1$.

EXAMPLE 1 The linear system $A\mathbf{x} = \mathbf{b}$ given by

$$\begin{aligned}
E_1:&\quad 10x_1 - x_2 + 2x_3 = 6,\\
E_2:&\quad -x_1 + 11x_2 - x_3 + 3x_4 = 25,\\
E_3:&\quad 2x_1 - x_2 + 10x_3 - x_4 = -11,\\
E_4:&\quad 3x_2 - x_3 + 8x_4 = 15
\end{aligned}$$

has solution $\mathbf{x} = (1, 2, -1, 1)^t$. To convert $A\mathbf{x} = \mathbf{b}$ to the form $\mathbf{x} = T\mathbf{x} + \mathbf{c}$, solve equation E_i for x_i obtaining

$$\begin{aligned}
x_1 &= \frac{1}{10}x_2 - \frac{1}{5}x_3 + \frac{3}{5},\\
x_2 &= \frac{1}{11}x_1 + \frac{1}{11}x_3 - \frac{3}{11}x_4 + \frac{25}{11},\\
x_3 &= -\frac{1}{5}x_1 + \frac{1}{10}x_2 + \frac{1}{10}x_4 - \frac{11}{10},\\
x_4 &= -\frac{3}{8}x_2 + \frac{1}{8}x_3 + \frac{15}{8}.
\end{aligned}$$

Then $A\mathbf{x} = \mathbf{b}$ has the form $\mathbf{x} = T\mathbf{x} + \mathbf{c}$, with

$$T = \begin{bmatrix} 0 & \frac{1}{10} & -\frac{1}{5} & 0 \\ \frac{1}{11} & 0 & \frac{1}{11} & -\frac{3}{11} \\ -\frac{1}{5} & \frac{1}{10} & 0 & \frac{1}{10} \\ 0 & -\frac{3}{8} & \frac{1}{8} & 0 \end{bmatrix} \quad \text{and} \quad \mathbf{c} = \begin{bmatrix} \frac{3}{5} \\ \frac{25}{11} \\ -\frac{11}{10} \\ \frac{15}{8} \end{bmatrix}.$$

For an initial approximation, suppose $\mathbf{x}^{(0)} = (0, 0, 0, 0)^t$. Then $\mathbf{x}^{(1)}$ is given by

$$\begin{aligned}
x_1^{(1)} &= \frac{1}{10}x_2^{(0)} - \frac{1}{5}x_3^{(0)} + \frac{3}{5} = 0.6000,\\
x_2^{(1)} &= \frac{1}{11}x_1^{(0)} + \frac{1}{11}x_3^{(0)} - \frac{3}{11}x_4^{(0)} + \frac{25}{11} = 2.2727,
\end{aligned}$$

$$x_3^{(1)} = -\frac{1}{5}x_1^{(0)} + \frac{1}{10}x_2^{(0)} \qquad + \frac{1}{10}x_4^{(0)} - \frac{11}{10} = -1.1000,$$

$$x_4^{(1)} = \qquad -\frac{3}{8}x_2^{(0)} + \frac{1}{8}x_3^{(0)} \qquad + \frac{15}{8} = 1.8750.$$

Additional iterates, $\mathbf{x}^{(k)} = (x_1^{(k)}, x_2^{(k)}, x_3^{(k)}, x_4^{(k)})^t$, are generated in a similar manner and are presented in Table 7.1. The decision to stop after 10 iterations was based on the criterion

$$\|\mathbf{x}^{(10)} - \mathbf{x}^{(9)}\|_\infty = 8.0 \times 10^{-4} < 10^{-3}.$$

In fact, $\|\mathbf{x}^{(10)} - \mathbf{x}\|_\infty \approx 0.0002$. ◆ ◆ ◆

Table 7.1

k	0	1	2	3	4	5	6	7	8	9	10
$x_1^{(k)}$	0.000	0.6000	1.0473	0.9326	1.0152	0.9890	1.0032	0.9981	1.0006	0.9997	1.0001
$x_2^{(k)}$	0.0000	2.2727	1.7159	2.053	1.9537	2.0114	1.9922	2.0023	1.9987	2.0004	1.9998
$x_3^{(k)}$	0.0000	−1.1000	−0.8052	−1.0493	−0.9681	−1.0103	−0.9945	−1.0020	−0.9990	−1.0004	−0.9998
$x_4^{(k)}$	0.0000	1.8750	0.8852	1.1309	0.9739	1.0214	0.9944	1.0036	0.9989	1.0006	0.9998

The method of Example 1 is called the **Jacobi** iterative method. It consists of solving the ith equation in $A\mathbf{x} = \mathbf{b}$ for x_i to obtain, provided $a_{ii} \neq 0$,

$$x_i = \sum_{\substack{j=1 \\ j\neq i}}^{n} \left(-\frac{a_{ij}x_j}{a_{ii}}\right) + \frac{b_i}{a_{ii}}, \qquad \text{for } i = 1, 2, \ldots, n,$$

and generating each $x_i^{(k)}$ from components of $\mathbf{x}^{(k-1)}$ for $k \geq 1$ by

$$x_i^{(k)} = \frac{\sum_{\substack{j=1 \\ j\neq i}}^{n} \left(-a_{ij}x_j^{(k-1)}\right) + b_i}{a_{ii}}, \qquad \text{for } i = 1, 2, \ldots, n. \tag{7.1}$$

The method is written in the form $\mathbf{x}^{(k)} = T\mathbf{x}^{(k-1)} + \mathbf{c}$ by splitting A into its diagonal and off-diagonal parts. To see this, let D be the diagonal matrix whose diagonal entries are those of A, $-L$ be the strictly lower-triangular part of A, and $-U$ be the strictly upper triangular part of A. With this notation,

$$A = \begin{bmatrix} a_{11} & a_{12} & \cdots & a_{1n} \\ a_{21} & a_{22} & \cdots & a_{2n} \\ \vdots & \vdots & & \vdots \\ a_{n1} & a_{n2} & \cdots & a_{nn} \end{bmatrix}$$

is split into

$$A = \begin{bmatrix} a_{11} & 0 & \cdots & 0 \\ 0 & a_{22} & \ddots & \vdots \\ \vdots & \ddots & \ddots & 0 \\ 0 & \cdots & 0 & a_{nn} \end{bmatrix} - \begin{bmatrix} 0 & \cdots & \cdots & 0 \\ -a_{21} & \ddots & & \vdots \\ \vdots & \ddots & \ddots & \vdots \\ -a_{n1} & \cdots & -a_{n,n-1} & 0 \end{bmatrix}$$

$$- \begin{bmatrix} 0 & -a_{12} & \cdots & -a_{1n} \\ \vdots & \ddots & \ddots & \vdots \\ \vdots & & \ddots & -a_{n-1,n} \\ 0 & \cdots & \cdots & 0 \end{bmatrix}$$

$$= D - L - U.$$

The equation $A\mathbf{x} = \mathbf{b}$ or $(D - L - U)\mathbf{x} = \mathbf{b}$ is then transformed into

$$D\mathbf{x} = (L + U)\mathbf{x} + \mathbf{b},$$

and, if D^{-1} exists—that is, if $a_{ii} \neq 0$ for each i—then

$$\mathbf{x} = D^{-1}(L + U)\mathbf{x} + D^{-1}\mathbf{b}.$$

This results in the matrix form of the Jacobi iterative technique:

$$\mathbf{x}^{(k)} = T_j\mathbf{x}^{(k-1)} + \mathbf{c}_j,$$

where $T_j = D^{-1}(L + U)$ and $\mathbf{c}_j = D^{-1}\mathbf{b}$.

The program JACITR71 implements the Jacobi method. If $a_{ii} \neq 0$ for some i and the system is nonsingular, a reordering of the equations is performed so that no $a_{ii} = 0$. To speed convergence, the equations should be arranged so that a_{ii} is as large as possible.

A possible improvement on the Jacobi method can be seen by reconsidering Eq.(7.1). In this equation the components of $\mathbf{x}^{(k-1)}$ are used to compute $x_i^{(k)}$. Since $x_1^{(k)}, \ldots, x_{i-1}^{(k)}$ have already been computed and are probably better approximations to the actual solutions $x_1, \ldots, x_{i-1}$ than $x_1^{(k-1)}, \ldots, x_{i-1}^{(k-1)}$, we can compute $x_i^{(k)}$ using these most recently calculated values. That is, we can use

$$x_i^{(k)} = \frac{-\sum_{j=1}^{i-1}(a_{ij}x_j^{(k)}) - \sum_{j=i+1}^{n}(a_{ij}x_j^{(k-1)}) + b_i}{a_{ii}}, \tag{7.2}$$

for each $i = 1, 2, \ldots, n$. This modification is called the **Gauss-Seidel** iterative technique and is illustrated in the following example.

EXAMPLE 2 The linear system given by

$$\begin{aligned} 10x_1 - x_2 + 2x_3 \phantom{{}+ 3x_4} &= 6, \\ -x_1 + 11x_2 - x_3 + 3x_4 &= 25, \\ 2x_1 - x_2 + 10x_3 - x_4 &= -11, \\ 3x_2 - x_3 + 8x_4 &= 15 \end{aligned}$$

was solved in Example 1 by the Jacobi iterative method. Using Eq. (7.2) gives the equations

$$
\begin{aligned}
x_1^{(k)} &= \frac{1}{10}x_2^{(k-1)} - \frac{1}{5}x_3^{(k-1)} + \frac{3}{5},\\
x_2^{(k)} &= \frac{1}{11}x_1^{(k)} + \frac{1}{11}x_3^{(k-1)} - \frac{3}{11}x_4^{(k-1)} + \frac{25}{11},\\
x_3^{(k)} &= -\frac{1}{5}x_1^{(k)} + \frac{1}{10}x_2^{(k)} + \frac{1}{10}x_4^{(k-1)} - \frac{11}{10},\\
x_4^{(k)} &= -\frac{3}{8}x_2^{(k)} + \frac{1}{8}x_3^{(k)} + \frac{15}{8}.
\end{aligned}
$$

Letting $\mathbf{x}^{(0)} = (0,0,0,0)^t$, we generate the iterates in Table 7.2. Since

$$\|\mathbf{x}^{(5)} - \mathbf{x}^{(4)}\|_\infty = 0.0008 < 10^{-3},$$

$\mathbf{x}^{(5)}$ is accepted as a reasonable approximation to the solution. Note that Jacobi's method in Example 1 required twice the iterations for the same accuracy. ◆ ◆ ◆

Table 7.2

k	0	1	2	3	4	5
$x_1^{(k)}$	0.0000	0.6000	1.030	1.0065	1.0009	1.0001
$x_2^{(k)}$	0.0000	2.3272	2.037	2.0036	2.0003	2.0000
$x_3^{(k)}$	0.0000	−0.9873	−1.014	−1.0025	−1.0003	−1.0000
$x_4^{(k)}$	0.0000	0.8789	0.9844	0.9983	0.9999	1.0000

To write the Gauss-Seidel method in matrix form, multiply both sides of Eq.(7.2) by a_{ii} and collect all kth iterate terms, to give

$$a_{i1}x_1^{(k)} + a_{i2}x_2^{(k)} + \cdots + a_{ii}x_i^{(k)} = -a_{i,i+1}x_{i+1}^{(k-1)} - \cdots - a_{in}x_n^{(k-1)} + b_i,$$

for each $i = 1, 2, \ldots, n$. Writing all n equations gives

$$
\begin{aligned}
a_{11}x_1^{(k)} &= -a_{12}x_2^{(k-1)} - a_{13}x_3^{(k-1)} - \cdots - a_{1n}x_n^{(k-1)} + b_1,\\
a_{21}x_1^{(k)} + a_{22}x_2^{(k)} &= -a_{23}x_3^{(k-1)} - \cdots - a_{2n}x_n^{(k-1)} + b_2,\\
&\ \ \vdots\\
a_{n1}x_1^{(k)} + a_{n2}x_2^{(k)} + \cdots + a_{nn}x_n^{(k)} &= b_n.
\end{aligned}
$$

With the definitions of D, L, and U that we used previously, we have the Gauss-Seidel method represented by

$$(D - L)\mathbf{x}^{(k)} = U\mathbf{x}^{(k-1)} + \mathbf{b}$$

or, if $(D - L)^{-1}$ exists, by

$$\mathbf{x}^{(k)} = T_g\mathbf{x}^{(k-1)} + \mathbf{c}_g, \qquad \text{for each } k = 1, 2, \ldots,$$

where $T_g = (D - L)^{-1}U$ and $c_g = (D - L)^{-1}\mathbf{b}$. Since $\det(D - L) = \prod_{i=1}^{n} a_{ii}$, the lower-triangular matrix $D - L$ is nonsingular precisely when $a_{ii} \neq 0$ for each $i = 1, 2, \ldots, n$. The Gauss-Seidel method is performed by the program GSEITR72.

The preceding discussion and the results of Examples 1 and 2 seem to imply that the Gauss-Seidel method is superior to the Jacobi method. This is almost always true, but there are linear systems for which the Jacobi method converges and the Gauss-Seidel method does not. However, if A is strictly diagonally dominant, then for any $\mathbf{b}$ and any choice of $\mathbf{x}^{(0)}$, the Jacobi and Gauss-Seidel methods will both converge to the unique solution of $A\mathbf{x} = \mathbf{b}$.

EXERCISE SET 7.4

1. Find the first two iterations of the Jacobi method for the following linear systems, using $\mathbf{x}^{(0)} = \mathbf{0}$:

a.
$$\begin{aligned} 3x_1 - x_2 + x_3 &= 1,\\ 3x_1 + 6x_2 + 2x_3 &= 0,\\ 3x_1 + 3x_2 + 7x_3 &= 4. \end{aligned}$$

b.
$$\begin{aligned} 10x_1 - x_2 \phantom{{}-2x_3} &= 9,\\ -x_1 + 10x_2 - 2x_3 &= 7,\\ -2x_2 + 10x_3 &= 6. \end{aligned}$$

c.
$$\begin{aligned} 10x_1 + 5x_2 &= 6,\\ 5x_1 + 10x_2 - 4x_3 &= 25,\\ -4x_2 + 8x_3 - x_4 &= -11,\\ -x_3 + 5x_4 &= -11. \end{aligned}$$

d.
$$\begin{aligned} 4x_1 + x_2 - x_3 + x_4 &= -2,\\ x_1 + 4x_2 - x_3 - x_4 &= -1,\\ -x_1 - x_2 + 5x_3 + x_4 &= 0,\\ x_1 - x_2 + x_3 + 3x_4 &= 1. \end{aligned}$$

e.
$$\begin{aligned} 4x_1 + x_2 + x_3 + x_5 &= 6,\\ -x_1 - 3x_2 + x_3 + x_4 &= 6,\\ 2x_1 + x_2 + 5x_3 - x_4 - x_5 &= 6,\\ -x_1 - x_2 - x_3 + 4x_4 &= 6,\\ 2x_2 - x_3 + x_4 + 4x_5 &= 6. \end{aligned}$$

f.
$$\begin{aligned} 4x_1 - x_2 - x_4 &= 0,\\ -x_1 + 4x_2 - x_3 - x_5 &= 5,\\ -x_2 + 4x_3 - x_6 &= 0,\\ -x_1 + 4x_4 - x_5 &= 6,\\ -x_2 - x_4 + 4x_5 - x_6 &= -2,\\ -x_3 - x_5 + 4x_6 &= 6. \end{aligned}$$

2. Repeat Exercise 1 using the Gauss-Seidel method.

3. Use the Jacobi method to solve the linear systems in Exercise 1, with $TOL = 10^{-3}$ in the l_∞ norm.

4. Repeat Exercise 3 using the Gauss-Seidel method.

5. The linear system

$$\begin{aligned} x_1 \qquad\qquad - x_3 &= 0.2, \\ -\frac{1}{2}x_1 + x_2 - \frac{1}{4}x_3 &= -1.425, \\ x_1 - \frac{1}{2}x_2 + x_3 &= 2. \end{aligned}$$

has the solution $(0.9, -0.8, 0.7)^t$.

a. Is the coefficient matrix

$$A = \begin{bmatrix} 1 & 0 & -1 \\ -\frac{1}{2} & 1 & -\frac{1}{4} \\ 1 & -\frac{1}{2} & 1 \end{bmatrix}$$

strictly diagonally dominant?

b. Compute the spectral radius of the Jacobi matrix T_j.

c. Use the Jacobi iterative method to approximate the solution to the linear system with a tolerance of 10^{-2} and a maximum of 300 iterations.

d. What happens in part (c) when the system is changed to

$$\begin{aligned} x_1 \qquad\qquad - 2x_3 &= 0.2, \\ -\frac{1}{2}x_1 + x_2 - \frac{1}{4}x_3 &= -1.425, \\ x_1 - \frac{1}{2}x_2 + x_3 &= 2. \end{aligned}$$

6. Repeat Exercise 5 using the Gauss-Seidel method.

7. Show that if A is strictly diagonally dominant, then $\|T_j\|_\infty < 1$.

7.5 The SOR Method

The SOR method is similar to the Jacobi and Gauss-Seidel methods, but it uses a scaling factor to more rapidly reduce the approximation error. In contrast to the classic methods discussed in the previous section, the SOR technique is a more recent innovation.

The SOR technique is one of a class of *relaxation* methods that compute approximations $\mathbf{x}^{(k)}$ by the formula

$$x_i^{(k)} = (1-\omega)x_i^{(k-1)} + \frac{\omega}{a_{ii}}\left[b_i - \sum_{j=1}^{i-1} a_{ij}x_j^{(k)} - \sum_{j=i+1}^{n} a_{ij}x_j^{(k-1)}\right],$$

where ω is the scaling factor.

When $\omega = 1$, we have the Gauss-Seidel method. When $0 < \omega < 1$, the procedures are called **under-relaxation methods** and can be used to obtain convergence of some systems that are not convergent by the Gauss-Seidel method. When $1 < \omega$, the procedures are called **over-relaxation methods**, which are used to accelerate the convergence for systems that are convergent by the Gauss-Seidel technique. These methods are abbreviated **SOR** for **Successive Over-Relaxation** and are particularly useful for solving the linear systems that occur in the numerical solution of certain partial-differential equations.

To determine the matrix form of the SOR method, we rewrite the preceding equation as

$$a_{ii}x_i^{(k)} + \omega \sum_{j=1}^{i-1} a_{ij}x_j^{(k)} = (1-\omega)a_{ii}x_i^{(k-1)} - \omega \sum_{j=i+1}^{n} a_{ij}x_j^{(k-1)} + \omega b_i,$$

so that in vector form we have

$$(D - \omega L)\mathbf{x}^{(k)} = [(1-\omega)D + \omega U]\,\mathbf{x}^{(k-1)} + \omega \mathbf{b}.$$

If $(D - \omega L)^{-1}$ exists, then

$$\mathbf{x}^{(k)} = T_\omega \mathbf{x}^{(k-1)} + \mathbf{c}_\omega,$$

where $T_\omega = (D - \omega L)^{-1}[(1-\omega)D + \omega U]$ and $c_\omega = \omega(D - \omega L)^{-1}\mathbf{b}$.

The SOR technique can be applied using the program SORITR73.

EXAMPLE 1 The linear system $A\mathbf{x} = \mathbf{b}$ given by

$$\begin{aligned} 4x_1 + 3x_2 \qquad &= 24, \\ 3x_1 + 4x_2 - \ x_3 &= 30, \\ -x_2 + 4x_3 &= -24 \end{aligned}$$

has the solution $(3, 4, -5)^t$. The Gauss-Seidel method and the SOR method with $\omega = 1.25$ will be used to solve this system, using $\mathbf{x}^{(0)} = (1, 1, 1)^t$ for both methods. For each $k = 1, 2, \ldots$, the equations for the Gauss-Seidel method are

$$\begin{aligned} x_1^{(k)} &= -0.75x_2^{(k-1)} + 6, \\ x_2^{(k)} &= -0.75x_1^{(k)} + 0.25x_3^{(k-1)} + 7.5, \\ x_3^{(k)} &= 0.25x_2^{(k)} - 6, \end{aligned}$$

and the equations for the SOR method with $\omega = 1.25$ are

$$\begin{aligned} x_1^{(k)} &= -0.25x_1^{(k-1)} - 0.9375x_2^{(k-1)} + 7.5, \\ x_2^{(k)} &= -0.9375x_1^{(k)} - 0.25x_2^{(k-1)} + 0.3125x_3^{(k-1)} + 9.375, \\ x_3^{(k)} &= 0.3125x_2^{(k)} - 0.25x_3^{(k-1)} - 7.5. \end{aligned}$$

The first seven iterates for each method are listed in Tables 7.3 and 7.4. To be accurate to seven decimal places, the Gauss-Seidel method required 34 iterations, as opposed to 14 iterations for the SOR method with $\omega = 1.25$. ◆ ◆ ◆

Table 7.3

k	0	1	2	3	4	5	6	7
$x_1^{(k)}$	1	5.250000	3.1406250	3.0878906	3.0549316	3.0343323	3.0214577	3.0134110
$x_1^{(k)}$	1	3.812500	3.8828125	3.9267578	3.9542236	3.9713898	3.9821186	3.9888241
$x_1^{(k)}$	1	−5.046875	−5.0292969	−5.0183105	−5.0114441	−5.0071526	−5.0044703	−5.0027940

Table 7.4

k	0	1	2	3	4	5	6	7
$x_1^{(k)}$	1	6.312500	2.6223145	3.1333027	2.9570512	3.0037211	2.9963276	3.0000498
$x_2^{(k)}$	1	3.5195313	3.9585266	4.0102646	4.0074838	4.0029250	4.0009262	4.0002586
$x_3^{(k)}$	1	−6.6501465	−4.6004238	−5.0966863	−4.9734897	−5.0057135	−4.9982822	−5.0003486

The obvious question to ask is how the appropriate value of ω is chosen. Although no complete answer to this question is known for the general $n \times n$ linear system, the following result can be used in certain situations.

SOR Method Convergence

If A is a positive definite matrix and $0 < \omega < 2$, then the SOR method converges for any choice of initial approximate solution vector $\mathbf{x}^{(0)}$.

If, in addition, A is tridiagonal, then $\rho(T_g) = [\rho(T_j)]^2 < 1$, and the optimal choice of ω for the SOR method is

$$\omega = \frac{2}{1 + \sqrt{1 - [\rho(T_j)]^2}}.$$

With this choice of ω, $\rho(T_\omega) = \omega - 1$.

EXAMPLE 2 The matrix

$$A = \begin{bmatrix} 4 & 3 & 0 \\ 3 & 4 & -1 \\ 0 & -1 & 4 \end{bmatrix}$$

given in Example 1 is positive definite and tridiagonal. Since

$$T_j = D^{-1}(L+U) = \begin{bmatrix} \frac{1}{4} & 0 & 0 \\ 0 & \frac{1}{4} & 0 \\ 0 & 0 & \frac{1}{4} \end{bmatrix} \begin{bmatrix} 0 & -3 & 0 \\ -3 & 0 & 1 \\ 0 & 1 & 0 \end{bmatrix} = \begin{bmatrix} 0 & -0.75 & 0 \\ -0.75 & 0 & 0.25 \\ 0 & 0.25 & 0 \end{bmatrix},$$

we have

$$\det(T_j - \lambda I) = -\lambda(\lambda^2 - 0.625) \quad \text{and} \quad \rho(T_j) = \sqrt{0.625}.$$

Hence, the optimal choice of ω is

$$\omega = \frac{2}{1+\sqrt{1-[\rho(T_j)]^2}} = \frac{2}{1+\sqrt{1-0.625}} \approx 1.24.$$

This explains the rapid convergence obtained in Example 1 by using $\omega = 1.25$. ◆ ◆ ◆

EXERCISE SET 7.5

1. Find the first two iterations of the SOR method with $\omega = 1.1$ for the following linear systems, using $\mathbf{x}^{(0)} = \mathbf{0}$:

   a. $$\begin{aligned} 3x_1 - x_2 + x_3 &= 1, \\ 3x_1 + 6x_2 + 2x_3 &= 0, \\ 3x_1 + 3x_2 + 7x_3 &= 4. \end{aligned}$$

   b. $$\begin{aligned} 10x_1 - x_2 &= 9, \\ -x_1 + 10x_2 - 2x_3 &= 7, \\ -2x_2 + 10x_3 &= 6. \end{aligned}$$

   c. $$\begin{aligned} 10x_1 + 5x_2 &= 6, \\ 5x_1 + 10x_2 - 4x_3 &= 25, \\ -4x_2 + 8x_3 - x_4 &= -11, \\ -x_3 + 5x_4 &= -11. \end{aligned}$$

   d. $$\begin{aligned} 4x_1 + x_2 - x_3 + x_4 &= -2, \\ x_1 + 4x_2 - x_3 - x_4 &= -1, \\ -x_1 - x_2 + 5x_3 + x_4 &= 0, \\ x_1 - x_2 + x_3 + 3x_4 &= 1. \end{aligned}$$

   e. $$\begin{aligned} 4x_1 + x_2 + x_3 + x_5 &= 6, \\ -x_1 - 3x_2 + x_3 + x_4 &= 6, \\ 2x_1 + x_2 + 5x_3 - x_4 - x_5 &= 6, \\ -x_1 - x_2 - x_3 + 4x_4 &= 6, \\ 2x_2 - x_3 + x_4 + 4x_5 &= 6. \end{aligned}$$

   f. $$\begin{aligned} 4x_1 - x_2 - x_4 &= 0, \\ -x_1 + 4x_2 - x_3 - x_5 &= 5, \\ -x_2 + 4x_3 - x_6 &= 0, \\ -x_1 + 4x_4 - x_5 &= 6, \\ -x_2 - x_4 + 4x_5 - x_6 &= -2, \\ -x_3 - x_5 + 4x_6 &= 6. \end{aligned}$$

2. Use the SOR method with $\omega = 1.2$ to solve the linear systems in Exercise 1 with a tolerance $TOL = 10^{-3}$ in the l_∞ norm.

3. For those matrices in Exercise 1 that are both tridiagonal and positive definite, use the SOR method with the optimal choice of ω.

4. Suppose that an object can be at any one of $n+1$ equally spaced points $x_0, x_1, \ldots, x_n$. When an object is at location x_i, it is equally likely to move to either x_{i-1} or x_{i+1} and cannot directly move to any other location. Consider the probabilities $\{P_i\}_{i=0}^n$ that an object starting at location x_i will reach the left endpoint x_0 before reaching the right endpoint x_n. Clearly, $P_0 = 1$ and $P_n = 0$. Since the object can move to x_i only from x_{i-1} or x_{i+1} and does so with probability $\frac{1}{2}$ for each of these locations,

$$P_i = \frac{1}{2}P_{i-1} + \frac{1}{2}P_{i+1}, \quad \text{for each } i = 1, 2, \ldots, n-1.$$

a. Show that

$$\begin{bmatrix} 1 & -\frac{1}{2} & 0 & \cdots & \cdots & 0 \\ -\frac{1}{2} & 1 & -\frac{1}{2} & \ddots & & \vdots \\ 0 & -\frac{1}{2} & 1 & \ddots & \ddots & \vdots \\ \vdots & \ddots & \ddots & \ddots & \ddots & 0 \\ \vdots & & \ddots & -\frac{1}{2} & 1 & -\frac{1}{2} \\ 0 & \cdots & \cdots & 0 & -\frac{1}{2} & 1 \end{bmatrix} \begin{bmatrix} P_1 \\ P_2 \\ \vdots \\ P_{n-1} \end{bmatrix} = \begin{bmatrix} \frac{1}{2} \\ 0 \\ \vdots \\ 0 \end{bmatrix}.$$

b. Solve this system using $n = 10$, 50, and 100.

c. Change the probabilities to α and $1-\alpha$ for movement to the left and right, respectively, and derive the linear system similar to the one in part (a).

d. Repeat part (b) with $\alpha = \frac{1}{3}$.

5. The forces on the bridge truss shown here satisfy the equations in the following table:

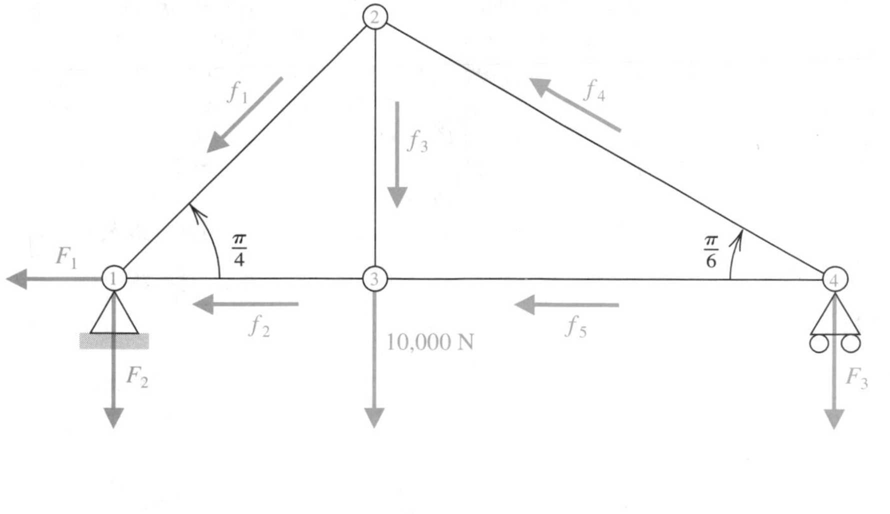

Joint	Horizontal Component	Vertical Component
①	$-F_1 + \frac{\sqrt{2}}{2} f_1 + f_2 = 0$	$\frac{\sqrt{2}}{2} f_1 - F_2 = 0$
②	$-\frac{\sqrt{2}}{2} f_1 + \frac{\sqrt{3}}{2} f_4 = 0$	$-\frac{\sqrt{2}}{2} f_1 - f_3 + \frac{1}{2} f_4 = 0$
③	$-f_2 + f_5 = 0$	$f_3 - 10,000 = 0$
④	$-\frac{\sqrt{3}}{2} f_4 - f_5 = 0$	$\frac{1}{2} f_4 - F_3 = 0$

This linear system can be placed in the matrix form

$$\begin{bmatrix} -1 & 0 & 0 & \frac{\sqrt{2}}{2} & 1 & 0 & 0 & 0 \\ 0 & -1 & 0 & \frac{\sqrt{2}}{2} & 0 & 0 & 0 & 0 \\ 0 & 0 & -1 & 0 & 0 & 0 & \frac{1}{2} & 0 \\ 0 & 0 & 0 & -\frac{\sqrt{2}}{2} & 0 & -1 & \frac{1}{2} & 0 \\ 0 & 0 & 0 & 0 & -1 & 0 & 0 & 1 \\ 0 & 0 & 0 & 0 & 0 & 1 & 0 & 0 \\ 0 & 0 & 0 & -\frac{\sqrt{2}}{2} & 0 & 0 & \frac{\sqrt{3}}{2} & 0 \\ 0 & 0 & 0 & 0 & 0 & 0 & -\frac{\sqrt{3}}{2} & -1 \end{bmatrix} \begin{bmatrix} F_1 \\ F_2 \\ F_3 \\ f_1 \\ f_2 \\ f_3 \\ f_4 \\ f_5 \end{bmatrix} = \begin{bmatrix} 0 \\ 0 \\ 0 \\ 0 \\ 0 \\ 10{,}000 \\ 0 \\ 0 \end{bmatrix}.$$

a. Explain why the system of equations was reordered.

b. Approximate the solution of the resulting linear system to within 10^{-2} in the l_∞ norm using as initial approximation the vector all of whose entries are 1s and (i) the Gauss-Seidel method, (ii) the Jacobi method, and (iii) the SOR method with $\omega = 1.25$.

7.6 Error Bounds and Iterative Refinement

This section considers convergence concerns that should be addressed when applying either an iterative or direct method to a linear system. There is no universally superior technique for approximating the solution to these systems, but some methods will commonly give better results than others, particularly when certain conditions are known about the matrix and the likely range of the solution.

It seems intuitively reasonable that if $\tilde{\mathbf{x}}$ is an approximation to the solution $\mathbf{x}$ of $A\mathbf{x} = \mathbf{b}$ and the **residual** vector, defined by $\mathbf{b} - A\tilde{\mathbf{x}}$, has the property that $\|\mathbf{b} - A\tilde{\mathbf{x}}\|$ is small, then $\|\mathbf{x} - \tilde{\mathbf{x}}\|$ should be small as well. This is often the case, but certain systems, which occur quite often in practice, fail to have this property.

EXAMPLE 1 The linear system $A\mathbf{x} = \mathbf{b}$ given by

$$\begin{bmatrix} 1 & 1 \\ 1.0001 & 2 \end{bmatrix} \begin{bmatrix} x_1 \\ x_2 \end{bmatrix} = \begin{bmatrix} 3 \\ 3.0001 \end{bmatrix}$$

has the unique solution $\mathbf{x} = (1, 1)^t$. The poor approximation $\tilde{\mathbf{x}} = (3, 0)^t$ has the residual vector

$$\mathbf{b} - A\tilde{\mathbf{x}} = \begin{bmatrix} 3 \\ 3.0001 \end{bmatrix} - \begin{bmatrix} 1 & 2 \\ 1.0001 & 2 \end{bmatrix} \begin{bmatrix} 3 \\ 0 \end{bmatrix} = \begin{bmatrix} 0 \\ -0.0002 \end{bmatrix},$$

so $\|\mathbf{b} - A\tilde{\mathbf{x}}\|_\infty = 0.0002$. Although the norm of the residual vector is small, the approximation $\tilde{\mathbf{x}} = (3, 0)^t$ is obviously quite poor; in fact, $\|\mathbf{x} - \tilde{\mathbf{x}}\|_\infty = 2$.

◆ ◆ ◆

The difficulty in Example 1 is explained quite simply by noting that the solution to the system represents the intersection of the lines

$$l_1: \quad x_1 + 2x_2 = 3 \quad \text{and} \quad l_2: \quad 1.0001x_1 + 2x_2 = 3.0001.$$

The point $(3, 0)$ lies on l_1, and the lines are nearly parallel. This implies that $(3, 0)$ also lies close to l_2, even though it differs significantly from the solution of the system, which is the intersection point $(1, 1)$. (See Figure 7.6.)

Figure 7.6

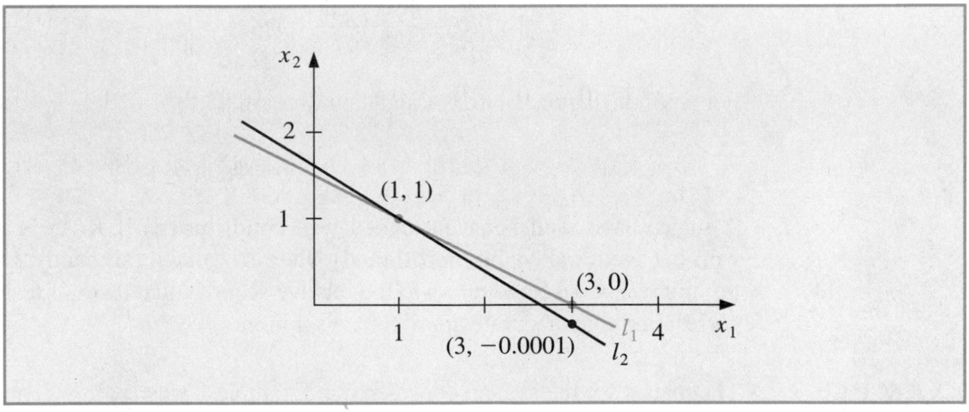

Example 1 was clearly constructed to show the difficulties that might—and, in fact, do—arise. Had the lines not been nearly coincident, we would expect a small residual vector to imply an accurate approximation. In the general situation, we cannot rely on the geometry of the system to give an indication of when problems might occur. We can, however, obtain this information by considering the norms of the matrix and its inverse.

Residual Vector Error Bounds

If $\tilde{\mathbf{x}}$ is an approximation to the solution of $A\mathbf{x} = \mathbf{b}$ and A is a nonsingular matrix, then for any natural norm,

$$\|\mathbf{x} - \tilde{\mathbf{x}}\| \le \|\mathbf{b} - A\tilde{\mathbf{x}}\| \cdot \|A^{-1}\|$$

and

$$\frac{||\mathbf{x}-\tilde{\mathbf{x}}||}{||\mathbf{x}||} \le ||A|| \cdot ||A^{-1}|| \frac{||\mathbf{b}-A\tilde{\mathbf{x}}||}{||\mathbf{b}||}, \quad \text{provided } \mathbf{x} \ne \mathbf{0} \text{ and } \mathbf{b} \ne \mathbf{0}.$$

This implies that $||A^{-1}||$ and $||A|| \cdot ||A^{-1}||$ provide an indication of the connection between the residual vector and the accuracy of the approximation. In general, the relative error $||\mathbf{x}-\tilde{\mathbf{x}}||/||\mathbf{x}||$ is of most interest, and this error is bounded by the product of $||A||\cdot||A^{-1}||$ with the relative residual for this approximation, $||\mathbf{b}-A\tilde{\mathbf{x}}||/||\mathbf{b}||$. Any convenient norm can be used for this approximation; the only requirement is that it be used consistently throughout.

The **condition number**, $K(A)$, of the nonsingular matrix A relative to a norm $||\cdot||$ is

$$K(A) = ||A|| \cdot ||A^{-1}||.$$

With this notation, we can reexpress the inequalities in the previous result by

$$||\mathbf{x}-\tilde{\mathbf{x}}|| \le K(A)\frac{||\mathbf{b}-A\tilde{\mathbf{x}}||}{||A||} \quad \text{and} \quad \frac{||\mathbf{x}-\tilde{\mathbf{x}}||}{||\mathbf{x}||} \le K(A)\frac{||\mathbf{b}-A\tilde{\mathbf{x}}||}{||\mathbf{b}||}.$$

For any nonsingular matrix A and natural norm $||\cdot||$,

$$1 = ||I|| = ||A \cdot A^{-1}|| \le ||A|| \cdot ||A^{-1}|| = K(A).$$

A matrix A is well-behaved (called **well-conditioned**) if $K(A)$ is close to 1 and is not well-behaved (called **ill-conditioned**) when $K(A)$ is significantly greater than 1. Conditioning in this instance refers to the relative security that a small residual vector implies a correspondingly accurate approximate solution.

EXAMPLE 2 The matrix for the system considered in Example 1 was

$$A = \begin{bmatrix} 1 & 2 \\ 1.0001 & 2 \end{bmatrix},$$

which has $||A||_\infty = 3.0001$. This norm would not be considered large. However,

$$A^{-1} = \begin{bmatrix} -10000 & 10000 \\ 5000.5 & -5000 \end{bmatrix}, \quad \text{so} \quad ||A^{-1}||_\infty = 20000,$$

and for the infinity norm, $K(A) = (20000)(3.0001) = 60002$. The size of the condition number for this example should certainly keep us from making hasty accuracy decisions based on the residual of an approximation. ◆ ◆ ◆

In Maple the condition number K_∞ for the matrix in Example 2 can be computed as follows:

```
>with(linalg);
>A:=matrix(2,2,[1,2,1.0001,2]);
>cond(A);
```

$$60002.00000$$

The residual of an approximation can also be used to improve the accuracy of the approximation. Suppose that $\tilde{\mathbf{x}}$ is an approximation to the solution of the linear system $A\mathbf{x} = \mathbf{b}$ and that $\mathbf{r} = \mathbf{b} - A\mathbf{x}$ is the residual associated with $\tilde{\mathbf{x}}$. Consider $\tilde{\mathbf{y}}$ the approximate solution to the system $A\mathbf{y} = \mathbf{r}$. Then

$$\tilde{\mathbf{y}} \approx A^{-1}\mathbf{r} = A^{-1}(\mathbf{b} - A\tilde{\mathbf{x}}) = A^{-1}\mathbf{b} - A^{-1}A\tilde{\mathbf{x}} = \mathbf{x} - \tilde{\mathbf{x}}.$$

So

$$\mathbf{x} \approx \tilde{\mathbf{x}} + \tilde{\mathbf{y}}.$$

This new approximation $\tilde{\mathbf{x}} + \tilde{\mathbf{y}}$ is often much closer to the solution of $A\mathbf{x} = \mathbf{b}$ than is $\tilde{\mathbf{x}}$ and $\tilde{\mathbf{y}}$ is easy to determine since it involves the same matrix, A, as the original system. This technique is called **iterative refinement**, or *iterative improvement*, and is illustrated in the following example. To increase accuracy, the residual vector is computed using double-digit arithmetic. The method can also be implemented with the program ITREF74.

EXAMPLE 3 The linear system given by

$$\begin{bmatrix} 3.3330 & 15920 & -10.333 \\ 2.2220 & 16.710 & 9.6120 \\ 1.5611 & 5.1791 & 1.6852 \end{bmatrix} \begin{bmatrix} x_1 \\ x_2 \\ x_3 \end{bmatrix} = \begin{bmatrix} 15913 \\ 28.544 \\ 8.4254 \end{bmatrix}$$

has the exact solution $\mathbf{x} = (1, 1, 1)^t$.

Using Gaussian elimination and five-digit rounding arithmetic leads successively to the augmented matrices

$$\left[\begin{array}{ccc:c} 3.3330 & 15920 & -10.333 & 15913 \\ 0 & -10596 & 16.501 & 10580 \\ 0 & -7451.4 & 6.5250 & -7444.9 \end{array}\right]$$

and

$$\left[\begin{array}{ccc:c} 3.3330 & 15920 & -10.333 & 15913 \\ 0 & -10596 & 16.501 & -10580 \\ 0 & 0 & -5.0790 & -4.7000 \end{array}\right].$$

The approximate solution to this system is

$$\tilde{\mathbf{x}} = (1.2001, 0.99991, 0.92538)^t.$$

When computed in 10-digit arithmetic, the residual vector corresponding to $\tilde{\mathbf{x}}$ is

$$\begin{aligned}\mathbf{r} &= \mathbf{b} - A\tilde{\mathbf{x}}\\ &= \begin{bmatrix} 15913 \\ 28.544 \\ 8.4254 \end{bmatrix} - \begin{bmatrix} 3.3330 & 15920 & -10.333 \\ 2.2220 & 16.710 & 9.6120 \\ 1.5611 & 5.1791 & 1.6852 \end{bmatrix} \begin{bmatrix} 1.20001 \\ 0.99991 \\ 0.92538 \end{bmatrix}\\ &= \begin{bmatrix} -0.00518 \\ 0.27413 \\ -0.18616 \end{bmatrix}.\end{aligned}$$

◆ ◆ ◆

Using five-digit arithmetic and Gaussian elimination, the approximate solution $\tilde{\mathbf{y}}$ to the equation $A\mathbf{y} = \mathbf{r}$ is

$$\tilde{\mathbf{y}} = (-0.20008, 8.9987 \times 10^{-5}, 0.074607)^t$$

and

$$\begin{aligned}\tilde{\mathbf{x}} + \tilde{\mathbf{y}} &= (1.2001, 0.99991, 0.92538)^t + (-0.20008, 8.9987 \times 10^{-5}, 0.074607)^t\\ &= (1.0000, 1.0000, 0.99999)^t\end{aligned}$$

is a much better approximation to the system $A\mathbf{x} = \mathbf{b}$ than is $\tilde{\mathbf{x}} = (1.2001, 0.99991, 0.92538)^t$. If we were continuing the iteration processes, we would, of course, use $\tilde{\mathbf{x}} + \tilde{\mathbf{y}}$ as our starting values rather than $\tilde{\mathbf{x}}$.

EXERCISE SET 7.6

1. Compute the condition numbers of the following matrices relative to $\|\cdot\|_\infty$.

   a. $\begin{bmatrix} \frac{1}{2} & \frac{1}{3} \\ \frac{1}{3} & \frac{1}{4} \end{bmatrix}$

   b. $\begin{bmatrix} 3.9 & 1.6 \\ 6.8 & 2.9 \end{bmatrix}$

   c. $\begin{bmatrix} 1 & 2 \\ 1.0001 & 2 \end{bmatrix}$

   d. $\begin{bmatrix} 1.003 & 58.09 \\ 5.550 & 321.8 \end{bmatrix}$

   e. $\begin{bmatrix} 1 & -1 & -1 \\ 0 & 1 & -1 \\ 0 & 0 & -1 \end{bmatrix}$

   f. $\begin{bmatrix} 0.04 & 0.01 & -0.01 \\ 0.2 & 0.5 & -0.2 \\ 1 & 2 & 4 \end{bmatrix}$

2. The following linear systems $A\mathbf{x} = \mathbf{b}$ have $\mathbf{x}$ as the actual solution and $\tilde{\mathbf{x}}$ as an approximate solution. Using the results of Exercise 1, compute $\|\mathbf{x} - \tilde{\mathbf{x}}\|_\infty$ and

$$K_\infty(A)\frac{\|\mathbf{b} - A\tilde{\mathbf{x}}\|_\infty}{\|A\|_\infty}.$$

a. $\frac{1}{2}x_1 + \frac{1}{3}x_2 = \frac{1}{63},$

$\frac{1}{3}x_1 + \frac{1}{4}x_2 = \frac{1}{168},$

$\mathbf{x} = \left(\frac{1}{7}, -\frac{1}{6}\right)^t,$

$\tilde{\mathbf{x}} = (0.142, -0.166)^t.$

b. $3.9x_1 + 1.6x_2 = 5.5,$

$6.8x_1 + 2.9x_2 = 9.7,$

$\mathbf{x} = (1, 1)^t,$

$\tilde{\mathbf{x}} = (0.98, 1.1)^t.$

c. $x_1 + 2x_2 = 3,$

$1.0001x_1 + 2x_2 = 3.0001,$

$\mathbf{x} = (1, 1)^t,$

$\tilde{\mathbf{x}} = (0.96, 1.02)^t$

d. $1.003x_1 + 58.09x_2 = 68.12,$

$5.550x_1 + 321.8x_2 = 377.3,$

$\mathbf{x} = (10, 1)^t,$

$\tilde{\mathbf{x}} = (-10, 1)^t$

e. $x_1 - x_2 - x_3 = 2\pi,$

$x_2 - x_3 = 0,$

$-x_3 = \pi.$

$\mathbf{x} = (0, -\pi, -\pi)^t,$

$\tilde{\mathbf{x}} = (-0.1, -3.15, -3.14)^t$

f. $0.04x_1 + 0.01x_2 - 0.01x_3 = 0.06,$

$0.2x_1 + 0.5x_2 - 0.2x_3 = 0.3,$

$x_1 + 2x_2 + 4x_3 = 11,$

$\mathbf{x} = (1.827586, 0.6551724, 1.965517)^t,$

$\tilde{\mathbf{x}} = (1.8, 0.64, 1.9)^t$

3. The linear system

$$\begin{bmatrix} 1 & 2 \\ 1.0001 & 2 \end{bmatrix} \begin{bmatrix} x_1 \\ x_2 \end{bmatrix} = \begin{bmatrix} 3 \\ 3.0001 \end{bmatrix}$$

has solution $(1, 1)^t$. Change A slightly to

$$\begin{bmatrix} 1 & 2 \\ 0.9999 & 2 \end{bmatrix}$$

and consider the linear system

$$\begin{bmatrix} 1 & 2 \\ 0.9999 & 2 \end{bmatrix} \begin{bmatrix} x_1 \\ x_2 \end{bmatrix} = \begin{bmatrix} 3 \\ 3.0001 \end{bmatrix}.$$

Compute the new solution using five-digit rounding arithmetic, and compare the change in A to the change in $\mathbf{x}$.

4. The linear system $A\mathbf{x} = \mathbf{b}$ given by

$$\begin{bmatrix} 1 & 2 \\ 1.00001 & 2 \end{bmatrix} \begin{bmatrix} x_1 \\ x_2 \end{bmatrix} = \begin{bmatrix} 3 \\ 3.00001 \end{bmatrix}$$

has solution $(1,1)^t$. Use seven-digit rounding arithmetic to find the solution of the perturbed system

$$\begin{bmatrix} 1 & 2 \\ 1.000011 & 2 \end{bmatrix} \begin{bmatrix} x_1 \\ x_2 \end{bmatrix} = \begin{bmatrix} 3.00001 \\ 3.00003 \end{bmatrix},$$

and compare the change in A and $\mathbf{b}$ to the change in $\mathbf{x}$.

5. (i) Use Gaussian elimination and three-digit rounding arithmetic to approximate the solutions to the following linear systems. (ii) Then use one iteration of iterative refinement to improve the approximation, and compare the approximations to the actual solutions.

a.
$$\begin{aligned} 0.03x_1 + 58.9x_2 &= 59.2 \\ 5.31x_1 - 6.10x_2 &= 47.0 \end{aligned}$$
Actual solution $(10,1)^t$.

b.
$$\begin{aligned} 3.3330x_1 + 15920x_2 + 10.333x_3 &= 7953 \\ 2.2220x_1 + 16.710x_2 + 9.6120x_3 &= 0.965 \\ -1.5611x_1 + 5.1792x_2 - 1.6855x_3 &= 2.714 \end{aligned}$$
Actual solution $(1, 0.5, -1)^t$.

c.
$$\begin{aligned} 1.19x_1 + 2.11x_2 - 100x_3 + x_4 &= 1.12 \\ 14.2x_1 - 0.122x_2 + 12.2x_3 - x_4 &= 3.44 \\ 100x_2 - 99.9x_3 + x_4 &= 2.15 \\ 15.3x_1 + 0.110x_2 - 13.1x_3 - x_4 &= 4.16 \end{aligned}$$
Actual solution $(0.17682530, 0.01269269, -0.02065405, -1.18260870)^t$.

d.
$$\begin{aligned} \pi x_1 - e x_2 + \sqrt{2}x_3 - \sqrt{3}x_4 &= \sqrt{11} \\ \pi^2 x_1 + e x_2 - e^2 x_3 + \frac{3}{7}x_4 &= 0 \\ \sqrt{5}x_1 - \sqrt{6}x_2 + x_3 - \sqrt{2}x_4 &= \pi \\ \pi^3 x_1 + e^2 x_2 - \sqrt{7}x_3 + \frac{1}{9}x_4 &= \sqrt{2} \end{aligned}$$
Actual solution $(0.78839378, -3.12541367, 0.16759660, 4.55700252)^t$.

6. Repeat Exercise 5 using four-digit rounding arithmetic.

7. **a.** Use single precision on a computer to solve the following linear system using Gaussian elimination.

$$\begin{aligned} \frac{1}{3}x_1 - \frac{1}{3}x_2 - \frac{1}{3}x_3 - \frac{1}{3}x_4 - \frac{1}{3}x_5 &= 1 \\ \frac{1}{3}x_2 - \frac{1}{3}x_3 - \frac{1}{3}x_4 - \frac{1}{3}x_5 &= 0 \\ \frac{1}{3}x_3 - \frac{1}{3}x_4 - \frac{1}{3}x_5 &= -1 \end{aligned}$$

$$\frac{1}{3}x_4 - \frac{1}{3}x_5 = 2$$
$$\frac{1}{3}x_5 = 7$$

b. Compute the condition number of the matrix for the system relative to $\|\cdot\|_\infty$.
c. Find the exact solution to the linear system.

8. The $n \times n$ *Hilbert* matrix, $H^{(n)}$, defined by

$$H_{ij}^{(n)} = \frac{1}{i+j-1}, \qquad 1 \le i, j \le n$$

is an ill-conditioned matrix that arises in solving for the coefficients of least squares polynomials (see Section 8.3).

a. Show that

$$[H^{(4)}]^{-1} = \begin{bmatrix} 16 & -120 & 240 & -140 \\ -120 & 1200 & -2700 & 1680 \\ 240 & -2700 & 6480 & -4200 \\ -140 & 1680 & -4200 & 2800 \end{bmatrix},$$

and compute $K_\infty(H^{(4)})$.

b. Show that

$$[H^{(5)}]^{-1} = \begin{bmatrix} 25 & -300 & 1050 & -1400 & 630 \\ -300 & 4800 & -18900 & 26880 & -12600 \\ 1050 & -18900 & 79380 & -117600 & 56700 \\ -1400 & 26880 & -117600 & 179200 & -88200 \\ 630 & -12600 & 56700 & -88200 & 44100 \end{bmatrix}$$

and compute $K_\infty(H^{(5)})$.

c. Solve the linear system

$$H^{(4)} \begin{bmatrix} x_1 \\ x_2 \\ x_3 \\ x_4 \end{bmatrix} = \begin{bmatrix} 1 \\ 0 \\ 0 \\ 1 \end{bmatrix}$$

using three-digit rounding arithmetic, and compare the actual error to the residual vector error bound.

9. **a.** Use four-digit rounding arithmetic to compute the inverse H^{-1} of the 3×3 Hilbert matrix H.
b. Use four-digit rounding arithmetic to compute $\hat{H} = (H^{-1})^{-1}$.
c. Determine $\|H - \hat{H}\|_\infty$.

7.7 Survey of Methods and Software

In this chapter we have studied iterative techniques to approximate the solution of linear systems. We began with the Jacobi method and the Gauss-Seidel method to introduce the iterative methods. Both methods require an arbitrary initial approximation $\mathbf{x}^{(0)}$ and generate a sequence of vectors $\mathbf{x}^{(i+1)}$ using an equation of the form

$$\mathbf{x}^{(i+1)} = T\mathbf{x}^{(i)} + \mathbf{c}.$$

The method will converge if and only if $\rho(T) < 1$, and the smaller the spectral radius, the faster the convergence. Analysis of the residual vectors of the Gauss-Seidel technique leads to the SOR iterative method, which uses a parameter ω to speed convergence.

These iterative methods and modifications are used in the solution of linear systems that arise in the numerical solution of boundary value problems and partial-differential equations (see Chapters 11 and 12). These systems are often very large and are sparse with their nonzero entries in predictable positions. The iterative methods are also useful for other large sparse systems and are easily adapted for efficient use on parallel computers.

The package LAPACK contains only direct methods for the solution of linear systems. Neither the IMSL Library nor the NAG Library contains subroutines for the iterative solution of linear systems since neither library focuses on methods for boundary value problems or partial-differential equations that require the solution of large sparse systems. However, the public domain packages ITPACK, SLAP, and SPARSPAK contain iterative methods.

The concepts of condition number and poorly conditioned matrices were introduced in the last section of the chapter. Many of the subroutines for solving a linear system or for factoring a matrix into an LU factorization include checks for ill-conditioned matrices and also give an estimate of the condition number. This is generally true for the subroutines in LAPACK, ISML, and NAG.

LAPACK, the IMSL Library, and the NAG Library have subroutines that improve on a solution to a linear system that is poorly conditioned. The subroutines test the condition number and then use iterative refinement to obtain the most accurate solution possible given the precision of the computer.

More information on the use of iterative methods for solving linear systems can be found in Varga [Var], Young [Y], Hageman and Young [HY], and Axelsson [Ax]. Iterative methods for large sparse systems are discussed in Barrett et al. [Ba], Hackbush [Ha], and Saad [Sa2].

CHAPTER

8 Approximation Theory

8.1 Introduction

Approximation theory involves two types of problems. One arises when a function is given explicitly, but we wish to find a "simpler" type of function, such as a polynomial, for representation. The other problem concerns fitting functions to given data and finding the "best" function in a certain class that can be used to represent the data. We will begin the chapter with this problem.

8.2 Discrete Least Squares Approximation

Consider the problem of estimating the values of a function at nontabulated points, given the experimental data in Table 8.1.

Table 8.1

x_i	y_i	x_i	y_i
1	1.3	6	8.8
2	3.5	7	10.1
3	4.2	8	12.5
4	5.0	9	13.0
5	7.0	10	15.6

Interpolation requires a function that assumes the value of y_i at x_i for each $i = 1, 2, \ldots, 10$. Figure 8.1 on the following page shows a graph of the values in Table 8.1. From this graph, it appears that the actual relationship between x and y is linear. It is likely that no line precisely fits the data, because of errors in the data. In this case, it is unreasonable to require that the approximating function agree exactly with the given data; in fact, such

Figure 8.1

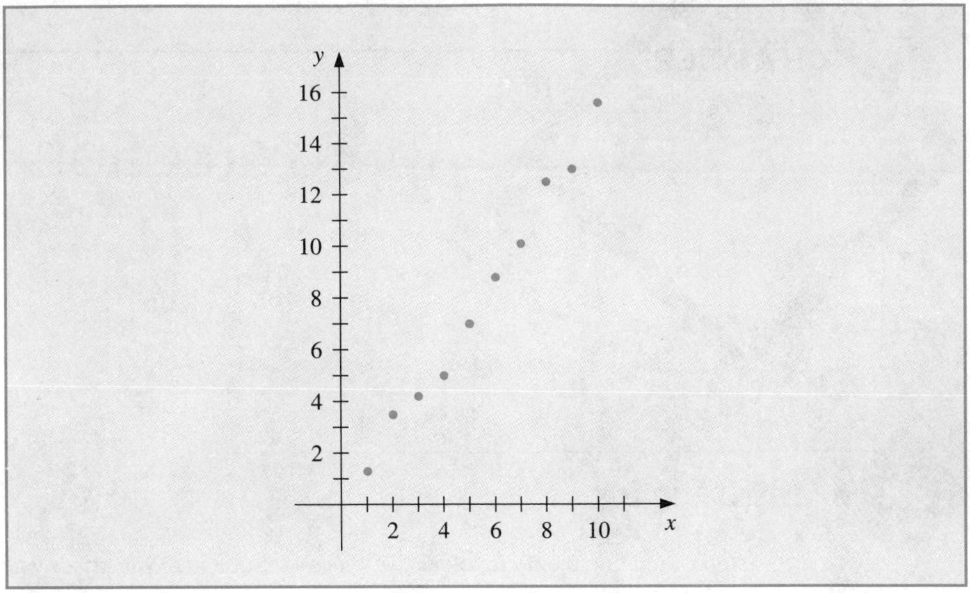

a function would introduce oscillations that should not be present. For example, the ninth degree interpolating polynomial on the data shown in Figure 8.2 is obtained in Maple using the commands

```
>p:=interp([1,2,3,4,5,6,7,8,9,10],[1.3,3.5,4.2,5.0,7.0,8.8,10.1,
 12.5,13.0,15.6],x);
>plot({p},x=1..10);
```

Figure 8.2

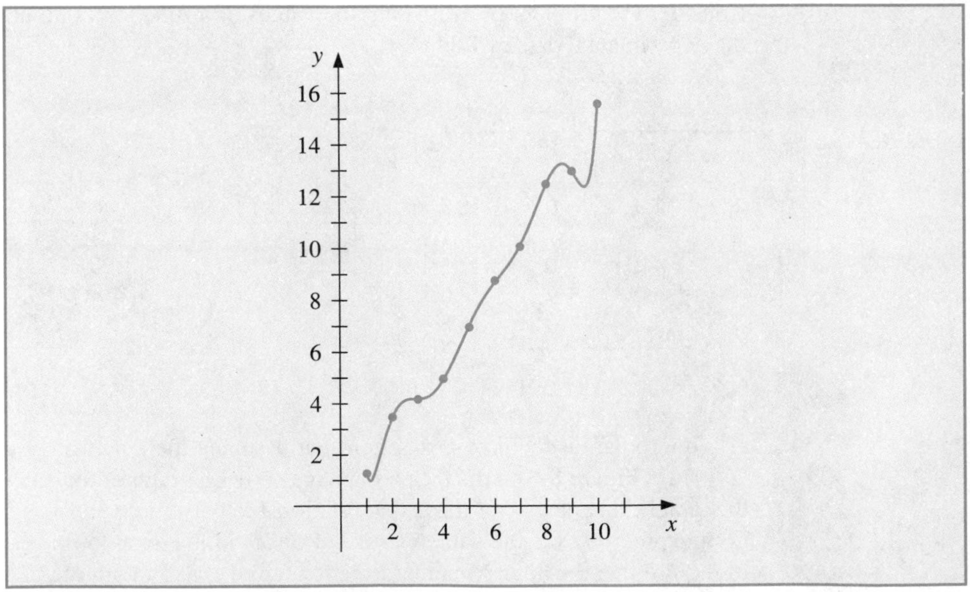

This polynomial is clearly a poor predictor of information between a number of the data points.

A better approach for a problem of this type would be to find the "best" (in some sense) approximating line, even if it did not agree precisely with the data at any point.

Let $a_1x_i + a_0$ denote the ith value on the approximating line and y_i the ith given y-value. The problem of finding the equation of the best linear approximation in the absolute sense requires that values of a_0 and a_1 be found to minimize

$$E_\infty(a_0, a_1) = \max_{1\le i\le 10} \{|y_i - (a_1x_i + a_0)|\}.$$

This is commonly called a **minimax** problem and cannot be handled by elementary techniques. Another approach to determining the best linear approximation involves finding values of a_0 and a_1 to minimize

$$E_1(a_0, a_1) = \sum_{i=1}^{10} |y_i - (a_1x_i + a_0)|.$$

This quantity is called the **absolute deviation**. To minimize a function of two variables, we need to set its partial derivatives to zero and simultaneously solve the resulting equations. In the case of the absolute deviation, we would need to find a_0 and a_1 with

$$0 = \frac{\partial}{\partial a_0} \sum_{i=1}^{10} |y_i - (a_1x_i + a_0)| \quad \text{and} \quad 0 = \frac{\partial}{\partial a_1} \sum_{i=1}^{10} |y_i - (a_1x_i + a_0)|.$$

The difficulty with this procedure is that the absolute-value function is not differentiable at zero, and solutions to this pair of equations cannot necessarily be obtained.

The **least squares** approach to this problem involves determining the best approximating line when the error involved is the sum of the squares of the differences between the y-values on the approximating line and the given y-values. Hence, constants a_0 and a_1 must be found that minimize the *total least squares error*:

$$E_2(a_0, a_1) = \sum_{i=1}^{10} \left[y_i - (a_1x_i + a_0)\right]^2.$$

The least squares method is the most convenient procedure for determining best linear approximations, and there are also important theoretical considerations that favor this method. The minimax approach generally assigns too much weight to a bit of data that is badly in error, whereas the absolute deviation method does not give sufficient weight to a point that is badly out of line. The least squares approach puts substantially more weight on a point that is out of line with the rest of the data but will not allow that point to dominate the approximation.

The general problem of fitting the best least squares line to a collection of data $\{(x_i, y_i)\}_{i=1}^m$ involves minimizing the total error $E_2(a_0, a_1) = \sum_{i=1}^m [y_i - (a_1x_i + a_0)]^2$ with respect to the parameters a_0 and a_1. For a minimum to occur, we need

$$0 = \frac{\partial}{\partial a_0} \sum_{i=1}^{m} \left[y_i - (a_1x_i - a_0)\right]^2 = 2\sum_{i=1}^{m} (y_i - a_1x_i - a_0)(-1)$$

and

$$0 = \frac{\partial}{\partial a_1} \sum_{i=1}^{m} \left[y_i - (a_1 x_i + a_0)\right]^2 = 2\sum_{i=1}^{m} (y_i - a_1 x_i - a_0)(-x_i)$$

These equations simplify to the **normal equations**

$$a_0 \sum_{i=1}^{m} x_i + a_1 \sum_{i=1}^{m} x_i^2 = \sum_{i=1}^{m} x_i y_i \quad \text{and} \quad a_0 \cdot m + a_1 \sum_{i=1}^{m} x_i = \sum_{i=1}^{m} y_i.$$

The solution to this system is as follows.

Linear Least Squares

The linear least squares solution for a given collection of data $\{(x_i, y_i)\}_{i=1}^{m}$ has the form $y = a_1 x + a_0$, where

$$a_0 = \frac{(\sum_{i=1}^{m} x_i^2)(\sum_{i=1}^{m} y_i) - (\sum_{i=1}^{m} x_i y_i)(\sum_{i=1}^{m} x_i)}{m(\sum_{i=1}^{m} x_i^2) - (\sum_{i=1}^{m} x_i)^2}$$

and

$$a_1 = \frac{m\left(\sum_{i=1}^{m} x_i y_i\right) - \left(\sum_{i=1}^{m} x_i\right)\left(\sum_{i=1}^{m} y_i\right)}{m\left(\sum_{i=1}^{m} x_i^2\right) - \left(\sum_{i=1}^{m} x_i\right)^2}.$$

EXAMPLE 1 Consider the data presented in Table 8.1. To find the least squares line approximating this data, extend the table as shown in the third and fourth columns of Table 8.2, and sum the columns.

Table 8.2

x_i	y_i	x_i^2	$x_i y_i$	$P(x_i) = 1.538x_i - 0.360$
1	1.3	1	1.3	1.18
2	3.5	4	7.0	2.72
3	4.2	9	12.6	4.25
4	5.0	16	20.0	5.79
5	7.0	25	35.0	7.33
6	8.8	36	52.8	8.87
7	10.1	49	70.7	10.41
8	12.5	64	100.0	11.94
9	13.0	81	117.0	13.48
10	15.6	100	156.0	15.02
55	81.0	385	572.4	$E_2 = \sum_{i=1}^{10} (y_i - P(x_i))^2 \approx 2.34$

Solving the normal equations produces

$$a_0 = \frac{385(81) - 55(572.4)}{10(385) - (55)^2} = -0.360 \quad \text{and} \quad a_1 = \frac{10(572.4) - 55(81)}{10(385) - (55)^2} = 1.538.$$

So

$$P(x) = 1.538x - 0.360.$$

The graph of this line and the data points are shown in Figure 8.3. The approximate values given by the least squares technique at the data points are in the final column in Table 8.2. ◆ ◆ ◆

Figure 8.3

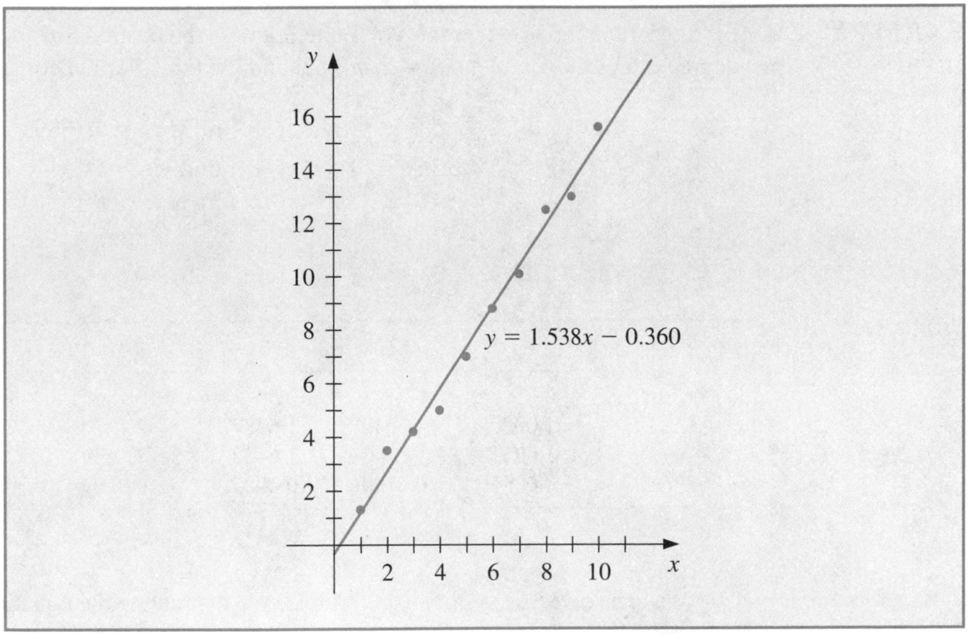

The problem of approximating a set of data, $\{(x_i, y_i) \mid i = 1, 2, \ldots, m\}$, with an algebraic polynomial

$$P_n(x) = a_n x^n + a_{n-1}x^{n-1} + \cdots + a_1 x + a_0$$

of degree $n < m - 1$ using least squares is handled in a similar manner. It requires choosing the constants $a_0, a_1, \ldots, a_n$ to minimize the *total least squares error*:

$$E_2 = \sum_{i=1}^{m} (y_i - P_n(x_i))^2.$$

For E_2 to be minimized, it is necessary that $\partial E_2/\partial a_j = 0$ for each $j = 0, 1, \ldots, n$. This gives $n + 1$ **normal equations** in the $n + 1$ unknowns, a_j,

$$a_0 \sum_{i=1}^{m} x_i^0 + a_1 \sum_{i=1}^{m} x_i^1 + a_2 \sum_{i=1}^{m} x_i^2 + \cdots + a_n \sum_{i=1}^{m} x_i^n = \sum_{i=1}^{m} y_i x_i^0,$$

$$a_0 \sum_{i=1}^{m} x_i^1 + a_1 \sum_{i=1}^{m} x_i^2 + a_2 \sum_{i=1}^{m} x_i^3 + \cdots + a_n \sum_{i=1}^{m} x_i^{n+1} = \sum_{i=1}^{m} y_i x_i^1,$$

$$\vdots$$

$$a_0 \sum_{i=1}^{m} x_i^n + a_1 \sum_{i=1}^{m} x_i^{n+1} + a_2 \sum_{i=1}^{m} x_i^{n+2} + \cdots + a_n \sum_{i=1}^{m} x_i^{2n} = \sum_{i=1}^{m} y_i x_i^n.$$

The normal equations will have a unique solution, provided that the x_i are distinct.

EXAMPLE 2 Fit the data in the first two rows of Table 8.3 with the discrete least squares polynomial of degree 2. For this problem, $n = 2, m = 5$, and the three normal equations are

$$\begin{aligned} 5a_0 + \quad 2.5a_1 + \quad 1.875a_2 &= 8.7680, \\ 2.5a_0 + \quad 1.875a_1 + 1.5625a_2 &= 5.4514, \\ 1.875a_0 + 1.5625a_1 + 1.3828a_2 &= 4.4015. \end{aligned}$$

Table 8.3

i	1	2	3	4	5
x_i	0	0.25	0.50	0.75	1.00
y_i	1.0000	1.2840	1.6487	2.1170	2.7183
$P(x_i)$	1.0051	1.2740	1.6482	2.1279	2.7130
$y_i - P(x_i)$	−0.0051	0.0100	0.0005	−0.0109	0.0053

We can solve this system using Maple. We first define the equations

```
>eq1:=5*a0+2.5*a1+1.875*a2=8.7680;
>eq2:=2.5*a0+1.875*a1+1.5625*a2=5.4514;
>eq3:=1.875*a0+1.5625*a1+1.3828*a2=4.4015;
```

To solve the system we enter

```
>solve({eq1,eq2,eq3},{a0,a1,a2});
```

which gives, with `Digits:=5;`

$$a_0 = 1.0051, \quad a_1 = 0.86468, \quad \text{and} \quad a_2 = 0.84316.$$

Thus, the least squares polynomial of degree 2 fitting the preceding data is $P_2(x) = 1.0051 + 0.86468x + 0.84316x^2$, whose graph is shown in Figure 8.4. At the given values of x_i, we have the approximations shown in Table 8.3.

Figure 8.4

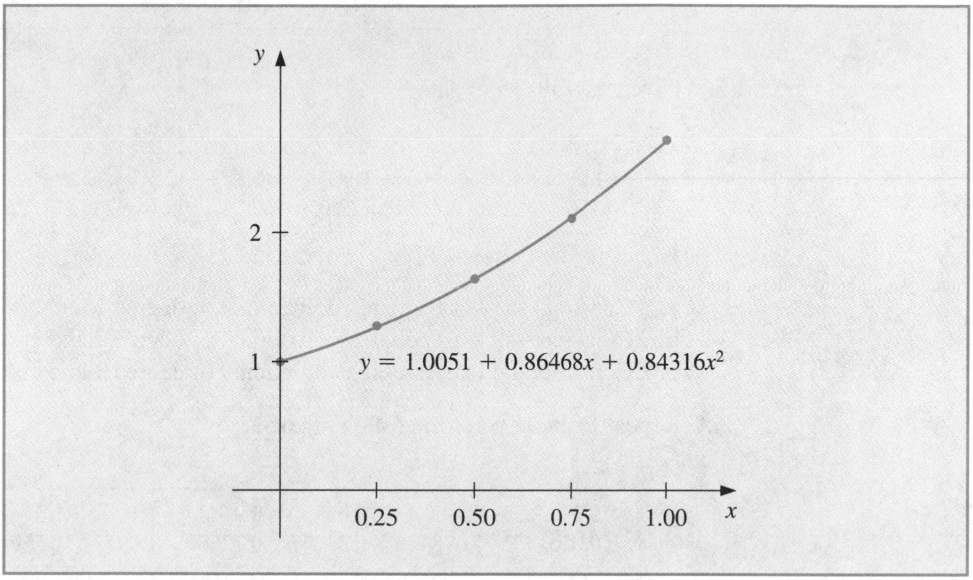

The total error,

$$\sum_{i=1}^{5}(y_i - P(x_i))^2 = 2.76 \times 10^{-4},$$

is the least that can be obtained by using a polynomial of degree at most 2. ◆ ◆ ◆

EXERCISE SET 8.2

1. Compute the linear least squares polynomial for the data of Example 2.
2. Compute the least squares polynomial of degree 2 for the data of Example 1 and compare the total error E_2 for the two polynomials.
3. Find the least squares polynomials of degrees 1, 2, and 3 for the data in the following table. Compute the error E_2 in each case. Graph the data and the polynomials.

x_i	1.0	1.1	1.3	1.5	1.9	2.1
y_i	1.84	1.96	2.21	2.45	2.94	3.18

4. Find the least squares polynomials of degrees 1, 2, and 3 for the data in the following table. Compute the error E_2 in each case. Graph the data and the polynomials.

x_i	0	0.15	0.31	0.5	0.6	0.75
y_i	1.0	1.004	1.031	1.117	1.223	1.422

5. Given the following data

x_i	4.0	4.2	4.5	4.7	5.1	5.5	5.9	6.3	6.8	7.1
y_i	102.56	113.18	130.11	142.05	167.53	195.14	224.87	256.73	299.50	326.72

a. Construct the least squares polynomial of degree 1 and compute the error.
b. Construct the least squares polynomial of degree 2 and compute the error.
c. Construct the least squares polynomial of degree 3 and compute the error.

6. Repeat Exercise 5 for the following data.

x_i	0.2	0.3	0.6	0.9	1.1	1.3	1.4	1.6
y_i	0.050446	0.098426	0.33277	0.72660	1.0972	1.5697	1.8487	2.5015

7. Hooke's law states that when a force is applied to a spring constructed of uniform material, the length of the spring is a linear function of the force that is applied, as shown in the accompanying figure.

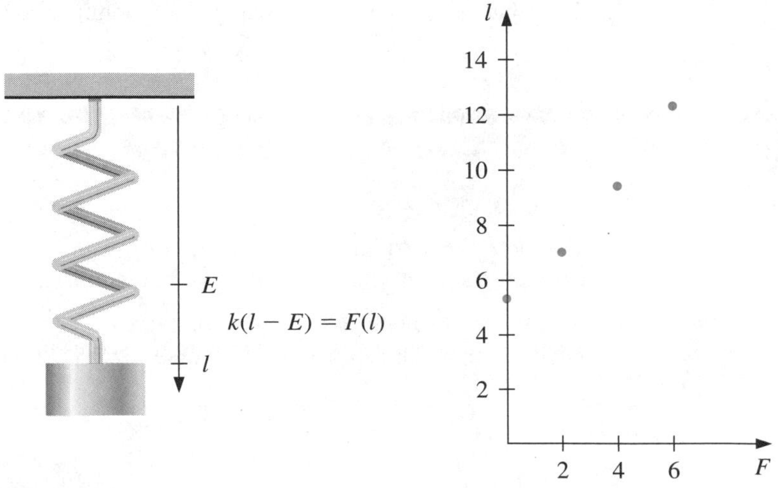

a. Suppose that $E = 5.3$ in. and that measurements are made of the length l in inches for applied weights $F(l)$ in pounds, as given in the following table. Find the least squares approximation for k.

$F(l)$	l
2	7.0
4	9.4
6	12.3

b. Additional measurements are made, giving the following additional data. Use these data to compute a new least squares approximation for k. Which of (a) or (b) best fits the total experimental data?

$F(l)$	l
3	8.3
5	11.3
8	14.4
10	15.9

8. To determine a relationship between the number of fish and the number of species of fish in samples taken for a portion of the Great Barrier Reef, P. Sale and R. Dybdahl [SD] fit a linear least squares polynomial to the following collection of data, which were collected in samples over a 2-year period. Let x be the number of fish in the sample and y be the number of species in the sample and determine the linear least squares polynomial for these data.

x	y	x	y	x	y
13	11	29	12	60	14
15	10	30	14	62	21
16	11	31	16	64	21
21	12	36	17	70	24
22	12	40	13	72	17
23	13	42	14	100	23
25	13	55	22	130	34

9. The following table lists the college grade-point averages of 20 mathematics and computer science majors, together with the scores that these students received on the mathematics portion of the ACT (American College Testing Program) test while in high school. Plot these data, and find the equation of the least squares line for this data. Do you think that the ACT scores are a reasonable predictor of college grade-point averages?

ACT score	Grade-point average	ACT score	Grade-point average
28	3.84	29	3.75
25	3.21	28	3.65
28	3.23	27	3.87
27	3.63	29	3.75
28	3.75	21	1.66
33	3.20	28	3.12
28	3.41	28	2.96
29	3.38	26	2.92
23	3.53	30	3.10
27	2.03	24	2.81

8.3 Continuous Least Squares Approximation

Suppose $f \in C[a, b]$ and we would like to determine a polynomial of degree at most n,

$$P_n(x) = a_n x^n + a_{n-1}x^{n-1} + \cdots + a_1 x + a_0 = \sum_{k=0}^{n} a_k x^k,$$

to minimize the error

$$E(a_0, a_1, \ldots, a_n) = \int_a^b (f(x) - P_n(x))^2\, dx = \int_a^b \left(f(x) - \sum_{k=0}^{n} a_k x^k \right)^2 dx.$$

A necessary condition for the numbers $a_0, a_1, \ldots, a_n$ to minimize the total error E is that

$$\frac{\partial E}{\partial a_j} = 0 \quad \text{for each } j = 0, 1, \ldots, n.$$

We can expand the integrand in this expression to

$$E = \int_a^b [f(x)]^2\, dx - 2\sum_{k=0}^{n} a_k \int_a^b x^k f(x)\, dx + \int_a^b \left(\sum_{k=0}^{n} a_k x^k \right)^2 dx,$$

so

$$\frac{\partial E}{\partial a_j} = -2\int_a^b x^j f(x)\, dx + 2\sum_{k=0}^{n} a_k \int_a^b x^{j+k}\, dx.$$

for each $j = 0, 1, \ldots, n$. Setting these to zero and rearranging, we obtain the $(n+1)$ linear **normal equations**

$$\sum_{k=0}^{n} a_k \int_a^b x^{j+k}\,dx = \int_a^b x^j f(x)\,dx, \quad \text{for each } j = 0, 1, \ldots, n,$$

which must be solved for the $n+1$ unknowns a_j. The normal equations have a unique solution provided that $f \in C[a,b]$.

EXAMPLE 1 Find the least squares approximating polynomial of degree 2 for the function $f(x) = \sin \pi x$ on the interval $[0, 1]$.

The normal equations for $P_2(x) = a_2x^2 + a_1x + a_0$ are

$$a_0 \int_0^1 1\,dx + a_1 \int_0^1 x\,dx + a_2 \int_0^1 x^2\,dx = \int_0^1 \sin \pi x\,dx,$$
$$a_0 \int_0^1 x\,dx + a_1 \int_0^1 x^2\,dx + a_2 \int_0^1 x^3\,dx = \int_0^1 x \sin \pi x\,dx,$$
$$a_0 \int_0^1 x^2\,dx + a_1 \int_0^1 x^3\,dx + a_2 \int_0^1 x^4\,dx = \int_0^1 x^2 \sin \pi x\,dx.$$

Performing the integration yields

$$a_0 + \frac{1}{2}a_1 + \frac{1}{3}a_2 = \frac{2}{\pi}, \quad \frac{1}{2}a_0 + \frac{1}{3}a_1 + \frac{1}{4}a_2 = \frac{1}{\pi}, \quad \frac{1}{3}a_0 + \frac{1}{4}a_1 + \frac{1}{5}a_2 = \frac{\pi^2 - 4}{\pi^3}.$$

These three equations in three unknowns can be solved to obtain

$$a_0 = \frac{12\pi^2 - 120}{\pi^3} \approx -0.050465 \quad \text{and} \quad a_1 = -a_2 = \frac{720 - 60\pi^2}{\pi^3} \approx 4.12251.$$

Consequently, the least squares polynomial approximation of degree 2 for $f(x) = \sin \pi x$ on $[0, 1]$ is $P_2(x) = -4.12251x^2 + 4.12251x - 0.050465$. (See Figure 8.5 on the following page.) ◆ ◆ ◆

Example 1 illustrates the difficulty in obtaining a least squares polynomial approximation. An $(n+1) \times (n+1)$ linear system for the unknowns $a_0, \ldots, a_n$ must be solved, and the coefficients in the linear system are of the form

$$\int_a^b x^{j+k}\,dx = \frac{b^{j+k+1} - a^{j+k+1}}{j+k+1}.$$

The matrix in the linear system is known as a *Hilbert matrix*, which is a classic example for demonstrating round-off error difficulties.

Another disadvantage to the technique used in Example 1 is similar to the situation that occurred when the Lagrange polynomials were first introduced in Section 3.2. The calculations that were performed in obtaining the best nth-degree polynomial do not lessen the amount of work required to obtain the best polynomials of higher degree. Both disad-

Figure 8.5

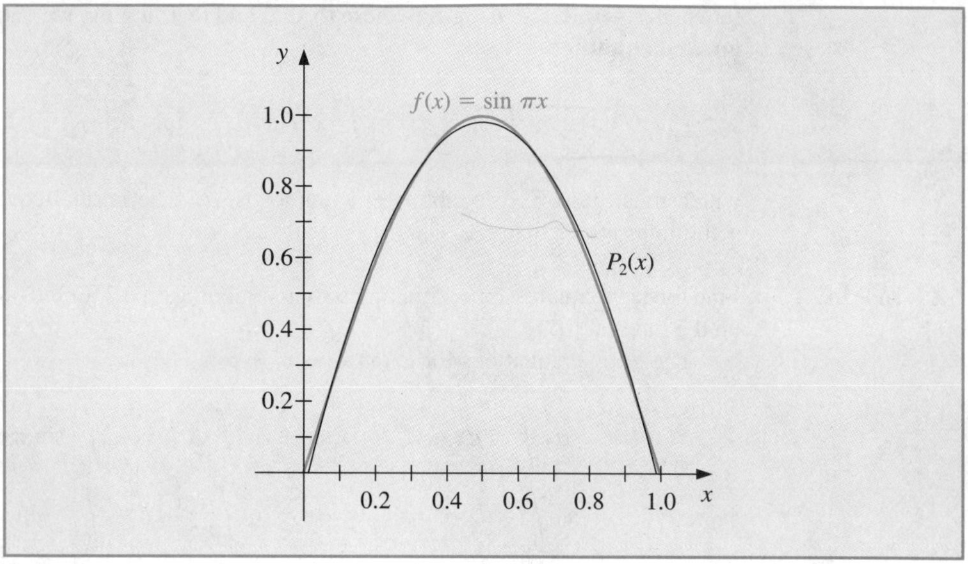

vantages are overcome by resorting to a technique that reduces the $n + 1$ equations in $n + 1$ unknowns to $n + 1$ equations, each of which contains only one unknown. This simplifies the problem to one that can be easily solved, but the technique requires some new concepts.

The set of functions $\{\phi_0, \phi_1, \ldots, \phi_n\}$ is said to be **linearly independent** on $[a, b]$ if, whenever

$$c_0\phi_0(x) + c_1\phi_1(x) + \cdots + c_n\phi_n(x) = 0 \quad \text{for all } x \in [a, b],$$

we have $c_0 = c_1 = \cdots = c_n = 0$. Otherwise the set of functions is said to be **linearly dependent**.

Linearly independent sets of functions are basic to our discussion and, since the functions we are using for approximations are polynomials, the following result is fundamental.

Linearly Independent Sets of Polynomials

If ϕ_j is a polynomial of degree j, for each $j = 0, 1, \ldots, n$, then $\{\phi_0, \ldots, \phi_n\}$ is linearly independent on any interval $[a, b]$.

The situation illustrated in the following example demonstrates a fact that holds in a much more general setting. Let $\prod_n$ be the **set of all polynomials of degree at most *n***. If $\{\phi_0(x), \phi_1(x), \ldots, \phi_n(x)\}$ is any collection of linearly independent polynomials in $\prod_n$, then each polynomial in $\prod_n$ can be written uniquely as a linear combination of $\{\phi_0(x), \phi_1(x), \ldots, \phi_n(x)\}$.

EXAMPLE 2 Let $\phi_0(x) = 2$, $\phi_1(x) = x - 3$, and $\phi_2(x) = x^2 + 2x + 7$. Then $\{\phi_0, \phi_1, \phi_2\}$ is linearly independent on any interval $[a, b]$. Suppose $Q(x) = a_0 + a_1x + a_2x^2$. We will show that there exist constants c_0, c_1, and c_2 such that $Q(x) = c_0\phi_0(x) + c_1\phi_1(x) + c_2\phi_2(x)$. Note first that

$$1 = \frac{1}{2}\phi_0(x), \quad x = \phi_1(x) + 3 = \phi_1(x) + \frac{3}{2}\phi_0(x),$$

and

$$\begin{aligned} x^2 &= \phi_2(x) - 2x - 7 = \phi_2(x) - 2\left[\phi_1(x) + \frac{3}{2}\phi_0(x)\right] - 7\left[\frac{1}{2}\phi_0(x)\right] \\ &= \phi_2(x) - 2\phi_1(x) - \frac{13}{2}\phi_0(x). \end{aligned}$$

Hence,

$$\begin{aligned} Q(x) &= a_0\left[\frac{1}{2}\phi_0(x)\right] + a_1\left[\phi_1(x) + \frac{3}{2}\phi_0(x)\right] + a_2\left[\phi_2(x) - 2\phi_1(x) - \frac{13}{2}\phi_0(x)\right] \\ &= \left[\frac{1}{2}a_0 + \frac{3}{2}a_1 - \frac{13}{2}a_2\right]\phi_0(x) + [a_1 - 2a_2]\phi_1(x) + a_2\phi_2(x), \end{aligned}$$

so any quadratic polynomial can be expressed as a linear combination of $\phi_0(x)$, $\phi_1(x)$, and $\phi_2(x)$. ◆ ◆ ◆

To discuss general function approximation requires the introduction of the notions of weight functions and orthogonality.

An integrable function w is called a **weight function** on the interval I if $w(x) \geq 0$ for all x in I, but w is not identically zero on any subinterval of I.

The purpose of a weight function is to assign varying degrees of importance to approximations on certain portions of the interval. For example, the weight function

$$w(x) = \frac{1}{\sqrt{1 - x^2}}$$

places less emphasis near the center of the interval $(-1, 1)$ and more emphasis when $|x|$ is near 1 (see Figure 8.6). This weight function will be used in the next section.

Figure 8.6

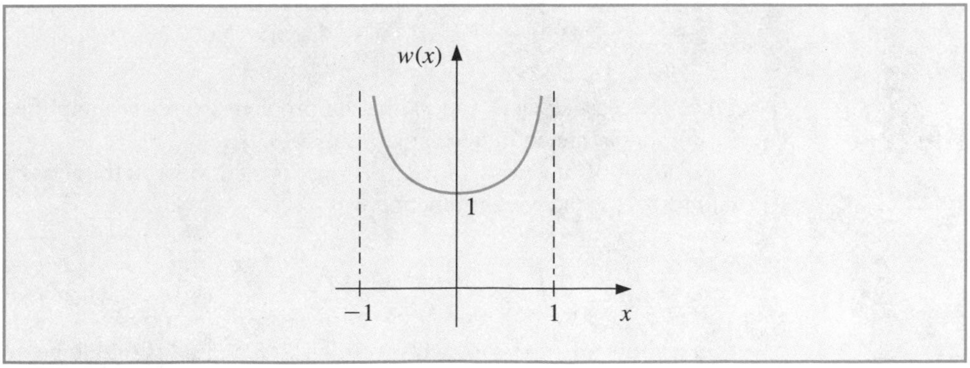

Suppose $\{\phi_0, \phi_1, \ldots, \phi_n\}$ is a set of linearly independent functions on $[a, b]$, w is a weight function for $[a, b]$, and, for $f \in C[a, b]$, a linear combination

$$P(x) = \sum_{k=0}^{n} a_k \phi_k(x)$$

is sought to minimize the error

$$E(a_0, \ldots, a_n) = \int_a^b w(x) \left[f(x) - \sum_{k=0}^{n} a_k \phi_k(x) \right]^2 dx.$$

This problem reduces to the situation considered at the beginning of this section in the special case when $w(x) \equiv 1$ and $\phi_k(x) = x^k$ for each $k = 0, 1, \ldots, n$.

The normal equations associated with this problem are derived from the fact that for each $j = 0, 1, \ldots, n$,

$$0 = \frac{\partial E}{\partial a_j} = 2 \int_a^b w(x) \left[f(x) - \sum_{k=0}^{n} a_k \phi_k(x) \right] \phi_j(x)\, dx.$$

The system of normal equations can be written

$$\int_a^b w(x) f(x) \phi_j(x)\, dx = \sum_{k=0}^{n} a_k \int_a^b w(x) \phi_k(x) \phi_j(x)\, dx, \quad \text{for each } j = 0, 1, \ldots, n.$$

If the functions $\phi_0, \phi_1, \ldots, \phi_n$ can be chosen so that

$$\int_a^b w(x) \phi_k(x) \phi_j(x)\, dx = \begin{cases} 0, & \text{when } j \neq k, \\ \alpha_k > 0, & \text{when } j = k, \end{cases} \tag{8.1}$$

then the normal equations reduce to

$$\int_a^b w(x) f(x) \phi_j(x)\, dx = a_j \int_a^b w(x) [\phi_j(x)]^2\, dx = a_j \alpha_j$$

for each $j = 0, 1, \ldots, n$, and are easily solved as

$$a_j = \frac{1}{\alpha_j} \int_a^b w(x) f(x) \phi_j(x)\, dx.$$

Hence the least squares approximation problem is greatly simplified when the functions $\phi_0, \phi_1, \ldots, \phi_n$ are chosen to satisfy Eq. (8.1).

The set of functions $\{\phi_0, \phi_1, \ldots, \phi_n\}$ is said to be **orthogonal** for the interval $[a, b]$ with respect to the weight function w if

$$\int_a^b w(x) \phi_j(x) \phi_k(x)\, dx = \begin{cases} 0, & \text{when } j \neq k, \\ \alpha_k > 0, & \text{when } j = k. \end{cases}$$

If, in addition, $\alpha_k = 1$ for each $k = 0, 1, \ldots, n$, the set is said to be **orthonormal**.

This definition, together with the remarks preceding it, implies the following.

Least Squares for Orthogonal Functions

If $\{\phi_0, \phi_1, \dots, \phi_n\}$ is an orthogonal set of functions on an interval $[a, b]$ with respect to the weight function w, then the least squares approximation to f on $[a, b]$ with respect to w is

$$P(x) = \sum_{k=0}^{n} a_k \phi_k(x),$$

where

$$a_k = \frac{\int_a^b w(x)\phi_k(x) f(x)\,dx}{\int_a^b w(x)[\phi_k(x)]^2\,dx} = \frac{1}{\alpha_k}\int_a^b w(x)\phi_k(x) f(x)\,dx.$$

The next result, which is based on the *Gram-Schmidt* process, describes a recursive procedure for constructing orthogonal polynomials on $[a, b]$ with respect to a weight function w.

Recursive Generation of Orthogonal Polynomials

The set of polynomials $\{\phi_0(x), \phi_1(x), \dots, \phi_n(x)\}$ defined in the following way is linearly independent and orthogonal on $[a, b]$ with respect to the weight function w.

$$\phi_0(x) \equiv 1, \qquad \phi_1(x) = x - B_1,$$

where

$$B_1 = \frac{\int_a^b x w(x)[\phi_0(x)]^2\,dx}{\int_a^b w(x)[\phi_0(x)]^2\,dx},$$

and when $k \geq 2$,

$$\phi_k(x) = (x - B_k)\phi_{k-1}(x) - C_k \phi_{k-2}(x),$$

where

$$B_k = \frac{\int_a^b x w(x)[\phi_{k-1}(x)]^2\,dx}{\int_a^b w(x)[\phi_{k-1}(x)]^2\,dx} \quad \text{and} \quad C_k = \frac{\int_a^b x w(x)\phi_{k-1}(x)\phi_{k-2}(x)\,dx}{\int_a^b w(x)[\phi_{k-2}(x)]^2\,dx}.$$

Moreover, for any polynomial $Q_k(x)$ of degree $k < n$,

$$\int_a^b w(x)\phi_n(x) Q_k(x)\,dx = 0.$$

EXAMPLE 3 The set of **Legendre polynomials**, $\{P_n(x)\}$, is orthogonal on $[-1, 1]$ with respect to the weight function $w(x) \equiv 1$. The classical definition of the Legendre polynomials requires that $P_n(1) = 1$ for each n, and a recursive relation can be used to generate the polynomials when $n \geq 2$. This normalization will not be needed in our discussion, and the least squares approximating polynomials generated in either case are essentially the same. Using the recursive procedure, $P_0(x) \equiv 1$, so

$$B_1 = \frac{\int_{-1}^{1} x\,dx}{\int_{-1}^{1} dx} = 0 \quad \text{and} \quad P_1(x) = (x - B_1)P_0(x) = x.$$

Also,

$$B_2 = \frac{\int_{-1}^{1} x^3\,dx}{\int_{-1}^{1} x^2\,dx} = 0 \quad \text{and} \quad C_2 = \frac{\int_{-1}^{1} x^2\,dx}{\int_{-1}^{1} 1\,dx} = \frac{1}{3},$$

so

$$P_2(x) = (x - B_2)P_1(x) - C_2P_0(x) = (x - 0)x - \frac{1}{3}\cdot 1 = x^2 - \frac{1}{3}.$$

Higher-degree Legendre polynomials are derived in the same manner. The Maple command `int` can be used to compute B_k and C_k. For example,

```
>B3:=int(x*(x^2-1/3)^2,x=-1..1)/int((x^2-1/3)^2,x=-1..1);
>C3:=int(x*(x^2-1/3)*x,x=-1..1)/int(x^2,x=-1..1);
```

Figure 8.7

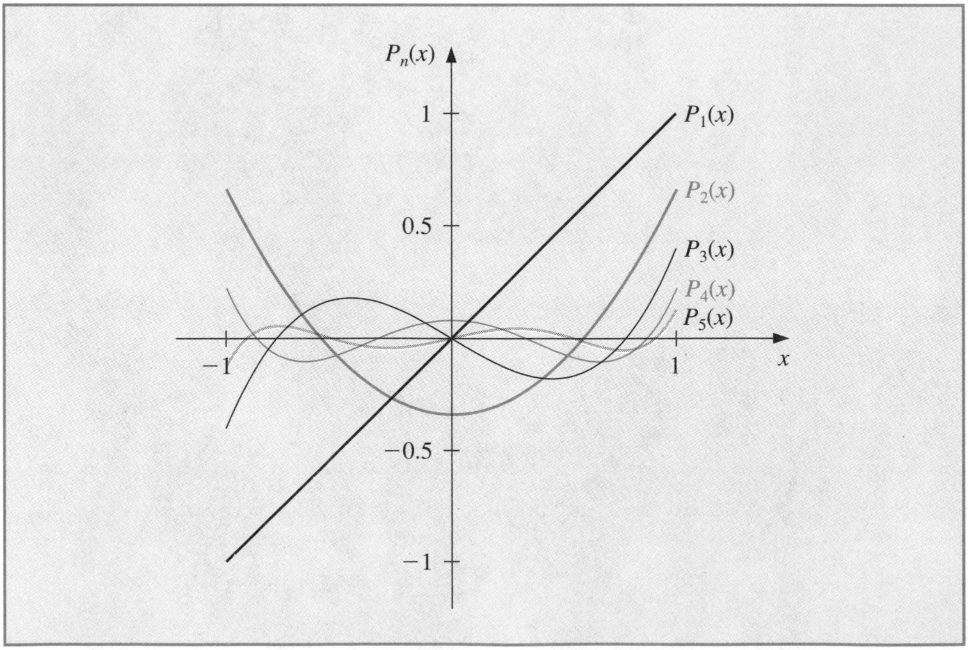

gives $B_3 = 0$ and $C_3 = \frac{4}{15}$. Thus,

$$P_3(x) = xP_2(x) - \frac{4}{15}P_1(x) = x^3 - \frac{1}{3}x - \frac{4}{15}x = x^3 - \frac{3}{5}x.$$

The next two Legendre polynomials are $P_4(x) = x^4 - \frac{6}{7}x^2 + \frac{3}{35}$ and $P_5(x) = x^5 - \frac{10}{9}x^3 + \frac{5}{21}x$. Figure 8.7 shows the graphs of these polynomials. ◆ ◆ ◆

EXERCISE SET 8.3

1. Find the linear least squares polynomial approximation to $f(x)$ on the indicated interval in each case.

a. $f(x) = x^2 + 3x + 2$, $[0, 1]$; **b.** $f(x) = x^3$, $[0, 2]$;

c. $f(x) = \dfrac{1}{x}$, $[1, 3]$; **d.** $f(x) = e^x$, $[0, 2]$;

e. $f(x) = \dfrac{1}{2}\cos x + \dfrac{1}{3}\sin 2x$, $[0, 1]$; **f.** $f(x) = x \ln x$, $[1, 3]$.

2. Find the least squares polynomial approximation of degree 2 to the functions and intervals in Exercise 1.

3. Find the linear least squares polynomial approximation on the interval $[-1, 1]$ for the following functions.

a. $f(x) = x^2 - 2x + 3$ **b.** $f(x) = x^3$

c. $f(x) = \dfrac{1}{x+2}$ **d.** $f(x) = e^x$

e. $f(x) = \dfrac{1}{2}\cos x + \dfrac{1}{3}\sin 2x$ **f.** $f(x) = \ln(x + 2)$

4. Find the least squares polynomial approximation of degree 2 on the interval $[-1, 1]$ for the functions in Exercise 3.

5. Compute the error E for the approximations in Exercise 3.

6. Compute the error E for the approximations in Exercise 4.

7. Use the Gram-Schmidt process to construct $\phi_0(x)$, $\phi_1(x)$, $\phi_2(x)$, and $\phi_3(x)$ for the following intervals.

a. $[0, 1]$ **b.** $[0, 2]$ **c.** $[1, 3]$

8. Repeat Exercise 1 using the results of Exercise 7.

9. Repeat Exercise 2 using the results of Exercise 7.

10. Use the Gram-Schmidt procedure to calculate L_1, L_2, and L_3, where $\{L_0(x), L_1(x), L_2(x), L_3(x)\}$ is an orthogonal set of polynomials on $(0, \infty)$ with respect to the weight functions $w(x) = e^{-x}$ and $L_0(x) \equiv 1$. The polynomials obtained from this procedure are called the **Laguerre polynomials**.

11. Use the Laguerre polynomials calculated in Exercise 10 to compute the least squares polynomials of degree 1, 2, and 3 on the interval $(0, \infty)$ with respect to the weight function $w(x) = e^{-x}$ for the following functions.

a. $f(x) = x^2$

b. $f(x) = e^{-x}$

c. $f(x) = x^3$

d. $f(x) = e^{-2x}$

12. Show that if $\{\phi_0, \phi_1, \ldots, \phi_n\}$ is an orthogonal set of functions on $[a, b]$ with respect to the weight function w, then $\{\phi_0, \phi_1, \ldots, \phi_n\}$ is a linearly independent set.

13. Suppose $\{\phi_0(x), \phi_1(x), \ldots, \phi_n(x)\}$ is a linearly independent set in $\prod_n$. Show that for any element $Q(x) \in \prod_n$ there exist unique constants $c_0, c_1, \ldots, c_n$ such that

$$Q(x) = \sum_{k=o}^{n} c_k \phi_k(x).$$

8.4 Chebyshev Polynomials

The **Chebyshev polynomials** $\{T_n(x)\}$ are orthogonal on $(-1, 1)$ with respect to the weight function $w(x) = (1 - x^2)^{-1/2}$. Although they can be derived by the method in the previous section, it is easier to give their definition and then show that they satisfy the required orthogonality properties.

For $x \in [-1, 1]$, define

$$T_n(x) = \cos(n \arccos x) \quad \text{for each } n \geq 0.$$

It is not obvious from this definition that for each n, $T_n(x)$ is a polynomial in x, but we will now show that it is. First note that

$$T_0(x) = \cos 0 = 1 \quad \text{and} \quad T_1(x) = \cos(\arccos x) = x.$$

For $n \geq 1$ we introduce the substitution $\theta = \arccos x$ to change this equation to

$$T_n(\theta(x)) \equiv T_n(\theta) = \cos(n\theta), \quad \text{where } \theta \in [0, \pi].$$

A recurrence relation is derived by noting that

$$T_{n+1}(\theta) = \cos(n\theta + \theta) = \cos(n\theta)\cos\theta - \sin(n\theta)\sin\theta$$

and

$$T_{n-1}(\theta) = \cos(n\theta - \theta) = \cos(n\theta)\cos\theta + \sin(n\theta)\sin\theta.$$

Adding these equations gives

$$T_{n+1}(\theta) = 2\cos(n\theta)\cos\theta - T_{n-1}(\theta).$$

Returning to the variable x we have, for $n \geq 1$,

$$T_{n+1} = 2\cos(n \arccos x) \cdot x - T_{n-1}(x) = 2T_n(x) \cdot x - T_{n-1}(x).$$

Since $T_0(x)$ and $T_1(x)$ are both polynomials in x, $T_{n+1}(x)$ will be a polynomial in x for each n.

Chebyshev Polynomials

$$T_0(x) = 1, \quad T_1(x) = x,$$

and, for $n \geq 1$, $T_{n+1}(x)$ is the polynomial of degree $n+1$ given by

$$T_{n+1}(x) = 2xT_n(x) - T_{n-1}(x).$$

The recurrence relation implies that $T_n(x)$ is a polynomial of degree n, and it has leading coefficient 2^{n-1}, when $n \geq 1$. The next three Chebyshev polynomials are

$$T_2(x) = 2xT_1(x) - T_0(x) = 2x^2 - 1,$$
$$T_3(x) = 2xT_2(x) - T_1(x) = 4x^3 - 3x,$$

and

$$T_4(x) = 2xT_3(x) - T_2(x) = 8x^4 - 8x^2 + 1.$$

The graphs of T_1, T_2, T_3, and T_4 are shown in Figure 8.8.

Figure 8.8

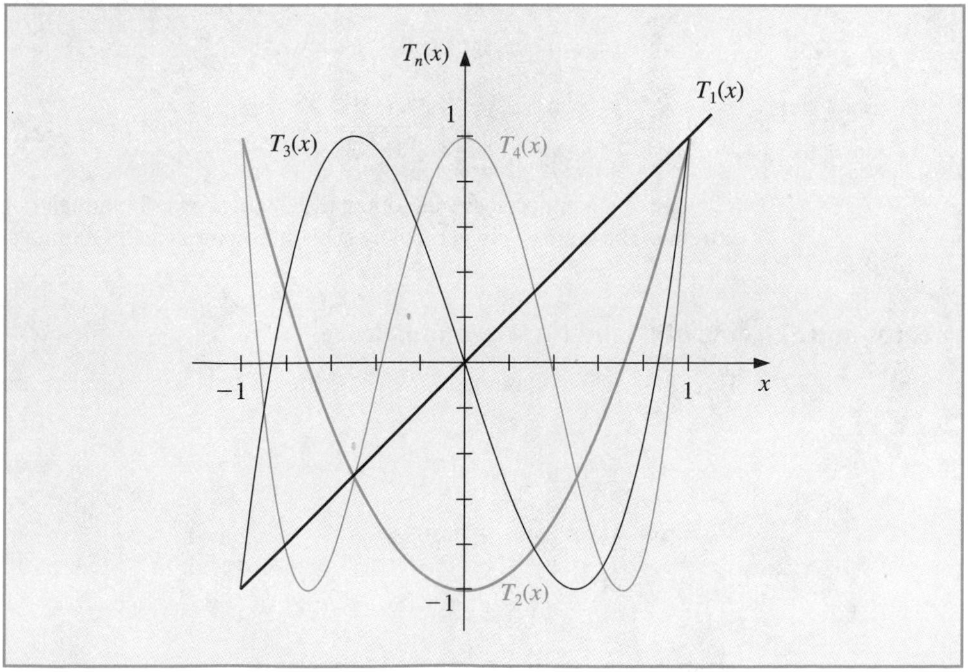

To show the orthogonality of the Chebyshev polynomials, consider

$$\int_{-1}^{1} \frac{T_n(x)T_m(x)}{\sqrt{1-x^2}}\,dx = \int_{-1}^{1} \frac{\cos(n \arccos x)\cos(m \arccos x)}{\sqrt{1-x^2}}\,dx.$$

Reintroducing the substitution $\theta = \arccos x$ gives

$$d\theta = -\frac{1}{\sqrt{1-x^2}}\,dx$$

and

$$\int_{-1}^{1} \frac{T_n(x)T_m(x)}{\sqrt{1-x^2}}\,dx = -\int_{\pi}^{0} \cos(n\theta)\cos(m\theta)\,d\theta = \int_{0}^{\pi} \cos(n\theta)\cos(m\theta)\,d\theta.$$

Suppose $n \neq m$. Since

$$\cos(n\theta)\cos(m\theta) = \frac{1}{2}\left[\cos(n+m)\theta + \cos(n-m)\theta\right],$$

we have

$$\begin{aligned}\int_{-1}^{1} \frac{T_n(x)T_m(x)}{\sqrt{1-x^2}}\,dx &= \frac{1}{2}\int_{0}^{\pi} \cos((n+m)\theta)\,d\theta + \frac{1}{2}\int_{0}^{\pi} \cos((n-m)\theta)\,d\theta \\ &= \left[\frac{1}{2(n+m)}\sin\left((n+m)\theta\right) + \frac{1}{2(n-m)}\sin(n-m)\theta\right]_{0}^{\pi} = 0.\end{aligned}$$

By a similar technique, it can also be shown that

$$\int_{-1}^{1} \frac{[T_n(x)]^2}{\sqrt{1-x^2}}\,dx = \frac{\pi}{2} \quad \text{for each } n \geq 1.$$

One of the important results about the Chebyshev polynomials concerns their zeros and extrema. These are easily verified by substitution into the polynomials and their derivatives.

Zeros and Extrema of Chebyshev Polynomials

The Chebyshev polynomial $T_n(x)$, of degree $n \geq 1$, has n simple zeros in $[-1, 1]$ at

$$\bar{x}_k = \cos\left(\frac{2k-1}{2n}\pi\right) \quad \text{for each } k = 1, 2, \ldots, n.$$

Moreover, T_n assumes its absolute extrema at

$$\bar{x}'_k = \cos\left(\frac{k\pi}{n}\right) \quad \text{with} \quad T_n\left(\bar{x}'_k\right) = (-1)^k \quad \text{for each } k = 0, 1, \ldots, n.$$

The *monic Chebyshev polynomial* (polynomial with leading coefficient 1), $\tilde{T}_n(x)$, is derived from the Chebyshev polynomial, $T_n(x)$, by dividing by the leading coefficient, 2^{n-1}, when $n \geq 1$. So

$$\tilde{T}_0(x) = 1, \quad \text{and} \quad \tilde{T}_n(x) = 2^{1-n}T_n(x) \quad \text{for each } n \geq 1.$$

These polynomials satisfy the recurrence relation

$$\tilde{T}_2(x) = x\tilde{T}_1(x) - \frac{1}{2}\tilde{T}_0(x); \quad \tilde{T}_{n+1}(x) = x\tilde{T}_n(x) - \frac{1}{4}\tilde{T}_{n-1}(x),$$

for each $n \geq 2$. Because $\tilde{T}_n$ is a multiple of T_n, the zeros of $\tilde{T}_n$ also occur at

$$\bar{x}_k = \cos\left(\frac{2k-1}{2n}\pi\right) \quad \text{for each } k = 1, 2, \ldots, n,$$

and the extreme values of $\tilde{T}_n$ occur at

$$\bar{x}'_k = \cos\left(\frac{k\pi}{n}\right) \quad \text{with} \quad \tilde{T}_n(\bar{x}'_k) = \frac{(-1)^k}{2^{n-1}} \quad \text{for each } k = 0, 1, 2, \ldots, n.$$

Let $\tilde{\Pi}_n$ denote the **set of all monic polynomials of degree *n***. We have an important minimization property that distinguishes the polynomials $\tilde{T}_n(x)$ from the other members of $\tilde{\Pi}_n$.

Minimum Property of Monic Chebyshev Polynomials

The polynomial $\tilde{T}_n(x)$, when $n \geq 1$, has the property that

$$\frac{1}{2^{n-1}} = \max_{x\in[-1,1]} |\tilde{T}_n(x)| \leq \max_{x\in[-1,1]} |P_n(x)| \quad \text{for all } P_n \in \tilde{\Pi}_n.$$

Moreover, equality can occur only if $P_n = \tilde{T}_n$.

This result is used to answer the question of where to place interpolating nodes to minimize the error in Lagrange interpolation. The error form for the Lagrange polynomial applied to the interval $[-1, 1]$ states that if $x_0, \ldots, x_n$ are distinct numbers in the interval $[-1, 1]$ and if $f \in C^{n+1}[-1, 1]$, then, for each $x \in [-1, 1]$, a number $\xi(x)$ exists in $(-1, 1)$ with

$$f(x) - P(x) = \frac{f^{(n+1)}(\xi(x))}{(n+1)!}(x - x_0)\cdots(x - x_n),$$

where $P(x)$ is the Lagrange interpolating polynomial. There is no control over $\xi(x)$, so to minimize the error by shrewd placement of the nodes $x_0, \ldots, x_n$ is equivalent to choosing $x_0, \ldots, x_n$ to minimize the quantity

$$|(x - x_0)(x - x_1)\cdots(x - x_n)|$$

throughout the interval $[-1, 1]$.

Since $(x-x_0)(x-x_1)\cdots(x-x_n)$ is a monic polynomial of degree $n+1$, the minimum is obtained when

$$(x-x_0)(x-x_1)\cdots(x-x_n)=\tilde{T}_{n+1}(x).$$

When x_k is chosen to be the $(k+1)$st zero of $\tilde{T}_{n+1}$, that is, when x_k is

$$\overline{x}_{k+1}=\cos\frac{2k+1}{2(n+1)}\pi,$$

the maximum value of $|(x-x_0)(x-x_1)\cdots(x-x_n)|$ is minimized. Since

$$\max_{x\in[-1,1]}\left|\tilde{T}_{n+1}(x)\right|=\frac{1}{2^n}$$

this also implies that

$$\begin{aligned}\frac{1}{2^n}&=\max_{x\in[-1,1]}|(x-\overline{x}_1)(x-\overline{x}_2)\cdots(x-\overline{x}_{n+1})|\\&\le\max_{x\in[-1,1]}|(x-x_0)(x-x_1)\cdots(x-x_n)|,\end{aligned}$$

for any choice of $x_0, x_1, \ldots, x_n$ in the interval $[-1,1]$.

Minimizing Lagrange Interpolation Error

If $P(x)$ is the interpolating polynomial of degree at most n with nodes at the roots of $T_{n+1}(x)$, then, for any $f\in C^{n+1}[-1,1]$,

$$\max_{x\in[-1,1]}|f(x)-P(x)|\le\frac{1}{2^n(n+1)!}\max_{x\in[-1,1]}\left|f^{(n+1)}(x)\right|.$$

The technique for choosing points to minimize the interpolating error can be easily extended to a general closed interval $[a,b]$ by using the change of variable

$$\tilde{x}=\frac{1}{2}[(b-a)x+a+b]$$

to transform the numbers $\overline{x}_k$ in the interval $[-1,1]$ into the corresponding numbers in the interval $[a,b]$, as shown in the next example.

EXAMPLE 1 Let $f(x)=xe^x$ on $[0,1.5]$. Two interpolation polynomials of degree at most 3 will be constructed. First, the equally spaced nodes $x_0=0$, $x_1=0.5$, $x_2=1$, and $x_3=1.5$ are used. The methods in Section 3.2 give this interpolating polynomial as

$$P_3(x)=1.3875x^3+0.057570x^2+1.2730x.$$

For the second interpolating polynomial, shift the zeros $\overline{x}_k=\cos((2k+1)/8)\pi$, for $k=0,1,2,3$, of $\tilde{T}_4$ from $[-1,1]$ to $[0,1.5]$, using the linear transformation

$$\tilde{x}_k = \frac{1}{2}[(1.5 - 0)\bar{x}_k + (1.5 + 0)] = 0.75 + 0.75\bar{x}_k$$

to obtain

$$\tilde{x}_0 = 1.44291, \quad \tilde{x}_1 = 1.03701, \quad \tilde{x}_2 = 0.46299, \quad \text{and} \quad \tilde{x}_3 = 0.05709.$$

For these nodes, the interpolation polynomial of degree at most 3 is

$$\tilde{P}_3(x) = 1.3811x^3 + 0.044652x^2 + 1.3031x - 0.014352.$$

Table 8.4

x	$f(x) = xe^x$	$P_3(x)$	$\lvert xe^x - P_3(x)\rvert$	$\tilde{P}_3(x)$	$\lvert xe^x - \tilde{P}_3(x)\rvert$
0.15	0.1743	0.1969	0.0226	0.1868	0.0125
0.25	0.3210	0.3435	0.0225	0.3358	0.0148
0.35	0.4967	0.5121	0.0154	0.5064	0.0097
0.65	1.245	1.233	0.0120	1.231	0.0140
0.75	1.588	1.572	0.0160	1.571	0.0170
0.85	1.989	1.976	0.0130	1.974	0.0150
1.15	3.632	3.650	0.0180	3.644	0.0120
1.25	4.363	4.391	0.0280	4.382	0.0190
1.35	5.208	5.237	0.0290	5.224	0.0160

For comparison, Table 8.4 lists various values of x, together with the values of $f(x)$, $P_3(x)$, and $\tilde{P}_3(x)$. Although the error using P_3 is less than using $\tilde{P}_3$ near the middle of the table, the maximum error involved with using $\tilde{P}_3$, 0.019, is considerably less than using P_3, 0.029. (See Figure 8.9.) ◆ ◆ ◆

Figure 8.9

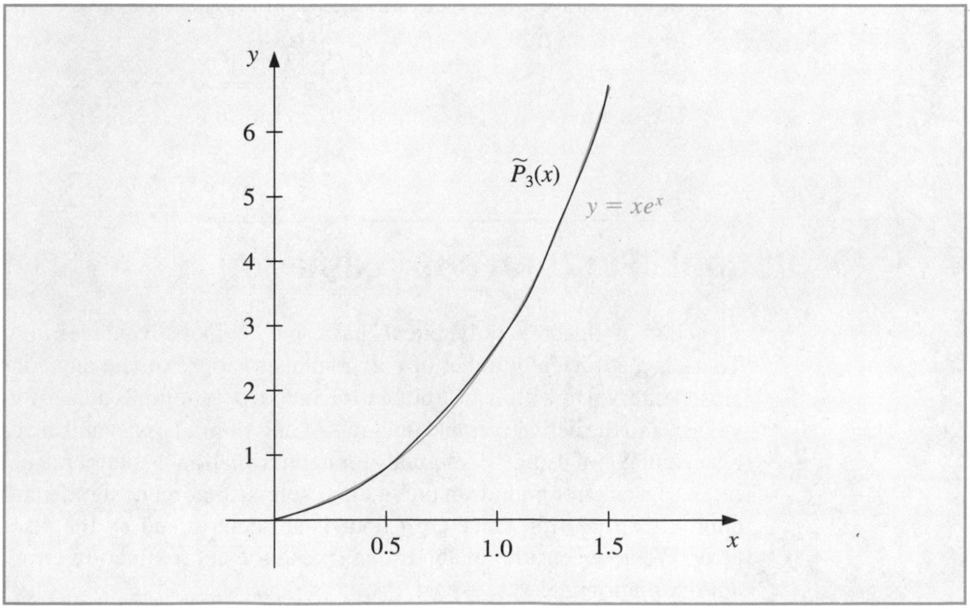

EXERCISE SET 8.4

1. Use the zeros of $\tilde{T}_3$ to construct an interpolating polynomial of degree 2 for the following functions on the interval $[-1, 1]$.
   a. $f(x) = e^x$
   b. $f(x) = \sin x$
   c. $f(x) = \ln(x + 2)$
   d. $f(x) = x^4$
2. Find a bound for the maximum error of the approximation in Exercise 1 on the interval $[-1, 1]$.
3. Use the zeros of $\tilde{T}_4$ to construct an interpolating polynomial of degree 3 for the functions in Exercise 1.
4. Repeat Exercise 2 for the approximations computed in Exercise 3.
5. Use the zeros of $\tilde{T}_3$ and transformations of the given interval to construct an interpolating polynomial of degree 2 for the following functions.
   a. $f(x) = \frac{1}{x}$, $[1, 3]$
   b. $f(x) = e^{-x}$, $[0, 2]$
   c. $f(x) = \frac{1}{2}\cos x + \frac{1}{3}\sin 2x$, $[0, 1]$
   d. $f(x) = x \ln x$, $[1, 3]$
6. Use the zeros of $\tilde{T}_4$ to construct an interpolating polynomial of degree 3 for the functions in Exercise 5.
7. Show that for any positive integers i and j with $i > j$ we have
$$T_i(x)T_j(x) = \frac{1}{2}[T_{i+j}(x) + T_{i-j}(x)].$$
8. Show that for each Chebyshev polynomial $T_n(x)$ we have
$$\int_{-1}^{1} \frac{[T_n(x)]^2}{\sqrt{1 - x^2}}\, dx = \frac{\pi}{2}.$$

8.5 Rational Function Approximation

The class of algebraic polynomials has some distinct advantages for use in approximation. There is a sufficient number of polynomials to approximate any continuous function on a closed interval to within an arbitrary tolerance, polynomials are easily evaluated at arbitrary values, and the derivatives and integrals of polynomials exist and are easily determined. The disadvantage of using polynomials for approximation is their tendency for oscillation. This often causes error bounds in polynomial approximation to significantly exceed the average approximation error, since error bounds are determined by the maximum approximation error. We now consider methods that spread the approximation error more evenly over the approximation interval.

A **rational function** r of degree N has the form

$$r(x) = \frac{p(x)}{q(x)},$$

where $p(x)$ and $q(x)$ are polynomials whose degrees sum to N.

Since every polynomial is a rational function (simply let $q(x) \equiv 1$), approximation by rational functions gives results with no greater error bounds than approximation by polynomials. However, rational functions whose numerator and denominator have the same or nearly the same degree generally produce approximation results superior to polynomial methods for the same amount of computational effort. Rational functions have the added advantage of permitting efficient approximation of functions with infinite discontinuities near the interval of approximation. Polynomial approximation is generally unacceptable in this situation.

Suppose r is a rational function of degree $N = n + m$ of the form

$$r(x) = \frac{p(x)}{q(x)} = \frac{p_0 + p_1 x + \cdots + p_n x^n}{q_0 + q_1 x + \cdots + q_m x^m}$$

that is used to approximate a function f on a closed interval I containing zero. For r to be defined at zero requires that $q_0 \neq 0$. In fact, we can assume that $q_0 = 1$, for if this is not the case we simply replace $p(x)$ by $p(x)/q_0$ and $q(x)$ by $q(x)/q_0$. Consequently, there are $N + 1$ parameters $q_1, q_2, \ldots, q_m, p_0, p_1, \ldots, p_n$ available for the approximation of f by r.

The **Padé approximation technique** chooses the $N + 1$ parameters so that $f^{(k)}(0) = r^{(k)}(0)$ for each $k = 0, 1, \ldots, N$. Padé approximation is the extension of Taylor polynomial approximation to rational functions. In fact, when $n = N$ and and $m = 0$, the Padé approximation is the Nth Taylor polynomial expanded about zero—that is, the Nth Maclaurin polynomial.

Consider the difference

$$f(x) - r(x) = f(x) - \frac{p(x)}{q(x)} = \frac{f(x)q(x) - p(x)}{q(x)} = \frac{f(x)\sum_{i=0}^{m} q_i x^i - \sum_{i=0}^{n} p_i x^i}{q(x)}$$

and suppose f has the Maclaurin series expansion $f(x) = \sum_{i=0}^{\infty} a_i x^i$. Then

$$f(x) - r(x) = \frac{\sum_{i=0}^{\infty} a_i x^i \sum_{i=0}^{m} q_i x^i - \sum_{i=0}^{n} p_i x^i}{q(x)}. \tag{8.2}$$

The object is to choose the constants $q_1, q_2, \ldots, q_m$ and $p_0, p_1, \ldots, p_n$ so that

$$f^{(k)}(0) - r^{(k)}(0) = 0 \quad \text{for each } k = 0, 1, \ldots, N.$$

This is equivalent to $f - r$ having a root of multiplicity $N + 1$ at zero. As a consequence, we choose $q_1, q_2, \ldots, q_m$ and $p_0, p_1, \ldots, p_n$ so that the numerator on the right side of Eq. (8.2),

$$(a_0 + a_1 x + \cdots)(1 + q_1 x + \cdots + q_m x^m) - (p_0 + p_1 x + \cdots + p_n x^n),$$

has no terms of degree less than or equal to N.

To simplify notation, we define $p_{n+1} = p_{n+2} = \cdots = p_N = 0$ and $q_{m+1} = q_{m+2} = \cdots = q_N = 0$. The coefficient of x^k is then

$$\left(\sum_{i=0}^{k} a_i q_{k-i}\right) - p_k,$$

and the rational function for Padé approximation results from the solution of the $N+1$ linear equations

$$\sum_{i=0}^{k} a_i q_{k-i} = p_k, \quad k = 0, 1, \ldots, N$$

in the $N+1$ unknowns $q_1, q_2, \ldots, q_m, p_0, p_1, \ldots, p_n$.

The Padé technique can be implemented using the program PADEMD81.

EXAMPLE 1 The Maclaurin series expansion for e^{-x} is

$$\sum_{i=0}^{\infty} \frac{(-1)^i}{i!} x^i.$$

To find the Padé approximation to e^{-x} of degree 5 with $n = 3$ and $m = 2$ requires choosing p_0, p_1, p_2, p_3, q_1, and q_2 so that the coefficients of x^k for $k = 0, 1, \ldots, 5$ are zero in the expression

$$\left(1 - x + \frac{x^2}{2} - \frac{x^3}{6} + \cdots\right)\left(1 + q_1 x + q_2 x^2\right) - \left(p_0 + p_1 x + p_2 x^2 + p_3 x^3\right).$$

Expanding and collecting terms produces

$$\begin{aligned}
x^5&: \quad -\frac{1}{120} + \frac{1}{24}q_1 - \frac{1}{6}q_2 = 0; & x^2&: \quad \frac{1}{2} - q_1 + q_2 = p_2;\\
x^4&: \quad \frac{1}{24} - \frac{1}{6}q_1 + \frac{1}{2}q_2 = 0; & x^1&: \quad -1 + q_1 = p_1;\\
x^3&: \quad -\frac{1}{6} + \frac{1}{2}q_1 - q_2 = p_3; & x^0&: \quad 1 = p_0.
\end{aligned}$$

To solve the system in Maple we use the following:

```
>eq1:=-1+q1=p1;
>eq2:=1/2-q1+q2=p2;
>eq3:=-1/6+1/2*q1-q2=p3;
>eq4:=1/24-1/6*q1+1/2*q2=0;
>eq5:=-1/120+1/24*q1-1/6*q2=0;
>solve({eq1,eq2,eq3,eq4,eq5},{q1,q2,p1,p2,p3});
```

The solution to this system is

$$p_0 = 1, \quad p_1 = -\frac{3}{5}, \quad p_2 = \frac{3}{20}, \quad p_3 = -\frac{1}{60}, \quad q_1 = \frac{2}{5}, \quad \text{and} \quad q_2 = \frac{1}{20},$$

so the Padé approximation is

$$r(x) = \frac{1 - \frac{3}{5}x + \frac{3}{20}x^2 - \frac{1}{60}x^3}{1 + \frac{2}{5}x + \frac{1}{20}x^2}.$$

Maple can be used to compute a Padé approximation directly. We first compute the Maclaurin series with the call

```
>series(exp(-x),x);
```

to obtain

$$1 - x + \frac{1}{2}x^2 - \frac{1}{6}x^3 + \frac{1}{24}x^4 - \frac{1}{120}x^5 + O(x^6)$$

The Padé approximation with $n = 3$ and $m = 2$ is computed using the command

```
>g:=convert(",ratpoly,3,2);
```

where the double quote refers to the result of the preceding calculation, that is, the Maclaurin series. The result is

$$g := \frac{1 - \frac{3}{5}x + \frac{3}{20}x^2 - \frac{1}{60}x^3}{1 + \frac{2}{5}x + \frac{1}{20}x^2}.$$

We can then compute the approximations, such as $g(0.8)$, by entering

```
>evalf(subs(x=0.8,g));
```

to get .4493096647.

Table 8.5 lists values of $r(x)$ and $P_5(x)$, the fifth Maclaurin polynomial. The Padé approximation is clearly superior in this example. ◆ ◆ ◆

Table 8.5

x	e^{-x}	$P_5(x)$	$\lvert e^{-x} - P_5(x)\rvert$	$r(x)$	$\lvert e^{-x} - r(x)\rvert$
0.2	0.81873075	0.81873067	8.64×10^{-8}	0.81873075	7.55×10^{-9}
0.4	0.67032005	0.67031467	5.38×10^{-6}	0.67031963	4.11×10^{-7}
0.6	0.54881164	0.54875200	5.96×10^{-5}	0.54880763	4.00×10^{-6}
0.8	0.44932896	0.44900267	3.26×10^{-4}	0.44930966	1.93×10^{-5}
1.0	0.36787944	0.36666667	1.21×10^{-3}	0.36781609	6.33×10^{-5}

It is interesting to compare the number of arithmetic operations required for calculations of $P_5(x)$ and $r(x)$ in Example 1. Using nested multiplication, $P_5(x)$ can be expressed as

$$P_5(x) = \left(\left(\left(\left(-\frac{1}{120}x + \frac{1}{24}\right)x - \frac{1}{6}\right)x + \frac{1}{2}\right)x - 1\right)x + 1.$$

Assuming that the coefficients of $1, x, x^2, x^3, x^4$, and x^5 are represented as decimals, a single calculation of $P_5(x)$ in nested form requires five multiplications and five additions/subtractions. Using nested multiplication, $r(x)$ is expressed as

$$r(x) = \frac{\left(\left(-\frac{1}{60}x + \frac{3}{20}\right)x - \frac{3}{5}\right)x + 1}{\left(\frac{1}{20}x + \frac{2}{5}\right)x + 1},$$

so a single calculation of $r(x)$ requires five multiplications, five additions/subtractions, and one division. Hence, computational effort appears to favor the polynomial approximation. However, by reexpressing $r(x)$ by continued division, we can write

$$\begin{aligned} r(x) &= \frac{1 - \frac{3}{5}x + \frac{3}{20}x^2 - \frac{1}{60}x^3}{1 + \frac{2}{5}x + \frac{1}{20}x^2} \\ &= \frac{-\frac{1}{3}x^3 + 3x^2 - 12x + 20}{x^2 + 8x + 20} \\ &= -\frac{1}{3}x + \frac{17}{3} + \frac{\left(-\frac{152}{3}x - \frac{280}{3}\right)}{x^2 + 8x + 20} \\ &= -\frac{1}{3}x + \frac{17}{3} + \frac{-\frac{152}{3}}{\left(\frac{x^2+8x+20}{x+\frac{35}{19}}\right)} \end{aligned}$$

or

$$r(x) = -\frac{1}{3}x + \frac{17}{3} - \frac{\frac{152}{3}}{\left(x + \frac{117}{19} + \frac{\frac{3125}{361}}{\left(x + \frac{35}{19}\right)}\right)}.$$

Written in this form, a single calculation of $r(x)$ requires one multiplication, five addition/subtractions, and two divisions. If the amount of computation required for division is approximately the same as for multiplication, the computational effort required for an evaluation of $P_5(x)$ significantly exceeds that required for an evaluation of $r(x)$. Expressing a rational function approximation in this form is called **continued-fraction** approximation. This is a classical approximation technique of current interest because of the computational efficiency of this representation. It is, however, a specialized technique—one we will not discuss further.

Although the rational function approximation in Example 1 gave results superior to the polynomial approximation of the same degree, the approximation has a wide variation in accuracy; the approximation at 0.2 is accurate to within 8×10^{-9}, whereas, at 1.0, the approximation and the function agree only to within 7×10^{-5}. This accuracy variation is expected, because the Padé approximation is based on a Taylor polynomial representation of e^{-x}, and the Taylor representation has a wide variation of accuracy in [0.2, 1.0].

To obtain more uniformly accurate rational function approximations, we use the set of Chebyshev polynomials, a class that exhibits more uniform behavior. The general Chebyshev rational function approximation method proceeds in the same manner as Padé approximation, except that each x^k-term in the Padé approximation is replaced by the kth-degree Chebyshev polynomial T_k. An introduction to this technique and more detailed references can be found in Burden and Faires [BF], pp. 512–517.

EXERCISE SET 8.5

1. Determine all Padé approximations for $f(x) = e^{2x}$ of degree 2. Compare the results at $x_i = 0.2i$, for $i = 1, 2, 3, 4, 5$, with the actual values $f(x_i)$.

2. Determine all Padé approximations for $f(x) = x\ln(x + 1)$ of degree 3. Compare the results at $x_i = 0.2i$, for $i = 1, 2, 3, 4, 5$, with the actual values $f(x_i)$.

3. Determine the Padé approximation of degree 5 with $n = 2$ and $m = 3$ for $f(x) = e^x$. Compare the results at $x_i = 0.2i$, for $i = 1, 2, 3, 4, 5$, with those from the fifth Maclaurin polynomial.

4. Repeat Exercise 3 using instead the Padé approximation of degree 5 with $n = 3$ and $m = 2$.

5. Determine the Padé approximation of degree 6 with $n = m = 3$ for $f(x) = \sin x$. Compare the results at $x_i = 0.1i$, for $i = 0, 1, \ldots, 5$, with the exact results and with the results of the sixth Maclaurin polynomial.

6. Determine the Padé approximations of degree 6 with (a) $n = 2, m = 4$ and (b) $n = 4, m = 2$ for $f(x) = \sin x$. Compare the results at each x_i to those obtained in Exercise 5.

7. Table 8.5 lists results of the Padé approximation of degree 5 with $n = 3$ and $m = 2$, the fifth Maclaurin polynomial, and the exact values of $f(x) = e^{-x}$ when $x_i = 0.2i$, for $i = 1, 2, 3, 4$, and 5. Compare these results with those produced from the other Padé approximations of degree 5.

   a. $n = 0, m = 5$ **b.** $n = 1, m = 4$

   c. $n = 3, m = 2$ **d.** $n = 4, m = 1$

8. Express the following rational functions in continued-fraction form.

   a. $\dfrac{x^2 + 3x + 2}{x^2 - x + 1}$ **b.** $\dfrac{4x^2 + 3x - 7}{2x^3 + x^2 - x + 5}$

   c. $\dfrac{2x^3 - 3x^2 + 4x - 5}{x^2 + 2x + 4}$ **d.** $\dfrac{2x^3 + x^2 - x + 3}{3x^3 + 2x^2 - x + 1}$

9. To accurately approximate $\sin x$ and $\cos x$ for inclusion in a mathematical library, we first restrict their domains. Given a real number x, divide by π to obtain the relation

$$|x| = M\pi + s, \quad \text{where } M \text{ is an integer and } |s| \le \frac{\pi}{2}.$$

a. Show that $\sin x = \text{sgn}(x)(-1)^M \sin s$.

b. Construct a rational approximation to $\sin s$ using $n = m = 4$. Estimate the error when $0 \le |s| \le \frac{\pi}{2}$.

c. Design an implementation of $\sin x$ using parts (a) and (b).

d. Repeat part (c) for $\cos x$ using the fact that $\cos x = \sin(x + \frac{\pi}{2})$.

10. To accurately approximate $f(x) = e^x$ for inclusion in a mathematical library, we first restrict the domain of f. Given a real number x, divide by $\ln\sqrt{10}$ to obtain the relation

$$x = M \ln\sqrt{10} + s,$$

where M is an integer and s is a real number satisfying $|s| \le \frac{1}{2}\ln\sqrt{10}$.

a. Show that $e^x = e^s 10^{M/2}$.

b. Construct a rational function approximation for e^s using $n = m = 3$. Estimate the error when $0 \le |s| \le \frac{1}{2}\ln\sqrt{10}$.

c. Design an implementation of e^x using the results of part (a) and (b) and the approximations

$$\frac{1}{\ln\sqrt{10}} = 0.8685889638 \quad \text{and} \quad \sqrt{10} = 3.162277660.$$

8.6 Trigonometric Polynomial Approximation

Trigonometric functions are used to approximate functions that have periodic behavior, functions with the property that for some constant T, $f(x + T) = f(x)$ for all x. We can generally transform the problem so that $T = 2\pi$ and restrict the approximation to the interval $[-\pi, \pi]$.

For each positive integer n, the set $\mathcal{T}_n$ of **trigonometric polynomials** of degree less than or equal to n is the set of all linear combinations of $\{\phi_0, \phi_1, \ldots, \phi_{2n-1}\}$, where

$$\phi_0(x) = \frac{1}{\sqrt{2\pi}},$$

$$\phi_k(x) = \frac{1}{\sqrt{\pi}} \cos kx \quad \text{for each } k = 1, 2, \ldots, n,$$

and

$$\phi_{n+k}(x) = \frac{1}{\sqrt{\pi}} \sin kx \quad \text{for each } k = 1, 2, \ldots, n-1.$$

(Some sources include an additional function in the set, $\phi_{2n}(x) = (1/\sqrt{\pi}) \sin nx$.)

The set $\{\phi_0, \phi_1, \ldots, \phi_{2n-1}\}$ is orthonormal on $[-\pi, \pi]$ with respect to the weight function $w(x) \equiv 1$. This follows from a demonstration similar to that which shows that the

Chebyshev polynomials are orthogonal on $[-1, 1]$. For example, if $k \neq j$ and $j \neq 0$,

$$\int_{-\pi}^{\pi} \phi_{n+k}(x)\phi_j(x)\,dx = \int_{-\pi}^{\pi} \frac{1}{\sqrt{\pi}} \sin kx \frac{1}{\sqrt{\pi}} \cos jx\,dx = \frac{1}{\pi}\int_{-\pi}^{\pi} \sin kx \cos jx\,dx.$$

The trigonometric identity

$$\sin kx \cos jx = \frac{1}{2}\sin(k+j)x + \frac{1}{2}\sin(k-j)x$$

can now be used to give

$$\begin{aligned}\int_{-\pi}^{\pi} \phi_{n+k}(x)\phi_j(x)\,dx &= \frac{1}{2\pi}\int_{-\pi}^{\pi} [\sin(k+j)x + \sin(k-j)x]\,dx \\ &= \frac{1}{2\pi}\left[\frac{-\cos(k+j)x}{k+j} - \frac{\cos(k-j)x}{k-j}\right]_{-\pi}^{\pi} = 0,\end{aligned}$$

since $\cos(k+j)\pi = \cos(k+j)(-\pi)$ and $\cos(k-j)\pi = \cos(k-j)(-\pi)$. The result also holds when $k = j$, for in this case we have $\sin(k-j)x = \sin 0 = 0$.

Showing orthogonality for the other possibilities from $\{\phi_0, \phi_1, \ldots, \phi_{2n-1}\}$ is similar and uses the appropriate trigonometric identities from the collection

$$\begin{aligned}\sin jx \cos kx &= \frac{1}{2}[\sin(j+k)x + \sin(j-k)x], \\ \sin jx \sin kx &= \frac{1}{2}[\cos(j-k)x - \cos(j+k)x], \\ \cos jx \cos kx &= \frac{1}{2}[\cos(j+k)x + \cos(j-k)x],\end{aligned}$$

to convert the products into sums.

Given $f \in C[-\pi, \pi]$, the continuous least squares approximation by functions in $\mathcal{T}_n$ is defined by

$$S_n(x) = \sum_{k=0}^{2n-1} a_k \phi_k(x),$$

where

$$a_k = \int_{-\pi}^{\pi} f(x)\phi_k(x)\,dx \quad \text{for each } k = 0, 1, \ldots, 2n-1.$$

The limit of $S_n(x)$ as $n \to \infty$ is called the **Fourier series** of f. Fourier series are used to describe the solution of various ordinary and partial-differential equations that occur in physical situations.

EXAMPLE 1 To determine the trigonometric polynomial from $\mathcal{T}_n$ that approximates

$$f(x) = |x| \quad \text{for } -\pi < x < \pi$$

requires finding

$$
\begin{aligned}
a_0 &= \int_{-\pi}^{\pi} |x| \frac{1}{\sqrt{2\pi}}\, dx = -\frac{1}{\sqrt{2\pi}} \int_{-\pi}^{0} x\, dx + \frac{1}{\sqrt{2\pi}} \int_{0}^{\pi} x\, dx \\
&= \frac{2}{\sqrt{2\pi}} \int_0^{\pi} x\, dx = \frac{\sqrt{2}\pi^2}{2\sqrt{\pi}}, \\
a_k &= \frac{1}{\sqrt{\pi}} \int_{-\pi}^{\pi} |x| \cos kx\, dx = \frac{2}{\sqrt{\pi}} \int_0^{\pi} x \cos kx\, dx \\
&= \frac{2}{\sqrt{\pi}k^2}[(-1)^k - 1], \quad \text{for each } k = 1, 2, \ldots, n,
\end{aligned}
$$

and the coefficients a_{n+k}. The coefficients a_{n+k} in the Fourier expansion are commonly denoted b_k; that is, $b_k = a_{n+k}$ for $k = 1, 2, \ldots, n-1$. In our example, we have

$$
b_k = \frac{1}{\sqrt{\pi}} \int_{-\pi}^{\pi} |x| \sin kx\, dx = 0, \quad \text{for each } k = 1, 2, \ldots, n-1,
$$

since the integrand is an odd function. The trigonometric polynomial from $\mathcal{T}_n$ approximating f is, therefore,

$$
S_n(x) = \frac{\pi}{2} + \frac{2}{\pi} \sum_{k=0}^{n} \frac{(-1)^k - 1}{k^2} \cos kx.
$$

The first few trigonometric polynomials for $f(x) = |x|$ are shown in Figure 8.10.

Figure 8.10

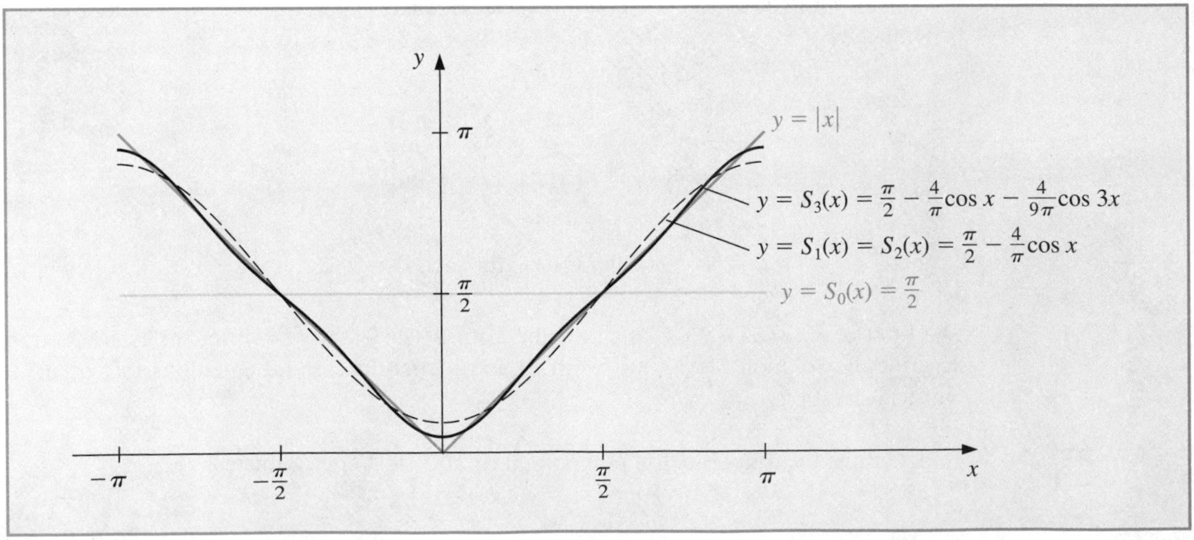

The Fourier series for f is

$$S(x) = \lim_{n\to\infty} S_n(x) = \frac{\pi}{2} + \frac{2}{\pi}\sum_{k=1}^{\infty} \frac{(-1)^k - 1}{k^2} \cos kx.$$

Since $|\cos kx| \le 1$ for every k and x, the series converges and $S(x)$ exists for all real numbers x. ◆ ◆ ◆

There is a discrete analog to Fourier series that is useful for the least squares approximation and interpolation of large amounts of data when the data are given at equally spaced points. Suppose that a collection of $2m$ paired data points $\{(x_j, y_j)\}_{j=0}^{2m-1}$ is given, with the first elements in the pairs equally partitioning a closed interval. For convenience, we assume that the interval is $[-\pi, \pi]$ and that, as shown in Figure 8.11,

$$x_j = -\pi + \left(\frac{j}{m}\right)\pi \quad \text{for each } j = 0, 1, \ldots, 2m-1.$$

If this were not the case, a linear transformation could be used to change the data into this form.

Figure 8.11

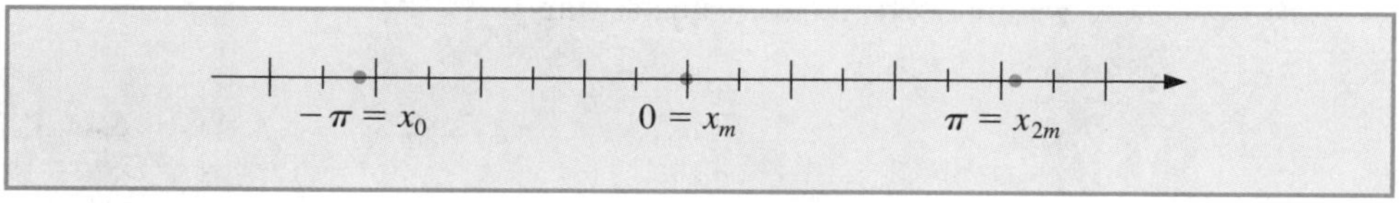

For a fixed $n < m$, consider the set of functions $\hat{\mathcal{T}}_n$ consisting of all the linear combinations of

$$\hat{\phi}_0(x) = \frac{1}{2},$$

$$\hat{\phi}_k(x) = \cos kx, \quad \text{for each } k = 1, 2, \ldots, n,$$

and

$$\hat{\phi}_{n+k}(x) = \sin kx, \quad \text{for each } k = 1, 2, \ldots, n-1.$$

The goal is to determine the linear combination of these functions for which

$$E(S_n) = \sum_{j=0}^{2m-1} \{y_j - S_n(x_j)\}^2$$

is minimized. That is, we want to choose the constants $a_0, a_1, \ldots, a_n$ and $b_1, b_2, \ldots, b_{n-1}$ to minimize the total error

$$E(S_n) = \sum_{j=0}^{2m-1} \left\{ y_j - \left[\frac{a_0}{2} + a_n \cos nx_j + \sum_{k=1}^{n-1}(a_k \cos kx_j + b_k \sin kx_j)\right]\right\}^2.$$

The determination of the constants is simplified by the fact that the set is orthogonal with respect to summation over the equally spaced points $\{x_j\}_{j=0}^{2m-1}$ in $[-\pi, \pi]$. By this we mean that, for each $k \neq l$,

$$\sum_{j=0}^{2m-1} \hat{\phi}_k(x_j)\hat{\phi}_l(x_j) = 0.$$

The orthogonality follows from the fact that if r and m are positive integers with $r < 2m$, we have (See Burden and Faires [BF], p. 521, for a verification)

$$\sum_{j=0}^{2m-1} \cos rx_j = 0 \quad \text{and} \quad \sum_{j=0}^{2m-1} \sin rx_j = 0.$$

To obtain the constants a_k for $k = 0, 1, \ldots, n$ and b_k for $k = 1, 2, \ldots, n-1$ in the summation

$$S_n(x) = \frac{a_0}{2} + a_n \cos nx + \sum_{k=1}^{n-1} (a_k \cos kx + b_k \sin kx),$$

we minimize the least squares sum

$$E(a_0, \ldots, a_n, b_1, \ldots, b_{n-1}) = \sum_{j=0}^{2m-1} [y_j - S_n(x_j)]^2$$

by setting to zero the partial derivatives of E with respect to the a_k's and the b_k's. This implies that

$$a_k = \frac{1}{m} \sum_{j=0}^{2m-1} y_j \cos kx_j, \quad \text{for each } k = 0, 1, \ldots, n,$$

and

$$b_k = \frac{1}{m} \sum_{j=0}^{2m-1} y_j \sin kx_j, \quad \text{for each } k = 1, 2, \ldots, n-1.$$

EXAMPLE 2 Let $f(x) = x^4 - 3x^3 + 2x^2 - \tan x(x-2)$. To find the discrete least squares approximation S_3 for the data $\{(x_j, y_j)\}_{j=0}^{9}$, where $x_j = j/5$ and $y_j = f(x_j)$, requires a transformation from $[0, 2]$ to $[-\pi, \pi]$. The required linear transformation is

$$z_j = \pi(x_j - 1),$$

and the transformed data are of the form

$$\left\{\left(z_j, f\left(1 + \frac{z_j}{\pi}\right)\right)\right\}_{j=0}^{9}.$$

Consequently, the least squares trigonometric polynomial is

$$S_3(z) = \frac{a_0}{2} + a_3 \cos 3z + \sum_{k=1}^{2} (a_k \cos kz + b_k \sin kz),$$

where

$$a_k = \frac{1}{5} \sum_{j=0}^{9} f\left(1 + \frac{z_j}{\pi}\right) \cos kz_j, \quad \text{for } k = 0, 1, 2, 3,$$

and

$$b_k = \frac{1}{5} \sum_{j=0}^{9} f\left(1 + \frac{z_j}{\pi}\right) \sin kz_j, \quad \text{for } k = 1, 2.$$

Evaluating these sums produces the approximation

$$\begin{aligned} S_3(z) = {} & 0.76201 + 0.77177 \cos z + 0.017423 \cos 2z + 0.0065673 \cos 3z \\ & - 0.38676 \sin z + 0.047806 \sin 2z. \end{aligned}$$

Converting back to the variable x gives

$$\begin{aligned} S_3(x) = {} & 0.76201 + 0.77177 \cos \pi(x - 1) + 0.017423 \cos 2\pi(x - 1) \\ & + 0.0065673 \cos 3\pi(x - 1) \\ & - 0.38676 \sin \pi(x - 1) + 0.047806 \sin 2\pi(x - 1). \end{aligned}$$

Table 8.6 lists values of $f(x)$ and $S_3(x)$. ◆ ◆ ◆

Table 8.6

x	$f(x)$	$S_3(x)$	$\lvert f(x) - S_3(x) \rvert$
0.125	0.26440	0.24060	2.38×10^{-2}
0.375	0.84081	0.85154	1.07×10^{-2}
0.625	1.36150	1.36248	9.74×10^{-4}
0.875	1.61282	1.60406	8.75×10^{-3}
1.125	1.36672	1.37566	8.94×10^{-3}
1.375	0.71697	0.71545	1.52×10^{-3}
1.625	0.07909	0.06929	9.80×10^{-3}
1.875	−0.14576	−0.12302	2.27×10^{-2}

EXERCISE SET 8.6

1. Find the continuous least squares trigonometric polynomial $S_2(x)$ for $f(x) = x^2$ on $[-\pi, \pi]$.
2. Find the continuous least squares trigonometric polynomial $S_n(x)$ for $f(x) = x$ on $[-\pi, \pi]$.

3. Find the continuous least squares trigonometric polynomial $S_3(x)$ for $f(x) = e^x$ on $[-\pi, \pi]$.

4. Find the general continuous least squares trigonometric polynomial $S_n(x)$ for $f(x) = e^x$ on $[-\pi, \pi]$.

5. Find the general continuous least squares trigonometric polynomial $S_n(x)$ for

$$f(x) = \begin{cases} 0, & \text{if } -\pi < x \leq 0, \\ 1, & \text{if } 0 < x < \pi. \end{cases}$$

6. Find the general continuous least squares trigonometric polynomial $S_n(x)$ for

$$f(x) = \begin{cases} -1, & \text{if } -\pi < x < 0. \\ 1, & \text{if } 0 \leq x \leq \pi. \end{cases}$$

7. Determine the discrete least squares trigonometric polynomial $S_n(x)$ on the interval $[-\pi, \pi]$ for the following functions, using the given values of m and n:

   a. $f(x) = \cos 2x, \quad m = 4, n = 2$ **b.** $f(x) = \cos 3x, \quad m = 4, n = 2$

   c. $f(x) = \sin \frac{1}{2}x + 2\cos \frac{1}{3}x, \quad m = 6, n = 3$

   d. $f(x) = x^2 \cos x, \quad m = 6, n = 3$

8. Compute the error $E(S_n)$ for each of the functions in Exercise 7.

9. Determine the discrete least squares trigonometric polynomial $S_3(x)$, using $m = 4$ for $f(x) = e^x \cos 2x$ on the interval $[-\pi, \pi]$. Compute the error $E(S_3)$.

10. Repeat Exercise 9 using $m = 8$. Compare the values of the approximating polynomials with the values of f at the points $\xi_j = -\pi + 0.2j\pi$, for $0 \leq j \leq 10$. Which approximation is better?

11. Show that for any continuous odd function f defined on the interval $[-a, a]$, we have $\int_{-a}^{a} f(x)\,dx = 0$.

12. Show that for any continuous even function f defined on the interval $[-a, a]$, we have $\int_{-a}^{a} f(x)\,dx = 2\int_0^a f(x)\,dx$.

13. Show that the functions $\phi_0(x) = 1/2, \phi_1(x) = \cos x, \ldots, \phi_n(x) = \cos nx, \phi_{n+1}(x) = \sin x, \ldots, \phi_{2n-1}(x) = \sin(n-1)x$,are orthogonal on $[-\pi, \pi]$ with respect to $w(x) \equiv 1$.

14. In Example 1 the Fourier series was determined for $f(x) = |x|$. Use this series and the assumption that it represents f at zero to find the value of the convergent infinite series

$$\sum_{k=0}^{\infty} \frac{1}{(2k+1)^2}.$$

8.7 Fast Fourier Transforms

The *interpolatory* trigonometric polynomial on the $2m$ data points $\{(x_j, y_j)\}_{j=0}^{2m-1}$ is the least squares polynomial from $\tilde{\mathcal{T}}_m$ for this collection of points. This least squares trigonometric polynomial was found in Section 8.6 to have the form

$$S_n(x) = \frac{a_0}{2} + a_n \cos nx + \sum_{k=1}^{n-1}(a_k \cos kx + b_k \sin kx),$$

where

$$a_k = \frac{1}{m}\sum_{j=0}^{2m-1} y_j \cos kx_j \quad \text{for each } k = 0, 1, \ldots, n \tag{8.3}$$

and

$$b_k = \frac{1}{m}\sum_{j=0}^{2m-1} y_j \sin kx_j \quad \text{for each } k = 1, 2, \ldots, n-1. \tag{8.4}$$

We can use this form with $n = m$ for interpolation if we make a minor modification. For interpolation, we balance the system by replacing the term a_m with $a_m/2$. The interpolatory polynomial then has the form

$$S_m(x) = \frac{a_0 + a_m \cos mx}{2} + \sum_{k=1}^{m-1}[a_k \cos kx + b_k \sin kx],$$

where the coefficients a_k and b_k are given in Eqs. (8.3) and (8.4).

The interpolation of large amounts of equally spaced data by trigonometric polynomials can produce very accurate results. It is the appropriate approximation technique in areas involving digital filters, antenna field patterns, quantum mechanics, optics, and certain simulation problems. Until the middle of the 1960s, the method had not been extensively applied due to the number of arithmetic calculations required for the determination of the constants in the approximation. The interpolation of $2m$ data points requires approximately $(2m)^2$ multiplications and $(2m)^2$ additions by the direct calculation technique. The approximation of many thousands of data points is not unusual in areas requiring trigonometric interpolation, so the direct methods for evaluating the constants require multiplication and addition operations numbering in the millions. The computation time for this many calculations is prohibitive, and the round-off error would generally dominate the approximation.

In 1965 a paper by Cooley and Tukey [CT] described an alternative method of calculating the constants in the interpolating trigonometric polynomial. This method requires only $O(m \log_2 m)$ multiplications and $O(m \log_2 m)$ additions, provided m is chosen in an appropriate manner. For a problem with thousands of data points, this reduces the number of calculations from millions to thousands. The method had actually been discovered a number of years before the Cooley-Tukey paper appeared but had gone unnoticed by most researchers until that time.

The method described by Cooley and Tukey is generally known as the **Fast Fourier Transform (FFT) method** and has led to a revolution in the use of interpolatory trigonometric polynomials. The method consists of organizing the problem so that the number of data points being used can be easily factored, particularly into powers of 2.

The relationship between the number of data points $2m$ and the degree of the trigonometric polynomial used in the fast Fourier transform procedure allows some notational simplification. The nodes are given, for each $j = 0, 1, \ldots, 2m-1$, by

$$x_j = -\pi + \left(\frac{j}{m}\right)\pi$$

and the coefficients, for each $k = 0, 1, \ldots, m$, as

$$a_k = \frac{1}{m}\sum_{j=0}^{2m-1} y_j \cos kx_j \quad \text{and} \quad b_k = \frac{1}{m}\sum_{j=0}^{2m-1} y_j \sin kx_j.$$

For notational convenience, b_0 and b_m have been added to the collection, but both are zero and do not contribute to the sum.

Instead of directly evaluating the constants a_k and b_k, the fast Fourier transform procedure computes the complex coefficients c_k in the formula

$$F(x) = \frac{1}{m}\sum_{k=0}^{2m-1} c_k e^{ikx},$$

where

$$c_k = \sum_{j=0}^{2m-1} y_j e^{\pi i jk/m}, \quad \text{for each } k = 0, 1, \ldots, 2m-1. \tag{8.5}$$

Once the constants c_k have been determined, a_k and b_k can be recovered. To do this we need *Euler's formula*, which states that for all real numbers z we have

$$e^{iz} = \cos z + i \sin z,$$

where the complex number i satisfies $i^2 = -1$. Then, for each $k = 0, 1, \ldots, m$,

$$\frac{1}{m}c_k e^{-i\pi k} = \frac{1}{m}\sum_{j=0}^{2m-1} y_j e^{\pi i jk/m} e^{-i\pi k} = \frac{1}{m}\sum_{j=0}^{2m-1} y_j e^{ik(-\pi+(\pi j/m))}$$

$$= \frac{1}{m}\sum_{j=0}^{2m-1} y_j(\cos kx_j + i \sin kx_j),$$

so

$$\frac{1}{m}c_k e^{-i\pi k} = a_k + ib_k.$$

The operation reduction feature of the fast Fourier transform results from calculating the coefficients c_k in clusters. The following example gives a simple illustration of the technique.

EXAMPLE 1 Consider the construction of the trigonometric interpolating polynomial of degree 2 for the data $\{(x_0, y_0), (x_1, y_1), (x_2, y_2), (x_3, y_3)\}$, where $m = 2$ and $x_j = -\pi + (j/2)\pi$ for $j = 0, 1, 2, 3$. The polynomial is given by

$$S_2(x) = \frac{a_0 + a_2 \cos 2x}{2} + a_1 \cos x + b_1 \sin x,$$

where the coefficients are

$$\begin{aligned}
a_0 &= \frac{1}{2}[y_0 \cos 0 + y_1 \cos 0 + y_2 \cos 0 + y_3 \cos 0] = \frac{1}{2}[y_0 + y_1 + y_2 + y_3],\\
a_1 &= \frac{1}{2}[y_0 \cos x_0 + y_1 \cos x_1 + y_2 \cos x_2 + y_3 \cos x_3],\\
a_2 &= \frac{1}{2}[y_0 \cos 2x_0 + y_1 \cos 2x_1 + y_2 \cos 2x_2 + y_3 \cos 2x_3],\\
b_1 &= \frac{1}{2}[y_0 \sin x_0 + y_1 \sin x_1 + y_2 \sin x_2 + y_3 \sin x_3].
\end{aligned}$$

Introducing the complex coefficients defined in Eq. (8.5) we have

$$\begin{aligned}
c_0 &= y_0 e^0 + y_1 e^0 + y_2 e^0 + y_3 e^0,\\
c_1 &= y_0 e^0 + y_1 e^{\pi i/2} + y_2 e^{\pi i} + y_3 e^{3\pi i/2},\\
c_2 &= y_0 e^0 + y_1 e^{\pi i} + y_2 e^{2\pi i} + y_3 e^{3\pi i},\\
c_3 &= y_0 e^0 + y_1 e^{3\pi i/2} + y_2 e^{3\pi i} + y_3 e^{9\pi i/2},
\end{aligned}$$

and $a_k + ib_k = \frac{1}{2} e^{-k\pi i} c_k$ for $k = 0, 1, 2, 3$. Thus

$$\begin{aligned}
a_0 &= \frac{1}{2}\text{Re}(c_0),\\
a_1 &= \frac{1}{2}\text{Re}(c_1 e^{-\pi i}),\\
a_2 &= \frac{1}{2}\text{Re}(c_2 e^{-2\pi i}),\\
b_1 &= \frac{1}{2}\text{Im}(c_1 e^{-\pi i}).
\end{aligned}$$

If we let $\delta = e^{\pi i/2}$ we can rewrite the equations as the matrix equation

$$\begin{bmatrix} c_0 \\ c_1 \\ c_2 \\ c_3 \end{bmatrix} = \begin{bmatrix} 1 & 1 & 1 & 1 \\ 1 & \delta & \delta^2 & \delta^3 \\ 1 & \delta^2 & \delta^4 & \delta^6 \\ 1 & \delta^3 & \delta^6 & \delta^9 \end{bmatrix} \begin{bmatrix} y_0 \\ y_1 \\ y_2 \\ y_3 \end{bmatrix}.$$

Since $\delta^4 = 1$, $\delta^5 = \delta$, $\delta^6 = \delta^2$, $\delta^7 = \delta^3$, $\delta^8 = \delta^4 = 1$, $\delta^9 = \delta^5 = \delta$, we have

$$\begin{bmatrix} c_0 \\ c_1 \\ c_2 \\ c_3 \end{bmatrix} = \begin{bmatrix} 1 & 1 & 1 & 1 \\ 1 & \delta & \delta^2 & \delta^3 \\ 1 & \delta^2 & 1 & \delta^2 \\ 1 & \delta^3 & \delta^2 & \delta \end{bmatrix} \begin{bmatrix} y_0 \\ y_1 \\ y_2 \\ y_3 \end{bmatrix}.$$

Factoring the matrix product gives

$$\begin{bmatrix} c_0 \\ c_1 \\ c_2 \\ c_3 \end{bmatrix} = \begin{bmatrix} 1 & 1 & 0 & 0 \\ 0 & 0 & 1 & 1 \\ 1 & \delta^2 & 0 & 0 \\ 0 & 0 & 1 & \delta^2 \end{bmatrix} \begin{bmatrix} 1 & 1 & 0 & 0 \\ 0 & 0 & \delta^3 & \delta^3 \\ 1 & \delta^2 & 0 & 0 \\ 0 & 0 & 1 & \delta^2 \end{bmatrix} \begin{bmatrix} 1 & 0 & 0 & 0 \\ 0 & 0 & 1 & 0 \\ 0 & \delta & 0 & 0 \\ 0 & 0 & 0 & \delta \end{bmatrix} \begin{bmatrix} y_0 \\ y_1 \\ y_2 \\ y_3 \end{bmatrix},$$

and since $\delta^2 = e^{\pi i} = \cos\pi + i\sin\pi = -1$ we have

$$\begin{bmatrix} c_0 \\ c_1 \\ c_2 \\ c_3 \end{bmatrix} = \begin{bmatrix} 1 & 1 & 0 & 0 \\ 0 & 0 & 1 & 1 \\ 1 & -1 & 0 & 0 \\ 0 & 0 & 1 & -1 \end{bmatrix} \begin{bmatrix} 1 & 1 & 0 & 0 \\ 0 & 0 & \delta^3 & \delta^3 \\ 1 & -1 & 0 & 0 \\ 0 & 0 & 1 & -1 \end{bmatrix} \begin{bmatrix} 1 & 0 & 0 & 0 \\ 0 & 0 & 1 & 0 \\ 0 & \delta & 0 & 0 \\ 0 & 0 & 0 & \delta \end{bmatrix} \begin{bmatrix} y_0 \\ y_1 \\ y_2 \\ y_3 \end{bmatrix}.$$

Computing $(c_0, c_1, c_2, c_3)^t$ using the factored form requires 2 multiplications by δ, 2 by δ^3, and 4 by $\delta^2 = -1$. In addition, each matrix multiplication requires 4 additions. Hence we have a total of 8 multiplications and 4 additions. The original product required 16 multiplications and 12 additions. ◆ ◆ ◆

The program FFTRNS82 performs the fast Fourier transform when $m = 2^p$ for some positive integer p. Modifications to the technique can be made when m takes other forms.

EXAMPLE 2 Let $f(x) = x^4 - 3x^3 + 2x^2 - \tan x(x-2)$. To determine the trigonometric interpolating polynomial of degree 4 for the data $\{(x_j, y_j)\}_{j=0}^{7}$ where $x_j = j/4$ and $y_j = f(x_j)$ requires a transformation of the interval $[0, 2]$ to $[-\pi, \pi]$. The linear transformation is given by

$$z_j = \pi(x_j - 1)$$

so that the input data to the fast Fourier transform method is

$$\left\{ z_j, f\left(1 + \frac{z_j}{\pi}\right) \right\}_{j=0}^{7}.$$

The interpolating polynomial in z is

$$\begin{aligned} S_4(z) = {} & 0.761979 + 0.771841\cos z + 0.0173037\cos 2z \\ & + 0.00686304\cos 3z - 0.000578545\cos 4z \\ & - 0.386374\sin z + 0.0468750\sin 2z - 0.0113738\sin 3z. \end{aligned}$$

The trigonometric polynomial $S_4(x)$ on $[0, 2]$ is obtained by substituting $z = \pi(x-1)$ into $S_4(z)$. The graphs of $y = f(x)$ and $y = S_4(x)$ are shown in Figure 8.12. Values of $f(x)$ and $S_4(x)$ are given in Table 8.7. ◆ ◆ ◆

Table 8.7

x	$f(x)$	$S_4(x)$	$\lvert f(x) - S_4(x)\rvert$
0.125	0.26440	0.25001	1.44×10^{-2}
0.375	0.84081	0.84647	5.66×10^{-3}
0.625	1.36150	1.35824	3.27×10^{-3}
0.875	1.61282	1.61515	2.33×10^{-3}
1.125	1.36672	1.36471	2.02×10^{-3}
1.375	0.71697	0.71931	2.33×10^{-3}
1.625	0.07909	0.07496	4.14×10^{-3}
1.875	−0.14576	−0.13301	1.27×10^{-2}

Figure 8.12

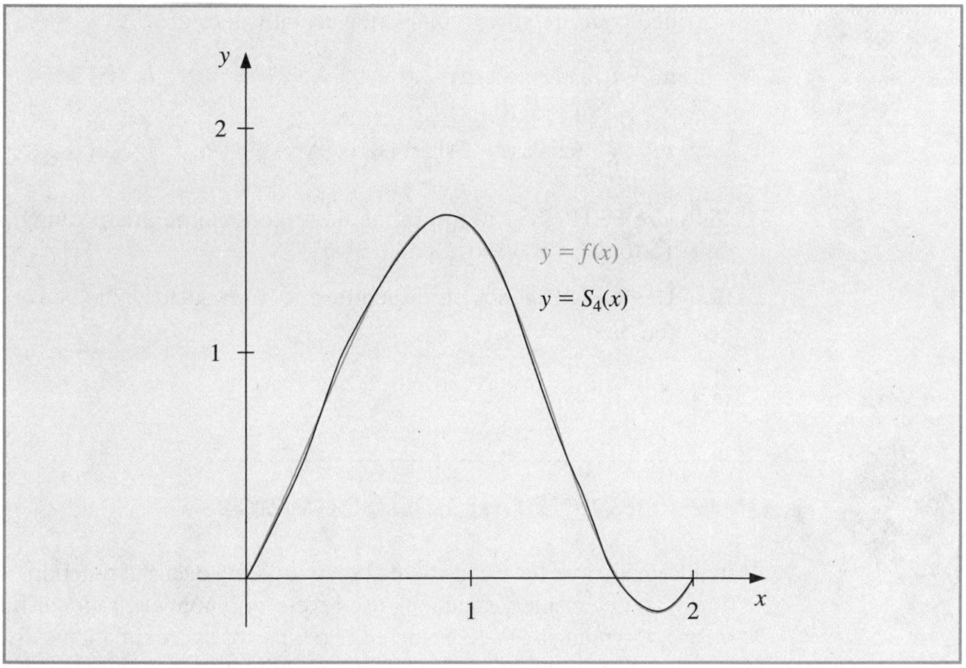

EXERCISE SET 8.7

1. Determine the trigonometric interpolating polynomial $S_2(x)$ of degree 2 on $[-\pi, \pi]$ for the following functions, and graph $f(x) - S_2(x)$.

   a. $f(x) = \pi(x - \pi)$

   b. $f(x) = x(\pi - x)$

   c. $f(x) = |x|$

   d. $f(x) = \begin{cases} -1, & -\pi \le x \le 0 \\ 1, & 0 < x \le \pi \end{cases}$

2. Determine the trigonometric interpolating polynomial of degree 4 for $f(x) = x(\pi - x)$ on the interval $[-\pi, \pi]$.
   a. Use direct calculation.
   b. Use fast Fourier transforms.

3. Use the fast Fourier transform method to compute the trigonometric interpolating polynomial of degree 4 on $[-\pi, \pi]$ for the following functions.
   a. $f(x) = \pi(x - \pi)$
   b. $f(x) = |x|$
   c. $f(x) = \cos \pi x - 2 \sin \pi x$
   d. $f(x) = x \cos x^2 + e^x \cos e^x$

4. a. Determine the trigonometric interpolating polynomial $S_4(x)$ of degree 4 for $f(x) = x^2 \sin x$ on the interval $[0, 1]$.
   b. Compute $\int_0^1 S_4(x)\, dx$.
   c. Compare the integral in part (b) to $\int_0^1 x^2 \sin x\, dx$.

5. Use the approximations obtained in Exercise 3 to approximate the following integrals, and compare your results to the actual values.
   a. $\displaystyle\int_{-\pi}^{\pi} \pi(x - \pi)\, dx$
   b. $\displaystyle\int_{-\pi}^{\pi} |x|\, dx$
   c. $\displaystyle\int_{-\pi}^{\pi} (\cos \pi x - 2 \sin \pi x)\, dx$
   d. $\displaystyle\int_{-\pi}^{\pi} (x \cos x^2 + e^x \cos e^x)\, dx$

6. Use FFTRNS82 to determine the trigonometric interpolating polynomial of degree 16 for $f(x) = x^2 \cos x$ on $[-\pi, \pi]$.

7. Use basic trigonometric identities to show that $\sum_{j=0}^{2m-1} (\sin mx_j)^2 = 0$ and $\sum_{j=0}^{2m-1} (\cos mx_j)^2 = 2m$.

8.8 Survey of Methods and Software

In this chapter we have considered approximating data and functions with elementary functions. The elementary functions used were polynomials, rational functions, and trigonometric polynomials. We considered two types of approximations, discrete and continuous. Discrete approximations arise when approximating a finite set of data with an elementary function. Continuous approximations are used when the function to be approximated is known.

Discrete least squares techniques are recommended when the function is specified by giving a set of data that may not exactly represent the function. Least squares fit of data can take the form of a linear or other polynomial approximation. These approximations are computed by solving sets of normal equations, as given in Section 8.2. If the data are periodic, a trigonometric least squares fit may be appropriate. Because of the orthonormality of the trigonometric basis functions, the least squares trigonometric approximation does not require the solution of a linear system. For large amounts of periodic data, interpolation by trigonometric polynomials is also recommended. An efficient method of computing the trigonometric interpolating polynomial is given by the fast Fourier transform.

When the function to be approximated can be evaluated at any required argument, the approximations seek to minimize an integral instead of a sum. The continuous least squares polynomial approximations were considered in Section 8.3. Efficient computation of least squares polynomials lead to orthonormal sets of polynomials, such as the Legendre and Chebyshev polynomials. Approximation by rational functions was studied in Section 8.5, where Padé approximation as a generalization of the Maclaurin polynomial was presented. This method allows a more uniform method of approximation than polynomials. Continuous least squares approximation by trigonometric functions was discussed in Section 8.6, especially as it relates to Fourier series.

The IMSL Library and the NAG Libraries provide subroutines for least squares approximation of data. The approximation can be by polynomials, cubic splines, or a users' choice of basis functions. Chebyshev polynomials are also used in the constructions to minimize round-off error and enhance accuracy, and Fast Fourier transformations are available.

For further information on the general theory of approximation theory see Powell [Po], Davis [Da], or Cheney [Ch]. A good reference for methods of least squares is Lawson and Hanson [LH], and information about Fourier transforms can be found in Van Loan [Van] and in Briggs and Hanson [BH].

CHAPTER

9 Approximating Eigenvalues

9.1 Introduction

Eigenvalues and eigenvectors were introduced in Chapter 7 in connection with the convergence of iterative methods for approximating the solution to a linear system. To determine the eigenvalues of an $n \times n$ matrix A, we construct the characteristic polynomial

$$p(\lambda) = \det(A - \lambda I)$$

and then determine its zeros. Finding the determinant of an $n \times n$ matrix is computationally expensive, and finding good approximations to the roots of $p(\lambda)$ is also difficult. In this chapter we will explore other means for approximating the eigenvalues of a matrix.

9.2 Isolating Eigenvalues

In Chapter 7 we found that an iterative technique for solving a linear system will converge if all the eigenvalues associated with the problem have magnitude less than 1 but cannot be expected to converge if there are eigenvalues with magnitude greater than 1. The exact values of the eigenvalues in this case are not of primary importance—only the region of the complex plane in which they lie.

Even when we need to know the eigenvalues, the fact that many of the techniques for their approximation are iterative implies that determining regions in which they lie is a first step in the direction of determining the approximation, since it provides us with the initial approximation that iterative methods need.

Before proceeding with the techniques for approximating eigenvalues, we need some more results from linear algebra to help us isolate them. We will give here all the additional linear algebra that will be used in this chapter so that it is in one place for ease of reference. For more complete results we recommend *Applied Linear Algebra* by Noble and Daniel [ND]. The first definitions and results parallel those in Section 8.3 for sets of polynomials.

The set of nonzero vectors $\{\mathbf{v}^{(1)}, \mathbf{v}^{(2)}, \mathbf{v}^{(3)}, \ldots, \mathbf{v}^{(k)}\}$ is said to be **linearly independent** if whenever

$$\mathbf{0} = \alpha_1 \mathbf{v}^{(1)} + \alpha_2 \mathbf{v}^{(2)} + \alpha_3 \mathbf{v}^{(3)} + \cdots + \alpha_k \mathbf{v}^{(k)},$$

we have $\alpha_1 = \alpha_2 = \alpha_3 = \cdots = \alpha_k = 0$. A set of vectors that is not linearly independent is called **linearly dependent**.

Unique Representation of Vectors in $\mathbb{R}^n$

If $\{\mathbf{v}^{(1)}, \mathbf{v}^{(2)}, \mathbf{v}^{(3)}, \ldots, \mathbf{v}^{(n)}\}$ is a set of n linearly independent vectors in $\mathbb{R}^n$, then any vector $\mathbf{x} \in \mathbb{R}^n$ can be written uniquely as

$$\mathbf{x} = \beta_1 \mathbf{v}^{(1)} + \beta_2 \mathbf{v}^{(2)} + \beta_3 \mathbf{v}^{(3)} + \cdots + \beta_n \mathbf{v}^{(n)}$$

for some collection of constants $\beta_1, \beta_2, \ldots, \beta_n$.

Any collection of n linearly independent vectors in $\mathbb{R}^n$ is called a **basis** for $\mathbb{R}^n$.

EXAMPLE 1 Let $\mathbf{v}^{(1)} = (1, 0, 0)^t$, $\mathbf{v}^{(2)} = (-1, 1, 1)^t$, and $\mathbf{v}^{(3)} = (0, 4, 2)^t$. If α_1, α_2, and α_3 are numbers with

$$\mathbf{0} = \alpha_1 \mathbf{v}^{(1)} + \alpha_2 \mathbf{v}^{(2)} + \alpha_3 \mathbf{v}^{(3)},$$

then

$$\begin{aligned}(0,0,0)^t &= \alpha_1(1,0,0)^t + \alpha_2(-1,1,1)^t + \alpha_3(0,4,2)^t \\ &= (\alpha_1 - \alpha_2, \alpha_2 + 4\alpha_3, \alpha_2 + 2\alpha_3)^t,\end{aligned}$$

so

$$\alpha_1 - \alpha_2 = 0, \qquad \alpha_2 + 4\alpha_3 = 0, \qquad \text{and} \qquad \alpha_2 + 2\alpha_3 = 0.$$

Since the only solution to this system is $\alpha_1 = \alpha_2 = \alpha_3 = 0$, the set $\{\mathbf{v}^{(1)}, \mathbf{v}^{(2)}, \mathbf{v}^{(3)}\}$ is linearly independent in $\mathbb{R}^3$ and is a basis for $\mathbb{R}^3$. A vector $\mathbf{x} = (x_1, x_2, x_3)^t$ in $\mathbb{R}^3$ can be written as

$$\mathbf{x} = \beta_1 \mathbf{v}^{(1)} + \beta_2 \mathbf{v}^{(2)} + \beta_3 \mathbf{v}^{(3)}$$

by choosing

$$\beta_1 = x_1 - x_2 + 2x_3, \qquad \beta_2 = 2x_3 - x_2, \qquad \text{and} \qquad \beta_3 = \frac{1}{2}(x_2 - x_3).$$ ◆ ◆ ◆

The next result will be used in the following section to develop the Power method for approximating eigenvalues.

Linear Independence of Eigenvectors

If A is a matrix and $\lambda_1, \ldots, \lambda_k$ are distinct eigenvalues of A with associated eigenvectors $\mathbf{x}^{(1)}, \mathbf{x}^{(2)}, \ldots, \mathbf{x}^{(k)}$, then $\{\mathbf{x}^{(1)}, \mathbf{x}^{(2)}, \ldots, \mathbf{x}^{(k)}\}$ is linearly independent.

A set of vectors $\{\mathbf{v}^{(1)}, \mathbf{v}^{(2)}, \ldots, \mathbf{v}^{(n)}\}$ is **orthogonal** if $(\mathbf{v}^{(i)})^t \mathbf{v}^{(j)} = 0$ for all $i \neq j$. If, in addition, $(\mathbf{v}^{(i)})^t \mathbf{v}^{(i)} = 1$ for all $i = 1, 2, \ldots, n$, then the set is **orthonormal**.

Since $\mathbf{x}^t\mathbf{x} = \|\mathbf{x}\|_2^2$, a set of orthogonal vectors $\{\mathbf{v}^{(1)}, \mathbf{v}^{(2)}, \ldots, \mathbf{v}^{(n)}\}$ is orthonormal if and only if

$$\|\mathbf{v}^{(i)}\|_2 = 1 \qquad \text{for each } i = 1, 2, \ldots, n.$$

Linear Independence of Orthogonal Vectors

Any orthogonal set of vectors that does not contain the zero vector is linearly independent.

EXAMPLE 2 The vectors $\mathbf{v}^{(1)} = (0, 4, 2)^t$, $\mathbf{v}^{(2)} = (-1, -\frac{1}{5}, \frac{2}{5})^t$, and $\mathbf{v}^{(3)} = (\frac{1}{6}, -\frac{1}{6}, \frac{1}{3})^t$ form an orthogonal set. For these vectors we have

$$\|\mathbf{v}^{(1)}\|_2 = 2\sqrt{5}, \qquad \|\mathbf{v}^{(2)}\|_2 = \frac{\sqrt{30}}{5}, \quad \text{and} \quad \|\mathbf{v}^{(3)}\|_2 = \frac{\sqrt{6}}{6}.$$

The vectors

$$\mathbf{u}^{(1)} = \frac{\mathbf{v}^{(1)}}{\|\mathbf{v}^{(1)}\|_2} = \left(0, \frac{2\sqrt{5}}{5}, \frac{\sqrt{5}}{5}\right)^t, \qquad \mathbf{u}^{(2)} = \frac{\mathbf{v}^{(2)}}{\|\mathbf{v}^{(2)}\|_2} = \left(-\frac{\sqrt{30}}{6}, -\frac{\sqrt{30}}{30}, \frac{\sqrt{30}}{15}\right)^t,$$

and

$$\mathbf{u}^{(3)} = \frac{\mathbf{v}^{(3)}}{\|\mathbf{v}^{(3)}\|_2} = \left(\frac{\sqrt{6}}{6}, -\frac{\sqrt{6}}{6}, \frac{\sqrt{6}}{3}\right)^t$$

form an orthonormal set, since they inherit orthogonality from $\mathbf{v}^{(1)}, \mathbf{v}^{(2)}$, and $\mathbf{v}^{(3)}$; in addition,

$$\|\mathbf{u}^{(1)}\|_2 = \|\mathbf{u}^{(2)}\|_2 = \|\mathbf{u}^{(3)}\|_2 = 1.$$

◆ ◆ ◆

An $n \times n$ matrix P is **orthogonal** if $P^{-1} = P^t$. This terminology follows from the fact that the columns of an orthogonal matrix form an orthogonal—in fact, orthonormal—set of vectors.

EXAMPLE 3 The orthogonal matrix P formed from the orthonormal set of vectors found in Example 2 is

$$P = [\mathbf{u}^{(1)}, \mathbf{u}^{(2)}, \mathbf{u}^{(3)}] = \begin{bmatrix} 0 & -\frac{\sqrt{30}}{6} & \frac{\sqrt{6}}{6} \\ \frac{2\sqrt{5}}{5} & -\frac{\sqrt{30}}{30} & -\frac{\sqrt{6}}{6} \\ \frac{\sqrt{5}}{5} & \frac{\sqrt{30}}{15} & \frac{\sqrt{6}}{3} \end{bmatrix}.$$

Note that

$$PP^t = \begin{bmatrix} 0 & \frac{-\sqrt{30}}{6} & \frac{\sqrt{6}}{6} \\ \frac{2\sqrt{5}}{5} & -\frac{\sqrt{30}}{30} & -\frac{\sqrt{6}}{6} \\ \frac{\sqrt{5}}{5} & \frac{\sqrt{30}}{15} & \frac{\sqrt{6}}{3} \end{bmatrix} \cdot \begin{bmatrix} 0 & \frac{2\sqrt{5}}{5} & \frac{\sqrt{5}}{5} \\ -\frac{\sqrt{30}}{6} & -\frac{\sqrt{30}}{30} & \frac{\sqrt{30}}{15} \\ \frac{\sqrt{6}}{6} & -\frac{\sqrt{6}}{6} & \frac{\sqrt{6}}{3} \end{bmatrix} = \begin{bmatrix} 1 & 0 & 0 \\ 0 & 1 & 0 \\ 0 & 0 & 1 \end{bmatrix}.$$

It is also true that $P^tP = I$, so $P^t = P^{-1}$. ◆ ◆ ◆

Two $n \times n$ matrices A and B are **similar** if a matrix S exists with $A = S^{-1}BS$. The important feature of similar matrices is that they have the same eigenvalues. The next result follows from observing that if $\lambda\mathbf{x} = A\mathbf{x} = S^{-1}BS\mathbf{x}$, then $BS\mathbf{x} = \lambda S\mathbf{x}$. Also, if $\mathbf{x} \neq \mathbf{0}$ and S is nonsingular, then $S\mathbf{x} \neq \mathbf{0}$, so $S\mathbf{x}$ is an eigenvector of B corresponding to its eigenvalue λ.

Eigenvalues and Eigenvectors of Similar Matrices

Suppose A and B are similar $n \times n$ matrices and λ is an eigenvalue of A with associated eigenvector $\mathbf{x}$. Then λ is also an eigenvalue of B, and if $A = S^{-1}BS$, then $S\mathbf{x}$ is an eigenvector associated with λ and the matrix B.

The Maple command `issimilar(A,B)` returns *true* if A and B are similar and *false* otherwise.

The determination of eigenvalues is easy for a triangular matrix A, because in this case λ is a solution to the equation

$$0 = \det(A - \lambda I) = \prod_{i=1}^{n}(a_{ii} - \lambda)$$

if and only if $\lambda = a_{ii}$ for some i. The next result provides a relationship, called a **similarity transformation**, between arbitrary matrices and triangular matrices.

Schur's Theorem

Let A be an arbitrary $n \times n$ matrix. A nonsingular matrix U exists with the property that

$$T = U^{-1}AU,$$

where T is an upper-triangular matrix whose diagonal entries consist of the eigenvalues of A.

Schur's Theorem is an existence theorem that ensures that the triangular matrix T exists, but the proof of the theorem does not provide a constructive means for determining T. In most instances, the similarity transformation is difficult to determine. The restriction to symmetric matrices reduces the complication since in this case the transformation matrix is orthogonal.

Eigenvalues of Symmetric Matrices

Suppose that A is an $n \times n$ symmetric matrix.

a. If D is the diagonal matrix whose diagonal entries are the eigenvalues of A, then there exists an orthogonal matrix P such that $D = P^{-1}AP = P^tAP$.
b. There exist n eigenvectors of A that form an orthonormal set and are the columns of the matrix P described in (a).
c. The eigenvalues are all real numbers.
d. A is positive definite if and only if all the eigenvalues of A are positive.

The final result of the section concerns bounds for the approximation of eigenvalues.

Gerschgorin Circle Theorem

Let A be an $n \times n$ matrix and R_i denote the circle in the complex plane with center a_{ii} and radius $\sum_{\substack{j=1\\ j\neq i}}^{n} |a_{ij}|$; that is,

$$R_i = \left\{ z \in C \;\middle|\; |z - a_{ii}| \le \sum_{\substack{j=1\\ j\neq i}}^{n} |a_{ij}| \right\},$$

where C denotes the complex plane. The eigenvalues of A are contained within $R = \cup_{i=1}^{n} R_i$. Moreover, the union of any k of these circles that do not intersect the remaining $n - k$ contain precisely k (counting multiplicities) of the eigenvalues.

EXAMPLE 4 For the matrix

$$A = \begin{bmatrix} 4 & 1 & 1 \\ 0 & 2 & 1 \\ -2 & 0 & 9 \end{bmatrix},$$

the circles in the Gerschgorin Theorem are (see Figure 9.1):

$$R_1 = \left\{ z \in C \;\middle|\; |z - 4| \le 2 \right\}, \qquad R_2 = \left\{ z \in C \;\middle|\; |z - 2| \le 1 \right\},$$

and

$$R_3 = \left\{ z \in C \;\middle|\; |z - 9| \le 2 \right\}.$$

Since R_1 and R_2 are disjoint from R_3, there are two eigenvalues within $R_1 \cup R_2$ and one within R_3. Moreover, since $\rho(A) = \max_{1 \le i \le 3} |\lambda_i|$, we have $7 \le \rho(A) \le 11$. ◆ ◆ ◆

Figure 9.1

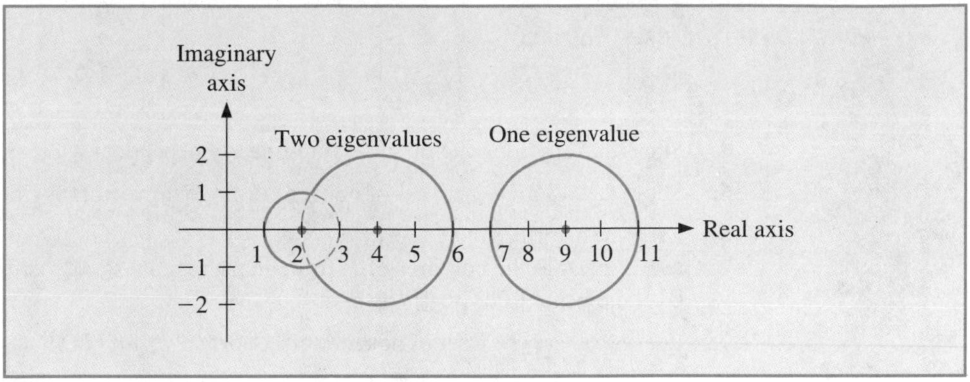

EXERCISE SET 9.2

1. Find the eigenvalues and associated eigenvectors for the following 3×3 matrices. Is there a set of three linearly independent eigenvectors?

   a. $A = \begin{bmatrix} 2 & -3 & 6 \\ 0 & 3 & -4 \\ 0 & 2 & -3 \end{bmatrix}$ b. $A = \begin{bmatrix} 1 & 0 & 0 \\ -1 & 0 & 1 \\ -1 & -1 & 2 \end{bmatrix}$

   c. $A = \begin{bmatrix} 2 & 0 & 1 \\ 0 & 2 & 0 \\ 1 & 0 & 2 \end{bmatrix}$ d. $A = \begin{bmatrix} 2 & -1 & -1 \\ -1 & 2 & -1 \\ -1 & -1 & 2 \end{bmatrix}$

   e. $A = \begin{bmatrix} 1 & 1 & 1 \\ 1 & 1 & 0 \\ 1 & 0 & 1 \end{bmatrix}$ f. $A = \begin{bmatrix} 2 & 1 & 1 \\ 1 & 2 & 1 \\ 1 & 1 & 2 \end{bmatrix}$

2. The matrices in Exercise 1(c), (d), (e), and (f) are symmetric.
   a. Are any positive definite?
   b. Consider the positive definite matrices in part (a). Construct an orthogonal matrix P for which $P^t AP = D$, a diagonal matrix, using the eigenvectors found in Exercise 1.

3. Use the Gerschgorin Circle Theorem to determine bounds for the eigenvalues of the following matrices.

   a. $\begin{bmatrix} 1 & 0 & 0 \\ -1 & 0 & 1 \\ -1 & -1 & 2 \end{bmatrix}$ b. $\begin{bmatrix} 4 & -1 & 0 \\ -1 & 4 & -1 \\ -1 & -1 & 4 \end{bmatrix}$

   c. $\begin{bmatrix} 3 & 2 & 1 \\ 2 & 3 & 0 \\ 1 & 0 & 3 \end{bmatrix}$ d. $\begin{bmatrix} 4.75 & 2.25 & -0.25 \\ 2.25 & 4.75 & 1.25 \\ -0.25 & 1.25 & 4.75 \end{bmatrix}$

e. $\begin{bmatrix} -4 & 0 & 1 & 3 \\ 0 & -4 & 2 & 1 \\ 1 & 2 & -2 & 0 \\ 3 & 1 & 0 & -4 \end{bmatrix}$ f. $\begin{bmatrix} 1 & 0 & -1 & 1 \\ 2 & 2 & -1 & 1 \\ 0 & 1 & 3 & -2 \\ 1 & 0 & 1 & 4 \end{bmatrix}$

4. Show that any four vectors in $\mathbb{R}^3$ are linearly dependent.

5. Show that a set $\{\mathbf{v}_1, \ldots, \mathbf{v}_k\}$ of nonzero orthogonal vectors is linearly independent.

6. Let P be an orthogonal matrix.

a. Show that the columns of P form an orthogonal set of vectors.

b. Show that $\|P\|_2 = \|P^t\|_2 = 1$.

7. Let $\{\mathbf{v}_1, \ldots, \mathbf{v}_n\}$ be a set of nonzero orthonormal vectors in $\mathbb{R}^n$ and $\mathbf{x} \in \mathbb{R}^n$. Show that

$$\mathbf{x} = \sum_{k=1}^{n} c_k \mathbf{v}_k, \qquad \text{where } c_k = \mathbf{v}_k^t \mathbf{x}.$$

8. Show that if A is an $n \times n$ matrix with n distinct eigenvalues, then A has n linearly independent eigenvectors.

9. In Exercise 11 of Section 6.6, a symmetric matrix

$$A = \begin{bmatrix} 1.59 & 1.69 & 2.13 \\ 1.69 & 1.31 & 1.72 \\ 2.13 & 1.72 & 1.85 \end{bmatrix}$$

was used to describe the average wing lengths of fruit flies that were offspring resulting from the mating of three mutants of the flies. The entry a_{ij} represents the average wing length of a fly that is the offspring of a male fly of type i and a female fly of type j.

a. Find the eigenvalues and associated eigenvectors of this matrix.

b. Use a result in this section to answer the question posed in part (b) of Exercise 11, Section 6.6: Is this matrix positive definite?

9.3 The Power Method

The Power method is an iterative technique used to determine the dominant eigenvalue of a matrix—that is, the eigenvalue with the largest magnitude. By modifying the method slightly it can also be used to determine other eigenvalues. One useful feature of the Power method is that it produces not only an eigenvalue, but an associated eigenvector. In fact, the Power method is often applied to find an eigenvector for an eigenvalue that is determined by some other means.

To apply the Power method, we must assume that the $n \times n$ matrix A has n eigenvalues, $\lambda_1, \lambda_2, \ldots, \lambda_n$, with an associated collection of linearly independent eigenvectors, $\{\mathbf{v}^{(1)}, \mathbf{v}^{(2)}, \mathbf{v}^{(3)}, \ldots, \mathbf{v}^{(n)}\}$. Moreover, we assume that A has precisely one eigenvalue, λ_1, that is largest in magnitude, and that $|\lambda_1| > |\lambda_2| \geq |\lambda_3| \geq \cdots \geq |\lambda_n| \geq 0$.

If $\mathbf{x}$ is any vector in $\mathbb{R}^n$, the fact that $\{\mathbf{v}^{(1)}, \mathbf{v}^{(2)}, \mathbf{v}^{(3)}, \ldots, \mathbf{v}^{(n)}\}$ is linearly independent implies that constants $\beta_1, \beta_2, \ldots, \beta_n$ exist with

$$\mathbf{x} = \sum_{j=1}^{n} \beta_j \mathbf{v}^{(j)}.$$

Multiplying both sides of this equation successively by $A, A^2, \ldots, A^k$ gives

$$\begin{aligned}
A\mathbf{x} &= \sum_{j=1}^{n} \beta_j A\mathbf{v}^{(j)} = \sum_{j=1}^{n} \beta_j \lambda_j \mathbf{v}^{(j)}, \\
A^2\mathbf{x} &= \sum_{j=1}^{n} \beta_j \lambda_j A\mathbf{v}^{(j)} = \sum_{j=1}^{n} \beta_j \lambda_j^2 \mathbf{v}^{(j)}, \\
&\vdots \\
A^k\mathbf{x} &= \sum_{j=1}^{n} \beta_j \lambda_j^k \mathbf{v}^{(j)}.
\end{aligned}$$

If λ_1^k is factored from each term on the right side of the last equation, then

$$A^k\mathbf{x} = \lambda_1^k \sum_{j=1}^{n} \beta_j \left(\frac{\lambda_j}{\lambda_1}\right)^k \mathbf{v}^{(j)}.$$

Since $|\lambda_1| > |\lambda_j|$ for all $j = 2, 3, \ldots, n$, we have $\lim_{k\to\infty} (\lambda_j/\lambda_1)^k = 0$, and

$$\lim_{k\to\infty} A^k\mathbf{x} = \lim_{k\to\infty} \lambda_1^k \beta_1 \mathbf{v}^{(1)}.$$

This sequence converges to zero if $|\lambda_1| < 1$ and diverges if $|\lambda_1| \geq 1$, provided, of course, that $\beta_1 \neq 0$. Advantage can be made of this relationship by scaling the powers of $A^k\mathbf{x}$ in an appropriate manner to ensure that the limit is finite and nonzero. The scaling begins by choosing $\mathbf{x}$ to be a unit vector $\mathbf{x}^{(0)}$ relative to $\|\cdot\|_\infty$ and a component $x_{p_0}^{(0)}$ of $\mathbf{x}^{(0)}$ with

$$x_{p_0}^{(0)} = 1 = \|\mathbf{x}^{(0)}\|_\infty.$$

Let $\mathbf{y}^{(1)} = A\mathbf{x}^{(0)}$ and define $\mu^{(1)} = y_{p_0}^{(1)}$. With this notation,

$$\mu^{(1)} = y_{p_0}^{(1)} = \frac{y_{p_0}^{(1)}}{x_{p_0}^{(0)}} = \frac{\beta_1 \lambda_1 v_{p_0}^{(1)} + \sum_{j=2}^{n} \beta_j \lambda_j v_{p_0}^{(j)}}{\beta_1 v_{p_0}^{(1)} + \sum_{j=2}^{n} \alpha_j v_{p_0}^{(j)}} = \lambda_1 \left[\frac{\beta_1 v_{p_0}^{(1)} + \sum_{j=2}^{n} \beta_j (\lambda_j/\lambda_1) v_{p_0}^{(j)}}{\beta_1 v_{p_0}^{(1)} + \sum_{j=2}^{n} \beta_j v_{p_0}^{(j)}}\right].$$

Then let p_1 be the least integer such that

$$\left|y_{p_1}^{(1)}\right| = \|\mathbf{y}^{(1)}\|_\infty$$

and define

$$\mathbf{x}^{(1)} = \frac{1}{y_{p_1}^{(1)}} \mathbf{y}^{(1)} = \frac{1}{y_{p_1}^{(1)}} A\mathbf{x}^{(0)}.$$

This continues with each iteration, calculating $\mathbf{x}^{(m-1)}$, $\mathbf{y}^{(m)}$, $\mu^{(m)}$, and $y_{p_{m-1}}^{(m)}$.

The Power method has the disadvantage that it is unknown at the outset whether the matrix has a single dominant eigenvalue. Nor is it known how $\mathbf{x}^{(0)}$ should be chosen to ensure that its representation in terms of the eigenvectors of the matrix will contain a nonzero contribution from the eigenvector associated with the dominant eigenvalue, should it exist.

Choosing the least integer, p_m, for which $|y^{(m)}_{p_m}| = \|\mathbf{y}^{(m)}\|_\infty$ will generally ensure that this index eventually becomes invariant. The rate at which $\{\mu^{(m)}\}_{m=1}^{\infty}$ converges to λ_1 is determined by the ratios $|\lambda_j/\lambda_1|^m$ for $j = 2, 3, \ldots, n$ and, in particular, by $|\lambda_2/\lambda_1|^m$; that is, the convergence is of order $O(|\lambda_2/\lambda_1|^m)$. There is a constant k such that for large m

$$|\mu^{(m)} - \lambda_1| \approx k \left|\frac{\lambda_2}{\lambda_1}\right|^m,$$

which implies that

$$\lim_{m\to\infty} \frac{|\mu^{(m+1)} - \lambda_1|}{|\mu^{(m)} - \lambda_1|} \approx \left|\frac{\lambda_2}{\lambda_1}\right| < 1.$$

The sequence $\{\mu^{(m)}\}$ converges linearly to λ_1, so Aitken's Δ^2 procedure can be used to speed the convergence.

In actuality, it is not necessary for the matrix to have n distinct eigenvalues for the Power method to converge. If the matrix has a unique dominant eigenvalue, λ_1, with multiplicity r greater than 1 and $\mathbf{v}^{(1)}, \mathbf{v}^{(2)}, \ldots, \mathbf{v}^{(r)}$ are linearly independent eigenvectors associated with λ_1, the procedure will still converge to λ_1. The sequence of vectors $\{\mathbf{x}^{(m)}\}_{m=0}^{\infty}$ will in this case converge to an eigenvector of λ_1 of norm 1 that depends on the choice of the initial vector $\mathbf{x}^{(0)}$ and is a linear combination of $\mathbf{v}^{(1)}, \mathbf{v}^{(2)}, \ldots, \mathbf{v}^{(r)}$.

The program POWERM91 implements the Power method.

EXAMPLE 1 The matrix

$$A = \begin{bmatrix} -4 & 14 & 0 \\ -5 & 13 & 0 \\ -1 & 0 & 2 \end{bmatrix}$$

has eigenvalues $\lambda_1 = 6$, $\lambda_2 = 3$, and $\lambda_3 = 2$, so the Power method will converge. In this example we present some Maple commands to illustrate the Power method. To access the linear algebra library, enter

```
>with(linalg);
```

We define the matrix A and the vector $\mathbf{x}^{(0)} = (1, 1, 1)^t$ by

```
>A:=matrix(3,3,[-4,14,0,-5,13,0,-1,0,2]);
```

and

```
>x0:=vector(3,[1,1,1]);
```

The command

```
>y1:=multiply(A,x);
```

gives $\mathbf{y}^{(1)} = A\mathbf{x}^{(0)} = (10, 8, 1)^t$. Thus,

$$\|\mathbf{y}^{(1)}\|_\infty = 10 \quad \text{and} \quad \mu^{(1)} = y_1^{(1)} = 10.$$

We compute $\mathbf{x}^{(1)}$ with

```
>x1:=0.1*y1;
```

We continue with computing $\mathbf{y}^{(2)}$ by

```
>y2:=multiply(A,x1);
```

to obtain

$$\mathbf{y}^{(2)} = A\mathbf{x}^{(1)} = A(1, 0.8, 0.1)^t = (7.2, 5.4, -0.8)^t,$$

$$\mu^{(2)} = y_1^{(2)} = 7.2, \quad \text{and} \quad \mathbf{x}^{(2)} = \frac{1}{7.2}\mathbf{y}^{(2)} = (1, 0.75, -0.\overline{1})^t.$$

Continuing in this manner leads to the values in Table 9.1, where $\hat{\mu}^{(m)}$ represents the sequence generated by the Aitken's Δ^2 procedure.

Table 9.1

m	$(\mathbf{x}^{(m)})^t$	$\mu^{(m)}$	$\hat{\mu}^{(m)}$
0	$(1, 1, 1)$		
1	$(1, 0.8, 0.1)$	10	6.266667
2	$(1, 0.75, -0.111)$	7.2	6.062473
3	$(1, 0.730769, -0.188034)$	6.5	6.015054
4	$(1, 0.722200, -0.220850)$	6.230769	6.004202
5	$(1, 0.718182, -0.235915)$	6.111000	6.000855
6	$(1, 0.716216, -0.243095)$	6.054546	6.000240
7	$(1, 0.715247, -0.246588)$	6.027027	6.000058
8	$(1, 0.714765, -0.248306)$	6.013453	6.000017
9	$(1, 0.714525, -0.249157)$	6.006711	6.000003
10	$(1, 0.714405 - 0.249579)$	6.003352	6.000000
11	$(1, 0.714346, -0.249790)$	6.001675	
12	$(1, 0.714316, -0.249895)$	6.000837	

An approximation to the dominant eigenvalue, 6, at this stage is $\hat{\mu}^{(10)} = 6.000000$ with approximate unit eigenvector $(1, 0.714316, -0.249895)^t$. Although the approximation to the eigenvalue is correct to the places listed, the eigenvector approximation is considerably less accurate to the true eigenvector, $(1, 0.714286, -0.25)^t$. ◆ ◆ ◆

When A is symmetric, a variation in the choice of the vectors $\mathbf{x}^{(m)}$ and $\mathbf{y}^{(m)}$ and scalars $\mu^{(m)}$ can be made to significantly improve the rate of convergence of the sequence $\{\mu^{(m)}\}_{m=1}^{\infty}$ to the dominant eigenvalue λ_1. First, select $\mathbf{x}^{(0)}$ with $\|\mathbf{x}^{(0)}\|_2 = 1$. For each $m = 1, 2, \ldots,$ define

$$\mu^{(m)} = \left(\mathbf{x}^{(m-1)}\right)^t A\mathbf{x}^{(m-1)} \quad \text{and} \quad \mathbf{x}^{(m)} = \frac{1}{\|A\mathbf{x}^{(m-1)}\|_2} A\mathbf{x}^{(m-1)}.$$

The rate of convergence of the Power method is $O((\lambda_2/\lambda_1)^m)$, but with this modification the rate of convergence for symmetric matrices is $O((\lambda_2/\lambda_1)^{2m})$. The program SYMPWR92 implements the Symmetric Power method in this way.

EXAMPLE 2 The matrix

$$A = \begin{bmatrix} 4 & -1 & 1 \\ -1 & 3 & -2 \\ 1 & -2 & 3 \end{bmatrix}$$

is symmetric with eigenvalues $\lambda_1 = 6, \lambda_2 = 3$, and $\lambda_3 = 1$. Tables 9.2 and 9.3 list, respectively, the first ten iterations of the Power method and the Symmetric Power method, assuming, in each case, that $\mathbf{y}^{(0)} = \mathbf{x}^{(0)} = (1, 0, 0)^t$.

Table 9.2

m	$\mathbf{y}^{(m)}$	$\mu^{(m)}$	$\mathbf{x}^{(m)}$
0			$(1, 0, 0)^t$
1	$(4, -1, 1)^t$	4	$(1, -0.25, 0.25)^t$
2	$(4.5, -2.25, 2.25)^t$	4.5	$(1, -0.5, 0.5)^t$
3	$(5, -3.5, 3.5)^t$	5	$(1, -0.7, 0.7)^t$
4	$(5.4, -4.5, 4.5)^t$	5.4	$(1, -0.83\overline{3}, 0.83\overline{3})^t$
5	$(5.66\overline{6}, -5.166\overline{6}, 5.166\overline{6})^t$	$5.66\overline{6}$	$(1, -0.911765, 0.911765)^t$
6	$(5.823529, -5.558824, 5.558824)^t$	5.823529	$(1, -0.954545, 0.954545)^t$
7	$(5.909091, -5.772727, 5.772727)^t$	5.909091	$(1, -0.976923, 0.976923)^t$
8	$(5.953846, -5.884615, 5.884615)^t$	5.953846	$(1, -0.988372, 0.988372)^t$
9	$(5.976744, -5.941861, 5.941861)^t$	5.976744	$(1, -0.994163, 0.994163)^t$
10	$(5.988327, -5.970817, 5.970817)^t$	5.988327	$(1, -0.997076, 0.997076)^t$

Notice the significant improvement that the Symmetric Power method provides. The approximations to the eigenvectors produced in the Power method are converging to $(1, -1, 1)^t$, a vector with $\|(1, -1, 1)^t\|_\infty = 1$. In the Symmetric Power method the convergence is to the parallel vector $(\sqrt{3}/3, -\sqrt{3}/3, \sqrt{3}/3)^t$, since we have $\|\sqrt{3}/3, -\sqrt{3}/3, \sqrt{3}/3)^t\|_2 = 1$. ◆ ◆ ◆

The following Maple commands are needed to compute the first two rows of Table 9.3:

```
>with(linalg):
>A:=matrix(3,3,[4,-1,1,-1,3,-2,1,-2,3]);
```

```
>x0:=vector(3,[1,0,0]);
>y1:=multiply(A,x0);
>n1:=evalf(norm(y1,2));
>x1:=1/n1*y1: evalm(x1);
>mu1:=innerprod(x0,y1);
>y2:=multiply(A,x1);
```

This gives the vector $\mathbf{y}^{(2)} = (4.242640690, -2.121320345, 2.121320345)^t$. To find $\mathbf{x}^{(2)}$ we use the Maple commands

```
>n2:=norm(y2,2);
>x2:=1/n2*y2: evalm(x2);
```

which gives $\mathbf{x}^{(2)} = (0.8164965809, -0.4082482905, 0.4082482905)^t$. $\mu^{(2)}$ is found by

```
>mu2:=innerprod(x1,y2);
```

Table 9.3

m	$\mathbf{y}^{(m)}$	$\mu^{(m)}$	$\mathbf{x}^{(m)}$
0	$(1, 0, 0)^t$		$(1, 0, 0)^t$
1	$(4, -1, 1)^t$	4	$(0.942809, -0.235702, 0.235702)^t$
2	$(4.242641, -2.121320, 2.121320)^t$	5	$(0.816497, -0.408248, 0.408248)^t$
3	$(4.082483, -2.857738, 2.857738)^t$	5.666667	$(0.710669, -0.497468, 0.497468)^t$
4	$(3.837613, -3.198011, 3.198011)^t$	5.909091	$(0.646997, -0.539164, 0.539164)^t$
5	$(3.666314, -3.342816, 3.342816)^t$	5.976744	$(0.612836, -0.558763, 0.558763)^t$
6	$(3.568871, -3.406650, 3.406650)^t$	5.994152	$(0.595247, -0.568190, 0.568190)^t$
7	$(3.517370, -3.436200, 3.436200)^t$	5.998536	$(0.586336, -0.572805, 0.572805)^t$
8	$(3.490952, -3.450359, 3.450359)^t$	5.999634	$(0.581852, -0.575086, 0.575086)^t$
9	$(3.477580, -3.457283, 3.457283)^t$	5.999908	$(0.579603, -0.576220, 0.576220)^t$
10	$(3.470854, -3.460706, 3.460706)^t$	5.999977	$(0.578477, -0.576786, 0.576786)^t$

The **Inverse Power Method** is a modification of the Power method that is used to determine the eigenvalue of A closest to a specified number q.

Suppose that the matrix A has eigenvalues $\lambda_1, \ldots, \lambda_n$ with linearly independent eigenvectors $\mathbf{v}^{(1)}, \mathbf{v}^{(2)}, \ldots, \mathbf{v}^{(n)}$. The eigenvalues of $(A - qI)^{-1}$, where $q \neq \lambda_i$ for each $i = 1, 2, \ldots, n$, are

$$\frac{1}{\lambda_1 - q}, \frac{1}{\lambda_2 - q}, \ldots, \frac{1}{\lambda_n - q}$$

with eigenvectors $\mathbf{v}^{(1)}, \mathbf{v}^{(2)}, \ldots, \mathbf{v}^{(n)}$. Applying the Power method to $(A - qI)^{-1}$ gives

$$\mathbf{y}^{(m)} = (A - qI)^{-1}\mathbf{x}^{(m-1)},$$

$$\mu^{(m)} = y_{p_{m-1}}^{(m)} = \frac{y_{p_{m-1}}^{(m)}}{x_{p_{m-1}}^{(m-1)}} = \frac{\sum_{j=1}^{n} \beta_j \dfrac{1}{(\lambda_j - q)^m} v_{p_{m-1}}^{(j)}}{\sum_{j=1}^{n} \beta_j \dfrac{1}{(\lambda_j - q)^{m-1}} v_{p_{m-1}}^{(j)}},$$

and

$$\mathbf{x}^{(m)} = \frac{\mathbf{y}^{(m)}}{y_{p_m}^{(m)}},$$

where, at each step, p_m represents the smallest integer for which $|y_{p_m}^{(m)}| = \|\mathbf{y}^{(m)}\|_\infty$. The sequence $\{\mu^{(m)}\}$ converges to $1/(\lambda_k - q)$, where

$$\frac{1}{|\lambda_k - q|} = \max_{1 \le i \le n} \frac{1}{|\lambda_i - q|},$$

and $\lambda_k \approx q + 1/\mu^{(m)}$ is the eigenvalue of A closest to q.

The choice of q determines the convergence, provided that $1/(\lambda_k - q)$ is a unique dominant eigenvalue of $(A - qI)^{-1}$ (although it may be a multiple eigenvalue). The closer q is to an eigenvalue λ_k, the faster the convergence since the convergence is of order

$$O\left(\left|\frac{(\lambda - q)^{-1}}{(\lambda_k - q)^{-1}}\right|^m\right) = O\left(\left|\frac{(\lambda_k - q)}{(\lambda - q)}\right|^m\right),$$

where λ represents the eigenvalue of A that is second closest to q.

The vector $\mathbf{y}^{(m)}$ is obtained from the equation

$$(A - qI)\mathbf{y}^{(m)} = \mathbf{x}^{(m-1)}.$$

Gaussian elimination with pivoting can be used to solve this system.

Although the Inverse Power method requires the solution of an $n \times n$ system at each step, the multipliers can be saved to reduce the computation. The selection of q can be based on the Gerschgorin Theorem or on another means of localizing an eigenvalue.

One choice of q comes from the initial approximation to the eigenvector $\mathbf{x}^{(0)}$:

$$q = \frac{\mathbf{x}^{(0)t} A \mathbf{x}^{(0)}}{\mathbf{x}^{(0)t} \mathbf{x}^{(0)}}.$$

This choice of q results from the observation that if $\mathbf{x}$ is an eigenvector of A with respect to the eigenvalue λ, then $A\mathbf{x} = \lambda\mathbf{x}$. So $\mathbf{x}^t A\mathbf{x} = \lambda \mathbf{x}^t\mathbf{x}$ and

$$\lambda = \frac{\mathbf{x}^t A \mathbf{x}}{\mathbf{x}^t \mathbf{x}} = \frac{\mathbf{x}^t A \mathbf{x}}{\|\mathbf{x}\|_2^2}$$

If q is close to an eigenvalue, the convergence will be quite rapid. In fact, this method is often used to approximate an eigenvector when an approximate eigenvalue q is obtained.

Since the convergence of the Inverse Power method is linear, Aitken's Δ^2 procedure can be used to speed convergence. The following example illustrates the fast convergence of the Inverse Power method if q is close to an eigenvalue.

EXAMPLE 3 The matrix

$$A = \begin{bmatrix} -4 & 14 & 0 \\ -5 & 13 & 0 \\ -1 & 0 & 2 \end{bmatrix}$$

was considered in Example 1. The Power method gave the approximation $\mu^{(12)} = 6.000837$ using $\mathbf{x}^{(0)} = (1, 1, 1)^t$.

The following Maple commands generate the first two rows of Table 9.4 for the Inverse Power method:

```
>with(linalg);
>A:=matrix(3,3,[-4,14,0,-5,13,0,-1,0,2]);
>x0:=vector(3,[1,1,1]);
```

To compute

$$q = \frac{\mathbf{x}^{(0)t} A\mathbf{x}^{(0)}}{\mathbf{x}^{(0)t}\mathbf{x}^{(0)}} = \frac{19}{3} = 6.333333,$$

we use the command

```
>q:=evalf(innerprod(x0,A,x0)/innerprod(x0,x0));
```

The identity matrix I_3 is given by

```
>I3:=matrix(3,3,[1,0,0,0,1,0,0,0,1]):
```

We form $A - qI$ with

```
>AQ:=A-q*I3;
```

giving the matrix

$$\begin{bmatrix} -10.33333333 & 14 & 0 \\ -5 & 6.666666667 & 0 \\ -1 & 0 & -4.333333333 \end{bmatrix}$$

We form the augmented matrix $M = [AQ : \mathbf{x}^{(0)}]$ using

```
>M:=augment(AQ,x0);
```

and perform Gaussian elimination with

```
>N:=gausselim(M);
```

We obtain $\mathbf{y}^{(1)} = (-6.599999864, -4.799999898, 1.292307661)^t$ with the command

```
>y1:=backsub(N);
```

Then we find $\mathbf{x}^{(1)}$ and $\mu^{(1)}$:

```
>x1:=1/(-6.599999864)*y1;
>mu1:=q+1/(-6.599999864);
```

Table 9.4

m	$\mathbf{x}^{(m)}$	$\mu^{(m)}$	$\hat{\mu}^{(m)}$
0	$(1, 1, 1)^t$		
1	$(1, 0.720727, -0.194042)^t$	6.183183	6.000116
2	$(1, 0.715518, -0.245052)^t$	6.017244	6.000004
3	$(1, 0.714409, -0.249522)^t$	6.001719	6.000004
4	$(1, 0.714298, -0.249953)^t$	6.000175	6.000003
5	$(1, 0.714287, -0.250000)^t$	6.000021	
6	$(1, 0.714286, -0.249999)^t$	6.000005	

The results of applying the program INVPOW93 for the Inverse Power method with this value of q are listed in Table 9.4, and the right column lists the results of Aitken's Δ^2 method applied to the $\mu^{(m)}$. ◆ ◆ ◆

Numerous techniques are available for obtaining approximations to the other eigenvalues of a matrix once an approximation to the dominant eigenvalue has been computed. We will restrict our presentation to **deflation techniques**. These techniques involve forming a new matrix B whose eigenvalues are the same as those of A, except that the dominant eigenvalue of A is replaced by the eigenvalue 0 in B. The following result justifies the procedure.

Eigenvalues and Eigenvectors of Deflated Matrices

Suppose $\lambda_1, \lambda_2, \ldots, \lambda_n$ are eigenvalues of A with associated eigenvectors $\mathbf{v}^{(1)}, \mathbf{v}^{(2)}, \ldots, \mathbf{v}^{(n)}$ and that λ_1 has multiplicity 1. If $\mathbf{x}$ is a vector with $\mathbf{x}^t\mathbf{v}^{(1)} = 1$, then

$$B = A - \lambda_1 \mathbf{v}^{(1)}\mathbf{x}^t$$

has eigenvalues $0, \lambda_2, \lambda_3, \ldots, \lambda_n$ with associated eigenvectors $\mathbf{v}^{(1)}, \mathbf{w}^{(2)}, \mathbf{w}^{(3)}, \ldots, \mathbf{w}^{(n)}$, where $\mathbf{v}^{(i)}$ and $\mathbf{w}^{(i)}$ are related by the equation

$$\mathbf{v}^{(i)} = (\lambda_i - \lambda_1)\mathbf{w}^{(i)} + \lambda_1(\mathbf{x}^t\mathbf{w}^{(i)})\mathbf{v}^{(1)},$$

for each $i = 2, 3, \ldots, n$.

Wielandt's deflation results from defining

$$\mathbf{x} = \frac{1}{\lambda_1 v_i^{(1)}}(a_{i1}, a_{i2}, \ldots, a_{in})^t,$$

where $v_i^{(1)}$ is a nonzero coordinate of the eigenvector $\mathbf{v}^{(1)}$ and the values $a_{i1}, a_{i2}, \ldots, a_{in}$ are the entries in the ith row of A. With this definition,

$$\mathbf{x}^t\mathbf{v}^{(1)} = \frac{1}{\lambda_1 v_i^{(1)}}(a_{i1}, a_{i2}, \ldots, a_{in})(v_1^{(1)}, v_2^{(1)}, \ldots, v_n^{(1)})^t = \frac{1}{\lambda_1 v_i^{(1)}}\sum_{j=1}^{n} a_{ij}v_j^{(1)},$$

where the sum is the ith coordinate of the product $A\mathbf{v}^{(1)}$. Since $A\mathbf{v}^{(1)} = \lambda_1\mathbf{v}^{(1)}$, we have

$$\sum_{j=1}^{n} a_{ij}v_j^{(1)} = \lambda_1 v_i^{(1)}$$

which implies that

$$\mathbf{x}^t\mathbf{v}^{(1)} = \frac{1}{\lambda_1 v_i^{(1)}}(\lambda_1 v_i^{(1)}) = 1.$$

So $\mathbf{x}$ satisfies the hypotheses of the result concerning the eigenvalues of deflated matrices. Moreover, the ith row of $B = A - \lambda_1\mathbf{v}^{(1)}\mathbf{x}^t$ consists entirely of zero entries.

If $\lambda \neq 0$ is an eigenvalue with associated eigenvector $\mathbf{w}$, the relation $B\mathbf{w} = \lambda\mathbf{w}$ implies that the ith coordinate of $\mathbf{w}$ must also be zero. Consequently, the ith column of the matrix B makes no contribution to the product $B\mathbf{w} = \lambda\mathbf{w}$. Thus, the matrix B can be replaced by an $(n-1)\times(n-1)$ matrix B', obtained by deleting the ith row and column from B. The matrix B' has eigenvalues $\lambda_2, \lambda_3, \ldots, \lambda_n$. If $|\lambda_2| > |\lambda_3|$, the Power method is reapplied to the matrix B' to determine this new dominant eigenvalue and an eigenvector, $\mathbf{w}^{(2)'}$, associated with λ_2, with respect to the matrix B'. To find the associated eigenvector $\mathbf{w}^{(2)}$ for the matrix B, insert a zero coordinate between the coordinates $w_{i-1}^{(2)'}$ and $w_i^{(2)'}$ of the $(n-1)$-dimensional vector $\mathbf{w}^{(2)'}$ and then calculate $\mathbf{v}^{(2)}$ by using the result for deflated matrices. This deflation technique can be performed using the program WIEDEF94.

EXAMPLE 4 From Example 2, we know that the matrix

$$A = \begin{bmatrix} 4 & -1 & 1 \\ -1 & 3 & -2 \\ 1 & -2 & 3 \end{bmatrix}$$

has eigenvalues $\lambda_1 = 6$, $\lambda_2 = 3$, and $\lambda_3 = 1$. Assuming that the dominant eigenvalue $\lambda_1 = 6$ and associated unit eigenvector $\mathbf{v}^{(1)} = (1, -1, 1)^t$ have been calculated, the procedure just outlined for obtaining λ_2 proceeds as follows:

$$\mathbf{x} = \frac{1}{6}\begin{bmatrix} 4 \\ -1 \\ 1 \end{bmatrix} = \left(\frac{2}{3}, -\frac{1}{6}, \frac{1}{6}\right)^t,$$

$$\mathbf{v}^{(1)}\mathbf{x}^t = \begin{bmatrix} 1 \\ -1 \\ 1 \end{bmatrix}\left[\tfrac{2}{3}, \ -\tfrac{1}{6}, \ \tfrac{1}{6}\right] = \begin{bmatrix} \frac{2}{3} & -\frac{1}{6} & \frac{1}{6} \\ -\frac{2}{3} & \frac{1}{6} & -\frac{1}{6} \\ \frac{2}{3} & -\frac{1}{6} & \frac{1}{6} \end{bmatrix},$$

and

$$B = A - \lambda_1 \mathbf{v}^{(1)}\mathbf{x}^t = \begin{bmatrix} 4 & -1 & 1 \\ -1 & 3 & -2 \\ 1 & -2 & 3 \end{bmatrix} - 6 \begin{bmatrix} \frac{2}{3} & -\frac{1}{6} & \frac{1}{6} \\ -\frac{2}{3} & \frac{1}{6} & -\frac{1}{6} \\ \frac{2}{3} & -\frac{1}{6} & \frac{1}{6} \end{bmatrix}$$

$$= \begin{bmatrix} 0 & 0 & 0 \\ 3 & 2 & -1 \\ -3 & -1 & 2 \end{bmatrix}.$$

Deleting the first row and column gives

$$B' = \begin{bmatrix} 2 & -1 \\ -1 & 2 \end{bmatrix},$$

which has eigenvalues $\lambda_2 = 3$ and $\lambda_3 = 1$. For $\lambda_2 = 3$ the eigenvector $\mathbf{w}^{(2)'}$ can be obtained by solving the second-order linear sytem

$$(B' - 3I)\mathbf{w}^{(2)'} = \mathbf{0},$$

resulting in

$$\mathbf{w}^{(2)'} = (1, -1)^t.$$

Adding a zero for the first component gives $\mathbf{w}^{(2)} = (0, 1, -1)^t$ and

$$\mathbf{v}^{(2)} = (3-6)(0, 1, -1)^t + 6\left[\left(\frac{2}{3}, -\frac{1}{6}, \frac{1}{6}\right)(0, 1, -1)^t\right](1, -1, 1)^t = (-2, -1, 1)^t.$$

◆ ◆ ◆

Although deflation can be used to find approximations to all the eigenvalues and eigenvectors of a matrix, the process is susceptible to round-off error. Techniques based on similarity transformations are presented in the next two sections. These methods are generally preferable when approximations to all the eigenvalues are needed.

EXERCISE SET 9.3

1. Find the first three iterations obtained by the Power method applied to the following matrices.

   a. $\begin{bmatrix} 2 & 1 & 1 \\ 1 & 2 & 1 \\ 1 & 1 & 2 \end{bmatrix}$

   Use $\mathbf{x}^{(0)} = (1, -1, 2)^t$.

   b. $\begin{bmatrix} 1 & 1 & 1 \\ 1 & 1 & 0 \\ 1 & 0 & 1 \end{bmatrix}$

   Use $\mathbf{x}^{(0)} = (-1, 0, 1)^t$.

c. $\begin{bmatrix} 1 & -1 & 0 \\ -2 & 4 & -2 \\ 0 & -1 & 2 \end{bmatrix}$
Use $\mathbf{x}^{(0)} = (-1, 2, 1)^t$.

d. $\begin{bmatrix} 4 & 1 & 1 & 1 \\ 1 & 3 & -1 & 1 \\ 1 & -1 & 2 & 0 \\ 1 & 1 & 0 & 2 \end{bmatrix}$
Use $\mathbf{x}^{(0)} = (1, -2, 0, 3)^t$.

e. $\begin{bmatrix} 5 & -2 & -\frac{1}{2} & \frac{3}{2} \\ -2 & 5 & \frac{3}{2} & -\frac{1}{2} \\ -\frac{1}{2} & \frac{3}{2} & 5 & -2 \\ \frac{3}{2} & -\frac{1}{2} & -2 & 5 \end{bmatrix}$
Use $\mathbf{x}^{(0)} = (1, 1, 0, -3)^t$.

f. $\begin{bmatrix} -4 & 0 & \frac{1}{2} & \frac{1}{2} \\ \frac{1}{2} & -2 & 0 & \frac{1}{2} \\ \frac{1}{2} & \frac{1}{2} & 0 & 0 \\ 0 & 1 & 1 & 4 \end{bmatrix}$
Use $\mathbf{x}^{(0)} = (0, 0, 0, 1)^t$.

2. Repeat Exercise 1 using the Inverse Power method.
3. Find the first three iterations obtained by the Symmetric Power method applied to the following matrices.

a. $\begin{bmatrix} 2 & 1 & 1 \\ 1 & 2 & 1 \\ 1 & 1 & 2 \end{bmatrix}$
Use $\mathbf{x}^{(0)} = (1, -1, 2)^t$.

b. $\begin{bmatrix} 1 & 1 & 1 \\ 1 & 1 & 0 \\ 1 & 0 & 1 \end{bmatrix}$
Use $\mathbf{x}^{(0)} = (-1, 0, 1)^t$.

c. $\begin{bmatrix} 4.75 & 2.25 & -0.25 \\ 2.25 & 4.75 & 1.25 \\ -0.25 & 1.25 & 4.75 \end{bmatrix}$
Use $\mathbf{x}^{(0)} = (0, 1, 0)^t$.

d. $\begin{bmatrix} 4 & 1 & -1 & 0 \\ 1 & 3 & -1 & 0 \\ -1 & -1 & 5 & 2 \\ 0 & 0 & 2 & 4 \end{bmatrix}$
Use $\mathbf{x}^{(0)} = (0, 1, 0, 0)^t$.

e. $\begin{bmatrix} 4 & 1 & 1 & 1 \\ 1 & 3 & -1 & 1 \\ 1 & -1 & 2 & 0 \\ 1 & 1 & 0 & 2 \end{bmatrix}$
Use $\mathbf{x}^{(0)} = (1, 0, 0, 0)^t$.

f. $\begin{bmatrix} 5 & -2 & -\frac{1}{2} & \frac{3}{2} \\ -2 & 5 & \frac{3}{2} & -\frac{1}{2} \\ -\frac{1}{2} & \frac{3}{2} & 5 & -2 \\ \frac{3}{2} & -\frac{1}{2} & -2 & 5 \end{bmatrix}$
Use $\mathbf{x}^{(0)} = (1, 1, 0, -3)^t$.

4. Use the Power method and Wielandt deflation to approximate the second most dominant eigenvalues for the matrices in Exercise 1, iterating until $\|\mathbf{x}^{(m)} - \mathbf{x}^{(m-1)}\|_\infty < 10^{-4}$ or until the number of iterations exceeds 25.
5. Use the Power method and Aitken's Δ^2 technique to approximate the dominant eigenvalue for the matrices in Exercise 1, iterating until $\|\mathbf{x}^{(m)} - \mathbf{x}^{(m-1)}\|_\infty < 10^{-4}$ or until the number of iterations exceeds 25.
6. Use the Symmetric Power method to approximate the dominant eigenvalue for the matrices given in Exercise 3, iterating until $\|\mathbf{x}^{(m)} - \mathbf{x}^{(m-1)}\|_2 < 10^{-4}$ or until the number of iterations exceeds 25.
7. Use the Inverse Power method on the matrices in Exercise 1, iterating until $\|\mathbf{x}^{(m)} - \mathbf{x}^{(m-1)}\|_\infty < 10^{-4}$ or until the number of iterations exceeds 25.
8. Show that the ith row of $B = A - \lambda_1 \mathbf{v}^{(1)}\mathbf{x}^t$ is zero, where λ_1 is the largest value of A in absolute value, $\mathbf{v}^{(1)}$ is the associated eigenvector of A for λ_1, and $\mathbf{x} = 1/\lambda_1 v_i^{(1)} (a_{i1}, a_{i2}, \ldots, a_{in})^t$.

9. Following along the line of Exercise 8 in Section 6.4 and Exercise 9 in Section 7.3, suppose that a species of beetle has a life span of 4 years and that a female in the first year has a survival rate of $\frac{1}{2}$, in the second year a survival rate of $\frac{1}{4}$, and in the third year a survival rate of $\frac{1}{8}$. Suppose additionally that a female gives birth, on the average, to two new females in the third year and to four new females in the fourth year. The matrix describing a single female's contribution in one year to the female population in the succeeding year is

$$A = \begin{bmatrix} 0 & 0 & 2 & 4 \\ \frac{1}{2} & 0 & 0 & 0 \\ 0 & \frac{1}{4} & 0 & 0 \\ 0 & 0 & \frac{1}{8} & 0 \end{bmatrix},$$

where again the entry in the ith row and jth column denotes the probabilistic contribution that a female of age j makes on the next year's female population of age i.

a. Use the Gerschgorin Circle Theorem to determine a region in the complex plane containing all the eigenvalues of A.

b. Use the Power method to determine the dominant eigenvalue of the matrix and its associated eigenvector.

c. Use deflation to determine any remaining eigenvalues and eigenvectors of A.

d. Find the eigenvalues of A by using the characteristic polynomial of A and the Newton-Raphson method.

e. What is your long-range prediction for the population of these beetles?

10. A linear dynamical system can be represented by the equations

$$\frac{d\mathbf{x}}{dt} = A(t)\mathbf{x}(t) + B(t)\mathbf{u}(t), \qquad \mathbf{y}(t) = C(t)\mathbf{x}(t) + D(t)\mathbf{u}(t),$$

where A is an $n \times n$ variable matrix, B is an $n \times r$ variable matrix, C is an $m \times n$ variable matrix, D is an $m \times r$ variable matrix, $\mathbf{x}$ is an n-dimensional vector variable, $\mathbf{y}$ is an m-dimensional vector variable, and $\mathbf{u}$ is an r-dimensional vector variable. For the system to be stable, the matrix A must have all its eigenvalues with nonpositive real part for all t.

a. Is the system stable if

$$A(t) = \begin{bmatrix} -1 & 2 & 0 \\ -2.5 & -7 & 4 \\ 0 & 0 & -5 \end{bmatrix}?$$

b. Is the system stable if

$$A(t) = \begin{bmatrix} -1 & 1 & 0 & 0 \\ 0 & -2 & 1 & 0 \\ 0 & 0 & -5 & 1 \\ -1 & -1 & -2 & -3 \end{bmatrix}?$$

9.4 Householder's Method

In the next section we use the QR method to reduce a symmetric tridiagonal matrix to a nearly diagonal matrix to which it is similar. The diagonal entries of the reduced matrix are approximations to the eigenvalues of both matrices. In this section, we consider the associated problem of reducing an arbitrary symmetric matrix to a similar tridiagonal matrix using a method devised by Alton Householder. Although there is a clear connection between the problems we are solving in these two sections, Householder's method has wide application in areas other than eigenvalue approximation.

Householder's method is used to find a symmetric tridiagonal matrix B that is similar to a given symmetric matrix A. Schur's Theorem ensures that a symmetric matrix A is similar to a diagonal matrix D, since an orthogonal matrix Q exists with the property that $D = Q^{-1}AQ = Q^tAQ$. However, the matrix Q (and consequently D) is generally difficult to compute, and Householder's method offers a compromise.

Let $\mathbf{w} \in \mathbb{R}^n$ with $\mathbf{w}^t\mathbf{w} = 1$. The $n \times n$ matrix

$$P = I - 2\mathbf{w}\mathbf{w}^t$$

is called a **Householder transformation**.

Householder transformations are used to selectively zero out blocks of entries in vectors or in columns of matrices in a manner that is extremely stable with respect to round-off error. An important property of Householder transformations follows.

Householder Transformations

If $P = I - 2\mathbf{w}\mathbf{w}^t$ is a Householder transformation, then P is symmetric and orthogonal, so $P^{-1} = P$.

Householder's method begins by determining a transformation $P^{(1)}$ with the property that $A^{(2)} = P^{(1)}AP^{(1)}$ will have

$$a_{j1}^{(2)} = 0 \qquad \text{for each } j = 3, 4, \ldots, n.$$

By symmetry, this also implies that $a_{1j}^{(2)} = 0$ for each $j = 3, 4, \ldots, n$.

The vector $\mathbf{w} = (w_1, w_2, \ldots, w_n)^t$ is chosen so that $\mathbf{w}^t\mathbf{w} = 1$, and in the matrix

$$A^{(2)} = P^{(1)}AP^{(1)} = (I - 2\mathbf{w}\mathbf{w}^t)A(I - 2\mathbf{w}\mathbf{w}^t)$$

we have $a_{11}^{(2)} = a_{11}$ and $a_{j1}^{(2)} = 0$ for each $j = 3, 4, \ldots, n$. This imposes n conditions on the n unknowns $w_1, w_2, \ldots, w_n$. Setting $w_1 = 0$ ensures that $a_{11}^{(2)} = a_{11}$. We want

$$P^{(1)} = I - 2\mathbf{w}\mathbf{w}^t$$

to satisfy

$$P^{(1)}(a_{11}, a_{21}, a_{31}, \ldots, a_{n1})^t = (a_{11}, \alpha, 0, \ldots, 0)^t, \tag{9.1}$$

where α will be chosen later. To simplify notation, let

$$\hat{\mathbf{w}} = (w_2, w_3, \ldots, w_n)^t \in \mathbb{R}^{n-1}, \hat{\mathbf{y}} = (a_{21}, a_{31}, \ldots, a_{n1})^t \in \mathbb{R}^{n-1},$$

and $\hat{P}$ be the $(n-1) \times (n-1)$ Householder transformation

$$\hat{P} = I_{n-1} - 2\hat{\mathbf{w}}\hat{\mathbf{w}}^t.$$

Equation (9.1) then becomes

$$P^{(1)} \begin{bmatrix} a_{11} \\ a_{21} \\ a_{31} \\ \vdots \\ a_{n1} \end{bmatrix} = \left[\begin{array}{c|ccc} 1 & 0 & \cdots & 0 \\ \hline 0 & & & \\ \vdots & & \hat{P} & \\ 0 & & & \end{array}\right] \cdot \begin{bmatrix} a_{11} \\ \\ \mathbf{y} \\ \\ \end{bmatrix} = \begin{bmatrix} a_{11} \\ \\ \hat{P}\hat{\mathbf{y}} \\ \\ \end{bmatrix} = \begin{bmatrix} a_{11} \\ \alpha \\ 0 \\ \vdots \\ 0 \end{bmatrix}$$

with

$$\hat{P}\hat{\mathbf{y}} = (I_{n-1} - 2\hat{\mathbf{w}}\hat{\mathbf{w}}^t)\hat{\mathbf{y}} = \hat{\mathbf{y}} - 2\hat{\mathbf{w}}^t\hat{\mathbf{y}}\hat{\mathbf{w}} = (\alpha, 0, \ldots, 0)^t. \tag{9.2}$$

Let $r = \hat{\mathbf{w}}^t\hat{\mathbf{y}}$. Then

$$(\alpha, 0, \ldots, 0)^t = (a_{21} - 2rw_2, a_{31} - 2rw_3, \ldots, a_{n1} - 2rw_n)^t.$$

Equating components gives

$$\alpha = a_{21} - 2rw_2 \qquad 0 = a_{j1} - 2rw_j, \qquad \text{for each } j = 3, \ldots, n.$$

Thus,

$$2rw_2 = a_{21} - \alpha \quad \text{and} \quad 2rw_j = a_{j1}, \qquad \text{for each } j = 3, \ldots, n. \tag{9.3}$$

Squaring both sides of each of the equations and summing gives

$$4r^2 \sum_{j=2}^{n} w_j^2 = (a_{21} - \alpha)^2 + \sum_{j=3}^{n} a_{j1}^2.$$

Since $\mathbf{w}^t\mathbf{w} = 1$ and $w_1 = 0$, we have $\sum_{j=2}^{n} w_j^2 = 1$ and

$$4r^2 = \sum_{j=2}^{n} a_{j1}^2 - 2\alpha a_{21} + \alpha^2. \tag{9.4}$$

Equation (9.2) and the fact that P is orthogonal imply that

$$\alpha^2 = (\alpha, 0, \ldots, 0)(\alpha, 0, \ldots, 0)^t = (\hat{P}\hat{\mathbf{y}})^t\hat{P}\hat{\mathbf{y}} = \hat{\mathbf{y}}^t\hat{P}^t\hat{P}\hat{\mathbf{y}} = \hat{\mathbf{y}}^t\hat{\mathbf{y}}.$$

Thus,

$$\alpha^2 = \sum_{j=2}^{n} a_{j1}^2,$$

which, when substituted into Eq. (9.4), gives

$$2r^2 = \sum_{j=2}^{n} a_{j1}^2 - \alpha a_{21}.$$

To ensure that $2r^2 = 0$ only if $a_{21} = a_{31} = \cdots = a_{n1} = 0$, we choose

$$\alpha = -(\text{sign } a_{21}) \left(\sum_{j=2}^{n} a_{j1}^2 \right)^{1/2}$$

which implies that

$$2r^2 = \sum_{j=2}^{n} a_{j1}^2 + |a_{21}| \left(\sum_{j=2}^{n} a_{j1}^2 \right)^{1/2}.$$

With this choice of α and $2r^2$, we solve the equations in (9.3) to obtain

$$w_2 = \frac{a_{21} - \alpha}{2r} \quad \text{and} \quad w_j = \frac{a_{j1}}{2r}, \qquad \text{for each } j = 3, \ldots, n.$$

To summarize the choice of $P^{(1)}$, we have

$$\begin{aligned}
\alpha &= -(\text{sign } a_{21}) \left(\sum_{j=2}^{n} a_{j1}^2 \right)^{1/2}, \\
r &= \left(\frac{1}{2}\alpha^2 - \frac{1}{2} a_{21}\alpha \right)^{1/2}, \\
w_1 &= 0, \\
w_2 &= \frac{a_{21} - \alpha}{2r},
\end{aligned}$$

and

$$w_j = \frac{a_{j1}}{2r}, \qquad \text{for each } j = 3, \ldots, n.$$

With this choice,

$$A^{(2)} = P^{(1)} A P^{(1)} = \begin{bmatrix} a_{11}^{(2)} & a_{12}^{(2)} & 0 & \cdots & 0 \\ a_{21}^{(2)} & a_{22}^{(2)} & a_{23}^{(2)} & \cdots & a_{2n}^{(2)} \\ 0 & a_{32}^{(2)} & a_{33}^{(2)} & \cdots & a_{3n}^{(2)} \\ \vdots & \vdots & \vdots & & \vdots \\ 0 & a_{n2}^{(2)} & a_{n3}^{(2)} & \cdots & a_{nn}^{(2)} \end{bmatrix}.$$

Having found $P^{(1)}$ and computed $A^{(2)}$, the process is repeated for $k = 2, 3, \ldots, n-2$ as follows:

$$\begin{aligned}
\alpha &= -(\text{sign } a_{k+1,k}^{(k)}) \left(\sum_{j=k+1}^{n} (a_{jk}^{(k)})^2 \right)^{1/2}, \\
r &= \left(\frac{1}{2}\alpha^2 - \frac{1}{2}\alpha a_{k+1,k}^{(k)} \right)^{1/2}, \\
w_1^{(k)} &= w_2^{(k)} = \cdots = w_k^{(k)} = 0, \\
w_{k+1}^{(k)} &= \frac{a_{k+1,k}^{(k)} - \alpha}{2r} \\
w_j^{(k)} &= \frac{a_{jk}^{(k)}}{2r}, \qquad \text{for each } j = k+2, k+3, \ldots, n, \\
P^{(k)} &= I - 2\mathbf{w}^{(k)} \cdot \left(\mathbf{w}^{(k)}\right)^t,
\end{aligned}$$

and

$$A^{(k+1)} = P^{(k)} A^{(k)} P^{(k)},$$

where

$$A^{(k+1)} = \begin{bmatrix}
a_{11}^{(k+1)} & a_{12}^{(k+1)} & 0 & \cdots & \cdots & \cdots & 0 \\
a_{21}^{(k+1)} & \ddots & \ddots & \ddots & & & \vdots \\
0 & \ddots & \ddots & \ddots & 0 & \cdots & 0 \\
\vdots & & \ddots & a_{k+1,k}^{(k+1)} & a_{k+1,k+1}^{(k+1)} & a_{k+1,k+2}^{(k+1)} \quad \cdots & a_{k+1,n}^{(k+1)} \\
\vdots & & & 0 & \vdots & & \vdots \\
\vdots & & & \vdots & \vdots & & \vdots \\
0 & \cdots & \cdots & 0 & a_{n,k+1}^{(k+1)} & \cdots & a_{nn}^{(k+1)}
\end{bmatrix}.$$

Continuing in this manner, the tridiagonal and symmetric matrix $A^{(n-1)}$ is formed, where

$$A^{(n-1)} = P^{(n-2)} P^{(n-3)} \cdots P^{(1)} A P^{(1)} \cdots P^{(n-3)} P^{(n-2)}.$$

The program HSEHLD95 performs Householder's method in this manner on a symmetric matrix.

EXAMPLE 1 The 4×4 matrix

$$A = \begin{bmatrix} 4 & 1 & -2 & 2 \\ 1 & 2 & 0 & 1 \\ -2 & 0 & 3 & -2 \\ 2 & 1 & -2 & -1 \end{bmatrix}$$

is symmetric. For the first application of a Householder transformation:

$$\alpha = -(1)\left(\sum_{j=2}^{4} a_{j1}^2\right)^{1/2} = -3, \qquad r = \left(\frac{1}{2}(-3)^2 - \frac{1}{2}(1)(-3)\right)^{1/2} = \sqrt{6},$$

$$\mathbf{w} = \left(0, \frac{\sqrt{6}}{3}, -\frac{\sqrt{6}}{6}, \frac{\sqrt{6}}{6}\right),$$

$$P^{(1)} = \begin{bmatrix} 1 & 0 & 0 & 0 \\ 0 & 1 & 0 & 0 \\ 0 & 0 & 1 & 0 \\ 0 & 0 & 0 & 1 \end{bmatrix} - 2\left(\frac{\sqrt{6}}{6}\right)^2 \begin{bmatrix} 0 \\ 2 \\ -1 \\ 1 \end{bmatrix} \cdot (0, 2, -1, 1)$$

$$= \begin{bmatrix} 1 & 0 & 0 & 0 \\ 0 & -\frac{1}{3} & \frac{2}{3} & -\frac{2}{3} \\ 0 & \frac{2}{3} & \frac{2}{3} & \frac{1}{3} \\ 0 & -\frac{2}{3} & \frac{1}{3} & \frac{2}{3} \end{bmatrix},$$

and

$$A^{(2)} = \begin{bmatrix} 4 & -3 & 0 & 0 \\ -3 & \frac{10}{3} & 1 & \frac{4}{3} \\ 0 & 1 & \frac{5}{3} & -\frac{4}{3} \\ 0 & \frac{4}{3} & -\frac{4}{3} & -1 \end{bmatrix}.$$

Continuing to the second iteration,

$$\alpha = -\frac{5}{3}, \qquad r = \frac{2\sqrt{5}}{3}, \qquad \mathbf{w} = \left(0, 0, 2\sqrt{5}, \frac{\sqrt{5}}{5}\right)^t,$$

$$P^{(2)} = \begin{bmatrix} 1 & 0 & 0 & 0 \\ 0 & 1 & 0 & 0 \\ 0 & 0 & -\frac{3}{5} & -\frac{4}{5} \\ 0 & 0 & -\frac{4}{5} & \frac{3}{5} \end{bmatrix}.$$

and the symmetric tridiagonal matrix is

$$A^{(3)} = \begin{bmatrix} 4 & -3 & 0 & 0 \\ -3 & \frac{10}{3} & -\frac{5}{3} & 0 \\ 0 & -\frac{5}{3} & -\frac{33}{25} & \frac{68}{75} \\ 0 & 0 & \frac{68}{75} & \frac{149}{75} \end{bmatrix}.$$

◆ ◆ ◆

In the next section, we will examine how the QR method can then be applied to $A^{(n-1)}$ to determine its eigenvalues, which are the same as those of the original matrix A.

EXERCISE SET 9.4

1. Use Householder's method to place the following matrices in tridiagonal form.

a. $\begin{bmatrix} 12 & 10 & 4 \\ 10 & 8 & -5 \\ 4 & -5 & 3 \end{bmatrix}$ **b.** $\begin{bmatrix} 2 & -1 & -1 \\ -1 & 2 & -1 \\ -1 & -1 & 2 \end{bmatrix}$

c. $\begin{bmatrix} 1 & 1 & 1 \\ 1 & 1 & 0 \\ 1 & 0 & 1 \end{bmatrix}$ **d.** $\begin{bmatrix} 4.75 & 2.25 & -0.25 \\ 2.25 & 4.75 & 1.25 \\ -0.25 & 1.25 & 4.75 \end{bmatrix}$

2. Use Householder's Method to place the following matrices in tridiagonal form.

a. $\begin{bmatrix} 4 & -1 & -1 & 0 \\ -1 & 4 & 0 & -1 \\ -1 & 0 & 4 & -1 \\ 0 & -1 & -1 & 4 \end{bmatrix}$ **b.** $\begin{bmatrix} 5 & -2 & -0.5 & 1.5 \\ -2 & 5 & 1.5 & -0.5 \\ -0.5 & 1.5 & 5 & -2 \\ 1.5 & -0.5 & -2 & 5 \end{bmatrix}$

c. $\begin{bmatrix} 8 & 0.25 & 0.5 & 2 & -1 \\ 0.25 & -4 & 0 & 1 & 2 \\ 0.5 & 0 & 5 & 0.75 & -1 \\ 2 & 1 & 0.75 & 5 & -0.5 \\ -1 & 2 & -1 & -0.5 & 6 \end{bmatrix}$

d. $\begin{bmatrix} 2 & -1 & -1 & 0 & 0 \\ -1 & 3 & 0 & -2 & 0 \\ -1 & 0 & 4 & 2 & 1 \\ 0 & -2 & 2 & 8 & 3 \\ 0 & 0 & 1 & 3 & 9 \end{bmatrix}$

9.5 The *QR* Method

To apply the *QR* method we begin with a symmetric matrix in tridiagonal form; that is, the only nonzero entries in the matrix lie either on the diagonal or on the subdiagonals directly above or below the diagonal. If this is not the form of the symmetric matrix, the first step is to apply Householder's method to compute a symmetric, tridiagonal matrix similar to the given matrix.

If A is a symmetric tridiagonal matrix, we can simplify the notation somewhat by labeling the entries of A as follows:

$$A = \begin{bmatrix} a_1 & b_2 & 0 & \cdots & 0 \\ b_2 & a_2 & b_3 & \ddots & \vdots \\ 0 & b_3 & a_3 & \ddots & 0 \\ \vdots & \ddots & \ddots & \ddots & b_n \\ 0 & \cdots & 0 & b_n & a_n \end{bmatrix}.$$

If $b_2 = 0$ or $b_n = 0$, then the 1×1 matrix $[a_1]$ or $[a_n]$ immediately produces an eigenvalue a_1 or a_n of A. When $b_j = 0$ for some j, where $2 < j < n$, the problem can be reduced to considering, instead of A, the smaller matrices

$$\begin{bmatrix} a_1 & b_2 & 0 & \cdots & 0 \\ b_2 & a_2 & b_3 & \ddots & \vdots \\ 0 & b_3 & a_3 & \ddots & 0 \\ \vdots & \ddots & \ddots & \ddots & b_{j-1} \\ 0 & \cdots & 0 & b_{j-1} & a_{j-1} \end{bmatrix} \quad \text{and} \quad \begin{bmatrix} a_j & b_{j+1} & 0 & \cdots & 0 \\ b_{j+1} & a_{j+1} & b_{j+2} & \ddots & \vdots \\ 0 & b_{j+2} & a_{j+2} & \ddots & 0 \\ \vdots & \ddots & \ddots & \ddots & b_n \\ 0 & \cdots & 0 & b_n & a_n \end{bmatrix}.$$

If none of the b_j are zero, the QR method proceeds by forming a sequence of matrices $A = A^{(1)}, A^{(2)}, A^{(3)}, \ldots,$ as follows:

1. $A^{(1)} = A$ is factored as a product $A^{(1)} = Q^{(1)}R^{(1)}$, where $Q^{(1)}$ is orthogonal and $R^{(1)}$ is upper triangular.
2. $A^{(2)}$ is defined as $A^{(2)} = R^{(1)}Q^{(1)}$.

In general, for $i \geq 2$, $A^{(i)}$ is factored as a product $A^{(i)} = Q^{(i)}R^{(i)}$ of an orthogonal matrix $Q^{(i)}$ and an upper triangular matrix $R^{(i)}$. Then $A^{(i+1)}$ is defined by the product of $R^{(i)}$ and $Q^{(i)}$ in the reverse direction $A^{(i+1)} = R^{(i)}Q^{(i)}$. Since $Q^{(i)}$ is orthogonal,

$$A^{(i+1)} = R^{(i)}Q^{(i)} = (Q^{(i)t}A^{(i)})Q^{(i)} = Q^{(i)t}A^{(i)}Q^{(i)},$$

and $A^{(i+1)}$ is symmetric with the same eigenvalues as $A^{(i)}$. By the way we define $R^{(i)}$ and $Q^{(i)}$, we can also ensure that $A^{(i+1)}$ is tridiagonal.

Continuing by induction, $A^{(i+1)}$ has the same eigenvalues as the original matrix A. The success of the procedure is a result of the fact that $A^{(i+1)}$ tends to a diagonal matrix with the eigenvalues of A along the diagonal.

To describe the construction of the factoring matrices $Q^{(i)}$ and $R^{(i)}$, we need the notion of a *rotation matrix*.

A **rotation matrix** P is an orthogonal matrix that differs from the identity matrix in at most four elements. These four elements are of the form

$$p_{ii} = p_{jj} = \cos\theta \quad \text{and} \quad p_{ij} = -p_{ji} = \sin\theta$$

for some θ and some $i \neq j$.

For any rotation matrix P, the matrix AP differs from A only in the ith and jth columns and the matrix PA differs from A only in the ith and jth rows. For any $i \neq j$, the angle θ can be chosen so that the product PA has a zero entry for $(PA)_{ij}$. In addition, P is orthogonal, since by its definition $PP^t = I$.

The factorization of $A^{(1)}$ into $A^{(1)} = Q^{(1)}R^{(1)}$ uses a product of $n-1$ rotation matrices of this type to construct

$$R^{(1)} = P_nP_{n-1}\cdots P_2A^{(1)}.$$

We first choose the rotation matrix P_2 to have

$$p_{11} = p_{22} = \cos\theta_2 \quad \text{and} \quad p_{12} = -p_{21} = \sin\theta_2,$$

where

$$\sin \theta_2 = \frac{b_2}{\sqrt{b_2^2 + a_1^2}} \quad \text{and} \quad \cos \theta_2 = \frac{a_1}{\sqrt{b_2^2 + a_1^2}}.$$

Then the matrix

$$A_2^{(1)} = P_2 A^{(1)}$$

has a zero in the (2, 1) position, that is, in the second row and first column, since the (2, 1) entry in $A_2^{(1)}$ is

$$(-\sin \theta_2)a_1 + (\cos \theta_2)b_2 = \frac{-b_2 a_1}{\sqrt{b_2^2 + a_1^2}} + \frac{a_1 b_2}{\sqrt{b_2^2 + a_1^2}} = 0.$$

Since the multiplication $P_2 A^{(1)}$ affects both rows 1 and 2 of $A^{(1)}$, the new matrix does not necessarily retain zero entries in positions $(1,3)$, $(1,4)$, ... , and $(1,n)$. However, $A^{(1)}$ is tridiagonal, so the $(1,4), \ldots, (1,n)$ entries of $A_2^{(1)}$ are zero. Only the $(1,3)$-entry, the one in the first row and third column, can become nonzero.

In general, the matrix P_k is chosen so that the $(k, k-1)$ entry in $A_k^{(1)} = P_k A_{k-1}^{(1)}$ is zero, which results in the $(k-1, k+1)$ entry becoming nonzero. The matrix $A_k^{(1)}$ has the form

$$A_k^{(1)} = \begin{bmatrix} z_1 & q_1 & r_1 & 0 & \cdots & \cdots & \cdots & \cdots & 0 \\ 0 & \ddots & \ddots & \ddots & \ddots & & & & \vdots \\ 0 & \ddots & \ddots & \ddots & \ddots & \ddots & & & \vdots \\ \vdots & & 0 & z_{k-1} & q_{k-1} & r_{k-1} & & & \vdots \\ \vdots & & & 0 & x_k & y_k & 0 & \ddots & \vdots \\ \vdots & & & & b_{k+1} & a_{k+1} & b_{k+2} & \ddots & 0 \\ \vdots & & & & & \ddots & \ddots & \ddots & 0 \\ \vdots & & & & & & \ddots & \ddots & b_n \\ 0 & \cdots & \cdots & \cdots & \cdots & \cdots & 0 & b_n & a_n \end{bmatrix}$$

and P_{k+1} has the form

$$P_{k+1} = \left[\begin{array}{c|cc|c} I_{k-1} & O & & O \\ \hline & c_{k+1} & s_{k+1} & \\ O & & & O \\ & -s_{k+1} & c_{k+1} & \\ \hline O & O & & I_{n-k-1} \end{array}\right] \begin{matrix} \\ \leftarrow \text{Row } k \\ \\ \\ \\ \end{matrix}$$

$\uparrow$
Column k

where O denotes the appropriately dimensioned matrix with all zero entries.

The constants $c_{k+1} = \cos\theta_{k+1}$ and $s_{k+1} = \sin\theta_{k+1}$ in P_{k+1} are chosen so that the $(k+1, k)$ entry in $A_{k+1}^{(1)}$ is zero; that is,

$$s_{k+1}x_k - c_{k+1}b_{k+1} = 0.$$

Since $c_{k+1}^2 + s_{k+1}^2 = 1$, the solution to this equation is

$$s_{k+1} = \frac{b_{k+1}}{\sqrt{b_{k+1}^2 + x_k^2}} \quad \text{and} \quad c_{k+1} = \frac{x_k}{\sqrt{b_{k+1}^2 + x_k^2}},$$

and $A_{k+1}^{(1)}$ has the form

$$A_{k+1}^{(1)} = \begin{bmatrix} z_1 & q_1 & r_1 & 0 & \cdots & \cdots & \cdots & \cdots & 0 \\ 0 & \ddots & \ddots & \ddots & \ddots & & & & \vdots \\ 0 & \ddots & \ddots & \ddots & \ddots & \ddots & & & \vdots \\ \vdots & & 0 & z_k & q_k & r_k & \ddots & & \vdots \\ \vdots & & & 0 & x_{k+1} & y_{k+1} & 0 & \ddots & \vdots \\ \vdots & & & & b_{k+2} & a_{k+2} & b_{k+3} & \ddots & 0 \\ \vdots & & & & & \ddots & \ddots & \ddots & 0 \\ \vdots & & & & & & \ddots & \ddots & b_n \\ 0 & \cdots & \cdots & \cdots & \cdots & \cdots & 0 & b_n & a_n \end{bmatrix}$$

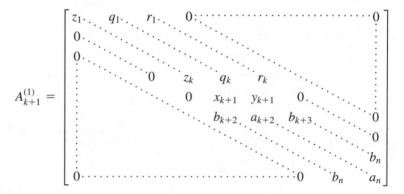

Proceeding with this construction in the sequence $P_2, \ldots, P_n$ produces the upper triangular matrix

$$R^{(1)} \equiv A_n^{(1)} = \begin{bmatrix} z_1 & q_1 & r_1 & 0 & \cdots & 0 \\ 0 & \ddots & \ddots & \ddots & \ddots & \vdots \\ \vdots & & \ddots & \ddots & \ddots & 0 \\ \vdots & & & \ddots & \ddots & r_{n-2} \\ \vdots & & & & z_{n-1} & q_{n-1} \\ 0 & \cdots & \cdots & \cdots & 0 & x_n \end{bmatrix}.$$

The orthogonal matrix $Q^{(1)}$ is defined as $Q^{(1)} = P_2^t P_3^t \cdots P_n^t$, so

$$A^{(2)} = R^{(1)}Q^{(1)} = R^{(1)}P_2^t P_3^t \cdots P_{n-1}^t = P_n P_{n-1} \cdots P_2 A P_2^t P_3^t \cdots P_n^t.$$

The matrix $A^{(2)}$ is tridiagonal, and, in general, the entries off the diagonal will be smaller in magnitude than the corresponding entries in $A^{(1)}$. The process is repeated to construct $A^{(3)}$, $A^{(4)}$, and so on.

If the eigenvalues of A have distinct moduli with $|\lambda_1| > |\lambda_2| > \cdots > |\lambda_n|$, then the rate of convergence of the entry $b_{j+1}^{(i+1)}$ to zero in the matrix $A^{(i+1)}$ depends on the ratio $|\lambda_{j+1}/\lambda_j|$. The rate of convergence of $b_{j+1}^{(i+1)}$ to zero determines the rate at which the entry $a_j^{(i+1)}$ converges to the jth eigenvalue λ_j. Thus, the rate of convergence can be slow if $|\lambda_{j+1}/\lambda_j|$ is close to unity.

To accelerate the convergence, a shifting technique is employed similar to that used with the Inverse Power method. A constant s is selected close to an eigenvalue of A. This modifies the factorization to choosing $Q^{(i)}$ and $R^{(i)}$ so that

$$A^{(i)} - sI = Q^{(i)}R^{(i)},$$

and correspondingly, the matrix $A^{(i+1)}$ is defined to be

$$A^{(i+1)} = R^{(i)}Q^{(i)} + sI.$$

With this modification, the rate of convergence of $b_{j+1}^{(i+1)}$ to zero depends on the ratio $|(\lambda_{j+1} - s)/\lambda_j - s|$, which can result in a significant improvement over the original rate of convergence of $a_j^{(i+1)}$ to λ_j if s is close to λ_{j+1} but not close to λ_j.

The shift s is computed at each step as the eigenvalue closest to $a_n^{(i)}$ of the 2×2 matrix

$$\begin{bmatrix} a_{n-1}^{(i)} & b_n^{(i)} \\ b_n^{(i)} & a_n^{(i)} \end{bmatrix}.$$

With this shift, $b_n^{(i+1)}$ converges to zero and $a_n^{(i+1)}$ converges to the eigenvalue λ_n. If $b_j^{(i+1)}$ converges to zero for some $j \neq n$, the splitting technique is employed. The shifts are accumulated and added to the approximation after convergence is obtained. The program QRSYMT96 performs the QR method for symmetric tridiagonal matrices in this way.

EXAMPLE 1 Let

$$A = \begin{bmatrix} 3 & 1 & 0 \\ 1 & 3 & 1 \\ 0 & 1 & 3 \end{bmatrix} = \begin{bmatrix} a_1^{(1)} & b_2^{(1)} & 0 \\ b_2^{(1)} & a_2^{(1)} & b_3^{(1)} \\ 0 & b_3^{(1)} & a_3^{(1)} \end{bmatrix}.$$

To find the acceleration parameter for shifting requires finding the eigenvalues of

$$\begin{bmatrix} a_2^{(1)} & b_3^{(1)} \\ b_3^{(1)} & a_3^{(1)} \end{bmatrix} = \begin{bmatrix} 3 & 1 \\ 1 & 3 \end{bmatrix},$$

which are $\mu_1 = 4$ and $\mu_2 = 2$. The choice of eigenvalue closest to $a_3^{(1)} = 3$ is arbitrary, and we choose $\mu_2 = 2$ and shift by this amount. Then $s_1 = 2$ and

$$\begin{bmatrix} d_1 & b_2^{(1)} & 0 \\ b_2^{(1)} & d_2 & b_3^{(1)} \\ 0 & b_3^{(1)} & d_3 \end{bmatrix} = \begin{bmatrix} 1 & 1 & 0 \\ 1 & 1 & 1 \\ 0 & 1 & 1 \end{bmatrix}.$$

Continuing the computation gives

$$x_1 = 1, \quad y_1 = 1, \quad z_1 = \sqrt{2}, \quad c_2 = \frac{\sqrt{2}}{2}, \quad s_2 = \frac{\sqrt{2}}{2},$$
$$q_1 = \sqrt{2}, \quad x_2 = 0, \quad r_1 = \frac{\sqrt{2}}{2}, \quad \text{and} \quad y_2 = \frac{\sqrt{2}}{2},$$

so

$$A_2^{(1)} = \begin{bmatrix} \sqrt{2} & \sqrt{2} & \frac{\sqrt{2}}{2} \\ 0 & 0 & \sqrt{2} \\ 0 & 1 & 1 \end{bmatrix}.$$

Further,

$$z_2 = 1, \quad c_3 = 0, \quad s_3 = 1, \quad q_2 = 1, \quad \text{and} \quad x_3 = -\frac{\sqrt{2}}{2},$$

so

$$R^{(1)} = A_3^{(1)} = \begin{bmatrix} \sqrt{2} & \sqrt{2} & \frac{\sqrt{2}}{2} \\ 0 & 1 & 1 \\ 0 & 0 & -\frac{\sqrt{2}}{2} \end{bmatrix}.$$

To compute $A^{(2)}$, we have

$$z_3 = -\frac{\sqrt{2}}{2}, \quad a_1^{(2)} = 2, \quad b_2^{(2)} = \frac{\sqrt{2}}{2}, \quad a_2^{(2)} = 1, \quad b_3^{(2)} = -\frac{\sqrt{2}}{2}, \quad \text{and} \quad a_3^{(2)} = 0,$$

so

$$A^{(2)} = R^{(1)}Q^{(1)} = \begin{bmatrix} 2 & \frac{\sqrt{2}}{2} & 0 \\ \frac{\sqrt{2}}{2} & 1 & -\frac{\sqrt{2}}{2} \\ 0 & -\frac{\sqrt{2}}{2} & 0 \end{bmatrix}.$$

One iteration of the *QR* method is complete. Since neither $b_2^{(2)} = \sqrt{2}/2$ nor $b_3^{(2)} = -\sqrt{2}/2$ is small, another iteration of the *QR* method will be performed. For this iteration we calculate the eigenvalues $\frac{1}{2} \pm \frac{1}{2}\sqrt{3}$ of the matrix

$$\begin{bmatrix} a_2^{(2)} & b_3^{(2)} \\ b_3^{(2)} & a_3^{(2)} \end{bmatrix} = \begin{bmatrix} 1 & -\frac{\sqrt{2}}{2} \\ -\frac{\sqrt{2}}{2} & 0 \end{bmatrix},$$

and choose $s_2 = \frac{1}{2} - \frac{1}{2}\sqrt{3}$, the closest eigenvalue to $a_3^{(2)} = 0$. Completing the calculations gives

$$A^{(3)} = \begin{bmatrix} 2.6720277 & 0.37597448 & 0 \\ 0.37597448 & 1.4736080 & 0.030396964 \\ 0 & 0.030396964 & -0.047559530 \end{bmatrix}.$$

If $b_3^{(3)} = 0.030396964$ is sufficiently small, then the approximation to the eigenvalue λ_3 is 1.5864151, the sum of $a_3^{(3)}$ and the shifts $s_1 + s_2 = 2 + (1 - \sqrt{3})/2$. Deleting the third row and column gives

$$A^{(3)} = \begin{bmatrix} 2.6720277 & 0.37597448 \\ 0.37597448 & 1.4736080 \end{bmatrix},$$

which has eigenvalues $\mu_1 = 2.7802140$ and $\mu_2 = 1.3654218$. Adding the shifts gives the approximations

$$\lambda_1 \approx 4.4141886 \quad \text{and} \quad \lambda_2 \approx 2.9993964.$$

Since the actual eigenvalues of the matrix A are 4.41420, 3.00000, and 1.58579, the QR method gave four significant digits of accuracy in only two iterations. ◆ ◆ ◆

EXERCISE SET 9.5

1. Apply two iterations of the QR method to the following matrices.

   a. $\begin{bmatrix} 2 & -1 & 0 \\ -1 & 2 & -1 \\ 0 & -1 & 2 \end{bmatrix}$ b. $\begin{bmatrix} 3 & 1 & 0 \\ 1 & 4 & 2 \\ 0 & 2 & 1 \end{bmatrix}$

   c. $\begin{bmatrix} 4 & -1 & 0 \\ -1 & 3 & -1 \\ 0 & -1 & 2 \end{bmatrix}$ d. $\begin{bmatrix} 1 & 1 & 0 & 0 \\ 1 & 2 & -1 & 0 \\ 0 & -1 & 3 & 1 \\ 0 & 0 & 1 & 4 \end{bmatrix}$

   e. $\begin{bmatrix} -2 & 1 & 0 & 0 \\ 1 & -3 & -1 & 0 \\ 0 & -1 & 1 & 1 \\ 0 & 0 & 1 & 3 \end{bmatrix}$ f. $\begin{bmatrix} 0.5 & 0.25 & 0 & 0 \\ 0.25 & 0.8 & 0.4 & 0 \\ 0 & 0.4 & 0.6 & 0.1 \\ 0 & 0 & 0.1 & 1 \end{bmatrix}$

2. Use the QR method to determine all the eigenvalues of the following matrices. Iterate until all the off-diagonal elements have magnitude less than 10^{-5}.

   a. $\begin{bmatrix} 2 & -1 & 0 \\ -1 & -1 & -2 \\ 0 & -2 & 3 \end{bmatrix}$ b. $\begin{bmatrix} 3 & 1 & 0 \\ 1 & 4 & 2 \\ 0 & 2 & 3 \end{bmatrix}$

   c. $\begin{bmatrix} 4 & 2 & 0 & 0 & 0 \\ 2 & 4 & 2 & 0 & 0 \\ 0 & 2 & 4 & 2 & 0 \\ 0 & 0 & 2 & 4 & 2 \\ 0 & 0 & 0 & 2 & 4 \end{bmatrix}$ d. $\begin{bmatrix} 5 & -1 & 0 & 0 & 0 \\ -1 & 4.5 & 0.2 & 0 & 0 \\ 0 & 0.2 & 1 & -0.4 & 0 \\ 0 & 0 & -0.4 & 3 & 1 \\ 0 & 0 & 0 & 1 & 3 \end{bmatrix}$

3. Use the QR method given in QRSYMT96 with $TOL = 10^{-5}$ to determine all the eigenvalues for the matrices given in Exercise 1.

4. Use the Inverse Power method given in INVPWR93 with $TOL = 10^{-5}$ to determine the eigenvectors of the matrices in Exercise 1.

5. a. Show that the rotation matrix $\begin{bmatrix} \cos\theta & -\sin\theta \\ \sin\theta & \cos\theta \end{bmatrix}$ applied to the vector $\mathbf{x} = (x_1, x_2)^t$ has the geometric effect of rotating $\mathbf{x}$ through the angle θ without changing its magnitude with respect to $\|\cdot\|_2$.

   b. Show that the magnitude of $\mathbf{x}$ with respect to $\|\cdot\|_\infty$ can be changed by a rotation matrix.

6. Let P be the rotation matrix with $p_{ii} = p_{jj} = \cos\theta$, $p_{ij} = -p_{ji} = \sin\theta$ for $j < i$. Show that for any $n \times n$ matrix A:

$$(AP)_{pq} = \begin{cases} a_{pq}, & \text{if } q \neq i, j, \\ (\cos\theta)a_{pj} + (\sin\theta)a_{pi}, & \text{if } q = j, \\ (\cos\theta)a_{pi} - (\sin\theta)a_{pj}, & \text{if } q = i, \end{cases}$$

$$(PA)_{pq} = \begin{cases} a_{pq}, & \text{if } p \neq i, j, \\ (\cos\theta)a_{jq} - (\sin\theta)a_{iq}, & \text{if } p = j, \\ (\sin\theta)a_{jq} + (\cos\theta)a_{iq}, & \text{if } p = i. \end{cases}$$

9.6 Survey of Methods and Software

This chapter discussed the approximation of eigenvalues and eigenvectors of a matrix. The Gerschgorin circles give a crude approximation to the location of the eigenvalues. The Power method can be used to find the dominant eigenvalue and an associated eigenvector for an arbitrary matrix A. If A is symmetric, the Symmetric Power method gives faster convergence to the dominant eigenvalue and an associated eigenvector. The Inverse Power method will find the eigenvalue closest to a given value and an associated eigenvector. This method is often used to refine an approximate eigenvalue and to compute an eigenvector once an eigenvalue has been found by some other technique.

Deflation methods, such as Wielandt deflation, obtain other eigenvalues once the dominant eigenvalue is known. These methods are used if only a few eigenvalues are required since they are susceptible to round-off error. The Inverse Power method should be used to improve the accuracy of approximate eigenvalues obtained from a deflation technique.

Methods based on similarity transformations, such as Householder's method, are used to convert a symmetric matrix into a similar matrix that is tridiagonal (or *upper Hessenberg* if the matrix is not symmetric). Techniques such as the *QR* method can then be applied to the tridiagonal (or *upper Hessenberg*) matrix to obtain approximations to all the eigenvalues. The associated eigenvectors can be found by using an iterative method, such as the Inverse Power method, or by applying the transformations to eigenvectors obtained directly from the *QR* method. We restricted our study to symmetric matrices and presented the *QR* method only to compute eigenvalues for the symmetric case.

The subroutines in the IMSL and NAG libraries are based on those contained in LAPACK. In general, the subroutines transform a matrix into the appropriate form for the *QR* method or one of its modifications, such as the *QL* method. The subroutines approximate all the eigenvalues and can approximate an associated eigenvector for each eigenvalue. There are special routines that find all the eigenvalues within an interval or region or find only the largest or smallest eigenvalue. Subroutines are also available to approximate the accuracy of the eigenvalue and the sensitivity of the process to round-off error.

The Maple procedure `Eigenvals(A)` computes the eigenvalues of A by first balancing and then transforming A to *upper Hessenberg* form. The QR method is then applied to obtain all eigenvalues and eigenvectors. The tridiagonal form is used for a symmetric matrix. Maple has the command `QRdecomp` to compute the QR factorization of a matrix. If the matrix A has been created the function call

```
>R:=QRdecomp(A,Q='G');
```

returns the upper triangular matrix R as the value of the function and returns the orthonormal matrix Q in G.

The books by Wilkinson [Wi2] and Wilkinson and Reinsch [WR] are classics in the study of eigenvalue problems. Stewart [Ste] is also a good source of information on the general problem and Parlett [Pa] considers the symmetric problem. A study of the nonsymmetric problem can be found in Saad [Sa1].

CHAPTER 10

Solutions of Systems of Nonlinear Equations

10.1 Introduction

A large part of the material in this book has involved the solution of systems of equations. Even so, to this point, the methods have been appropriate only for systems of *linear* equations, equations in the variables $x_1, x_2, \ldots, x_n$ of the form

$$a_{i1}x_1 + a_{i2}x_2 + \cdots + a_{in}x_n = b_i$$

for $i = 1, 2, \ldots, n$. If you have wondered why we have not considered more general systems of equations, the reason is simple. It is much harder to approximate the solutions to a system of general, or *nonlinear*, equations.

Solving a system of nonlinear equations is a problem that is avoided when possible, customarily by approximating the nonlinear system by a system of linear equations. When this is unsatisfactory, the problem must be tackled directly. The most straightforward method of approach is to adapt the methods from Chapter 2 that approximate the solutions of a single nonlinear equation in one variable to apply when the single-variable problem is replaced by a vector problem that incorporates all the variables.

The principal tool in Chapter 2 was Newton's method, a technique that is generally quadratically convergent. This is the first technique we modify to solve systems of nonlinear equations. Newton's method, as modified for systems of equations, is quite costly to apply, so in Section 10.3 we describe how a modified Secant method can be used to obtain approximations more easily, although with a loss of the extremely rapid convergence that Newton's method provides.

Section 10.4 describes the method of Steepest Descent. This technique is only linearly convergent, but it does not require the accurate starting approximations needed for more rapidly converging techniques. It is often used to find a good initial approximation for Newton's method or one of its modifications.

A system of nonlinear equations has the form

$$
\begin{aligned}
f_1(x_1, x_2, \dots, x_n) &= 0,\\
f_2(x_1, x_2, \dots, x_n) &= 0,\\
\vdots \qquad\quad &\quad \vdots\\
f_n(x_1, x_2, \dots, x_n) &= 0,
\end{aligned}
$$

where each function f_i can be thought of as mapping a vector $\mathbf{x} = (x_1, x_2, \dots, x_n)^t$ of the n-dimensional space $\mathbb{R}^n$ into the real line $\mathbb{R}$. A geometric representation of a nonlinear system when $n = 2$ is given in Figure 10.1.

Figure 10.1

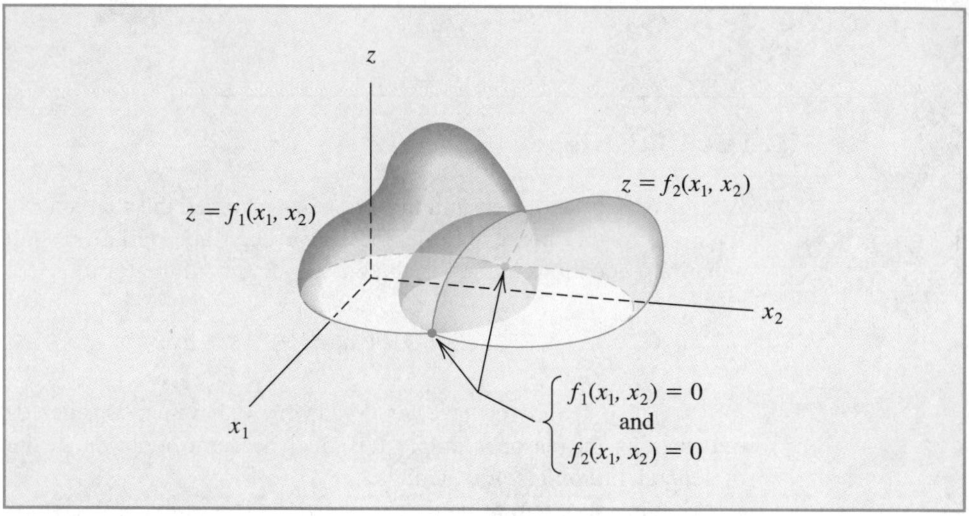

A general system of n nonlinear equations in n unknowns can be represented alternatively by defining a function $\mathbf{F}$, mapping $\mathbb{R}^n$ into $\mathbb{R}^n$, by

$$\mathbf{F}(x_1, x_2, \dots, x_n) = (f_1(x_1, x_2, \dots, x_n), f_2(x_1, x_2, \dots, x_n), \dots, f_n(x_1, x_2, \dots, x_n))^t .$$

If vector notation is used to represent the variables $x_1, x_2, \dots, x_n$, the nonlinear system assumes the form

$$\mathbf{F}(\mathbf{x}) = \mathbf{0}.$$

The functions $f_1, f_2, \dots, f_n$ are the **coordinate functions** of $\mathbf{F}$.

EXAMPLE 1 The 3×3 nonlinear system

$$3x_1 - \cos(x_2x_3) - \frac{1}{2} = 0,$$
$$x_1^2 - 81(x_2 + 0.1)^2 + \sin x_3 + 1.06 = 0,$$
$$e^{-x_1x_2} + 20x_3 + \frac{10\pi - 3}{3} = 0$$

can be placed in the form $\mathbf{F}(\mathbf{x}) = \mathbf{0}$ by first defining the three coordinate functions f_1, f_2, and f_3 from $\mathbb{R}^3$ to $\mathbb{R}$ as

$$f_1(x_1, x_2, x_3) = 3x_1 - \cos(x_2x_3) - \frac{1}{2},$$
$$f_2(x_1, x_2, x_3) = x_1^2 - 81(x_2 + 0.1)^2 + \sin x_3 + 1.06,$$
$$f_3(x_1, x_2, x_3) = e^{-x_1x_2} + 20x_3 + \frac{10\pi - 3}{3}$$

and then defining $\mathbf{F}$ from $\mathbb{R}^3 \to \mathbb{R}^3$ by

$$\begin{aligned}\mathbf{F}(\mathbf{x}) &= \mathbf{F}(x_1, x_2, x_3)\\ &= (f_1(x_1, x_2, x_3), f_2(x_1, x_2, x_3), f_3(x_1, x_2, x_3))^t\\ &= \left(3x_1 - \cos(x_1x_2) - \frac{1}{2}, x_1^2 - 81(x_2 + 0.1)^2 + \sin x_3 + 1.06,\right.\\ &\qquad \left. e^{-x_1x_2} + 20x_3 + \frac{10\pi - 3}{3}\right)^t.\end{aligned}$$

◆ ◆ ◆

Before discussing the solution of a system of nonlinear equations, we need some results concerning continuity and differentiability of functions from $\mathbb{R}^n$ into $\mathbb{R}$ and $\mathbb{R}^n$ into $\mathbb{R}^n$. These results parallel those given in Section 1.2 for a function from $\mathbb{R}$ into $\mathbb{R}$.

Let f be a function defined on a set $D \subset \mathbb{R}^n$ and mapping $\mathbb{R}^n$ into $\mathbb{R}$. The function f has the **limit** L at $\mathbf{x}_0$, written

$$\lim_{\mathbf{x} \to \mathbf{x}_0} f(\mathbf{x}) = L,$$

if, given any number $\varepsilon > 0$, a number $\delta > 0$ exists with the property that

$$|f(\mathbf{x}) - L| < \varepsilon \qquad \text{whenever } \mathbf{x} \in D \quad \text{and} \quad 0 < \|\mathbf{x} - \mathbf{x}_0\| < \delta.$$

Any convenient norm can be used to satisfy the condition in this definition. The specific value of δ will depend on the norm chosen, but the existence of δ is independent of the norm.

The function f from $\mathbb{R}^n$ into $\mathbb{R}$ is **continuous** at $\mathbf{x}_0 \in D$ provided $\lim_{\mathbf{x}\to\mathbf{x}_0} f(\mathbf{x})$ exists and is $f(\mathbf{x}_0)$. In addition, f is **continuous on a set** D provided f is continuous at every point of D. This is expressed by writing $f \in C(D)$.

We define the limit and continuity concepts for functions from $\mathbb{R}^n$ into $\mathbb{R}^n$ by considering the coordinate functions from $\mathbb{R}^n$ into $\mathbb{R}$.

Let $\mathbf{F}$ be a function from $D \subset \mathbb{R}^n$ into $\mathbb{R}^n$ and suppose $\mathbf{F}$ has the representation

$$\mathbf{F}(\mathbf{x}) = (f_1(\mathbf{x}), f_2(\mathbf{x}), \ldots, f_n(\mathbf{x}))^t.$$

We define the limit of $\mathbf{F}$ from $\mathbb{R}^n$ to $\mathbb{R}^n$ as

$$\lim_{\mathbf{x}\to\mathbf{x}_0} \mathbf{F}(\mathbf{x}) = \mathbf{L} = (L_1, L_2, \ldots, L_n)^t$$

if and only if $\lim_{\mathbf{x}\to\mathbf{x}_0} f_i(\mathbf{x}) = L_i$ for each $i = 1, 2, \ldots, n$.

The function $\mathbf{F}$ is **continuous** at $\mathbf{x}_0 \in D$ provided $\lim_{\mathbf{x}\to\mathbf{x}_0} \mathbf{F}(\mathbf{x})$ exists and is $\mathbf{F}(\mathbf{x}_0)$. In addition, $\mathbf{F}$ is **continuous on the set** D if $\mathbf{F}$ is continuous at each $\mathbf{x}$ in D.

Maple provides the function `fsolve` to solve systems of equations. For example, the nonlinear system

$$\begin{aligned} x_1 + \cos(x_1x_2x_3) &= 1, \\ (1 - x_1)^{1/4} + x_2 + 0.05x_3^2 - 0.15x_3 &= 1, \\ -x_1^2 - 0.1x_2^2 + 0.01x_2 + x_3 &= 1 \end{aligned}$$

has a solution $x_1 = 0$, $x_2 = 0.1$, and $x_3 = 1.0$. The system can be solved with the following commands:

```
>f1:=x1+cos(x1*x2*x3)=1;
>f2:=(1-x1)^0.25+x2+0.05*x3^2-0.15*x3=1;
>f3:=-x1^2-0.1*x2^2+0.01*x2+x3=1;
>fsolve({f1,f2,f3},{x1,x2,x3},{x1=-1..1,x2=0..1,x3=0.5..1.5});
```

The first three commands define the system, and the last command invokes the procedure `fsolve`, which gives the answer

$$x_1 = -0.162638 \times 10^{-17},\ x_2 = 0.1,\ \text{and } x_3 = 1.$$

In general, `fsolve(eqns,vars,options)` solves the system of equations represented by the parameter `eqns` for the variables represented by the parameter `vars` under optional parameters represented by `options`. Under `options` we specify a region in which the routine is required to search for a solution. This specification is not mandatory, and Maple determines its own search space if the options are omitted.

10.2 Newton's Method for Systems

Newton's method for approximating the solution p to the single nonlinear equation

$$f(x) = 0$$

requires an initial approximation p_0 to p and generates a sequence defined by

$$p_k = p_{k-1} - \frac{f(p_{k-1})}{f'(p_{k-1})}, \quad \text{for } k \geq 1.$$

To modify this technique to find the vector solution $\mathbf{p}$ to the vector equation

$$\mathbf{F}(\mathbf{x}) = \mathbf{0},$$

we first need to determine an initial approximation vector $\mathbf{p}^{(0)}$. We must then decide how to modify the single-variable Newton's method to a vector function method that will have the same convergence properties but not require division, since this operation is not defined for vectors. We must also replace the derivative of f in the single-variable version of Newton's method with something that is appropriate for the vector function $\mathbf{F}$.

The derivative f' of the single-variable function f describes how the values of the function change relative to a change in its independent variable. The vector function $\mathbf{F}$ has n different variables, $x_1, x_2, \ldots, x_n$, and n different component functions, $f_1, f_2, \ldots, f_n$, each of which can change as any one of the variables change. The appropriate derivative modification from the single-variable Newton's method to the vector form must involve all these n^2 possible changes. The natural way to represent n^2 items is by an $n \times n$ matrix. Each change in a component function f_i with respect to the change in the variable x_j is described by the partial derivative

$$\frac{\partial f_i}{\partial x_j},$$

and the $n \times n$ matrix that replaces the derivative that occurs in the single-variable case is

$$J(\mathbf{x}) = \begin{bmatrix} \frac{\partial f_1(\mathbf{x})}{\partial x_1} & \frac{\partial f_1(\mathbf{x})}{\partial x_2} & \cdots & \frac{\partial f_1(\mathbf{x})}{\partial x_n} \\ \frac{\partial f_2(\mathbf{x})}{\partial x_1} & \frac{\partial f_2(\mathbf{x})}{\partial x_2} & \cdots & \frac{\partial f_2(\mathbf{x})}{\partial x_n} \\ \vdots & \vdots & & \vdots \\ \frac{\partial f_n(\mathbf{x})}{\partial x_1} & \frac{\partial f_n(\mathbf{x})}{\partial x_2} & \cdots & \frac{\partial f_n(\mathbf{x})}{\partial x_n} \end{bmatrix}.$$

The matrix $J(\mathbf{x})$ is called the **Jacobian** matrix and has a number of applications in analysis. It might, in particular, be familiar due to its application in the multiple integration of a function of several variables over a region that requires a change of variables to be performed.

Newton's method for systems replaces the derivative in the single-variable case with the $n \times n$ Jacobian matrix in the vector situation and substitutes the division by the derivative with the inversion of the Jacobian matrix. As a consequence, Newton's method for finding the solution $\mathbf{p}$ to the nonlinear system of equations represented by the vector equation $\mathbf{F}(\mathbf{x}) = \mathbf{0}$ has the form

$$\mathbf{p}^{(k)} = \mathbf{p}^{(k-1)} - [J(\mathbf{p}^{(k-1)})]^{-1}\mathbf{F}(\mathbf{p}^{(k-1)}), \quad \text{for } k \geq 1,$$

given the initial approximation $\mathbf{p}^{(0)}$ to the solution $\mathbf{p}$.

A weakness in Newton's method for systems arises from the necessity of inverting the matrix $J(\mathbf{p}^{(k-1)})$ at each iteration. In practice, explicit computation of the inverse of $J(\mathbf{p}^{(k-1)})$ is avoided by performing the operation in a two-step manner. First, a vector $\mathbf{y}^{(k-1)}$ is found that will satisfy

$$J(\mathbf{p}^{(k-1)})\mathbf{y}^{(k-1)} = -\mathbf{F}(\mathbf{p}^{(k-1)}).$$

After this has been accomplished, the new approximation, $\mathbf{p}^{(k)}$, is obtained by adding $\mathbf{y}^{(k-1)}$ to $\mathbf{p}^{(k-1)}$. The general implementation of Newton's method for systems of nonlinear equations can be performed using the program NWTSY101.

EXAMPLE 1 The nonlinear system

$$\begin{aligned} 3x_1 - \cos(x_2x_3) - \frac{1}{2} &= 0, \\ x_1^2 - 81(x_2+0.1)^2 + \sin x_3 + 1.06 &= 0, \\ e^{-x_1x_2} + 20x_3 + \frac{10\pi - 3}{3} &= 0 \end{aligned}$$

has an approximate solution at $(0.5, 0, -0.52359877)^t$. Newton's method will be used to obtain this approximation when the initial approximation is $\mathbf{p}^{(0)} = (0.1, 0.1, -0.1)^t$, and

$$\mathbf{F}(x_1, x_2, x_3) = (f_1(x_1, x_2, x_3), f_2(x_1, x_2, x_3), f_3(x_1, x_2, x_3))^t$$

where

$$\begin{aligned} f_1(x_1, x_2, x_3) &= 3x_1 - \cos(x_2x_3) - \frac{1}{2}, \\ f_2(x_1, x_2, x_3) &= x_1^2 - 81(x_2+0.1)^2 + \sin x_3 + 1.06, \end{aligned}$$

and

$$f_3(x_1, x_2, x_3) = e^{-x_1x_2} + 20x_3 + \frac{10\pi - 3}{3}.$$

The Jacobian matrix, $J(\mathbf{x})$, for this system is

$$J(x_1, x_2, x_3) = \begin{bmatrix} 3 & x_3 \sin x_2x_3 & x_2 \sin x_2x_3 \\ 2x_1 & -162(x_2+0.1) & \cos x_3 \\ -x_2e^{-x_1x_2} & -x_1e^{-x_1x_2} & 20 \end{bmatrix}$$

and

$$\begin{bmatrix} p_1^{(k)} \\ p_2^{(k)} \\ p_3^{(k)} \end{bmatrix} = \begin{bmatrix} p_1^{(k-1)} \\ p_2^{(k-1)} \\ p_3^{(k-1)} \end{bmatrix} + \begin{bmatrix} y_1^{(k-1)} \\ y_2^{(k-1)} \\ y_3^{(k-1)} \end{bmatrix},$$

where

$$\begin{bmatrix} y_1^{(k-1)} \\ y_2^{(k-1)} \\ y_3^{(k-1)} \end{bmatrix} = -\left(J(p_1^{(k-1)}, p_2^{(k-1)}, p_3^{(k-1)})\right)^{-1} \mathbf{F}\left(p_1^{(k-1)}, p_2^{(k-1)}, p_3^{(k-1)}\right).$$

Thus, at the kth step, the linear system $J(\mathbf{p}^{(k-1)})\mathbf{y}^{(k-1)} = -\mathbf{F}(\mathbf{p}^{(k-1)})$ must be solved, where

$$J(\mathbf{p}^{(k-1)}) = \begin{bmatrix} 3 & p_3^{(k-1)} \sin p_2^{(k-1)} p_3^{(k-1)} & p_2^{(k-1)} \sin p_2^{(k-1)} p_3^{(k-1)} \\ 2p_1^{(k-1)} & -162(p_2^{(k-1)} + 0.1) & \cos p_3^{(k-1)} \\ -p_2^{(k-1)} e^{-p_1^{(k-1)} p_2^{(k-1)}} & -p_1^{(k-1)} e^{-p_1^{(k-1)} p_2^{(k-1)}} & 20 \end{bmatrix}$$

and

$$\mathbf{F}(\mathbf{p}^{(k-1)}) = \begin{bmatrix} 3p_1^{(k-1)} - \cos p_2^{(k-1)} p_3^{(k-1)} - \frac{1}{2} \\ (p_1^{(k-1)})^2 - 81(p_2^{(k-1)} + 0.1)^2 + \sin p_3^{(k-1)} + 1.06 \\ e^{-p_1^{(k-1)} p_2^{(k-1)}} + 20p_3^{(k-1)} + \frac{10\pi - 3}{3} \end{bmatrix}.$$

To define the nonlinear system in Maple we use the commands

```
>f1:=3*x1-cos(x2*x3)-0.5;
>f2:=x1^2-81*(x2+0.1)^2+sin(x3)+1.06;
>f3:=exp(-x1*x2)+20*x3+(10*Pi-3)/3;
```

We compute the entries in the Jacobian matrix by

```
>j11:=diff(f1,x1); j12:=diff(f1,x2); j13:=diff(f1,x3);
>j21:=diff(f2,x1); j22:=diff(f2,x2); j23:=diff(f2,x3);
>j31:=diff(f3,x1); j32:=diff(f3,x2); j33:=diff(f3,x3);
```

The vector $\mathbf{p}^{(0)}$ is defined by

```
x:=vector(3,[0.1,0.1,-0.1]);
```

We evaluate the entries in the Jacobian with the commands

```
>b1:=evalf(subs(x1=x[1],x2=x[2],x3=x[3],j11));
>b2:=evalf(subs(x1=x[1],x2=x[2],x3=x[3],j12));
>b3:=evalf(subs(x1=x[1],x2=x[2],x3=x[3],j13));
>b4:=evalf(subs(x1=x[1],x2=x[2],x3=x[3],j21));
>b5:=evalf(subs(x1=x[1],x2=x[2],x3=x[3],j22));
>b6:=evalf(subs(x1=x[1],x2=x[2],x3=x[3],j23));
>b7:=evalf(subs(x1=x[1],x2=x[2],x3=x[3],j31));
>b8:=evalf(subs(x1=x[1],x2=x[2],x3=x[3],j32));
>b9:=evalf(subs(x1=x[1],x2=x[2],x3=x[3],j33));
```

Collecting the entries in the matrix $A = J\mathbf{p}^{(0)}$ is accomplished by

```
>A:=matrix(3,3,[b1,b2,b3,b4,b5,b6,b7,b8,b9]);
```

to produce

$$A = \begin{bmatrix} 3 & 0.0009999833334 & -0.0009999833334 \\ 0.2 & -32.4 & 0.9950041653 \\ -0.9900498337 & -0.09900498337 & 20 \end{bmatrix}.$$

We obtain $\mathbf{v} = -F(\mathbf{p}^{(0)})$ with

```
>c1:=evalf(subs(x1=x[1],x2=x[2],x3=x[3],f1));
>c2:=evalf(subs(x1=x[1],x2=x[2],x3=x[3],f2));
>c3:=evalf(subs(x1=x[1],x2=x[2],x3=x[3],f3));
>v:=vector(3,[-c1,-c2,-c3]);
```

giving

$$\mathbf{v} = (1.19995, 2.269833417, -8.462025344)^t.$$

To solve the linear system $J(\mathbf{p}^{(0)})\mathbf{y}^{(0)} = -F(\mathbf{p}^{(0)})$ requires the commands

```
>M:=augment(A,v);
>N:=gausselim(M);
>y:=backsub(N);
```

which produces

$$\mathbf{y} = (0.3998696727, -0.08053315145, -0.4215204719)^t.$$

We obtain $\mathbf{x}^{(1)}$ by adding $\mathbf{y}$ to $\mathbf{x}^{(0)}$, giving

$$\mathbf{x}^{(1)} = (0.4998696727, 0.01946684855, -0.5215204719)^t.$$

◆ ◆ ◆

The results using this iterative procedure are shown in Table 10.1.

Table 10.1

k	$p_1^{(k)}$	$p_2^{(k)}$	$p_3^{(k)}$	$\|\mathbf{p}^{(k)} - \mathbf{p}^{(k-1)}\|_\infty$
0	0.10000000	0.10000000	−0.10000000	
1	0.49986967	0.01946684	−0.52152047	0.422
2	0.50001424	0.00158859	−0.52355696	1.79×10^{-2}
3	0.50000011	0.00001244	−0.52359845	1.58×10^{-3}
4	0.50000000	0.00000000	−0.52359877	1.24×10^{-5}
5	0.50000000	0.00000000	−0.52359877	8.04×10^{-10}

The previous example illustrates that Newton's method can converge very rapidly once an approximation is obtained that is near the true solution. However, it is not always easy to determine starting values that will lead to a solution, and the method is computationally expensive. In the next section, we consider a method for overcoming the latter weakness. Good starting values can usually be found by the method discussed in Section 10.4.

Initial approximation to the solutions of 2×2 and often 3×3 nonlinear systems can also be obtained using the graphing facilities of Maple. The nonlinear system

$$\begin{aligned} x_1^2 - x_2^2 + 2x_2 &= 0, \\ 2x_1 + x_2^2 - 6 &= 0. \end{aligned}$$

has two solutions, $(0.625204094, 2.179355825)^t$ and $(2.109511920, -1.334532188)^t$. To use Maple we first define the two equations

```
>eq1:=x1^2-x2^2+2*x2=0;
>eq2:=2*x1+x2^2-6=0;
```

To obtain a graph of the two equations for $-3 \leq x_1, x_2 \leq 3$, enter the commands

```
>with(plots);
>implicitplot({eq1,eq2},x1=-3..3,x2=-3..3);
```

From the graph shown in Figure 10.2, we are able to estimate that there are solutions near $(0.64, 2.2)^t$ and $(2.1, -1.3)^t$. This gives us good starting values for Newton's method.

Figure 10.2

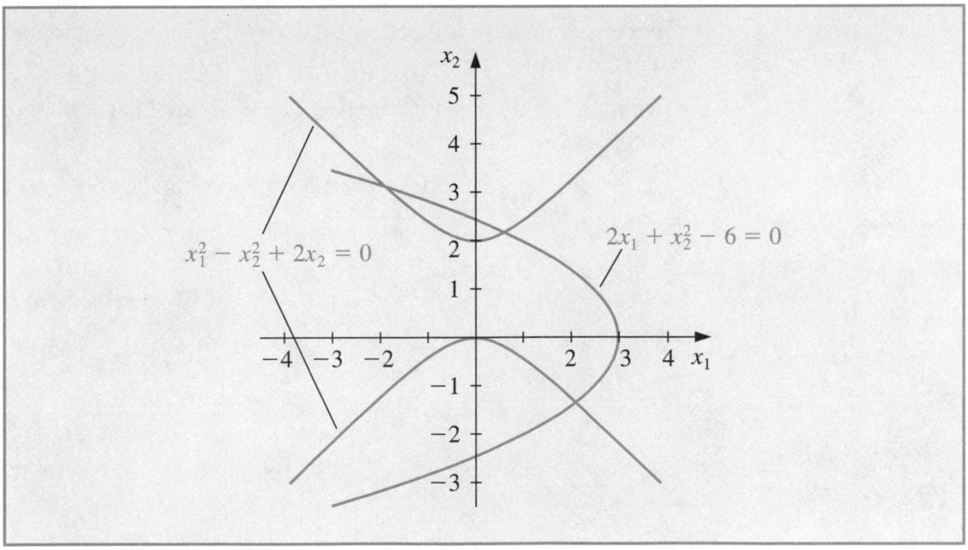

The problem is more difficult in three dimensions. Consider the nonlinear system

$$\begin{aligned} 2x_1 - 3x_2 + x_3 - 4 &= 0, \\ 2x_1 + x_2 - x_3 + 4 &= 0, \\ x_1^2 + x_2^2 + x_3^2 - 4 &= 0. \end{aligned}$$

Define three equations using the Maple commands

```
>eq1:=2*x1-3*x2+x3-4=0;
>eq2:=2*x1+x2-x3+4=0;
>eq3:=x1^2+x2^2+x3^2-4=0;
```

The third equation describes a sphere of radius 2 and center $(0, 0, 0)$, so `x1,x2,` and `x3` are in $[-2, 2]$. The Maple commands to obtain the graph in this case are

```
>with(plots);
>implicitplot3d({eq1,eq2,eq3},x1=-2..2,x2=-2..2,x3=-2..2);
```

Various three-dimensional plotting options are available in Maple for isolating a solution to the nonlinear system. For example, we can rotate the graph to better view the sections of the surfaces. Then we can zoom into regions where the intersections lie and alter the display form of the axes for a more accurate view of the intersection's coordinates. For this problem a reasonable initial approximation is $(x_1, x_2, x_3) = (-0.5, -1.5, 1.5)$.

EXERCISE SET 10.2

1. Give an example of a function $\mathbf{F} : \mathbb{R}^2 \to \mathbb{R}^2$ that is continuous at each point of $\mathbb{R}^2$ except at $(1, 0)$.

2. Give an example of a function $\mathbf{F} : \mathbb{R}^3 \to \mathbb{R}^3$ that is continuous at each point of $\mathbb{R}^3$ except at $(1, 2, 3)$.

3. Use the graphing facilities in Maple to approximate solutions to the following nonlinear systems.

   a. $$x_1(1 - x_1) + 4x_2 = 12,$$ $$(x_1 - 2)^2 + (2x_2 - 3)^2 = 25.$$

   b. $$5x_1^2 - x_2^2 = 0,$$ $$x_2 - 0.25(\sin x_1 + \cos x_2) = 0.$$

   c. $$15x_1 + x_2^2 - 4x_3 = 13,$$ $$x_1^2 + 10x_2 - x_3 = 11,$$ $$x_2^3 - 25x_3 = -22.$$

   d. $$10x_1 - 2x_2^2 + x_2 - 2x_3 - 5 = 0,$$ $$8x_2^2 + 4x_3^2 - 9 = 0,$$ $$8x_2 \quad x_3 + 4 = 0.$$

4. Use Newton's method with $\mathbf{x}^{(0)} = \mathbf{0}$ to compute $\mathbf{x}^{(2)}$ for each of the following nonlinear systems.

   a. $$4x_1^2 - 20x_1 + \frac{1}{4}x_2^2 + 8 = 0,$$ $$\frac{1}{2}x_1x_2^2 + 2x_1 - 5x_2 + 8 = 0.$$

   b. $$\sin(4\pi x_1 x_2) - 2x_2 - x_1 = 0,$$ $$\left(\frac{4\pi - 1}{4\pi}\right)(e^{2x_1} - e) + 4ex_2^2 - 2ex_1 = 0.$$

c.
$$3x_1 - \cos(x_2x_3) - \frac{1}{2} = 0,$$
$$4x_1^2 - 625x_2^2 + 2x_2 - 1 = 0,$$
$$e^{-x_1x_2} + 20x_3 + \frac{10\pi - 3}{3} = 0.$$

d.
$$x_1^2 + x_2 - 37 = 0,$$
$$x_1 - x_2^2 - 5 = 0,$$
$$x_1 + x_2 + x_3 - 3 = 0.$$

5. Use the answers obtained in Exercise 3 as initial approximations to Newton's method, iterating until $\|\mathbf{x}^{(k)} - \mathbf{x}^{(k-1)}\|_\infty < 10^{-6}$.

6. The nonlinear system

$$3x_1 - \cos(x_2x_3) - \frac{1}{2} = 0,$$
$$x_1^2 - 625x_2^2 - \frac{1}{4} = 0,$$
$$e^{-x_1x_2} + 20x_3 + \frac{10\pi - 3}{3} = 0$$

has a singular Jacobian matrix at the solution. Apply Newton's method with $\mathbf{x}^{(0)} = (1, 1 - 1)^t$. Note that convergence may be slow or may not occur within a reasonable number of iterations.

7. The nonlinear system

$$4x_1 - x_2 + x_3 = x_1x_4,$$
$$-x_1 + 3x_2 - 2x_3 = x_2x_4,$$
$$x_1 - 2x_2 + 3x_3 = x_3x_4,$$
$$x_1^2 + x_2^2 + x_3^2 = 1$$

has three solutions. Use Newton's method to approximate all three solutions, iterating until $\|\mathbf{x}^{(k)} - \mathbf{x}^{(k-1)}\|_\infty < 10^{-5}$.

8. In Exercise 6 of Section 5.7, we considered the problem of predicting the population of two species that compete for the same food supply. In the problem, we made the assumption that the populations could be predicted by solving the system of equations

$$\frac{dx_1(t)}{dt} = x_1(t)\,(4 - 0.0003x_1(t) - 0.0004x_2(t))$$

and

$$\frac{dx_2(t)}{dt} = x_2(t)\,(2 - 0.0002x_1(t) - 0.0001x_2(t))\,.$$

In this exercise, we would like to consider the problem of determining equilibrium populations of the two species. The mathematical criteria that must be satisfied in order for the populations to be at equilibrium is that, simultaneously,

$$\frac{dx_1(t)}{dt} = 0 \quad \text{and} \quad \frac{dx_2(t)}{dt} = 0.$$

This occurs when the first species is extinct and the second species has a population of 20,000 or when the second species is extinct and the first species has a population of 13,333. Can an equilibrium occur in any other situation?

9. The amount of pressure required to sink a large, heavy object in a soft homogeneous soil that lies above a hard base soil can be predicted by the amount of pressure required to sink smaller objects in the same soil. Specifically, the amount of pressure p required to sink a circular plate of radius r a distance d in the soft soil, where the hard base soil lies a distance $D > d$ below the surface, can be approximated by an equation of the form

$$p = k_1 e^{k_2 r} + k_3 r,$$

where k_1, k_2, and k_3 are constants, with $k_2 > 0$, depending on d and the consistency of the soil but not on the radius of the plate. (See [Bek], pp. 89–94.)

a. Find the values of k_1, k_2, and k_3 if we assume that a plate of radius 1 in. requires a pressure of 10 lb/in.^{2} to sink 1 ft in a muddy field, a plate of radius 2 in. requires a pressure of 12 lb/in.^{2} to sink 1 ft, and a plate of radius 3 in. requires a pressure of 15 lb/in.^{2} to sink this distance (assuming that the mud is more than 1 ft deep).

b. Use your calculations from part (a) to predict the minimal size of circular plate that would be required to sustain a load of 500 lb on this field with sinkage of less than 1 ft.

10.3 Quasi-Newton Methods

A significant weakness of Newton's method for solving systems of nonlinear equations lies in the requirement that, at each iteration, a Jacobian matrix be computed and an $n \times n$ linear system solved that involves this matrix. To illustrate the magnitude of this weakness, let us consider the amount of computation associated with one iteration of Newton's method. The Jacobian matrix associated with a system of n nonlinear equations written in the form $\mathbf{F}(\mathbf{x}) = \mathbf{0}$ requires that the n^2 partial derivatives of the n component functions of $\mathbf{F}$ be determined and evaluated. In most situations, the exact evaluation of the partial derivatives is inconvenient, and in many applications it is impossible. This difficulty can be overcome by using finite-difference approximations to the partial derivatives. For example,

$$\frac{\partial f_j}{\partial x_k}\left(\mathbf{x}^{(i)}\right) \approx \frac{f_j\left(\mathbf{x}^{(i)} + h\mathbf{e}_k\right) - f_j\left(\mathbf{x}^{(i)}\right)}{h},$$

where h is small in absolute value and $\mathbf{e}_k$ is the vector whose only nonzero entry is a 1 in the kth coordinate. This approximation, however, still requires that at least n^2 scalar functional evaluations be performed to approximate the Jacobian and does not decrease the amount of calculation, in general $O(n^3)$, required for solving the linear system involving this approximate Jacobian. The total computational effort for just one iteration of Newton's method is, consequently, at least $n^2 + n$ scalar functional evaluations (n^2 for the evaluation of the Jacobian matrix and n for the evaluation of $\mathbf{F}$) together with $O(n^3)$ arithmetic operations to solve the linear system. This amount of computational effort is prohibitive except for relatively small values of n and easily evaluated scalar functions.

In this section, we consider a generalization of the Secant method to systems of nonlinear equations, in particular, a technique known as **Broyden's method**. The method requires only n scalar functional evaluations per iteration and also reduces the number of arithmetic calculations to $O(n^2)$. It belongs to a class of methods known as *least-change secant updates* that produce algorithms called *quasi-Newton*. These methods replace the Jacobian matrix in Newton's method with an approximation matrix that is updated at each iteration. The disadvantage to the method is that the quadratic convergence of Newton's method is lost. It is replaced in general by a convergence called *superlinear*, which implies that

$$\lim_{i\to\infty} \frac{\|\mathbf{p}^{(i+1)} - \mathbf{p}\|}{\|\mathbf{p}^{(i)} - \mathbf{p}\|} = 0,$$

where $\mathbf{p}$ denotes the solution to $\mathbf{F}(\mathbf{x}) = \mathbf{0}$, and $\mathbf{p}^{(i)}$ and $\mathbf{p}^{(i+1)}$ are consecutive approximations to $\mathbf{p}$. In most applications, the reduction to superlinear convergence is a more than acceptable trade-off for the decrease in the amount of computation.

An additional disadvantage of quasi-Newton methods is that, unlike Newton's method, they are not self-correcting. Newton's method, for example, will generally correct for round-off error with successive iterations, but unless special safeguards are incorporated, Broyden's method will not.

To describe Broyden's method, suppose that an initial approximation $\mathbf{p}^{(0)}$ is given to the solution $\mathbf{p}$ of $\mathbf{F}(\mathbf{x}) = \mathbf{0}$. We calculate the next approximation $\mathbf{p}^{(1)}$ in the same manner as Newton's method, or, if it is inconvenient to determine $J(\mathbf{p}^{(0)})$ exactly, we can use difference equations to approximate the partial derivatives. To compute $\mathbf{p}^{(2)}$, however, we depart from Newton's method and examine the Secant method for a single nonlinear equation. The Secant method uses the approximation

$$f'(p_1) \approx \frac{f(p_1) - f(p_0)}{p_1 - p_0}$$

as a replacement for $f'(p_1)$ in Newton's method. For nonlinear systems, $\mathbf{p}^{(1)} - \mathbf{p}^{(0)}$ is a vector, and the corresponding quotient is undefined. However, the method proceeds similarly in that we replace the matrix $J(\mathbf{p}^{(1)})$ in Newton's method by a matrix A_1 with the property that

$$A_1\left(\mathbf{p}^{(1)} - \mathbf{p}^{(0)}\right) = \mathbf{F}\left(\mathbf{p}^{(1)}\right) - \mathbf{F}\left(\mathbf{p}^{(0)}\right).$$

Any nonzero vector in $\mathbb{R}^n$ can be written as the sum of a multiple of $\mathbf{p}^{(1)} - \mathbf{p}^{(0)}$ and a multiple of a vector in the orthogonal complement of $\mathbf{p}^{(1)} - \mathbf{p}^{(0)}$. So, to uniquely define the matrix A_1, we need to specify how it acts on the orthogonal complement of $\mathbf{p}^{(1)} - \mathbf{p}^{(0)}$. Since no information is available about the change in $\mathbf{F}$ in a direction orthogonal to $\mathbf{p}^{(1)} - \mathbf{p}^{(0)}$, we require that

$$A_1\mathbf{z} = J\left(\mathbf{p}^{(0)}\right)\mathbf{z} \quad \text{whenever} \quad \left(\mathbf{p}^{(1)} - \mathbf{p}^{(0)}\right)^t \mathbf{z} = 0.$$

Thus, any vector orthogonal to $\mathbf{p}^{(1)} - \mathbf{p}^{(0)}$ is unaffected by the update from $J(\mathbf{p}^{(0)})$, which was used to compute $\mathbf{p}^{(1)}$, to A_1, which is used in the determination of $\mathbf{p}^{(2)}$.

These conditions uniquely define A_1 as

$$A_1 = J\left(\mathbf{p}^{(0)}\right) + \frac{\left[\mathbf{F}\left(\mathbf{p}^{(1)}\right) - \mathbf{F}\left(\mathbf{p}^{(0)}\right) - J\left(\mathbf{p}^{(0)}\right)\left(\mathbf{p}^{(1)} - \mathbf{p}^{(0)}\right)\right]\left(\mathbf{p}^{(1)} - \mathbf{p}^{(0)}\right)^t}{\|\mathbf{p}^{(1)} - \mathbf{p}^{(0)}\|_2^2}.$$

It is this matrix that is used in place of $J(\mathbf{p}^{(1)})$ to determine $\mathbf{p}^{(2)}$:

$$\mathbf{p}^{(2)} = \mathbf{p}^{(1)} - A_1^{-1}\mathbf{F}\left(\mathbf{p}^{(1)}\right).$$

Once $\mathbf{p}^{(2)}$ has been determined, the method is repeated to determine $\mathbf{p}^{(3)}$, with A_1 used in place of $A_0 \equiv J(\mathbf{p}^{(0)})$ and with $\mathbf{p}^{(2)}$ and $\mathbf{p}^{(1)}$ in place of $\mathbf{p}^{(1)}$ and $\mathbf{p}^{(0)}$, respectively. In general, once $\mathbf{p}^{(i)}$ has been determined, $\mathbf{p}^{(i+1)}$ is computed by

$$A_i = A_{i-1} + \frac{\mathbf{y}_i - A_{i-1}\mathbf{s}_i}{\|\mathbf{s}_i\|_2^2}\mathbf{s}_i^t$$

and

$$\mathbf{p}^{(i+1)} = \mathbf{p}^{(i)} - A_i^{-1}\mathbf{F}\left(\mathbf{p}^{(i)}\right),$$

where the notation $\mathbf{s}_i = \mathbf{p}^{(i)} - \mathbf{p}^{(i-1)}$ and $\mathbf{y}_i = \mathbf{F}(\mathbf{p}^{(i)}) - \mathbf{F}(\mathbf{p}^{(i-1)})$ is introduced to simplify the equations.

If the method is performed as outlined, the number of scalar functional evaluations is reduced from $n^2 + n$ to n (those required for evaluating $\mathbf{F}(\mathbf{p}^{(i)})$), but the method still requires $O(n^3)$ calculations to solve the associated $n \times n$ linear system

$$A_i\mathbf{y}_i = -\mathbf{F}\left(\mathbf{p}^{(i)}\right).$$

Employing the method in this form would not be justified because of the reduction to superlinear convergence from the quadratic convergence of Newton's method. A significant improvement can be incorporated by employing a matrix-inversion formula.

Sherman-Morrison Formula

If A is a nonsingular matrix and $\mathbf{x}$ and $\mathbf{y}$ are vectors with $\mathbf{y}^t A^{-1}\mathbf{x} \neq -1$, then $A + \mathbf{x}\mathbf{y}^t$ is nonsingular and

$$\left(A + \mathbf{x}\mathbf{y}^t\right)^{-1} = A^{-1} - \frac{A^{-1}\mathbf{x}\mathbf{y}^t A^{-1}}{1 + \mathbf{y}^t A^{-1}\mathbf{x}}.$$

This formula permits A_i^{-1} to be computed directly from A_{i-1}^{-1}, eliminating the need for a matrix inversion with each iteration. By letting $A = A_{i-1}$, $\mathbf{x} = (\mathbf{y}_i - A_{i-1}\mathbf{s}_i)/\|\mathbf{s}_i\|_2^2$, and

$\mathbf{y} = \mathbf{s}_i$, the Sherman-Morrison formula implies that

$$A_i^{-1} = \left(A_{i-1} + \frac{\mathbf{y}_i - A_{i-1}\mathbf{s}_i}{\|\mathbf{s}_i\|_2^2}\mathbf{s}_i^t\right)^{-1}$$

$$= A_{i-1}^{-1} - \frac{A_{i-1}^{-1}\left(\dfrac{\mathbf{y}_i - A_{i-1}\mathbf{s}_i}{\|\mathbf{s}_i\|_2^2}\mathbf{s}_i^t\right)A_{i-1}^{-1}}{1 + \mathbf{s}_i^t A_{i-1}^{-1}\left(\dfrac{\mathbf{y}_i - A_{i-1}\mathbf{s}_i}{\|\mathbf{s}_i\|_2^2}\right)}$$

$$= A_{i-1}^{-1} - \frac{(A_{i-1}^{-1}y_i - \mathbf{s}_i)\mathbf{s}_i^t A_{i-1}^{-1}}{\|\mathbf{s}_i\|_2^2 + \mathbf{s}_i^t A_{i-1}^{-1}\mathbf{y}_i - \|\mathbf{s}_i\|_2^2}$$

$$= A_{i-1}^{-1} + \frac{(\mathbf{s}_i - A_{i-1}^{-1}\mathbf{y}_i)\,\mathbf{s}_i^t A_{i-1}^{-1}}{\mathbf{s}_i^t A_{i-1}^{-1}\mathbf{y}_i}.$$

This computation involves only matrix-vector multiplication at each step and therefore requires only $O(n^2)$ arithmetic calculations. The calculation of A_i is bypassed, as is the necessity of solving the linear system. The program BROYM102 is used to implement this technique.

EXAMPLE 1 The nonlinear system

$$3x_1 - \cos(x_2x_3) - \frac{1}{2} = 0,$$
$$x_1^2 - 81(x_2 + 0.1)^2 + \sin x_3 + 1.06 = 0,$$
$$e^{-x_1x_2} + 20x_3 + \frac{10\pi - 3}{3} = 0$$

was solved by Newton's method in Example 1 of Section 10.2. The Jacobian matrix for this system is

$$J(x_1, x_2, x_3) = \begin{bmatrix} 3 & x_3 \sin x_2x_3 & x_2 \sin x_2x_3 \\ 2x_1 & -162(x_2 + 0.1) & \cos x_3 \\ -x_2e^{-x_1x_2} & -x_1e^{-x_1x_2} & 20 \end{bmatrix}.$$

Let $\mathbf{p}^{(0)} = (0.1, 0.1, -0.1)^t$, and

$$\mathbf{F}(x_1, x_2, x_3) = (f_1(x_1, x_2, x_3), f_2(x_1, x_2, x_3), f_3(x_1, x_2, x_3))^t$$

where

$$f_1(x_1, x_2, x_3) = 3x_1 - \cos(x_2x_3) - \frac{1}{2},$$
$$f_2(x_1, x_2, x_3) = x_1^2 - 81(x_2 + 0.1)^2 + \sin x_3 + 1.06,$$

and

$$f_3(x_1, x_2, x_3) = e^{-x_1 x_2} + 20x_3 + \frac{10\pi - 3}{3}.$$

Then

$$\mathbf{F}\left(\mathbf{p}^{(0)}\right) = \begin{bmatrix} -1.199950 \\ -2.269833 \\ 8.462025 \end{bmatrix}.$$

Since

$$\begin{aligned} A_0 &= J\left(p_1^{(0)}, p_2^{(0)}, p_3^{(0)}\right) \\ &= \begin{bmatrix} 3 & 9.999836 \times 10^{-4} & -9.999833 \times 10^{-4} \\ 0.2 & -32.4 & 0.9950042 \\ -9.900498 \times 10^{-2} & -9.900498 \times 10^{-2} & 20 \end{bmatrix}, \end{aligned}$$

we have

$$\begin{aligned} A_0^{-1} &= J\left(p_1^{(0)}, p_2^{(0)}, p_3^{(0)}\right)^{-1} \\ &= \begin{bmatrix} 0.33333312 & 1.023852 \times 10^{-5} & 1.615701 \times 10^{-5} \\ 2.108607 \times 10^{-3} & -3.086883 \times 10^{-2} & 1.535836 \times 10^{-3} \\ 1.660520 \times 10^{-3} & -1.527577 \times 10^{-4} & 5.000768 \times 10^{-2} \end{bmatrix}, \\ \mathbf{p}^{(1)} &= \mathbf{p}^{(0)} - A_0^{-1}\mathbf{F}\left(\mathbf{p}^{(0)}\right) = \begin{bmatrix} 0.4998697 \\ 1.946685 \times 10^{-2} \\ -0.5215205 \end{bmatrix}, \\ \mathbf{F}\left(\mathbf{p}^{(1)}\right) &= \begin{bmatrix} -3.394465 \times 10^{-4} \\ -0.3443879 \\ 3.188238 \times 10^{-2} \end{bmatrix}, \\ \mathbf{y}_1 &= \mathbf{F}\left(\mathbf{p}^{(1)}\right) - \mathbf{F}\left(\mathbf{p}^{(0)}\right) = \begin{bmatrix} 1.199611 \\ 1.925445 \\ -8.430143 \end{bmatrix}, \\ \mathbf{s}_1 &= \begin{bmatrix} 0.3998697 \\ -8.053315 \times 10^{-2} \\ -0.4215204 \end{bmatrix}, \\ \mathbf{s}_1^t A_0^{-1}\mathbf{y}_1 &= 0.3424604, \\ A_1^{-1} &= A_0^{-1} + \left(\frac{1}{0.3424604}\right)\left[\left(\mathbf{s}_1 - A_0^{-1}\mathbf{y}_1\right)\mathbf{s}_1^t A_0^{-1}\right] \\ &= \begin{bmatrix} 0.3333781 & 1.11050 \times 10^{-5} & 8.967344 \times 10^{-6} \\ -2.021270 \times 10^{-3} & -3.094849 \times 10^{-2} & 2.196906 \times 10^{-3} \\ 1.022214 \times 10^{-3} & -1.650709 \times 10^{-4} & 5.010986 \times 10^{-2} \end{bmatrix}, \end{aligned}$$

and

$$\mathbf{p}^{(2)} = \mathbf{p}^{(1)} - A_1^{-1}\mathbf{F}\left(\mathbf{p}^{(1)}\right) = \begin{bmatrix} 0.4999863 \\ 8.737833 \times 10^{-3} \\ -0.5231746 \end{bmatrix}.$$

Additional iterations are listed in Table 10.2. The fifth iteration of Broyden's method is slightly less accurate than was the fourth iteration of Newton's method given in the example at the end of the preceding section. ◆ ◆ ◆

Table 10.2

k	$p_1^{(k)}$	$p_2^{(k)}$	$p_3^{(k)}$	$\lVert\mathbf{p}^{(k)} - \mathbf{p}^{(k-1)}\rVert_2$
3	0.5000066	8.672215×10^{-4}	−0.5236918	7.88×10^{-3}
4	0.5000005	6.087473×10^{-5}	−0.5235954	8.12×10^{-4}
5	0.5000002	-1.445223×10^{-6}	−0.5235989	6.24×10^{-5}

EXERCISE SET 10.3

1. Use Broyden's method with $\mathbf{x}^{(0)} = \mathbf{0}$ to compute $\mathbf{x}^{(2)}$ for each of the following nonlinear systems.

   a.
   $$4x_1^2 - 20x_1 + \frac{1}{4}x_2^2 + 8 = 0,$$
   $$\frac{1}{2}x_1x_2^2 + 2x_1 - 5x_2 + 8 = 0.$$

   b.
   $$\sin(4\pi x_1x_2) - 2x_2 - x_1 = 0,$$
   $$\left(\frac{4\pi - 1}{4\pi}\right)(e^{2x_1} - e) + 4ex_2^2 - 2ex_1 = 0.$$

   c.
   $$3x_1 - \cos(x_2x_3) - \frac{1}{2} = 0,$$
   $$4x_1^2 - 625x_2^2 + 2x_2 - 1 = 0,$$
   $$e^{-x_1x_2} + 20x_3 + \frac{10\pi - 3}{3} = 0.$$

   d.
   $$x_1^2 + x_2 - 37 = 0,$$
   $$x_1 - x_2^2 - 5 = 0,$$
   $$x_1 + x_2 + x_3 - 3 = 0.$$

2. Use Broyden's method to approximate solutions to the nonlinear systems in Exercise 1, iterating until $\lVert\mathbf{x}^{(k)} - \mathbf{x}^{(k-1)}\rVert_\infty < 10^{-6}$. The initial approximations $\mathbf{x}^{(0)}$ in Exercise 1 may not lead to convergence. If not, use a different value of $\mathbf{x}^{(0)}$.

3. Use Broyden's method to find a solution to the following nonlinear systems, iterating until $\lVert\mathbf{x}^{(k)} - \mathbf{x}^{(k-1)}\rVert_\infty < 10^{-6}$.

   a.
   $$3x_1^2 - x_2^2 = 0$$
   $$3x_1x_2^2 - x_1^3 - 1 = 0$$
   Use $\mathbf{x}^{(0)} = (1, 1)^t$.

b. $\ln(x_1^2 + x_2^2) - \sin(x_1 x_2) = \ln 2 + \ln \pi$

$e^{x_1 - x_2} + \cos(x_1 x_2) = 0$

Use $\mathbf{x}^{(0)} = (2, 2)^t$.

c. $x_1^3 + x_1^2 x_2 - x_1 x_3 + 6 = 0$

$e^{x_1} + e^{x_2} - x_3 = 0$

$x_2^2 - 2x_1 x_3 = 4$

Use $\mathbf{x}^{(0)} = (-1, -2, 1)^t$.

d. $6x_1 - 2\cos(x_2 x_3) - 1 = 0$

$9x_2 + \sqrt{x_1^2 + \sin x_3 + 1.06} + 0.9 = 0$

$60x_3 + 3e^{-x_1 x_2} + 10\pi - 3 = 0$

Use $\mathbf{x}^{(0)} = (0, 0, 0)^t$.

4. The nonlinear system

$$3x_1 - \cos(x_2 x_3) - \frac{1}{2} = 0,$$

$$x_1^2 - 625x_2^2 - \frac{1}{4} = 0,$$

$$e^{-x_1 x_2} + 20x_3 + \frac{10\pi - 3}{3} = 0$$

has a singular Jacobian matrix at the solution. Apply Broyden's method with $\mathbf{x}^{(0)} = (1, 1 - 1)^t$. Note that convergence may be slow or may not occur within a reasonable number of iterations.

5. The nonlinear system

$$\begin{aligned} 4x_1 - x_2 + x_3 &= x_1 x_4, \\ -x_1 + 3x_2 - 2x_3 &= x_2 x_4, \\ x_1 - 2x_2 + 3x_3 &= x_3 x_4, \\ x_1^2 + x_2^2 + x_3^2 &= 1 \end{aligned}$$

has three solutions. Use Broyden's method to approximate all three solutions. Iterate until $\|\mathbf{x}^{(k)} - \mathbf{x}^{(k-1)}\|_\infty < 10^{-5}$.

6. Show that if $\mathbf{0} \neq \mathbf{y} \in \mathbb{R}^n$ and $\mathbf{z} \in \mathbb{R}^n$, then $\mathbf{z} = \mathbf{z}_1 + \mathbf{z}_2$, where

$$\mathbf{z}_1 = \frac{\mathbf{y}^t \mathbf{z}}{\|\mathbf{y}\|_2^2}\mathbf{y}$$

is parallel to $\mathbf{y}$ and $\mathbf{z}_2 = \mathbf{z} - \mathbf{z}_1$ is orthogonal to $\mathbf{y}$.

7. It can be shown that if A^{-1} exists and $\mathbf{x}, \mathbf{y} \in \mathbb{R}^n$, then $(A + \mathbf{x}\mathbf{y}^t)^{-1}$ exists if and only if $\mathbf{y}^t A^{-1}\mathbf{x} \neq -1$. Use this result to show the Sherman-Morrison formula: If A^{-1} exists and $\mathbf{y}^t A^{-1}\mathbf{x} \neq -1$, then $(A + \mathbf{x}\mathbf{y}^t)^{-1}$ exists, and

$$\left(A + \mathbf{x}\mathbf{y}^t\right)^{-1} = A^{-1} - \frac{A^{-1}\mathbf{x}\mathbf{y}^t A^{-1}}{1 + \mathbf{y}^t A^{-1}\mathbf{x}}.$$

10.4 The Steepest Descent Method

The advantage of the Newton and quasi-Newton methods for solving systems of nonlinear equations is their speed of convergence once a sufficiently accurate approximation is known. A weakness of these methods is that an accurate initial approximation to the solution is needed to ensure convergence. The **method of Steepest Descent** will generally converge only linearly to the solution, but it is global in nature. It is often used to find sufficiently accurate starting approximations for the Newton-based techniques.

The method of Steepest Descent determines a local minimum for a multivariable function of the form $g : \mathbb{R}^n \to \mathbb{R}$. The method is valuable quite apart from the application as a starting method for solving nonlinear systems.

The connection between the minimization of a function from $\mathbb{R}^n$ to $\mathbb{R}$ and the solution of a system of nonlinear equations is due to the fact that a system of the form

$$\begin{aligned} f_1(x_1, x_2, \ldots, x_n) &= 0, \\ f_2(x_1, x_2, \ldots, x_n) &= 0, \\ \vdots \qquad & \quad \vdots \\ f_n(x_1, x_2, \ldots, x_n) &= 0, \end{aligned}$$

has a solution at $\mathbf{x} = (x_1, x_2, \ldots, x_n)^t$ precisely when the function g defined by

$$g(x_1, x_2, \ldots, x_n) = \sum_{i=1}^{n} \left[f_i(x_1, x_2, \ldots, x_n)\right]^2$$

has the minimal value zero.

The method of Steepest Descent for finding a local minimum for an arbitrary function g from $\mathbb{R}^n$ into $\mathbb{R}$ can be intuitively described as follows:

1. Evaluate g at an initial approximation $\mathbf{p}^{(0)} = (p_1^{(0)}, p_2^{(0)}, \ldots, p_n^{(0)})^t$.
2. Determine a direction from $\mathbf{p}^{(0)}$ that results in a decrease in the value of g.
3. Move an appropriate amount in this direction and call the new value $\mathbf{p}^{(1)}$.
4. Repeat the steps with $\mathbf{p}^{(0)}$ replaced by $\mathbf{p}^{(1)}$.

Before describing how to choose the correct direction and the appropriate distance to move in this direction, we need to review some results from calculus. The Extreme Value Theorem implies that a differentiable single-variable function can have a relative minimum only when the derivative is zero. To extend this result to multivariable functions, we need the following definition.

If $g : \mathbb{R}^n \to \mathbb{R}$, the **gradient** of g at $\mathbf{x} = (x_1, x_2, \ldots, x_n)^t$ is denoted $\nabla g(\mathbf{x})$ and defined by

$$\nabla g(\mathbf{x}) = \left(\frac{\partial g}{\partial x_1}(\mathbf{x}), \frac{\partial g}{\partial x_2}(\mathbf{x}), \ldots, \frac{\partial g}{\partial x_n}(\mathbf{x}) \right)^t.$$

The gradient for a multivariable function is analogous to the derivative of a single variable function in the sense that a differentiable multivariable function can have a relative minimum at $\mathbf{x}$ only when the gradient at $\mathbf{x}$ is the zero vector.

The gradient has another important property connected with the minimization of multivariable functions. Suppose that $\mathbf{v} = (v_1, v_2, \ldots, v_n)^t$ is a vector in $\mathbb{R}^n$ with

$$\|\mathbf{v}\|_2^2 = \sum_{i=1}^{n} v_i^2 = 1.$$

The **directional derivative** of g at $\mathbf{x}$ in the direction of $\mathbf{v}$ is defined by

$$D_{\mathbf{v}} g(\mathbf{x}) = \lim_{h \to 0} \frac{1}{h} \left[g(\mathbf{x} + h\mathbf{v}) - g(\mathbf{x}) \right] = \mathbf{v} \cdot \nabla g(\mathbf{x}).$$

The directional derivative of g at $\mathbf{x}$ in the direction of $\mathbf{v}$ measures the change in the value of the function g relative to the change in the variable in the direction of $\mathbf{v}$.

A standard result from the calculus of multivariable functions states that the direction that produces the maximum magnitude for the directional derivative occurs when $\mathbf{v}$ is chosen to be parallel to $\nabla g(\mathbf{x})$, provided that $\nabla g(\mathbf{x}) \neq \mathbf{0}$. The direction of greatest decrease in the value of g at $\mathbf{x}$ is the direction given by $-\nabla g(\mathbf{x})$. (See Figure 10.3 for an illustration when g is a function of two variables.)

Figure 10.3

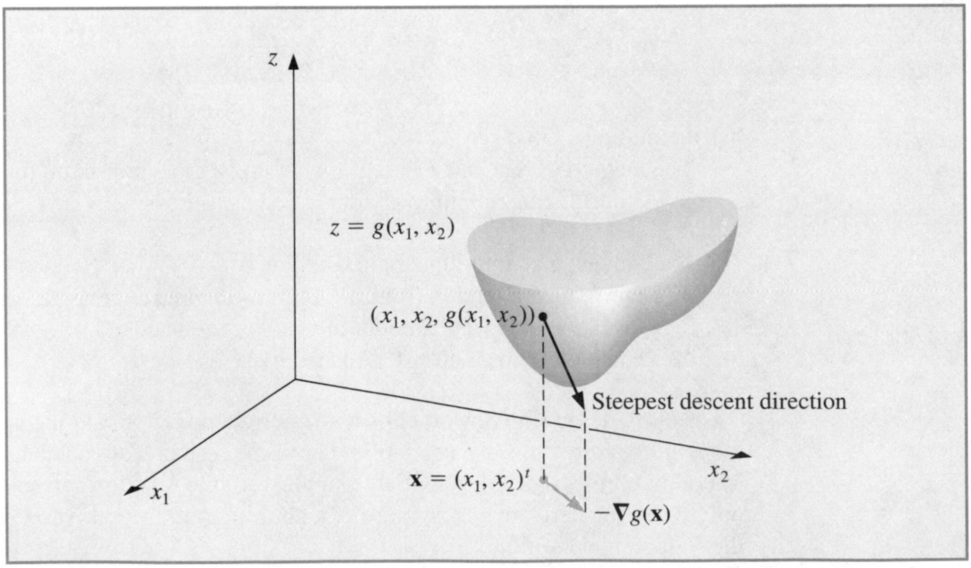

The object is to reduce $g(\mathbf{x})$ to its minimal value of zero, so given the initial approximation $\mathbf{p}^{(0)}$, we choose

$$\mathbf{p}^{(1)} = \mathbf{p}^{(0)} - \alpha \nabla g\left(\mathbf{p}^{(0)}\right) \tag{10.1}$$

for some constant $\alpha > 0$.

The problem now reduces to choosing α so that $g(\mathbf{p}^{(1)})$ will be significantly less than $g(\mathbf{p}^{(0)})$. To determine an appropriate choice for the value α, we consider the single-variable function

$$h(\alpha) = g\left(\mathbf{p}^{(0)} - \alpha \nabla g(\mathbf{p}^{(0)})\right).$$

The value of α that minimizes h is the value needed for Eq. (10.1).

Finding a minimal value for h directly would require differentiating h and then solving a root-finding problem to determine the critical points of h. This procedure is generally too costly. Instead, we choose three numbers $\alpha_1 < \alpha_2 < \alpha_3$ that, we hope, are close to where the minimum value of $h(\alpha)$ occurs. Then we construct the quadratic polynomial $P(x)$ that interpolates h at α_1, α_2, and α_3. We define $\hat{\alpha}$ in $[\alpha_1, \alpha_3]$ so that $P(\hat{\alpha})$ is a minimum in $[\alpha_1, \alpha_3]$ and use $P(\hat{\alpha})$ to approximate the minimal value of $h(\alpha)$. Then $\hat{\alpha}$ is used to determine the new iterate for approximating the minimal value of g:

$$\mathbf{p}^{(1)} = \mathbf{p}^{(0)} - \hat{\alpha} \nabla g\left(\mathbf{p}^{(0)}\right).$$

Since $g(\mathbf{p}^{(0)})$ is available, we first choose $\alpha_1 = 0$ to minimize the computation. Next a number α_3 is found with $h(\alpha_3) < h(\alpha_1)$. (Since α_1 does not minimize h, such a number α_3 does exist.) Finally, α_2 is chosen to be $\alpha_3/2$.

The minimum value $\hat{\alpha}$ of P on $[\alpha_1, \alpha_3]$ occurs either at the only critical point of P or at the right endpoint α_3, because, by assumption, $P(\alpha_3) = h(\alpha_3) < h(\alpha_1) = P(\alpha_1)$. The critical point is easily determined since P is a quadratic polynomial.

The program STPDC103 applies the method of Steepest Descent to approximate the minimal value of $g(\mathbf{x})$. To begin each iteration, the value 0 is assigned to α_1, and the value 1 is assigned to α_3. If $h(\alpha_3) \geq h(\alpha_1)$, then successive divisions of α_3 by 2 are performed and the value of α_3 is reassigned until $h(\alpha_3) \leq h(\alpha_1)$.

To employ the method to approximate the solution to the system

$$\begin{aligned} f_1(x_1, x_2, \ldots, x_n) &= 0, \\ f_2(x_1, x_2, \ldots, x_n) &= 0, \\ \vdots \qquad & \quad \vdots \\ f_n(x_1, x_2, \ldots, x_n) &= 0, \end{aligned}$$

we simply replace the function g with $\sum_{i=1}^{n} f_i^2$.

EXAMPLE 1 To find a reasonable starting approximation to the solution of the nonlinear system

$$f_1(x_1, x_2, x_3) = 3x_1 - \cos(x_2 x_3) - \frac{1}{2} = 0,$$
$$f_2(x_1, x_2, x_3) = x_1^2 - 81(x_2 + 0.1)^2 + \sin x_3 + 1.06 = 0,$$
$$f_3(x_1, x_2, x_3) = e^{-x_1 x_2} + 20x_3 + \frac{10\pi - 3}{3} = 0,$$

we use the Steepest Descent method with $\mathbf{p}^{(0)} = (0, 0, 0)^t$.

Let $g(x_1, x_2, x_3) = [f_1(x_1, x_2, x_3)]^2 + [f_2(x_1, x_2, x_3)]^2 + [f_3(x_1, x_2, x_3)]^2$. Then

$$\begin{aligned}\nabla g(x_1, x_2, x_3) \equiv \nabla g(\mathbf{x}) = &\Big(2f_1(\mathbf{x})\frac{\partial f_1}{\partial x_1}(\mathbf{x}) + 2f_2(\mathbf{x})\frac{\partial f_2}{\partial x_1}(\mathbf{x}) + 2f_3(\mathbf{x})\frac{\partial f_3}{\partial x_1}(\mathbf{x}),\\ &2f_1(\mathbf{x})\frac{\partial f_1}{\partial x_2}(\mathbf{x}) + 2f_2(\mathbf{x})\frac{\partial f_2}{\partial x_2}(\mathbf{x}) + 2f_3(\mathbf{x})\frac{\partial f_3}{\partial x_2}(\mathbf{x}),\\ &2f_1(\mathbf{x})\frac{\partial f_1}{\partial x_3}(\mathbf{x}) + 2f_2(\mathbf{x})\frac{\partial f_2}{\partial x_3}(\mathbf{x}) + 2f_3(\mathbf{x})\frac{\partial f_3}{\partial x_3}(\mathbf{x})\Big)\\ = &\,2\mathbf{J}(\mathbf{x})^t\mathbf{F}(\mathbf{x}).\end{aligned}$$

For $\mathbf{p}^{(0)} = (0, 0, 0)^t$, we have

$$g\left(\mathbf{p}^{(0)}\right) = 111.975 \quad \text{and} \quad z_0 = \|\nabla g\left(\mathbf{p}^{(0)}\right)\|_2 = 419.554.$$

Let

$$\mathbf{z} = \frac{1}{z_0}\nabla g\left(\mathbf{p}^{(0)}\right) = (-0.0214514, -0.0193062, 0.999583)^t.$$

With $\alpha_1 = 0$, we have $g_1 = g(\mathbf{p}^{(0)} - \alpha_1\mathbf{z}) = g(\mathbf{p}^{(0)}) = 111.975$. We arbitrarily let $\alpha_3 = 1$ so that

$$g_3 = g\left(\mathbf{p}^{(0)} - \alpha_3\mathbf{z}\right) = 93.5649.$$

Since $g_3 < g_1$ we accept α_3 as is and set $\alpha_2 = 0.5$. Thus,

$$g_2 = g\left(\mathbf{p}^{(0)} - \alpha_2\mathbf{z}\right) = 2.53557.$$

We now form the Newton's forward divided-difference interpolating polynomial

$$P(\alpha) = g_1 + h_1\alpha + h_3\alpha(\alpha - \alpha_2)$$

that interpolates

$$g\left(\mathbf{p}^{(0)} - \alpha\nabla g(\mathbf{p}^{(0)})\right) = g\left(\mathbf{p}^{(0)} - \alpha\mathbf{z}\right)$$

at $\alpha_1 = 0$, $\alpha_2 = 0.5$, and $\alpha_3 = 1$ as follows:

$$\alpha_1 = 0, \quad g_1 = 111.975,$$

$$\alpha_2 = 0.5, \quad g_2 = 2.53557, \quad h_1 = \frac{g_2 - g_1}{\alpha_2 - \alpha_1} = -218.878,$$

$$\alpha_3 = 1, \quad g_3 = 93.5649, \quad h_2 = \frac{g_3 - g_2}{\alpha_3 - \alpha_2} = 182.059, \quad h_3 = \frac{h_2 - h_1}{\alpha_3 - \alpha_1} = 400.937.$$

Thus,

$$P(\alpha) = 111.975 - 218.878\alpha + 400.937\alpha(\alpha - 0.5).$$

Now $P'(\alpha) = 0$ when $\alpha = \alpha_0 = 0.522959$. Since $g_0 = g(\mathbf{p}^{(0)} - \alpha_0\mathbf{z}) = 2.32762$ is smaller than g_1 and g_3, we set

$$\mathbf{p}^{(1)} = \mathbf{p}^{(0)} - \alpha_0\mathbf{z} = \mathbf{p}^{(0)} - 0.522959\mathbf{z} = (0.0112182, 0.0100964, -0.522741)^t,$$

and

$$g\left(\mathbf{p}^{(1)}\right) = 2.32762.$$

Table 10.3 contains the remainder of the results. A true solution to the nonlinear system is $(0.5, 0, -0.5235988)^t$, so $\mathbf{p}^{(2)}$ would likely be adequate as an initial approximation for Newton's method or Broyden's method. ◆ ◆ ◆

Table 10.3

k	$p_1^{(k)}$	$p_2^{(k)}$	$p_3^{(k)}$	$g(p_1^{(k)}, p_2^{(k)}, p_3^{(k)})$
0	0.0	0.0	0.0	111.975
1	0.0112182	0.0100964	−0.522741	2.32762
2	0.137860	−0.205453	−0.522059	1.27406
3	0.266959	0.00551102	−0.558494	1.06813
4	0.272734	−0.00811751	−0.522006	0.468309

EXERCISE SET 10.4

1. Use the method of Steepest Descent to approximate the solutions of the following nonlinear systems, iterating until $\|\mathbf{x}^{(k)} - \mathbf{x}^{(k-1)}\|_\infty < 0.05$.

   a. $$4x_1^2 - 20x_1 + \frac{1}{4}x_2^2 + 8 = 0$$
   $$\frac{1}{2}x_1x_2^2 + 2x_1 - 5x_2 + 8 = 0$$

   b. $$3x_1^2 - x_2^2 = 0$$
   $$3x_1x_2^2 - x_1^3 - 1 = 0$$

   c. $$\ln(x_1^2 + x_2^2) - \sin(x_1x_2) = \ln 2 + \ln \pi$$
   $$e^{x_1 - x_2} + \cos(x_1x_2) = 0$$

d.
$$\sin(4\pi x_1 x_2) - 2x_2 - x_1 = 0$$
$$\left(\frac{4\pi - 1}{4\pi}\right)(e^{2x_1} - e) + 4ex_2^2 - 2ex_1 = 0$$

2. Use the results in Exercise 1 and Newton's method to approximate the solutions of the nonlinear systems in Exercise 1, iterating until $\|\mathbf{x}^{(k)} - \mathbf{x}^{(k-1)}\|_\infty < 10^{-6}$.

3. Use the method of Steepest Descent to approximate the solutions of the following nonlinear systems, iterating until $\|\mathbf{x}^{(k)} - \mathbf{x}^{(k-1)}\|_\infty < 0.05$.

a.
$$15x_1 + x_2^2 - 4x_3 = 13$$
$$x_1^2 + 10x_2 - x_3 = 11$$
$$x_2^3 - 25x_3 = -22$$

b.
$$10x_1 - 2x_2^2 + x_2 - 2x_3 - 5 = 0$$
$$8x_2^2 + 4x_3^2 - 9 = 0$$
$$8x_2x_3 + 4 = 0$$

c.
$$x_1^3 + x_1^2x_2 - x_1x_3 + 6 = 0$$
$$e^{x_1} + e^{x_2} - x_3 = 0$$
$$x_2^2 - 2x_1x_3 = 4$$

d.
$$x_1 + \cos(x_1x_2x_3) - 1 = 0$$
$$(1 - x_1)^{1/4} + x_2 + 0.05x_3^2 - 0.15x_3 - 1 = 0$$
$$-x_1^2 - 0.1x_2^2 + 0.01x_2 + x_3 - 1 = 0$$

4. Use the results of Exercise 3 and Newton's method to approximate the solutions of the nonlinear systems in Exercise 3, iterating until $\|\mathbf{x}^{(k)} - \mathbf{x}^{(k-1)}\|_\infty < 10^{-6}$.

5. Use the method of Steepest Descent to approximate minima for the following functions, iterating until $\|\mathbf{x}^{(k)} - \mathbf{x}^{(k-1)}\|_\infty < 0.005$.

a. $g(x_1, x_2) = \cos(x_1 + x_2) + \sin x_1 + \cos x_2$

b. $g(x_1, x_2) = 100(x_1^2 - x_2)^2 + (1 - x_1)^2$

c. $g(x_1, x_2, x_3) = x_1^2 + 2x_2^2 + x_3^2 - 2x_1x_2 + 2x_1 - 2.5x_2 - x_3 + 2$

d. $g(x_1, x_2, x_3) = x_1^4 + 2x_2^4 + 3x_3^4 + 1.01$

6. a. Show that the quadratic polynomial that interpolates the function

$$h(\alpha) = g\left(\mathbf{p}^{(0)} - \alpha\nabla g(\mathbf{p}^{(0)})\right)$$

at $\alpha = 0$, α_2, and α_3 is

$$P(\alpha) = g\left(\mathbf{p}^{(0)}\right) + h_1\alpha + h_3\alpha(\alpha - \alpha_2)$$

where

$$h_1 = \frac{g\left(\mathbf{p}^{(0)} - \alpha_2\mathbf{z}\right) - g\left(\mathbf{p}^{(0)}\right)}{\alpha_2},$$

$$h_2 = \frac{g\left(\mathbf{p}^{(0)} - \alpha_3\mathbf{z}\right) - g\left(\mathbf{p}^{(0)} - \alpha_2\mathbf{z}\right)}{\alpha_3 - \alpha_2}, \quad \text{and} \quad h_3 = \frac{h_2 - h_1}{\alpha_3}.$$

b. Show that the only critical point of P occurs at

$$\alpha_0 = 0.5\left(\alpha_2 - \frac{h_1}{h_3}\right).$$

10.5 Survey of Methods and Software

In this chapter we considered methods to approximate solutions to nonlinear systems

$$\begin{aligned} f_1(x_1, x_2, \ldots, x_n) &= 0, \\ f_2(x_1, x_2, \ldots, x_n) &= 0, \\ \vdots \quad & \quad \vdots \\ f_n(x_1, x_2, \ldots, x_n) &= 0, \end{aligned}$$

Newton's method for systems requires a good initial approximation $(p_1^{(0)}, p_2^{(0)}, \ldots, p_n^{(0)})^t$ and generates a sequence

$$\mathbf{p}^{(k)} = \mathbf{p}^{(k-1)} - J\left(\mathbf{p}^{(k-1)}\right)^{-1} \mathbf{F}\left(\mathbf{p}^{(k-1)}\right)$$

that converges rapidly to a solution $\mathbf{p}$ if $\mathbf{p}^{(0)}$ is sufficiently close to $\mathbf{p}$. However, Newton's method requires evaluating, or approximating, n^2 partial derivatives and solving an $n \times n$ linear system at each step.

Broyden's method reduces the amount of computation at each step without significantly degrading the speed of convergence. This technique replaces the Jacobian matrix J, with a matrix A_{k-1} whose inverse is directly determined at each step. This reduces the arithmetic computations from $O(n^3)$ to $O(n^2)$. Moreover, the only scalar function evaluations required are in evaluating the f_i, saving n^2 scalar function evaluations per step. Broyden's method also requires a good initial approximation.

The Steepest Descent method was presented as a way to obtain good initial approximations for Newton and Broyden's methods. Although Steepest Descent does not give a rapidly convergent sequence, it does not require a good initial approximation. The Steepest Descent method approximates a minimum of a multivariable function g. For our application we chose

$$g(x_1, x_2, \ldots, x_n) = \sum_{i=1}^{n} \left[f_i(x_1, x_2, \ldots, x_n)\right]^2.$$

The minimum of g is zero, which occurs when the functions f_i are simultaneously zero.

Homotopy and continuation methods are also used for nonlinear systems and are the subject of current research. In these methods, a given problem

$$\mathbf{F}(\mathbf{x}) = \mathbf{0}$$

is embedded in a one-parameter family of problems using a parameter λ, assuming values in [0, 1]. The original problem corresponds to $\lambda = 1$, and a problem with a known solution corresponds to $\lambda = 0$. For example, the set of problems

$$G(\mathbf{x}, \lambda) = \lambda \mathbf{F}(\mathbf{x}) + (1 - \lambda)\mathbf{F}(\mathbf{p}_0) = \mathbf{0}, \qquad 0 \le \lambda \le 1,$$

for fixed $\mathbf{p}_0 \in \mathbb{R}^n$ forms a homotopy. When $\lambda = 0$, the solution is $\mathbf{x}(\lambda = 0) = \mathbf{p}_0$. The solution to the original problem corresponds to $\mathbf{x}(\lambda = 1)$. A continuation method attempts to determine $\mathbf{x}(\lambda = 1)$ by solving the sequence of problems corresponding to $\lambda_0 = 0 < \lambda_1 < \lambda_2 < \cdots < \lambda_m = 1$. The initial approximation to the solution of

$$\lambda_i \mathbf{F}(\mathbf{x}) + (1 - \lambda_i)\mathbf{F}(\mathbf{p}_0) = \mathbf{0}$$

is the solution $\mathbf{x}(\lambda = \lambda_{i-1})$ to the problem

$$\lambda_{i-1} \mathbf{F}(\mathbf{x}) + (1 - \lambda_{i-1})\mathbf{F}(\mathbf{p}_0) = \mathbf{0}.$$

The methods in the IMSL and NAG libraries are based on two subroutines contained in MINPACK, a public-domain package. Both methods use the Levenberg-Marquardt method, which is a weighted average of Newton's method and the Steepest Descent method. The weight is biased toward the Steepest Descent method until convergence is detected, at which time the weight is shifted toward the more rapidly convergent Newton's method. One subroutine uses a finite-difference approximation to the Jacobian, and the other requires a user-supplied subroutine to compute the Jacobian.

A comprehensive treatment of methods for solving nonlinear systems of equations can be found in Ortega and Rheinbolt [OR] and in Dennis and Schnabel [DS]. Recent developments on iterative methods can be found in Argyros and Szidarovszky [AS], and information on the use of homotopy and continuation methods is available in Allgower and Georg [AG].

CHAPTER

11 Boundary-Value Problems for Ordinary Differential Equations

11.1 Introduction

The differential equations in Chapter 5 are of first order and have one initial condition to satisfy. Later in the chapter we saw that the techniques could be extended to systems of equations and then to higher-order equations, but all the specified conditions must be on the same endpoint. These are initial-value problems. In this chapter we show how to approximate the solution to **two-point boundary-value** problems, differential equations where conditions are imposed at different points. For first-order differential equations only one condition is specified, so there is no distinction between initial-value and boundary-value problems.

The differential equations whose solutions we will approximate are of second order, specifically of the form

$$y'' = f(x, y, y'), \qquad \text{for } a \le x \le b,$$

with the boundary conditions on the solution prescribed by

$$y(a) = \alpha \quad \text{and} \quad y(b) = \beta,$$

for some constants α and β. Such a problem has a unique solution provided that f and its partial derivatives with respect to y and y' are continuous and that the partial derivative with respect to y is positive and with respect to y' is bounded. These are all reasonable conditions for boundary-value problems representing physical problems.

11.2 The Linear Shooting Method

A boundary-value problem is **linear** when the function f has the form

$$f(x, y, y') = p(x)y' + q(x)y + r(x).$$

Linear problems occur frequently in applications and are much easier to solve than nonlinear equations since adding any solution to the **inhomogeneous** differential equation

$$y'' - p(x)y' - q(x)y = r(x)$$

to the complete solution of the **homogeneous** differential equation

$$y'' - p(x)y' - q(x)y = 0$$

gives all the solutions to the inhomogeneous problem. The solutions of the homogeneous problem are easier to determine than are those of the inhomogeneous. Moreover, to show that a linear problem has a unique solution, we need only show that p, q, and r are continuous and that the values of q are positive.

To approximate the unique solution to the linear boundary-value problem, let us first consider the two initial-value problems

$$y'' = p(x)y' + q(x)y + r(x), \quad \text{for } a \le x \le b, \quad \text{where } y(a) = \alpha \text{ and } y'(a) = 0, \qquad \textbf{(11.1)}$$

and

$$y'' = p(x)y' + q(x)y, \quad \text{for } a \le x \le b, \quad \text{where } y(a) = 0 \text{ and } y'(a) = 1, \qquad \textbf{(11.2)}$$

both of which have unique solutions. Suppose that $y_1(x)$ is the solution to Eq. (11.1) and $y_2(x)$ is the solution to Eq. (11.2). Then

$$y(x) = y_1(x) + \frac{\beta - y_1(b)}{y_2(b)} y_2(x) \qquad \textbf{(11.3)}$$

is the unique solution to the linear boundary-value problem

$$y'' = p(x)y' + q(x)y + r(x), \quad \text{for } a \le x \le b, \quad \text{with } y(a) = \alpha \text{ and } y(b) = \beta. \qquad \textbf{(11.4)}$$

To verify this, first note that

$$\begin{aligned} y'' - p(x)y' - q(x)y &= y_1'' - p(x)y_1' - q(x)y_1 + \frac{\beta - y_1(b)}{y_2(b)}[y_2'' - p(x)y_2' - q(x)y_2] \\ &= r(x) + \frac{\beta - y_1(b)}{y_2(b)} \cdot 0 = r(x). \end{aligned}$$

Moreover,

$$y(a) = y_1(a) + \frac{\beta - y_1(b)}{y_2(b)} y_2(a) = y_1(a) + \frac{\beta - y_1(b)}{y_2(b)} \cdot 0 = \alpha$$

and

$$y(b) = y_1(b) + \frac{\beta - y_1(b)}{y_2(b)} y_2(b) = y_1(b) + \beta - y_1(b) = \beta.$$

The situation when $y_2(b) = 0$ is discussed in Exercise 8.

The Linear Shooting method is based on the replacement of the boundary-value problem by the two initial-value problems, (11.1) and (11.2). Numerous methods are available from Chapter 5 for approximating the solutions $y_1(x)$ and $y_2(x)$, and once these approximations are available, the solution to the boundary-value problem is approximated using the weighted sum in Eq. (11.3). Graphically, the method has the appearance shown in Figure 11.1.

Figure 11.1

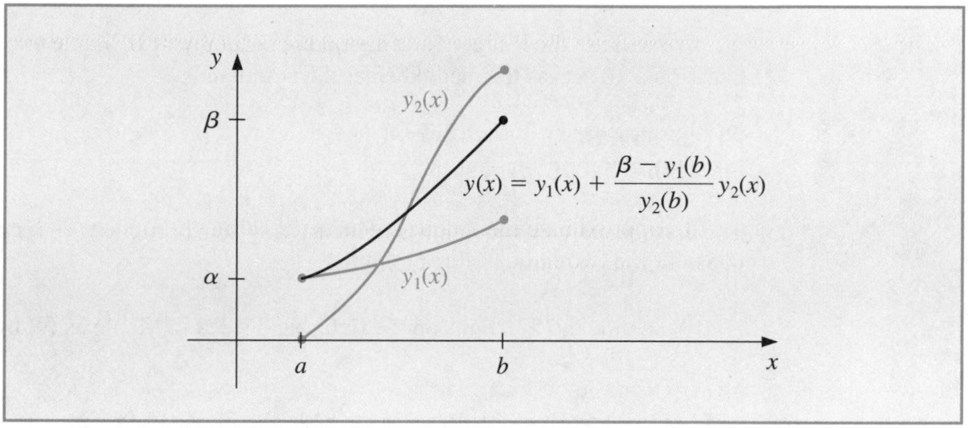

The program LINST111 incorporates the Runge-Kutta method of order 4 to find the approximations to $y_1(x)$ and $y_2(x)$, but any technique for approximating the solutions to initial-value problems can be substituted. The program has the additional feature of obtaining approximations for the derivative of the solution to the boundary-value problem as well as to the solution of the problem itself.

EXAMPLE 1 The boundary-value problem

$$y'' = -\frac{2}{x}y' + \frac{2}{x^2}y + \frac{\sin(\ln x)}{x^2}, \quad \text{for } 1 \le x \le 2, \quad \text{where} \quad y(1) = 1 \quad \text{and} \quad y(2) = 2,$$

has the exact solution

$$y = c_1 x + \frac{c_2}{x^2} - \frac{3}{10}\sin(\ln x) - \frac{1}{10}\cos(\ln x),$$

where

$$c_2 = \frac{1}{70}\left[8 - 12\sin(\ln 2) - 4\cos(\ln 2)\right]$$

and

$$c_1 = \frac{11}{10} - c_2.$$

Applying the Linear Shooting method to this problem requires approximating the solutions to the initial-value problems

$$y_1'' = -\frac{2}{x}y_1' + \frac{2}{x^2}y_1 + \frac{\sin(\ln x)}{x^2}, \quad \text{for } 1 \le x \le 2, \quad \text{where} \quad y_1(1) = 1 \quad \text{and} \quad y_1'(1) = 0,$$

and

$$y_2'' = -\frac{2}{x}y_2' + \frac{2}{x^2}y_2, \quad \text{for } 1 < x < 2, \quad \text{where} \quad y_2(1) = 0 \quad \text{and} \quad y_2'(1) = 1.$$

We will use the Runge-Kutta method of order 4 within Maple to solve both differential equations. To access the method enter the commands

```
>with(share);
>readshare(ODE,plots);
```

The first second-order equation is written as a system of two first-order differential equations $u_1' = f_1(x, u_1, u_2)$ and $u_2' = f_2(x, u_1, u_2)$.

We define the functions on the right-hand sides in Maple as follows:

```
>f1:=(x,u1,u2)->u2;
>f2:=(x,u1,u2)->-2*u2/x+2*u1/x^2+sin(ln(x))/x^2;
```

The Runge-Kutta method of order 4 is invoked with the command

```
>rhp:=rungekutta([f1,f2],[1,1,0],0.1,10);
```

and gives the values of $u_{1,i}$ in the second column of Table 11.1.

Table 11.1

x_i	$u_{1,i}$	$v_{1,i}$	w_i	$y(x_i)$	$\lvert y(x_i) - w_i \rvert$
1.0	1.00000000	0.00000000	1.00000000	1.00000000	
1.1	1.00896058	0.09117986	1.09262916	1.09262930	1.43×10^{-7}
1.2	1.03245472	0.16851175	1.18708471	1.18708484	1.34×10^{-7}
1.3	1.06674375	0.23608704	1.28338227	1.28338236	9.78×10^{-8}
1.4	1.10928795	0.29659067	1.38144589	1.38144595	6.02×10^{-8}
1.5	1.15830000	0.35184379	1.48115939	1.48115942	3.06×10^{-8}
1.6	1.21248371	0.40311695	1.58239245	1.58239246	1.08×10^{-8}
1.7	1.27087454	0.45131840	1.68501396	1.68501396	5.43×10^{-10}
1.8	1.33273851	0.49711137	1.78889854	1.78889853	5.05×10^{-9}
1.9	1.39750618	0.54098928	1.89392951	1.89392951	4.41×10^{-9}
2.0	1.46472815	0.58332538	2.00000000	2.00000000	

The next second-order differential equation as a system of two first-order differential equations uses the function f_1 defined previously and g_2 defined by

```
>g2:=(x,u1,u2)->-2*u2/x+2*u1/x^2;
```

The Runge-Kutta method produces the values of $v_{1,i}$ in the third column of Table 11.1 with the command

```
>rhp1:=rungekutta([f1,g2],[1,0,1],0.1,10);
```

We form the combination

$$y(x) = y_1(x) + \frac{2 - y_1(2)}{y_2(2)} y_2(x)$$

using the following code

```
>for i from 1 to 10 do
> w[i]:=rhp[i][2]+(2-rhp[10][2])/rhp1[10][2]*rhp1[i][2];
>od;
```

This gives the results presented in the fourth column of Table 11.1.

The value listed as $u_{1,i}$ approximates $y_1(x_i)$, the value of $v_{1,i}$ approximates $y_2(x_i)$, and w_i approximates $y(x_i)$. ◆ ◆ ◆

The accurate results in this example are due to the fact that the Runge-Kutta method of order 4 gives $O(h^4)$ approximations to the solutions of the initial-value problems. Unfortunately, there can be round-off error problems hidden in this technique. If $y_1(x)$ rapidly increases as x goes from a to b, then $u_{1,N} \approx y_1(b)$ will be large. Should β be small in magnitude compared to $u_{1,N}$, the term $(\beta - u_{1,N})/v_{1,N}$ will be approximately $-u_{1,N}/v_{1,N}$. So the approximations

$$y(x_i) \approx w_i = u_{1,i} - \left(\frac{\beta - u_{1,N}}{v_{1,N}}\right) v_{1,i}, \approx u_{1,i} - \left(\frac{u_{1,N}}{v_{1,N}}\right) v_{1,i}$$

allow the possibility of a loss of significant digits due to cancellation. However, since $u_{1,i}$ is an approximation to $y_1(x_i)$, the behavior of y_1 can be easily monitored, and if $u_{1,i}$ increases rapidly from a to b, the shooting technique can be employed in the other direction—that is, solving instead the initial-value problems

$$y'' = p(x)y' + q(x)y + r(x), \qquad \text{for } a \le x \le b, \quad \text{where} \quad y(b) = \beta \quad \text{and} \quad y'(b) = 0,$$

and

$$y'' = p(x)y' + q(x)y, \qquad \text{for } a \le x \le b, \quad \text{where} \quad y(b) = 0 \quad \text{and} \quad y'(b) = 1.$$

If the reverse shooting technique still gives cancellation of significant digits and if increased precision does not yield greater accuracy, other techniques must be employed. In general, however, if $u_{1,i}$ and $v_{1,i}$ are $O(h^n)$ approximations to $y_1(x_i)$ and $y_2(x_i)$, respectively, for each $i = 0, 1, \ldots, N$, then $w_{1,i}$ will be an $O(h^n)$ approximation to $y(x_i)$.

EXERCISE SET 11.2

1. The boundary-value problem

$$y'' = 4(y - x), \qquad \text{for } 0 \le x \le 1 \quad \text{with } y(0) = 0 \text{ and } y(1) = 2$$

has the solution $y(x) = e^2(e^4 - 1)^{-1}(e^{2x} - e^{-2x}) + x$. Use the Linear Shooting method to approximate the solution and compare the results to the actual solution.

a. With $h = \frac{1}{3}$ **b.** With $h = \frac{1}{4}$

2. The boundary-value problem

$$y'' = y' + 2y + \cos x, \qquad \text{for } 0 \le x \le \frac{\pi}{2} \quad \text{with } y(0) = -0.3 \text{ and } y\left(\frac{\pi}{2}\right) = -0.1$$

has the solution $y(x) = -\frac{1}{10}(\sin x + 3\cos x)$. Use the Linear Shooting method to approximate the solution and compare the results to the actual solution.

a. With $h = \dfrac{\pi}{4}$ **b.** With $h = \dfrac{\pi}{6}$.

3. Use the Linear Shooting method to approximate the solution to the following boundary-value problems.

a. $y'' = -3y' + 2y + 2x + 3$, for $0 \le x \le 1$ with $y(0) = 2$ and $y(1) = 1$; use $h = 0.1$.

b. $y'' = -\dfrac{4}{x}y' - \dfrac{2}{x^2}y - \dfrac{2\ln x}{x^2}$, for $1 \le x \le 2$ with $y(1) = -\dfrac{1}{2}$ and $y(2) = \ln 2$; use $h = 0.05$.

c. $y'' = -(x + 1)y' + 2y + (1 - x^2)e^{-x}$, for $0 \le x \le 1$ with $y(0) = -1$ and $y(1) = 0$; use $h = 0.1$.

d. $y'' = \dfrac{y'}{x} + \dfrac{3y}{x^2} + \dfrac{\ln x}{x} - 1$, for $1 \le x \le 2$ with $y(1) = y(2) = 0$; use $h = 0.1$.

4. Although $q(x) < 0$ in the following boundary-value problems, unique solutions exist and are given. Use the Linear Shooting method to approximate the solutions to the following problems and compare the results to the actual solutions.

a. $y'' + y = 0$, for $0 \le x \le \dfrac{\pi}{4}$ with $y(0) = 1$ and $y\left(\dfrac{\pi}{4}\right) = 1$; use $h = \dfrac{\pi}{20}$;

actual solution $y(x) = \cos x + \dfrac{2 - \sqrt{2}}{\sqrt{2}}\sin x$.

b. $y'' + 4y = \cos x$, for $0 \le x \le \dfrac{\pi}{4}$ with $y(0) = 0$ and $y\left(\dfrac{\pi}{4}\right) = 0$; use $h = \dfrac{\pi}{20}$;

actual solution $y(x) = -\dfrac{1}{3}\cos 2x - \dfrac{\sqrt{2}}{6}\sin 2x + \dfrac{1}{3}\cos x$.

c. $y'' = -\dfrac{4}{x}y' - \dfrac{2}{x^2}y + \dfrac{2}{x^2}\ln x$, for $1 \le x \le 2$ with $y(1) = \dfrac{1}{2}$ and $y(2) = \ln 2$;

use $h = 0.05$; actual solution $y(x) = \dfrac{4}{x} - \dfrac{2}{x^2} + \ln x - \dfrac{3}{2}$.

d. $y'' = 2y' - y + xe^x - x$, for $0 \le x \le 2$ with $y(0) = 0$ and $y(2) = -4$; use $h = 0.2$; actual solution $y(x) = \frac{1}{6}x^3e^x - \frac{5}{3}xe^x + 2e^x - x - 2$.

5. Use the Linear Shooting method to approximate the solution $y = e^{-10x}$ to the boundary-value problem

$$y'' = 100y, \qquad \text{for } 0 \le x \le 1 \quad \text{with} \quad y(0) = 1 \quad \text{and} \quad y(1) = e^{-10}.$$

Use $h = 0.1$ and 0.05.

6. Write the second-order initial-value problems (11.1) and (11.2) as first-order systems, and derive the equations necessary to solve the systems using the fourth-order Runge-Kutta method for systems.

7. Let u represent the electrostatic potential between two concentric metal spheres of radii R_1 and R_2 ($R_1 < R_2$), such that the potential of the inner sphere is kept constant at V_1 volts and the potential of the outer sphere is 0 volts. The potential in the region between the two spheres is governed by Laplace's equation, which, in this particular application, reduces to

$$\frac{d^2u}{dr^2} + \frac{2}{r}\frac{du}{dr} = 0, \qquad \text{for } R_1 \le r \le R_2 \quad \text{with} \quad u(R_1) = V_1 \quad \text{and} \quad u(R_2) = 0.$$

Suppose $R_1 = 2$ in., $R_2 = 4$ in., and $V_1 = 110$ volts.

a. Approximate $u(3)$ using the Linear Shooting method.

b. Compare the results of part (a) with the actual potential $u(3)$, where

$$u(r) = \frac{V_1R_1}{r}\left(\frac{R_2 - r}{R_2 - R_1}\right).$$

8. Show that if y_2 is the solution to $y'' = p(x)y' + q(x)y$ and $y_2(a) = y_2(b) = 0$, then $y_2 \equiv 0$.

9. Consider the boundary-value problem

$$y'' + y = 0, \qquad \text{for } 0 \le x \le b \quad \text{with} \quad y(0) = 0 \quad \text{and} \quad y(b) = B.$$

Find choices for b and B so that the boundary-value problem has

a. No solution;

b. Exactly one solution;

c. Infinitely many solutions.

10. Explain what happens when you attempt to apply the instructions in Exercise 9 to the boundary-value problem

$$y'' - y = 0, \qquad \text{for } 0 \le x \le b \quad \text{with} \quad y(0) = 0 \quad \text{and} \quad y(b) = B.$$

11.3 Linear Finite Difference Methods

The Shooting method often has round-off error difficulties. The methods we present in this section have better rounding characteristics, but they generally require more computation to obtain a specified accuracy.

Methods involving finite differences for solving boundary-value problems replace each of the derivatives in the differential equation with an appropriate difference-quotient approximation of the type considered in Section 4.9. The particular difference quotient is chosen to maintain a specified order of error.

The finite-difference method for the linear second-order boundary-value problem,

$$y'' = p(x)y' + q(x)y + r(x), \qquad \text{for } a \le x \le b, \quad \text{where} \quad y(a) = \alpha \quad \text{and} \quad y(b) = \beta,$$

requires that difference-quotient approximations be used for approximating both y' and y''. First, we select an integer $N > 0$ and divide the interval $[a, b]$ into $(N + 1)$ equal subintervals, whose endpoints are the mesh points $x_i = a + ih$, for $i = 0, 1, \ldots, N + 1$, where $h = (b - a)/(N + 1)$.

At the interior mesh points, x_i, for $i = 1, 2, \ldots, N$, the differential equation to be approximated is

$$y''(x_i) = p(x_i)y'(x_i) + q(x_i)y(x_i) + r(x_i). \tag{11.5}$$

Expanding y in a third-degree Taylor polynomial about x_i evaluated at x_{i+1} and x_{i-1}, we have, assuming that $y \in C^4[x_{i-1}, x_{i+1}]$,

$$y(x_{i+1}) = y(x_i + h) = y(x_i) + hy'(x_i) + \frac{h^2}{2}y''(x_i) + \frac{h^3}{6}y'''(x_i) + \frac{h^4}{24}y^{(4)}(\xi_i^+),$$

for some ξ_i^+ in (x_i, x_{i+1}), and

$$y(x_{i-1}) = y(x_i - h) = y(x_i) - hy'(x_i) + \frac{h^2}{2}y''(x_i) - \frac{h^3}{6}y'''(x_i) + \frac{h^4}{24}y^{(4)}(\xi_i^-),$$

for some ξ_i^- in (x_{i-1}, x_i). If these equations are added, we have

$$y(x_{i+1}) + y(x_{i-1}) = 2y(x_i) + h^2y''(x_i) + \frac{h^4}{24}\left[y^{(4)}\left(\xi_i^+\right) + y^{(4)}\left(\xi_i^-\right)\right],$$

and a simple algebraic manipulation gives

$$y''(x_i) = \frac{1}{h^2}\left[y(x_{i+1}) - 2y(x_i) + y(x_{i-1})\right] - \frac{h^2}{24}\left[y^{(4)}\left(\xi_i^+\right) + y^{(4)}\left(\xi_i^-\right)\right].$$

The Intermediate Value Theorem can be used to simplify this even further.

Centered-Difference Formula for $y''(x_i)$

$$y''(x_i) = \frac{1}{h^2}\left[y(x_{i+1}) - 2y(x_i) + y(x_{i-1})\right] - \frac{h^2}{12}y^{(4)}(\xi_i),$$

for some ξ_i in (x_{i-1}, x_{i+1}).

A centered-difference formula for $y'(x_i)$ is obtained in a similar manner.

Centered-Difference Formula for $y'(x_i)$

$$y'(x_i) = \frac{1}{2h}[y(x_{i+1}) - y(x_{i-1})] - \frac{h^2}{6}y'''(\eta_i),$$

for some η_i in (x_{i-1}, x_{i+1}).

The use of these centered-difference formulas in Eq. (11.5) results in the equation

$$\frac{y(x_{i+1}) - 2y(x_i) + y(x_{i-1})}{h^2} = p(x_i)\left[\frac{y(x_{i+1}) - y(x_{i-1})}{2h}\right] + q(x_i)y(x_i)$$
$$+ r(x_i) - \frac{h^2}{12}[2p(x_i)y'''(\eta_i) - y^{(4)}(\xi_i)].$$

A Finite-Difference method with truncation error of order $O(h^2)$ results from using this equation together with the boundary conditions $y(a) = \alpha$ and $y(b) = \beta$ to define

$$w_0 = \alpha, \qquad w_{N+1} = \beta,$$

and

$$\left(\frac{2w_i - w_{i+1} - w_{i-1}}{h^2}\right) + p(x_i)\left(\frac{w_{i+1} - w_{i-1}}{2h}\right) + q(x_i)w_i = -r(x_i)$$

for each $i = 1, 2, \ldots, N$.

In the form we will consider, the equation is rewritten as

$$-\left(1 + \frac{h}{2}p(x_i)\right)w_{i-1} + \left(2 + h^2q(x_i)\right)w_i - \left(1 - \frac{h}{2}p(x_i)\right)w_{i+1} = -h^2r(x_i),$$

and the resulting system of equations is expressed in the tridiagonal $N \times N$ matrix form $A\mathbf{w} = \mathbf{b}$, where

$$A = \begin{bmatrix} 2 + h^2 q(x_1) & -1 + \frac{h}{2}p(x_1) & 0 & \cdots & \cdots & 0 \\ -1 - \frac{h}{2}p(x_2) & 2 + h^2 q(x_2) & -1 + \frac{h}{2}p(x_2) & \ddots & & \vdots \\ 0 & \ddots & \ddots & \ddots & \ddots & 0 \\ \vdots & & \ddots & \ddots & \ddots & -1 + \frac{h}{2}p(x_{N-1}) \\ 0 & \cdots & \cdots & 0 & -1 - \frac{h}{2}p(x_N) & 2 + h^2 q(x_N) \end{bmatrix},$$

$$\mathbf{w} = \begin{bmatrix} w_1 \\ w_2 \\ \vdots \\ w_{N-1} \\ w_N \end{bmatrix}, \quad \text{and} \quad \mathbf{b} = \begin{bmatrix} -h^2 r(x_1) + \left(1 + \frac{h}{2}p(x_1)\right) w_0 \\ -h^2 r(x_2) \\ \vdots \\ -h^2 r(x_{N-1}) \\ -h^2 r(x_N) + \left(1 - \frac{h}{2}p(x_N)\right) w_{N+1} \end{bmatrix}.$$

This system has a unique solution provided that p, q, and r are continuous on $[a, b]$, that $q(x) \geq 0$ on $[a, b]$, and that $h < 2/L$, where $L = \max_{a \leq x \leq b} |p(x)|$.

The program LINFD112 implements the Linear Finite-Difference method.

EXAMPLE 1 The Linear Finite-Difference method will be used to approximate the solution to the linear boundary-value problem

$$y'' = -\frac{2}{x}y' + \frac{2}{x^2}y + \frac{\sin(\ln x)}{x^2}, \quad \text{for } 1 \leq x \leq 2, \quad \text{where} \quad y(1) = 1 \quad \text{and} \quad y(2) = 2,$$

which was also approximated by the Shooting method in Example 1 of Section 11.2. For this example, we use the same spacing as in Example 1 of Section 11.2.

To use Maple to apply the Linear Finite-Difference method, first we need to access the linear algebra library with the command

```
>with(linalg);
```

Then we define the endpoints of the interval, the boundary conditions, N, and h.

```
>a:=1; b:=2; alpha:=1; beta:=2; N:=9; h:=(b-a)/(N+1);
```

The mesh points are defined in the following loop:

```
>For i from 1 to N do
>x[i]:=a+i*h;
>od;
```

The functions $p(x)$, $q(x)$, and $r(x)$ are defined by

```
>p:=x->-2/x;
>q:=x->2/x^2;
>r:=x->sin(ln(x))/x^2;
```

We initialize the 9×10 array A as the zero matrix.

```
>A:=matrix(9,10,0);
```

Then we generate the nonzero entries with the following statements:

```
>A[1,1]:=2+h*h*evalf(q(x[1]));
>A[1,2]:=-1+h*evalf(p(x[1]))/2;
>A[1,N+1]:=-h*h*evalf(r(x[1]))+(1+h*p(x[1])/2)*alpha;
>for i from 2 to N-1 do
>A[i,i-1]:=-1-h*evalf(p(x[i]))/2;
>A[i,i]:=2+h*h*evalf(q(x[i]));
>A[i,i+1]:=-1+h*evalf(p(x[i]))/2;
>A[i,N+1]:=-h*h*evalf(r(x[i]));
>od;
>A[N,N-1]:=-1-h*evalf(p(x[N]))/2;
>A[N,N]:=2+h*h*evalf(q(x[N]));
>A[N,N+1]:=-h*h*evalf(r(x[N]))+(1-h*p(x[N])/2)*beta;
```

We now apply Gaussian elimination to solve the 9×9 linear system for the values w_1, $w_2, \ldots, w_9$.

```
>C:=gausselim(A);
>w:=backsub(C);
```

The complete results are presented in Table 11.2. ◆ ◆ ◆

Table 11.2

x_i	w_i	$y(x_i)$	$\lvert w_i - y(x_i)\rvert$
1.0	1.00000000	1.00000000	
1.1	1.09260052	1.09262930	2.88×10^{-5}
1.2	1.18704313	1.18708484	4.17×10^{-5}
1.3	1.28333687	1.28338236	4.55×10^{-5}
1.4	1.38140204	1.38144595	4.39×10^{-5}
1.5	1.48112026	1.48115942	3.92×10^{-5}
1.6	1.58235990	1.58239246	3.26×10^{-5}
1.7	1.68498902	1.68501396	2.49×10^{-5}
1.8	1.78888175	1.78889853	1.68×10^{-5}
1.9	1.89392110	1.89392951	8.41×10^{-6}
2.0	2.00000000	2.00000000	

Note that these results are considerably less accurate than those obtained in Example 1 of Section 11.2 and listed in Table 11.1. This is because the method used in that example involved a Runge-Kutta technique with error of order $O(h^4)$, whereas the difference method used here has error of order $O(h^2)$.

To obtain a difference method with greater accuracy, we can proceed in a number of ways. Using fifth-order Taylor series for approximating $y''(x_i)$ and $y'(x_i)$ results in an error term involving h^4. However, this requires using multiples not only of $y(x_{i+1})$ and $y(x_{i-1})$, but also $y(x_{i+2})$ and $y(x_{i-2})$ in the approximation formulas for $y''(x_i)$ and $y'(x_i)$. This leads to difficulty at $i = 0$ and $i = N$. Moreover, the resulting system of equations is not in tridiagonal form, and the solution to the system requires many more calculations.

Instead of obtaining a difference method with a higher-order error term in this manner, it is generally more satisfactory to consider a reduction in step size. In addition, the Richardson's extrapolation technique can be used effectively for this method, since the error term is expressed in even powers of h with coefficients independent of h, provided y is sufficiently differentiable.

EXAMPLE 2 Richardson's extrapolation for approximating the solution to the boundary-value problem

$$y'' = -\frac{2}{x}y' + \frac{2}{x^2}y + \frac{\sin(\ln x)}{x^2}, \qquad \text{for } 1 \le x \le 2, \quad \text{where} \quad y(1) = 1 \quad \text{and} \quad y(2) = 2,$$

with $h = 0.1, 0.05$, and 0.025, gives the results listed in Table 11.3. The first extrapolation is

$$\text{Ext}_{1i} = \frac{4w_i(h = 0.05) - w_i(h = 0.1)}{3};$$

the second extrapolation is

$$\text{Ext}_{2i} = \frac{4w_i(h = 0.025) - w_i(h = 0.05)}{3};$$

Table 11.3

x_i	$w_i(h = 0.1)$	$w_i(h = 0.05)$	$w_i(h = 0.025)$	Ext_{1i}	Ext_{2i}	Ext_{3i}
1.0	1.00000000	1.00000000	1.00000000	1.00000000	1.00000000	1.00000000
1.1	1.09260052	1.09262207	1.09262749	1.09262925	1.09262930	1.09262930
1.2	1.18704313	1.18707436	1.18708222	1.18708477	1.18708484	1.18708484
1.3	1.28333687	1.28337094	1.28337950	1.28338230	1.28338236	1.28338236
1.4	1.38140204	1.38143493	1.38144319	1.38144589	1.38144595	1.38144595
1.5	1.48112026	1.48114959	1.48115696	1.48115937	1.48115941	1.48115942
1.6	1.58235990	1.58238429	1.58239042	1.58239242	1.58239246	1.58239246
1.7	1.68498902	1.68500770	1.68501240	1.68501393	1.68501396	1.68501396
1.8	1.78888175	1.78889432	1.78889748	1.78889852	1.78889853	1.78889853
1.9	1.89392110	1.89392740	1.89392898	1.89392950	1.89392951	1.89392951
2.0	2.00000000	2.00000000	2.00000000	2.00000000	2.00000000	2.00000000

and the final extrapolation is

$$\text{Ext}_{3i} = \frac{16\text{Ext}_{2i} - \text{Ext}_{1i}}{15}.$$

All the results of Ext_{3i} are correct to the decimal places listed. In fact, if sufficient digits are maintained, this approximation gives results that agree with the exact solution with a maximum error of 6.3×10^{-11}. ◆ ◆ ◆

EXERCISE SET 11.3

1. The boundary-value problem

$$y'' = 4(y - x), \qquad \text{for } 0 \le x \le 1 \text{ with } y(0) = 0 \text{ and } y(1) = 2$$

has the solution $y(x) = e^2(e^4 - 1)^{-1}(e^{2x} - e^{-2x}) + x$. Use the Linear Finite-Difference method to approximate the solution and compare the results to the actual solution.

a. With $h = \frac{1}{3}$ **b.** With $h = \frac{1}{4}$.

2. The boundary-value problem

$$y'' = y' + 2y + \cos x, \text{ for } 0 \le x \le \frac{\pi}{2} \text{ with } y(0) = -0.3 \text{ and } y\left(\frac{\pi}{2}\right) = -0.1$$

has the solution $y(x) = -\frac{1}{10}(\sin x + 3\cos x)$. Use the Linear Finite-Difference method to approximate the solution and compare the results to the actual solution.

a. With $h = \dfrac{\pi}{4}$ **b.** With $h = \dfrac{\pi}{6}$

3. Use the Linear Finite-Difference method to approximate the solution to the following boundary-value problems.

a. $y'' = -3y' + 2y + 2x + 3$, for $0 \le x \le 1$ with $y(0) = 2$ and $y(1) = 1$; use $h = 0.1$.

b. $y'' = -\dfrac{4}{x}y' - \dfrac{2}{x^2}y + \dfrac{2}{x^2}\ln x$, for $1 \le x \le 2$ with $y(1) = -\dfrac{1}{2}$ and $y(2) = \ln 2$; use $h = 0.05$.

c. $y'' = -(x + 1)y' + 2y + (1 - x^2)e^{-x}$, for $0 \le x \le 1$ with $y(0) = -1$ and $y(1) = 0$; use $h = 0.1$.

d. $y'' = \dfrac{y'}{x} + \dfrac{3y}{x^2} + \dfrac{\ln x}{x} - 1$, for $1 \le x \le 2$ for $y(1) = y(2) = 0$; use $h = 0.1$.

4. Although $q(x) < 0$ in the following boundary-value problems, unique solutions exist and are given. Use the Linear Finite-Difference method to approximate the solutions and compare the results to the actual solutions.

a. $y'' + y = 0, \quad \text{for } 0 \le x \le \frac{\pi}{4}$ with $y(0) = 1$ and $y\left(\frac{\pi}{4}\right) = 1$; use $h = \frac{\pi}{20}$; actual solution $y(x) = \cos x + (\sqrt{2} - 1)\sin x$.

b. $y'' + 4y = \cos x, \quad \text{for } 0 \le x \le \frac{\pi}{4}$ with $y(0) = 0$ and $y\left(\frac{\pi}{4}\right) = 0$; use $h = \frac{\pi}{20}$; actual solution $y(x) = -\frac{1}{3}\cos 2x - \frac{\sqrt{2}}{6}\sin 2x + \frac{1}{3}\cos x$.

c. $y'' = -\frac{4}{x}y' - \frac{2}{x^2}y + \frac{2\ln x}{x^2}, \quad \text{for } 1 \le x \le 2$ with $y(1) = \frac{1}{2}$ and $y(2) = \ln 2$; use $h = 0.05$; actual solution $y(x) = \frac{4}{x} - \frac{2}{x^2} + \ln x - \frac{3}{2}$.

d. $y'' = 2y' - y + xe^x - x, \quad \text{for } 0 \le x \le 2$ with $y(0) = 0$ and $y(2) = -4$; use $h = 0.2$; actual solution $y(x) = \frac{1}{6}x^3e^x - \frac{5}{3}xe^x + 2e^x - x - 2$.

5. Use the Linear Finite-Difference method to approximate the solution $y = e^{-10x}$ to the boundary-value problem

$$y'' = 100y, \qquad \text{for } 0 \le x \le 1 \quad \text{with} \quad y(0) = 1 \quad \text{and} \quad y(1) = e^{-10}.$$

Use $h = 0.1$ and 0.05. Can you explain the consequences?

6. Repeat Exercise 3(a) and (b) using the extrapolation discussed in Example 2.

7. The deflection of a uniformly loaded, long rectangular plate under an axial tension force is governed by a second-order differential equation. Let S represent the axial force and q, the intensity of the uniform load. The deflection w along the elemental length is given by

$$w''(x) - \frac{S}{D}w(x) = \frac{-ql}{2D}x + \frac{q}{2D}x^2, \qquad \text{for } 0 \le x \le l \quad \text{with} \quad w(0) = w(l) = 0,$$

where l is the length of the plate and D is the flexual rigidity of the plate. Let $q = 200$ lb/in.^{2}, $S = 100$ lb/in., $D = 8.8 \times 10^7$ lb/in., and $l = 50$ in. Approximate the deflection at 1-in. intervals.

8. The boundary-value problem governing the deflection of a beam with supported ends subject to uniform loading is

$$\frac{d^2w}{dx^2} = \frac{S}{EI}w + \frac{qx}{2EI}(x - l), \qquad \text{for } 0 < x < l \quad \text{with} \quad w(0) = 0 \quad \text{and} \quad w(l) = 0.$$

Suppose the beam is a W10-type steel I-beam with the following characteristics: length $l = 120$ in., intensity of uniform load $q = 100$ lb/ft, modulus of elasticity $E = 3.0 \times 10^7$ lb/in.^{2}, stress at ends $S = 1000$ lb, and central moment of inertia $I = 625$ in.^{4}.

a. Approximate the deflection $w(x)$ of the beam every 6 in.

b. The actual relationship is given by

$$w(x) = c_1 e^{ax} + c_2 e^{-ax} + b(x - l)x + c,$$

where $c_1 = 7.7042537 \times 10^4$, $c_2 = 7.9207462 \times 10^4$, $a = 2.3094010 \times 10^{-4}$, $b = -4.1666666 \times 10^{-3}$, and $c = -1.5625 \times 10^5$. Is the maximum error on the interval within 0.2 in.?

c. State law requires that $\max_{0<x<l} w(x) < 1/300$. Does this beam meet state code?

11.4 The Nonlinear Shooting Method

The shooting technique for the nonlinear second-order boundary-value problem

$$y'' = f(x, y, y'), \qquad \text{for } a \le x \le b, \quad \text{where} \quad y(a) = \alpha \quad \text{and} \quad y(b) = \beta, \tag{11.6}$$

is similar to the linear method, except that the solution to a nonlinear problem cannot be expressed as a linear combination of the solutions to two initial-value problems. Instead, we approximate the solution to the boundary-value problem by using the solutions to a sequence of initial-value problems involving a parameter t. These problems have the form

$$y'' = f(x, y, y'), \quad \text{for } a \le x \le b, \quad \text{where} \quad y(a) = \alpha \quad \text{and} \quad y'(a) = t. \tag{11.7}$$

We do this by choosing the parameters $t = t_k$ in a manner to ensure that

$$\lim_{k \to \infty} y(b, t_k) = y(b) = \beta,$$

where $y(x, t_k)$ denotes the solution to the initial-value problem (11.7) with $t = t_k$ and $y(x)$ denotes the solution to the boundary-value problem (11.6).

This technique is called a *shooting* method, by analogy to the procedure of firing objects at a stationary target. (See Figure 11.2.) We start with a parameter t_0 that determines the

Figure 11.2

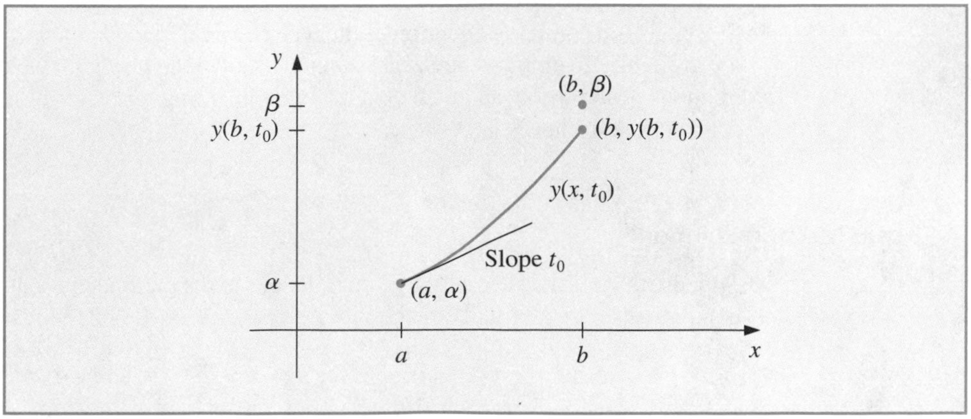

initial elevation at which the object is fired from the point (a, α) along the curve described by the solution to the initial-value problem:

$$y'' = f(x, y, y'), \quad \text{for } a \leq x \leq b, \quad \text{where} \quad y(a) = \alpha \quad \text{and} \quad y'(a) = t_0.$$

If $y(b, t_0)$ is not sufficiently close to β, we correct our approximation by choosing elevations t_1, t_2, and so on, until $y(b, t_k)$ is sufficiently close to "hitting" β. (See Figure 11.3.)

Figure 11.3

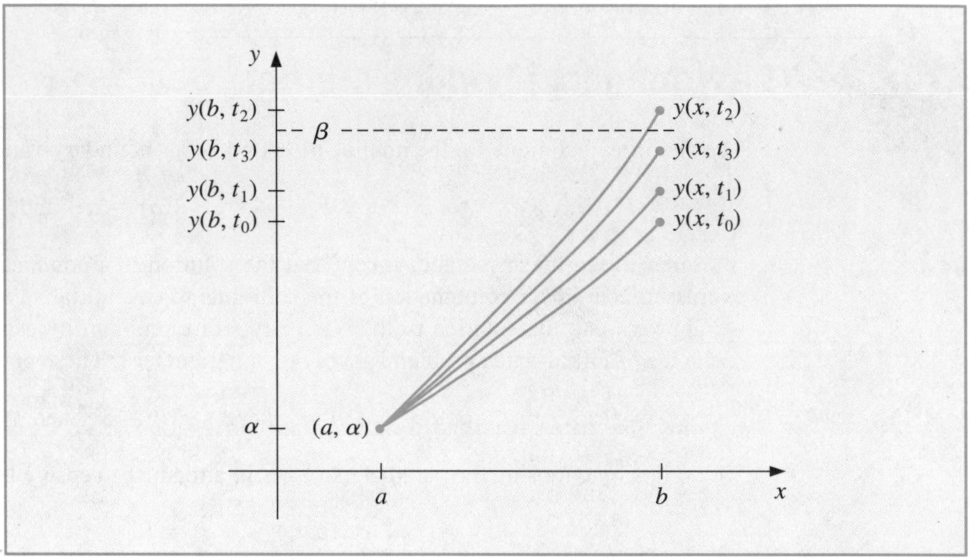

The problem is to determine the parameter t in the initial-value problem so that

$$y(b, t) - \beta = 0.$$

Since this is a nonlinear equation of the type considered in Chapter 2, a number of methods are available. To employ the Secant method to solve the problem, we choose initial approximations t_0 and t_1 to t and then generate the remaining terms of the sequence by using the following procedure.

Secant Method Solution

Suppose that t_0 and t_1 are initial approximations to the parameter t that solves the nonlinear equation $y(b, t) - \beta = 0$. For each successive $k = 2, 3, \ldots$, solve the initial-value problem

$$y'' = f(x, y, y'), \quad \text{for } a \leq x \leq b, \quad \text{where} \quad y(a) = \alpha,$$

with $y'(a) = t_{k-2}$ to find $y(b, t_{k-2})$ and with $y'(a) = t_{k-1}$ to find $y(b, t_{k-1})$. Then define

$$t_k = t_{k-1} - \frac{(y(b, t_{k-1}) - \beta)(t_{k-1} - t_{k-2})}{y(b, t_{k-1}) - y(b, t_{k-2})}$$

and repeat the process with t_{k-1} replacing t_{k-2} and t_k replacing t_{k-1}.

To use the more powerful Newton's method to generate the sequence $\{t_k\}$, only one initial approximation, t_0, is needed. However, the iteration has the form

$$t_k = t_{k-1} - \frac{y(b, t_{k-1}) - \beta}{(dy/dt)(b, t_{k-1})},$$

and requires the knowledge of $(dy/dt)(b, t_{k-1})$. This presents a difficulty, since an explicit representation for $y(b, t)$ is not known; we know only the values $y(b, t_0)$, $y(b, t_1), \ldots, y(b, t_{k-1})$.

To overcome this difficulty, first rewrite the initial-value problem, emphasizing that the solution depends on both x and t:

$$y''(x, t) = f(x, y(x, t), y'(x, t)), \quad \text{for } a \le x \le b, \text{ where } y(a, t) = \alpha \text{ and } y'(a, t) = t,$$

retaining the prime notation to indicate differentiation with respect to x. Since we are interested in determining $(dy/dt)(b, t)$ when $t = t_{k-1}$, we take the partial derivative with respect to t. This implies that

$$\begin{aligned}
\frac{\partial y''}{\partial t}(x, t) &= \frac{\partial f}{\partial t}(x, y(x, t), y'(x, t)) \\
&= \frac{\partial f}{\partial x}(x, y(x, t), y'(x, t))\frac{\partial x}{\partial t} + \frac{\partial f}{\partial y}(x, y(x, t), y'(x, t))\frac{\partial y}{\partial t}(x, t) \\
&\quad + \frac{\partial f}{\partial y'}(x, y(x, t), y'(x, t))\frac{\partial y'}{\partial t}(x, t)
\end{aligned}$$

or, since x and t are independent,

$$\frac{\partial y''}{\partial t}(x, t) = \frac{\partial f}{\partial y}\left(x, y(x, t), y'(x, t)\right)\frac{\partial y}{\partial t}(x, t) + \frac{\partial f}{\partial y'}\left(x, y(x, t), y'(x, t)\right)\frac{\partial y'}{\partial t}(x, t) \qquad \textbf{(11.8)}$$

for $a \le x \le b$. The initial conditions give

$$\frac{\partial y}{\partial t}(a, t) = 0 \quad \text{and} \quad \frac{\partial y'}{\partial t}(a, t) = 1.$$

If we simplify the notation by using $z(x, t)$ to denote $(\partial y/\partial t)(x, t)$ and assume that the order of differentiation of x and t can be reversed, Eq. (11.8) becomes the linear initial-value problem

$$z''(x,t) = \frac{\partial f}{\partial y}(x,y,y')z(x,t) + \frac{\partial f}{\partial y'}(x,y,y')z'(x,t), \qquad \text{for } a \le x \le b,$$

where $z(a,t) = 0$ and $z'(a,t) = 1$. Newton's method therefore requires that two initial-value problems be solved for each iteration.

Newton's Method Solution

Suppose that t_0 is an initial approximation to the parameter t that solves the nonlinear equation $y(b,t) - \beta = 0$. For each successive $k = 1, 2, \ldots$, solve the initial-value problems

$$y'' = f(x,y,y'), \qquad \text{for } a \le x \le b, \quad \text{where} \quad y(a) = \alpha \quad \text{and} \quad y'(a) = t_{k-1}$$

and

$$z'' = f_y(x,y,y')z + f_{y'}(x,y,y')z', \quad \text{for } a \le x \le b, \quad \text{where} \quad z(a,t) = 0 \quad \text{and} \quad z'(a,t) = 1.$$

Then define

$$t_k = t_{k-1} - \frac{y(b,t_{k-1}) - \beta}{z(b,t_{k-1})}$$

and repeat the process with t_k replacing t_{k-1}.

In practice, none of these initial-value problems is likely to be solved exactly; instead the solutions are approximated by one of the methods discussed in Chapter 5. The program NLINS113 uses the Runge-Kutta method of order 4 to approximate both solutions required for Newton's method.

EXAMPLE 1 Consider the boundary-value problem

$$y'' = \frac{1}{8}(32 + 2x^3 - yy'), \qquad \text{for } 1 \le x \le 3, \quad \text{where} \quad y(1) = 17 \quad \text{and} \quad y(3) = \frac{43}{3},$$

which has the exact solution $y(x) = x^2 + 16/x$.

Applying the Shooting method to this problem requires approximating the solutions to the initial-value problems

$$y'' = \frac{1}{8}(32 + 2x^3 - yy'), \quad \text{for } 1 \le x \le 3, \quad \text{where} \quad y(1) = 17 \quad \text{and} \quad y'(1) = t_k,$$

and

$$z'' = \frac{\partial f}{\partial y}z + \frac{\partial f}{\partial y'}z' = -\frac{1}{8}(y'z + yz'), \quad \text{for } 1 \leq x \leq 3, \quad \text{where } z(1) = 0 \text{ and } z'(1) = 1,$$

at each step in the iteration.

We will use the Runge-Kutta method of order 4 within Maple to solve both differential equations. To access the method enter the commands

```
>with(share);
>readshare(ODE,plots);
```

We write the second order equation as a system of two first order diffential equations,

```
>f1:=(x,u1,u2)->u2;
>f2:=(x,u1,u2)->(32+2*x^3-u1*u2)/8;
```

We define a, b, N, α, and β by

```
>a:=1; b:=3; N:=20; h:=(b-a)/N; alpha:=17; beta:=43/3;
```

For the initial value of t_0 we use

```
>tk:=(beta-alpha)/(b-a);
```

The Runge-Kutta method of order 4 for solving this system is invoked by

```
>r:=rungekutta([f1,f2],[1,alpha,tk],h,N);
```

We use the values returned from the Runge-Kutta method to solve the second initial value problem written as a system of two first-order differential equations. We set up the system as follows:

```
>fy:=(x,v1,v2)->-v2/8;
>fyp:=(x,v1,v2)->-v1/8;
>u1:=0;
>u2:=1;
```

Since we need to use the approximations to y and y_1 at t_i, the Runge-Kutta method cannot be called from Maple. We must generate our own code for the computation. We will include only the values from $z(x_i, t_0)$.

```
>for i from 1 to N do
>k11:=h*u2;
```

```
>k12:=h*fy(x,r[i-1][2],r[i-1][3])*u1+h*fyp(x,r[i-1][2],r[i-1][3])
 *u2;
>k21:=h*(u2+k12/2);
>k22:=h*fy(x+h/2,r[i-1][2],r[i-1][3])*(u1+k11/2)+h*fyp(x+h/2,r[i-1]
 [2],r[i-1][3])*(u2+k12/2);
>k31:=h*(u2+k22/2);
>k32:=h*fy(x+h/2,r[i-1][2],r[i-1][3])*(u1+k21/2)+h*fyp(x+h/2,
 r[i-1][2],r[i-1][3])*(u2+k22/2);
>k41:=h*(u2+k32);
>k42:=h*fy(x+h,r[i-1][2],r[i-1][3])*(u1+k31)+h*fyp(x+h,r[i-1][2],
 r[i-1][3])*(u2+k32);
>uu1:=u1+(k11+2*k21+2*k31+k41)/6;
>uu2:=u2+(k12+2*k22+2*k32+k42)/6;
>u1:=uu1;
>u2:=uu2;
>od;
```

We make our test for convergence on $|y(b, t_0) - \beta|$

```
>abs(r[20][2]-beta);
```

which gives 6.14586912. Clearly, we need at least one more iteration, so we compute $t_1 = -16.20583517$ with

```
>tk1:=tk-(r[20][2]-beta)/u1;
```

and repeat the entire process.

If the stopping technique requires $|w_{1,N}(t_k) - y(3)| \leq 10^{-5}$, this problem requires four iterations and $t_4 = -14.000203$. The results obtained for this value of t are shown in Table 11.4. ◆ ◆ ◆

Table 11.4

x_i	$w_{1,i}$	$y(x_i)$	$\|w_{1,i} - y(x_i)\|$	x_i	$w_{1,i}$	$y(x_i)$	$\|w_{1,i} - y(x_i)\|$
1.0	17.000000	17.000000		2.0	12.000023	12.000000	2.32×10^{-5}
1.1	15.755495	15.755455	4.06×10^{-5}	2.1	12.029066	12.029048	1.84×10^{-5}
1.2	14.773389	14.773333	5.60×10^{-5}	2.2	12.112741	12.112727	1.40×10^{-5}
1.3	13.997752	13.997692	5.94×10^{-5}	2.3	12.246532	12.246522	1.01×10^{-5}
1.4	13.388629	13.388571	5.71×10^{-5}	2.4	12.426673	12.426667	6.68×10^{-6}
1.5	12.916719	12.916667	5.23×10^{-5}	2.5	12.650004	12.650000	3.61×10^{-6}
1.6	12.560046	12.560000	4.64×10^{-5}	2.6	12.913847	12.913845	9.17×10^{-7}
1.7	12.301805	12.301765	4.02×10^{-5}	2.7	13.215924	13.215926	1.43×10^{-6}
1.8	12.128923	12.128889	3.14×10^{-5}	2.8	13.554282	13.554286	3.46×10^{-6}
1.9	12.031081	12.031053	2.84×10^{-5}	2.9	13.927236	13.927241	5.21×10^{-6}
				3.0	14.333327	14.333333	6.69×10^{-6}

Although Newton's method used with the shooting technique requires the solution of an additional initial-value problem, it will generally be faster than the Secant method. Both methods are only locally convergent, since they require good initial approximations.

EXERCISE SET 11.4

1. Use the Nonlinear Shooting method with $h = 0.5$ to approximate the solution to the boundary-value problem

$$y'' = -(y')^2 - y + \ln x, \qquad \text{for } 1 \le x \le 2 \quad \text{with} \quad y(1) = 0 \quad \text{and} \quad y(2) = \ln 2.$$

   Compare your results to the actual solution $y = \ln x$.

2. Use the Nonlinear Shooting method with $h = 0.25$ to approximate the solution to the boundary-value problem

$$y'' = 2y^3, \qquad \text{for } -1 \le x \le 0 \quad \text{with } y(-1) = \frac{1}{2} \text{ and } y(0) = \frac{1}{3}.$$

   Compare your results to the actual solution $y(x) = 1/(x + 3)$.

3. Use the Nonlinear Shooting method to approximate the solution to the following boundary-value problems, iterating until $|w_{1,n} - \beta| \le 10^{-4}$. The actual solution is given for comparison to your results.

   a. $y'' = y^3 - yy'$, for $1 \le x \le 2$ with $y(1) = \frac{1}{2}$ and $y(2) = \frac{1}{3}$; use $h = 0.1$ and compare the results to $y(x) = (x + 1)^{-1}$.

   b. $y'' = 2y^3 - 6y - 2x^3$, for $1 \le x \le 2$ with $y(1) = 2$ and $y(2) = \frac{5}{2}$; use $h = 0.1$ and compare the results to $y(x) = x + x^{-1}$.

   c. $y'' = y' + 2(y - \ln x)^3 - x^{-1}$, for $1 \le x \le 2$ with $y(1) = 1$ and $y(2) = \frac{1}{2} + \ln 2$; use $h = 0.1$ and compare the results to $y(x) = x^{-1} + \ln x$.

   d. $y'' = \left[x^2(y')^2 - 9y^2 + 4x^6\right]/x^5$, for $1 \le x \le 3$ with $y(1) = 0$ and $y(3) = 27 \ln 3$; use $h = 0.1$ and compare the results to $y(x) = x^3 \ln x$.

4. Use the Secant method with $t_0 = (\beta - \alpha)/(b - a)$ and $t_1 = t_0 + (\beta - y(b, t_0))/(b - a)$ to solve the problems in Exercises 3(a) and 3(c) and compare the number of iterations required with that of Newton's method.

5. The Van der Pol equation,

$$y'' - \mu(y^2 - 1)y' + y = 0, \qquad \text{for } \mu > 0,$$

   governs the flow of current in a vacuum tube with three internal elements. Let $\mu = \frac{1}{2}$, $y(0) = 0$, and $y(2) = 1$. Approximate the solution $y(t)$ for $t = 0.2i$, where $1 \le i \le 9$.

11.5 Nonlinear Finite-Difference Methods

The difference method for the general nonlinear boundary-value problem

$$y'' = f(x, y, y'), \qquad \text{for } a \le x \le b, \quad \text{where} \quad y(a) = \alpha \quad \text{and} \quad y(b) = \beta,$$

is similar to the method applied to linear problems in Section 11.3. Here, however, the system of equations will not be linear, so an iterative process is required to solve it.

As in the linear case, we divide $[a, b]$ into $(N + 1)$ equal subintervals whose endpoints are at $x_i = a + ih$ for $i = 0, 1, \ldots, N + 1$. Assuming that the exact solution has a bounded fourth derivative allows us to replace $y''(x_i)$ and $y'(x_i)$ in each of the equations by the appropriate centered-difference formula to obtain, for each $i = 1, 2, \ldots, N$,

$$\frac{y(x_{i+1}) - 2y(x_i) + y(x_{i-1})}{h^2} = f\left(x_i, y(x_i), \frac{y(x_{i+1}) - y(x_{i-1})}{2h} - \frac{h^2}{6}y'''(\eta_i)\right) + \frac{h^2}{12}y^{(4)}(\xi_i),$$

for some ξ_i and η_i in the interval (x_{i-i}, x_{i+1}).

The difference method results when the error terms are deleted and the boundary conditions are added. This produces the $N \times N$ nonlinear system

$$\begin{aligned}
2w_1 - w_2 + h^2 f\left(x_1, w_1, \frac{w_2 - \alpha}{2h}\right) - \alpha &= 0,\\
-w_1 + 2w_2 - w_3 + h^2 f\left(x_2, w_2, \frac{w_3 - w_1}{2h}\right) &= 0,\\
&\vdots\\
-w_{N-2} + 2w_{N-1} - w_N + h^2 f\left(x_{N-1}, w_{N-1}, \frac{w_N - w_{N-2}}{2h}\right) &= 0,\\
-w_{N-1} + 2w_N + h^2 f\left(x_N, w_N, \frac{\beta - w_{N-1}}{2h}\right) - \beta &= 0.
\end{aligned}$$

To approximate the solution to this system, we use Newton's method for nonlinear systems, as discussed in Section 10.2. A sequence of iterates $\{(w_1^{(k)}, w_2^{(k)}, \ldots, w_N^{(k)})^t\}$ is generated that converges to the solution of system, provided that the initial approximation $(w_1^{(0)}, w_2^{(0)}, \ldots, w_N^{(0)})^t$ is sufficiently close to the true solution, $(w_1, w_2, \ldots, w_N)^t$.

Newton's method for nonlinear systems requires solving, at each iteration, an $N \times N$ linear system involving the Jacobian matrix. In our case, the Jacobian matrix is tridiagonal, and Crout factorization can be applied. The initial approximations $w_i^{(0)}$ to w_i for each $i = 1, 2, \ldots, N$, are obtained by passing a straight line through (a, α) and (b, β) and evaluating at x_i.

Since a good initial approximation may be required, an upper bound for k should be specified and, if exceeded, a new initial approximation or a reduction in step size considered.

The program NLFDM114 can be used to employ the Nonlinear Finite-Difference method.

EXAMPLE 1 We apply the Nonlinear Finite-Difference method, with $h = 0.1$, to the nonlinear boundary-value problem

$$y'' = \frac{1}{8}\left(32 + 2x^3 - yy'\right), \qquad \text{for } 1 \le x \le 3, \quad \text{where } y(1) = 17 \text{ and } y(3) = \frac{43}{3}.$$

To use Maple, we first access the linear algebra library

```
>with(linalg);
```

and then we define a, b, α, β, N, h, and $\mathbf{w}$ by

```
>a:=1; b:=3; alpha:=17; beta:=43/3; N:=19; h:=(b-a)/(N+1);
>w:=vector(19,0);
```

We define x_i and initialize w_i by passing a straight line through (a, α) and (b, β) and evaluating at x_i. We also define and initialize the 19×19 matrix A and the 19-dimensional vector $\mathbf{u}$.

```
>for i from 1 to N do
>x[i]:=a+i*h;
>w[i]:=alpha+i*(beta-alpha)/(b-a)*h;
>od;
>A:=matrix(19,19,0);
>u:=vector(19,0);
```

The functions $f(x, y, y')$, $f_y(x, y, y')$, and $f_{y'}(x, y, y')$ are defined by

```
>f:=(x,y,yp)->(32+2*x^3-y*yp)/8;
>fy:=(x,y,yp)->-yp/8;
>fyp:=(x,y,yp)->-y/8;
```

The nonzero entries of A and the right-hand side, u, of the linear system are generated as follows:

```
>A[1,1]:=2+h*h*evalf(fy(x[1],w[1],(w[2]-alpha)/(2*h)));
>A[1,2]:=-1+h*evalf(fyp(x[1],w[1],(w[2]-alpha)/(2*h)))/2;
>u[1]:=-(2*w[1]-w[2]-alpha+h*h*evalf(f(x[1],w[1],
 (w[2]-alpha)/(2*h))));
>for i from 2 to N-1 do
>A[i,i-1]:=-1-h*evalf(fyp(x[i],w[i],(w[i+1]-w[i-1])/(2*h)))/2;
>A[i,i+1]:=-1+h*evalf(fyp(x[i],w[i],(w[i+1]-w[i-1])/(2*h)))/2;
>A[i,i]:=2+h*h*evalf(fy(x[i],w[i],(w[i+1]-w[i-1])/(2*h)));
```

```
>u[i]:=-(-w[i-1]+2*w[i]-w[i+1]+h*h*evalf(f(x[i],w[i],
 (w[i+1]-w[i-1])/(2*h))));
>od;
>A[N,N-1]:=-1-h*evalf(fyp(x[N],w[N],(beta-w[N-1])/(2*h)))/2;
>A[N,N]:=2+h*h*evalf(fy(x[N],w[N],(beta-w[N-1])/(2*h)));
>u[N]:=-(-w[N-1]+2*w[N]-beta+h*h*evalf(f(x[N],w[N],
 (beta-w[N-1])/(2*h))));
```

The augmented matrix for the linear system is formed by

```
>B:=augment(A,u);
```

Then we use the Gaussian elimination of Maple with the command

```
>M:=gausselim(B);
```

We obtain the solution to the linear system with the command

```
>v:=backsub(M);
```

The first iteration for $w(x_i)$ is obtained by

```
>z:=w+v;
>w:=evalm(z);
```

We continue to iterate the process and generate the results shown in Table 11.5.

The stopping procedure used in this example was to iterate until values of successive iterates differed by less than 10^{-8}. This was accomplished with four iterations. The problem in this example is the same as that considered for the Nonlinear Shooting method, Example 1 of Section 11.4. ◆ ◆ ◆

Table 11.5

x_i	w_i	$y(x_i)$	$\|w_i - y(x_i)\|$	x_i	w_i	$y(x_i)$	$\|w_i - y(x_i)\|$
1.0	17.000000	17.000000		2.0	11.997915	12.000000	2.085×10^{-3}
1.1	15.754503	15.755455	9.520×10^{-4}	2.1	12.027142	12.029048	1.905×10^{-3}
1.2	14.771740	14.773333	1.594×10^{-3}	2.2	12.111020	12.112727	1.707×10^{-3}
1.3	13.995677	13.997692	2.015×10^{-3}	2.3	12.245025	12.246522	1.497×10^{-3}
1.4	13.386297	13.388571	2.275×10^{-3}	2.4	12.425388	12.426667	1.278×10^{-3}
1.5	12.914252	12.916667	2.414×10^{-3}	2.5	12.648944	12.650000	1.056×10^{-3}
1.6	12.557538	12.560000	2.462×10^{-3}	2.6	12.913013	12.913846	8.335×10^{-4}
1.7	12.299326	12.301765	2.438×10^{-3}	2.7	13.215312	13.215926	6.142×10^{-4}
1.8	12.126529	12.128889	2.360×10^{-3}	2.8	13.553885	13.554286	4.006×10^{-4}
1.9	12.028814	12.031053	2.239×10^{-3}	2.9	13.927046	13.927241	1.953×10^{-4}
				3.0	14.333333	14.333333	

Richardson's extrapolation can also be used for the Nonlinear Finite-Difference method. Table 11.6 lists the results when this method is applied to our example using $h = 0.1, 0.05$, and 0.025, with four iterations in each case. The notation is the same as in Example 2 of Section 11.3, and the values of Ext_{3i} are all accurate to the places listed, with an actual maximum error of 3.68×10^{-10}. The values of $w_i(h = 0.1)$ are omitted from the table since they were listed previously.

Table 11.6

x_i	$w_i(h = 0.05)$	$w_i(h = 0.025)$	Ext_{1i}	Ext_{2i}	Ext_{3i}
1.0	17.00000000	17.00000000	17.00000000	17.00000000	17.00000000
1.1	15.75521721	15.75539525	15.75545543	15.75545460	15.75545455
1.2	14.77293601	14.77323407	14.77333479	14.77333342	14.77333333
1.3	13.99718996	13.99756690	13.99769413	13.99769242	13.99769231
1.4	13.38800424	13.38842973	13.38857346	13.38857156	13.38857143
1.5	12.91606471	12.91651628	12.91666881	12.91666680	12.91666667
1.6	12.55938618	12.55984665	12.56000217	12.56000014	12.56000000
1.7	12.30115670	12.30161280	12.30176684	12.30176484	12.30176471
1.8	12.12830042	12.12874287	12.12899094	12.12888902	12.12888889
1.9	12.03049438	12.03091316	12.03105457	12.03105275	12.03105263
2.0	11.99948020	11.99987013	12.00000179	12.00000011	12.00000000
2.1	12.02857252	12.02892892	12.02902924	12.02904772	12.02904762
2.2	12.11230149	12.11262089	12.11272872	12.11272736	12.11272727
2.3	12.24614846	12.24642848	12.24652299	12.24652182	12.24652174
2.4	12.42634789	12.42658702	12.42666773	12.42666673	12.42666667
2.5	12.64973666	12.64993420	12.65000086	12.65000005	12.65000000
2.6	12.91362828	12.91379422	12.91384683	12.91384620	12.91384615
2.7	13.21577275	13.21588765	13.21592641	13.21592596	13.21592593
2.8	13.55418579	13.55426075	13.55428603	13.55428573	13.55428571
2.9	13.92719268	13.92722921	13.92724153	13.92724139	13.92724138
3.0	14.33333333	14.33333333	14.33333333	14.33333333	14.33333333

EXERCISE SET 11.5

1. Use the Nonlinear Finite-Difference method with $h = 0.5$ to approximate the solution to the boundary-value problem

$$y'' = -(y')^2 - y + \ln x, \qquad \text{for } 1 \le x \le 2 \quad \text{with} \quad y(1) = 0 \quad \text{and} \quad y(2) = \ln 2.$$

Compare your results to the actual solution $y = \ln x$.

2. Use the Nonlinear Finite-Difference method with $h = 0.25$ to approximate the solution to the boundary-value problem

$$y'' = 2y^3, \qquad \text{for } -1 \le x \le 0 \quad \text{with} \quad y(-1) = \tfrac{1}{2} \quad \text{and} \quad y(0) = \tfrac{1}{3}.$$

Compare your results to the actual solution $y(x) = 1/(x + 3)$.

3. Use the Nonlinear Finite-Difference method to approximate the solution to the following boundary-value problems, iterating until successive iterations differ by less than 10^{-4}. The actual solution is given for comparison to your results.
   a. $y'' = y^3 - yy'$, for $1 \le x \le 2$ with $y(1) = \frac{1}{2}$ and $y(2) = \frac{1}{3}$; use $h = 0.1$ and compare the results to $y(x) = (x+1)^{-1}$.
   b. $y'' = 2y^3 - 6y - 2x^3$, for $1 \le x \le 2$ with $y(1) = 2$ and $y(2) = \frac{5}{2}$; use $h = 0.1$ and compare the results to $y(x) = x + x^{-1}$.
   c. $y'' = y' + 2(y - \ln x)^3 - x^{-1}$, for $1 \le x \le 2$ with $y(1) = 1$ and $y(2) = \frac{1}{2} + \ln 2$; use $h = 0.1$ and compare the results to $y(x) = x^{-1} + \ln x$.
   d. $y'' = (x^2(y')^2 - 9y^2 + 4x^6)/x^5$, for $1 \le x \le 3$ with $y(1) = 0$ and $y(3) = 27 \ln 3$; use $h = 0.1$ and compare the results to $y(x) = x^3 \ln x$.
4. Repeat Exercise 3(a) and (b) using extrapolation.
5. In Exercise 8 of Section 11.3 the deflection of beam with supported ends subject to uniform loading was approximated. Using a more appropriate representation of curvature gives the differential equation

$$\left[1 + (w'(x))^2\right]^{-3/2} w''(x) = \frac{S}{EI} w(x) + \frac{qx}{2EI}(x - l), \qquad \text{for } 0 < x < l.$$

   Approximate the deflection $w(x)$ of the beam every 6 in. and compare the results to those of Exercise 8 of Section 11.3.

11.6 Variational Techniques

The Shooting method for approximating the solution to a boundary-value problem replaced it with initial-value problems. The finite-difference approach replaces the continuous operation of differentiation with the discrete operation of finite differences. The Rayleigh-Ritz method is a variational technique that attacks the problem from a third approach. The boundary-value problem is first reformulated as a problem of choosing, from the set of all sufficiently differentiable functions satisfying the boundary conditions, the function to minimize a certain integral. Then the set of feasible functions is reduced in size, to result in an approximation to the solution to the minimization problem and (as a consequence) an approximation to the solution to the boundary-value problem.

To describe the Rayleigh-Ritz method, we consider approximating the solution to a linear two-point boundary-value problem from beam stress analysis. This boundary-value problem is described by the differential equation

$$-\frac{d}{dx}\left(p(x)\frac{dy}{dx}\right) + q(x)y = f(x) \qquad \text{for } 0 \le x \le 1,$$

with the boundary conditions

$$y(0) = y(1) = 0.$$

The differential equation describes the deflection $y(x)$ on a beam of length 1 with variable cross section given by $q(x)$. The deflection is due to the added stresses $p(x)$ and $f(x)$.

As is the case with many boundary-value problems that describe physical phenomena, the solution to the beam equation satisfies a variational property. The solution to the beam equation is the function that minimizes a certain integral over all functions in $C_0^2[0,1]$, the set of those functions u on $[0,1]$ that have two continuous derivatives and satisfy $u(0) = u(1) = 0$. Details concerning this connection can be found in *Spline Analysis* by Schultz [Schu], pp. 88–89.

Variational Property for the Beam Equation

The function $y \in C_0^2[0,1]$ is the unique solution to the boundary-value problem

$$-\frac{d}{dx}\left(p(x)\frac{dy}{dx}\right) + q(x)y = f(x), \qquad \text{for } 0 \le x \le 1,$$

if and only if y is the unique function in $C_0^2[0,1]$ that minimizes the integral

$$I[u] = \int_0^1 \left\{p(x)\left[u'(x)\right]^2 + q(x)\left[u(x)\right]^2 - 2f(x)u(x)\right\} dx.$$

The Rayleigh-Ritz method approximates the solution y by minimizing the integral, not over all the functions in $C_0^2[0,1]$, but over a smaller set of functions consisting of linear combinations of certain basis functions $\phi_1, \phi_2, \ldots, \phi_n$. The basis functions are linearly independent and satisfy

$$\phi_i(0) = \phi_i(1) = 0, \qquad \text{for each } i = 1, 2, \ldots, n.$$

An approximation $\phi(x) = \sum_{i=1}^n c_i\phi_i(x)$ to the solution $y(x)$ is obtained by finding constants $c_1, c_2, \ldots, c_n$ to minimize $I[\sum_{i=1}^n c_i\phi_i]$.

From the variational property,

$$\begin{aligned} I[\phi] &= I\left[\sum_{i=1}^n c_i\phi_i\right] \\ &= \int_0^1 \left[p(x)\left[\sum_{i=1}^n c_i\phi_i'(x)\right]^2 + q(x)\left[\sum_{i=1}^n c_i\phi_i(x)\right]^2 - 2f(x)\sum_{i=1}^n c_i\phi_i(x)\right] dx, \end{aligned}$$

and, for a minimum to occur, it is necessary to have

$$\frac{\partial I}{\partial c_j} = 0, \qquad \text{for each } j = 1, 2, \ldots, n.$$

Differentiating with respect to the coefficients gives

$$\frac{\partial I}{\partial c_j} = \int_0^1 \left[2p(x) \sum_{i=1}^{n} c_i \phi_i'(x)\phi_j'(x) + 2q(x) \sum_{i=1}^{n} c_i \phi_i(x)\phi_j(x) - 2f(x)\phi_j(x) \right] dx,$$

so

$$0 = \sum_{i=1}^{n} \left[\int_0^1 \{p(x)\phi_i'(x)\phi_j'(x) + q(x)\phi_i(x)\phi_j(x)\}\, dx \right] c_i - \int_0^1 f(x)\phi_j(x)\, dx,$$

for each $j = 1, 2, \ldots, n$. These *normal equations* produce an $n \times n$ linear system $A\mathbf{c} = \mathbf{b}$ in the variables $c_1, c_2, \ldots, c_n$, where the symmetric matrix A is given by

$$a_{ij} = \int_0^1 [p(x)\phi_i'(x)\phi_j'(x) + q(x)\phi_i(x)\phi_j(x)]\, dx,$$

and $\mathbf{b}$ has coordinates

$$b_i = \int_0^1 f(x)\phi_i(x)\, dx.$$

The most elementary choice of basis functions involves piecewise linear polynomials. The first step is to form a partition of $[0, 1]$ by choosing points $x_0, x_1, \ldots, x_{n+1}$ with

$$0 = x_0 < x_1 < \cdots < x_n < x_{n+1} = 1.$$

Let $h_i = x_{i+1} - x_i$ for each $i = 0, 1, \ldots, n$, and define the basis functions $\phi_1(x), \phi_2(x), \ldots, \phi_n(x)$ by

$$\phi_i(x) = \begin{cases} 0, & \text{for } 0 \le x \le x_{i-1}, \\ \dfrac{x - x_{i-1}}{h_{i-1}}, & \text{for } x_{i-1} < x \le x_i, \\ \dfrac{x_{i+1} - x}{h_i}, & \text{for } x_i < x \le x_{i+1}, \\ 0, & \text{for } x_{i+1} < x \le 1, \end{cases} \tag{11.9}$$

for each $i = 1, 2, \ldots, n$. (See Figure 11.4.)

Figure 11.4

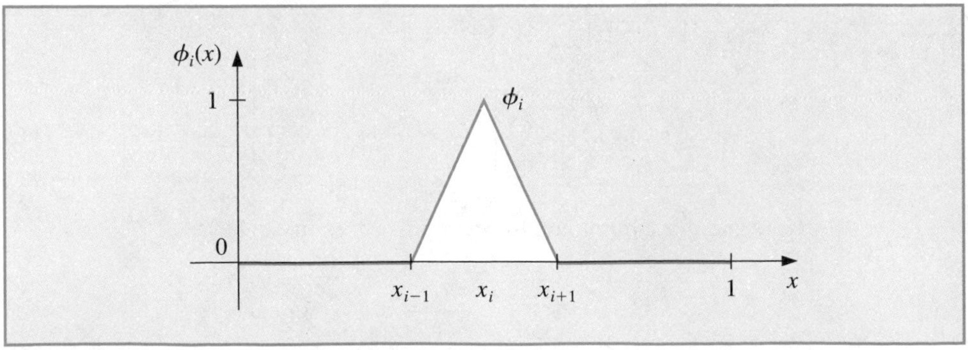

Since the functions ϕ_i are piecewise linear, the derivatives ϕ_i', while not continuous, are constant on the open subinterval (x_j, x_{j+1}) for each $j = 0, 1, \ldots, n$. Thus, we have

$$\phi_i'(x) = \begin{cases} 0, & \text{for } 0 < x < x_{i-1}, \\ \dfrac{1}{h_{i-1}}, & \text{for } x_{i-1} < x < x_i, \\ -\dfrac{1}{h_i}, & \text{for } x_i < x < x_{i+1}, \\ 0, & \text{for } x_{i+1} < x < 1, \end{cases}$$

for each $i = 1, 2, \ldots, n$. Because ϕ_i and ϕ_i' are nonzero only on (x_{i-1}, x_{i+1}),

$$\phi_i(x)\phi_j(x) \equiv 0 \quad \text{and} \quad \phi_i'(x)\phi_j'(x) \equiv 0,$$

except when j is $i-1$, i, or $i+1$. As a consequence, the linear system reduces to an $n \times n$ tridiagonal linear system. The nonzero entries in A are

$$\begin{aligned} a_{ii} &= \int_0^1 \{p(x)[\phi_i'(x)]^2 + q(x)[\phi_i(x)]^2\}\,dx \\ &= \int_{x_{i-1}}^{x_i} \left(\frac{1}{h_{i-1}}\right)^2 p(x)\,dx + \int_{x_i}^{x_{i+1}} \left(\frac{-1}{h_i}\right)^2 p(x)\,dx \\ &\quad + \int_{x_{i-1}}^{x_i} \left(\frac{1}{h_{i-1}}\right)^2 (x - x_{i-1})^2 q(x)\,dx + \int_{x_i}^{x_{i+1}} \left(\frac{1}{h_i}\right)^2 (x_{i+1} - x)^2 q(x)\,dx, \end{aligned}$$

for each $i = 1, 2, \ldots, n$;

$$\begin{aligned} a_{i,i+1} &= \int_0^1 \{p(x)\phi_i'(x)\phi_{i+1}'(x) + q(x)\phi_i(x)\phi_{i+1}(x)\}\,dx \\ &= \int_{x_i}^{x_{i+1}} -\left(\frac{1}{h_i}\right)^2 p(x)\,dx + \int_{x_i}^{x_{i+1}} \left(\frac{1}{h_i}\right)^2 (x_{i+1} - x)(x - x_i) q(x)\,dx, \end{aligned}$$

for each $i = 1, 2, \ldots, n-1$; and

$$\begin{aligned} a_{i,i-1} &= \int_0^1 \{p(x)\phi_i'(x)\phi_{i-1}'(x) + q(x)\phi_i(x)\phi_{i-1}(x)\}\,dx \\ &= \int_{x_{i-1}}^{x_i} -\left(\frac{1}{h_{i-1}}\right)^2 p(x)\,dx + \int_{x_{i-1}}^{x_i} \left(\frac{1}{h_{i-1}}\right)^2 (x_i - x)(x - x_{i-1}) q(x)\,dx, \end{aligned}$$

for each $i = 2, \ldots, n$. The entries in $\mathbf{b}$ are

$$\begin{aligned} b_i &= \int_0^1 f(x)\phi_i(x)\,dx \\ &= \int_{x_{i-1}}^{x_i} \frac{1}{h_{i-1}}(x - x_{i-1}) f(x)\,dx + \int_{x_i}^{x_{i+1}} \frac{1}{h_i}(x_{i+1} - x) f(x)\,dx, \end{aligned}$$

for each $i = 1, 2, \ldots, n$.

There are six types of integrals to be evaluated

$$Q_{1,i} = \left(\frac{1}{h_i}\right)^2 \int_{x_i}^{x_{i+1}} (x_{i+1} - x)(x - x_i)q(x)\,dx, \qquad \text{for each } i = 1, 2, \ldots, n-1,$$

$$Q_{2,i} = \left(\frac{1}{h_{i-1}}\right)^2 \int_{x_{i-1}}^{x_i} (x - x_{i-1})^2 q(x)\,dx, \qquad \text{for each } i = 1, 2, \ldots, n,$$

$$Q_{3,i} = \left(\frac{1}{h_i}\right)^2 \int_{x_i}^{x_{i+1}} (x_{i+1} - x)^2 q(x)\,dx, \qquad \text{for each } i = 1, 2, \ldots, n,$$

$$Q_{4,i} = \left(\frac{1}{h_{i-1}}\right)^2 \int_{x_{i-1}}^{x_i} p(x)\,dx, \qquad \text{for each } i = 1, 2, \ldots, n+1,$$

$$Q_{5,i} = \frac{1}{h_{i-1}} \int_{x_{i-1}}^{x_i} (x - x_{i-1})f(x)\,dx, \qquad \text{for each } i = 1, 2, \ldots, n,$$

and

$$Q_{6,i} = \frac{1}{h_i} \int_{x_i}^{x_{i+1}} (x_{i+1} - x)f(x)\,dx, \qquad \text{for each } i = 1, 2, \ldots, n.$$

Then

$$\begin{aligned} a_{i,i} &= Q_{4,i} + Q_{4,i+1} + Q_{2,i} + Q_{3,i}, \qquad \text{for each } i = 1, 2, \ldots, n, \\ a_{i,i+1} &= -Q_{4,i+1} + Q_{1,i}, \qquad \text{for each } i = 1, 2, \ldots, n-1, \\ a_{i,i-1} &= -Q_{4,i} + Q_{1,i-1}, \qquad \text{for each } i = 2, 3, \ldots, n, \end{aligned}$$

and

$$b_i = Q_{5,i} + Q_{6,i}, \qquad \text{for each } i = 1, 2, \ldots, n.$$

The entries in **c** are the unknown coefficients $c_1, c_2, \ldots, c_n$, from which the Rayleigh-Ritz approximation ϕ, given by $\phi(x) = \sum_{i=1}^{n} c_i\phi_i(x)$, is constructed.

A practical difficulty with this method is the necessity of evaluating $6n$ integrals. The integrals can be evaluated either directly or by a quadrature formula such as Simpson's method. An alternative approach for the integral evaluation is to approximate each of the functions p, q, and f with its piecewise linear interpolating polynomial and then integrate the approximation. Consider, for example, the integral $Q_{1,i}$. The piecewise linear interpolation of q is

$$P_q(x) = \sum_{i=0}^{n+1} q(x_i)\phi_i(x),$$

where $\phi_1, \ldots, \phi_n$ are defined in Eq. (11.9) and

$$\phi_0(x) = \begin{cases} \dfrac{x_1 - x}{x_1}, & 0 \le x \le x_1, \\ 0, & \text{elsewhere,} \end{cases} \qquad \text{and} \qquad \phi_{n+1}(x) = \begin{cases} \dfrac{x - x_n}{1 - x_n}, & x_n \le x \le 1, \\ 0, & \text{elsewhere.} \end{cases}$$

Since the interval of integration is $[x_i, x_{i+1}]$, the piecewise polynomial $P_q(x)$ reduces to

$$P_q(x) = q(x_i)\phi_i(x) + q(x_{i+1})\phi_{i+1}(x).$$

This is the first-degree interpolating polynomial studied in Section 3.2, with error

$$|q(x) - P_q(x)| = O(h_i^2), \qquad \text{when } x_i \le x \le x_{i+1},$$

if $q \in C^2[x_i, x_{i+1}]$. For $i = 1, 2, \ldots, n-1$, the approximation to $Q_{1,i}$ is obtained by integrating the approximation to the integrand

$$\begin{aligned} Q_{1,i} &= \left(\frac{1}{h_i}\right)^2 \int_{x_i}^{x_{i+1}} (x_{i+1} - x)(x - x_i)q(x)\,dx \\ &\approx \left(\frac{1}{h_i}\right)^2 \int_{x_i}^{x_{i+1}} (x_{i+1} - x)(x - x_i)\left[\frac{q(x_i)(x_{i+1} - x)}{h_i} + \frac{q(x_{i+1})(x - x_i)}{h_i}\right] dx \\ &= \frac{h_i}{12}[q(x_i) + q(x_{i+1})], \end{aligned}$$

with

$$\left|Q_{1,i} - \frac{h_i}{12}\left[q(x_i) + q(x_{i+1})\right]\right| = O(h_i^3).$$

Approximations to the other integrals are derived in a similar manner and given by

$$Q_{2,i} \approx \frac{h_{i-1}}{12}\left[3q(x_i) + q(x_{i-1})\right], \qquad Q_{3,i} \approx \frac{h_i}{12}\left[3q(x_i) + q(x_{i+1})\right],$$

$$Q_{4,i} \approx \frac{h_{i-1}}{2}\left[p(x_i) + p(x_{i-1})\right], \qquad Q_{5,i} \approx \frac{h_{i-1}}{6}\left[2f(x_i) + f(x_{i-1})\right],$$

and

$$Q_{6,i} \approx \frac{h_i}{6}[2f(x_i) + f(x_{i+1})].$$

The program PLRRG115 sets up the tridiagonal linear system and incorporates Crout factorization for tridiagonal systems to solve the system. The integrals $Q_{1,i}, \ldots, Q_{6,i}$ can be computed by one of the methods just discussed. Because of the elementary nature of the following example, the integrals were found directly.

EXAMPLE 1 Consider the boundary-value problem

$$-y'' + \pi^2 y = 2\pi^2 \sin(\pi x), \quad \text{for } 0 \le x \le 1, \quad \text{where } y(0) = y(1) = 0.$$

Let $h_i = h = 0.1$, so that $x_i = 0.1i$ for each $i = 0, 1, \ldots, 9$. The integrals are

$$Q_{1,i} = 100 \int_{0.1i}^{0.1i+0.1} (0.1i + 0.1 - x)(x - 0.1i)\pi^2\, dx = \frac{\pi^2}{60},$$

$$Q_{2,i} = 100 \int_{0.1i-0.1}^{0.1i} (x - 0.1i + 0.1)^2 \pi^2\, dx = \frac{\pi^2}{30},$$

$$Q_{3,i} = 100 \int_{0.1i}^{0.1i+0.1} (0.1i + 0.1 - x)^2 \pi^2\, dx = \frac{\pi^2}{30},$$

$$Q_{4,i} = 100 \int_{0.1i-0.1}^{0.1i} dx = 10,$$

$$Q_{5,i} = 10 \int_{0.1i-0.1}^{0.1i} (x - 0.1i + 0.1)2\pi^2 \sin \pi x\, dx$$
$$= -2\pi \cos 0.1\pi i + 20\left[\sin(0.1\pi i) - \sin\left((0.1i - 0.1)\pi\right)\right],$$

and

$$Q_{6,i} = 10 \int_{0.1i}^{0.1i+0.1} (0.1i + 0.1 - x)2\pi^2 \sin \pi x\, dx$$
$$= 2\pi \cos 0.1\pi i - 20\left[\sin\left((0.1i + 0.1)\pi\right) - \sin(0.1\pi i)\right].$$

The linear system $A\mathbf{c} = \mathbf{b}$ has

$$a_{i,i} = 20 + \frac{\pi^2}{15}, \quad \text{for each } i = 1, 2, \ldots, 9,$$

$$a_{i,i+1} = -10 + \frac{\pi^2}{60}, \quad \text{for each } i = 1, 2, \ldots, 8,$$

$$a_{i,i-1} = -10 + \frac{\pi^2}{60}, \quad \text{for each } i = 2, 3, \ldots, 9,$$

and

$$b_i = 40 \sin(0.1\pi i)[1 - \cos 0.1\pi], \qquad \text{for each } i = 1, 2, \ldots, 9.$$

The solution to the tridiagonal linear system is

$$\begin{array}{lll} c_9 = 0.3102866742, & c_6 = 0.9549641893, & c_3 = 0.8123410598, \\ c_8 = 0.5902003271, & c_5 = 1.004108771, & c_2 = 0.5902003271, \\ c_7 = 0.8123410598, & c_4 = 0.9549641893, & c_1 = 0.3102866742. \end{array}$$

The piecewise linear approximation is

$$\phi(x) = \sum_{i=1}^{9} c_i \phi_i(x).$$

The actual solution to the boundary-value problem is

$$y(x) = \sin \pi x.$$

Table 11.7 lists the error in the approximation at x_i for each $i = 1, \ldots, 9$. ◆ ◆ ◆

Table 11.7

i	x_i	$\phi(x_i)$	$y(x_i)$	$\lvert\phi(x_i) - y(x_i)\rvert$
1	0.1	0.3102866742	0.3090169943	0.00127
2	0.2	0.5902003271	0.5877852522	0.00242
3	0.3	0.8123410598	0.8090169943	0.00332
4	0.4	0.9549641896	0.9510565162	0.00391
5	0.5	1.0041087710	1.0000000000	0.00411
6	0.6	0.9549641893	0.9510565162	0.00391
7	0.7	0.8123410598	0.8090169943	0.00332
8	0.8	0.5902003271	0.5877852522	0.00242
9	0.9	0.3102866742	0.3090169943	0.00127

The tridiagonal matrix A given by the piecewise linear basis functions is positive definite, so the linear system is stable with respect to round-off error and

$$|\phi(x) - y(x)| = O(h^2), \qquad \text{when } 0 \le x \le 1.$$

The use of piecewise-linear basis functions results in an approximate solution that is continuous but not differentiable on $[0, 1]$. A more complicated set of basis functions is required to construct an approximation that belongs to $C_0^2[0, 1]$. These basis functions are similar to the cubic interpolatory splines discussed in Section 3.5.

Recall that the cubic *interpolatory* spline S on the five nodes x_0, x_1, x_2, x_3, and x_4 for a function f is defined as follows:

(a) S is a cubic polynomial, denoted by S_j, on $[x_j, x_{j+1}]$, for $j = 0, 1, 2, 3$. (*This gives* 16 *selectable constants for* S, 4 *for each cubic*.)
(b) $S(x_j) = f(x_j)$, for $j = 0, 1, 2, 3, 4$ (5 *specified conditions*).
(c) $S_{j+1}(x_{j+1}) = S_j(x_{j+1})$, for $j = 0, 1, 2$ (3 *specified conditions*).
(d) $S'_{j+1}(x_{j+1}) = S'_j(x_{j+1})$, for $j = 0, 1, 2$ (3 *specified conditions*).
(e) $S''_{j+1}(x_{j+1}) = S''_j(x_{j+1})$, for $j = 0, 1, 2$ (3 *specified conditions*).
(f) One of the following boundary conditions is satisfied:
  (i) Free: $S''(x_0) = S''(x_4) = 0$ (2 *specified conditions*).
  (ii) Clamped: $S'(x_0) = f'(x_0)$ and $S'(x_4) = f'(x_4)$ (2 *specified conditions*).

Since uniqueness of solution requires the number of constants in (a), 16, to equal the number of conditions in (b) through (f), only one of the boundary conditions in (f) can be specified for the interpolatory cubic splines.

The cubic spline functions we will use for our basis functions are called **B-splines**, or *bell-shaped splines*. These differ from interpolatory splines in that both sets of boundary conditions in (f) are satisfied. This requires the relaxation of two of the conditions in (b) through (e). Since the spline must have two continuous derivatives on $[x_0, x_4]$, we delete from the description of the interpolatory splines two of the interpolation conditions. In particular, we modify condition (b) to

b. $S(x_j) = f(x_j)$ for $j = 0, 2, 4$.

The basic B-spline S defined next and shown in Figure 11.5 uses the equally spaced nodes $x_0 = -2$, $x_1 = -1$, $x_2 = 0$, $x_3 = 1$, and $x_4 = 2$. It satisfies the interpolatory conditions

b′. $S(x_0) = 0$, $S(x_2) = 1$, $S(x_4) = 0$;

as well as both sets of conditions

$$\text{(i)}\quad S''(x_0) = S''(x_4) = 0 \quad \text{and} \quad \text{(ii)}\quad S'(x_0) = S'(x_4) = 0.$$

Figure 11.5

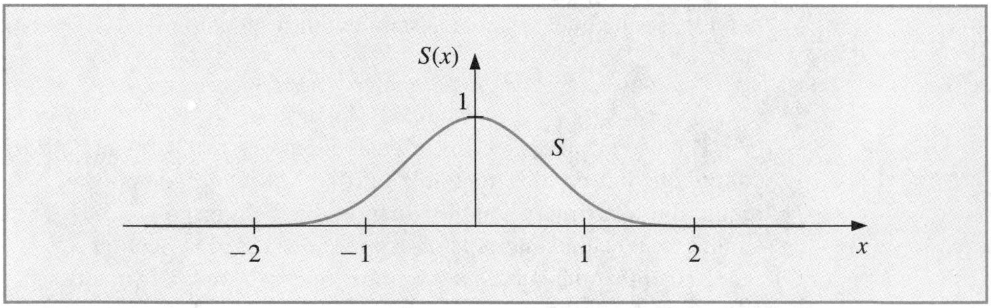

As a consequence, $S \in C^2(-\infty, \infty)$.

$$S(x) = \begin{cases} 0, & \text{for } x \le -2, \\ \frac{1}{4}(2+x)^3, & \text{for } -2 \le x \le -1, \\ \frac{1}{4}\left[(2+x)^3 - 4(1+x)^3\right], & \text{for } -1 < x \le 0, \\ \frac{1}{4}\left[(2-x)^3 - 4(1-x)^3\right], & \text{for } 0 < x \le 1, \\ \frac{1}{4}(2-x)^3, & \text{for } 1 < x \le 2, \\ 0, & \text{for } 2 < x. \end{cases} \tag{11.10}$$

To construct the basis functions ϕ_i in $C_0^2[0, 1]$ we first partition $[0, 1]$ by choosing a positive integer n and defining $h = 1/(n+1)$. This produces the equally spaced nodes $x_i = ih$, for each $i = 0, 1, \ldots, n+1$. We then define the basis functions $\{\phi_i\}_{i=0}^{n+1}$ as

$$\phi_i(x) = \begin{cases} S\left(\frac{x}{h}\right) - 4S\left(\frac{x+h}{h}\right), & \text{for } i = 0, \\ S\left(\frac{x-h}{h}\right) - S\left(\frac{x+h}{h}\right), & \text{for } i = 1, \\ S\left(\frac{x-ih}{h}\right), & \text{for } 2 \le i \le n-1, \\ S\left(\frac{x-nh}{h}\right) - S\left(\frac{x-(n+2)h}{h}\right), & \text{for } i = n, \\ S\left(\frac{x-(n+1)h}{h}\right) - 4S\left(\frac{x-(n+2)h}{h}\right), & \text{for } i = n+1. \end{cases}$$

It is not difficult to show that $\{\phi_i\}_{i=0}^{n+1}$ is a linearly independent set of cubic splines satisfying $\phi_i(0) = \phi_i(1) = 0$, for each $i = 0, 1, \ldots, n, n+1$. The graphs of ϕ_i, for $2 \le i \le n-1$, are shown in Figure 11.6 and the graphs of ϕ_0, ϕ_1, ϕ_n, and ϕ_{n+1} are in Figure 11.7.

Figure 11.6

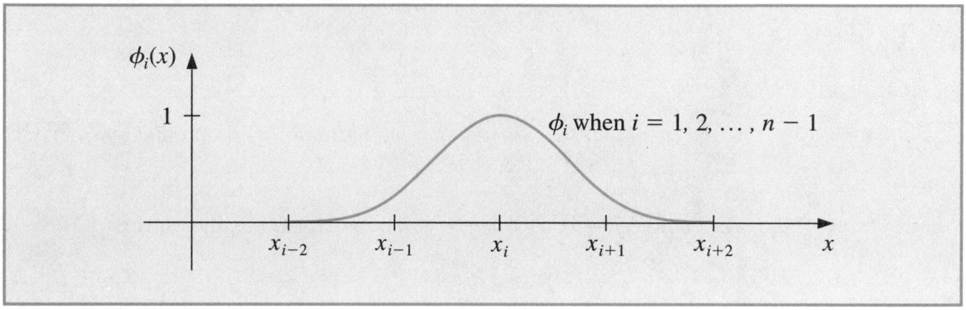

Figure 11.7

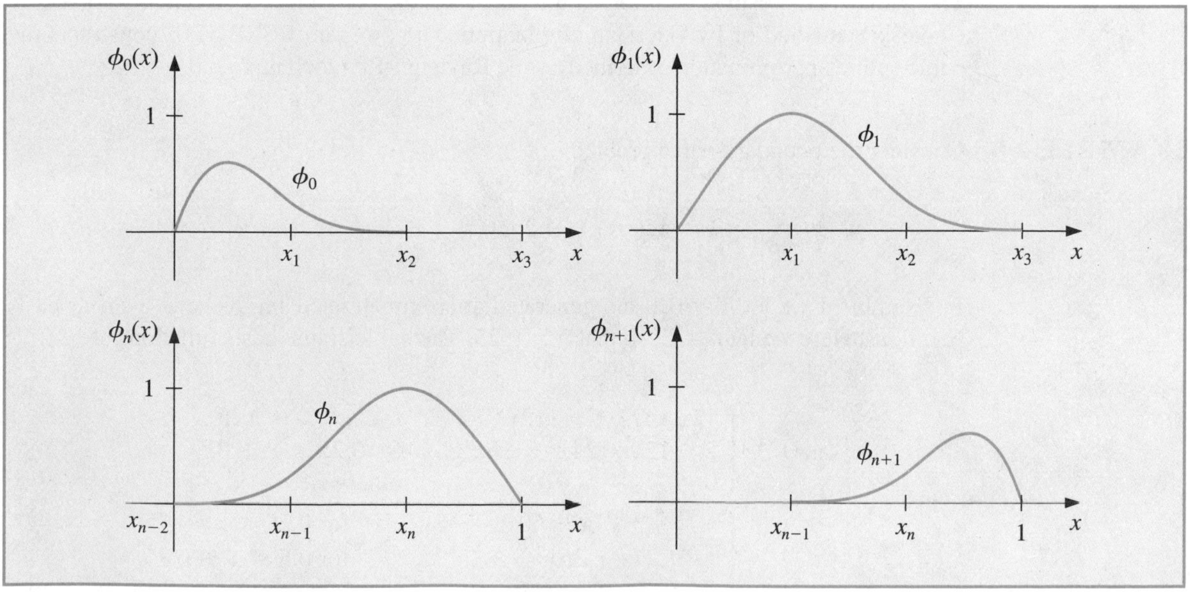

Since $\phi_i(x)$ and $\phi_i'(x)$ are nonzero only for $x_{i-2} \le x \le x_{i+2}$, the matrix in the Rayleigh-Ritz approximation is a band matrix with bandwidth at most seven:

$$A = \begin{bmatrix} a_{00} & a_{01} & a_{02} & a_{03} & 0 & \cdots & \cdots & \cdots & 0 \\ a_{10} & a_{11} & a_{12} & a_{13} & a_{14} & \ddots & & & \vdots \\ a_{20} & a_{21} & a_{22} & a_{23} & a_{24} & a_{25} & \ddots & & \vdots \\ a_{30} & a_{31} & a_{32} & a_{33} & a_{34} & a_{35} & a_{36} & \ddots & \vdots \\ 0 & \ddots & \ddots & \ddots & \ddots & \ddots & \ddots & \ddots & 0 \\ \vdots & \ddots & \ddots & \ddots & \ddots & \ddots & \ddots & \ddots & a_{n-2,n+1} \\ \vdots & & \ddots & \ddots & \ddots & \ddots & \ddots & \ddots & a_{n-1,n+1} \\ \vdots & & & \ddots & \ddots & \ddots & \ddots & \ddots & a_{n,n+1} \\ 0 & \cdots & \cdots & \cdots & 0 & a_{n+1,n-2} & a_{n+1,n-1} & a_{n+1,n} & a_{n+1,n+1} \end{bmatrix}, \qquad \textbf{(11.11)}$$

where

$$a_{ij} = \int_0^1 \{p(x)\phi_i'(x)\phi_j'(x) + q(x)\phi_i(x)\phi_j(x)\}\,dx,$$

for each $i, j = 0, 1, \ldots, n+1$. The vector $\mathbf{b}$ has the entries

$$b_i = \int_0^1 f(x)\phi_i(x)\,dx.$$

The matrix A is positive definite, so the linear system $A\mathbf{c} = \mathbf{b}$ can be quickly solved by Choleski's method or by Gaussian elimination. The program CSRRG116 constructs the cubic spline approximation described by the Rayleigh-Ritz technique.

EXAMPLE 2 Consider the boundary-value problem

$$-y'' + \pi^2 y = 2\pi^2 \sin(\pi x), \qquad \text{for } 0 \le x \le 1, \quad \text{where } y(0) = y(1) = 0.$$

In Example 1 we let $h = 0.1$ and generated approximations using piecewise-linear basis functions. Here we let $n = 3$, so that $h = 0.25$. The cubic spline basis functions are

$$\phi_0(x) = \begin{cases} 12x - 72x^2 + 112x^3, & \text{for } 0 \le x \le 0.25 \\ 2 - 12x + 24x^2 - 16x^3, & \text{for } 0.25 < x \le 0.5 \\ 0, & \text{otherwise} \end{cases}$$

$$\phi_1(x) = \begin{cases} -\frac{1665}{1332} + 21x - 60x^2 + 48x^3, & \text{for } 0.25 < x \le 0.5 \\ \frac{27}{4} - 27x + 36x^2 - 16x^3, & \text{for } 0.5 < x \le 0.75 \\ 0, & \text{otherwise} \end{cases}$$

$$\phi_2(x) = \begin{cases} 16x^3, & \text{for } 0 \le x \le 0.25 \\ 1 - 12x + 48x^2 - 48x^3, & \text{for } 0 < x \le 0.5 \\ -11 + 60x - 96x^2 + 48x^3, & \text{for } 0.5 < x \le 0.75 \\ 16 - 48x + 48x^2 - 16x^3, & \text{for } 0.75 < x \le 1 \\ 0, & \text{otherwise} \end{cases}$$

$$\phi_3(x) = \begin{cases} -\frac{1}{4} + 3x - 12x^2 + 16x^3, & \text{for } 0.25 < x \le 0.5 \\ \frac{3441}{444} - 45x + 84x^2 - 48x^3, & \text{for } 0.5 < x \le 0.75 \\ -26 + 90x - 96x^2 + 32x^3, & \text{for } 0.75 < x \le 1 \\ 0, & \text{otherwise} \end{cases}$$

$$\phi_4(x) = \begin{cases} -2 + 12x - 24x^2 + 16x^3, & \text{for } 0.5 < x \le 0.75 \\ 52 - 204x + 264x^2 - 112x^3, & \text{for } 0.75 \le x \le 1 \\ 0, & \text{otherwise} \end{cases}$$

The entries in the matrix A are generated as follows

$$\begin{aligned} a_{00} &= \int_0^1 \left\{ p(x) \left[\phi_0'(x)\right]^2 + q(x) \left[\phi_0(x)\right]^2 \right\} dx \\ &= \int_0^{0.5} \left\{ \left[\phi_0'(x)\right]^2 + \pi^2 \left[\phi_0(x)\right]^2 \right\} dx = 6.5463531 \end{aligned}$$

$$\begin{aligned} a_{01} = a_{10} &= \int_0^1 \{p(x)\phi_0'(x)\phi_1'(x) + q(x)\phi_0(x)\phi_1(x)\} \, dx \\ &= \int_0^{0.5} \{\phi_0'(x)\phi_1'(x) + \pi^2\phi_0(x)\phi_1(x)\} \, dx = 4.0764737 \end{aligned}$$

$$\begin{aligned} a_{02} = a_{20} &= \int_0^1 \{p(x)\phi_0'(x)\phi_2'(x) + q(x)\phi_0(x)\phi_2(x)\} \, dx \\ &= \int_0^{0.5} \{\phi_0'(x)\phi_2'(x) + \pi^2\phi_0(x)\phi_2(x)\} \, dx = -1.3722239 \end{aligned}$$

$$\begin{aligned} a_{03} = a_{30} &= \int_0^1 \{p(x)\phi_0'(x)\phi_3'(x) + q(x)\phi_0(x)\phi_3(x)\} \, dx \\ &= \int_{0.25}^{0.5} \{\phi_0'(x)\phi_3'(x) + \pi^2\phi_0(x)\phi_3(x)\} \, dx = -0.73898482 \end{aligned}$$

$$\begin{aligned} a_{11} &= \int_0^1 \left\{ p(x) \left[\phi_1'(x)\right]^2 + q(x) \left[\phi_1(x)\right]^2 \right\} dx \\ &= \int_0^{0.75} \left\{ \left[\phi_1'(x)\right]^2 + \pi^2 \left[\phi_1(x)\right]^2 \right\} dx = 10.329086 \end{aligned}$$

$$\begin{aligned} a_{12} = a_{21} &= \int_0^1 \{p(x)\phi_1'(x)\phi_2'(x) + q(x)\phi_1(x)\phi_2(x)\} \, dx \\ &= \int_0^{0.75} \{\phi_1'(x)\phi_2'(x) + \pi^2\phi_1(x)\phi_2(x)\} \, dx = 0.26080684 \end{aligned}$$

$$a_{13} = a_{31} = \int_0^1 \{p(x)\phi_1'(x)\phi_3'(x) + q(x)\phi_1(x)\phi_3(x)\} \, dx$$

$$= \int_{0.25}^{0.75} \{\phi_1'(x)\phi_3'(x) + \pi^2\phi_1(x)\phi_3(x)\}\, dx = -1.6678178$$

$$a_{14} = a_{41} = \int_0^1 \{p(x)\phi_1'(x)\phi_4'(x) + q(x)\phi_1(x)\phi_4(x)\}\, dx$$

$$= \int_{0.5}^{0.75} \{\phi_1'(x)\phi_4'(x) + \pi^2\phi_1(x)\phi_4(x)\}\, dx = -0.73898482$$

$$a_{22} = \int_0^1 \left\{p(x)\left[\phi_2'(x)\right]^2 + q(x)\left[\phi_2(x)\right]^2\right\} dx$$

$$= \int_0^1 \left\{\left[\phi_2'(x)\right]^2 + \pi^2\left[\phi_2(x)\right]^2\right\} dx = 8.6612683$$

$$a_{23} = a_{32} = \int_0^1 \{p(x)\phi_2'(x)\phi_3'(x) + q(x)\phi_2(x)\phi_3(x)\}\, dx$$

$$= \int_{0.25}^1 \{\phi_2'(x)\phi_3'(x) + \pi^2\phi_2(x)\phi_3(x)\}\, dx = 0.26080684$$

$$a_{24} = a_{42} = \int_0^1 \{p(x)\phi_2'(x)\phi_4'(x) + q(x)\phi_2(x)\phi_4(x)\}\, dx$$

$$= \int_{0.5}^1 \{\phi_2'(x)\phi_4'(x) + \pi^2\phi_2(x)\phi_4(x)\}\, dx = -1.3722239$$

$$a_{33} = \int_0^1 \left\{p(x)\left[\phi_3'(x)\right]^2 + q(x)\left[\phi_3(x)\right]^2\right\} dx$$

$$= \int_{0.25}^1 \left\{\left[\phi_3'(x)\right]^2 + \pi^2\left[\phi_3(x)\right]^2\right\} dx = 10.329086$$

$$a_{34} = a_{43} = \int_0^1 \{p(x)\phi_3'(x)\phi_4'(x) + q(x)\phi_3(x)\phi_4(x)\}\, dx$$

$$= \int_{0.5}^1 \{\phi_3'(x)\phi_4'(x) + \pi^2\phi_3(x)\phi_4(x)\}\, dx = 4.0764737$$

$$a_{44} = \int_0^1 \left\{p(x)\left[\phi_4'(x)\right]^2 + q(x)\left[\phi_4(x)\right]^2\right\} dx$$

$$= \int_{0.5}^1 \left\{\left[\phi_4'(x)\right]^2 + \pi^2\left[\phi_4(x)\right]^2\right\} dx = 6.5463531$$

and

$$b_0 = \int_0^1 f(x)\phi_0(x)\, dx = \int_0^{0.5} \left(2\pi^2 \sin \pi x\right) \phi_0(x)\, dx = 1.0803542$$

$$b_1 = \int_0^1 f(x)\phi_1(x)\, dx = \int_0^{0.75} \left(2\pi^2 \sin \pi x\right) \phi_1(x)\, dx = 4.7202512$$

$$b_2 = \int_0^1 f(x)\phi_2(x)\, dx = \int_0^1 \left(2\pi^2 \sin \pi x\right) \phi_2(x)\, dx = 6.6754433$$

$$b_3 = \int_0^1 f(x)\phi_3(x)\,dx = \int_{0.25}^1 (2\pi^2 \sin \pi x)\,\phi_3(x)\,dx = 4.7202512$$

$$b_4 = \int_0^1 f(x)\phi_4(x)\,dx = \int_{0.5}^1 (2\pi^2 \sin \pi x)\,\phi_4(x)\,dx = 1.08035418$$

The solution to the system $A\mathbf{c} = \mathbf{b}$ is $c_0 = 0.00060266150$, $c_1 = 0.52243908$, $c_2 = 0.73945127$, $c_3 = 0.52243908$, and $c_4 = 0.00060264906$. Evaluating $\phi(x) = \sum_{i=0}^{4} c_i\phi_i(x)$ at each x_j, for $0 \le j \le 4$, gives the results in Table 11.8. ◆ ◆ ◆

Table 11.8

i	x_i	$\phi(x_i)$	$y(x_i)$	$\lvert\phi(x_i) - y(x_i)\rvert$
0	0	0	0	0
1	0.25	0.70745256	0.70710678	0.00034578
2	0.5	1.0006708	1	0.0006708
3	0.75	0.70745256	0.70710678	0.00034578
4	1	0	0	0

The integrations should be performed in two steps, as was done in the Piecewise Linear method. First, construct cubic spline interpolatory polynomials for p, q, and f using the methods presented in Section 3.5. Then approximate the integrands by products of cubic splines or derivatives of cubic splines. Since these integrands are piecewise polynomials, they can be integrated exactly on each subinterval and then summed.

In general, this technique produces approximations $\phi(x)$ to $y(x)$ that satisfy

$$\left[\int_0^1 |y(x) - \phi(x)|^2 dx\right]^{1/2} = O(h^4), \qquad 0 \le x \le 1.$$

EXERCISE SET 11.6

1. Use the Piecewise Linear method to approximate the solution to the boundary-value problem

$$y'' + \frac{\pi^2}{4}y = \frac{\pi^2}{16}\cos\frac{\pi}{4}x, \qquad \text{for } 0 \le x \le 1 \quad \text{with } y(0) = y(1) = 0$$

using $x_0 = 0, x_1 = 0.3, x_2 = 0.7, x_3 = 1$ and compare the results to the actual solution $y(x) = -\frac{1}{3}\cos\frac{\pi}{2}x - \frac{\sqrt{2}}{6}\sin\frac{\pi}{2}x + \frac{1}{3}\cos\frac{\pi}{4}x$.

2. Use the Piecewise Linear method to approximate the solution to the boundary-value problem

$$-\frac{d}{dx}(xy') + 4y = 4x^2 - 8x + 1, \qquad \text{for } 0 \le x \le 1 \quad \text{with } y(0) = y(1) = 0$$

using $x_0 = 0$, $x_1 = 0.4$, $x_2 = 0.8$, $x_3 = 1$ and compare the results to the actual solution $y(x) = x^2 - x$.

3. Use the Piecewise Linear method to approximate the solutions to the following boundary-value problems and compare the results to the actual solution:

a. $-x^2y'' - 2xy' + 2y = -4x^2$, for $0 \le x \le 1$ with $y(0) = y(1) = 0$; use $h = 0.1$; actual solution $y(x) = x^2 - x$.

b. $-\frac{d}{dx}(e^x y') + e^x y = x + (2 - x)e^x$, for $0 \le x \le 1$ with $y(0) = y(1) = 0$; use $h = 0.1$; actual solution $y(x) = (x - 1)(e^{-x} - 1)$.

c. $-\frac{d}{dx}(e^{-x} y') + e^{-x} y = (x - 1) - (x + 1)e^{-(x-1)}$, for $0 \le x \le 1$ with $y(0) = y(1) = 0$; use $h = 0.05$; actual solution $y(x) = x(e^x - e)$.

d. $-(x + 1)y'' - y' + (x + 2)y = [2 - (x + 1)^2]e \ln 2 - 2e^x$, for $0 \le x \le 1$ with $y(0) = y(1) = 0$; use $h = 0.05$; actual solution $y(x) = e^x \ln(x + 1) - (e \ln 2)x$.

4. Use the Cubic Spline method with $n = 3$ to approximate the solution to each of the following boundary-value problems and compare the results to the actual solutions.

a. $y'' + \frac{\pi^2}{4}y = \frac{\pi^2}{16}\cos\frac{\pi}{4}x$, for $0 \le x \le 1$ with $y(0) = 0$ and $y(1) = 0$

b. $-\frac{d}{dx}(xy') + 4y = 4x^2 - 8x + 1$, for $0 \le x \le 1$ with $y(0) = 0$ and $y(1) = 0$.

5. Repeat Exercise 3 using the Cubic Spline method.

6. Show that the boundary-value problem

$$-\frac{d}{dx}(p(x)y') + q(x)y = f(x), \qquad \text{for } 0 \le x \le 1 \quad \text{with } y(0) = \alpha \text{ and } y(1) = \beta,$$

can be transformed by the change of variable

$$z = y - \beta x - (1 - x)\alpha$$

into the form

$$-\frac{d}{dx}(p(x)z') + q(x)z = F(x), \quad 0 \le x \le 1, \qquad z(0) = 0, \qquad z(1) = 0.$$

7. Use Exercise 6 and the Piecewise Linear method with $n = 9$ to approximate the solution to the boundary-value problem

$$-y'' + y = x, \quad \text{for } 0 \le x \le 1 \quad \text{with} \quad y(0) = 1 \quad \text{and} \quad y(1) = 1 + e^{-1}.$$

8. Repeat Exercise 7 using the Cubic Spline method.

9. Show that the boundary-value problem

$$-\frac{d}{dx}(p(x)y') + q(x)y = f(x), \qquad \text{for } a \leq x \leq b \quad \text{with } y(a) = \alpha \text{ and } y(b) = \beta,$$

can be transformed into the form

$$-\frac{d}{dw}(p(w)z') + q(w)z = F(w), \qquad \text{for } 0 \leq w \leq 1 \quad \text{with } z(0) = 0 \text{ and } z(1) = 0,$$

by a method similar to that given in Exercise 6.

10. Show that the set of piecewise linear basis functions is linearly independent on [0, 1].

11. Use the definition of positive definite to show that the matrix given by the piecewise linear basis functions satisfies this condition.

11.7 Survey of Methods and Software

In this chapter we discussed methods for approximating solutions to boundary-value problems. For the linear boundary-value problem

$$y'' = p(x)y' + q(x)y + r(x), \quad \text{for } a \leq x \leq b, \quad \text{where} \quad y(a) = \alpha \quad \text{and} \quad y(b) = \beta,$$

we considered both a Linear Shooting method and a Finite-Difference method to approximate the solution. The Shooting method uses an initial-value technique to solve the problems

$$y'' = p(x)y' + q(x)y + r(x), \qquad \text{for } a \leq x \leq b, \quad \text{where} \quad y(a) = \alpha \quad \text{and} \quad y'(a) = 0,$$

and

$$y'' = p(x)y' + q(x)y, \qquad \text{for } a \leq x \leq b, \quad \text{where} \quad y(a) = 0 \quad \text{and} \quad y'(a) = 1.$$

A weighted average of these solutions produces a solution to the linear boundary-value problem. In the Finite-Difference method, we replaced y'' and y' with difference approximations and solved a linear system. Although the approximations may not be as accurate as the Shooting method, there is less sensitivity to round-off error. Higher-order difference methods are available, or extrapolation can be used to improve accuracy.

For the nonlinear boundary problem

$$y'' = f(x, y, y'), \qquad \text{for } a \leq x \leq b, \quad \text{where} \quad y(a) = \alpha \quad \text{and} \quad y(b) = \beta,$$

we also presented two methods. The Nonlinear Shooting method requires the solution of the initial-value problem

$$y'' = f(x, y, y'), \qquad \text{for } a \le x \le b, \quad \text{where} \quad y(a) = \alpha \quad \text{and} \quad y'(a) = t,$$

for an initial choice of t. We improved the choice by using Newton's method to approximate the solution, t, to $y(b, t) = \beta$. This method required solving two initial-value problems at each iteration. The accuracy is dependent on the choice of method for solving the initial-value problems.

The Finite-Difference method for the nonlinear equation requires the replacement of y'' and y' by difference quotients, which results in a nonlinear system. This system is solved using Newton's method. Higher-order differences or extrapolation can be used to improve accuracy. Finite-Difference methods tend to be less sensitive to round-off error than shooting methods.

The Rayleigh-Ritz-Galerkin method was illustrated by approximating the solution to the boundary-value problem

$$-\frac{d\left(p(x)\dfrac{dy}{dx}\right)}{dx} + q(x)y = f(x), \qquad \text{for } 0 \le x \le 1, \quad \text{where } y(0) = y(1) = 0.$$

A piecewise linear approximation or a cubic spline approximation can be obtained.

Most of the material concerning second-order boundary-value problems can be extended to problems with boundary conditions of the form

$$\alpha_1 y(a) + \beta_1 y'(a) = \alpha \quad \text{and} \quad \alpha_2 y(b) + \beta_2 y'(b) = \beta,$$

where $|\alpha_1| + |\beta_1| \ne 0$ and $|\alpha_2| + |\beta_2| \ne 0$, but some of the techniques become quite complicated. The reader who is interested in problems of this type is advised to consider a book specializing in boundary-value problems, such as Keller [K, H].

The IMSL and NAG libraries contain methods for boundary-value problems. There are Finite-Difference methods and Shooting methods that are based on their adaptations of variable-step-size Runge-Kutta methods.

Further information on the general problems involved with the numerical solution to two-point boundary-value problems can be found in Keller [K, H] and Bailey, Shampine, and Waltman [BSW]. The book by Ascher, Mattheij, and Russell [AMR] has a comprehensive presentation of Multiple Shooting and Parallel Shooting methods.

CHAPTER 12

Numerical Methods for Partial-Differential Equations

12.1 Introduction

Physical problems involving more than one variable are often expressed using equations involving partial derivatives. In this chapter, we present a brief introduction to some of the techniques available for approximating the solution to partial-differential equations involving two variables by showing how these techniques can be applied to certain standard physical problems.

The partial-differential equation we will consider in Section 12.2 is an **elliptic** equation known as the **Poisson equation**:

$$\frac{\partial^2 u}{\partial x^2}(x, y) + \frac{\partial^2 u}{\partial y^2}(x, y) = f(x, y).$$

In this equation we assume that f describes the input to the problem on a plane region R with boundary S. Equations of this type arise in the study of various time-independent physical problems such as the steady-state distribution of heat in a plane region, the potential energy of a point in a plane acted on by gravitational forces in the plane, and two-dimensional steady-state problems involving incompressible fluids.

Additional constraints must be imposed to obtain a unique solution to the Poisson equation. For example, the study of the steady-state distribution of heat in a plane region requires that $f(x, y) \equiv 0$, resulting in a simplification to

$$\frac{\partial^2 u}{\partial x^2}(x, y) + \frac{\partial^2 u}{\partial y^2}(x, y) = 0,$$

which is called **Laplace's equation**. If the temperature within the region is determined by the temperature distribution on the boundary of the region, the constraints are called the

Dirichlet boundary conditions, given by

$$u(x, y) = g(x, y)$$

for all (x, y) on S, the boundary of the region R. (See Figure 12.1.)

Figure 12.1

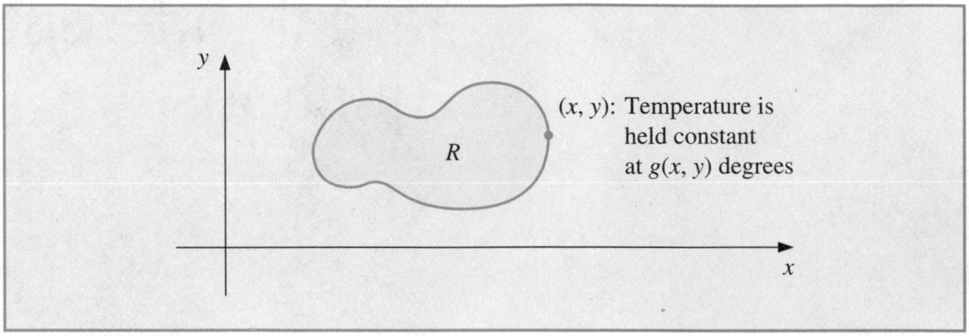

In Section 12.3 we consider the numerical solution to a problem involving a **parabolic** partial-differential equation of the form

$$\frac{\partial u}{\partial t}(x, t) - \alpha^2 \frac{\partial^2 u}{\partial x^2}(x, t) = 0.$$

The physical problem considered here concerns the flow of heat along a rod of length l (see Figure 12.2), which has a uniform temperature within each cross-sectional element. This requires the rod to be perfectly insulated on its lateral surface. The constant α is independent of the position in the rod and is determined by the heat-conductive properties of the material of which the rod is composed.

Figure 12.2

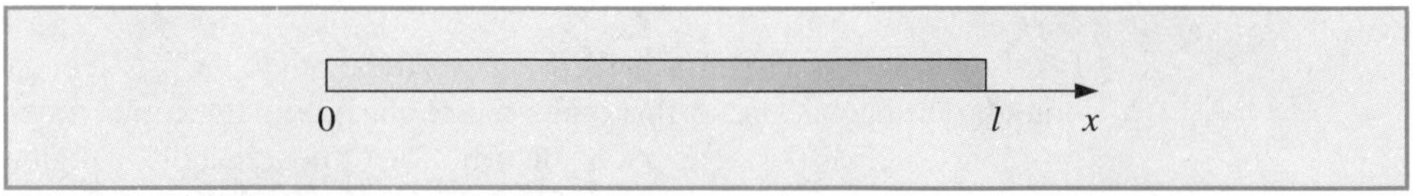

One of the typical sets of constraints for a heat-flow problem of this type is to specify the initial heat distribution in the rod,

$$u(x, 0) = f(x),$$

and to describe the behavior at the ends of the rod. For example, if the ends are held at constant temperatures U_1 and U_2, the boundary conditions have the form

$$u(0, t) = U_1 \quad \text{and} \quad u(l, t) = U_2,$$

and the heat distribution approaches the limiting temperature distribution

$$\lim_{t\to\infty} u(x,t) = U_1 + \frac{U_2 - U_1}{l}x.$$

If, instead, the rod is insulated so that no heat flows through the ends, the boundary conditions are

$$\frac{\partial u}{\partial x}(0,t) = 0 \quad \text{and} \quad \frac{\partial u}{\partial x}(l,t) = 0,$$

resulting in a constant temperature in the rod as the limiting case. The parabolic partial-differential equation is also of importance in the study of gas diffusion; in fact, it is known in some circles as the **diffusion equation**.

The problem studied in Section 12.4 is the one-dimensional **wave equation** and is an example of a **hyperbolic** partial-differential equation. Suppose an elastic string of length l is stretched between two supports at the same horizontal level (see Figure 12.3).

Figure 12.3

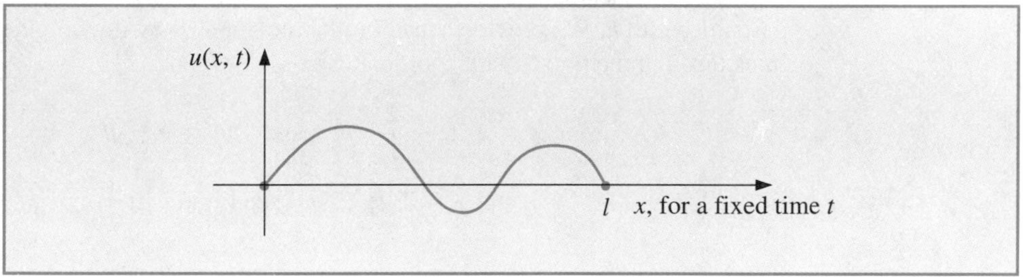

If the string is set to vibrate in a vertical plane, the vertical displacement $u(x,t)$ of a point x at time t satisfies the partial-differential equation

$$\alpha^2\frac{\partial^2 u}{\partial x^2}(x,t) = \frac{\partial^2 u}{\partial t^2}(x,t), \quad \text{for } 0 < x < l \quad \text{and} \quad 0 < t,$$

provided that damping effects are neglected and the amplitude is not too large. To impose constraints on this problem, assume that the initial position and velocity of the string are given by

$$u(x,0) = f(x) \quad \text{and} \quad \frac{\partial u}{\partial t}(x,0) = g(x), \qquad \text{for } 0 \le x \le l.$$

If the endpoints are fixed, we also have $u(0,t) = 0$ and $u(l,t) = 0$.

Other physical problems involving the hyperbolic partial-differential equation occur in the study of vibrating beams with one or both ends clamped, and in the transmission of electricity on a long line where there is some leakage of current to the ground.

12.2 Finite-Difference Methods for Elliptic Problems

The *elliptic* partial-differential equation we consider is the Poisson equation,

$$\nabla^2 u(x, y) \equiv \frac{\partial^2 u}{\partial x^2}(x, y) + \frac{\partial^2 u}{\partial y^2}(x, y) = f(x, y)$$

on $R = \{(x, y) \mid a < x < b, c < y < d\}$, with

$$u(x, y) = g(x, y) \qquad \text{for } (x, y) \in S,$$

where S denotes the boundary of R. If f and g are continuous on their domains, then there is a unique solution to this equation.

The method used here is an adaptation of the Finite-Difference method for linear boundary-value problems, which was discussed in Section 11.3. The first step is to choose integers n and m and define step sizes h and k by $h = (b - a)/n$ and $k = (d - c)/m$. Partition the interval $[a, b]$ into n equal parts of width h and the interval $[c, d]$ into m equal parts of width k. We provide a grid on the rectangle R by drawing vertical and horizontal lines through the points with coordinates (x_i, y_j), where

$$x_i = a + ih \quad \text{and} \quad y_j = c + jk$$

for each $i = 0, 1, \ldots, n$ and $j = 0, 1, \ldots, m$ (see Figure 12.4).

Figure 12.4

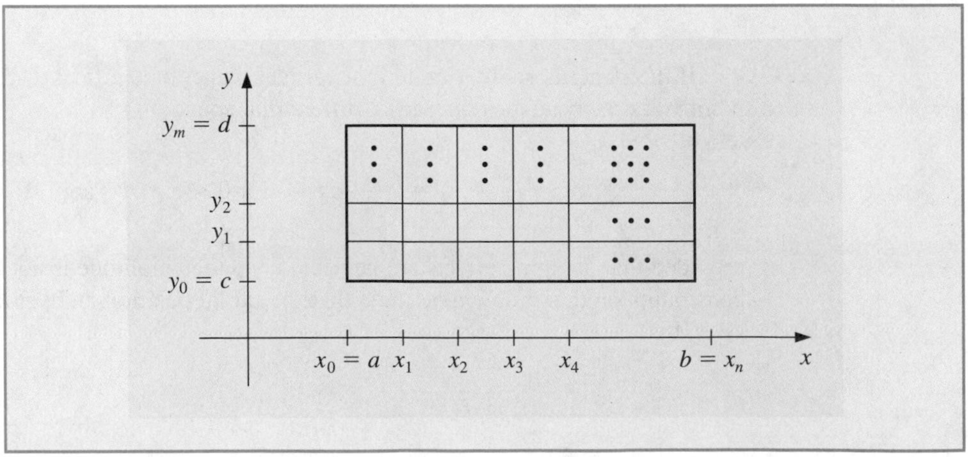

The lines $x = x_i$ and $y = y_j$ are **grid lines**, and their intersections are the **mesh points** of the grid. For each mesh point in the interior of the grid, we use the Taylor polynomial in the variable x about x_i to generate the centered-difference formula

$$\frac{\partial^2 u}{\partial x^2}(x_i, y_j) = \frac{u(x_{i+1}, y_j) - 2u(x_i, y_j) + u(x_{i-1}, y_j)}{h^2} - \frac{h^2}{12}\frac{\partial^4 u}{\partial x^4}(\xi_i, y_j),$$

for some ξ_i in (x_{i-1}, x_{i+1}) and the Taylor polynomial in the variable y about y_j to generate the centered-difference formula

$$\frac{\partial^2 u}{\partial y^2}(x_i, y_j) = \frac{u(x_i, y_{j+1}) - 2u(x_i, y_j) + u(x_i, y_{j-1})}{k^2} - \frac{k^2}{12}\frac{\partial^4 u}{\partial y^4}(x_i, \eta_j),$$

for some η_j in (y_{j-1}, y_{j+1}).

Using these formulas in the Poisson equation produces the following equations

$$\frac{u(x_{i+1}, y_j) - 2u(x_i, y_j) + u(x_{i-1}, y_j)}{h^2} + \frac{u(x_i, y_{j+1}) - 2u(x_i, y_j) + u(x_i, y_{j-1})}{k^2}$$
$$= f(x_i, y_j) + \frac{h^2}{12}\frac{\partial^4 u}{\partial x^4}(\xi_i, y_j) + \frac{k^2}{12}\frac{\partial^4 u}{\partial y^4}(x_i, \eta_j),$$

for each $i = 1, 2, \ldots, n-1$ and $j = 1, 2, \ldots, m-1$, and the boundary conditions give

$$u(x_i, y_0) = g(x_i, y_0) \quad \text{and} \quad u(x_i, y_m) = g(x_i, y_m),$$

for each $i = 1, 2, \ldots, n-1$, and

$$u(x_0, y_j) = g(x_0, y_j) \quad \text{and} \quad u(x_n, y_j) = g(x_n, y_j),$$

for each $j = 0, 1, \ldots, m$.

In difference-equation form, this results in the *Finite-Difference* method for the Poisson equation, with error of order $O(h^2 + k^2)$.

Finite-Difference Method

$$2\left[\left(\frac{h}{k}\right)^2 + 1\right] w_{ij} - (w_{i+1,j} + w_{i-1,j}) - \left(\frac{h}{k}\right)^2 (w_{i,j+1} + w_{i,j-1}) = -h^2 f(x_i, y_j),$$

for each $i = 1, 2, \ldots, n-1$ and $j = 1, 2, \ldots, m-1$, and

$$w_{0j} = g(x_0, y_j), \quad w_{nj} = g(x_n, y_j), \quad w_{i0} = g(x_i, y_0), \quad \text{and} \quad w_{im} = g(x_i, y_m)$$

for each $i = 1, 2, \ldots, n-1$ and $j = 0, 1, \ldots, m$, where w_{ij} approximates $u(x_i, y_j)$.

The typical equation involves approximations to $u(x, y)$ at the points

$$(x_{i-1}, y_j), \quad (x_i, y_j), \quad (x_{i+1}, y_j), \quad (x_i, y_{j-1}), \quad \text{and} \quad (x_i, y_{j+1}).$$

Reproducing the portion of the grid where these points are located (see Figure 12.5) shows that each equation involves approximations in a star-shaped region about (x_i, y_j).

Figure 12.5

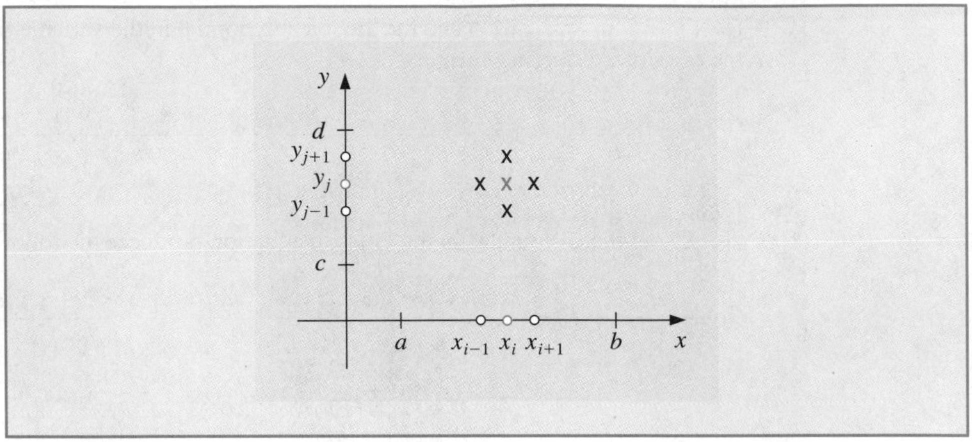

If we use the information from the boundary conditions, whenever appropriate, in the system given by the Finite-Difference method (that is, at all points (x_i, y_j) that are adjacent to a boundary mesh point), we have an $(n-1)(m-1)\times(n-1)(m-1)$ linear system with the unknowns being the approximations w_{ij} to $u(x_i, y_j)$ at the interior mesh points. The linear system involving these unknowns is expressed for matrix calculations more efficiently if

Figure 12.6

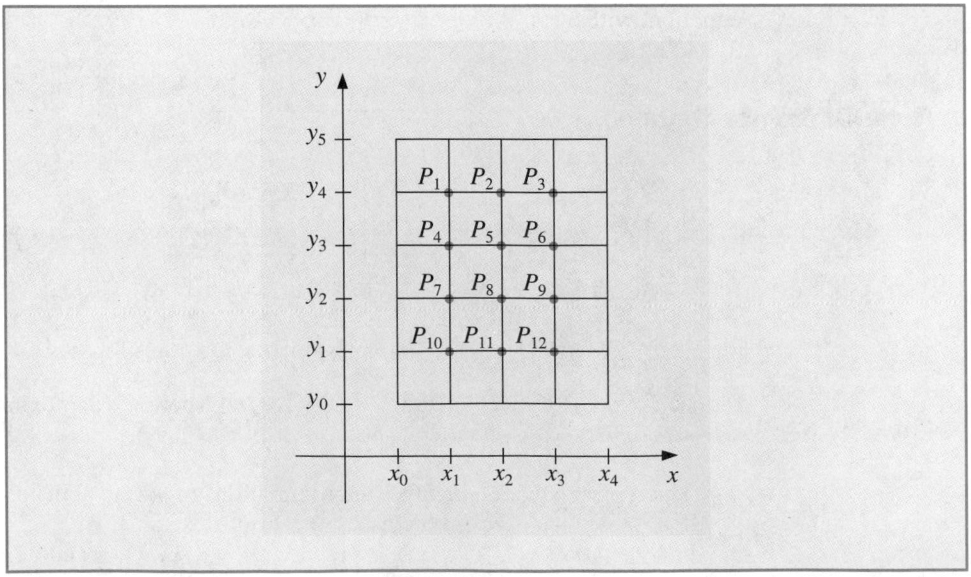

the interior mesh points are relabeled by letting

$$P_l = (x_i, y_j) \quad \text{and} \quad w_l = w_{ij},$$

where $l = i + (m - 1 - j)(n - 1)$ for each $i = 1, 2, \ldots, n - 1$ and $j = 1, 2, \ldots, m - 1$. This, in effect, labels the mesh points consecutively from left to right and top to bottom. For example, with $n = 4$ and $m = 5$, the relabeling results in a grid whose points are shown in Figure 12.6 on the preceding page. Labeling the points in this manner ensures that the system needed to determine the w_{ij} is a banded matrix with band width at most $2n - 1$.

EXAMPLE 1 Consider the problem of determining the steady-state heat distribution in a thin square metal plate 0.5 meters on a side. Two adjacent boundaries are held at 0°C, and the heat on the other boundaries increases linearly from 0°C at one corner to 100°C where the sides meet. If we place the sides with the zero boundary conditions along the x- and y-axes, the problem is expressed as

$$\frac{\partial^2 u}{\partial x^2}(x, y) + \frac{\partial^2 u}{\partial y^2}(x, y) = 0,$$

for (x, y) in the set $R = \{(x, y) \mid 0 < x < 0.5; 0 < y < 0.5\}$, with the boundary conditions

$$u(0, y) = 0, \qquad u(x, 0) = 0, \qquad u(x, 0.5) = 200x, \qquad u(0.5, y) = 200y.$$

Figure 12.7

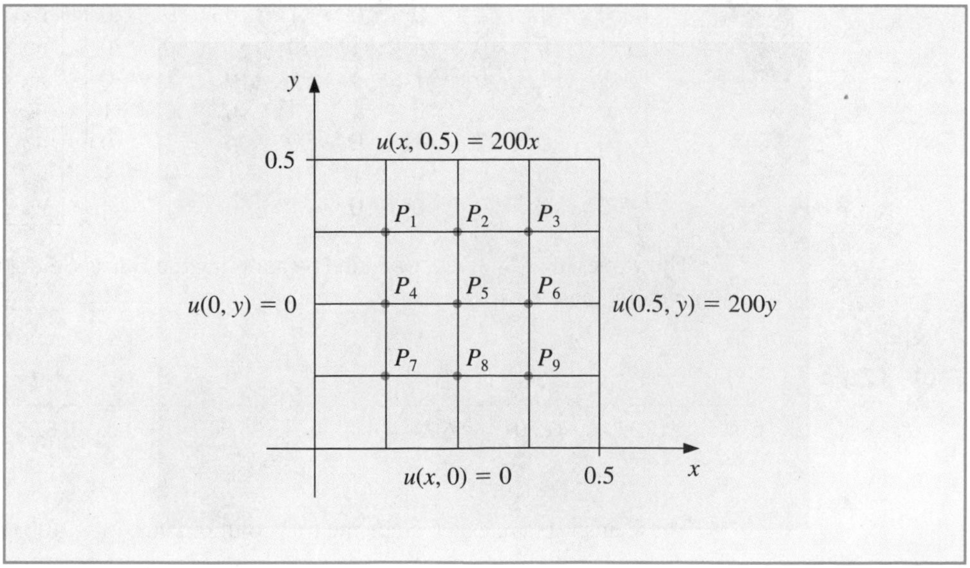

If $n = m = 4$, the problem has the grid given in Figure 12.7, and the difference equation is

$$4w_{i,j} - w_{i+1,j} - w_{i-1,j} - w_{i,j-1} - w_{i,j+1} = 0,$$

for each $i = 1, 2, 3$, and $j = 1, 2, 3$.

Expressing this in terms of the relabeled interior grid points $w_i = u(P_i)$ implies that the equations at the points P_i are

$$\begin{aligned}
P_1:&\quad 4w_1 - w_2 - w_4 = w_{0,3} + w_{1,4},\\
P_2:&\quad 4w_2 - w_3 - w_1 - w_5 = w_{2,4},\\
P_3:&\quad 4w_3 - w_2 - w_6 = w_{4,3} + w_{3,4},\\
P_4:&\quad 4w_4 - w_5 - w_1 - w_7 = w_{0,2},\\
P_5:&\quad 4w_5 - w_6 - w_4 - w_2 - w_8 = 0,\\
P_6:&\quad 4w_6 - w_5 - w_3 - w_9 = w_{4,2},\\
P_7:&\quad 4w_7 - w_8 - w_4 = w_{0,1} + w_{1,0},\\
P_8:&\quad 4w_8 - w_9 - w_7 - w_5 = w_{2,0},\\
P_9:&\quad 4w_9 - w_8 - w_6 = w_{3,0} + w_{4,1},
\end{aligned}$$

where the right sides of the equations are obtained from the boundary conditions. In fact, the boundary conditions imply that $w_{1,0} = w_{2,0} = w_{3,0} = w_{0,1} = w_{0,2} = w_{0,3} = 0$, $w_{1,4} = w_{4,1} = 25$, $w_{2,4} = w_{4,2} = 50$, and $w_{3,4} = w_{4,3} = 75$.

The linear system associated with this problem has the form

$$\begin{bmatrix}
4 & -1 & 0 & -1 & 0 & 0 & 0 & 0 & 0\\
-1 & 4 & -1 & 0 & -1 & 0 & 0 & 0 & 0\\
0 & -1 & 4 & 0 & 0 & -1 & 0 & 0 & 0\\
-1 & 0 & 0 & 4 & -1 & 0 & -1 & 0 & 0\\
0 & -1 & 0 & -1 & 4 & -1 & 0 & -1 & 0\\
0 & 0 & -1 & 0 & -1 & 4 & 0 & 0 & -1\\
0 & 0 & 0 & -1 & 0 & 0 & 4 & -1 & 0\\
0 & 0 & 0 & 0 & -1 & 0 & -1 & 4 & -1\\
0 & 0 & 0 & 0 & 0 & -1 & 0 & -1 & 4
\end{bmatrix}
\begin{bmatrix} w_1\\ w_2\\ w_3\\ w_4\\ w_5\\ w_6\\ w_7\\ w_8\\ w_9 \end{bmatrix}
=
\begin{bmatrix} 25\\ 50\\ 150\\ 0\\ 0\\ 50\\ 0\\ 0\\ 25 \end{bmatrix}.$$

The values of $w_1, w_2, \ldots, w_9$ found by applying the Gauss-Seidel method to this matrix are given in Table 12.1.

Table 12.1

i	1	2	3	4	5	6	7	8	9
w_i	18.75	37.50	56.25	12.50	25.00	37.50	6.25	12.50	18.75

These answers are exact, since the true solution, $u(x, y) = 400xy$, has

$$\frac{\partial^4 u}{\partial x^4} = \frac{\partial^4 u}{\partial y^4} = 0,$$

so the error is zero at each step. ◆ ◆ ◆

For simplicity, the Gauss-Seidel iterative procedure is used in the program POIFD121, but it is generally advisable to use a direct technique such as Gaussian elimination when the system is small, on the order of 100 or less, since the positive definiteness ensures stability with respect to round-off errors. For large systems, an iterative method should be used. If the SOR method is used, the choice of the optimal ω in this situation comes from the fact that when A is decomposed into its diagonal D and upper- and lower-triangular parts U and L,

$$A = D - L - U,$$

and B is the matrix for the Jacobi method,

$$B = D^{-1}(L + U),$$

then the spectral radius of B is

$$\rho(B) = \frac{1}{2}\left[\cos\left(\frac{\pi}{m}\right) + \cos\left(\frac{\pi}{n}\right)\right].$$

The value of ω is, consequently,

$$\omega = \frac{2}{1 + \sqrt{1 - [\rho(B)]^2}} = \frac{4}{2 + \sqrt{4 - \left[\cos\left(\frac{\pi}{m}\right) + \cos\left(\frac{\pi}{n}\right)\right]^2}}.$$

EXAMPLE 2 Consider Poisson's equation

$$\frac{\partial^2 u}{\partial x^2}(x, y) + \frac{\partial^2 u}{\partial y^2}(x, y) = xe^y, \quad \text{for} \quad 0 < x < 2 \quad \text{and} \quad 0 < y < 1,$$

with the boundary conditions

$$\begin{aligned} u(0, y) = 0, u(2, y) &= 2e^y, \qquad \text{for} \quad 0 \le y \le 1, \\ u(x, 0) = x, u(x, 1) &= ex, \qquad \text{for} \quad 0 \le x \le 2. \end{aligned}$$

We will use the Finite-Difference method to approximate the exact solution $u(x, y) = xe^y$ with $n = 6$ and $m = 5$. The stopping criterion for the Gauss-Seidel method in the program POIFD121 requires that

$$\left| w_{ij}^{(l)} - w_{ij}^{(l-1)} \right| \le 10^{-10},$$

for each $i = 1, \ldots, 5$ and $j = 1, \ldots, 4$. So, the solution to the difference equation was accurately obtained, and the procedure stopped at $l = 61$. The results, along with the correct values, are presented in Table 12.2. ◆ ◆ ◆

Table 12.2

i	j	x_i	y_i	$w_{i,j}^{(61)}$	$u(x_i, y_i)$	$\lvert u(x_i, y_j) - w_{i,j}^{(61)}\rvert$
1	1	0.3333	0.2000	0.40726	0.40713	1.30×10^{-4}
1	2	0.3333	0.4000	0.49748	0.49727	2.08×10^{-4}
1	3	0.3333	0.6000	0.60760	0.60737	2.23×10^{-4}
1	4	0.3333	0.8000	0.74201	0.74185	1.60×10^{-4}
2	1	0.6667	0.2000	0.81452	0.81427	2.55×10^{-4}
2	2	0.6667	0.4000	0.99496	0.99455	4.08×10^{-4}
2	3	0.6667	0.6000	1.2152	1.2147	4.37×10^{-4}
2	4	0.6667	0.8000	1.4840	1.4837	3.15×10^{-4}
3	1	1.0000	0.2000	1.2218	1.2214	3.64×10^{-4}
3	2	1.0000	0.4000	1.4924	1.4918	5.80×10^{-4}
3	3	1.0000	0.6000	1.8227	1.8221	6.24×10^{-4}
3	4	1.0000	0.8000	2.2260	2.2255	4.51×10^{-4}
4	1	1.3333	0.2000	1.6290	1.6285	4.27×10^{-4}
4	2	1.3333	0.4000	1.9898	1.9891	6.79×10^{-4}
4	3	1.3333	0.6000	2.4302	2.4295	7.35×10^{-4}
4	4	1.3333	0.8000	2.9679	2.9674	5.40×10^{-4}
5	1	1.6667	0.2000	2.0360	2.0357	3.71×10^{-4}
5	2	1.6667	0.4000	2.4870	2.4864	5.84×10^{-4}
5	3	1.6667	0.6000	3.0375	3.0369	6.41×10^{-4}
5	4	1.6667	0.8000	3.7097	3.7092	4.89×10^{-4}

EXERCISE SET 12.2

1. Use the Finite-Difference method to approximate the solution to the elliptic partial-differential equation

$$\frac{\partial^2 u}{\partial x^2} + \frac{\partial^2 u}{\partial y^2} = 4, \quad 0 < x < 1, \quad 0 < y < 2;$$
$$u(x, 0) = x^2, \quad u(x, 2) = (x - 2)^2, \quad 0 \le x \le 1;$$
$$u(0, y) = y^2, \quad u(1, y) = (y - 1)^2, \quad 0 \le y \le 2.$$

Use $h = k = \frac{1}{2}$ and compare the results to the actual solution $u(x, y) = (x - y)^2$.

2. Use the Finite-Difference method to approximate the solution to the elliptic partial-differential equation

$$\frac{\partial^2 u}{\partial x^2} + \frac{\partial^2 u}{\partial y^2} = 0, \quad 1 < x < 2, \quad 0 < y < 1;$$
$$u(x, 0) = 2 \ln x, \quad u(x, 1) = \ln(x^2 + 1), \quad 1 \le x \le 2;$$
$$u(1, y) = \ln(y^2 + 1), \quad u(2, y) = \ln(y^2 + 4), \quad 0 \le y \le 1.$$

Use $h = k = \frac{1}{3}$ and compare the results to the actual solution $u(x, y) = \ln(x^2 + y^2)$.

3. Use the Finite-Difference method to approximate the solutions to the following elliptic partial-differential equations:

   a. $$\frac{\partial^2 u}{\partial x^2} + \frac{\partial^2 u}{\partial y^2} = 0, \quad 0 < x < 1, \quad 0 < y < 1;$$
   $$u(x, 0) = 0, \quad u(x, 1) = x, \quad 0 \le x \le 1;$$
   $$u(0, y) = 0, \quad u(1, y) = y, \quad 0 \le y \le 1.$$
   Use $h = k = 0.2$ and compare the results with the solution $u(x, y) = xy$.

   b. $$\frac{\partial^2 u}{\partial x^2} + \frac{\partial^2 u}{\partial y^2} = -(\cos(x + y) + \cos(x - y)), \quad 0 < x < \pi, \quad 0 < y < \frac{\pi}{2};$$
   $$u(0, y) = \cos y, \quad u(\pi, y) = -\cos y, \quad 0 \le y \le \frac{\pi}{2},$$
   $$u(x, 0) = \cos x, \quad u\left(x, \frac{\pi}{2}\right) = 0, \quad 0 \le x \le \pi.$$
   Use $h = \pi/5$ and $k = \pi/10$ and compare the results with the solution $u(x, y) = \cos x \cos y$.

   c. $$\frac{\partial^2 u}{\partial x^2} + \frac{\partial^2 u}{\partial y^2} = (x^2 + y^2)e^{xy}, \quad 0 < x < 2, \quad 0 < y < 1;$$
   $$u(0, y) = 1, \quad u(2, y) = e^{2y}, \quad 0 \le y \le 1;$$
   $$u(x, 0) = 1, \quad u(x, 1) = e^x, \quad 0 \le x \le 2.$$
   Use $h = 0.2$ and $k = 0.1$, and compare the results with the solution $u(x, y) = e^{xy}$.

   d. $$\frac{\partial^2 u}{\partial x^2} + \frac{\partial^2 u}{\partial y^2} = \frac{x}{y} + \frac{y}{x}, \quad 1 < x < 2, \quad 1 < y < 2;$$
   $$u(x, 1) = x \ln x, \quad u(x, 2) = x \ln(4x^2), \quad 1 \le x \le 2;$$
   $$u(1, y) = y \ln y, \quad u(2, y) = 2y \ln(2y), \quad 1 \le y \le 2.$$
   Use $h = k = 0.1$ and compare the results with the solution $u(x, y) = xy \ln xy$.

4. Repeat Exercise 3(a) using extrapolation with $h_0 = 0.2$, $h_1 = h_0/2$, and $h_2 = h_0/4$.

5. A coaxial cable is made of a 0.1-in.-square inner conductor and a 0.5-in.-square outer conductor. The potential at a point in the cross section of the cable is described by Laplace's equation. Suppose the inner conductor is kept at 0 volts and the outer conductor is kept at 110 volts. Find the potential between the two conductors by placing a grid with horizontal mesh spacing $h = 0.1$ in. and vertical mesh spacing $k = 0.1$ in. on the region

   $$D = \{(x, y) | 0 \le x, y \le 0.5\}.$$

   Approximate the solution to Laplace's equation at each grid point, and use the two sets of boundary conditions to derive a linear system to be solved by the Gauss-Seidel method.

6. A 6-cm $\times$ 5-cm rectangular silver plate has heat being uniformly generated at each point at the rate $q = 1.5$ cal/cm$^3 \cdot$ s. Let x represent the distance along the edge of the plate of length 6 cm and y be the distance along the edge of the plate of length 5 cm.

Suppose the temperature u along the edges is kept at the following temperatures:

$$u(x, 0) = x(6 - x), \qquad u(x, 5) = 0, \qquad 0 \leq x \leq 6,$$
$$u(0, y) = y(5 - y), \qquad u(6, y) = 0, \qquad 0 \leq y \leq 5,$$

where the origin lies at a corner of the plate with coordinates $(0, 0)$ and the edges lie along the positive x- and y-axes. The steady-state temperature $u = u(x, y)$ satisfies Poisson's equation:

$$\frac{\partial^2 u}{\partial x^2}(x, y) + \frac{\partial^2 u}{\partial y^2}(x, y) = -\frac{q}{K}, \qquad 0 < x < 6, \qquad 0 < y < 5,$$

where K, the thermal conductivity, is 1.04 cal/cm · deg · s. Use the Finite-Difference method with $h = 0.4$ and $k = \frac{1}{3}$ to approximate the temperature $u(x, y)$.

12.3 Finite-Difference Methods for Parabolic Problems

The *parabolic* partial-differential equation we will study is the heat or diffusion equation

$$\frac{\partial u}{\partial t}(x, t) = \alpha^2 \frac{\partial^2 u}{\partial x^2}(x, t), \qquad \text{for } 0 < x < l \quad \text{and} \quad t > 0,$$

subject to the conditions

$$u(0, t) = u(l, t) = 0 \qquad \text{for } t > 0, \quad \text{and} \quad u(x, 0) = f(x) \qquad \text{for } 0 \leq x \leq l.$$

The approach we use to approximate the solution to this problem involves finite differences similar to those in Section 12.2. First select an integer $m > 0$ and define $h = l/m$. Then select a time-step size k. The grid points for this situation are (x_i, t_j), where $x_i = ih$, for $i = 0, 1, \ldots, m$, and $t_j = jk$, for $j = 0, 1, \ldots$.

We obtain the Difference method by using the Taylor polynomial in t to form the difference quotient

$$\frac{\partial u}{\partial t}(x_i, t_j) = \frac{u(x_i, t_j + k) - u(x_i, t_j)}{k} - \frac{k}{2}\frac{\partial^2 u}{\partial t^2}(x_i, \mu_j),$$

for some μ_j in (t_j, t_{j+1}), and the Taylor polynomial in x to form the difference quotient

$$\frac{\partial^2 u}{\partial x^2}(x_i, t_j) = \frac{u(x_i + h, t_j) - 2u(x_i, t_j) + u(x_i - h, t_j)}{h^2} - \frac{h^2}{12}\frac{\partial^4 u}{\partial x^4}(\xi_i, t_j),$$

for some ξ_i in (x_{i-1}, x_{i+1}).

The parabolic partial-differential equation implies that at the interior gridpoint (x_i, t_j) we have

$$\frac{\partial u}{\partial t}(x_i, t_j) - \alpha^2 \frac{\partial^2 u}{\partial x^2}(x_i, t_j) = 0,$$

so the Difference method using the two difference quotients is

$$\frac{w_{i,j+1} - w_{ij}}{k} - \alpha^2 \frac{w_{i+1,j} - 2w_{ij} + w_{i-1,j}}{h^2} = 0,$$

where w_{ij} approximates $u(x_i, t_j)$. The error for this difference equation is

$$\tau_{ij} = \frac{k}{2}\frac{\partial^2 u}{\partial t^2}(x_i, \mu_j) - \alpha^2 \frac{h^2}{12}\frac{\partial^4 u}{\partial x^4}(\xi_i, t_j).$$

Solving the difference equation for $w_{i,j+1}$ gives

$$w_{i,j+1} = \left(1 - \frac{2\alpha^2 k}{h^2}\right) w_{ij} + \alpha^2 \frac{k}{h^2}(w_{i+1,j} + w_{i-1,j}),$$

for each $i = 1, 2, \ldots, m-1$ and $j = 1, 2, \ldots$. Since the initial condition $u(x, 0) = f(x)$ implies that $w_{i0} = f(x_i)$, for each $i = 0, 1, \ldots, m$, these values can be used in the difference equation to find the value of w_{i1} for each $i = 1, 2, \ldots, m-1$. The additional conditions $u(0, t) = 0$ and $u(l, t) = 0$ imply that $w_{01} = w_{m1} = 0$, so all the entries of the form w_{i1} can be determined. If the procedure is reapplied once all the approximations w_{i1}, are known, the values of $w_{i2}, w_{i3}, \ldots, w_{i,m-1}$ can be obtained in a similar manner.

The explicit nature of the Difference method implies that the $(m-1) \times (m-1)$ matrix associated with this system can be written in the tridiagonal form

$$A = \begin{bmatrix} (1-2\lambda) & \lambda & 0 & \cdots & 0 \\ \lambda & (1-2\lambda) & \lambda & \ddots & \vdots \\ 0 & \ddots & \ddots & \ddots & 0 \\ \vdots & & \ddots & \ddots & \lambda \\ 0 & \cdots & 0 & \lambda & (1-2\lambda) \end{bmatrix},$$

where $\lambda = \alpha^2(k/h^2)$. If we let

$$\mathbf{w}^{(0)} = (f(x_1), f(x_2), \ldots, f(x_{m-1}))^t$$

and

$$\mathbf{w}^{(j)} = (w_{1j}, w_{2j}, \ldots, w_{m-1,j})^t, \qquad \text{for each } j = 1, 2, \ldots,$$

then the approximate solution is given by

$$\mathbf{w}^{(j)} = A\mathbf{w}^{(j-1)}, \qquad \text{for each } j = 1, 2, \ldots,$$

so $\mathbf{w}^{(j)}$ is obtained from $\mathbf{w}^{(j-1)}$ by a simple matrix multiplication. This is known as the **Forward-Difference method**, and it is of order $O(k + h^2)$.

EXAMPLE 1 Consider the heat equation

$$\frac{\partial u}{\partial t}(x,t) - \frac{\partial^2 u}{\partial x^2}(x,t) = 0, \qquad \text{for } 0 < x < 1 \quad \text{and} \quad 0 \le t,$$

with boundary conditions

$$u(0,t) = u(1,t) = 0, \qquad \text{for } 0 < t,$$

and initial conditions

$$u(x,0) = \sin(\pi x), \qquad \text{for } 0 \le x \le 1.$$

It is easily verified that the solution to this problem is $u(x,t) = e^{-\pi^2 t}\sin(\pi x)$. The solution at $t = 0.5$ will be approximated using the Forward-Difference method, first with $h = 0.1$, $k = 0.0005$, and $\lambda = 0.05$ and then with $h = 0.1$, $k = 0.01$, and $\lambda = 1$.

To use Maple for the calculations we first access the linear algebra library.

```
>with(linalg);
```

We define l, α, m, h, and the function $f(x)$ with the commands

```
>l:=1; alpha:=1; m:=10; h:=l/m; k:=0.0005;
>f:=x->sin(Pi*x);
```

We then define and initialize the $(m-1) \times (m-1)$ matrix A and $(m-1)$-dimensional vectors $\mathbf{w}$ and $\mathbf{u}$ with

```
>A:=matrix(m-1,m-1,0);
>w:=vector(m-1,0);
>u:=vector(m-1,0);
```

The vector $\mathbf{w}^{(0)}$ is defined by

```
>for i from 1 to m-1 do
>w[i]:=evalf(f(i*h));
>od;
```

and we define λ and the nonzero entries of A with the following code:

```
>lambda:=alpha^2*k/h^2;
>A[1,1]:=1-2*lambda;
>A[1,2]:=lambda;
>for i from 2 to m-2 do
```

```
>A[i,i-1]:=lambda;
>A[i,i+1]:=lambda;
>A[i,i]:=1-2*lambda;
>od;
>A[m-1,m-2]:=lambda;
>A[m-1,m-1]:=1-2*lambda;
```

To obtain $\mathbf{w}^{(1000)}$ we raise A to the 1000th power and then multiply $\mathbf{w}^{(0)}$ by B.

```
>B:=A^1000;
>u:=multiply(B,w);
```

The results are summarized in Table 12.3. ◆ ◆ ◆

Table 12.3

x_i	$u(x_i, 0.5)$	$w_{i,1000}$ $k = 0.0005$	$\lvert u(x_i, 0.5) - w_{i,1000}\rvert$	$w_{i,50}$ $k = 0.01$	$\lvert u(x_i, 0.5) - w_{i,50}\rvert$
0.0	0	0		0	
0.1	0.00222241	0.00228652	6.411×10^{-5}	8.19876×10^{7}	8.199×10^{7}
0.2	0.00422728	0.00434922	1.219×10^{-4}	-1.55719×10^{8}	1.557×10^{8}
0.3	0.00581836	0.00598619	1.678×10^{-4}	2.13833×10^{8}	2.138×10^{8}
0.4	0.00683989	0.00703719	1.973×10^{-4}	-2.50642×10^{8}	2.506×10^{8}
0.5	0.00719188	0.00739934	2.075×10^{-4}	2.62685×10^{8}	2.627×10^{8}
0.6	0.00683989	0.00703719	1.973×10^{-4}	-2.49015×10^{8}	2.490×10^{8}
0.7	0.00581836	0.00598619	1.678×10^{-4}	2.11200×10^{8}	2.112×10^{8}
0.8	0.00422728	0.00434922	1.219×10^{-4}	-1.53086×10^{8}	1.531×10^{8}
0.9	0.00222241	0.00228652	6.511×10^{-5}	8.03604×10^{7}	8.036×10^{7}
1.0	0	0		0	

An error of order $O(k + h^2)$ is expected in Example 1. This is obtained with $h = 0.1$ and $k = 0.0005$, but it is certainly not obtained when $h = 0.1$ and $k = 0.01$. To explain the difficulty, we must look at the stability of the Forward-Difference method.

If the error $\mathbf{e}^{(0)} = (e_1^{(0)}, e_2^{(0)}, \ldots, e_{m-1}^{(0)})^t$ is made in representing the initial data $\mathbf{w}^{(0)} = (f(x_1), f(x_2), \ldots, f(x_{m-1}))^t$, or in any particular step (the choice of the initial step is simply for convenience), an error of $A\mathbf{e}^{(0)}$ propagates in $\mathbf{w}^{(1)}$ since

$$\mathbf{w}^{(1)} = A(\mathbf{w}^{(0)} + \mathbf{e}^{(0)}) = A\mathbf{w}^{(0)} + A\mathbf{e}^{(0)}.$$

This process continues. At the nth time step, the error in $\mathbf{w}^{(n)}$ due to $\mathbf{e}^{(0)}$ is $A^n\mathbf{e}^{(0)}$. The method is consequently conditionally stable, that is, stable precisely when these errors do not grow as n increases. But this is true if and only if $\|A^n\mathbf{e}^{(0)}\| \leq \|\mathbf{e}^{(0)}\|$ for all n. Hence we must have $\|A^n\| \leq 1$, a condition that requires that the spectral radius $\rho(A^n) = (\rho(A))^n \leq 1$. The Forward-Difference method is therefore stable only if $\rho(A) \leq 1$.

The eigenvalues of A are

$$\mu_i = 1 - 4\lambda\left(\sin\left(\frac{i\pi}{2m}\right)\right)^2, \qquad \text{for each } i = 1, 2, \ldots, m-1,$$

so the condition for stability consequently reduces to determining whether

$$\rho(A) = \max_{1 \le i \le m-1} \left| 1 - 4\lambda\left(\sin\left(\frac{i\pi}{2m}\right)\right)^2 \right| \le 1,$$

which simplifies to

$$0 \le \lambda\left(\sin\left(\frac{i\pi}{2m}\right)\right)^2 \le \frac{1}{2}, \qquad \text{for each } i = 1, 2, \ldots, m-1.$$

Since stability requires that this inequality condition hold as $h \to 0$ or, equivalently, as $m \to \infty$, the fact that

$$\lim_{m\to\infty}\left[\sin\left(\frac{(m-1)\pi}{2m}\right)\right]^2 = 1$$

means that stability will occur only if $0 \le \lambda \le \frac{1}{2}$. Since $\lambda = \alpha^2(k/h^2)$, this inequality requires that h and k be chosen so that

$$\alpha^2 \frac{k}{h^2} \le \frac{1}{2}.$$

In Example 1 we have $\alpha = 1$, so the condition is satisfied when $h = 0.1$ and $k = 0.0005$. But when k was increased to 0.01 with no corresponding increase in h, the ratio was

$$\frac{0.01}{(0.1)^2} = 1 > \frac{1}{2},$$

and stability problems became apparent.

To obtain a more stable method, we consider an implicit-difference method that results from using the backward-difference quotient for $(\partial u/\partial t)(x_i, t_j)$ in the form

$$\frac{\partial u}{\partial t}(x_i, t_j) = \frac{u(x_i, t_j) - u(x_i, t_{j-1})}{k} + \frac{k}{2}\frac{\partial^2 u}{\partial t^2}(x_i, \mu_j),$$

for some μ_j in (t_{j-1}, t_j). Substituting this equation and the centered-difference formula for $\partial^2 u/\partial x^2$ into the partial-differential equation gives

$$\frac{u(x_i, t_j) - u(x_i, t_{j-1})}{k} - \alpha^2\frac{u(x_{i+1}, t_j) - 2u(x_i, t_j) + u(x_{i-1}, t_j)}{h^2}$$
$$= -\frac{k}{2}\frac{\partial^2 u}{\partial t^2}(x_i, \mu_j) - \alpha^2\frac{h^2}{12}\frac{\partial^4 u}{\partial x^4}(\xi_i, t_j),$$

for some ξ_i in (x_{i-1}, x_{i+1}). The difference method this produces is called the *Backward-Difference method* for the heat equation.

Backward-Difference Method

$$\frac{w_{ij} - w_{i,j-1}}{k} - \alpha^2 \frac{w_{i+1,j} - 2w_{ij} + w_{i-1,j}}{h^2} = 0,$$

for each $i = 1, 2, \ldots, m-1$, and $j = 1, 2, \ldots$.

This method involves, at a typical step, the mesh points

$$(x_i, t_j), \qquad (x_i, t_{j-1}), \qquad (x_{i-1}, t_j), \quad \text{and} \quad (x_{i+1}, t_j),$$

and involves approximations at the points marked with $\times$'s in Figure 12.8.

Figure 12.8

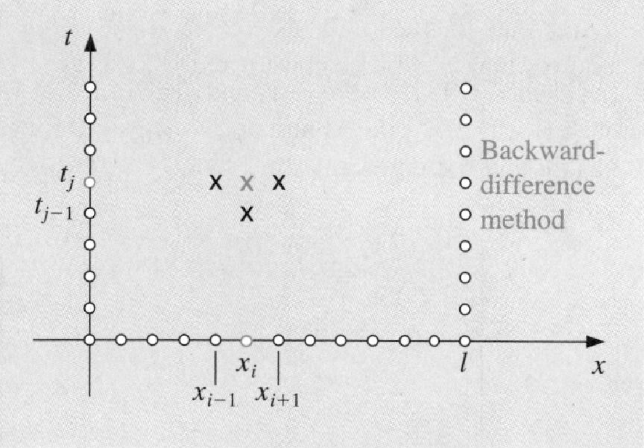

Since the boundary and initial conditions associated with the problem give information at the circled mesh points, the figure shows that no explicit procedures can be used to find the approximations in the Backward-Difference method.

Recall that in the Forward-Difference method (see Figure 12.9 on the following page), approximations at

$$(x_{i-1}, t_j), \qquad (x_i, t_j), \qquad (x_i, t_{j+1}), \quad \text{and} \quad (x_{i+1}, t_j)$$

were used, so an explicit method for finding the approximations, based on the information from the initial and boundary conditions, was available.

Figure 12.9

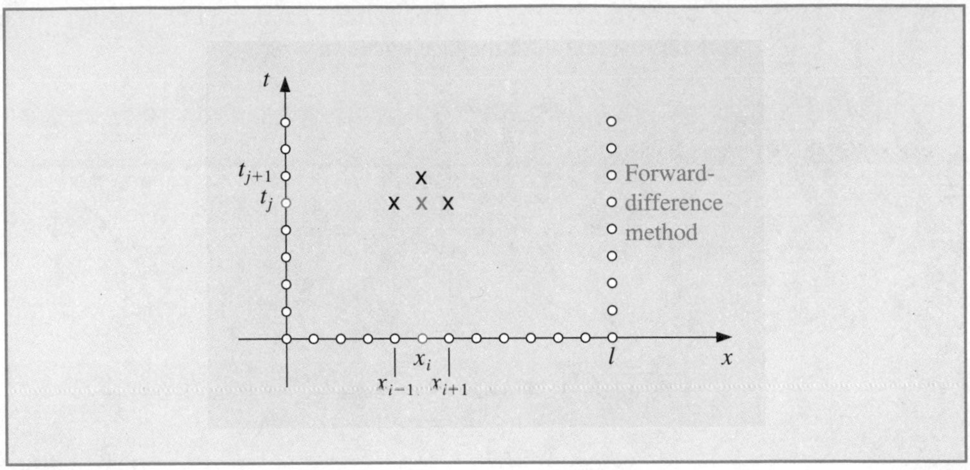

If we again let λ denote the quantity $\alpha^2(k/h^2)$, the Backward-Difference method becomes

$$(1+2\lambda)w_{ij} - \lambda w_{i+1,j} - \lambda w_{i-1,j} = w_{i,j-1},$$

for each $i = 1, 2, \ldots, m-1$, and $j = 1, 2, \ldots$. Using the knowledge that $w_{i0} = f(x_i)$ for each $i = 1, 2, \ldots, m-1$ and $w_{mj} = w_{0j} = 0$ for each $j = 1, 2, \ldots$, this Difference method has the matrix representation

$$\begin{bmatrix} (1+2\lambda) & -\lambda & 0 & \cdots & 0 \\ -\lambda & \ddots & \ddots & \ddots & \vdots \\ 0 & \ddots & \ddots & \ddots & 0 \\ \vdots & \ddots & \ddots & \ddots & -\lambda \\ 0 & \cdots & 0 & -\lambda & (1+2\lambda) \end{bmatrix} \begin{bmatrix} w_{1j} \\ w_{2j} \\ \vdots \\ w_{m-1,j} \end{bmatrix} = \begin{bmatrix} w_{1,j-1} \\ w_{2,j-1} \\ \vdots \\ w_{m-1,j-1} \end{bmatrix},$$

or $A\mathbf{w}^{(j)} = \mathbf{w}^{(j-1)}$.

Hence we must now solve a linear system to obtain $\mathbf{w}^{(j)}$ from $\mathbf{w}^{(j-1)}$. Since $\lambda > 0$, the matrix A is positive definite and strictly diagonally dominant, as well as being tridiagonal. We can use either Crout factorization for tridiagonal linear systems or the SOR method to solve this system. The program HEBDM122 uses Crout factorization, which is an acceptable method unless m is large.

EXAMPLE 2 The Backward-Difference method with $h = 0.1$ and $k = 0.01$ will be used to approximate the solution to the heat equation

$$\frac{\partial u}{\partial t}(x,t) - \frac{\partial^2 u}{\partial x^2}(x,t) = 0, \qquad 0 < x < 1, \qquad 0 < t,$$

subject to the constraints

$$u(0,t) = u(1,t) = 0, \quad 0 < t, \qquad u(x,0) = \sin \pi x, \qquad 0 \leq x \leq 1,$$

which was considered in Example 1. To demonstrate the stability of the Backward-Difference method, we again compare $w_{i,50}$ to $u(x_i, 0.5)$, where $i = 0, 1, \ldots, 10$.

To use Maple to obtain $\mathbf{w}^{(1)}$ from $\mathbf{w}^{(0)}$, we first access the linear algebra library.

```
>with(linalg);
```

We define l, α, m, h, k and the function $f(x)$ with the commands

```
>l:=1; alpha:=1; m:=10; h:=l/m; k:=0.01;
>f:=x->sin(Pi*x);
```

and we define and initialize the $(m-1) \times (m-1)$ matrix A and the $(m-1)$-dimensional vectors $\mathbf{w}$ and $\mathbf{u}$ by

```
>A:=matrix(m-1,m-1,0);
>w:=vector(m-1,0);
>u:=vector(m-1,0);
```

We then compute $\mathbf{w}^{(0)}$ with

```
>for i from 1 to m-1 do
>w[i]:=evalf(f(i*h));
>od;
```

and we define λ and generate the nonzero entries of A.

```
>lambda:=alpha^2*k/h^2;
>A[1,1]:=1+2*lambda;
>A[1,2]:=-lambda;
>for i from 2 to m-2 do
>A[i,i-1]:=-lambda;
>A[i,i+1]:=-lambda;
>A[i,i]:=1+2*lambda;
>od;
>A[m-1,m-2]:=-lambda;
>A[m-1,m-1]:=1+2*lambda;
```

We augment A with $\mathbf{w}^{(0)}$ and use Gaussian elimination to compute $\mathbf{v} = \mathbf{w}^{(1)}$ by solving the linear system $\mathbf{w}^{(1)} = A\mathbf{w}^{(0)}$.

```
>B:=augment(A,w);
>M:=gausselim(B);
>v:=backsub(M);
```

The final results are listed in Table 12.4. ◆ ◆ ◆

Table 12.4

x_i	$w_{i,50}$	$u(x_i, 0.5)$	$\lvert w_{i,50} - u(x_i, 0.5)\rvert$
0.0	0	0	
0.1	0.00289802	0.00222241	6.756×10^{-4}
0.2	0.00551236	0.00422728	1.285×10^{-3}
0.3	0.00758711	0.00581836	1.769×10^{-3}
0.4	0.00891918	0.00683989	2.079×10^{-3}
0.5	0.00937818	0.00719188	2.186×10^{-3}
0.6	0.00891918	0.00683989	2.079×10^{-3}
0.7	0.00758711	0.00581836	1.769×10^{-3}
0.8	0.00551236	0.00422728	1.285×10^{-3}
0.9	0.00289802	0.00222241	6.756×10^{-4}
1.0	0	0	

The results listed in the second column of Table 12.4 should be compared with those in the third and fifth columns of Table 12.3.

The reason that the Backward-Difference method does not have the stability problems of the Forward-Difference method can be seen by analyzing the eigenvalues of the matrix A. For the Backward-Difference method the eigenvalues are

$$\mu_i = 1 + 4\lambda\left[\sin\left(\frac{i\pi}{2m}\right)\right]^2 \qquad \text{for each } i = 1, 2, \ldots, m-1;$$

and since $\lambda > 0$, we have $\mu_i > 1$ for all $i = 1, 2, \ldots, m-1$. This implies that A^{-1} exists, since zero is not an eigenvalue of A. An error $\mathbf{e}^{(0)}$ in the initial data produces an error $(A^{-1})^n\mathbf{e}^{(0)}$ at the nth step. Since the eigenvalues of A^{-1} are the reciprocals of the eigenvalues of A, the spectral radius of A^{-1} is bounded above by 1 and the method is unconditionally stable, that is, stable independent of the choice of $\lambda = \alpha^2(k/h^2)$. The error for the method is of order $O(k + h^2)$, provided the solution of the differential equation satisfies the usual differentiability conditions.

The weakness in the Backward-Difference method results from the fact that the error has a portion with order $O(k)$, requiring that time intervals be made much smaller than spatial intervals. It would clearly be desirable to have a procedure with error of order $O(k^2 + h^2)$. A method with this error term is derived by averaging the Forward-Difference method at the jth step in t,

$$\frac{w_{i,j+1} - w_{ij}}{k} - \alpha^2\frac{w_{i+1,j} - 2w_{ij} + w_{i-1,j}}{h^2} = 0,$$

which has error $(k/2)(\partial^2 u/\partial t^2)(x_i, \mu_j) + O(h^2)$, and the Backward-Difference method at the $(j+1)$st step in t,

$$\frac{w_{i,j+1} - w_{ij}}{k} - \alpha^2\frac{w_{i+1,j+1} - 2w_{i,j+1} + w_{i-1,j+1}}{h^2} = 0,$$

which has error $-(k/2)(\partial^2 u/\partial t^2)(x_i, \hat{u}_j) + O(h^2)$. If we assume that

$$\frac{\partial^2 u}{\partial t^2}(x_i, \hat{\mu}_j) \approx \frac{\partial^2 u}{\partial t^2}(x_i, \mu_j),$$

then the averaged-difference method,

$$\frac{w_{i,j+1} - w_{ij}}{k} - \frac{\alpha^2}{2}\left[\frac{w_{i+1,j} - 2w_{ij} + w_{i-1,j}}{h^2} + \frac{w_{i+1,j+1} - 2w_{i,j+1} + w_{i-1,j+1}}{h^2}\right] = 0,$$

has error of order $O(k^2 + h^2)$, provided, of course, that the usual differentiability conditions are satisfied. This is known as the **Crank-Nicolson method** and is represented in the matrix form $A\mathbf{w}^{(j+1)} = B\mathbf{w}^{(j)}$ for each $j = 0, 1, 2, \ldots$, where

$$\lambda = \alpha^2 \frac{k}{h^2}, \qquad \mathbf{w}^{(j)} = (w_{1j}, w_{2j}, \ldots, w_{m-1,j})^t,$$

and the matrices A and B are given by

$$A = \begin{bmatrix} (1+\lambda) & -\frac{\lambda}{2} & 0 & \cdots & \cdots & 0 \\ -\frac{\lambda}{2} & \ddots & \ddots & \ddots & & \vdots \\ 0 & \ddots & \ddots & \ddots & \ddots & \vdots \\ \vdots & \ddots & \ddots & \ddots & \ddots & 0 \\ \vdots & & \ddots & \ddots & \ddots & -\frac{\lambda}{2} \\ 0 & \cdots & \cdots & 0 & -\frac{\lambda}{2} & (1+\lambda) \end{bmatrix}$$

and

$$B = \begin{bmatrix} (1-\lambda) & \frac{\lambda}{2} & 0 & \cdots & \cdots & 0 \\ \frac{\lambda}{2} & \ddots & \ddots & \ddots & & \vdots \\ 0 & \ddots & \ddots & \ddots & \ddots & \vdots \\ \vdots & \ddots & \ddots & \ddots & \ddots & 0 \\ \vdots & & \ddots & \ddots & \ddots & \frac{\lambda}{2} \\ 0 & \cdots & \cdots & 0 & \frac{\lambda}{2} & (1-\lambda) \end{bmatrix}.$$

Since A is a positive definite, strictly diagonal dominant, and a tridiagonal matrix, it is nonsingular. Either Crout factorization for tridiagonal linear systems or the SOR method can be used to obtain $\mathbf{w}^{(j+1)}$ from $\mathbf{w}^{(j)}$, for each $j = 0, 1, 2, \ldots$. The program HECNM123 incorporates Crout factorization into the Crank-Nicolson technique.

EXAMPLE 3 The Crank-Nicolson method will be used to approximate the solution to the problem in Examples 1 and 2, consisting of the equation

$$\frac{\partial u}{\partial t}(x,t) - \frac{\partial^2 u}{\partial x^2}(x,t) = 0, \qquad \text{for } 0 < x < 1 \quad \text{and} \quad 0 < t,$$

subject to the conditions

$$u(0,t) = u(1,t) = 0, \quad 0 < t, \quad \text{and} \quad u(x,0) = \sin(\pi x), \quad 0 \le x \le 1.$$

The choices $m = 10$, $h = 0.1$, $k = 0.01$, and $\lambda = 1$ are used, as they were in the previous examples.

To use Maple to obtain $\mathbf{w}^{(1)}$ from $\mathbf{w}^{(0)}$. We first access the linear algebra library.

```
>with(linalg);
```

We define l, α, m, h, k and the function $f(x)$ with the commands

```
>l:=1; alpha:=1; m:=10; h:=l/m; k:=0.01;
>f:=x -> sin(Pi*x);
```

and we define and initialize the $(m-1) \times (m-1)$ matrices A and B and the $(m-1)$-dimensional vectors $\mathbf{w}$ and $\mathbf{u}$ by

```
>A:=matrix(m-1,m-1,0);
>B:=matrix(m-1,m-1,0);
>w:=vector(m-1,0);
>u:=vector(m-1,0);
```

We then compute $\mathbf{w}^{(0)}$ with

```
>for i from 1 to m-1 do
>w[i]:=evalf(f(i*h));
>od;
```

and we define λ and generate the nonzero entries of A and B by

```
>lambda:=alpha^2*k/h^2;
>A[1,1]:=1+lambda; B[1,1]:=1-lambda;
>A[1,2]:=-lambda/2; B[1,2]:=lambda/2;
>for i from 2 to m-2 do
>A[i,i-1]:=-lambda/2; B[i,i-1]:=lambda/2;
>A[i,i+1]:=-lambda/2; B[i,i+1]:=lambda/2;
>A[i,i]:=1+lambda; B[i,i]:=1-lambda;
```

```
>od;
>A[m-1,m-2]:=-lambda/2; B[m-1,m-2]:=lambda/2;
>A[m-1,m-1]:=1+lambda; B[m-1,m-1]:=1-lambda;
```

We need to form $\mathbf{u} = B\mathbf{w} = B\mathbf{w}^{(0)}$ and the augmented matrix $C = [A : \mathbf{u}]$

```
>u:=multiply(B,w);
>C:=augment(A,u);
```

Then we perform Gaussian elimination on the system $A\mathbf{w}^{(1)} = B\mathbf{w}^{(0)}$ to compute $\mathbf{w}^{(1)}$:

```
>M:=gausselim(C);
>z:=backsub(M);
```

The vector $\mathbf{z}$ is $\mathbf{w}^{(1)}$.

The results in Table 12.5 indicate the increase in accuracy of the Crank-Nicolson method over the Backward-Difference method, the best of the two previously discussed techniques. ◆ ◆ ◆

Table 12.5

x_i	$w_{i,50}$	$u(x_i, 0.5)$	$\lvert w_{i,50} - u(x_i, 0.5)\rvert$
0.0	0	0	
0.1	0.00230512	0.00222241	8.271×10^{-5}
0.2	0.00438461	0.00422728	1.573×10^{-4}
0.3	0.00603489	0.00581836	2.165×10^{-4}
0.4	0.00709444	0.00683989	2.546×10^{-4}
0.5	0.00745954	0.00719188	2.677×10^{-4}
0.6	0.00709444	0.00683989	2.546×10^{-4}
0.7	0.00603489	0.00581836	2.165×10^{-4}
0.8	0.00438461	0.00422728	1.573×10^{-4}
0.9	0.00230512	0.00222241	8.271×10^{-5}
1.0	0	0	

EXERCISE SET 12.3

1. Use the Backward-Difference method to approximate the solution to the following partial-differential equations.

   a. $$\frac{\partial u}{\partial t} - \frac{\partial^2 u}{\partial x^2} = 0, \qquad 0 < x < 2, \qquad 0 < t;$$
   $$u(0, t) = u(2, t) = 0, \qquad 0 < t,$$
   $$u(x, 0) = \sin \frac{\pi}{2} x, \qquad 0 \le x \le 2.$$
   Use $m = 4$, $T = 0.1$, and $N = 2$ and compare your answers to the actual solution $u(x, t) = e^{-(\pi^2/4)t} \sin(\pi x/2)$.

b. $\dfrac{\partial u}{\partial t} - \dfrac{1}{16}\dfrac{\partial^2 u}{\partial x^2} = 0, \quad 0 < x < 1, \quad 0 < t;$

$u(0, t) = u(1, t) = 0, \quad 0 < t,$

$u(x, 0) = 2 \sin 2\pi x, \quad 0 \le x \le 1.$

Use $m = 3$, $T = 0.1$, and $N = 2$ and compare your answers to the actual solution $u(x, t) = 2e^{-(\pi^2/4)t} \sin 2\pi x$.

2. Repeat Exercise 1 using the Crank-Nicolson method.

3. Use the Forward-Difference method to approximate the solution to the following parabolic partial-differential equations.

a. $\dfrac{\partial u}{\partial t} - \dfrac{\partial^2 u}{\partial x^2} = 0, \quad 0 < x < 2, \quad 0 < t;$

$u(0, t) = u(2, t) = 0, \quad 0 < t,$

$u(x, 0) = \sin 2\pi x, \quad 0 \le x \le 2.$

Use $h = 0.4$ and $k = 0.1$, and compare your answers at $t = 0.5$ to the actual solution $u(x, t) = e^{-4\pi^2 t} \sin 2\pi x$. Then use $h = 0.4$ and $k = 0.05$, and compare the answers.

b. $\dfrac{\partial u}{\partial t} - \dfrac{\partial^2 u}{\partial x^2} = 0, \quad 0 < x < \pi, \quad 0 < t;$

$u(0, t) = u(\pi, t) = 0, \quad 0 < t,$

$u(x, 0) = \sin x, \quad 0 \le x \le \pi.$

Use $h = \pi/10$ and $k = 0.05$ and compare your answers to the actual solution $u(x, t) = e^{-t} \sin x$ at $t = 0.5$.

c. $\dfrac{\partial u}{\partial t} - \dfrac{4}{\pi^2}\dfrac{\partial^2 u}{\partial x^2} = 0, \quad 0 < x < 4, \quad 0 < t;$

$u(0, t) = u(4, t) = 0, \quad 0 < t,$

$u(x, 0) = \sin \dfrac{\pi}{4} x \left(1 + 2 \cos \dfrac{\pi}{4} x\right), \quad 0 \le x \le 4.$

Use $h = 0.2$ and $k = 0.04$. Compare your answers to the actual solution $u(x, t) = e^{-t} \sin(\pi/2)x + e^{-t/4} \sin(\pi/4)x$ at $t = 0.4$.

d. $\dfrac{\partial u}{\partial t} - \dfrac{1}{\pi^2}\dfrac{\partial^2 u}{\partial x^2} = 0, \quad 0 < x < 1, \quad 0 < 1;$

$u(0, t) = u(1, t) = 0, \quad 0 < t,$

$u(x, 0) = \cos \pi \left(x - \dfrac{1}{2}\right), \quad 0 \le x \le 1.$

Use $h = 0.1$ and $k = 0.04$. Compare your answers to the actual solution $u(x, t) = e^{-t} \cos \pi(x - \frac{1}{2})$ at $t = 0.4$.

4. Repeat Exercise 3 using the Backward-Difference method.

5. Repeat Exercise 3 using the Crank-Nicolson method.

6. Modify the Backward-Difference method to accommodate the parabolic partial-differential equation

$$\frac{\partial u}{\partial t} - \frac{\partial^2 u}{\partial x^2} = F(x), \qquad 0 < x < l, \qquad 0 < t;$$
$$u(0, t) = u(l, t) = 0, \qquad 0 < t,$$
$$u(x, 0) = f(x), \qquad 0 \le x \le l.$$

7. Use the results of Exercise 6 to approximate the solution to

$$\frac{\partial u}{\partial t} - \frac{\partial^2 u}{\partial x^2} = 2, \qquad 0 < x < 1, \qquad 0 < t;$$
$$u(0, t) = u(1, t) = 0, \qquad 0 < t,$$
$$u(x, 0) = \sin \pi x + x(1 - x),$$

with $h = 0.1$ and $k = 0.01$. Compare your answer to the actual solution $u(x, t) = e^{-\pi^2 t} \sin \pi x + x(1 - x)$ at $t = 0.25$.

8. Modify the Backward-Difference method to accommodate the parabolic partial-differential equation

$$\frac{\partial u}{\partial t} - \alpha^2 \frac{\partial^2 u}{\partial x^2} = 0, \qquad 0 < x < l, \qquad 0 < t;$$
$$u(0, t) = \phi(t),\ u(l, t) = \Psi(t), \qquad 0 < t;$$
$$u(x, 0) = f(x), \qquad 0 \le x \le l,$$

where $f(0) = \phi(0)$ and $f(l) = \Psi(0)$.

9. The temperature $u(x, t)$ of a long, thin rod of constant cross section and homogeneous conducting material is governed by the one-dimensional heat equation. If heat is generated in the material, for example, by resistance to current or nuclear reaction, the heat equation becomes

$$\frac{\partial^2 u}{\partial x^2} + \frac{Kr}{\rho C} = K \frac{\partial u}{\partial t}, \qquad 0 < x < l, \qquad 0 < t,$$

where l is the length, ρ is the density, C is the specific heat, and K is the thermal diffusivity of the rod. The function $r = r(x, t, u)$ represents the heat generated per unit volume. Suppose that

$$l = 1.5 \text{ cm}, \qquad K = 1.04 \text{ cal/cm} \cdot \text{deg} \cdot \text{s},$$
$$\rho = 10.6 \text{ g/cm}^3, \qquad C = 0.056 \text{ cal/g} \cdot \text{deg},$$

and

$$r(x, t, u) = 5.0 \text{ cal/cm}^3 \cdot \text{s}.$$

If the ends of the rod are kept at 0° C, then

$$u(0, t) = u(l, t) = 0, \qquad t > 0.$$

Suppose the initial temperature distribution is given by

$$u(x,0) = \sin\frac{\pi x}{l}, \qquad 0 \le x \le l.$$

Use the results of Exercise 6 to approximate the temperature distribution with $h = 0.15$ and $k = 0.0225$.

10. Sagar and Payne [SP] analyze the stress-strain relationships and material properties of a cylinder subjected alternately to heating and cooling and consider the equation

$$\frac{\partial^2 T}{\partial r^2} + \frac{1}{r}\frac{\partial T}{\partial r} = \frac{1}{4K}\frac{\partial T}{\partial t}, \qquad \frac{1}{2} < r < 1, \qquad 0 < T,$$

where $T = T(r,t)$ is the temperature, r is the radial distance from the center of the cylinder, t is time, and K is a diffusivity coefficient.

a. Find approximations to $T(r, 10)$ for a cylinder with outside radius 1, given the initial and boundary conditions:

$$T(1,t) = 100 + 40t, \qquad 0 \le t \le 10,$$
$$T\left(\frac{1}{2}, t\right) = t, \qquad 0 \le t \le 10,$$
$$T(r,0) = 200(r - 0.5), \qquad 0.5 \le r \le 1.$$

Use a modification of the Backward-Difference method with $K = 0.1$, $k = 0.5$, and $h = \Delta r = 0.1$.

b. Using the temperature distribution of part (a), calculate the strain I by approximating the integral

$$I = \int_{0.5}^{1} \alpha T(r,t) r \, dr,$$

where $\alpha = 10.7$ and $t = 10$. Use the Composite Trapezoidal method with $n = 5$.

12.4 Finite-Difference Methods for Hyperbolic Problems

In this section we consider the numerical solution to the **wave equation**, an example of a *hyperbolic* partial-differential equation. The wave equation is given by the differential equation

$$\frac{\partial^2 u}{\partial t^2}(x,t) - \alpha^2 \frac{\partial^2 u}{\partial x^2}(x,t) = 0, \qquad \text{for } 0 < x < l \quad \text{and} \quad t > 0,$$

subject to the conditions

$$u(0,t) = u(l,t) = 0, \qquad \text{for } t > 0,$$

$$u(x,0) = f(x), \quad \text{and} \quad \frac{\partial u}{\partial t}(x,0) = g(x), \qquad \text{for } 0 \le x \le l,$$

where α is a constant. Select an integer $m > 0$ and time-step size $k > 0$. With $h = l/m$, the mesh points (x_i, t_j) are defined by

$$x_i = ih \quad \text{and} \quad t_j = jk,$$

for each $i = 0, 1, \ldots, m$ and $j = 0, 1, \ldots$. At any interior mesh point (x_i, t_j), the wave equation becomes

$$\frac{\partial^2 u}{\partial t^2}(x_i, t_j) - \alpha^2 \frac{\partial^2 u}{\partial x^2}(x_i, t_j) = 0.$$

The Difference method is obtained using the centered-difference quotient for the second partial derivatives given by

$$\frac{\partial^2 u}{\partial t^2}(x_i, t_j) = \frac{u(x_i, t_{j+1}) - 2u(x_i, t_j) + u(x_i, t_{j-1})}{k^2} - \frac{k^2}{12}\frac{\partial^4 u}{\partial t^4}(x_i, \mu_j),$$

for some μ_j in (t_{j-1}, t_{j+1}) and

$$\frac{\partial^2 u}{\partial x^2}(x_i, t_j) = \frac{u(x_{i+1}, t_j) - 2u(x_i, t_j) + u(x_{i-1}, t_j)}{h^2} - \frac{h^2}{12}\frac{\partial^4 u}{\partial x^4}(\xi_i, t_j),$$

for some ξ_i in (x_{i-1}, x_{i+1}). Substituting these into the wave equation gives

$$\frac{u(x_i, t_{j+1}) - 2u(x_i, t_j) + u(x_i, t_{j-1})}{k^2} - \alpha^2 \frac{u(x_{i+1}, t_j) - 2u(x_i, t_j) + u(x_{i-1}, t_j)}{h^2}$$
$$= \frac{1}{12}\left[k^2 \frac{\partial^4 u}{\partial t^4}(x_i, \mu_j) - \alpha^2 h^2 \frac{\partial^4 u}{\partial x^4}(\xi_i, t_j)\right].$$

Neglecting the error term

$$\tau_{ij} = \frac{1}{12}\left[k^2 \frac{\partial^4 u}{\partial t^4}(x_i, \mu_j) - \alpha^2 h^2 \frac{\partial^4 u}{\partial x^4}(\xi_i, t_j)\right]$$

leads to the difference equation

$$\frac{w_{i,j+1} - 2w_{ij} + w_{i,j-1}}{k^2} - \alpha^2 \frac{w_{i+1,j} - 2w_{ij} + w_{i-1,j}}{h^2} = 0.$$

With $\lambda = \alpha k/h$, we can solve for $w_{i,j+1}$, the most advanced time-step approximation, to obtain

$$w_{i,j+1} = 2(1 - \lambda^2)w_{ij} + \lambda^2(w_{i+1,j} + w_{i-1,j}) - w_{i,j-1}.$$

This equation holds for each $i = 1, 2, \ldots, (m-1)$ and $j = 1, 2, \ldots$. The boundary conditions give

$$w_{0j} = w_{mj} = 0, \qquad \text{for each } j = 1, 2, 3, \ldots,$$

and the initial condition implies that

$$w_{i0} = f(x_i), \qquad \text{for each } i = 1, 2, \ldots, m-1.$$

Writing this set of equations in matrix form gives

$$\begin{bmatrix} w_{1,j+1} \\ w_{2,j+1} \\ \vdots \\ w_{m-1,j+1} \end{bmatrix} = \begin{bmatrix} 2(1-\lambda^2) & \lambda^2 & 0 & \cdots & 0 \\ \lambda^2 & 2(1-\lambda^2) & \lambda^2 & \ddots & \vdots \\ 0 & \ddots & \ddots & \ddots & 0 \\ \vdots & & \ddots & \ddots & \lambda^2 \\ 0 & \cdots & 0 & \lambda^2 & 2(1-\lambda^2) \end{bmatrix} \begin{bmatrix} w_{1,j} \\ w_{2,j} \\ \vdots \\ w_{m-1,j} \end{bmatrix} - \begin{bmatrix} w_{1,j-1} \\ w_{2,j-1} \\ \vdots \\ w_{m-1,j-1} \end{bmatrix}.$$

To determine $w_{i,j+1}$ requires values from the jth and $(j-1)$st time steps. (See Figure 12.10.)

Figure 12.10

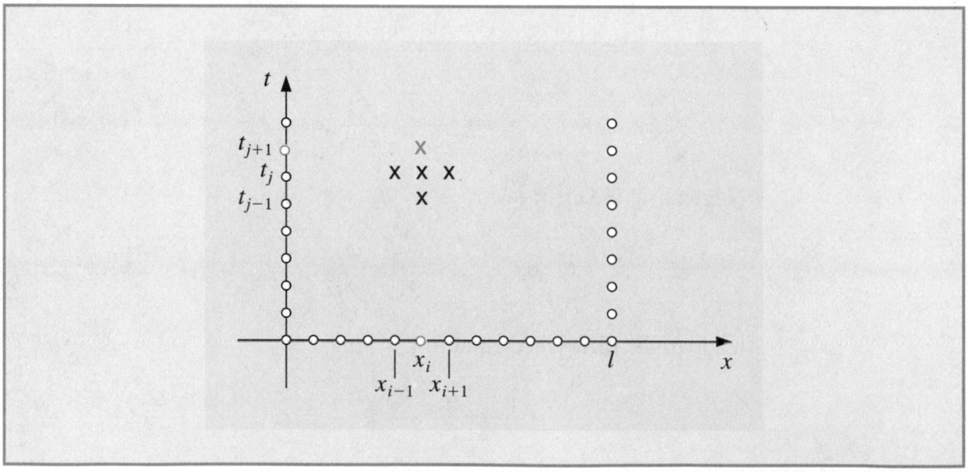

There is a minor starting problem since values for $j = 0$ are given in the initial conditions, but values for $j = 1$, which are needed to compute w_{i2}, must be obtained from the initial-velocity condition

$$\frac{\partial u}{\partial t}(x, 0) = g(x), \qquad \text{for } 0 \le x \le l.$$

One approach is to replace $\partial u/\partial t$ by a forward-difference approximation,

$$\frac{\partial u}{\partial t}(x_i, 0) = \frac{u(x_i, t_1) - u(x_i, 0)}{k} - \frac{k}{2}\frac{\partial^2 u}{\partial t^2}(x_i, \tilde{\mu}_i),$$

for some $\tilde{\mu}_i$ in $(0, t_1)$. Solving for $u(x_i, t_1)$ gives

$$\begin{aligned} u(x_i, t_1) &= u(x_i, 0) + k\frac{\partial u}{\partial t}(x_i, 0) + \frac{k^2}{2}\frac{\partial^2 u}{\partial t^2}(x_i, \tilde{\mu}_i) \\ &= u(x_i, 0) + kg(x_i) + \frac{k^2}{2}\frac{\partial^2 u}{\partial t^2}(x_i, \tilde{\mu}_i). \end{aligned}$$

As a consequence,

$$w_{i1} = w_{i0} + kg(x_i), \qquad \text{for each } i = 1, \ldots, m-1.$$

However, this gives an approximation that has error of only $O(k)$. A better approximation to $u(x_i, 0)$ can be obtained.

Consider the equation

$$u(x_i, t_1) = u(x_i, 0) + k\frac{\partial u}{\partial t}(x_i, 0) + \frac{k^2}{2}\frac{\partial^2 u}{\partial t^2}(x_i, 0) + \frac{k^3}{6}\frac{\partial^3 u}{\partial t^3}(x_i, \hat{u}_i)$$

for some $\hat{\mu}_i$ in $(0, t_1)$, which comes from expanding $u(x_i, t_1)$ in a second Maclaurin polynomial in t.

If f'' exists, then

$$\frac{\partial^2 u}{\partial t^2}(x_i, 0) = \alpha^2\frac{\partial^2 u}{\partial x^2}(x_i, 0) = \alpha^2\frac{d^2 f}{dx^2}(x_i) = \alpha^2 f''(x_i)$$

and

$$u(x_i, t_1) = u(x_i, 0) + kg(x_i) + \frac{\alpha^2 k^2}{2}f''(x_i) + \frac{k^3}{6}\frac{\partial^3 u}{\partial t^3}(x_i, \hat{\mu}_i),$$

producing an approximation with error $O(k^2)$:

$$w_{i1} = w_{i0} + kg(x_i) + \frac{\alpha^2 k^2}{2}f''(x_i).$$

If $f''(x_i)$ is not readily available, we can use a centered-difference quotient to write

$$f''(x_i) = \frac{f(x_{i+1}) - 2f(x_i) + f(x_{i-1})}{h^2} - \frac{h^2}{12}f^{(4)}(\tilde{\xi}_i),$$

for some $\tilde{\xi}_i$ in (x_{i-1}, x_{i+1}). The approximation then becomes

$$\frac{u(x_i, t_1) - u(x_i, 0)}{k} = g(x_i) + \frac{k\alpha^2}{2h^2}[f(x_{i+1}) - 2f(x_i) + f(x_{i-1})] + O(k^2 + h^2 k),$$

so

$$u(x_i, t_1) = u(x_i, 0) + kg(x_i) + \frac{k^2\alpha^2}{2h^2}[f(x_{i+1}) - 2f(x_i) + f(x_{i-1})] + O(k^3 + h^2k^2).$$

Letting $\lambda = k\alpha/h$ gives

$$\begin{aligned} u(x_i, t_1) &= u(x_i, 0) + kg(x_i) + \frac{\lambda^2}{2}[f(x_{i+1}) - 2f(x_i) + f(x_{i-1})] + O(k^3 + h^2k^2) \\ &= (1 - \lambda^2)f(x_i) + \frac{\lambda^2}{2}f(x_{i+1}) + \frac{\lambda^2}{2}f(x_{i-1}) + kg(x_i) + O(k^3 + h^2k^2). \end{aligned}$$

The difference equation for the wave equation

$$w_{i1} = (1 - \lambda^2)f(x_i) + \frac{\lambda^2}{2}f(x_{i+1}) + \frac{\lambda^2}{2}f(x_{i-1}) + kg(x_i)$$

can be used to find w_{i1} for each $i = 1, 2, \ldots, m-1$.

The program WVFDM124 uses this equation to find w_{i1}. It is assumed that there is an upper bound T for the value of t, to be used in the stopping technique, and that $k = T/N$, where N is also given.

EXAMPLE 1 Consider the hyperbolic problem

$$\frac{\partial^2 u}{\partial t^2}(x, t) - 4\frac{\partial^2 u}{\partial x^2}(x, t) = 0, \qquad \text{for } 0 < x < 1 \quad \text{and} \quad 0 < t,$$

with boundary conditions

$$u(0, t) = u(1, t) = 0, \qquad \text{for } 0 < t,$$

and initial conditions

$$u(x, 0) = \sin(\pi x), \quad 0 \le x \le 1, \quad \text{and} \quad \frac{\partial u}{\partial t}(x, 0) = 0, \quad \text{for } 0 \le x \le 1.$$

It is easily verified that the solution to this problem is

$$u(x, t) = \sin(\pi x)\cos(2\pi t).$$

The Finite-Difference method is used in this example with $m = 10$, $T = 1$, and $N = 20$, which implies that $h = 0.1$, $k = 0.05$, and $\lambda = 1$.

To use Maple for this example, we first need to access the linear algebra library.

```
>with(linalg);
```

We define l, α, m, N, h, k, and the functions $f(x)$ and $g(x)$ with the commands

```
>l:=1; alpha:=2; m:=10; N:=20; h:=l/m; k:=0.05;
>f:=x->sin(Pi*x); g:=x->0;
```

and we define and initialize the $m-1$ by $m-1$ matrix A and the vectors $\mathbf{w}^{(0)}$, $\mathbf{w}^{(1)}$, $\mathbf{w}^{(2)}$, and $\mathbf{u}$ by

```
>A:=matrix(m-1,m-1,0);
>w0:=vector(m-1,0);
>w1:=vector(m-1,0);
>w2:=vector(m-1,0);
>u:=vector(m-1,0);
```

We compute $\mathbf{w}^{(0)} = \mathbf{w}0$ with

```
>for i from 1 to m-1 do
>w0[i]:=evalf(f(i*h));
>od;
```

and define the constant λ by

```
>lambda:=alpha*k/h;
```

We use the $O(h^2)$ method to generate $\mathbf{w}^{(1)} = \mathbf{w}1$ with the commands

```
>for i from 1 to m-1 do
>w1[i]:=(1-lambda^2)*evalf(f(i*h))+(evalf(f((i+1)*h))+
 evalf(f((i-1)*h)))*lambda^2/2+k*evalf(g(i*h));
>od;
```

and generate the nonzero entries of A

```
>A[1,1]:=2*(1-lambda^2);
>A[1,2]:=lambda^2;
>for i from 2 to m-2 do
>A[i,i-1]:=lambda^2;
>A[i,i+1]:=lambda^2;
>A[i,i]:=2*(1-lambda^2);
>od;
>A[m-1,m-2]:=lambda^2;
>A[m-1,m-1]:=2*(1-lambda^2);
```

Each pass through the following loop gives $\mathbf{u} = A\mathbf{w}^{(i)} = A\mathbf{w}1$ and $\mathbf{w}2 = \mathbf{w}^{(i+1)} = A\mathbf{w}^{(i)} - \mathbf{w}^{(i-1)}$ and then prepares for the next pass.

```
>for i from 1 to N-1 do
>u:=multiply(A,w1);
```

```
>w2:=evalm(u - w0);
>w0:=evalm(w1);
>w1:=evalm(w2);
>od;
```

Table 12.6 lists the results of the approximation at the final time value, w_{iN}, for $i = 0, 1, \ldots, 10$, which are correct to the places given. ◆ ◆ ◆

Table 12.6

x_i	$w_{i,20}$
0.0	0.0000000000
0.1	0.3090169944
0.2	0.5877852523
0.3	0.8090169944
0.4	0.9510565163
0.5	1.0000000000
0.6	0.9510565163
0.7	0.8090169944
0.8	0.5877852523
0.9	0.3090169944
1.0	0.0000000000

EXERCISE SET 12.4

1. Use the Finite-Difference method with $m = 4$, $N = 4$, and $T = 1.0$ to approximate the solution to the wave equation

$$\frac{\partial^2 u}{\partial t^2} - \frac{\partial^2 u}{\partial x^2} = 0, \qquad 0 < x < 1, \qquad 0 < t;$$
$$u(0, t) = u(1, t) = 0, \qquad 0 < t,$$
$$u(x, 0) = \sin \pi x, \qquad 0 \le x \le 1,$$
$$\frac{\partial u}{\partial t}(x, 0) = 0, \qquad 0 \le x \le 1,$$

and compare your results to the actual solution $u(x, t) = \cos \pi t \sin \pi x$ at $t = 1.0$.

2. Use the Finite-Difference method with $m = 4$, $N = 4$, and $T = 0.5$ to approximate the solution to the wave equation

$$\frac{\partial^2 u}{\partial t^2} - \frac{1}{16\pi^2}\frac{\partial^2 u}{\partial x^2} = 0, \qquad 0 < x < 0.5, \qquad 0 < t;$$
$$u(0, t) = u(0.5, t) = 0, \qquad 0 < t,$$
$$u(x, 0) = 0, \qquad 0 \le x \le 0.5,$$
$$\frac{\partial u}{\partial t}(x, 0) = \sin 4\pi x, \qquad 0 \le x \le 0.5,$$

and compare your results to the actual solution $u(x, t) = \sin t \sin 4\pi x$ at $t = 0.5$.

3. Use the Finite-Difference method with

a. $h = \pi/10$ and $k = 0.05$,

b. $h = \pi/20$ and $k = 0.1$,

c. $h = \pi/20$ and $k = 0.05$

to approximate the solution to the wave equation

$$\frac{\partial^2 u}{\partial t^2} - \frac{\partial^2 u}{\partial x^2} = 0, \qquad 0 < x < \pi, \qquad 0 < t;$$
$$u(0,t) = u(\pi,t) = 0, \qquad 0 < t,$$
$$u(x,0) = \sin x, \qquad 0 \le x \le \pi,$$
$$\frac{\partial u}{\partial t}(x,0) = 0, \qquad 0 \le x \le \pi.$$

Compare your results to the actual solution $u(x,t) = \cos t \sin x$ at $t = 0.5$.

4. Use the Finite-Difference method with $h = k = 0.1$ to approximate the solution to the wave equation

$$\frac{\partial^2 u}{\partial t^2} - \frac{\partial^2 u}{\partial x^2} = 0, \qquad 0 < x < 1, \qquad 0 < t;$$
$$u(0,t) = u(1,t) = 0, \qquad 0 < t,$$
$$u(x,0) = \sin 2\pi x, \qquad 0 \le x \le 1,$$
$$\frac{\partial u}{\partial t}(x,0) = 2\pi \sin 2\pi x, \qquad 0 \le x \le 1,$$

and compare your results to the actual solution $u(x,t) = \sin 2\pi x(\cos 2\pi t + \sin 2\pi t)$, at $t = 0.3$.

5. Use the Finite-Difference method with $h = k = 0.1$ to approximate the solution to the wave equation

$$\frac{\partial^2 u}{\partial t^2} - \frac{\partial^2 u}{\partial x^2} = 0, \quad 0 < x < 1, 0 < t;$$
$$u(0,t) = u(1,t) = 0, \quad 0 < t,$$
$$u(x,0) = \begin{cases} 1, & 0 \le x \le \frac{1}{2}, \\ -1, & \frac{1}{2} < x \le 1, \end{cases}$$
$$\frac{\partial u}{\partial t}(x,0) = 0, \quad 0 \le x \le 1.$$

6. The air pressure $p(x,t)$ in an organ pipe is governed by the wave equation

$$\frac{\partial^2 p}{\partial x^2} = \frac{1}{c^2}\frac{\partial^2 p}{\partial t^2}, \qquad 0 < x < l, \qquad 0 < t,$$

where l is the length of the pipe and c is a physical constant. If the pipe is open, the boundary conditions are given by

$$p(0,t) = p_0 \quad \text{and} \quad p(l,t) = p_0.$$

If the pipe is closed at the end where $x = l$, the boundary conditions are

$$p(0,t) = p_0 \quad \text{and} \quad \frac{\partial p}{\partial x}(l,t) = 0.$$

Assume that $c = 1$, $l = 1$, and the initial conditions are

$$p(x,0) = p_0 \cos 2\pi x \quad \text{and} \quad \frac{\partial p}{\partial t}(x,0) = 0, \qquad 0 \le x \le 1.$$

a. Use the Finite-Difference method to approximate the pressure for an open pipe with $p_0 = 0.9$ at $x = \frac{1}{2}$ for $t = 0.5$ and $t = 1$, using $h = k = 0.1$.
b. Modify the Finite-Difference method for the closed-pipe problem with $p_0 = 0.9$, and approximate $p(0.5, 0.5)$ and $p(0.5, 1)$ using $h = k = 0.1$.

7. In an electric transmission line of length l that carries alternating current of high frequency (called a *lossless* line), the voltage V and current i are described by

$$\frac{\partial^2 V}{\partial x^2} = LC\frac{\partial^2 V}{\partial t^2}, \qquad 0 < x < l, \qquad 0 < t,$$
$$\frac{\partial^2 i}{\partial x^2} = LC\frac{\partial^2 i}{\partial t^2}, \qquad 0 < x < l, \qquad 0 < t,$$

where L is the inductance per unit length and C is the capacitance per unit length. Suppose the line is 200 ft long and the constants C and L are given by

$$C = 0.1 \text{ farads/ft} \quad \text{and} \quad L = 0.3 \text{ henries/ft}.$$

Suppose the voltage and current also satisfy

$$V(0,t) = V(200,t) = 0, \qquad 0 < t,$$
$$V(x,0) = 110 \sin\frac{\pi x}{200}, \qquad 0 \le x \le 200,$$
$$\frac{\partial V}{\partial t}(x,0) = 0, \qquad 0 \le x \le 200,$$
$$i(0,t) = i(200,t) = 0, \qquad 0 < t,$$
$$i(x,0) = 5.5 \cos\frac{\pi x}{200}, \qquad 0 \le x \le 200,$$

and

$$\frac{\partial i}{\partial t}(x,0) = 0, \qquad 0 \le x \le 200.$$

Use the Finite-Difference method to approximate the voltage and current at $t = 0.2$ and $t = 0.5$ using $h = 10$ and $k = 0.1$.

12.5 Introduction to the Finite-Element Method

The **Finite-Element method** for partial-differential equations is similar to the Rayleigh-Ritz method for approximating the solution to two-point boundary-value problems. It was originally developed for use in civil engineering, but it is now used for approximating the solutions to partial-differential equations that arise in all areas of applied mathematics.

One advantage of the Finite-Element method over finite-difference methods is the relative ease with which the boundary conditions of the problem are handled. Many physical problems have boundary conditions involving derivatives and irregularly shaped boundaries. Boundary conditions of this type are difficult to handle using finite-difference techniques, since each boundary condition involving a derivative must be approximated by a difference quotient at the grid points, and irregular shaping of the boundary makes placing the grid points difficult. The Finite-Element method includes the boundary conditions as integrals in a functional that is being minimized, so the construction procedure is independent of the particular boundary conditions of the problem.

In our discussion, we consider the partial-differential equation

$$\frac{\partial}{\partial x}\left(p(x,y)\frac{\partial u}{\partial x}\right) + \frac{\partial}{\partial y}\left(q(x,y)\frac{\partial u}{\partial y}\right) + r(x,y)u(x,y) = f(x,y),$$

with (x, y) in $\mathcal{D}$, where $\mathcal{D}$ is a plane region with boundary S.

Boundary conditions of the form

$$u(x,y) = g(x,y)$$

are imposed on a portion, S_1, of the boundary. On the remainder of the boundary, S_2, $u(x, y)$ is required to satisfy

$$p(x,y)\frac{\partial u}{\partial x}(x,y)\cos\theta_1 + q(x,y)\frac{\partial u}{\partial y}(x,y)\cos\theta_2 + g_1(x,y)u(x,y) = g_2(x,y),$$

where θ_1 and θ_2 are the direction angles of the outward normal to the boundary at the point (x, y). (See Figure 12.11 on the following page). Physical problems in the areas of solid mechanics and elasticity have associated partial-differential equations of this type. The solution to such a problem is typically the minimization of a certain functional, involving integrals, over a class of functions determined by the problem.

Suppose p, q, r, and f are all continuous in $\mathcal{D} \cup S$, p and q have continuous first partial derivatives, and g_1 and g_2 are continuous on S_2. Suppose, in addition, that $p(x, y) > 0$, $q(x, y) > 0$, $r(x, y) \le 0$, and $g_1(x, y) > 0$. Then a solution to our problem uniquely

Figure 12.11

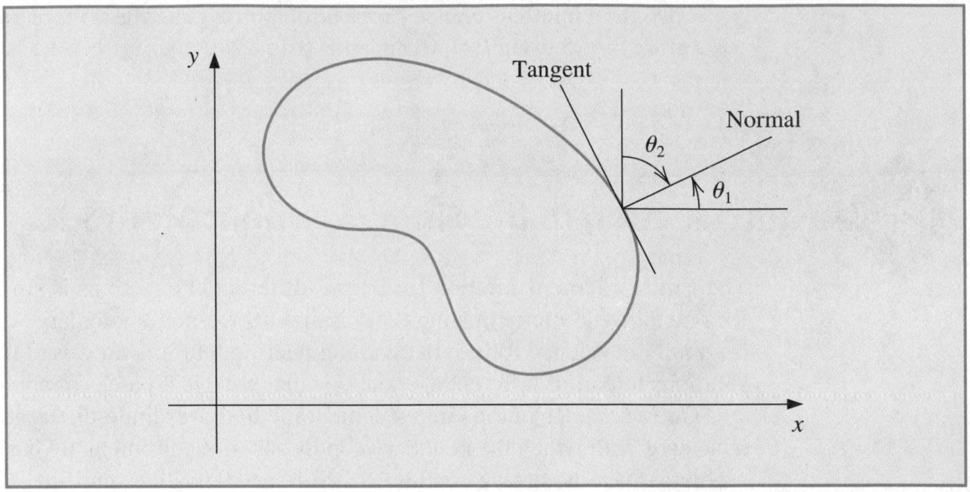

minimizes the functional

$$I[w] = \iint_D \left\{ \frac{1}{2}\left[p(x,y)\left(\frac{\partial w}{\partial x}\right)^2 + q(x,y)\left(\frac{\partial w}{\partial y}\right)^2 - r(x,y)w^2 \right] + f(x,y)w \right\} dx\,dy$$
$$+ \int_{S_2} \left\{ -g_2(x,y)w + \frac{1}{2}g_1(x,y)w^2 \right\} ds$$

over all twice continuously differentiable functions w satisfying $w(x, y) = g(x, y)$ on S_1. The Finite-Element method approximates this solution by minimizing the functional I over a smaller class of functions, just as the Rayleigh-Ritz method did for the boundary-value problem considered in Section 11.6.

The first step is to divide the region into a finite number of sections, or elements, of a regular shape, either rectangles or triangles. (See Figure 12.12.)

Figure 12.12

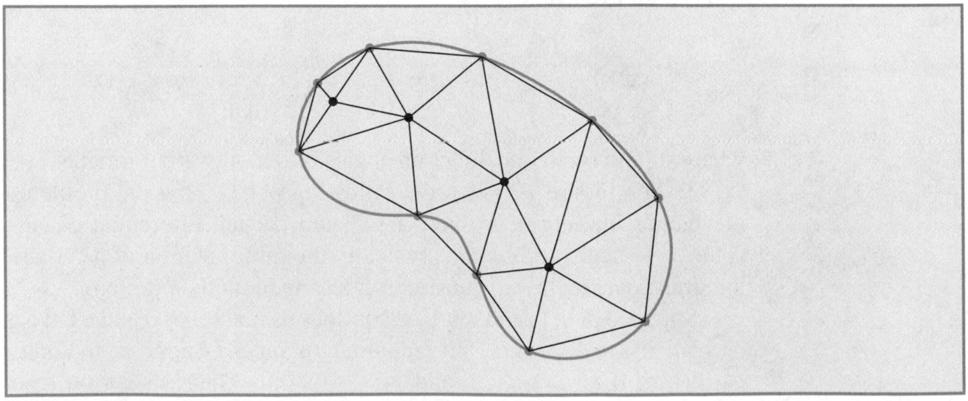

The set of functions used for approximation is generally a set of piecewise polynomials of fixed degree in x and y, and the approximation requires that the polynomials be pieced together in such a manner that the resulting function is continuous with an integrable or continuous first or second derivative on the entire region. Polynomials of linear type in x and y,

$$\phi(x, y) = a + bx + cy,$$

are commonly used with triangular elements, and polynomials of bilinear type in x and y,

$$\phi(x, y) = a + bx + cy + dxy,$$

are used with rectangular elements.

For our discussion, suppose that the region $\mathcal{D}$ has been subdivided into triangular elements. The collection of triangles is denoted D, and the vertices of these triangles are called **nodes**. The method seeks an approximation of the form

$$\phi(x, y) = \sum_{i=1}^{m} \gamma_i \phi_i(x, y),$$

where $\phi_1, \phi_2, \ldots, \phi_m$ are linearly independent piecewise-linear polynomials and $\gamma_1, \gamma_2, \ldots, \gamma_m$ are constants. Some of these constants, say, $\gamma_{n+1}, \gamma_{n+2}, \ldots, \gamma_m$, are used to ensure that the boundary condition

$$\phi(x, y) = g(x, y)$$

is satisfied on S_1, whereas the remaining constants $\gamma_1, \gamma_2, \ldots, \gamma_n$ are used to minimize the functional $I[\sum_{i=1}^{m} \gamma_i \phi_i]$.

Since the functional is of the form

$$\begin{aligned} I[\phi] &= I\left[\sum_{i=1}^{m} \gamma_i \phi_i\right] \\ &= \iint_D \left(\frac{1}{2}\left\{p(x, y)\left[\sum_{i=1}^{m} \gamma_i \frac{\partial \phi_i}{\partial x}(x, y)\right]^2 + q(x, y)\left[\sum_{i=1}^{m} \gamma_i \frac{\partial \phi_i}{\partial y}(x, y)\right]^2 \right.\right. \\ &\quad \left.\left. - r(x, y)\left[\sum_{i=1}^{m} \gamma_i \phi_i(x, y)\right]^2\right\} + f(x, y)\sum_{i=1}^{m} \gamma_i \phi_i(x, y)\right) dy\, dx \\ &\quad + \int_{S_2} \left\{-g_2(x, y)\sum_{i=1}^{m} \gamma_i \phi_i(x, y) + \frac{1}{2} g_1(x, y)\left[\sum_{i=1}^{m} \gamma_i \phi_i(x, y)\right]^2\right\} ds, \end{aligned}$$

for a minimum to occur, considering I as a function of $\gamma_1, \gamma_2, \ldots, \gamma_n$, it is necessary to have

$$\frac{\partial I}{\partial \gamma_j} = 0, \qquad \text{for each } j = 1, 2, \ldots, n.$$

Performing the partial differentiation allows us to write this set of equations as a linear system,

$$A\mathbf{c} = \mathbf{b},$$

where $\mathbf{c} = (\gamma_1, \ldots, \gamma_n)^t$ and where $A = (\alpha_{ij})$ and $\mathbf{b} = (\beta_1, \ldots, \beta_n)^t$ are defined by

$$\alpha_{ij} = \iint_{\mathcal{D}} \left[p(x,y)\frac{\partial \phi_i}{\partial x}(x,y)\frac{\partial \phi_j}{\partial x}(x,y) + q(x,y)\frac{\partial \phi_i}{\partial y}(x,y)\frac{\partial \phi_j}{\partial y}(x,y) \right.$$
$$\left. - r(x,y)\phi_i(x,y)\phi_j(x,y) \right] dx\,dy + \int_{S_2} g_1(x,y)\phi_i(x,y)\phi_j(x,y)\,ds,$$

for each $i = 1, 2, \ldots, n$, and $j = 1, 2, \ldots, m$, and

$$\beta_i = -\iint_{\mathcal{D}} f(x,y)\phi_i(x,y)\,dx\,dy + \int_{S_2} g_2(x,y)\phi_i(x,y)\,ds - \sum_{k=n+1}^{m} \alpha_{ik}\gamma_k,$$

for each $i = 1, \ldots, n$.

The particular choice of basis functions is important, since the appropriate choice can often make the matrix A positive definite and banded. For our second-order problem, we assume that $\mathcal{D}$ is polygonal and that S is a contiguous set of straight lines. In this case, $\mathcal{D} = D$.

To begin the procedure, we divide the region D into a collection of triangles $T_1, T_2, \ldots, T_M$, with the ith triangle having three vertices, or nodes, denoted

$$V_j^{(i)} = \left(x_j^{(i)}, y_j^{(i)} \right), \qquad \text{for } j = 1, 2, 3.$$

To simplify the notation, we write $V_j^{(i)}$ simply as $V_j = (x_j, y_j)$ when working with the fixed triangle T_i. With each vertex V_j we associate a linear polynomial

$$N_j^{(i)} \equiv N_j = a_j + b_j x + c_j y, \quad \text{where} \quad N_j^{(i)}(x_k, y_k) = \begin{cases} 1, & \text{if } j = k \\ 0, & \text{if } j \neq k. \end{cases}$$

This produces linear systems of the form

$$\begin{bmatrix} 1 & x_1 & y_1 \\ 1 & x_2 & y_2 \\ 1 & x_3 & y_3 \end{bmatrix} \begin{bmatrix} a_j \\ b_j \\ c_j \end{bmatrix} = \begin{bmatrix} 0 \\ 1 \\ 0 \end{bmatrix},$$

with the element one occurring in the jth row in the vector on the right.

Let $E_1, \ldots, E_n$ be a labeling of the nodes lying in $D \cup S$. With each node E_k, we associate a function ϕ_k that is linear on each triangle, has the value 1 at E_k, and is 0 at each of the other nodes. This choice makes ϕ_k identical to $N_j^{(i)}$ on triangle T_i when the node E_k is the vertex denoted $V_j^{(i)}$.

EXAMPLE 1 Suppose that a finite-element problem contains the triangles T_1 and T_2 shown in Figure 12.13. The linear function $N_1^{(1)}(x, y)$ that assumes the value 1 at $(1, 1)$ and 0 at both $(0, 0)$ and $(-1, 2)$ satisfies

$$\begin{aligned} a_1^{(1)} + b_1^{(1)}(1) + c_1^{(1)}(1) &= 1, \\ a_1^{(1)} + b_1^{(1)}(-1) + c_1^{(1)}(2) &= 0, \\ a_1^{(1)} + b_1^{(1)}(0) + c_1^{(1)}(0) &= 0, \end{aligned}$$

so $a_1^{(1)} = 0, b_1^{(1)} = \frac{2}{3}, c_1^{(1)} = \frac{1}{3}$, and

$$N_1^{(1)}(x, y) = \frac{2}{3}x + \frac{1}{3}y.$$

Figure 12.13

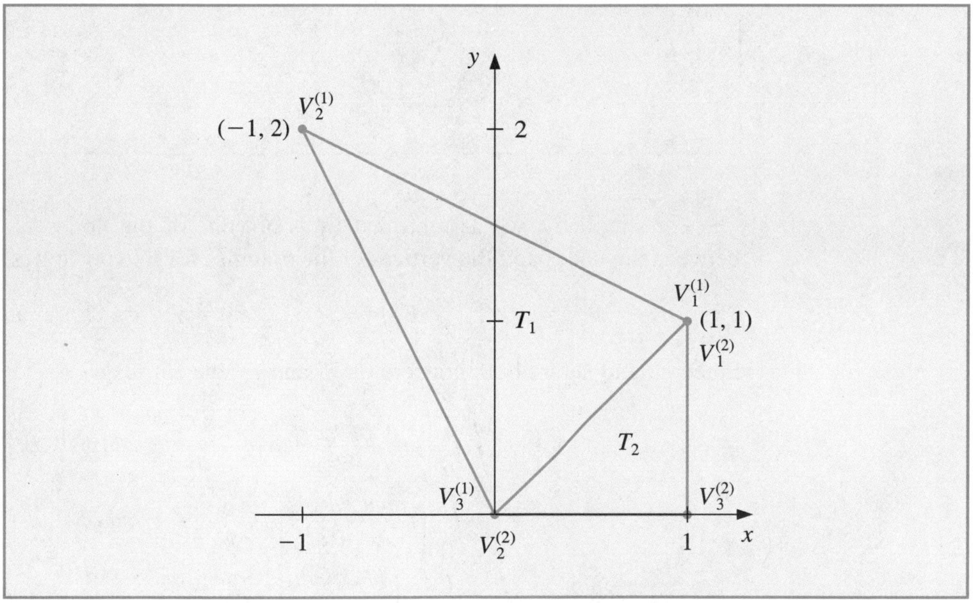

In a similar manner, the linear function $N_1^{(2)}(x, y)$ that assumes the value 1 at $(1, 1)$ and 0 at both $(0, 0)$ and $(1, 0)$ satisfies

$$\begin{aligned} a_1^{(2)} + b_1^{(2)}(1) + c_1^{(2)}(1) &= 1, \\ a_1^{(2)} + b_1^{(2)}(0) + c_1^{(2)}(0) &= 0, \\ a_1^{(2)} + b_1^{(2)}(1) + c_1^{(2)}(0) &= 0, \end{aligned}$$

so $a_1^{(2)} = 0, b_1^{(2)} = 0$, and $c_1^{(2)} = 1$. As a consequence, $N_1^{(2)}(x, y) = y$. Note that on the common boundary of T_1 and T_2, we have $N_1^{(1)}(x, y) = N_1^{(2)}(x, y)$, since $y = x$.

◆ ◆ ◆

Consider Figure 12.14, the upper left portion of the region shown in Figure 12.12. We will generate the entries in the matrix A that correspond to the nodes shown in this figure.

Figure 12.14

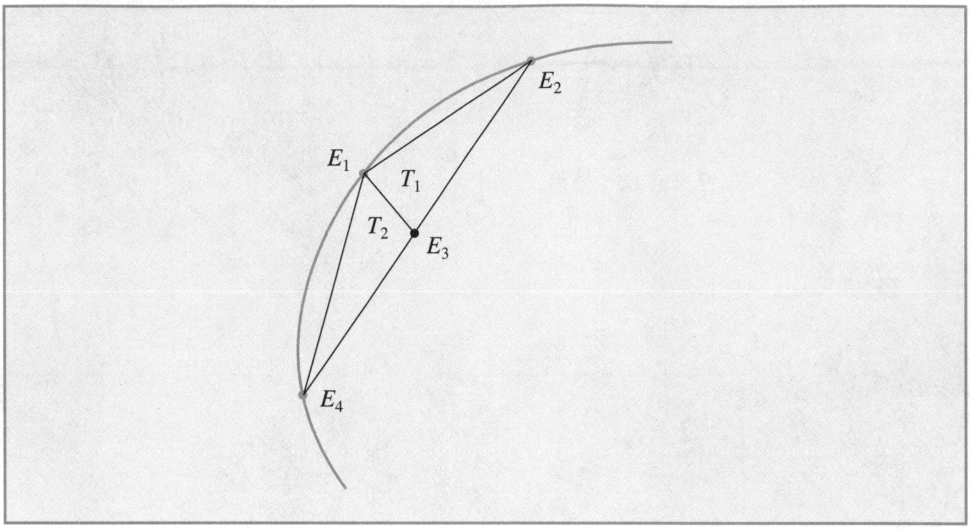

For simplicity, we assume that E_1 is not one of the nodes on S_2. The relationship between the nodes and the vertices of the triangles for this portion is

$$E_1 = V_3^{(1)} = V_1^{(2)}, \qquad E_4 = V_2^{(2)}, \qquad E_3 = V_2^{(1)} = V_3^{(2)}, \quad \text{and} \quad E_2 = V_1^{(1)}.$$

Since ϕ_1 and ϕ_3 are both nonzero on T_1 and T_2, the entries $\alpha_{1,3} = \alpha_{3,1}$ are computed by

$$\begin{aligned}
\alpha_{1,3} &= \iint_D \left[p\frac{\partial\phi_1}{\partial x}\frac{\partial\phi_3}{\partial x} + q\frac{\partial\phi_1}{\partial y}\frac{\partial\phi_3}{\partial y} - r\phi_1\phi_3 \right] dx\,dy \\
&= \iint_{T_1} \left[p\frac{\partial\phi_1}{\partial x}\frac{\partial\phi_3}{\partial x} + q\frac{\partial\phi_1}{\partial y}\frac{\partial\phi_3}{\partial y} - r\phi_1\phi_3 \right] dx\,dy \\
&\quad + \iint_{T_2} \left[p\frac{\partial\phi_1}{\partial x}\frac{\partial\phi_3}{\partial x} + q\frac{\partial\phi_1}{\partial y}\frac{\partial\phi_3}{\partial y} - r\phi_1\phi_3 \right] dx\,dy.
\end{aligned}$$

On triangle T_1,

$$\phi_1(x, y) = N_3^{(1)} = a_3^{(1)} + b_3^{(1)}x + c_3^{(1)}y$$

and

$$\phi_3(x, y) = N_2^{(1)} = a_2^{(1)} + b_2^{(1)}x + c_2^{(1)}y,$$

so

$$\frac{\partial\phi_1}{\partial x} = b_3^{(1)}, \qquad \frac{\partial\phi_1}{\partial y} = c_3^{(1)}, \qquad \frac{\partial\phi_3}{\partial x} = b_2^{(1)}, \quad \text{and} \quad \frac{\partial\phi_3}{\partial y} = c_2^{(1)}.$$

Similarly, on T_2,

$$\phi_1(x, y) = N_1^{(2)} = a_1^{(2)} + b_1^{(2)}x + c_1^{(2)}y$$

and

$$\phi_3(x, y) = N_3^{(2)} = a_3^{(2)} + b_3^{(2)}x + c_3^{(2)}y,$$

so

$$\frac{\partial \phi_1}{\partial x} = b_1^{(2)}, \quad \frac{\partial \phi_1}{\partial y} = c_1^{(2)}, \quad \frac{\partial \phi_3}{\partial x} = b_3^{(2)}, \quad \text{and} \quad \frac{\partial \phi_3}{\partial y} = c_3^{(2)}.$$

Thus,

$$\begin{aligned}
\alpha_{13} = {} & b_3^{(1)}b_2^{(1)} \iint_{T_1} p\,dx\,dy + c_3^{(1)}c_2^{(1)} \iint_{T_1} q\,dx\,dy \\
& - \iint_{T_1} r(a_3^{(1)} + b_3^{(1)}x + c_3^{(1)}y)(a_2^{(1)} + b_2^{(1)}x + c_2^{(1)}y)\,dx\,dy \\
& + b_1^{(2)}b_3^{(2)} \iint_{T_2} p\,dx\,dy + c_1^{(2)}c_3^{(2)} \iint_{T_2} q\,dx\,dy \\
& - \iint_{T_2} r(a_1^{(2)} + b_1^{(2)}x + c_1^{(2)}y)(a_3^{(2)} + b_3^{(2)}x + c_3^{(2)}y)\,dx\,dy.
\end{aligned}$$

All the double integrals over D reduce to double integrals over triangles. The usual procedure is to compute all possible integrals over the triangles and accumulate them into the correct entry α_{ij} in A.

Similarly, the double integrals of the form

$$\iint_D f(x, y)\phi_i(x, y)\,dx\,dy$$

are computed over triangles and then accumulated into the correct entry β_i of $\mathbf{b}$. For example, to determine β_1 we need

$$\begin{aligned}
-\iint_D f(x, y)\phi_1(x, y)\,dx\,dy = {} & -\iint_{T_1} f(x, y)\left[a_3^{(1)} + b_3^{(1)}x + c_3^{(1)}y\right]\,dx\,dy \\
& -\iint_{T_2} f(x, y)\left[a_1^{(2)} + b_1^{(2)}x + c_1^{(2)}y\right]\,dx\,dy.
\end{aligned}$$

Part of β_1 is contributed by ϕ_1 restricted to T_1 and the remainder by ϕ_1 restricted to T_2, since E_1 is a vertex of both T_1 and T_2. In addition, nodes that lie on S_2 have line integrals added to their entries in A and $\mathbf{b}$.

Consider the evaluation of the double integral

$$\iint_T F(x, y)\,dy\,dx$$

over the triangle T with vertices (x_0, y_0), (x_1, y_1) and (x_2, y_2). First, we define

$$\Delta = \frac{1}{2}\det\begin{bmatrix} 1 & x_1 & y_1 \\ 1 & x_2 & y_2 \\ 1 & x_3 & y_3 \end{bmatrix}.$$

Let (x_3, y_3), (x_4, y_4) and (x_5, y_5) be the midpoints of the sides of the triangle T and let (x_6, y_6) be the centroid as shown in Figure 12.15.

Figure 12.15

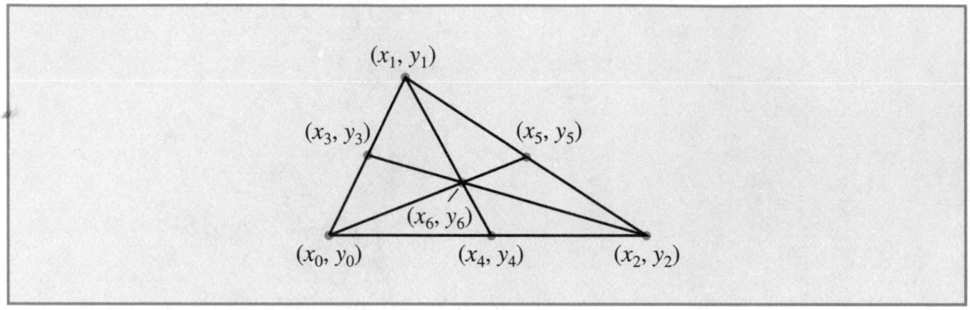

We have

$$x_3 = \frac{1}{2}(x_0 + x_1), \qquad y_3 = \frac{1}{2}(y_0 + y_1),$$

$$x_4 = \frac{1}{2}(x_0 + x_2), \qquad y_4 = \frac{1}{2}(y_0 + y_2),$$

$$x_5 = \frac{1}{2}(x_1 + x_2), \qquad y_5 = \frac{1}{2}(y_1 + y_2),$$

$$x_6 = \frac{1}{3}(x_0 + x_1 + x_2), \qquad y_6 = \frac{1}{3}(y_0 + y_1 + y_2).$$

An $O(h^2)$ formula for the double integral is

$$\iint_T F(x, y)\,dy\,dx = \frac{1}{2}|\Delta|\left\{\frac{1}{20}(F(x_0, y_0) + F(x_1, y_1) + F(x_2, y_2)) + \frac{2}{15}(F(x_3, y_3) + F(x_4, y_4) + F(x_5, y_5)) + \frac{9}{20}F(x_6, y_6)\right\}$$

Note that if $F(x, y) = 1$, then the double integral evaluates to $\frac{1}{2}|\Delta|$, which is the area of the triangle T.

Another integral that needs to be computed is the line integral

$$\int_L G(x, y)\,dS$$

where L is the line segment with endpoints (x_0, y_0) and (x_1, y_1). Let $x = x(t)$ and $y = y(t)$ be a parameterization of L with $(x_0, y_0) = (x(t_0), y(t_0))$ and $(x_1, y_1) = (x(t_1), y(t_1))$. Then

$$\int_L G(x, y)\, ds = \int_{t_0}^{t_1} G(x(t), y(t))\sqrt{[x'(t)]^2 + [y'(t)]^2}\, dt.$$

We could then evaluate the definite integral exactly or we could use a method presented in Chapter 4.

The program LINFE125 performs the Finite-Element method on a second-order elliptic differential equation. In the program, all values of the matrix A and vector $\mathbf{b}$ are initially set to zero and after all the integrations have been performed on all the triangles these values are added to the appropriate entries in A and $\mathbf{b}$.

EXAMPLE 2 The temperature, $u(x, y)$, in a two-dimensional region D satisfies Laplace's equation,

$$\frac{\partial^2 u}{\partial x^2}(x, y) + \frac{\partial^2 u}{\partial y^2}(x, y) = 0 \qquad \text{on } D.$$

Consider the region D shown in Figure 12.16 with boundary conditions given by

$$\begin{aligned} u(x, y) &= 4, && \text{for } (x, y) \text{ on } L_3 \text{ or } L_4, \\ \frac{\partial u}{\partial n}(x, y) &= x, && \text{for } (x, y) \text{ on } L_1, \\ \frac{\partial u}{\partial n}(x, y) &= \frac{x + y}{\sqrt{2}}, && \text{for } (x, y) \text{ on } L_2, \end{aligned}$$

where $\partial u/\partial n$ denotes the directional derivative in the direction of the normal to the boundary of the region D at the point (x, y).

Figure 12.16

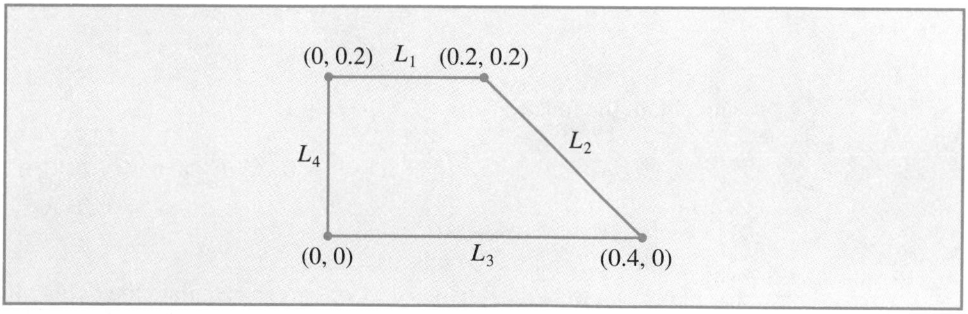

For this example, $S_1 = L_3 \cup L_4$ and $S_2 = L_1 \cup L_2$. Our functions are

$$p(x, y) = 1, \quad q(x, y) = 1, \quad r(x, y) = 0, \quad f(x, y) = 0$$

on D and its boundary. We also have

$$g(x, y) = 4, \qquad \text{on } S_1,$$

$$g_1(x,y) = 0, \quad \text{on } S_2,$$

$$g_2(x,y) = x, \quad \text{on} \quad L_1 \quad \text{and} \quad g_2(x,y) = \frac{x+y}{\sqrt{2}} \quad \text{on } L_2.$$

We first subdivide D into triangles with the labeling shown in Figure 12.17.

Figure 12.17

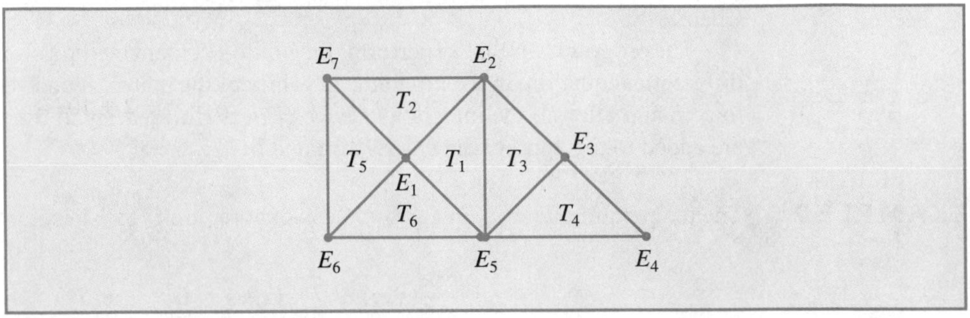

The nodes are given by $E_1 = (0.1, 0.1)$, $E_2 = (0.2, 0.2)$, $E_3 = (0.3, 0.1)$, $E_4 = (0.4, 0.0)$, $E_5 = (0.2, 0.0)$, $E_6 = (0.0, 0.0)$ and $E_7 = (0.0, 0.2)$. Since E_4, E_5, E_6, and E_7 are on S_1 and $g(x, y) = 4$, we have $\gamma_4 = \gamma_5 = \gamma_6 = \gamma_7 = 4$.

For each triangle T_i, when $1 \leq i \leq 6$, we assign the three vertices

$$V_j^{(i)} = (x_j^{(i)}, y_j^{(i)}) \qquad \text{for } j = 1, 2, 3$$

and calculate

$$\Delta_i = \frac{1}{2} \det \begin{bmatrix} 1 & x_1^{(i)} & y_1^{(i)} \\ 1 & x_2^{(i)} & y_2^{(i)} \\ 1 & x_3^{(i)} & y_3^{(i)} \end{bmatrix}$$

and obtain the functions

$$N_j^{(i)}(x, y) = a_j^{(i)} + b_j^{(i)} x + c_j^{(i)} y.$$

For each i, we have $\Delta_i = 0.02$.

Table 12.7 summarizes the assignments.

For each triangle T_i, when $1 \leq i \leq 6$, we calculate the double integrals

$$\begin{aligned} z_{j,k}^{(i)} &= b_j^{(i)} b_k^{(i)} \iint_{T_i} p(x,y)\,dy\,dx + c_j^{(i)} c_k^{(i)} \iint_{T_i} q(x,y)\,dy\,dx \\ &\quad - \iint_{T_i} r(x,y) N_j^{(i)}(x,y) N_k^{(i)}(x,y)\,dy\,dx \\ &= b_j^{(i)} b_k^{(i)} \frac{1}{2} |\Delta_i| + c_j^{(i)} c_k^{(i)} \frac{1}{2} |\Delta_i| - 0 = 0.01 \left[b_j^{(i)} b_k^{(i)} + c_j^{(i)} c_k^{(i)} \right] \end{aligned}$$

Table 12.7

$i = 1$	$j = 1$	$V_1^{(1)} = (0.1, 0.1) = E_1$	$N_1^{(1)} = 2 - 10x$
	$j = 2$	$V_2^{(1)} = (0.2, 0.2) = E_2$	$N_2^{(1)} = -1 + 5x + 5y$
	$j = 3$	$V_3^{(1)} = (0.2, 0.0) = E_5$	$N_3^{(1)} = 5x - 5y$
$i = 2$	$j = 1$	$V_1^{(2)} = (0.1, 0.1) = E_1$	$N_1^{(2)} = 2 - 10y$
	$j = 2$	$V_2^{(2)} = (0.0, 0.2) = E_7$	$N_2^{(2)} = -5x + 5y$
	$j = 3$	$V_3^{(2)} = (0.2, 0.2) = E_2$	$N_3^{(2)} = -1 + 5x + 5y$
$i = 3$	$j = 1$	$V_1^{(3)} = (0.2, 0.2) = E_2$	$N_1^{(3)} = 1 - 5x + 5y$
	$j = 2$	$V_2^{(3)} = (0.3, 0.1) = E_3$	$N_2^{(3)} = -2 + 10x$
	$j = 3$	$V_3^{(3)} = (0.2, 0.0) = E_5$	$N_3^{(3)} = 2 - 5x - 5y$
$i = 4$	$j = 1$	$V_1^{(4)} = (0.3, 0.1) = E_3$	$N_1^{(4)} = 10y$
	$j = 2$	$V_2^{(4)} = (0.4, 0.0) = E_4$	$N_2^{(4)} = -1 + 5x - 5y$
	$j = 3$	$V_3^{(4)} = (0.2, 0.0) = E_5$	$N_3^{(4)} = 2 - 5x - 5y$
$i = 5$	$j = 1$	$V_1^{(5)} = (0.0, 0.2) = E_7$	$N_1^{(5)} = -5x + 5y$
	$j = 2$	$V_2^{(5)} = (0.1, 0.1) = E_1$	$N_2^{(5)} = 10x$
	$j = 3$	$V_3^{(5)} = (0.0, 0.0) = E_6$	$N_3^{(5)} = 1 - 5x - 5y$
$i = 6$	$j = 1$	$V_1^{(6)} = (0.0, 0.0) = E_6$	$N_1^{(6)} = 1 - 5x - 5y$
	$j = 2$	$V_2^{(6)} = (0.2, 0.0) = E_5$	$N_2^{(6)} = 5x - 5y$
	$j = 3$	$V_3^{(6)} = (0.1, 0.1) = E_1$	$N_3^{(6)} = 10y$

for $j = 1, 2, 3$ and $k = 1, \ldots j$. Then

$$H_j^{(i)} = -\iint_{T_i} f(x, y) N_j^{(i)}(x, y)\, dy\, dx = 0 \qquad \text{for } j = 1, 2, 3.$$

We let l_1 be the line from $(0.4, 0.0)$ to $(0.3, 0.1)$, l_2 be the line from $(0.3, 0.1)$ to $(0.2, 0.2)$, and l_3 be the line from $(0.2, 0.2)$ to $(0.0, 0.2)$. We use the parametrization

$$\begin{aligned}
l_1&:\quad x = 0.4 - t, & y &= t, & &\text{for} & 0 &\le t \le 0.1,\\
l_2&:\quad x = 0.4 - t, & y &= t, & &\text{for} & 0.1 &\le t \le 0.2,\\
l_3&:\quad x = -t, & y &= 0.2, & &\text{for} & -0.2 &\le t \le 0.
\end{aligned}$$

The line integrals we calculate are over the edges of the triangle, which are on $S_2 = L_1 \cup L_2$. Suppose triangle T_i has an edge e_i on S_2 from the vertex $V_j^{(i)}$ to $V_k^{(i)}$. The line integrals we need are

$$J_{k,j}^{(i)} = J_{j,k}^{(i)} = \int_{e_i} g_1(x, y) N_j^{(i)}(x, y) N_k^{(i)}(x, y)\, ds$$

and

$$I_j^{(i)} = \int_{e_i} g_2(x, y) N_j^{(i)}(x, y)\, ds$$

$$I_k^{(i)} = \int_{e_i} g_2(x, y) N_k^{(i)}(x, y)\, ds.$$

Since $g_1(x, y) = 0$ on S_2, we need to consider only the line integrals involving $g_2(x, y)$.

Triangle T_4 has an edge on L_4 with vertices $V_2^{(4)} = (0.4, 0.0) = E_4$ and $V_1^{(4)} = (0.3, 0.1) = E_3$. Thus,

$$\begin{aligned}
I_1^{(4)} &= \int_{l_1} g_2(x, y) N_1^{(4)}(x, y)\, ds = \int_{l_1} \frac{x+y}{\sqrt{2}}(10y)\, ds \\
&= \int_0^{0.1} \frac{0.4}{\sqrt{2}}(10t)\sqrt{(-1)^2+1}\, dt = \int_0^{0.1} 4t\, dt = 0.02, \\
I_2^{(4)} &= \int_{l_1} g_2(x, y) N_2^{(4)}(x, y)\, ds = \int_{l_1} \frac{x+y}{\sqrt{2}}(-1+5x-5y)\, ds \\
&= \int_0^{0.1} \frac{0.4}{\sqrt{2}}(-1+2-5t-5t)\sqrt{2}\, dt = \int_0^{0.1} 0.4(1-10t)\, dt = 0.02.
\end{aligned}$$

Triangle T_3 has an edge on L_2 with vertices $V_2^{(3)} = (0.3, 0.1) = E_3$ and $V_1^{(3)} = (0.2, 0.2) = E_2$. Thus,

$$\begin{aligned}
I_2^{(3)} &= \int_{l_2} g_2(x, y) N_2^{(3)}(x, y)\, ds = \int_{l_2} \frac{x+y}{\sqrt{2}}(-2+10x)\, ds \\
&= \int_{0.1}^{0.2} \frac{0.4}{\sqrt{2}}(-2+4-10t)\sqrt{2}\, dt = \int_{0.1}^{0.2} (0.8-4t)\, dt = 0.02, \\
I_1^{(3)} &= \int_{l_2} g_2(x, y) N_1^{(3)}(x, y)\, ds = \int_{l_2} \frac{x+y}{\sqrt{2}}(1-5x+5y)\, ds \\
&= \int_{0.1}^{0.2} \frac{0.4}{\sqrt{2}}(1-2+5t+5t)\sqrt{2}\, dt = \int_{0.1}^{0.2} (-0.4+4t)\, dt = 0.02.
\end{aligned}$$

Triangle T_2 has an edge on L_1 with vertices $V_3^{(2)} = (0.2, 0.2) = E_2$ and $V_2^{(2)} = (0.0, 0.2) = E_7$. Thus,

$$\begin{aligned}
I_3^{(2)} &= \int_{l_3} g_2(x, y) N_3^{(2)}(x, y)\, ds = \int_{l_3} x(-1+5x+5y)\, ds \\
&= \int_{-0.2}^{0} (-t)(-1-5t+1)\sqrt{(-1)^2}\, dt = \int_{-0.2}^{0} 5t^2\, dt = 0.01\overline{3}, \\
I_2^{(2)} &= \int_{l_3} g_2(x, y) N_2^{(2)}(x, y)\, ds = \int_{l_3} x(-5x+5y)\, ds \\
&= \int_{-0.2}^{0} (-t)(5t+1)\, dt = \int_{-0.2}^{0} (-5t^2-t)\, dt = 0.00\overline{6}.
\end{aligned}$$

Assembling all the elements gives

$$\begin{aligned}
\alpha_{11} &= z_{1,1}^{(1)} + z_{1,1}^{(2)} + z_{2,2}^{(5)} + z_{3,3}^{(6)}\\
&= (-10)^2 0.01 + (-10)^2 0.01 + (10)^2 0.01 + (10)^2 0.01 = 4\\
\alpha_{12} &= \alpha_{21} = z_{2,1}^{(1)} + z_{3,1}^{(2)} = (-50)0.01 + (-50)0.01 = -1\\
\alpha_{13} &= \alpha_{31} = 0\\
\alpha_{22} &= z_{2,2}^{(1)} + z_{3,3}^{(2)} + z_{1,1}^{(3)} = (50)0.01 + (50)0.01 + (50)0.01 = 1.5\\
\alpha_{23} &= a_{32} = z_{2,1}^{(3)} = (-50)0.01 = -0.5\\
\alpha_{33} &= z_{2,2}^{(3)} + z_{1,1}^{(4)} = (100)0.01 + (100)0.01 = 2\\
\beta_1 &= -z_{3,1}^{(1)}\gamma_5 - z_{2,1}^{(2)}\gamma_7 - z_{2,1}^{(5)}\gamma_7 - z_{3,2}^{(5)}\gamma_6 - z_{3,1}^{(6)}\gamma_6 - z_{3,2}^{(6)}\gamma_5\\
&= -4(-50)0.01 - 4(-50)0.01 - 4(-50)0.01\\
&\quad - 4(-50)0.01 - 4(-50)0.01 - 4(-50)0.01 = 12\\
\beta_2 &= -z_{3,2}^{(1)}\gamma_5 - z_{3,2}^{(2)}\gamma_7 - z_{3,1}^{(3)}\gamma_5 + I_1^{(3)} + I_3^{(2)}\\
&= -4(25 - 25)0.01 - 4(-25 + 25)0.01 - 4(25 - 25)0.01\\
&\quad + 0.02 + 0.01\overline{3} = 0.0\overline{3}\\
\beta_3 &= -z_{3,2}^{(3)}\gamma_5 - z_{2,1}^{(4)}\gamma_4 - z_{3,1}^{(4)}\gamma_5 + I_1^{(4)} + I_2^{(3)}\\
&= -4(-50)0.01 - 4(-50)0.01 - 4(-50)0.01 + 0.02 + 0.02 = 6.04
\end{aligned}$$

Thus,

$$A = \begin{bmatrix} \alpha_{11} & \alpha_{12} & \alpha_{13}\\ \alpha_{21} & \alpha_{22} & \alpha_{23}\\ \alpha_{31} & \alpha_{32} & \alpha_{33} \end{bmatrix} = \begin{bmatrix} 4 & -1 & 0\\ -1 & 1.5 & -0.5\\ 0 & -0.5 & 2 \end{bmatrix} \quad \text{and} \quad \mathbf{b} = \begin{bmatrix} \beta_1\\ \beta_2\\ \beta_3 \end{bmatrix} = \begin{bmatrix} 12\\ 0.0\overline{3}\\ 6.04 \end{bmatrix}$$

The linear system $A\mathbf{c} = \mathbf{b}$, where

$$\mathbf{c} = \begin{bmatrix} \gamma_1\\ \gamma_2\\ \gamma_3 \end{bmatrix},$$

has solution $(4.00962963, 4.03851852, 4.02962963)^t$, which gives the approximate solution $\phi(x, y)$ on the triangles

$$\begin{aligned}
T_1&: \ \phi(x, y) = 4.00962963(2 - 10x) + 4.03851852(-1 + 5x + 5y) + 4(5x - 5y),\\
T_2&: \ \phi(x, y) = 4.00962963(2 - 10y) + 4(-5x + 5y) + 4.03851852(-1 + 5x + 5y),\\
T_3&: \ \phi(x, y) = 4.03851852(1 - 5x - 5y) + 4.02962963(-2 + 10x) + 4(2 - 5x - 5y),\\
T_4&: \ \phi(x, y) = 4.02962963(10y) + 4(-1 + 5x - 5y) + 4(2 - 5x - 5y),\\
T_5&: \ \phi(x, y) = 4(-5x + 5y) + 4.00962963(10x) + 4(1 - 5x - 5y),\\
T_6&: \ \phi(x, y) = 4(1 - 5x - 5y) + 4(5x - 5y) + 4.00962963(10y).
\end{aligned}$$

The actual solution to the boundary-value problem is $u(x, y) = xy + 4$. Table 12.8 compares the value of u to the value of ϕ at E_1, E_2, and E_3. ◆ ◆ ◆

Table 12.8

x	y	$\phi(x, y)$	$u(x, y)$	$\lvert\phi(x, y) - u(x, y)\rvert$
0.1	0.1	4.00962963	4.01	0.00037037
0.2	0.2	4.03851852	4.04	0.00148148
0.3	0.1	4.02962963	4.03	0.00037037

Typically, the error for elliptic second-order problems with smooth coefficient functions is $O(h^2)$, where h is the maximum diameter of the triangular elements. Piecewise bilinear basis functions on rectangular elements are also expected to give $O(h^2)$ results, where h is the maximum diagonal length of the rectangular elements. Other classes of basis functions can be used to give $O(h^4)$ results, but the construction is more complex. Efficient error theorems for finite-element methods are difficult to state and apply because the accuracy of the approximation depends on the continuity properties of the solution and the regularity of the boundary.

EXERCISE SET 12.5

1. Use the Finite-Element method to approximate the solution to the following partial-differential equation (see the figure):

$$\frac{\partial}{\partial x}\left(y^2\frac{\partial u}{\partial x}(x, y)\right) + \frac{\partial}{\partial y}\left(y^2\frac{\partial u}{\partial y}(x, y)\right) - yu(x, y) = -x, \qquad (x, y) \in D,$$

$$u(x, 0.5) = 2x, \quad 0 \le x \le 0.5, \quad u(0, y) = 0, \qquad 0.5 \le y \le 1,$$

$$y^2\frac{\partial u}{\partial x}(x, y)\cos\theta_1 + y^2\frac{\partial u}{\partial y}(x, y)\cos\theta_2 = \frac{\sqrt{2}}{2}(y - x) \quad \text{for } (x, y) \in S_2.$$

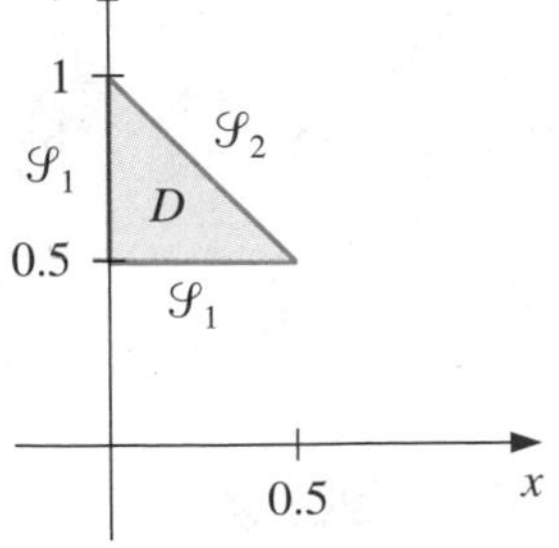

Let $M = 2$; T_1 have vertices $(0, 0.5)$, $(0.25, 0.75)$, $(0, 1)$; and T_2 have vertices $(0, 0.5)$, $(0.5, 0.5)$, and $(0.25, 0.75)$.

2. Repeat Exercise 1, using instead the triangles

$$\begin{aligned}
T_1&: \quad (0, 0.75), (0, 1), (0.25, 0.75);\\
T_2&: \quad (0.25, 0.5), (0.25, 0.75), (0.5, 0.5);\\
T_3&: \quad (0, 0.5), (0, 0.75), (0.25, 0.75);\\
T_4&: \quad (0, 0.5), (0.25, 0.5), (0.25, 0.75).
\end{aligned}$$

3. Use the Finite-Element method with the elements given in the accompanying figure to approximate the solution to the partial-differential equation

$$\frac{\partial^2 u}{\partial x^2}(x, y) + \frac{\partial^2 u}{\partial y^2}(x, y) - 12.5\pi^2 u(x, y) = -25\pi^2 \sin\frac{5\pi}{2}x \sin\frac{5\pi}{2}y,$$

for $0 < x < 0.4$ and $0 < y < 0.4$, subject to the Dirichlet boundary condition

$$u(x, y) = 0.$$

Compare the approximate solution to the actual solution

$$u(x, y) = \sin\frac{5\pi}{2}x \sin\frac{5\pi}{2}y$$

at the interior vertices and at the points $(0.125, 0.125)$, $(0.125, 0.25)$, $(0.25, 0.125)$, and $(0.25, 0.25)$.

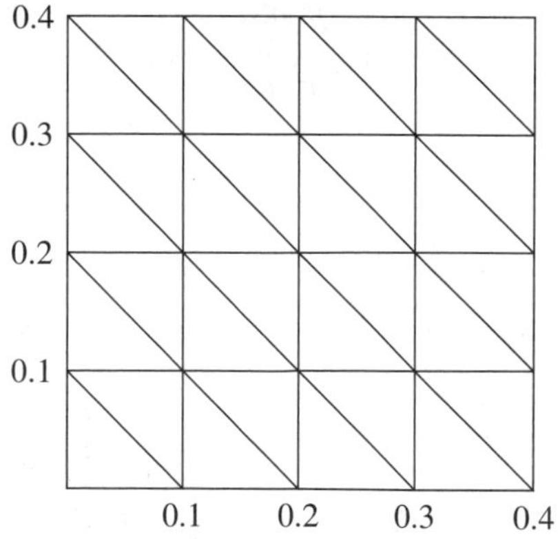

4. Repeat Exercise 3 with $f(x, y) = -25\pi^2 \cos\frac{5\pi}{2}x \cos\frac{5\pi}{2}y$, using the Neumann boundary condition

$$\frac{\partial u}{\partial n}(x, y) = 0.$$

The actual solution for this problem is

$$u(x, y) = \cos \frac{5\pi}{2} x \cos \frac{5\pi}{2} y.$$

5. A silver plate in the shape of a trapezoid (see the accompanying figure) has heat being uniformly generated at each point at the rate $q = 1.5\text{cal/cm}^3 \cdot \text{s}$. The steady-state temperature $u(x, y)$ of the plate satisfies the Poisson equation

$$\frac{\partial^2 u}{\partial x^2}(x, y) + \frac{\partial^2 u}{\partial y^2}(x, y) = \frac{-q}{k},$$

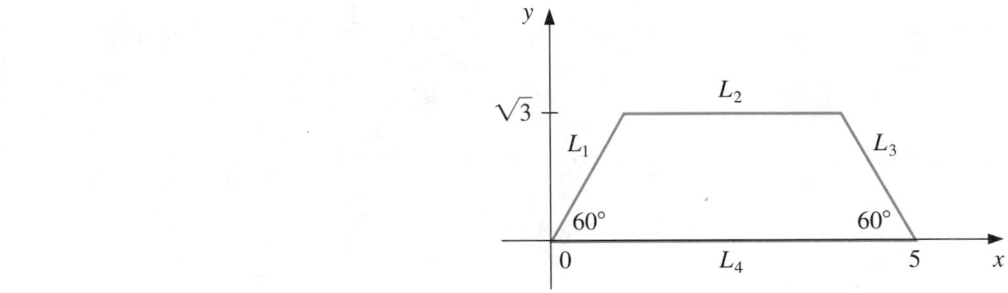

where k, the thermal conductivity, is $1.04\text{cal/cm} \cdot \text{deg} \cdot \text{s}$. Assume that the temperature is held at 15°C on L_2, that heat is lost on the slanted edges L_1 and L_3 according to the boundary condition $\partial u/\partial n = 4$, and that no heat is lost on L_4, that is, $\partial u/\partial n = 0$. Use the Finite-Element method to approximate the temperature of the plate at $(1, 0)$, $(4, 0)$, and $(\frac{5}{2}, \sqrt{3}/2)$.

12.6 Survey of Methods and Software

Methods to approximate solutions to partial-differential equations were considered in this chapter. We restricted our attention to Poisson's equation as an example of an elliptic partial-differential equation, the heat or diffusion equation as an example of a parabolic partial-differential equation, and the wave equation as an example of a hyperbolic partial-differential equation. Finite-difference approximations were discussed for these three examples.

Poisson's equation on a rectangle required the solution of a large sparse linear system, for which iterative techniques, such as the SOR method, are recommended. Three Finite-Difference methods were presented for the heat equation. The Forward-Difference method has stability problems, so the Backward-Difference method and the Crank-Nicolson method were introduced. Although a tridiagonal linear system must be solved at each time step with these implicit methods, they are more stable than the explicit Forward-Difference method. The Finite-Difference method for the wave equation is explicit and can also have stability problems for certain choice of time and space discretizations.

In the last section of the chapter we presented an introduction to the Finite-Element method for a self-adjoint elliptic partial-differential equation on a polygonal domain. Although our methods will work adequately for the problems and examples in the textbook, more powerful generalizations and modifications of these techniques are required for commercial applications.

We mention two subroutines from the IMSL Library. One subroutine is used to solve the partial-differential equation

$$\frac{\partial u}{\partial t} = F\left(x, t, u, \frac{\partial u}{\partial x}, \frac{\partial^2 u}{\partial x^2}\right)$$

with boundary conditions

$$\alpha(x,t)u(x,t) + \beta(x,t)\frac{\partial u}{\partial x}(x,t) = \gamma(x,t).$$

The method is based on collocation at Gaussian points on the x-axis for each value of t and uses cubic Hermite splines as basis functions.

The other subroutine is used to solve Poisson's equation on a rectangle. The method of solution is based on a choice of second- or fourth-order finite differences on a uniform mesh.

The NAG Library has a number of subroutines for partial-differential equations. One subroutine is used for Laplace's equation on an arbitrary domain in the xy-plane and another subroutine is used to solve a single parabolic partial-differential equation by the method of lines.

There are specialized packages, such as NASTRAN, consisting of codes for the Finite-Element method. These packages are popular in engineering applications. General codes for partial-differential equations are difficult to write because of the problem of specifying domains other than common geometrical figures. Research in the area of solution of partial-differential equations is currently very active.

We have presented only a small sample of the many techniques used for approximating the solutions to the problems involving partial-differential equations. Further information on the general topic can be found in Lapidus and Pinder [LP], Twizell [Tw], and the recent book by Morton and Mayers [MM].

Books that focus on Finite-Difference methods include Strikwerda [Str], Thomas [Th], and Shashkov and Steinberg [ShS]. Strang and Fix [SF] and Zienkiewicz and Morgan [ZM] are good sources for information on the finite-element method.

Multigrid methods use coarse grid approximations and iterative techniques to provide approximations on finer grids. References on these techniques include Briggs [Bri], McCormick [M], and Bramble [Bra].

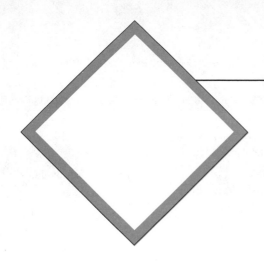

Bibliography

[AG] Allgower, E. and K. Georg, *Numerical continuation methods: an introduction*, Springer-Verlag, New York, 1990, 388 pp. QA377.A56.

[Am] Ames, W. F., *Numerical methods for partial differential equations* (Third edition), Academic Press, New York, 1992, 451 pp. QA374.A46.

[An] Anderson, E., et al., *LAPACK user's guide* (Second edition), SIAM Publications, Philadelphia, PA, 1995, 325 pp. QA76.73.F25 L36.

[AS] Argyros, I. K. and F. Szidarovszky, *The theory and applications of iteration methods*, CRC Press, Boca Raton, FL, 1993, 355 pp. QA297.8.A74.

[AMR] Ascher, U. M., R. M. M. Mattheij, and R. B. Russell, *Numerical solution of boundary value problems for ordinary differential equations*, Prentice-Hall, Englewood Cliffs, NJ, 1988, 595 pp. QA379.A83.

[Ax] Axelsson, O., *Iterative solution methods*, Cambridge University Press, New York, 1994, 654 pp. QA297.8.A94.

[AB] Axelsson, O. and V. A. Barker, *Finite element solution of boundary value problems: theory and computation*, Academic Press, Orlando, FL, 1984, 432 pp. QA379.A9.

[BSW] Bailey, P. B., L. F. Shampine, and P. E. Waltman, *Nonlinear two-point boundary-value problems*, Academic Press, New York, 1968, 171 pp. QA372.B27.

[Ba] Barrett, R., et al., *Templates for the solution of linear systems: building blocks for iterative methods,* SIAM Publications, Philadelphia, PA, 1994, 112 pp. QA297.8.T45.

[Bek] Bekker, M. G., *Introduction to terrain vehicle systems*, University of Michigan Press, Ann Arbor, MI, 1969, 846 pp. TL243.B39.

[Ber] Bernadelli, H., *Population waves*, Journal of the Burma Research Society **31** (1941), 1–18, DS527.B85.

[Bra] Bramble, J. H., *Multigrid methods*, John Wiley & Sons, New York, 1993, 161 pp. QA377.B73.

[Bre] Brent, R., *Algorithms for minimization without derivatives*, Prentice-Hall, Englewood Cliffs, NJ, 1973, 195 pp. QA402.5.B74.

[Bri] Briggs, W. L., *A multigrid tutorial*, SIAM Publications, Philadelphia, PA, 1987, 88 pp. QA377.B75.

[BH] Briggs, W. L. and V. E. Henson, *The DFT: an owner's manual for the discrete Fourier transform*, SIAM Publications, Philadelphia, PA, 1995, 434 pp. QA403.5.B75.

[BF] Burden, R. L., and J. D. Faires, *Numerical Analysis*, (Sixth edition) Brooks/Cole Publishing, Pacific Grove, CA, 1997, 811 pp. QA297.B84.

[Bur] Burrage, K., *Parallel and sequential methods for ordinary differential equations*, Oxford University Press, New York, 1995, 446 pp. QA372.B883.

[CF] Chaitin-Chatelin, F. and Fraysse, V., *Lectures on finite precision computations*, SIAM Publications, Philadelphia, PA, 1996, 235 pp. QA297.C417.

[Ch] Cheney, E. W., *Introduction to approximation theory*, McGraw-Hill, New York, 1966, 259 pp. QA221.C47.

[CC] Clenshaw, C. W. and C. W. Curtis, *A method for numerical integration on an automatic computer*, Numerische Mathematik **2** (1960), 197–205, QA241.N9.

[CT] Cooley, J. W. and J. W. Tukey, *An algorithm for the machine calculation of complex Fourier series*, Mathematics of Computation **19**, No. 90 (1965), 297–301, QA1.M4144.

[Co] Cowell, W. (ed.), *Sources and development of mathematical software*, Prentice-Hall, Englewood Cliffs, NJ, 1984, 404 pp. QA76.95.S68.

[Da] Davis, P. J., *Interpolation and Approximation*, Dover, New York, 1975, 393 pp. QA221.D33.

[DR] Davis, P. J. and P. Rabinowitz, *Methods of numerical integration*, Academic Press, New York, 1975, 459 pp. QA299.3.D28.

[De] De Boor, C., *A practical guide to splines*, Springer-Verlag, New York, 1978, 392 pp. QA1.A647 vol. 27.

[DS] Dennis, J. E., Jr. and R. B. Schnabel, *Numerical methods for unconstrained optimization and nonlinear equations*, Prentice-Hall, Englewood Cliffs, NJ, 1983, 378 pp. QA402.5.D44.

[Di] Dierckx, P., *Curve and surface fitting with splines*, Oxford University Press, New York, 1993, 285 pp. QA297.6.D54.

[Do] Dormand, J. R., *Numerical methods for differential equations: a computational approach*, CRC Press, Boca Raton, FL, 1996, 368 pp. QA372.D67.

[DB] Dorn, G. L. and A. B. Burdick, *On the recombinational structure of complementation relationships in the m-dy complex of the Drosophila melanogaster*, Genetics **47** (1962), 503–518, QH431.G43.

[E] Engels, H., *Numerical Quadrature and Cubature*, Academic Press, New York, 1980, 441 pp. QA299.3.E5.

[FM] Forsythe, G. E. and C. B. Moler, *Computer solution of linear algebraic systems*, Prentice-Hall, Englewood Cliffs, NJ, 1967, 148 pp. QA297.F57.

[G] Gear, C. W., *Numerical initial-value problems in ordinary differential equations*, Prentice-Hall, Englewood Cliffs, NJ, 1971, 253 pp. QA372.G4.

[GL] George, A. and J. W. Liu, *Computer solution of large sparse positive definite systems*, Prentice-Hall, Englewood Cliffs, NJ, 1981, 324 pp. QA188.G46.

[GO] Golub, G. H. and Ortega, J. M., *Scientific computing: an introduction with parallel computing*, Academic Press, Boston, MA, 1993, 442 pp. QA76.58.G64.

[GV] Golub, G. H. and C. F. Van Loan, *Matrix computations*, (Second edition), John Hopkins University Press, Baltimore, MD, 1989, 642 pp. QA188.G65.

[GKO] Gustafsson, B., H. Kreiss, and J. Oliger, *Time dependent problems and difference methods*, John Wiley & Sons, New York, 1995, 642 pp. QA374.G974.

[Ha] Hackbusch, W., *Iterative solution of large sparse systems of equations*, Springer-Verlag, New York, 1994, 429 pp. QA1.A647 vol. 95.

[HY] Hageman, L. A. and D. M. Young, *Applied iterative methods*, Academic Press, New York, 1981, 386 pp. QA297.8.H34.

[HNW1] Hairer, E., S. P. Nörsett, and G. Wanner, *Solving ordinary differential equations. Vol. 1: Nonstiff equations*, Springer-Verlag, New York, 1987, QA372.H16.

[HNW2] Hairer, E., S. P. Nörsett, and G. Wanner, *Solving ordinary differential equations. Vol. 2: Stiff and differential-algebraic problems*, Springer-Verlag, New York, 1991, QA372.H16.

[He] Henrici, P., *Discrete variable methods in ordinary differential equations* John Wiley & Sons, New York, 1962, 407 pp. QA372.H48.

[Hi] Hildebrand, F. B., *Introduction to numerical analysis*, (Second edition), McGraw-Hill, New York, 1974, 669 pp. QA297.H54.

[Ho] Householder, A. S., *The numerical treatment of a single nonlinear equation*, McGraw-Hill, New York, 1970, 216 pp. QA218.H68.

[IK] Issacson, E., and H. B. Keller, *Analysis of numerical methods*, John Wiley & Sons, New York, 1966, 541 pp. QA297.I8.

[K, H] Keller, H. B., *Numerical methods for two-point boundary-value problems*, Blaisdell, Waltham, MA, 1968, 184 pp. QA372.K42.

[K, J] Keller, J. B., *Probability of a shutout in racquetball*, SIAM Review **26**, No. 2 (1984), 267–268, QA1.S2.

[K] Kelley, C. T., *Iterative methods for linear and nonlinear equations*, SIAM Publications, Philadelphia, PA, 1995, 165 pp. QA297.8.K45.

[Ko] Köckler, N., *Numerical methods and scientific computing: using software libraries for problem solving*, Oxford University Press, New York, 328 pp. TA345.K653.

[LP] Lapidus, L. and G. F. Pinder, *Numerical solution of partial differential equations in science and engineering*, John Wiley & Sons, New York, 1982, 677 pp. Q172.L36.

[LH] Lawson, C. L. and Hanson, R. J., *Solving least squares problems*, SIAM Publications, Philadelphia, PA, 1995, 337 pp. QA275.L38.

[M] McCormick, S. F., *Multigrid methods*, SIAM Publications, Philadelphia, PA, 1987, 282 pp. QA374.M84.

[MM] Morton, K. W. and D. F. Mayers, *Numerical solution of partial differential equations: an introduction*, Cambridge University Press, New York, 1994, 227 pp. QA377.M69.

[ND] Noble, B. and J. W. Daniel, *Applied linear algebra*, (second edition), Prentice-Hall, Englewood Cliffs, NJ, 1977, 477 pp. QA184.N6.

[Or] Ortega, J. M., *Numerical analysis; a second course*, Academic Press, New York, 1972, 201 pp. QA297.O78.

[OR] Ortega, J. M. and W. C. Rheinboldt, *Iterative solution of nonlinear equations in several variables*, Academic Press, New York, 1970, 572 pp. QA297.8.O77.

[Os] Ostrowski, A. M., *Solution of equations and systems of equations*, (second edition), Academic Press, New York, 1966, 338 pp. QA3.P8 vol. 9.

[Pa] Parlett, B., *The symmetric eigenvalue problem*, Prentice-Hall, Englewood Cliffs, NJ, 1980, 348 pp. QA188.P37.

[PF] Phillips, C. and T. L. Freeman, *Parallel numerical algorithms*, Prentice-Hall, New York, 1992, 315 pp. QA76.9.A43 F74.

[PDUK] Piessens, R., E. de Doncker-Kapenga, C. W. Überhuber, and D. K. Kahaner, *QUADPACK: a subroutine package for automatic integration*, Springer-Verlag, New York, 1983, 301 pp. QA299.3.Q36.

[Pi] Pissanetzky, S., *Sparse matrix technology*, Academic Press, New York, 1984, 321 pp. QA188.P57.

[Po] Powell, M. J. D., *Approximation theory and methods*, Cambridge University Press, Cambridge, MA, 1981, 339 pp. QA221.P65.

[Ra] Rashevsky, N., *Looking at history through mathematics*, Massachusetts Institute of Technology Press, Cambridge, MA, 1968, 199 pp. D16.25.R3.

[Ri] Rice, J. R., *Numerical methods, software, and analysis: IMSL reference edition*, McGraw-Hill, New York, 1983, 661 pp. QA297.R49.

[Sa1] Saad, Y., *Numerical methods for large eigenvalue problems*, Halsted Press, New York, 1992, 346 pp. QA188.S18.

[Sa2] Saad, Y., *Iterative methods for sparse linear systems*, PWS-Kent Publishing, Boston, MA, 1996, 447 pp. QA188.S17.

[SP] Sagar, V. and D. J. Payne, *Incremental collapse of thick-walled circular cylinders under steady axial tension and torsion loads and cyclic transient heating*, Journal of the Mechanics and Physics of Solids **21**, No. 1 (1975), 39–54, TA350.J68.

[SD] Sale, P. F. and R. Dybdahl, *Determinants of community structure for coral-reef fishes in experimental habitat*, Ecology **56** (1975), 1343–1355, QH540.E3.

[Scho] Schoenberg, I. J., *Contributions to the problem of approximation of equidistant data by analytic functions*, Quarterly of Applied Mathematics **4**, (1946), Part A, 45–99; Part B, 112–141, QA1.A26.

[Schu] Schultz, M. H., *Spline analysis*, Prentice-Hall, Englewood Cliffs, NJ, 1973, 156 pp. QA211.S33.

[Sh] Shampine, L. F., *Numerical solution of ordinary differential equations*, Chapman & Hall, New York, 1994, 484 pp. QA372.S417.

[ShS] Shashkov, M. and S. Steinberg, *Conservative finite-difference methods on general grids*, CRC Press, Boca Raton, FL, 1996, 359 pp. QA431.S484.

[Ste] Stewart, G. W., *Introduction to matrix computations*, Academic Press, New York, 1973, 441 pp. QA188.S7.

[SF] Strang, W. G. and G. J. Fix, *An analysis of the finite element method*, Prentice-Hall, Englewood Cliffs, NJ, 1973, 306 pp. TA335.S77.

[Str] Strikwerda, J. C., *Finite difference schemes and partial differential equations*, Brooks/Cole Publishing, Pacific Grove, CA, 1989, 386 pp. QA374.S88.

[StS] Stroud, A. H. and D. Secrest, *Gaussian quadrature formulas*, Prentice-Hall, Englewood Cliffs, NJ, 1966, 374 pp. QA299.4.G4 S7.

[Sz] Szüsz, P., *Math bite*, Mathematics Magazine **68**, No. 2 (1995), 97, QA1.N28.

[Th] Thomas, J. W., *Numerical partial differential equations*, Springer-Verlag, New York, 1995, 445 pp. QA377.T495.

[Tr] Traub, J. F., *Iterative methods for the solution of equations*, Prentice-Hall, Englewood Cliffs, NJ, 1964, 310 pp. QA297.T7.

[Tw] Twizell, E. H., *Computational methods for partial differential equations*, Ellis Horwood Ltd., Chichester, West Sussex, England, 1984, 276 pp. QA377.T95.

[Van] Van Loan, C. F., *Computational frameworks for the fast Fourier transform*, SIAM Publications, Philadelphia, PA, 1992, 273 pp. QA403.5.V35.

[Var] Varga, R. S., *Matrix iterative analysis*, Prentice-Hall, Englewood Cliffs, NJ, 1962, 322 pp. QA263.V3.

[We] Wendroff, B., *Theoretical numerical analysis*, Academic Press, New York, 1966, 239 pp. QA297.W43.

[Wi1] Wilkinson, J. H., *Rounding errors in algebraic processes*, Prentice-Hall, Englewood Cliffs, NJ, 1963, 161 pp. QA76.5.W53.

[Wi2] Wilkinson, J. H., *The algebraic eigenvalue problem*, Clarendon Press, Oxford, 1965, 662 pp. QA218.W5.

[WR] Wilkinson, J. H. and C. Reinsch (eds.), *Handbook for automatic computation. Vol. 2: Linear algebra*, Springer-Verlag, New York, 1971, 439 pp. QA251.W67.

[Y] Young, D. M., *Iterative solution of large linear systems*, Academic Press, New York, 1971, 570 pp. QA195.Y68.

[ZM] Zienkiewicz, O. C. and K. Morgan, *Finite elements and approximation*, John Wiley & Sons, New York, 1983, 328 pp. QA297.5.Z53.

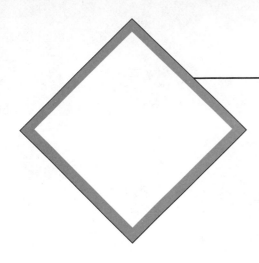

Answers for Numerical Methods

CHAPTER 1

EXERCISE SET 1.2 (PAGE 12)

1. For each part, $f \in C[a, b]$ on the given interval. Since $f(a)$ and $f(b)$ are of opposite sign, the Intermediate Value Theorem implies a number c exists with $f(c) = 0$.

3. For each part, $f \in C[a, b]$, f' exists on (a, b), and $f(a) = f(b) = 0$. Rolle's Theorem implies that a number c exists in (a, b) with $f'(c) = 0$. For part (d), we can use $[a, b] = [-1, 0]$ or $[a, b] = [0, 2]$.

5. **a.** $P_2(x) = 0$
b. $R_2(0.5) = 0.125$; actual error $= 0.125$
c. $P_2(x) = 1 + 3(x - 1) + 3(x - 1)^2$
d. $R_2(0.5) = -0.125$; actual error $= -0.125$

7. Since

$$P_2(x) = 1 + x \quad \text{and} \quad R_2(x) = \frac{-2e^{\xi}(\sin \xi + \cos \xi)}{6} x^3$$

for some ξ between x and 0, we have the following:
a. $P_2(0.5) = 1.5$ and $|f(0.5) - P_2(0.5)| \le 0.0532$
b. $|f(x) - P_2(x)| \le 1.252$
c. $\int_0^1 f(x)\,dx \approx 1.5$
d. $|\int_0^1 f(x)\,dx - \int_0^1 P_2(x)\,dx| \le \int_0^1 |R_2(x)|\,dx \le 0.313$, and the actual error is 0.122.

9. The error is approximately 8.86×10^{-7}.

11. **a.** $P_3(x) = \frac{1}{3}x + \frac{1}{6}x^2 + \frac{23}{648}x^3$
b. $f^{(4)}(x) = -\frac{119}{1296}e^{x/2}\sin\frac{x}{3} + \frac{5}{24}e^{x/2}\cos\frac{x}{3}$,
so

$$|f^{(4)}(x)| \le |f^{(4)}(0.60473891)| \le 0.09787176 \text{ for } 0 \le x \le 1$$

and

$$|f(x) - P_3(x)| \le \frac{|f^{(4)}(\xi)|}{4!}|x|^4 \le \frac{0.09787176}{24}(1)^4 = 0.004077990.$$

13. A bound for the maximum error is 0.0026.

15. **a.** $e^{-t^2} = \sum_{k=0}^{\infty} \frac{(-1)^k t^{2k}}{k!}$

Use this series to integrate

$$\frac{2}{\sqrt{\pi}} \int_0^x e^{-t^2}\, dt$$

and obtain the result.

b.
$$\frac{2}{\sqrt{\pi}} e^{-x^2} \sum_{k=0}^{\infty} \frac{2^k x^{2k+1}}{1 \cdot 3 \cdots (2k+1)} = \frac{2}{\sqrt{\pi}} \left[1 - x^2 + \frac{1}{2}x^4 - \frac{1}{6}x^7 + \frac{1}{24}x^8 + \cdots\right]$$
$$\cdot \left[x + \frac{2}{3}x^3 + \frac{4}{15}x^5 + \frac{8}{105}x^7 + \frac{16}{945}x^9 + \cdots\right]$$
$$= \frac{2}{\sqrt{\pi}} \left[x - \frac{1}{3}x^3 + \frac{1}{10}x^5 - \frac{1}{42}x^7 + \frac{1}{216}x^9 + \cdots\right]$$
$$= \operatorname{erf}(x)$$

c. 0.8427008 **d.** 0.8427069

e. The series in part (a) is alternating, so for any positive integer n and positive x we have the bound

$$\left|\operatorname{erf}(x) - \frac{2}{\sqrt{\pi}} \sum_{k=0}^{n} \frac{(-1)^k x^{2k+1}}{(2k+1)k!}\right| < \frac{x^{2n+3}}{(2n+3)(n+1)!}.$$

We have no such bound for the positive term series in part (b).

EXERCISE SET 1.3 (PAGE 18)

1.

	Absolute error	Relative error
a.	0.001264	4.025×10^{-4}
b.	7.346×10^{-6}	2.338×10^{-6}
c.	2.818×10^{-4}	1.037×10^{-4}
d.	2.136×10^{-4}	1.510×10^{-4}
e.	2.647×10^{1}	1.202×10^{-3}
f.	1.454×10^{1}	1.050×10^{-2}
g.	420	1.042×10^{-2}
h.	3.343×10^{3}	9.213×10^{-3}

3.

	Approximation	Absolute error	Relative error
a.	134	0.079	5.90×10^{-4}
b.	133	0.499	3.77×10^{-3}
c.	2.00	0.327	0.195
d.	1.67	0.003	1.79×10^{-3}
e.	1.80	0.154	0.0786
f.	−15.1	0.0546	3.60×10^{-3}
g.	0.286	2.86×10^{-4}	10^{-3}
h.	0.00	0.0215	1.00

5.

	Approximation	Absolute error	Relative error
a.	133.9	0.021	1.568×10^{-4}
b.	132.5	0.001	7.55×10^{-6}
c.	1.700	0.027	0.01614
d.	1.673	0	0
e.	1.986	0.03246	0.01662
f.	−15.16	0.005377	3.548×10^{-4}
g.	0.2857	1.429×10^{-5}	5×10^{-5}
h.	−0.01700	0.0045	0.2092

7.

	Approximation	Absolute error	Relative error
a.	3.14557613	3.983×10^{-3}	1.268×10^{-3}
b.	3.14162103	2.838×10^{-5}	9.032×10^{-6}

9. **b.** The first formula gives −0.00658 and the second formula gives −0.0100. The true three-digit value is −0.0116.

11. **a.** $39.375 \leq$ volume ≤ 86.625 **b.** $71.5 \leq$ surface area ≤ 119.5

EXERCISE SET 1.4 (PAGE 25)

1.

	x_1	Absolute error	Relative error
a.	92.26	0.01542	1.672×10^{-4}
b.	0.005421	1.264×10^{-6}	2.333×10^{-4}
c.	10.98	6.875×10^{-3}	6.257×10^{-4}
d.	−0.001149	7.566×10^{-8}	6.584×10^{-5}

	x_2	Absolute error	Relative error
a.	0.005419	6.273×10^{-7}	1.157×10^{-4}
b.	−92.26	4.580×10^{-3}	4.965×10^{-5}
c.	0.001149	7.566×10^{-8}	6.584×10^{-5}
d.	−10.98	6.875×10^{-3}	6.257×10^{-4}

3. **a.** −0.1000 **b.** −0.1010

c. Absolute error for part (a) is 2.331×10^{-3} with relative error 2.387×10^{-2}. Absolute error for part (b) is 3.331×10^{-3} with relative error 3.411×10^{-2}.

5.

	Approximation		Absolute error	Relative error
a.	3.743	b.	1.011×10^{-3}	2.694×10^{-3}
c.	3.755	d.	1.889×10^{-4}	5.033×10^{-4}

7. **a.** The approximate sums are 1.53 and 1.54, respectively. The actual value is 1.549. Significant round-off error occurs earlier with the first method.

9.

	Approximation	Absolute error	Relative error
a.	2.715	3.282×10^{-3}	1.207×10^{-3}
b.	2.716	2.282×10^{-3}	8.394×10^{-4}
c.	2.716	2.282×10^{-3}	8.394×10^{-4}
d.	2.718	2.818×10^{-4}	1.037×10^{-4}

11. The rates of convergence are as follows.

a. $O(h^2)$ **b.** $O(h)$

c. $O(h^2)$ **d.** $O(h)$

13. Since $\lim_{n\to\infty} x_n = \lim_{n\to\infty} x_{n+1} = x$ and $x_{n+1} = 1 + 1/x_n$, we have $x = 1 + 1/x$. This implies that $x = (1 + \sqrt{5})/2$. This number is called the *golden ratio*. It appears frequently in mathematics and the sciences.

15. **a.** $n = 50$ **b.** $n = 500$

CHAPTER 2

EXERCISE SET 2.2 (PAGE 34)

1. $p_3 = 0.625$

3. The Bisection method gives the following.

a. $p_7 = 0.5859$ **b.** $p_8 = 3.002$ **c.** $p_7 = 3.419$

5. **a.**

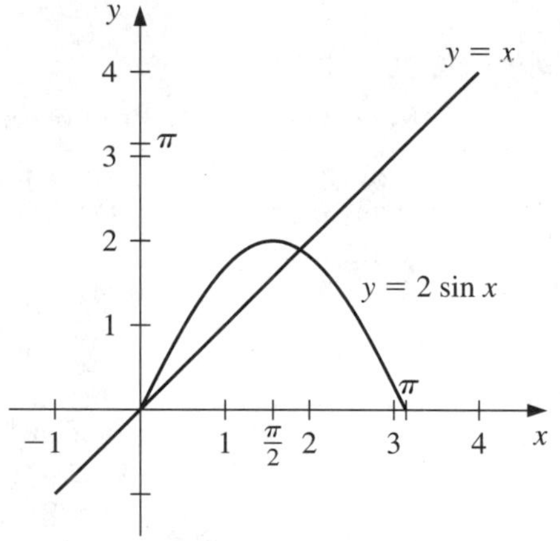

b. With $[1, 2]$, we have $p_7 = 1.8984$.

7. **a.** 2 **b.** -2

c. -1 **d.** 1

9. $\sqrt{3} \approx p_{14} = 1.7320$ using $[1, 2]$

11. A bound is $n \geq 12$, and $p_{12} = 1.3787$.

1. **a.** $[c, d]$ where $-1 < c < 0$, $2 < d < 3$, and $c + d < 2$; for example, $[-0.9, 2.1]$

b. $[c, d]$ where $-1 < c < 0$, $2 < d < 3$, and $c + d > 2$; for example, $[-0.1, 2.5]$

c. $[c, d]$ where $-2 < c < -1$, $1 < d < 2$ and $c + d > 0$; for example, $[-1.1, 1.9]$

EXERCISE SET 2.3 (PAGE 40)

1. **a.** $p_3 = 2.45454$ **b.** $p_3 = 2.44444$

3. Using the endpoints of the intervals as p_0 and p_1, we have the following.

a. $p_{11} = 2.69065$ **b.** $p_7 = -2.87939$

c. $p_6 = 0.73909$ **d.** $p_5 = 0.96433$

5. Using the endpoints of the intervals as p_0 and p_1, we have the following.
   a. $p_{16} = 2.69060$
   b. $p_6 = -2.87938$
   c. $p_7 = 0.73908$
   d. $p_6 = 0.96433$

7. For $p_0 = 0.1$ and $p_1 = 3$ we have $p_7 = 2.363171$.
   For $p_0 = 3$ and $p_1 = 4$ we have $p_7 = 3.817926$.
   For $p_0 = 5$ and $p_1 = 6$ we have $p_6 = 5.839252$.
   For $p_0 = 6$ and $p_1 = 7$ we have $p_9 = 6.603085$.

9. For $p_0 = 0$ and $p_1 = 1$, the Secant method gives $p_7 = 0.589755$. The closest point on the graph is $(0.589755, 0.347811)$.

11. a. For $p_0 = -1$ and $p_1 = 0$, we have $p_{17} = -0.04065850$, and for $p_0 = 0$ and $p_1 = 1$, we have $p_9 = 0.9623984$.
    b. For $p_0 = -1$ and $p_1 = 0$, we have $p_5 = -0.04065929$, and for $p_0 = 0$ and $p_1 = 1$, we have $p_{12} = -0.04065929$. The Secant method fails to find the zero in $[0, 1]$.

13. For $p_0 = \frac{1}{2}$, $p_1 = \pi/4$, and tolerance of 10^{-100}, the Secant method required 11 iterations, giving the 100-digit answer

$$p_{11} = .73908513321516064165531208767387340401341175890075746496568063577328465488354759459937610693176653 19.$$

15. For $p_0 = 0.1$ and $p_1 = 0.2$, the Secant method gives $p_3 = 0.16616$, so the depth of the water is $1 - p_3 = 0.83384$ ft.

EXERCISE SET 2.4 (PAGE 46)

1. $p_2 = 2.60714$

3. a. For $p_0 = 2$, we have $p_5 = 2.69065$.
   b. For $p_0 = -3$, we have $p_3 = -2.87939$.
   c. For $p_0 = 0$, we have $p_4 = 0.73909$.
   d. For $p_0 = 0$, we have $p_3 = 0.96434$.

5. Newton's method gives the following approximations:
   With $p_0 = 1.5, p_6 = 2.363171$; with $p_0 = 3.5, p_5 = 3.817926$;
   with $p_0 = 5.5, p_4 = 5.839252$; with $p_0 = 7, p_5 = 6.603085$.

7. Newton's method gives the following:
   a. For $p_0 = 0.5$ we have $p_{13} = 0.567135$.
   b. For $p_0 = -1.5$ we have $p_{23} = -1.414325$.
   c. For $p_0 = 0.5$ we have $p_{22} = 0.641166$.
   d. For $p_0 = -0.5$ we have $p_{23} = -0.183274$.

9. a. $p_{10} = 13.655776$
   b. $p_6 = 0.44743154$
   c. With $p_0 = 0$, Newton's method did not converge in 10 iterations. The initial approximation $p_0 = 0.48$ is sufficiently close to the solution for rapid convergence.

11. Newton's method gives $p_{15} = 1.895488$ for $p_0 = \pi/2$ and $p_{19} = 1.895489$ for $p_0 = 5\pi$. The sequence does not converge in 200 iterations for $p_0 = 10\pi$. The results do not indicate the fast convergence usually associated with Newton's method.

13. Using $p_0 = 0.75$, Newton's method gives $p_4 = 0.8423$.

15. The minimal interest rate is 6.67%.

17. a. $e/3$, $t = 3$ h
    b. 11 h and 5 min
    c. 21 h and 14 min

EXERCISE SET 2.5 (PAGE 52)

1. The results are listed in the following table.

	a.	b.	c.	d.
$\hat{p}_0$	0.258684	0.907859	0.548101	0.731385
$\hat{p}_1$	0.257613	0.909568	0.547915	0.736087
$\hat{p}_2$	0.257536	0.909917	0.547847	0.737653
$\hat{p}_3$	0.257531	0.909989	0.547823	0.738469
$\hat{p}_4$	0.257530	0.910004	0.547814	0.738798
$\hat{p}_5$	0.257530	0.910007	0.547810	0.738958

3. Newton's method gives $p_6 = -0.1828876$, and the improved value is $\hat{p}_6 = -0.183387$.

5. **a.** (i) Since $|p_{n+1} - 0| = \frac{1}{n+1} < \frac{1}{n} = |p_n - 0|$, the sequence $\{\frac{1}{n}\}$ converges linearly to 0. (ii) We need $\frac{1}{n} \leq 0.05$ or $n \geq 20$. (iii) Aitken's Δ^2 method gives $\hat{p}_{10} = 0.0\overline{45}$.

b. (i) Since $|p_{n+1} - 0| = \frac{1}{(n+1)^2} < \frac{1}{n^2} = |p_n - 0|$, the sequence $\{\frac{1}{n^2}\}$ converges linearly to 0. (ii) We need $\frac{1}{n^2} \leq 0.05$ or $n \geq 5$. (ii) Aitken's Δ^2 method gives $\hat{p}_2 = 0.0363$.

7. **a.** Since

$$\frac{|p_{n+1} - 0|}{|p_n - 0|^2} = \frac{10^{-2^{n+1}}}{\left(10^{-2^n}\right)^2} = \frac{10^{-2^{n+1}}}{10^{-2^{n+1}}} = 1,$$

the sequence is quadratically convergent.

b. Since

$$\frac{|p_{n+1} - 0|}{|p_n - 0|^2} = \frac{10^{-(n+1)^k}}{\left(10^{-n^k}\right)^2} = \frac{10^{-(n+1)^k}}{10^{-2n^k}} = 10^{2n^k-(n+1)^k}$$

diverges, the sequence $p_n = 10^{-n^k}$ does not converge quadratically.

EXERCISE SET 2.6 (PAGE 57)

1. **a.** For $p_0 = 1$, we have $p_{22} = 2.69065$.

b. For $p_0 = 1$, we have $p_5 = 0.53209$; for $p_0 = -1$, we have $p_3 = -0.65270$, and for $p_0 = -3$, we have $p_3 = -2.87939$.

c. For $p_0 = 1$, we have $p_4 = 1.12412$; and for $p_0 = 0$, we have $p_8 = -0.87605$.

d. For $p_0 = 0$, we have $p_{10} = 1.49819$.

3. The following table lists the initial approximation and the roots.

	p_0	p_1	p_2	Approximated roots	Complex conjugate roots
a.	-1	0	1	$p_7 = -0.34532 - 1.31873i$	$-0.34532 + 1.31873i$
	0	1	2	$p_6 = 2.69065$	
b.	0	1	2	$p_6 = 0.53209$	
	1	2	3	$p_9 = -0.65270$	
	-2	-3	-2.5	$p_4 = -2.87939$	
c.	0	1	2	$p_5 = 1.12412$	
	2	3	4	$p_{12} = -0.12403 + 1.74096i$	$-0.12403 - 1.74096i$
	-2	0	-1	$p_5 = -0.87605$	
d.	0	1	2	$p_6 = 1.49819$	
	-1	-2	-3	$p_{10} = -0.51363 - 1.09156i$	$-0.51363 + 1.09156i$
	1	0	-1	$p_8 = 0.26454 - 1.32837i$	$0.26454 + 1.32837i$

5. **a.** The roots are 1.244, 8.847, and -1.091, and the critical points are 0 and 6.

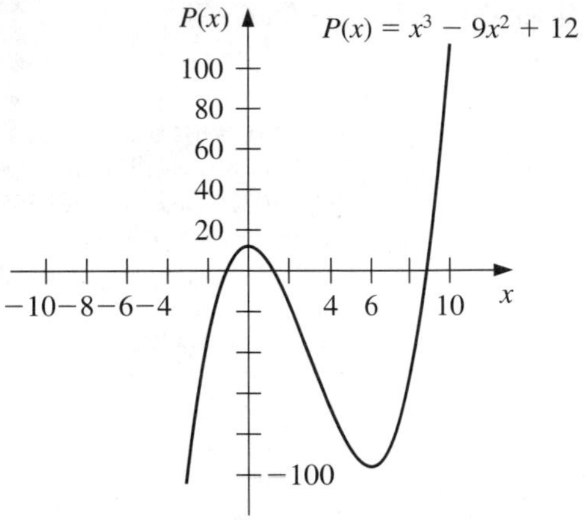

b. The roots are 0.5798, 1.521, 2.332, and -2.432, and the critical points are 1, 2.001, and -1.5.

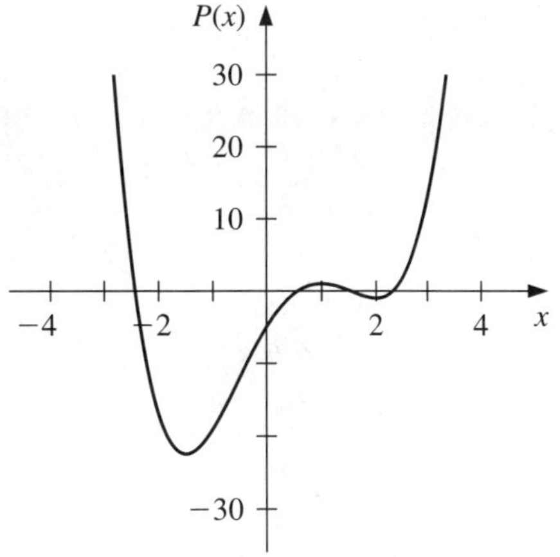

7. Let $c_1 = (2 + \frac{2}{9}\sqrt{129})^{-1/3}$ and $c_2 = (2 + \frac{2}{9}\sqrt{129})^{1/3}$. The roots are $c_2 - \frac{4}{3}c_1$, $-\frac{1}{2}c_2 + \frac{2}{3}c_1 + \frac{1}{2}\sqrt{3}(c_2 + \frac{4}{3}c_1)i$, and $-\frac{1}{2}c_2 + \frac{2}{3}c_1 - \frac{1}{2}\sqrt{3}(c_2 + \frac{4}{3}c_1)i$.

9. **a.** For $p_0 = 0.1$ and $p_1 = 1$ we have $p_{14} = 0.23233$.
b. For $p_0 = 0.55$ we have $p_6 = 0.23235$.
c. For $p_0 = 0.1$ and $p_1 = 1$ we have $p_8 = 0.23235$.
d. For $p_0 = 0.1$ and $p_1 = 1$ we have $p_{88} = 0.23035$.
e. For $p_0 = 0$, $p_1 = 0.25$, and $p_2 = 1$ we have $p_6 = 0.23235$.

11. The minimal material is approximately 573.64895 cm^2.

CHAPTER 3

EXERCISE SET 3.2 (PAGE 72)

1. a. (i) $P_1(x) = -0.29110731x + 1$; $P_1(0.45) = 0.86900171$; $|\cos 0.45 - P_1(0.45)| = 0.03144539$; (ii) $P_2(x) = -0.43108687x^2 - 0.03245519x + 1$; $P_2(0.45) = 0.89810007$; $|\cos 0.45 - P_2(0.45)| = 0.0023470$

b. (i) $P_1(x) = 0.44151844x + 1$; $P_1(0.45) = 1.1986833$; $|\sqrt{1.45} - P_1(0.45)| = 0.00547616$; (ii) $P_2(x) = -0.070228596x^2 + 0.483655598x + 1$; $P_2(0.45) = 1.20342373$; $|\sqrt{1.45} - P_2(0.45)| = 0.00073573$

c. (i) $P_1(x) = 0.78333938x$; $P_1(0.45) = 0.35250272$; $|\ln 1.45 - P_1(0.45)| = 0.01906083$; (ii) $P_2(x) = -0.23389466x^2 + 0.92367618x$; $P_2(0.45) = 0.36829061$; $|\ln 1.45 - P_2(0.45)| = 0.00327294$

d. (i) $P_1(x) = 1.14022801x$; $P_1(0.45) = 0.51310260$; $|\tan 0.45 - P_1(0.45)| = 0.03004754$; (ii) $P_2(x) = 0.86649261x^2 + 0.62033245x$; $P_2(0.45) = 0.45461436$; $|\tan 0.45 - P_2(0.45)| = 0.02844071$

3. a.

n	$x_0, x_1, \ldots, x_n$	$P_n(8.4)$
1	8.3, 8.6	17.87833
2	8.3, 8.6, 8.7	17.87716
3	8.3, 8.6, 8.7, 8.1	17.87714

b.

n	$x_0, x_1, \ldots, x_n$	$P_n(-\frac{1}{3})$
1	−0.5, −0.25	0.21504167
2	−0.5, −0.25, 0.0	0.16988889
3	−0.5, −0.25, 0.0, −0.75	0.17451852

c.

n	$x_0, x_1, \ldots, x_n$	$P_n(0.25)$
1	0.2, 0.3	−0.13869287
2	0.2, 0.3, 0.4	−0.13259734
3	0.2, 0.3, 0.4, 0.1	−0.13277477

d.

n	$x_0, x_1, \ldots, x_n$	$P_n(0.9)$
1	0.8, 1.0	0.44086280
2	0.8, 1.0, 0.7	0.43841352
3	0.8, 1.0, 0.7, 0.6	0.44198500

5. $\sqrt{3} \approx P_4(\frac{1}{2}) = 1.708\overline{3}$

7. a.

n	Actual error	Error bound
1	0.00118	0.00120
2	1.367×10^{-5}	1.452×10^{-5}

b.

n	Actual error	Error bound
1	4.0523×10^{-2}	4.5153×10^{-2}
2	4.6296×10^{-3}	4.6296×10^{-3}

c.

n	Actual error	Error bound
1	5.9210×10^{-3}	6.0971×10^{-3}
2	1.7455×10^{-4}	1.8128×10^{-4}

d.

n	Actual error	Error bound
1	2.7296×10^{-3}	1.4080×10^{-2}
2	5.1789×10^{-3}	9.2215×10^{-3}

9. $f(1.09) \approx 0.2826$. The actual error is 4.3×10^{-5}, and an error bound is 7.4×10^{-6}. The discrepancy is due to the fact that the data are given to only four decimal places and only four-digit arithmetic is used.

11. $y = 4.25$

13. The largest possible step size is 0.004291932, so 0.004 would be a reasonable choice.

15. The difference between the actual value and the computed value is $\frac{2}{3}$.

17. a.

x	erf(x)
0.0	0
0.2	0.2227
0.4	0.4284
0.6	0.6039
0.8	0.7421
1.0	0.8427

b. Linear interpolation with $x_0 = 0.2$ and $x_1 = 0.4$ gives $\text{erf}(\frac{1}{3}) \approx 0.3598$ and quadratic interpolation with $x_0 = 0.2, x_1 = 0.4$, and $x_2 = 0.6$ gives $\text{erf}(\frac{1}{3}) \approx 0.3632$. Since $\text{erf}(\frac{1}{3}) \approx 0.3626$, quadratic interpolation is more accurate.

EXERCISE SET 3.3 (PAGE 80)

1. Newton's interpolatory divided-difference formula gives the following:

a. $P_1(x) = 16.9441 + 3.1041(x - 8.1)$; $P_1(8.4) = 17.87533$
$P_2(x) = P_1(x) + 0.06(x - 8.1)(x - 8.3)$; $P_2(8.4) = 17.87713$
$P_3(x) = P_2(x) - 0.00208333(x - 8.1)(x - 8.3)(x - 8.6)$;
$P_3(8.4) = 17.87714$

b. $P_1(x) = -0.1769446 + 1.9069687(x - 0.6)$; $P_1(0.9) = 0.395146$
$P_2(x) = P_1(x) + 0.959224(x - 0.6)(x - 0.7)$; $P_2(0.9) = 0.4526995$
$P_3(x) = P_2(x) - 1.785741(x - 0.6)(x - 0.7)(x - 0.8)$; $P_3(0.9) = 0.4419850$

3. In the following equations we have $s = (1/h)(x - x_n)$.

a. $P_1(s) = 1.101 + 0.7660625s$; $f(-\frac{1}{3}) \approx P_1(-\frac{4}{3}) = 0.07958333$;
$P_2(s) = P_1(s) + 0.406375s(s + 1)/2$; $f(-\frac{1}{3}) \approx P_2(-\frac{4}{3}) = 0.1698889$;
$P_3(s) = P_2(s) + 0.09375s(s + 1)(s + 2)/6$; $f(-\frac{1}{3}) \approx P_3(-\frac{4}{3}) = 0.1745185$

b. $P_1(s) = 0.2484244 + 0.2418235s$; $f(0.25) \approx P_1(-1.5) = -0.1143108$
$P_2(s) = P_1(s) - 0.04876419s(s + 1)/2$; $f(0.25) \approx P_2(-1.5) = -0.1325973$
$P_3(s) = P_2(s) - 0.00283891s(s + 1)(s + 2)/6$; $f(0.25) \approx P_3(-1.5) = -0.1327748$

5. a. $f(0.05) \approx 1.05126$ **b.** $f(0.65) \approx 1.91555$

7. $\Delta^3 f(x_0) = -6$ and $\Delta^4 f(x_0) = \Delta^5 f(x_0) = 0$, so the interpolating polynomial has degree 3.

9. $\Delta^2 P(10) = 1140$.

11. The approximation to $f(0.3)$ should be increased by 5.9375.

13. $f[x_0] = f(x_0) = 1$, $f[x_1] = f(x_1) = 3$, $f[x_0, x_1] = 5$

EXERCISE SET 3.4 (PAGE 85)

1. The coefficients for the polynomials in divided-difference form are given in the following tables. For example, the polynomial in part (a) is

$$H_3(x) = 17.56492 + 3.116256(x - 8.3) + 0.05948(x - 8.3)^2 - 0.00202222(x - 8.3)^2(x - 8.6).$$

a.	b.	c.	d.
17.56492	0.022363362	−0.02475	−0.62049958
3.116256	2.1691753	0.751	3.5850208
0.05948	0.01558225	2.751	−2.1989182
−0.00202222	−3.2177925	1	−0.490447
		0	0.037205
		0	0.040475
			−0.0025277777
			0.0029629628

3. **a.** We have $\sin 0.34 \approx H_5(0.34) = 0.33349$.

b. The formula gives an error bound of 3.05×10^{-14}, but the actual error is 2.91×10^{-6}. The discrepancy is due to the fact that the data are given to only five decimal places.

c. We have $\sin 0.34 \approx H_7(0.34) = 0.33350$. Although the error bound is now 5.4×10^{-20}, the accuracy of the given data dominates the calculations. This result is actually less accurate than the approximation in part (b), since $\sin 0.34 = 0.333487$.

5. For 2(a) we have an error bound of 5.9×10^{-8}. The error bound for 2(c) is 0 since $f^{(n)}(x) \equiv 0$ for $n > 3$.

7. The Hermite polynomial generated from these data is

$$\begin{aligned} H_9(x) = {} & 75x + 0.222222x^2(x-3) - 0.0311111x^2(x-3)^2 \\ & - 0.00644444x^2(x-3)^2(x-5) + 0.00226389x^2(x-3)^2(x-5)^2 \\ & - 0.000913194x^2(x-3)^2(x-5)^2(x-8) \\ & + 0.000130527x^2(x-3)^2(x-5)^2(x-8)^2 \\ & - 0.0000202236x^2(x-3)^2(x-5)^2(x-8)^2(x-13). \end{aligned}$$

a. The Hermite polynomial predicts a position of $H_9(10) = 743$ ft and a speed of $H_9'(10) = 48$ ft/s. Although the position approximation is reasonable, the low-speed prediction is suspect.

b. To find the first time the speed exceeds 55 mi/h $= 80.\overline{6}$ ft/s, we solve for the smallest value of t in the equation $80.\overline{6} = H_9'(x)$. This gives $x \approx 5.6488092$.

c. The estimated maximum speed is $H_9'(12.37187) = 119.423$ ft/s ≈ 81.425 mi/h.

EXERCISE SET 3.5 (PAGE 96)

1. $S(x) = x$ on $[0, 2]$

3. The equations of the respective free cubic splines are given by

$$S(x) = S_i(x) = a_i + b_i(x - x_i) + c_i(x - x_i)^2 + d_i(x - x_i)^3,$$

for x in $[x_i, x_{i+1}]$ and the coefficients in the following tables.

a.

i	a_i	b_i	c_i	d_i
0	17.564920	3.13410000	0.00000000	0.00000000

b.

i	a_i	b_i	c_i	d_i
0	0.22363362	2.17229175	0.00000000	0.00000000

c.

i	a_i	b_i	c_i	d_i
0	−0.02475000	1.03237500	0.00000000	6.50200000
1	0.33493750	2.25150000	4.87650000	−6.50200000

d.

i	a_i	b_i	c_i	d_i
0	−0.62049958	3.45508693	0.00000000	−8.9957933
1	−0.28398668	3.18521313	−2.69873800	−0.94630333
2	0.00660095	2.61707643	−2.98262900	9.9420966

5. The equations of the respective clamped cubic splines are given by

$$s(x) = s_i(x) = a_i + b_i(x - x_i) + c_i(x - x_i)^2 + d_i(x - x_i)^3,$$

for x in $[x_i, x_{i+1}]$ and the coefficients in the following tables.

a.

i	a_i	b_i	c_i	d_i
0	17.564920	3.1162560	0.0600867	−0.00202222

b.

i	a_i	b_i	c_i	d_i
0	0.22363362	2.1691753	0.65914075	−3.2177925

c.

i	a_i	b_i	c_i	d_i
0	−0.02475000	0.75100000	2.5010000	1.0000000
1	0.33493750	2.18900000	3.2510000	1.0000000

d.

i	a_i	b_i	c_i	d_i
0	−0.62049958	3.5850208	−2.1498407	−0.49077413
1	−0.28398668	3.1403294	−2.2970730	−0.47458360
2	0.006600950	2.6666773	−2.4394481	−0.44980146

7. The equation of the spline is

$$S(x) = S_i(x) = a_i + b_i(x - x_i) + c_i(x - x_i)^2 + d_i(x - x_i)^3$$

on the interval $[x_i, x_{i+1}]$, where the coefficients are given in the following table.

x_i	a_i	b_i	c_i	d_i
0	1.0	−0.7573593	0.0	−6.627417
0.25	0.7071068	−2.0	−4.970563	6.627417
0.5	0.0	−3.242641	0.0	6.627417
0.75	−0.7071068	−2.0	4.970563	−6.627417

a. $\int_0^1 S(x)\,dx = 0.000000$

b. $S'(0.5) = -3.24264$, and $S''(0.5) = 0.0$

9. **a.** The equation of the spline is

$$s(x) = s_i(x) = a_i + b_i(x - x_i) + c_i(x - x_i)^2 + d_i(x - x_i)^3$$

on the interval $[x_i, x_{i+1}]$, where the coefficients are given in the following table.

x_i	a_i	b_i	c_i	d_i
0	1.0	0.0	−5.193321	2.028118
0.25	0.7071068	−2.216388	−3.672233	4.896310
0.5	0.0	−3.134447	0.0	4.896310
0.75	−0.7071068	−2.216388	3.672233	2.028118

b. $\int_0^1 s(x)\,dx = 0.000000$

c. $s(0.5) = -3.13445$, and $s''(0.5) = 0.0$.

11. $a = 4, b = -1, c = -3, d = 1$

13. $B = \frac{1}{4}, D = \frac{1}{4}, b = -\frac{1}{2}, d = \frac{1}{4}$

15. Let $f(x) = a + bx + cx^2 + dx^3$. Clearly, f satisfies properties (a), (c), (d), (e) of the definition and f interpolates itself for any choice of $x_0, \ldots, x_n$. Since (ii) of (f) in the definition holds, f must be its own clamped cubic spline. However, $f''(x) = 2c + 6dx$ can be zero only at $x = -c/3d$. Thus, part (i) of (f) in the definition cannot hold at two values x_0 and x_n, and f cannot be a natural cubic spline.

17.

x_i	a_i	b_i	c_i	d_i
1940	132,165	1651.85	0.00000	2.64248
1950	151,326	2444.59	79.2744	−4.37641
1960	179,323	2717.16	−52.0179	2.00918
1970	203,302	2279.55	8.25746	−0.381311
1980	226,542	2330.31	−3.18186	0.106062

$S(1930) = 113{,}004$, $S(1965) = 191{,}860$, and $S(2000) = 272{,}724$.

19. **a.** $S(x) = S_i(x) = a_i + b_i(x - x_i) + c_i(x - x_i)^2 + d_i(x - x_i)^3$ on $[x_i, x_{i+1}]$, where

x_i	a_i	b_i	c_i	d_i
0	0	88.8	0	12.8
0.25	22.4	91.2	9.6	0
0.5	45.8	96.0	9.6	−4.8
1.0	95.6	102.0	2.4	−3.2
1.25				

b. 1:10 $\frac{13}{40}$

c. Starting speed ≈ 40.54 mi/h. Ending speed ≈ 35.09 mi/h.

EXERCISE SET 3.6 (PAGE 104)

1. **a.** $x(t) = -10t^3 + 14t^2 + t, \quad y(t) = -2t^3 + 3t^2 + t$

b. $x(t) = -10t^3 + 14.5t^2 + 0.5t, \quad y(t) = -3t^3 + 4.5t^2 + 0.5t$

c. $x(t) = -10t^3 + 14t^2 + t, \quad y(t) = -4t^3 + 5t^2 + t$

d. $x(t) = -10t^3 + 13t^2 + 2t, \quad y(t) = 2t$

3. **a.** $x(t) = -11.5t^3 + 15t^2 + 1.5t + 1, \quad y(t) = -4.25t^3 + 4.5t^2 + 0.75t + 1$

b. $x(t) = -6.25t^3 + 10.5t^2 + 0.75t + 1, \quad y(t) = -3.5t^3 + 3t^2 + 1.5t + 1$

c. For t between $(0, 0)$ and $(4, 6)$ we have

$$x(t) = -5t^3 + 7.5t^2 + 1.5t, \qquad y(t) = -13.5t^3 + 18t^2 + 1.5t,$$

and for t between $(4, 6)$ and $(6, 1)$ we have

$$x(t) = -5.5t^3 + 6t^2 + 1.5t + 4, \qquad y(t) = 4t^3 - 6t^2 - 3t + 6.$$

d. For t between $(0, 0)$ and $(2, 1)$ we have

$$x(t) = -5.5t^3 + 6t^2 + 1.5t, \qquad y(t) = -0.5t^3 + 1.5t,$$

for t between $(2, 1)$ and $(4, 0)$ we have

$$x(t) = -4t^3 + 3t^2 + 3t + 2, \qquad y(t) = -t^3 + 1,$$

and for t between $(4, 0)$ and $(6, -1)$ we have

$$x(t) = -8.5t^3 + 13.5t^2 - 3t + 4, \qquad y(t) = -3.25t^3 + 5.25t^2 - 3t.$$

CHAPTER 4

EXERCISE SET 4.2 (PAGE 114)

1. The Midpoint Rule gives the following approximations.

a. 0.1582031 **b.** −0.2666667
c. 0.1743309 **d.** 0.1516327
e. −0.6753247 **f.** −0.5194805
g. 0.1180292 **h.** 1.8039148

3. The Trapezoidal Rule gives the following approximations.

a. 0.265625 **b.** −0.2678571
c. 0.2280741 **d.** 0.1839397
e. −0.8666667 **f.** −0.6166667
g. 0.2180895 **h.** 4.1432597

5. Simpson's Rule gives the following approximations.

a. 0.1940104 **b.** −0.2670635
c. 0.1922453 **d.** 0.16240168
e. −0.7391053 **f.** −0.5518759
g. 0.1513826 **h.** 2.5836964

7. Formula (i) gives the following approximations.

a. 0.19386574 **b.** −0.26706310
c. 0.19225309 **d.** 0.16140992
e. −0.73642770 **f.** −0.55053615
g. 0.15158524 **h.** 2.5857891

9. $f(1) = \frac{1}{2}$

11. $c_0 = \frac{1}{4}$, $c_1 = \frac{3}{4}$, and $x_1 = \frac{2}{3}$

13.

	(i) Midpoint Rule	(ii) Trapezoidal Rule	(iii) Simpson's Rule
a.	4.83393	5.43476	5.03420
b.	-7.2×10^{-7}	1.6×10^{-6}	5.3×10^{-8}

EXERCISE SET 4.3 (PAGE 123)

1. The Composite Trapezoidal Rule approximations are as follows.

a. 0.639900 **b.** 31.3653
c. 0.784241 **d.** -6.42872
e. -13.5760 **f.** 0.476977
g. 0.605498 **h.** 0.970926

3. The Composite Midpoint Rule approximations are as follows.

a. 0.633096 **b.** 11.1568
c. 0.786700 **d.** -6.11274
e. -14.9985 **f.** 0.478751
g. 0.602961 **h.** 0.947868

5. **a.** The Composite Trapezoidal Rule requires $h < 0.000922295$ and $n \geq 2168$.
b. The Composite Simpson's Rule requires $h < 0.037658$ and $n \geq 54$.
c. The Composite Midpoint Rule requires $h < 0.00065216$ and $n \geq 3066$.

7. **a.** The Composite Trapezoidal Rule requires $h < 0.04382$ and $n \geq 46$. The approximation is 0.405471.
b. The Composite Simpson's Rule requires $h < 0.44267$ and $n \geq 6$. The approximation is 0.405466.
c. The Composite Midpoint Rule requires $h < 0.03098$ and $n \geq 64$. The approximation is 0.405460.

9. $\alpha = 1.5$

11. **a.** 0.95449101, obtained using $n = 14$ in Composite Simpson's Rule.
b. 0.99728944, obtained using $n = 18$ in Composite Simpson's Rule.

13. The length of the track is approximately 9858 ft.

15. **a.** For $p_0 = 0.5$ we have $p_6 = 1.644854$ with $n = 20$.
b. For $p_0 = 0.5$ we have $p_6 = 1.645085$ with $n = 40$.

EXERCISE SET 4.4 (PAGE 132)

1. Romberg integration gives $R_{3,3}$ as follows:

a. 0.1922593 **b.** 0.1606105
c. -0.1768200 **d.** 0.08875677
e. 2.5879685 **f.** -0.7341567
g. 0.6362135 **h.** 0.6426970

3. Romberg integration gives the following values:

a. 0.19225936 with $n = 4$ **b.** 0.16060279 with $n = 5$
c. -0.17682002 with $n = 4$ **d.** 0.088755284 with $n = 5$
e. 2.5886286 with $n = 6$ **f.** -0.73396918 with $n = 6$
g. 0.63621335 with $n = 4$ **h.** 0.64269908 with $n = 5$

5. $R_{33} = 11.5246$

7. $f(2.5) \approx 0.43457$

9. $R_{31} = 5$

11. Let $N_2(h) = N\left(\frac{h}{3}\right) + \frac{1}{8}\left(N\left(\frac{h}{3}\right) - N(h)\right)$ and $N_3(h) = N_2\left(\frac{h}{3}\right) + \frac{1}{80}\left(2\left(\frac{h}{3}\right) - N_2(h)\right)$. Then $N_3(h)$ is an $O(h^6)$ approximation to M.

13. a. By L'Hôpital's rule, $\lim_{h\to 0} \frac{\ln(2+h)-\ln(2-h)}{h} = \lim_{h\to 0}\left[\frac{1}{2+h}+\frac{1}{2-h}\right] = 1$, so

$$\lim_{h\to 0}\left(\frac{2+h}{2-h}\right)^{1/h} = \lim_{h\to 0} e^{(1/h)[\ln(2+h)-\ln(2-h)]} = e^1 = e.$$

b. $N(0.04) = 2.718644377221219$, $N(0.02) = 2.718372444800607$, $N(0.01) = 2.718304481241685$

c. Let $N_2(h) = 2N\left(\frac{h}{2}\right) - N(h)$, $N_3(h) = N_2\left(\frac{h}{2}\right) + \frac{1}{3}[N_2\left(\frac{h}{2}\right) - N_2(h)]$. Then $N_2(0.04) = 2.718100512379995$, $N_2(0.02) = 2.718236517682763$, and $N_3(0.04) = 2.718281852783685$. $N_3(0.04)$ is at least an $O(h^3)$ approximation satisfying $|e - N_3(0.04)| \le 0.5 \times 10^{-7}$.

d. $N(-h) = \left(\frac{2-h}{2+h}\right)^{1/-h} = \left(\frac{2+h}{2-h}\right)^{1/h} = N(h)$

e. Let

$$e = N(h) + K_1 h + K_2 h^2 + K_3 h^3 + \cdots.$$

Replacing h by $-h$ gives

$$e = N(-h) - K_1 h + K_2 h^2 - K_3 h^3 + \cdots,$$

but $N(-h) = N(h)$, so that

$$e = N(h) - K_1 h + K_2 h^2 - K_3 h^3 + \cdots.$$

Thus,

$$K_1 h + K_3 h^3 + \cdots = -K_1 h - K_3 h^3 \cdots,$$

and it follows that $K_1 = K_3 = K_5 = \cdots = 0$ and

$$e = N(h) + K_2 h^2 + K_4 h^4 + \cdots.$$

f. Let

$$N_2(h) = N\left(\frac{h}{2}\right) + \frac{1}{3}\left(N\left(\frac{h}{2}\right) - N(h)\right)$$

and

$$N_3(h) = N_2\left(\frac{h}{2}\right) + \frac{1}{15}\left(N_2\left(\frac{h}{2}\right) - N_2(h)\right).$$

Then

$$N_2(0.04) = 2.718281800660402, \qquad N_2(0.02) = 2.718281826722043$$

and

$$N_3(0.04) = 2.718281828459487.$$

$N_3(0.04)$ is an $O(h^6)$ approximation satisfying $|e - N_3(0.04)| \le 0.5 \times 10^{-12}$.

EXERCISE SET 4.5 (PAGE 139)

1. Gaussian quadrature gives the following.

a. 0.1922687
b. 0.1594104
c. −0.1768190
d. 0.08926302
e. 2.5913247
f. −0.7307230
g. 0.6361966
h. 0.6423172

3. Gaussian quadrature gives the following.

a. 0.1922594 **b.** 0.1606028
c. −0.1768200 **d.** 0.08875529
e. 2.5886327 **f.** −0.7339604
g. 0.6362133 **h.** 0.6426991

5. $a = 1, b = 1, c = \frac{1}{3}, d = -\frac{1}{3}$

EXERCISE SET 4.6 (PAGE 144)

1. Simpson's Rule gives the following.

a. $S(1, 1.5) = 0.19224530$, $S(1, 1.25) = 0.039372434$, $S(1.25, 1.5) = 0.15288602$, and the actual value is 0.19225935.

b. $S(0, 1) = 0.16240168$, $S(0, 0.5) = 0.028861071$, $S(0.5, 1) = 0.13186140$, and the actual value is 0.16060279.

c. $S(0, 0.35) = -0.17682156$, $S(0, 0.175) = -0.087724382$, $S(0.175, 0.35) = -0.089095736$, and the actual value is −0.17682002.

d. $S(0, \pi/4) = 0.087995669$, $S(0, \pi/8) = 0.0058315797$, $S(\pi/8, \pi/4) = 0.082877624$, and the actual value is 0.088755285.

e. $S(0, \pi/4) = 2.5836964$, $S(0, \pi/8) = 0.33088926$, $S(\pi/8, \pi/4) = 2.2568121$, and the actual value is 2.5886286.

f. $S(1, 1.6) = -0.73910533$, $S(1, 1.3) = -0.26141244$, $S(1.3, 1.6) = -0.47305351$, and the actual value is −0.73396917.

g. $S(3, 3.5) = 0.63623873$, $S(3, 3.25) = 0.32567095$, $S(3.25, 3.5) = 0.31054412$, and the actual value is 0.63621334.

h. $S(0, \pi/4) = 0.64326905$, $S(0, \pi/8) = 0.37315002$, $S(\pi/8, \pi/4) = 0.26958270$, and the actual value is 0.64269908.

3. Adaptive quadrature gives the following.

a. 108.555281 **b.** −1724.966983
c. −15.306308 **d.** −18.945949

5. Adaptive quadrature gives the following.

$$\int_{0.1}^{2} \sin\frac{1}{x}\,dx = 1.1454 \quad\text{and}\quad \int_{0.1}^{2} \cos\frac{1}{x}\,dx = 0.67378.$$

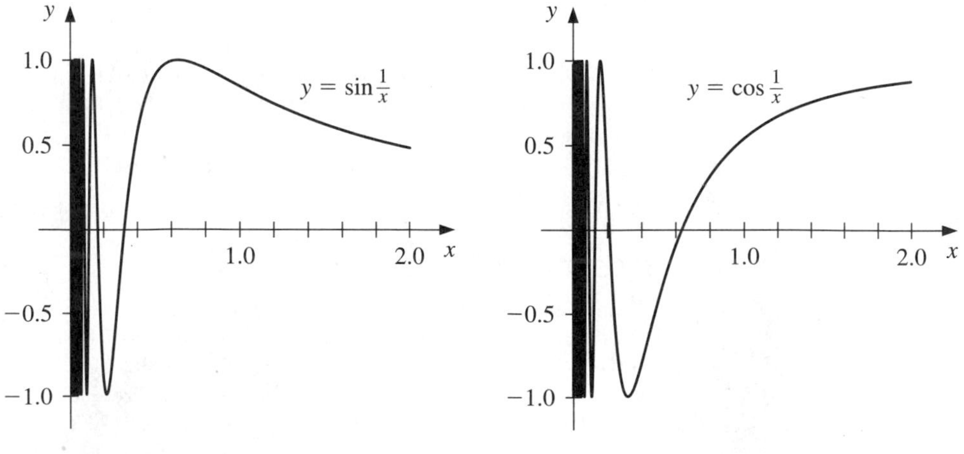

7. $\int_0^{2\pi} u(t)\,dt \approx 0.00001$

9.

t	$c(t)$	$s(t)$
0.1	0.0999975	0.000523589
0.2	0.199921	0.00418759
0.3	0.299399	0.0141166
0.4	0.397475	0.0333568
0.5	0.492327	0.0647203
0.6	0.581061	0.110498
0.7	0.659650	0.172129
0.8	0.722844	0.249325
0.9	0.764972	0.339747
1.0	0.779880	0.438245

EXERCISE SET 4.7 (PAGE 155)

1. Composite Simpson's rule with $n = m = 4$ gives these values.
   a. 0.3115733 **b.** 0.2552526
   c. 16.50864 **d.** 1.476684

3. Composite Simpson's rule with $n = 4$ and $m = 8$, $n = 8$ and $m = 4$, and $n = m = 6$ gives, respectively, the following.
   a. 0.5119875, 0.5118533, 0.5118722
   b. 1.718857, 1.718220, 1.718385
   c. 1.001953, 1.000122, 1.000386
   d. 0.7838542, 0.7833659, 0.7834362
   e. -1.985611, -1.999182, -1.997353
   f. 2.004596, 2.000879, 2.000980
   g. 0.3084277, 0.3084562, 0.3084323
   h. -22.61612, -19.85408, -20.14117

5. Gaussian quadrature with $n = m = 2$ gives the following.
   a. 0.3115733 **b.** 0.2552446
   c. 16.50863 **d.** 1.488875

7. Gaussian quadrature with $n = m = 6$, $n = 6$ and $m = 8$, $n = 8$ and $m = 6$, and $n = m = 8$ gives, respectively, the following.
   a. 0.5118655, 0.5118445, 0.5118655, 0.5118445, 2.1×10^{-5}, 1.3×10^{-7}, 2.1×10^{-5}, 1.3×10^{-7}
   b. 1.718163, 1.718302, 1.718139, 1.718277, 1.2×10^{-4}, 2.0×10^{-5}, 1.4×10^{-4}, 4.8×10^{-6}
   c. 1.000000, 1.000000, 1.0000000, 1.000000, 0, 0, 0, 0
   d. 0.7833333, 0.7833333, 0.7833333, 0.7833333, 0, 0, 0, 0
   e. -1.991878, -2.000124, -1.991878, -2.000124, 8.1×10^{-3}, 1.2×10^{-4}, 8.1×10^{-3}, 1.2×10^{-4}
   f. 2.001494, 2.000080, 2.001388, 1.999984, 1.5×10^{-3}, 8×10^{-5}, 1.4×10^{-3}, 1.6×10^{-5}
   g. 0.3084151, 0.3084145, 0.3084246, 0.3084245, 10^{-5}, 5.5×10^{-7}, 1.1×10^{-5}, 6.4×10^{-7}
   h. -12.74790, -21.21539, -11.83624, -20.30373, 7.0, 1.5, 7.9, 0.564

9. Gaussian quadrature with $n = m = p = 2$ gives the first listed value. The second is the exact result.
   a. 5.204036, $e(e^{0.5} - 1)(e - 1)^2$ **b.** 0.08429784, $\frac{1}{12}$
   c. 0.08641975, $\frac{1}{14}$ **d.** 0.09722222, $\frac{1}{12}$
   e. 7.103932, $2 + \frac{1}{2}\pi^2$ **f.** 1.428074, $\frac{1}{2}(e^2 + 1) - e$

11. Composite Simpson's Rule with $n = m = 14$ gives 0.1479103 and Gaussian quadrature with $n = m = 4$ gives 0.1506823.

13. The area approximations are (a) 1.040253 and (b) 1.040252.

15. Gaussian quadrature with $n = m = p = 4$ gives 3.0521250. The exact result is 3.0521249.

EXERCISE SET 4.8 (PAGE 162)

1. Composite Simpson's rule gives the following.
a. 0.5284163 **b.** 4.266654
c. 0.4329748 **d.** 0.8802210

3. Composite Simpson's rule gives the following.
a. 0.4112649 **b.** 0.2440679
c. 0.05501681 **d.** 0.2903746

5. Composite Simpson's rule gives the following.
a. 3.141569 **b.** 0.0
c. 1.178071 **d.** 2.221548

EXERCISE SET 4.9 (PAGE 171)

1. From the two-point formula we have the following approximations:
a. $f'(0.5) \approx 0.8520$, $f'(0.6) \approx 0.8520$, $f'(0.7) \approx 0.7960$
b. $f'(0.0) \approx 3.7070$, $f'(0.2) \approx 3.1520$, $f'(0.4) \approx 3.1520$

3. For the endpoints of the tables we use the three-point endpoint formula. The other approximations come from the three-point midpoint formula.
a. $f'(1.1) \approx 17.769705$, $f'(1.2) \approx 22.193635$, $f'(1.3) \approx 27.107350$, $f'(1.4) \approx 32.150850$
b. $f'(8.1) \approx 3.092050$, $f'(8.3) \approx 3.116150$, $f'(8.5) \approx 3.139975$, $f'(8.7) \approx 3.163525$
c. $f'(2.9) \approx 5.101375$, $f'(3.0) \approx 6.654785$, $f'(3.1) \approx 8.216330$, $f'(3.2) \approx 9.786010$
d. $f'(2.0) \approx 0.13533150$, $f'(2.1) \approx -0.09989550$, $f'(2.2) \approx -0.3298960$, $f'(2.3) \approx -0.5546700$

5. a. The five-point endpoint formula gives $f'(2.1) \approx 3.899344$, $f'(2.2) \approx 2.876876$, $f'(2.5) \approx 1.544210$, and $f'(2.6) \approx 1.355496$. The five-point midpoint formula gives $f'(2.3) \approx 2.249704$ and $f'(2.4) \approx 1.837756$.
b. The five-point endpoint formula gives $f'(3.0) \approx -5.877358$, $f'(-2.8) \approx -5.468933$, $f'(-2.2) \approx -4.239911$, and $f'(-2.0) \approx -3.828853$. The five-point midpoint formula gives $f'(-2.6) \approx -5.059884$ and $f'(-2.4) \approx -4.650223$.

7. The approximation is -4.8×10^{-9} and $f''(0.5) = 0$. The error bound is 0.35874. The method is very accurate since the function is symmetric about $x = 0.5$.

9. $f'(3) \approx \frac{1}{12}[f(1) - 8f(2) + 8f(4) - f(5)] = 0.21062$ with an error bound given by

$$\max_{1\le x\le 5} \frac{|f^{(5)}(x)|h^4}{30} \le \frac{23}{30} = 0.7\bar{6}.$$

11. The optimal $h = 2\sqrt{\varepsilon/M}$, where $M = \max |f''(x)|$.

13. Since $e'(h) = -\varepsilon/h^2 + hM/3$, we have $e'(h) = 0$ if and only if $h = \sqrt[3]{3\varepsilon/M}$. Also, $e'(h) < 0$ if $h < \sqrt[3]{\varepsilon/M}$ and $e'(h) > 0$ if $h > \sqrt[3]{\varepsilon/M}$, so an absolute minimum for $e(h)$ occurs at $h = \sqrt[3]{3\varepsilon/M}$.

15. Using three-point formulas gives the following table:

Time	0	3	5	8	10	13
Speed	79	82.4	74.2	76.8	69.4	71.2

CHAPTER 5

EXERCISE SET 5.2 (PAGE 187)

1. Euler's method gives the approximations in the following tables.

a.

i	t_i	w_i	$y(t_i)$
1	0.500	0.0000000	0.2836165
2	1.000	1.1204223	3.2190993

b.

i	t_i	w_i	$y(t_i)$
1	2.500	2.0000000	1.8333333
2	3.000	2.6250000	2.5000000

c.

i	t_i	w_i	$y(t_i)$
1	1.250	2.7500000	2.7789294
2	1.500	3.5500000	3.6081977
3	1.750	4.3916667	4.4793276
4	2.000	5.2690476	5.3862944

d.

i	t_i	w_i	$y(t_i)$
1	0.250	1.2500000	1.3291498
2	0.500	1.6398053	1.7304898
3	0.750	2.0242547	2.0414720
4	1.000	2.2364573	2.1179795

3. Euler's method gives the approximations in the following tables.

a.

i	t_i	w_i	$y(t_i)$
2	1.2	1.0082645	1.0149523
4	1.4	1.0385147	1.0475339
6	1.6	1.0784611	1.0884327
8	1.8	1.1232621	1.1336536
10	2.0	1.1706516	1.1812322

b.

i	t_i	w_i	$y(t_i)$
2	1.4	0.4388889	0.4896817
4	1.8	1.0520380	1.1994386
6	2.2	1.8842608	2.2135018
8	2.6	3.0028372	3.6784753
10	3.0	4.5142774	5.8741000

c.

i	t_i	w_i	$y(t_i)$
2	0.4	−1.6080000	−1.6200510
4	0.8	−1.3017370	−1.3359632
6	1.2	−1.1274909	−1.1663454
8	1.6	−1.0491191	−1.0783314
10	2.0	−1.0181518	−1.0359724

d.

i	t_i	w_i	$y(t_i)$
2	0.2	0.1083333	0.1626265
4	0.4	0.1620833	0.2051118
6	0.6	0.3455208	0.3765957
8	0.8	0.6213802	0.6461052
10	1.0	0.9803451	1.0022460

5. a.

i	t_i	w_i	$y(t_i)$
1	0.50	0.12500000	0.28361652
2	1.00	2.02323897	3.21909932

b.

i	t_i	w_i	$y(t_i)$
1	2.50	1.75000000	1.83333333
2	3.00	2.42578125	2.50000000

c.

i	t_i	w_i	$y(t_i)$
1	1.25	2.78125000	2.77892944
2	1.50	3.61250000	3.60819766
3	1.75	4.48541667	4.47932763
4	2.00	5.39404762	5.38629436

d.

i	t_i	w_i	$y(t_i)$
1	0.25	1.34375000	1.32914981
2	0.50	1.77218707	1.73048976
3	0.75	2.11067606	2.04147203
4	1.00	2.20164395	2.11797955

7. a.

i	t_i	w_i	$y(t_i)$
2	1.2	1.0149771	1.0149523
4	1.4	1.0475619	1.0475339
6	1.6	1.0884607	1.0884327
8	1.8	1.1336811	1.1336536
10	2.0	1.1812594	1.1812322

b.

i	t_i	w_i	$y(t_i)$
2	1.4	0.4896141	0.4896817
4	1.8	1.1993085	1.1994386
6	2.2	2.2132495	2.2135018
8	2.6	3.6779557	3.6784753
10	3.0	5.8729143	5.8741000

c.

i	t_i	w_i	$y(t_i)$
2	0.4	−1.6201137	−1.6200510
4	0.8	−1.3359853	−1.3359632
6	1.2	−1.1663295	−1.1663454
8	1.6	−1.0783171	−1.0783314
10	2.0	−1.0359674	−1.0359724

d.

i	t_i	w_i	$y(t_i)$
2	0.2	0.1627236	0.1626265
4	0.4	0.2051833	0.2051118
6	0.6	0.3766352	0.3765957
8	0.8	0.6461246	0.6461052
10	1.0	1.0022549	1.0022460

9. a.

i	t_i	w_i	$y(t_i)$
1	1.05	−0.9500000	−0.9523810
2	1.10	−0.9045353	−0.9090909
11	1.55	−0.6263495	−0.6451613
12	1.60	−0.6049486	−0.6250000
19	1.95	−0.4850416	−0.5128205
20	2.00	−0.4712186	−0.5000000

b. Linear interpolation gives (i) $y(1.052) \approx -0.9481814$, (ii) $y(1.555) \approx -0.6242094$, (iii) $y(1.978) \approx -0.4773007$.

The actual values are $y(1.052) = -0.9505703$, $y(1.555) = -0.6430868$, $y(1.978) = -0.5055612$.

c.

i	t_i	w_i	$y(t_i)$
1	1.05	−0.9525000	−0.9523810
2	1.10	−0.9093138	−0.9090909
11	1.55	−0.6459788	−0.6451613
12	1.60	−0.6258649	−0.6250000
19	1.95	−0.5139781	−0.5128205
20	2.00	−0.5011957	−0.5000000

d. Linear interpolation gives (i) $y(1.052) \approx -0.9507726$, (ii) $y(1.555) \approx -0.6439674$, (iii) $y(1.978) \approx -0.5068199$.

e.

i	t_i	w_i	$y(t_i)$
1	1.05	−0.9523813	−0.9523810
2	1.10	−0.9090914	−0.9090909
11	1.55	−0.6451629	−0.6451613
12	1.60	−0.6250017	−0.6250000
19	1.95	−0.5128226	−0.5128205
20	2.00	−0.5000022	−0.5000000

f. Hermite interpolation gives (i) $y(1.052) \approx -0.9505706$, (ii) $y(1.555) \approx -0.6430884$, (iii) $y(1.978) \approx -0.5055633$.

11. b. $w_{50} = 0.10430 \approx p(50)$

c. Since $p(t) = 1 - 0.99e^{-0.002t}$, $p(50) = 0.10421$.

EXERCISE SET 5.3 (PAGE 196)

1. a.

i	t_i	w_i	$y(t_i)$
1	0.5	0.2646250	0.2836165
2	1.0	3.1300023	3.2190993

b.

i	t_i	w_i	$y(t_i)$
1	2.5	1.7812500	1.8333333
2	3.0	2.4550638	2.5000000

c.

i	t_i	w_i	$y(t_i)$
1	1.25	2.7777778	2.7789294
2	1.50	3.6060606	3.6081977
3	1.75	4.4763015	4.4793276
4	2.00	5.3824398	5.3862944

d.

i	t_i	w_i	$y(t_i)$
1	0.25	1.3337962	1.3291498
2	0.50	1.7422854	1.7304898
3	0.75	2.0596374	2.0414720
4	1.00	2.1385560	2.1179795

3. **a.**

i	t_i	w_i	$y(t_i)$
1	0.5	0.5602111	0.2836165
2	1.0	5.3014898	3.2190993

b.

i	t_i	w_i	$y(t_i)$
1	2.5	1.8125000	1.8333333
2	3.0	2.4815531	2.5000000

c.

i	t_i	w_i	$y(t_i)$
1	1.25	2.7750000	2.7789294
2	1.50	3.6008333	3.6081977
3	1.75	4.4688294	4.4793276
4	2.00	5.3728586	5.3862944

d.

i	t_i	w_i	$y(t_i)$
1	0.25	1.3199027	1.3291498
2	0.50	1.7070300	1.7304898
3	0.75	2.0053560	2.0414720
4	1.00	2.0770789	2.1179795

5. **a.** $1.0221167 \approx y(1.25) = 1.0219569, 1.1640347 \approx y(1.93) = 1.1643901$
b. $1.9086500 \approx y(2.1) = 1.9249616, 4.3105913 \approx y(2.75) = 4.3941697$
c. $-1.1461434 \approx y(1.3) = -1.1382768, -1.0454854 \approx y(1.93) = -1.0412665$
d. $0.3271470 \approx y(0.54) = 0.3140018, 0.8967073 \approx y(0.94) = 0.8866318$

7. **a.** $1.0225530 \approx y(1.25) = 1.0219569, 1.1646155 \approx y(1.93) = 1.1643901$
b. $1.9132167 \approx y(2.1) = 1.9249616, 4.3246152 \approx y(2.75) = 4.3941697$
c. $-1.1441775 \approx y(1.3) = -1.1382768, -1.0447403 \approx y(1.93) = -1.0412665$
d. $0.3251049 \approx y(0.54) = 0.3140018, 0.8945125 \approx y(0.94) = 0.8866318$

9. **a.** $1.0227863 \approx y(1.25) = 1.0219569, 1.1649247 \approx y(1.93) = 1.1643901$
b. $1.9153749 \approx y(2.1) = 1.9249616, 4.3312939 \approx y(2.75) = 4.3941697$
c. $-1.1432070 \approx y(1.3) = -1.1382768, -1.0443743 \approx y(1.93) = -1.0412665$
d. $0.3240839 \approx y(0.54) = 0.3140018, 0.8934152 \approx y(0.94) = 0.8866318$

11. **a.** The Runge-Kutta method of order 4 gives the results in the following tables.

i	t_i	w_i	$y(t_i)$
2	1.2	1.0149520	1.0149523
4	1.4	1.0475336	1.0475339
6	1.6	1.0884323	1.0884327
8	1.8	1.1336532	1.1336536
10	2.0	1.1812319	1.1812322

b.

i	t_i	w_i	$y(t_i)$
2	1.4	0.4896842	0.4896817
4	1.8	1.1994320	1.1994386
6	2.2	2.2134693	2.2135018
8	2.6	3.6783790	3.6784753
10	3.0	5.8738386	5.8741000

c.

i	t_i	w_i	$y(t_i)$
2	0.4	−1.6200576	−1.6200510
4	0.8	−1.3359824	−1.3359632
6	1.2	−1.1663735	−1.1663454
8	1.6	−1.0783582	−1.0783314
10	2.0	−1.0359922	−1.0359724

d.

i	t_i	w_i	$y(t_i)$
2	0.2	0.1627655	0.1626265
4	0.4	0.2052405	0.2051118
6	0.6	0.3766981	0.3765957
8	0.8	0.6461896	0.6461052
10	1.0	1.0023207	1.0022460

13. With $f(t, y) = -y + t + 1$ we have

$$\begin{aligned} w_i + hf\left(t_i + \frac{h}{2}, w_i + \frac{h}{2}f(t_i, w_i)\right) &= w_i + \frac{h}{2}[f(t_i, w_i) + f(t_{i+1}, w_i + hf(t_i, w_i))] \\ &= w_i + \frac{h}{4}\left[f(t_i, w_i) + 3f\left(t_i + \frac{2}{3}h, w_i + \frac{2}{3}hf(t_i, w_i)\right)\right] \\ &= w_i\left(1 - h + \frac{h^2}{2}\right) + t_i\left(h - \frac{h^2}{2}\right) + h. \end{aligned}$$

15. In 0.2 s we have approximately 2099 units of KOH.

EXERCISE SET 5.4 (PAGE 204)

1. The Adams-Bashforth methods give the results in the following tables.

a.

i	t_i	2-step	3-step	4-step	5-step	$y(t_i)$
1	0.2	0.0268128	0.0268128	0.0268128	0.0268128	0.0268128
2	0.4	0.1200522	0.1507778	0.1507778	0.1507778	0.1507778
3	0.6	0.4153551	0.4613866	0.4960196	0.4960196	0.4960196
4	0.8	1.1462844	1.2512447	1.2961260	1.3308570	1.3308570
5	1.0	2.8241683	3.0360680	3.1461400	3.1854002	3.2190993

b.

i	t_i	2-step	3-step	4-step	5-step	$y(t_i)$
1	2.2	1.3666667	1.3666667	1.3666667	1.3666667	1.3666667
2	2.4	1.6750000	1.6857143	1.6857143	1.6857143	1.6857143
3	2.6	1.9632431	1.9794407	1.9750000	1.9750000	1.9750000
4	2.8	2.2323184	2.2488759	2.2423065	2.2444444	2.2444444
5	3.0	2.4884512	2.5051340	2.4980306	2.5011406	2.5000000

c.

i	t_i	2-step	3-step	4-step	5-step	$y(t_i)$
1	1.2	2.6187859	2.6187859	2.6187859	2.6187859	2.6187859
2	1.4	3.2734823	3.2710611	3.2710611	3.2710611	3.2710611
3	1.6	3.9567107	3.9514231	3.9520058	3.9520058	3.9520058
4	1.8	4.6647738	4.6569191	4.6582078	4.6580160	4.6580160
5	2.0	5.3949416	5.3848058	5.3866452	5.3862177	5.3862944

d.

i	t_i	2-step	3-step	4-step	5-step	$y(t_i)$
1	0.2	1.2529306	1.2529306	1.2529306	1.2529306	1.2529306
2	0.4	1.5986417	1.5712255	1.5712255	1.5712255	1.5712255
3	0.6	1.9386951	1.8827238	1.8750869	1.8750869	1.8750869
4	0.8	2.1766821	2.0844122	2.0698063	2.0789180	2.0789180
5	1.0	2.2369407	2.1115540	2.0998117	2.1180642	2.1179795

3. The Adams-Bashforth methods give the results in the following tables.

a.

i	t_i	2-step	3-step	4-step	5-step	$y(t_i)$
2	1.2	1.0161982	1.0149520	1.0149520	1.0149520	1.0149523
4	1.4	1.0497665	1.0468730	1.0477278	1.0475336	1.0475339
6	1.6	1.0910204	1.0875837	1.0887567	1.0883045	1.0884327
8	1.8	1.1363845	1.1327465	1.1340093	1.1334967	1.1336536
10	2.0	1.1840272	1.1803057	1.1815967	1.1810689	1.1812322

b.

i	t_i	2-step	3-step	4-step	5-step	$y(t_i)$
2	1.4	0.4867550	0.4896842	0.4896842	0.4896842	0.4896817
4	1.8	1.1856931	1.1982110	1.1990422	1.1994320	1.1994386
6	2.2	2.1753785	2.2079987	2.2117448	2.2134792	2.2135018
8	2.6	3.5849181	3.6617484	3.6733266	3.6777236	3.6784753
10	3.0	5.6491203	5.8268008	5.8589944	5.8706101	5.8741000

c.

i	t_i	2-step	3-step	4-step	5-step	$y(t_i)$
5	0.5	−1.5357010	−1.5381988	−1.5379372	−1.5378676	−1.5378828
10	1.0	−1.2374093	−1.2389605	−1.2383734	−1.2383693	−1.2384058
15	1.5	−1.0952910	−1.0950952	−1.0947925	−1.0948481	−1.0948517
20	2.0	−1.0366643	−1.0359996	−1.0359497	−1.0359760	−1.0359724

d.

i	t_i	2-step	3-step	4-step	5-step	$y(t_i)$
2	0.2	0.1739041	0.1627655	0.1627655	0.1627655	0.1626265
4	0.4	0.2144877	0.2026399	0.2066057	0.2052405	0.2051118
6	0.6	0.3822803	0.3747011	0.3787680	0.3765206	0.3765957
8	0.8	0.6491272	0.6452640	0.6487176	0.6471458	0.6461052
10	1.0	1.0037415	1.0020894	1.0064121	1.0073348	1.0022460

5. The Adams Fourth-order Predictor-Corrector Algorithm gives the results in the following tables.

a.

i	t_i	w_i	$y(t_i)$
2	1.2	1.0149520	1.0149523
4	1.4	1.0475227	1.0475339
6	1.6	1.0884141	1.0884327
8	1.8	1.1336331	1.1336536
10	2.0	1.1812112	1.1812322

b.

i	t_i	w_i	$y(t_i)$
2	1.4	0.4896842	0.4896817
4	1.8	1.1994245	1.1994386
6	2.2	2.2134701	2.2135018
8	2.6	3.6784144	3.6784753
10	3.0	5.8739518	5.8741000

c.

i	t_i	w_i	$y(t_i)$
5	0.5	−1.5378788	−1.5378828
10	1.0	−1.2384134	−1.2384058
15	1.5	−1.0948609	−1.0948517
20	2.0	−1.0359757	−1.0359724

d.

i	t_i	w_i	$y(t_i)$
2	0.2	0.1627655	0.1626265
4	0.4	0.2048557	0.2051118
6	0.6	0.3762804	0.3765957
8	0.8	0.6458949	0.6461052
10	1.0	1.0021372	1.0022460

7. Milne-Simpson's Predictor-Corrector method gives the results in the following tables.

a.

i	t_i	w_i	$y(t_i)$
2	1.2	1.01495200	1.01495231
5	1.5	1.06725997	1.06726235
7	1.7	1.11065221	1.11065505
10	2.0	1.18122584	1.18123222

b.

i	t_i	w_i	$y(t_i)$
2	1.4	0.48968417	0.48968166
5	2.0	1.66126150	1.66128176
7	2.4	2.87648763	2.87655142
10	3.0	5.87375555	5.87409998

c.

i	t_i	w_i	$y(t_i)$
5	0.5	−1.53788255	−1.53788284
10	1.0	−1.23840789	−1.23840584
15	1.5	−1.09485532	−1.09485175
20	2.0	−1.03597247	−1.03597242

d.

i	t_i	w_i	$y(t_i)$
2	0.2	0.16276546	0.16262648
5	0.5	0.27741080	0.27736167
7	0.7	0.50008713	0.50006579
10	1.0	1.00215439	1.00224598

EXERCISE SET 5.5 (PAGE 210)

1. $y_{22} = 0.14846014$ approximates $y(0.1) = 0.14846010$.

3. The Extrapolation method gives the results in the following tables.

a.

i	t_i	w_i	h_i	k	y_i
1	1.05	1.10385729	0.05	2	1.10385738
2	1.10	1.21588614	0.05	2	1.21588635
3	1.15	1.33683891	0.05	2	1.33683925
4	1.20	1.46756907	0.05	2	1.46756957

b.

i	t_i	w_i	h_i	k	y_i
1	0.25	0.25228680	0.25	3	0.25228680
2	0.50	0.51588678	0.25	3	0.51588678
3	0.75	0.79594460	0.25	2	0.79594458
4	1.00	1.09181828	0.25	3	1.09181825

c.

i	t_i	w_i	h_i	k	y_i
1	1.50	−1.50000055	0.50	5	−1.50000000
2	2.00	−1.33333435	0.50	3	−1.33333333
3	2.50	−1.25000074	0.50	3	−1.25000000
4	3.00	−1.20000090	0.50	2	−1.20000000

d.

i	t_i	w_i	h_i	k	y_i
1	0.25	1.08708817	0.25	3	1.08708823
2	0.50	1.28980537	0.25	3	1.28980528
3	0.75	1.51349008	0.25	3	1.51348985
4	1.00	1.70187009	0.25	3	1.70187005

5. $P(5) \approx 56,751$.

EXERCISE SET 5.6 (PAGE 219)

1. **a.** $w_1 = 0.4787456 \approx y(t_1) = y(0.2966446) = 0.4787309$
b. $w_4 = 0.31055852 \approx y(t_4) = y(0.2) = 0.31055897$

3. The Runge-Kutta-Fehlberg method gives the results in the following tables.

a.

i	t_i	w_i	h_i	y_i
1	1.05	1.1038574	0.05	1.1038574
2	1.10	1.2158864	0.05	1.2158863
3	1.15	1.3368393	0.05	1.3368393
4	1.20	1.4675697	0.05	1.4675696

b.

i	t_i	w_i	h_i	y_i
1	0.25	0.2522868	0.25	0.2522868
2	0.50	0.5158867	0.25	0.5158868
3	0.75	0.7959445	0.25	0.7959446
4	1.00	1.0918182	0.25	1.0918183

c.

i	t_i	w_i	h_i	y_i
1	1.1382206	−1.7834313	0.1382206	−1.7834282
3	1.6364797	−1.4399709	0.3071709	−1.4399551
5	2.6364797	−1.2340532	0.5000000	−1.2340298
6	3.0000000	−1.2000195	0.3635203	−1.2000000

d.

i	t_i	w_i	h_i	y_i
1	0.2	1.0571819	0.2	1.0571810
2	0.4	1.2014801	0.2	1.2014860
3	0.6	1.3809214	0.2	1.3809312
4	0.8	1.5550243	0.2	1.5550314
5	1.0	1.7018705	0.2	1.7018701

5. The Adams Variable Step-Size Predictor-Corrector method gives the results in the following tables.

a.

i	t_i	w_i	h_i	y_i
1	1.05	1.10385717	0.05	1.10385738
2	1.10	1.21588587	0.05	1.21588635
3	1.15	1.33683848	0.05	1.33683925
4	1.20	1.46756885	0.05	1.46756957

b.

i	t_i	w_i	h_i	y_i
1	0.2	0.20120278	0.2	0.20120267
2	0.4	0.40861919	0.2	0.40861896
3	0.6	0.62585310	0.2	0.62585275
4	0.8	0.85397394	0.2	0.85396433
5	1.0	1.09183759	0.2	1.09181825

c.

i	t_i	w_i	h_i	y_i
5	1.16289739	−1.75426113	0.03257948	−1.75426455
10	1.32579477	−1.60547206	0.03257948	−1.60547731
15	1.57235777	−1.46625721	0.04931260	−1.46626230
20	1.92943707	−1.34978308	0.07694168	−1.34978805
25	2.47170180	−1.25358275	0.11633076	−1.25358804
30	3.00000000	−1.19999513	0.10299186	−1.20000000

d.

i	t_i	w_i	h_i	y_i
1	0.0625	1.00583097	0.06250	1.00583095
5	0.3125	1.13099427	0.06250	1.13098105
10	0.6250	1.40361751	0.06250	1.40360196
12	0.8125	1.56515769	0.09375	1.56514800
14	1.0000	1.70186884	0.09375	1.70187005

7. The current after 2 s is approximately $i(2) = 8.693$ amperes.

EXERCISE SET 5.7 (PAGE 228)

1. The Runge-Kutta for Systems method gives the results in the following tables.

a.

i	t_i	w_{1i}	u_{1i}	w_{2i}	u_{2i}
1	0.2	2.12036583	2.12500839	1.50699185	1.51158743
2	0.4	4.44122776	4.46511961	3.24224021	3.26598528
3	0.6	9.73913329	9.83235869	8.16341700	8.25629549
4	0.8	22.67655977	23.00263945	21.34352778	21.66887674
5	1.0	55.66118088	56.73748265	56.03050296	57.10536209

b.

i	t_i	w_{1i}	u_{1i}	w_{2i}	u_{2i}
1	0.5	0.95671390	0.95672798	−1.08381950	−1.08383310
2	1.0	1.30654440	1.30655930	−0.83295364	−0.83296776
3	1.5	1.34416716	1.34418117	−0.56980329	−0.56981634
4	2.0	1.14332436	1.14333672	−0.36936318	−0.36937457

c.

i	t_i	w_{1i}	u_{1i}	w_{2i}
1	0.5	0.70787076	0.70828683	−1.24988663
2	1.0	−0.33691753	−0.33650854	−3.01764179
3	1.5	−2.41332734	−2.41345688	−5.40523279
4	2.0	−5.89479008	−5.89590551	−8.70970537

i	t_i	u_{2i}	w_{3i}	u_{3i}
1	0.5	−1.25056425	0.39884862	0.39815702
2	1.0	−3.01945051	−0.29932294	−0.30116868
3	1.5	−5.40844686	−0.92346873	−0.92675778
4	2.0	−8.71450036	−1.32051165	−1.32544426

d.

i	t_i	w_{1i}	u_{1i}	w_{2i}
2	0.2	1.38165297	1.38165325	1.00800000
5	0.5	1.90753116	1.90753184	1.12500000
7	0.7	2.25503524	2.25503620	1.34300000
10	1.0	2.83211921	2.83212056	2.00000000

i	t_i	u_{2i}	w_{3i}	u_{3i}
2	0.2	1.00800000	−0.61833075	−0.61833075
5	0.5	0.12500000	−0.09090565	−0.09090566
7	0.7	1.34000000	0.26343971	0.26343970
10	1.0	2.00000000	0.88212058	0.88212056

3. First use the Runge-Kutta method of order 4 for systems to compute all starting values:

$$w_{1,0}, w_{2,0}, \ldots, w_{m,0},$$
$$w_{1,1}, w_{2,1}, \ldots, w_{m,1},$$
$$w_{1,2}, w_{2,2}, \ldots, w_{m,2},$$
$$w_{1,3}, w_{2,3}, \ldots, w_{m,3}.$$

Then for each $j = 3, 4, \ldots N-1$, compute the predictor values

$$w_{i,j+1}^{(0)} = w_{i,j} + \frac{h}{24}[55 f_i(t_i, w_{1,j}, \ldots, w_{m,j}) - 59 f_i(t_{j-1}, w_{1,j-1}, \ldots, w_{m,j-1})$$
$$+ 37 f_i(t_{j-2}, w_{1,j-2}, \ldots, w_{m,j-2}) - 9 f_i(t_{j-3}, w_{1,j-3}, \ldots, w_{m,j-3})],$$

and for each $i = 1, \ldots, m$, compute the corrector values

$$w_{i,j+1} = w_{i,j} + \frac{h}{24}[9f_i(t_{j+1}, w^{(0)}_{i,j+1}, \ldots, w^{(0)}_{m,j+1}) + 19f_i(t_i, w_{1,j}, \ldots, w_{m,j})$$
$$- 5f_i(t_{j-1}, w_{1,j-1}, \ldots, w_{m,j-1}) + f_i(t_{j-2}, w_{1,j-2}, \ldots, w_{m,j-2})].$$

5. The predicted number of prey, x_{1i}, and predators, x_{2i}, are given in the following table.

i	t_i	x_{1i}	x_{2i}
10	1.0	4393	1512
20	2.0	288	3175
30	3.0	32	2042
40	4.0	25	1258

A stable solution is $x_1 = 833.\overline{3}$ and $x_2 = 1500$.

EXERCISE SET 5.8 (PAGE 234)

1. Euler's method gives the results in the following tables.

a.

i	t_i	w_i	$y(t_i)$
2	0.2	0.027182818	0.4493290
5	0.5	0.000027183	0.0301974
7	0.7	0.000000272	0.0049916
10	1.0	0.000000000	0.0003355

b.

i	t_i	w_i	$y(t_i)$
2	0.2	0.373333333	0.0461052
5	0.5	−0.933333333	0.2500151
7	0.7	0.146666667	0.4900003
10	1.0	1.333333333	1.0000000

c.

i	t_i	w_i	$y(t_i)$
2	0.5	16.47925	0.4794709
4	1.0	256.7930	0.8414710
6	1.5	4096.142	0.9974950
8	2.0	65,523.12	0.9092974

d.

i	t_i	w_i	$y(t_i)$
2	0.2	6.128259	1.000000001
5	0.5	−378.2574	1.000000000
7	0.7	−6052.063	1.000000000
10	1.0	387,332.0	1.000000000

3. The Adams Fourth-Order Predictor-Corrector method gives the results in the following tables.

a.

i	t_i	w_i	$y(t_i)$
2	0.2	0.4588119	0.4493290
5	0.5	−0.0112813	0.0301974
7	0.7	0.0013734	0.0049916
10	1.0	0.0023604	0.0003355

b.

i	t_i	w_i	$y(t_i)$
2	0.2	0.0792593	0.0461052
5	0.5	0.1554027	0.2500151
7	0.7	0.5507445	0.4900003
10	1.0	0.7278557	1.0000000

c.

i	t_i	w_i	$y(t_i)$
2	0.5	188.3082	0.4794709
4	1.0	38,932.03	0.8414710
6	1.5	9,073,607	0.9974950
8	2.0	2,115,741,299	0.9092974

d.

i	t_i	w_i	$y(t_i)$
2	0.2	−215.7459	1.000000000
5	0.5	−682,637.0	1.000000000
7	0.7	−159,172,736	1.000000000
10	1.0	−566,751,172,258	1.000000000

5. The following tables list the results of the Backward Euler method applied to the problems in Exercise 1.

a.

i	t_i	w_i	k	$y(t_i)$
2	0.2	0.75298666	2	0.44932896
5	0.5	0.10978082	2	0.03019738
7	0.7	0.03041020	2	0.00499159
10	1.0	0.00443362	2	0.00033546

b.

i	t_i	w_i	k	$y(t_i)$
2	0.2	0.08148148	2	0.04610521
5	0.5	0.25635117	2	0.25001513
7	0.7	0.49515013	2	0.49000028
10	1.0	1.00500556	2	1.00000000

c.

i	t_i	w_i	k	$y(t_i)$
2	0.5	0.50495522	2	0.47947094
4	1.0	0.83751817	2	0.84147099
6	1.5	0.99145076	2	0.99749499
8	2.0	0.90337560	2	0.90929743

d.

i	t_i	w_i	k	$y(t_i)$
2	0.2	1.00348713	3	1.00000001
5	0.5	1.00000262	2	1.00000000
7	0.7	1.00000002	1	1.00000000
10	1.0	1.00000000	1	1.00000000

CHAPTER 6

EXERCISE SET 6.2 (PAGE 245)

1. **a.** Intersecting lines with solution $x_1 = x_2 = 1$.
b. Intersecting lines with solution $x_1 = x_2 = 0$.
c. One line, so there are an infinite number of solutions with $x_2 = \frac{3}{2} - \frac{1}{2}x_1$.
d. Parallel lines, so there is no solution.
e. One line, so there are an infinite number of solutions with $x_2 = -\frac{1}{2}x_1$.
f. Three lines in the plane that do not intersect at a common point.
g. Intersecting lines with solution $x_1 = \frac{2}{7}$ and $x_2 = -\frac{11}{7}$.
h. Two planes in space that intersect in a line with $x_1 = -\frac{5}{4}x_2$ and $x_3 = \frac{3}{2}x_2 + 1$.

3. Gaussian elimination gives the following solutions.
a. $x_1 = 1.1875$, $x_2 = 1.8125$, $x_3 = 0.875$ with one row interchange required
b. $x_1 = -1$, $x_2 = 0$, $x_3 = 1$ with no interchange required
c. $x_1 = 1.5$, $x_2 = 2$, $x_3 = -1.2$, $x_4 = 3$ with no interchange required
d. $x_1 = \frac{22}{9}$, $x_2 = -\frac{4}{9}$, $x_3 = \frac{4}{3}$, $x_4 = 1$ with one row interchange required
e. No unique solution
f. $x_1 = -1$, $x_2 = 2$, $x_3 = 0$, $x_4 = 1$ with one row interchange required

5. **a.** When $\alpha = -\frac{1}{3}$, there is no solution.
b. When $\alpha = \frac{1}{3}$, there are an infinite number of solutions with $x_1 = x_2 + 1.5$, and x_2 is arbitrary.
c. If $\alpha \neq \pm\frac{1}{3}$, then the unique solution is

$$x_1 = \frac{3}{2(1 + 3\alpha)} \quad \text{and} \quad x_2 = \frac{-3}{2(1 + 3\alpha)}.$$

7. **a.** There is sufficient food to satisfy the average daily consumption.
b. We could add 200 of species 1, or 150 of species 2, or 100 of species 3, or 100 of species 4.
c. Assuming none of the increases indicated in part (b) was selected, species 2 could be increased by 650, or species 3 could be increased by 150, or species 4 could be increased by 150.
d. Assuming none of the increases indicated in parts (b) or (c) were selected, species 3 could be increased by 150, or species 4 could be increased by 150.

EXERCISE SET 6.3 (PAGE 255)

1. **a.** None **b.** Interchange rows 2 and 3.
c. None **d.** Interchange rows 1 and 2.

3. **a.** Interchange rows 1 and 3, then interchange rows 2 and 3.
b. Interchange rows 2 and 3.
c. Interchange rows 2 and 3.
d. Interchange rows 1 and 3, then interchange rows 2 and 3.

5. Gaussian elimination with three-digit chopping arithmetic gives the following results.
a. $x_1 = 30.0, x_2 = 0.990$
b. $x_1 = 1.00, x_2 = 9.98$
c. $x_1 = 0.00, x_2 = 10.0, x_3 = 0.142$
d. $x_1 = 12.0, x_2 = 0.492, x_3 = -9.78$
e. $x_1 = 0.206, x_2 = 0.0154, x_3 = -0.0156, x_4 = -0.716$
f. $x_1 = 0.828, x_2 = -3.32, x_3 = 0.153, x_4 = 4.91$

7. Gaussian elimination with partial pivoting and three-digit chopping arithmetic gives the following results.
a. $x_1 = 10.0, x_2 = 1.00$
b. $x_1 = 1.00, x_2 = 9.98$
c. $x_1 = -0.163, x_2 = 9.98, x_3 = 0.142$
d. $x_1 = 12.0, x_2 = 0.504, x_3 = -9.78$
e. $x_1 = 0.177, x_2 = -0.0072, x_3 = -0.0208, x_4 = -1.18$
f. $x_1 = 0.777, x_2 = -3.10, x_3 = 0.161, x_4 = 4.50$

9. **a.** $\alpha = 6$

EXERCISE SET 6.4 (PAGE 264)

1. **a.** $\begin{bmatrix} 1 & 0 & 0 \\ 1 & 2 & 0 \\ 9 & 5 & 1 \end{bmatrix}$ **b.** $\begin{bmatrix} 1 & -1 & 2 \\ 2 & -1 & 7 \\ -2 & 1 & -5 \end{bmatrix}$

c. $\begin{bmatrix} 1 & 0 & 0 \\ 2 & 1 & 0 \\ -7 & -2 & 1 \end{bmatrix}$ **d.** $\begin{bmatrix} 6 & -7 & 15 \\ 0 & -1 & 3 \\ 0 & 0 & 6 \end{bmatrix}$

3. **a.** Singular, $\det A = 0$ **b.** $\det A = -8$, $\det A^{-1} = -0.125$
c. Singular, $\det A = 0$ **d.** Singular, $\det A = 0$
e. $\det A = 28$, $\det A^{-1} = \frac{1}{28}$ **f.** $\det A = 3$, $\det A^{-1} = \frac{1}{3}$

5. **a.** Not true. For example, let

$$A = \begin{bmatrix} 2 & 1 \\ 1 & 0 \end{bmatrix} \quad \text{and} \quad B = \begin{bmatrix} 1 & -1 \\ -1 & 2 \end{bmatrix}.$$

Then

$$AB = \begin{bmatrix} 1 & 0 \\ 1 & -1 \end{bmatrix}$$

is not symmetric.

b. True. Let A be a nonsingular symmetric matrix. From the properties of transposes and inverses we have $(A^{-1})^t = (A^t)^{-1}$. Thus $(A^{-1})^t = (A^t)^{-1} = A^{-1}$, and A^{-1} is symmetric.

c. Not true. Use the matrices A and B from part (a).

7. **a.** The solution is $x_1 = 0$, $x_2 = 10$, and $x_3 = 26$.
   b. We have $D_1 = -1, D_2 = 3, D_3 = 7$, and $D = 0$, and there are no solutions.
   c. We have $D_1 = D_2 = D_3 = D = 0$, and there are infinitely many solutions.

9. **a.** For each $k = 1, 2, \dots, m$, the number a_{ik} represents the total number of plants of type v_i eaten by herbivores in the species h_k. The number of herbivores of types h_k eaten by species c_j is b_{kj}. Thus, the total number of plants of type v_i ending up in species c_j is $a_{i1}b_{1j} + a_{i2}b_{2j} + \cdots + a_{im}b_{mj} = (AB)_{ij}$.
   b. We first assume $n = m = k$ so that the matrices will have inverses. Let $x_1, \dots, x_n$ represent the vegetations of type $v_1, \dots, v_n$, let $y_1, \dots, y_n$ represent the number of herbivores of species $h_1, \dots, h_n$, and let $z_1, \dots, z_n$ represent the number of carnivores of species $c_1, \dots, c_n$.
   If

$$\begin{bmatrix} x_1 \\ x_2 \\ \vdots \\ x_n \end{bmatrix} = A \begin{bmatrix} y_1 \\ y_2 \\ \vdots \\ y_n \end{bmatrix}, \quad \text{then} \quad \begin{bmatrix} y_1 \\ y_2 \\ \vdots \\ y_n \end{bmatrix} = A^{-1} \begin{bmatrix} x_1 \\ x_2 \\ \vdots \\ x_n \end{bmatrix}.$$

Thus, $(A^{-1})_{i,j}$ represents the amount of type v_j plants eaten by a herbivore of species h_i. Similarly, if

$$\begin{bmatrix} y_1 \\ y_2 \\ \vdots \\ y_n \end{bmatrix} = B \begin{bmatrix} z_1 \\ z_2 \\ \vdots \\ z_n \end{bmatrix}, \quad \text{then} \quad \begin{bmatrix} z_1 \\ z_2 \\ \vdots \\ z_n \end{bmatrix} = B^{-1} \begin{bmatrix} y_1 \\ y_2 \\ \vdots \\ y_n \end{bmatrix}.$$

Thus $(B^{-1})_{i,j}$ represents the number of herbivores of species h_j eaten by a carnivore of species c_i. If $x = Ay$ and $y = Bz$, then $x = ABz$ and $z = (AB)^{-1}x$. But, $y = A^{-1}x$ and $z = B^{-1}y$, so $z = B^{-1}A^{-1}x$.

11. **a.** In component form:

$$(a_{11}x_1 - b_{11}y_1 + a_{12}x_2 - b_{12}y_2) + (b_{11}x_1 + a_{11}y_1 + b_{12}x_2 + a_{12}y_2)i = c_1 + id_1$$
$$(a_{21}x_1 - b_{21}y_1 + a_{22}x_2 - b_{22}y_2) + (b_{21}x_1 + a_{21}y_1 + b_{22}x_2 + a_{22}y_2)i = c_2 + id_2,$$

so

$$\begin{aligned} a_{11}x_1 + a_{12}x_2 - b_{11}y_1 - b_{12}y_2 &= c_1 \\ b_{11}x_1 + b_{12}x_2 + a_{11}y_1 + a_{12}y_2 &= d_1 \\ a_{21}x_1 + a_{22}x_2 - b_{21}y_1 - b_{22}y_2 &= c_2 \\ b_{21}x_1 + b_{22}x_2 + a_{21}y_1 + a_{22}y_2 &= d_2 \end{aligned}$$

   b. The system

$$\begin{bmatrix} 1 & 3 & 2 & -2 \\ -2 & 2 & 1 & 3 \\ 2 & 4 & -1 & -3 \\ 1 & 3 & 2 & 4 \end{bmatrix} \begin{bmatrix} x_1 \\ x_2 \\ y_1 \\ y_2 \end{bmatrix} = \begin{bmatrix} 5 \\ 2 \\ 4 \\ -1 \end{bmatrix}$$

has the solution $x_1 = -1.2$, $x_2 = 1$, $y_1 = 0.6$, and $y_2 = -1$.

EXERCISE SET 6.5 (PAGE 273)

1. **a.** $x_1 = -3, x_2 = 3, x_3 = 1$
   b. $x_1 = \frac{1}{2}, x_2 = \frac{-9}{2}, x_3 = \frac{7}{2}$

3. **a.** $P^tLU = \begin{bmatrix} 0 & 1 & 0 \\ 1 & 0 & 0 \\ 0 & 0 & 1 \end{bmatrix} \begin{bmatrix} 1 & 0 & 0 \\ 0 & 1 & 0 \\ 0 & -\frac{1}{2} & 1 \end{bmatrix} \begin{bmatrix} 1 & 1 & -1 \\ 0 & 2 & 3 \\ 0 & 0 & \frac{5}{2} \end{bmatrix}$

b. $P^tLU = \begin{bmatrix} 1 & 0 & 0 \\ 0 & 0 & 1 \\ 0 & 1 & 0 \end{bmatrix} \begin{bmatrix} 1 & 0 & 0 \\ 2 & 1 & 0 \\ 1 & 0 & 1 \end{bmatrix} \begin{bmatrix} 1 & 2 & -1 \\ 0 & -5 & 6 \\ 0 & 0 & 4 \end{bmatrix}$

c. $P^tLU = \begin{bmatrix} 1 & 0 & 0 & 0 \\ 0 & 0 & 0 & 1 \\ 0 & 1 & 0 & 0 \\ 0 & 0 & 1 & 0 \end{bmatrix} \begin{bmatrix} 1 & 0 & 0 & 0 \\ 2 & 1 & 0 & 0 \\ 1 & 0 & 1 & 0 \\ 3 & 0 & 0 & 1 \end{bmatrix} \begin{bmatrix} 1 & -2 & 3 & 0 \\ 0 & 5 & -2 & 1 \\ 0 & 0 & -1 & -2 \\ 0 & 0 & 0 & 3 \end{bmatrix}$

d. $P^tLU = \begin{bmatrix} 1 & 0 & 0 & 0 \\ 0 & 0 & 0 & 1 \\ 0 & 0 & 1 & 0 \\ 0 & 1 & 0 & 0 \end{bmatrix} \begin{bmatrix} 1 & 0 & 0 & 0 \\ 2 & 1 & 0 & 0 \\ 1 & 0 & 1 & 0 \\ 1 & 0 & 0 & 1 \end{bmatrix} \begin{bmatrix} 1 & -2 & 3 & 0 \\ 0 & 5 & -3 & -1 \\ 0 & 0 & -1 & -2 \\ 0 & 0 & 0 & 1 \end{bmatrix}$

EXERCISE SET 6.6 (PAGE 282)

1. **(i)** The symmetric matrices are in (a), (b), and (f).
(ii) The singular matrices are in (e) and (h).
(iii) The strictly diagonally dominant matrices are in (a), (b), (c), and (d).
(iv) The positive definite matrices are in (a) and (f).

3. Choleski factorization gives the following results.

a. $L = \begin{bmatrix} 1.414213 & 0 & 0 \\ -0.7071069 & 1.224743 & 0 \\ 0 & -0.8164972 & 1.154699 \end{bmatrix}$

b. $L = \begin{bmatrix} 2 & 0 & 0 & 0 \\ 0.5 & 1.658311 & 0 & 0 \\ 0.5 & -0.7537785 & 1.087113 & 0 \\ 0.5 & 0.4522671 & 0.08362442 & 1.240346 \end{bmatrix}$

c. $L = \begin{bmatrix} 2 & 0 & 0 & 0 \\ 0.5 & 1.658311 & 0 & 0 \\ -0.5 & -0.4522671 & 2.132006 & 0 \\ 0 & 0 & 0.9380833 & 1.766351 \end{bmatrix}$

d. $L = \begin{bmatrix} 2.449489 & 0 & 0 & 0 \\ 0.8164966 & 1.825741 & 0 & 0 \\ 0.4082483 & 0.3651483 & 1.923538 & 0 \\ -0.4082483 & 0.1825741 & -0.4678876 & 1.606574 \end{bmatrix}$

5. Crout factorization gives the following results.

a. $x_1 = 0.5, x_2 = 0.5, x_3 = 1$
b. $x_1 = -0.9999995, x_2 = 1.999999, x_3 = 1$
c. $x_1 = 1, x_2 = -1, x_3 = 0$
d. $x_1 = -0.09357798, x_2 = 1.587156, x_3 = -1.167431, x_4 = 0.5412844$

7. **a.** No, consider $\begin{bmatrix} 1 & 0 \\ 0 & 1 \end{bmatrix}$.
b. Yes, since $A = A^t$.
c. Yes, since $\mathbf{x}^t(A + B)\mathbf{x} = \mathbf{x}^tA\mathbf{x} + \mathbf{x}^tB\mathbf{x}$.

d. Yes, since $\mathbf{x}^t A^2 \mathbf{x} = \mathbf{x}^t A^t A \mathbf{x} = (A\mathbf{x})^t (A\mathbf{x}) \geq 0$ and because A is nonsingular, equality holds only if $\mathbf{x} = \mathbf{0}$.

e. No, consider $A = \begin{bmatrix} 1 & 0 \\ 0 & 1 \end{bmatrix}$ and $B = \begin{bmatrix} 2 & 0 \\ 0 & 2 \end{bmatrix}$.

9. a. Since $\det A = 3\alpha - 2\beta$, A is singular if and only if $\alpha = 2\beta/3$.
b. $|\alpha| > 1, |\beta| < 1$
c. $\beta = 1$
d. $\alpha > \frac{2}{3}, \beta = 1$

11. a. Mating male i with female j produces offspring with the same wing characteristics as mating male j with female i.
b. No. Consider, for example, $\mathbf{x} = (1, 0, -1)^t$.

CHAPTER 7

EXERCISE SET 7.2 (PAGE 294)

1. a. We have $\|\mathbf{x}\|_\infty = 4$ and $\|\mathbf{x}\|_2 = 5.220153$.
b. We have $\|\mathbf{x}\|_\infty = 4$ and $\|\mathbf{x}\|_2 = 5.477226$.
c. We have $\|\mathbf{x}\|_\infty = 2^k$ and $\|\mathbf{x}\|_2 = (1 + 4^k)^{1/2}$.
d. We have $\|\mathbf{x}\|_\infty = 4/(k+1)$ and $\|\mathbf{x}\|_2 = (16/(k+1)^2 + 4/k^4 + k^4 e^{-2k})^{1/2}$.

3. a. We have $\lim_{k\to\infty} \mathbf{x}^{(k)} = (0, 0, 0)^t$.
b. We have $\lim_{k\to\infty} \mathbf{x}^{(k)} = (0, 1, 3)^t$.
c. We have $\lim_{k\to\infty} \mathbf{x}^{(k)} = (0, 0, \frac{1}{2})^t$.
d. We have $\lim_{k\to\infty} \mathbf{x}^{(k)} = (1, -1, 1)^t$.

5. a. We have $\|\mathbf{x} - \hat{\mathbf{x}}\|_\infty = 8.57 \times 10^{-4}$ and $\|A\hat{\mathbf{x}} - \mathbf{b}\|_\infty = 2.06 \times 10^{-4}$.
b. We have $\|\mathbf{x} - \hat{\mathbf{x}}\|_\infty = 0.90$ and $\|A\hat{\mathbf{x}} - \mathbf{b}\|_\infty = 0.27$.
c. We have $\|\mathbf{x} - \hat{\mathbf{x}}\|_\infty = 0.5$ and $\|A\hat{\mathbf{x}} - \mathbf{b}\|_\infty = 0.3$.
d. We have $\|\mathbf{x} - \hat{\mathbf{x}}\|_\infty = 6.55 \times 10^{-2}$, and $\|A\hat{\mathbf{x}} - \mathbf{b}\|_\infty = 0.32$.

7. Let $A = \begin{bmatrix} 1 & 1 \\ 0 & 1 \end{bmatrix}$ and $B = \begin{bmatrix} 1 & 0 \\ 1 & 1 \end{bmatrix}$. Then $\|AB\|_{\otimes} = 2$, but $\|A\|_{\otimes} \cdot \|B\|_{\otimes} = 1$.

9. It is not difficult to show that (*i*) holds. If $\|A\| = 0$, then $\|A\mathbf{x}\| = 0$ for all vectors $\mathbf{x}$ with $\|\mathbf{x}\| = 1$. Using $\mathbf{x} = (1, 0, \ldots, 0)^t$, $\mathbf{x} = (0, 1, 0, \ldots, 0)^t, \ldots,$ and $\mathbf{x} = (0, \ldots, 0, 1)^t$ successively implies that each column of A is zero. Thus, $\|A\| = 0$ if and only if $A = 0$. Moreover,

$$\|\alpha A\| = \max_{\|\mathbf{x}\|=1} \|(\alpha A\mathbf{x})\| = |\alpha| \max_{\|\mathbf{x}\|=1} \|A\mathbf{x}\| = |\alpha| \cdot \|A\|,$$
$$\|A + B\| = \max_{\|\mathbf{x}\|=1} \|(A + B)\mathbf{x}\| \leq \max_{\|\mathbf{x}\|=1} (\|A\mathbf{x}\| + \|B\mathbf{x}\|),$$

so

$$\|A + B\| \leq \max_{\|\mathbf{x}\|=1} \|A\mathbf{x}\| + \max_{\|\mathbf{x}\|=1} \|B\mathbf{x}\| = \|A\| + \|B\|$$

and

$$\|AB\| = \max_{\|\mathbf{x}\|=1} \|(AB)\mathbf{x}\| = \max_{\|\mathbf{x}\|=1} \|A(B\mathbf{x})\|,$$

so

$$\|AB\| \leq \max_{\|\mathbf{x}\|=1} \|A\|\|B\mathbf{x}\| = \|A\| \max_{\|\mathbf{x}\|=1} \|B\mathbf{x}\| = \|A\|\|B\|.$$

EXERCISE SET 7.3 (PAGE 302)

1. a. The eigenvalue $\lambda_1 = 3$ has the eigenvector $\mathbf{x}_1 = (1, -1)^t$, and the eigenvalue $\lambda_2 = 1$ has the eigenvector $\mathbf{x}_2 = (1, 1)^t$.
   b. The eigenvalue $\lambda_1 = (1 + \sqrt{5})/2$ has the eigenvector $\mathbf{x}_1 = (1, (1 + \sqrt{5})/2)^t$, and the eigenvalue $\lambda_2 = (1 - \sqrt{5})/2$ has the eigenvector $\mathbf{x}_2 = (1, (1 - \sqrt{5})/2)^t$.
   c. The eigenvalue $\lambda_1 = \frac{1}{2}$ has the eigenvector $\mathbf{x}_1 = (1, 1)^t$ and the eigenvalue $\lambda_2 = -\frac{1}{2}$ has the eigenvector $\mathbf{x}_2 = (1, -1)^t$.
   d. The eigenvalue $\lambda_1 = 0$ has the eigenvector $\mathbf{x}_1 = (1, -1)^t$ and the eigenvalue $\lambda_2 = -1$ has the eigenvector $\mathbf{x}_2 = (1, -2)^t$.
   e. The eigenvalue $\lambda_1 = \lambda_2 = 3$ has the eigenvectors $\mathbf{x}_1 = (0, 0, 1)^t$ and $\mathbf{x}_2 = (1, 1, 0)^t$, and the eigenvalue $\lambda_3 = 1$ has the eigenvector $\mathbf{x}_3 = (1, -1, 0)^t$.
   f. The eigenvalue $\lambda_1 = 7$ has the eigenvector $\mathbf{x}_1 = (1, 4, 4)^t$, the eigenvalue $\lambda_2 = 3$ has the eigenvector $\mathbf{x}_2 = (1, 2, 0)^t$, and the eigenvalue $\lambda_3 = -1$ has the eigenvector $\mathbf{x}_3 = (1, 0, 0)^t$.
   g. The eigenvalue $\lambda_1 = \lambda_2 = 1$ has the eigenvectors $\mathbf{x}_1 = (-1, 1, 0)^t$ and $\mathbf{x}_2 = (-1, 0, 1)^t$, and the eigenvalue $\lambda_3 = 5$ has the eigenvector $\mathbf{x}_3 = (1, 2, 1)^t$.
   h. The eigenvalue $\lambda_1 = 3$ has the eigenvector $\mathbf{x}_1 = (-1, 1, 2)^t$, the eigenvalue $\lambda_2 = 4$ has the eigenvector $\mathbf{x}_2 = (0, 1, 2)^t$, and the eigenvalue $\lambda_3 = -2$ has the eigenvector $\mathbf{x}_3 = (-3, 8, 1)^t$.

3. Since

$$A_1^k = \begin{bmatrix} 1 & 0 \\ \frac{2^k-1}{2^{k+1}} & 2^{-k} \end{bmatrix}, \quad \text{we have} \quad \lim_{k\to\infty} A_1^k = \begin{bmatrix} 1 & 0 \\ \frac{1}{2} & 0 \end{bmatrix}.$$

   Also

$$A_2^k = \begin{bmatrix} 2^{-k} & 0 \\ \frac{16k}{2^{k-1}} & 2^{-k} \end{bmatrix}, \quad \text{so} \quad \lim_{k\to\infty} A_2^k = \begin{bmatrix} 0 & 0 \\ 0 & 0 \end{bmatrix}.$$

5. a. 3
   b. 1.618034
   c. 0.5
   d. 3.162278
   e. 3
   f. 8.224257
   g. 5.203527
   h. 5.601152

7. Let $A = \begin{bmatrix} 1 & 1 \\ 0 & 1 \end{bmatrix}$ and $B = \begin{bmatrix} 1 & 0 \\ 1 & 1 \end{bmatrix}$. Then $\rho(A) = \rho(B) = 1$ and $\rho(A + B) = 3$.

9. a. We have the real eigenvalue $\lambda = 1$ with the eigenvector $\mathbf{x} = (6, 3, 1)^t$.
   b. Choose any multiple of the vector $(6, 3, 1)^t$.

EXERCISE SET 7.4 (PAGE 308)

1. Two iterations of Jacobi's method give the following results.
   a. $(0.1428571, -0.3571429, 0.4285714)^t$
   b. $(0.97, 0.91, 0.74)^t$
   c. $(-0.65, 1.65, -0.4, -2.475)^t$
   d. $(-0.5208333, -0.04166667, -0.2166667, 0.4166667)^t$
   e. $(1.325, -1.6, 1.6, 1.675, 2.425)^t$
   f. $(0.6875, 1.125, 0.6875, 1.375, 0.5625, 1.375)^t$

3. Jacobi's method gives the following results.
   a. $\mathbf{x}^{(9)} = (0.03510079, -0.23663751, 0.65812732)^t$
   b. $\mathbf{x}^{(6)} = (0.99572500, 0.95777500, 0.79145000)^t$
   c. $\mathbf{x}^{(21)} = (-0.79710581, 2.79517067, -0.25939578, -2.25179299)^t$
   d. $\mathbf{x}^{(12)} = (-0.75205599, 0.04027028, -0.28025957, 0.69008536)^t$

e. $\mathbf{x}^{(12)} = (0.78708833, -1.00303576, 1.86604817, 1.91244923, 1.9857067)^t$
f. $\mathbf{x}^{(16)} = (0.99973534, 1.99925144, 0.99973534, 1.99947069, 0.99962572, 1.99947069)^t$

5. **a.** A is not strictly diagonally dominant.

b. $T_j = \begin{bmatrix} 0 & 0 & 1 \\ 0.5 & 0 & 0.25 \\ -1 & 0.5 & 0 \end{bmatrix}$ and $\rho(T_j) = 0.97210521$.

Since T_j is convergent, the Jacobi method will converge.

c. With $\mathbf{x}^{(0)} = (0, 0, 0)^t$, $\mathbf{x}^{(187)} = (0.90222655, -0.79595242, 0.69281316)^t$.

d. $\rho(T_j) = 1.39331779371$. Since T_j is not convergent, the Jacobi method will not converge.

7. $T_j = (t_{ik})$ has entries given by

$$t_{ik} = \begin{cases} 0, & i = k \text{ for } 1 \le i \le n \text{ and } 1 \le k \le n \\ -\dfrac{a_{ik}}{a_{ii}}, & i \ne k \text{ for } 1 \le i \le n \text{ and } 1 \le k \le n. \end{cases}$$

Thus,

$$\|T_j\|_\infty = \max_{1 \le i \le n} \sum_{\substack{k=1 \\ k \ne i}}^{n} \left| \frac{a_{ik}}{a_{ii}} \right| < 1,$$

since A is strictly diagonally dominant.

EXERCISE SET 7.5 (PAGE 312)

1. Two iterations of the SOR method give the following results.

a. $(0.05410079, -0.2115435, 0.6477159)^t$
b. $(0.9876790, 0.9784935, 0.7899328)^t$
c. $(-0.71885, 2.818822, -0.2809726, -2.235422)^t$
d. $(-0.6604902, 0.03700749, -0.2493513, 0.6561139)^t$
e. $(1.079675, -1.260654, 2.042489, 1.995373, 2.049536)^t$
f. $(0.8318750, 1.647766, 0.9189856, 1.791281, 0.8712129, 1.959155)^t$

3. **a.** The tridiagonal matrices are in parts (b) and (c).
b. For $\omega = 1.012823$ we have $\mathbf{x}^{(4)} = (0.9957846, 0.9578935, 0.7915788)^t$.
c. For $\omega = 1.153499$ we have $\mathbf{x}^{(7)} = (-0.7977651, 2.795343, -0.2588021, -2.251760)^t$.

5. **a.** The system was reordered so that the diagonal of the matrix had nonzero entries.

b. **(i)** The solution vector is $(-6.27212290601165 \times 10^{-3}, -2.36602456112022 \times 10^4, -1.36602492324141 \times 10^4, -3.34606444633457 \times 10^4, 2.36602456112022 \times 10^4, 1.00000000000000 \times 10^4, -2.73205026462435 \times 10^4, 2.36602492324141 \times 10^4)^t$, using 29 iterations with tolerance 1.00×10^{-2}.

(ii) The solution vector is $(-9.89308239877573 \times 10^{-3}, -2.36602492321617 \times 10^4, -1.36602492324141 \times 10^4, -3.34606444633457 \times 10^4, 2.36602456107651 \times 10^4, 1.00000000000000 \times 10^4, -2.73205026459521 \times 10^4, 2.36602456112022 \times 10^4)^t$, using 57 iterations with tolerance 1.00×10^{-2}.

(iii) The solution vector is $(-2.16147 \times 10^{-3}, -2.366025403900 \times 10^4, -1.366025404100 \times 10^4, -3.346065215000 \times 10^4, 2.366025411100 \times 10^4, 1.000000000000 \times 10^4, -2.732050807600 \times 10^4, 2.366025403600 \times 10^4)^t$, using 19 iterations with tolerance 1.00×10^{-2} and parameter 1.25.

EXERCISE SET 7.6 (PAGE 318)

1. The $\|\cdot\|_\infty$ condition number is as follows.

a. 50 **b.** 241.37 **c.** 60,002
d. 339,866 **e.** 12 **f.** 198.17

3. The matrix is ill-conditioned since $K_\infty = 60002$. For the new system we have $\tilde{\mathbf{x}} = (-1.0000, 2.0000)^t$.

5. **a.** (i) $(-10.0, 1.01)^t$, (ii) $(10.0, 1.00)^t$
b. (i) $(12.0, 0.499, -1.98)^t$, (ii) $(1.00, 0.500, -1.00)^t$
c. (i) $(0.185, 0.0103, -0.0200, -1.12)^t$, (ii) $(0.177, 0.0127, -0.0207, -1.18)^t$
d. (i) $(0.799, -3.12, 0.151, 4.56)^t$, (ii) $(0.758, -3.00, 0.159, 4.30)^t$

7. **a.** We have $\tilde{\mathbf{x}} = (188.9998, 92.99998, 45.00001, 27.00001, 21.00002)^t$.
b. The condition number is $K_\infty = 80$.
c. The exact solution is $\mathbf{x} = (189, 93, 45, 27, 21)^t$.

9. **a.** $$\hat{H}^{-1} = \begin{bmatrix} 8.968 & -35.77 & 29.77 \\ -35.77 & 190.6 & -178.6 \\ 29.77 & -178.6 & 178.6 \end{bmatrix}$$

b. $$\hat{H} = \begin{bmatrix} 0.9799 & 0.4870 & 0.3238 \\ 0.4860 & 0.3246 & 0.2434 \\ 0.3232 & 0.2433 & 0.1949 \end{bmatrix}$$

c. $\|H - \hat{H}\|_\infty = 0.04260$

CHAPTER 8

EXERCISE SET 8.2 (PAGE 329)

1. The linear least squares polynomial is $1.70784x + 0.89968$.

3. The least squares polynomials with their errors are:
$0.6208950 + 1.219621x$, with $E_2 = 2.719 \times 10^{-5}$;
$0.5965807 + 1.253293x - 0.01085343x^2$, with $E_2 = 1.801 \times 10^{-5}$;
$0.6290193 + 1.185010x + 0.03533252x^2 - 0.01004723x^3$, with $E_2 = 1.741 \times 10^{-5}$.

5. **a.** The linear least squares polynomial is $72.0845x - 194.138$, with $E_2 = 329$.
b. The least squares polynomial of degree 2 is $6.61821x^2 - 1.14352x + 1.23556$, with $E_2 = 1.44 \times 10^{-3}$.
c. The least squares polynomial of degree 3 is $-0.0136742x^3 + 6.84557x^2 - 2.37919x + 3.42904$, with $E_2 = 5.27 \times 10^{-4}$.

7. **a.** We have $k = 0.8996$, with $E_2(k) = 0.407$.
b. We have $k = 0.9069$, with $E_2(k) = 0.486$.
Part (b) best fits the total experimental data.

9. The predicted point average is 0.101(ACT score) +0.487.

EXERCISE SET 8.3 (PAGE 339)

1. The linear least squares approximations are as follows.
a. $P_1(x) = 1.833333 + 4x$
b. $P_1(x) = -1.600003 + 3.600003x$
c. $P_1(x) = 1.140981 - 0.2958375x$

d. $P_1(x) = 0.1945267 + 3.000001x$
e. $P_1(x) = 0.6109245 + 0.09167105x$
f. $P_1(x) = -1.861455 + 1.666667x$

3. The linear least squares approximations on $[-1, 1]$ are as follows.
a. $P_1(x) = 3.333333 - 2x$
b. $P_1(x) = 0.6000025x$
c. $P_1(x) = 0.5493063 - 0.2958375x$
d. $P_1(x) = 1.175201 + 1.103639x$
e. $P_1(x) = 0.4207355 + 0.4353975x$
f. $P_1(x) = 0.6479184 + 0.5281226x$

5. The errors for the approximations in Exercise 3 are as follows.
a. 0.177779 **b.** 0.0457206 **c.** 0.00484624
d. 0.0526541 **e.** 0.0153784 **f.** 0.00363453

7. The Gram-Schmidt process produces the following collections of polynomials;
a. $\phi_0(x) = 1$, $\phi_1(x) = x - 0.5$, $\phi_2(x) = x^2 - x + \frac{1}{6}$, and $\phi_3(x) = x^3 - 1.5x^2 + 0.6x - 0.05$
b. $\phi_0(x) = 1$, $\phi_1(x) = x - 1$, $\phi_2(x) = x^2 - 2x + \frac{2}{3}$, and $\phi_3(x) = x^3 - 3x^2 + \frac{12}{5}x - \frac{2}{5}$
c. $\phi_0(x) = 1$, $\phi_1(x) = x - 2$, $\phi_2(x) = x^2 - 4x + \frac{11}{3}$, and $\phi_3(x) = x^3 - 6x^2 + 11.4x - 6.8$

9. The least squares polynomials of degree 2 are as follows.
a. $P_2(x) = 3.833333\phi_0(x) + 4\phi_1(x) + 0.9999998\phi_2(x)$
b. $P_2(x) = 2\phi_0(x) + 3.6\phi_1(x) + 3\phi_2(x)$
c. $P_2(x) = 0.5493061\phi_0(x) - 0.2958369\phi_1(x) + 0.1588785\phi_2(x)$
d. $P_2(x) = 3.194528\phi_0(x) + 3\phi_1(x) + 1.458960\phi_2(x)$
e. $P_2(x) = 0.6567600\phi_0(x) + 0.09167105\phi_1(x) - 0.7375118\phi_2(x)$
f. $P_2(x) = 1.471878\phi_0(x) + 1.666667\phi_1(x) + 0.2597705\phi_2(x)$

11. **a.** $2L_0(x) + 4L_1(x) + L_2(x)$
b. $\frac{1}{2}L_0(x) - \frac{1}{4}L_1(x) + \frac{1}{16}L_2(x) - \frac{1}{96}L_3(x)$
c. $6L_0(x) + 18L_1(x) + 9L_2(x) + L_3(x)$
d. $\frac{1}{3}L_0(x) - \frac{2}{9}L_1(x) + \frac{2}{27}L_2(x) - \frac{4}{243}L_3(x)$

13. First let $\phi_i(x) = \sum_{k=0}^{n} b_{ik}x^k$ for each $i = 0, 1, \ldots, n$. Then $b_{ii} \neq 0$, since $\{\phi_0, \ldots, \phi_n\}$ is linearly independent. Now let $Q(x) = \sum_{k=0}^{n} a_k x^k$. Then $c_n = a_n/b_{nn}$ and $c_i = (a_i - \sum_{j=i+1}^{n} c_j b_{ji})/b_{ii}$ for each $i = n-1, n-2, \ldots, 0$.

EXERCISE SET 8.4 (PAGE 346)

1. The interpolating polynomials of degree 2 are as follows.
a. $P_2(x) = 2.377443 + 1.590534(x - 0.8660254) + 0.5320418(x - 0.8660254)x$
b. $P_2(x) = 0.7617600 + 0.8796047(x - 0.8660254)$
c. $P_2(x) = 1.052926 + 0.4154370(x - 0.8660254) - 0.1384262x(x - 0.8660254)$
d. $P_2(x) = 0.5625 + 0.649519(x - 0.8660254) + 0.75x(x - 0.8660254)$

3. The interpolating polynomials of degree 3 are as follows.
a.
$$P_3(x) = 2.519044 + 1.945377(x - 0.9238795) + 0.7047420(x - 0.9238795)(x - 0.3826834) + 0.1751757(x - 0.9238795)(x - 0.3826834)(x + 0.3826834)$$
b.
$$P_3(x) = 0.7979459 + 0.7844380(x - 0.9238795) - 0.1464394(x - 0.9238795)(x - 0.3826834) - 0.1585049(x - 0.9238795)(x - 0.3826834)(x + 0.3826834)$$

c. $P_3(x) = 1.072911 + 0.3782067(x - 0.9238795)$
$- 0.09799213(x - 0.9238795)(x - 0.3826834)$
$+ 0.04909073(x - 0.9238795)(x - 0.3826834)(x + 0.3826834)$

d. $P_3(x) = 0.7285533 + 1.306563(x - 0.9238795)$
$+ 0.9999999(x - 0.9238795)(x - 0.3826834)$

5. The zeros of $\tilde{T}_3$ produce the following interpolating polynomials of degree 2.

a. $P_2(x) = 0.3489153 - 0.1744576(x - 2.866025) + 0.1538462(x - 2.866025)(x - 2)$

b. $P_2(x) = 0.1547375 - 0.2461152(x - 1.866025) + 0.1957273(x - 1.866025)(x - 1)$

c. $P_2(x) = 0.6166200 - 0.2370869(x - 0.9330127) - 0.7427732(x - 0.9330127)(x - 0.5)$

d. $P_2(x) = 3.0177125 + 1.883800(x - 2.866025) + 0.2584625(x - 2.866025)(x - 2)$

7. If $i > j$, then

$$\tfrac{1}{2}(T_{i+j}(x) + T_{i-j}(x)) = \tfrac{1}{2}(\cos(i + j)\theta + \cos(i - j)\theta) = \cos i\theta \cos j\theta = T_i(x)T_j(x).$$

EXERCISE SET 8.5 (PAGE 351)

1. The Padé approximations of degree 2 for $f(x) = e^{2x}$ are
$n = 2, m = 0$: $r_{2,0}(x) = 1 + 2x + 2x^2$,
$n = 1, m = 1$: $r_{1,1}(x) = (1 + x)/(1 - x)$,
$n = 0, m = 2$: $r_{0,2}(x) = (1 - 2x + 2x^2)^{-1}$.

i	x_i	$f(x_i)$	$r_{2,0}(x_i)$	$r_{1,1}(x_i)$	$r_{0,2}(x_i)$
1	0.2	1.4918	1.4800	1.5000	1.4706
2	0.4	2.2255	2.1200	2.3333	1.9231
3	0.6	3.3201	2.9200	4.0000	1.9231
4	0.8	4.9530	3.8800	9.0000	1.4706
5	1.0	7.3891	5.0000	undefined	1.0000

3. $r_{2,3}(x) = (1 + \frac{2}{5}x + \frac{1}{20}x^2)/(1 - \frac{3}{5}x + \frac{3}{20}x^2 - \frac{1}{60}x^3)$

i	x_i	$f(x_i)$	$r_{2,3}(x_i)$
1	0.2	1.22140276	1.22140277
2	0.4	1.49182470	1.49182561
3	0.6	1.82211880	1.82213210
4	0.8	2.22554093	2.22563652
5	1.0	2.71828183	2.71875000

5. $r_{3,3}(x) = (x - \frac{7}{60}x^3)/(1 + \frac{1}{20}x^2)$

i	x_i	$f(x_i)$	6th Maclaurin polynomial	$r_{3,2}(x_i)$
0	0.0	0.00000000	0.00000000	0.00000000
1	0.1	0.09983342	0.09966675	0.09938640
2	0.2	0.19866933	0.19733600	0.19709571
3	0.3	0.29552021	0.29102025	0.29246305
4	0.4	0.38941834	0.37875200	0.38483660
5	0.5	0.47942554	0.45859375	0.47357724

7. The Padé approximations of degree 5 are as follows.

a. $r_{0,5}(x) = (1 + x + \frac{1}{2}x^2 + \frac{1}{6}x^3 + \frac{1}{24}x^4 + \frac{1}{120}x^5)^{-1}$

b. $r_{1,4}(x) = (1 - \frac{1}{5}x)/(1 + \frac{4}{5}x + \frac{3}{10}x^2 + \frac{1}{15}x^3 + \frac{1}{120}x^4)$

c. $r_{3,2}(x) = (1 - \frac{3}{5}x + \frac{3}{20}x^2 - \frac{1}{60}x^3)/(1 + \frac{2}{5}x + \frac{1}{20}x^2)$

d. $r_{4,1}(x) = (1 - \frac{4}{5}x + \frac{3}{10}x^2 - \frac{1}{15}x^3 + \frac{1}{120}x^4)/(1 + \frac{1}{5}x)$

i	x_i	$f(x_i)$	$r_{0,5}(x_i)$	$r_{1,4}(x_i)$	$r_{2,3}(x_i)$	$r_{4,1}(x_i)$
1	0.2	0.81873075	0.81873081	0.81873074	0.81873075	0.81873077
2	0.4	0.67032005	0.67032276	0.67031942	0.67031963	0.67032099
3	0.6	0.54881164	0.54883296	0.54880635	0.54880763	0.54882143
4	0.8	0.44932896	0.44941181	0.44930678	0.44930966	0.44937931
5	1.0	0.36787944	0.36809816	0.36781609	0.36781609	0.36805556

9. a. Since

$$\sin|x| = \sin(M\pi + s) = \sin M\pi \cos s + \cos M\pi \sin s = (-1)^M \sin s,$$

we have

$$\sin x = \operatorname{sgn} x \sin|x| = \operatorname{sgn}(x)(-1)^M \sin s.$$

b. $\sin x \approx \left(s - \dfrac{31}{294}s^3\right) \Big/ \left(1 + \dfrac{3}{49}s^2 + \dfrac{11}{5880}s^3\right)$ with $|\text{error}| \le 2.84 \times 10^{-4}$.

c. Set $M = \text{round}(|x|/\pi)$; $s = |x| - M\pi$; $f_1 = \left(s - \frac{31}{294}s^3\right) \Big/ \left(1 + \frac{3}{49}s^2 + \frac{11}{5880}s^4\right)$. Then $f = (-1)^M f_1 \cdot x/|x|$ is the approximation.

d. Set $y = x + \pi/2$ and repeat part (c) with y in place of x.

EXERCISE SET 8.6 (PAGE 357)

1. $S_2(x) = \pi^2/3 - 4\cos x + \cos 2x$

3. $S_3(x) = 3.676078 - 3.676078\cos x + 1.470431\cos 2x - 0.7352156\cos 3x + 3.676078\sin x - 2.940862\sin 2x$

5. $S_n(x) = \dfrac{1}{2} + \dfrac{1}{\pi}\displaystyle\sum_{k=1}^{n-1} \dfrac{1-(-1)^k}{k}\sin kx$

7. The trigonometric least squares polynomials are as follows.

a. $S_2(x) = \cos 2x$

b. $S_2(x) = 0$

c. $S_3(x) = 1.566453 + 0.5886815\cos x - 0.2700642\cos 2x + 0.2175679\cos 3x + 0.8341640\sin x - 0.3097866\sin 2x$

d. $S_3(x) = -2.046326 + 3.883872\cos x - 2.320482\cos 2x + 0.7310818\cos 3x$

9. The trigonometric least squares polynomial is $S_3(x) = -0.4968929 + 0.2391965\cos x + 1.515393\cos 2x + 0.2391965\cos 3x - 1.150649\sin x$ with error $E(S_3) = 7.271197$.

11. Let $f(-x) = -f(x)$. The integral $\int_{-a}^{0} f(x)\,dx$ under the change of variable $t = -x$ transforms to

$$-\int_a^0 f(-t)\,dt = \int_0^a f(-t)\,dt = -\int_0^a f(t)\,dt = -\int_0^a f(x)\,dx.$$

Thus,

$$\int_{-a}^{a} f(x)\,dx = \int_{-a}^{0} f(x)\,dx + \int_{0}^{a} f(x)\,dx = -\int_{0}^{a} f(x)\,dx + \int_{0}^{a} f(x)\,dx = 0.$$

13. Representative integrations that establish the orthogonality are:

$$\int_{-\pi}^{\pi} [\phi_0(x)]^2\,dx = \frac{1}{2}\int_{-\pi}^{\pi} dx = \pi,$$

$$\int_{-\pi}^{\pi} [\phi_k(x)]^2\,dx = \int_{-\pi}^{\pi} (\cos kx)^2\,dx = \int_{-\pi}^{\pi} \left[\frac{1}{2} + \frac{1}{2}\cos 2kx\right] dx$$

$$= \pi + \left[\frac{1}{4k}\sin 2kx\right]_{-\pi}^{\pi} = \pi,$$

$$\int_{-\pi}^{\pi} \phi_k(x)\phi_0(x)\,dx = \frac{1}{2}\int_{-\pi}^{\pi} \left[\cos kx\,dx = \frac{1}{2k}\sin kx\right]_{-\pi}^{\pi} = 0,$$

and

$$\int_{-\pi}^{\pi} \phi_k(x)\phi_{n+j}(x)\,dx = \int_{-\pi}^{\pi} \cos kx \sin jx\,dx$$

$$= \frac{1}{2}\int_{-\pi}^{\pi} [\sin(k+j)x - \sin(k-j)x]\,dx = 0.$$

EXERCISE SET 8.7 (PAGE 363)

1. The trigonometric interpolating polynomials are as follows.

a. $S_2(x) = -12.33701 + 4.934802\cos x - 2.467401\cos 2x + 4.934802\sin x$

b. $S_2(x) = -6.168503 + 9.869604\cos x - 3.701102\cos 2x + 4.934802\sin x$

c. $S_2(x) = 1.570796 - 1.570796\cos x$

d. $S_2(x) = -0.5 - 0.5\cos 2x + \sin x$

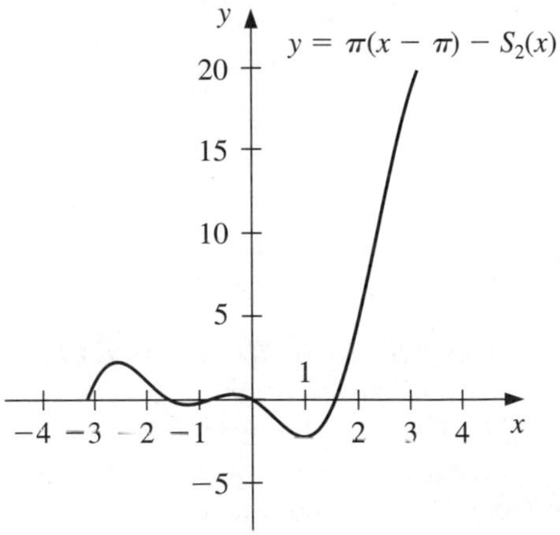

1(a)

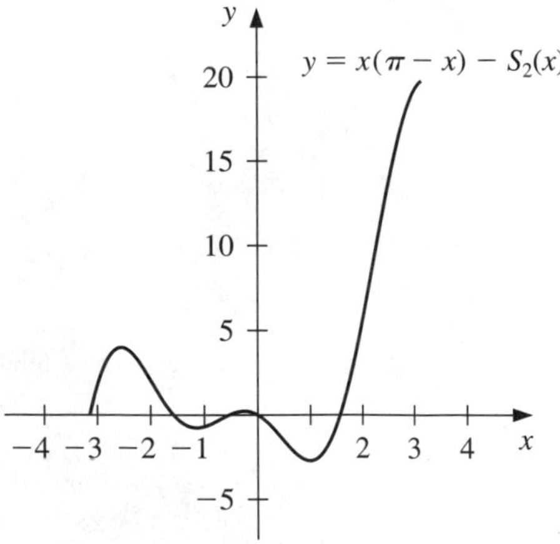

1(b)

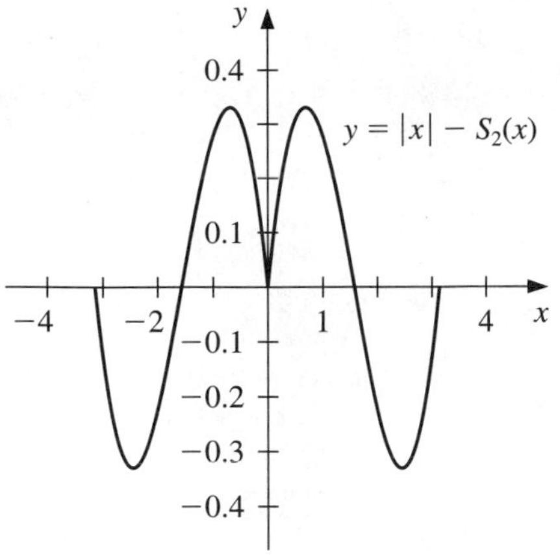

1(c)

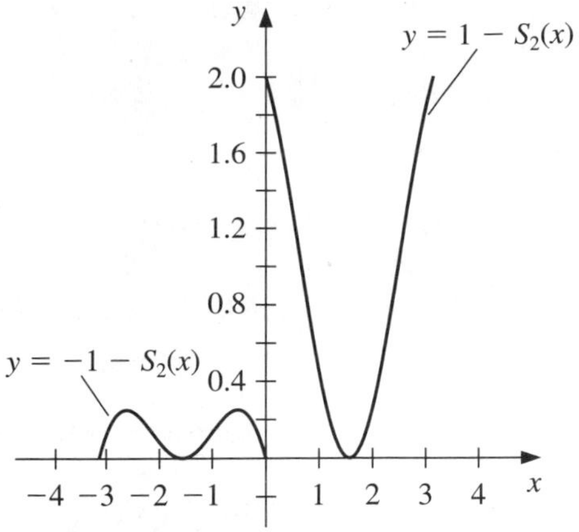

1(d)

3. The Fast Fourier Transform method gives the following trigonometric interpolating polynomials.

a. $S_4(x) = -11.10331 + 2.467401\cos x - 2.467401\cos 2x + 2.467401\cos 3x$
$\quad - 1.233701\cos 4x + 5.956833\sin x - 2.467401\sin 2x + 1.022030\sin 3x$

b. $S_4(x) = 1.570796 - 1.340759\cos x - 0.2300378\cos 3x$

c. $S_4(x) = -0.1264264 + 0.2602724\cos x - 0.3011140\cos 2x + 1.121372\cos 3x$
$\quad + 0.04589648\cos 4x - 0.1022190\sin x + 0.2754062\sin 2x - 2.052955\sin 3x$

d. $S_4(x) = -0.1526819 + 0.04754278\cos x + 0.6862114\cos 2x - 1.216913\cos 3x$
$\quad + 1.176143\cos 4x - 0.8179387\sin x + 0.1802450\sin 2x + 0.2753402\sin 3x$

5.

	Approximation	**Actual**
a.	−69.76415	−62.01255
b.	9.869602	9.869604
c.	−0.7943605	−0.2739383
d.	−0.9593287	−0.9557781

7. Since $(\sin x)^2 = \frac{1}{2} - \frac{1}{2}\cos 2x$ and $(\cos x)^2 = \frac{1}{2} + \frac{1}{2}\cos 2x$,

$$\sum_{j=0}^{2m-1}(\sin mx_j)^2 = \frac{1}{2}\sum_{j=0}^{2m-1}1 - \frac{1}{2}\sum_{j=0}^{2m-1}\cos 2mx_j = m - \frac{1}{2}\sum_{j=0}^{2m-1}\cos 2mx_j$$

and

$$\sum_{j=0}^{2m-1}(\cos mx_j)^2 = m + \frac{1}{2}\sum_{j=0}^{2m-1}\cos 2mx_j.$$

However,

$$\cos 2mx_j = \cos 2m\left(-\pi + \frac{j}{m}\pi\right)$$
$$= \cos(2j\pi - 2m\pi) = \cos(2j - 2m)\pi = (-1)^{2j-2m} = 1.$$

Thus,

$$\sum_{j=0}^{2m-1}(\sin mx_j)^2 = m - \frac{1}{2}\sum_{j=0}^{2m-1}1 = m - m = 0$$

and

$$\sum_{j=0}^{2m-1} (\cos mx_j)^2 = m + \frac{1}{2}\sum_{j=0}^{2m-1} 1 = m + m = 2m.$$

CHAPTER 9

EXERCISE SET 9.2 (PAGE 371)

1. **a.** The eigenvalues and associated eigenvectors are $\lambda_1 = 2$, $\mathbf{v}^{(1)} = (1, 0, 0)^t$; $\lambda_2 = 1$, $\mathbf{v}^{(2)} = (0, 2, 1)^t$; and $\lambda_3 = -1$, $\mathbf{v}^{(3)} = (-1, 1, 1)^t$. The set is linearly independent.

b. The eigenvalues and associated eigenvectors are $\lambda_1 = \lambda_2 = \lambda_3 = 1$, $\mathbf{v}^{(1)} = \mathbf{v}^{(2)} = (1, 0, 1)^t$ and $\mathbf{v}^{(3)} = (0, 1, 1)$. The set is linearly dependent.

c. The eigenvalues and associated eigenvectors are $\lambda_1 = 2$, $\mathbf{v}^{(1)} = (0, 1, 0)^t$; $\lambda_2 = 3$, $\mathbf{v}^{(2)} = (1, 0, 1)^t$; and $\lambda_3 = 1$, $\mathbf{v}^{(3)} = (1, 0, -1)^t$. The set is linearly independent.

d. The eigenvalues and associated eigenvectors are $\lambda_1 = \lambda_2 = 3$, $\mathbf{v}^{(1)} = (1, 0, -1)^t$, $\mathbf{v}^{(2)} = (0, 1, -1)^t$; and $\lambda_3 = 0$, $\mathbf{v}^{(3)} = (1, 1, 1)^t$. The set is linearly independent.

e. The eigenvalues and associated eigenvectors are $\lambda_1 = 1$, $\mathbf{v}^{(1)} = (0, -1, 1)^t$; $\lambda_2 = 1 + \sqrt{2}$, $\mathbf{v}^{(2)} = (\sqrt{2}, 1, 1)^t$; and $\lambda_3 = 1 - \sqrt{2}$, $\mathbf{v}^{(3)} = (-\sqrt{2}, 1, 1)^t$. The set is linearly independent.

f. The eigenvalues and associated eigenvectors are $\lambda_1 = 1$, $\mathbf{v}^{(1)} = (1, 0, -1)^t$; $\lambda_2 = 1$, $\mathbf{v}^{(2)} = (1, -1, 0)^t$; and $\lambda_3 = 4$, $\mathbf{v}^{(3)} = (1, 1, 1)^t$. The set is linearly independent.

3. **a.** The three eigenvalues are within $\{\lambda \mid |\lambda| \le 2\} \cup \{\lambda \mid |\lambda - 2| \le 2\}$.

b. The three eigenvalues are within $R_1 = \{\lambda \mid |\lambda - 4| \le 2\}$.

c. The three real eigenvalues satisfy $0 \le \lambda \le 6$.

d. The three real eigenvalues satisfy $1.25 \le \lambda \le 8.25$.

e. The four real eigenvalues satisfy $-8 \le \lambda \le 1$.

f. The four real eigenvalues are within $R_1 = \{\lambda \mid |\lambda - 2| \le 4\}$.

5. If $c_1\mathbf{v}_1 + \cdots + c_k\mathbf{v}_k = \mathbf{0}$, then for any j, with $1 \le j \le k$, we have $c_1\mathbf{v}_j^t\mathbf{v}_1 + \cdots + c_k\mathbf{v}_j^t\mathbf{v}_k = \mathbf{0}$. But orthogonality gives $c_i\mathbf{v}_j^t\mathbf{v}_i = 0$ for $i \ne j$, so $c_j\mathbf{v}_j^t\mathbf{v}_j = 0$ and $c_j = 0$.

7. Since $\{\mathbf{v}_i\}_{i=1}^n$ is linearly independent in $\mathbb{R}^n$, there exist numbers $c_1, \ldots, c_n$ with

$$\mathbf{x} = c_1\mathbf{v}_1 + \cdots + c_n\mathbf{v}_n.$$

Hence, for any j, with $1 \le j \le n$,

$$\mathbf{v}_j^t\mathbf{x} = c_1\mathbf{v}_j^t\mathbf{v}_1 + \cdots + c_n\mathbf{v}_j^t\mathbf{v}_n = c_j\mathbf{v}_j^t\mathbf{v}_j = c_j.$$

9. **a.** The eigenvalues are $\lambda_1 = 5.307857563$, $\lambda_2 = -0.4213112993$, $\lambda_3 = -0.1365462647$ with associated eigenvectors $(0.59020967, 0.51643129, 0.62044441)^t$, $(0.77264234, -0.13876278, -0.61949069)^t$, and $(0.23382978, -0.84501102, 0.48091581)^t$, respectively.

b. A is not positive definite, since $\lambda_2 < 0$ and $\lambda_3 < 0$.

EXERCISE SET 9.3 (PAGE 382)

1. The approximate eigenvalues and approximate eigenvectors are as follows.

a. $\mu^{(3)} = 3.666667$, $\mathbf{x}^{(3)} = (0.9772727, 0.9318182, 1)^t$

b. $\mu^{(3)} = 2.000000$, $\mathbf{x}^{(3)} = (1, 1, 0.5)^t$

c. $\mu^{(3)} = 5.000000$, $\mathbf{x}^{(3)} = (-0.2578947, 1, -0.2842105)^t$

d. $\mu^{(3)} = 5.038462$, $\mathbf{x}^{(3)} = (1, 0.2213741, 0.3893130, 0.4045802)^t$

e. $\mu^{(3)} = 7.531073$, $\mathbf{x}^{(3)} = (0.6886722, -0.6706677, -0.9219805, 1)^t$
f. $\mu^{(3)} = 4.106061$, $\mathbf{x}^{(3)} = (0.1254613, 0.08487085, 0.00922509, 1)^t$

3. The approximate eigenvalues and approximate eigenvectors are as follows.
a. $\mu^{(3)} = 3.959538$, $\mathbf{x}^{(3)} = (0.5816124, 0.5545606, 0.5951383)^t$
b. $\mu^{(3)} = 2.0000000$, $\mathbf{x}^{(3)} = (-0.6666667, -0.6666667, -0.3333333)^t$
c. $\mu^{(3)} = 7.189567$, $\mathbf{x}^{(3)} = (0.5995308, 0.7367472, 0.3126762)^t$
d. $\mu^{(3)} = 6.037037$, $\mathbf{x}^{(3)} = (0.5073714, 0.4878571, -0.6634857, -0.2536857)^t$
e. $\mu^{(3)} = 5.142562$, $\mathbf{x}^{(3)} = (0.8373051, 0.3701770, 0.1939022, 0.3525495)^t$
f. $\mu^{(3)} = 8.593142$, $\mathbf{x}^{(3)} = (-0.4134762, 0.4026664, 0.5535536, -0.6003962)^t$

5. The approximate eigenvalues and approximate eigenvectors are as follows.
a. $\mu^{(8)} = 4.000001$, $\mathbf{x}^{(8)} = (0.9999773, 0.99993134, 1)^t$
b. The method fails because of division by zero.
c. $\mu^{(7)} = 5.124890$, $\mathbf{x}^{(7)} = (-0.2425938, 1, -0.3196351)^t$
d. $\mu^{(15)} = 5.236112$, $\mathbf{x}^{(15)} = (1, 0.6125369, 0.1217216, 0.4978318)^t$
e. $\mu^{(10)} = 8.999890$, $\mathbf{x}^{(10)} = (0.9944137, -0.9942148, -0.9997991, 1)^t$
f. $\mu^{(11)} = 4.105317$, $\mathbf{x}^{(11)} = (0.11716540, 0.072853995, 0.01316655, 1)^t$

7. The approximate eigenvalues and approximate eigenvectors are as follows.
a. $\mu^{(9)} = 1.000015$, $\mathbf{x}^{(9)} = (-0.1999939, 1, -0.7999909)^t$
b. $\mu^{(12)} = -0.4142136$, $\mathbf{x}^{(12)} = (1, -0.7070918, -0.7071217)^t$
c. The method did not converge in 25 iterations. However, $\mu^{(42)} = 1.636636$, $\mathbf{x}^{(42)} = (-0.5706815, 0.3633636, 1)^t$.
d. $\mu^{(9)} = 1.381959$, $\mathbf{x}^{(9)} = (-0.3819400, -0.2361007, 0.2360191, 1)^t$
e. $\mu^{(6)} = 3.999997$, $\mathbf{x}^{(6)} = (0.9999939, 0.9999999, 0.9999940, 1)^t$
f. $\mu^{(3)} = 4.105293$, $\mathbf{x}^{(3)} = (0.06281419, 0.08704089, 0.01825213, 1)^t$

9. **a.** We have $|\lambda| \le 6$ for all eigenvalues λ.
b. The approximate eigenvalue is $\mu^{(133)} = 0.69766854$, with the approximate eigenvector $\mathbf{x}^{(133)} = (1, 0.7166727, 0.2568099, 0.04601217)^t$.
c. Wielandt's deflation fails because λ_2 and λ_3 are complex numbers.
d. The characteristic polynomial is $P(\lambda) = \lambda^4 - \frac{1}{4}\lambda - \frac{1}{16}$ and the eigenvalues are $\lambda_1 = 0.6976684972$, $\lambda_2 = -0.2301775942 + 0.56965884i$, $\lambda_3 = -0.2301775942 - 0.56965884i$, and $\lambda_4 = -0.237313308$.
e. The beetle population should approach zero since A is convergent.

EXERCISE SET 9.4 (PAGE 390)

1. Householder's method produces the following tridiagonal matrices.

a. $$\begin{bmatrix} 12.00000 & -10.77033 & 0.0 \\ -10.77033 & 3.862069 & 5.344828 \\ 0.0 & 5.344828 & 7.137931 \end{bmatrix}$$

b. $$\begin{bmatrix} 2.0000000 & 1.414214 & 0.0 \\ 1.414214 & 1.000000 & 0.0 \\ 0.0 & 0.0 & 3.0 \end{bmatrix}$$

c. $$\begin{bmatrix} 1.0000000 & -1.414214 & 0.0 \\ -1.414214 & 1.000000 & 0.0 \\ 0.0 & 0.0 & 1.000000 \end{bmatrix}$$

d. $$\begin{bmatrix} 4.750000 & -2.263846 & 0.0 \\ -2.263846 & 4.475610 & -1.219512 \\ 0.0 & -1.219512 & 5.024390 \end{bmatrix}$$

EXERCISE SET 9.5 (PAGE 396)

1. Two iterations of the QR method produce the following matrices.

a. $A^{(3)} = \begin{bmatrix} 0.6939977 & -0.3759745 & 0.0 \\ -0.3759745 & 1.892417 & -0.03039696 \\ 0.0 & -0.03039696 & 3.413585 \end{bmatrix}$

b. $A^{(3)} = \begin{bmatrix} 4.535466 & 1.212648 & 0.0 \\ 1.212648 & 3.533242 & 3.83 \times 10^{-7} \\ 0.0 & 3.83 \times 10^{-7} & -0.06870782 \end{bmatrix}$

c. $A^{(3)} = \begin{bmatrix} 4.679567 & -0.2969009 & 0.0 \\ -2.969009 & 3.052484 & -1.207346 \times 10^{-5} \\ 0.0 & -1.207346 \times 10^{-5} & 1.267949 \end{bmatrix}$

d. $A^{(3)} = \begin{bmatrix} 0.3862092 & 0.4423226 & 0.0 & 0.0 \\ 0.4423226 & 1.787694 & -0.3567744 & 0.0 \\ 0.0 & -0.3567744 & 3.080815 & 3.116382 \times 10^{-5} \\ 0.0 & 0.0 & 3.116382 \times 10^{-5} & 4.745281 \end{bmatrix}$

e. $A^{(3)} = \begin{bmatrix} -2.826365 & 1.130297 & 0.0 & 0.0 \\ 1.130297 & -2.429647 & -0.1734156 & 0.0 \\ 0.0 & -0.1734156 & 0.8172086 & 1.863997 \times 10^{-9} \\ 0.0 & 0.0 & 1.863997 \times 10^{-9} & 3.438803 \end{bmatrix}$

f. $A^{(3)} = \begin{bmatrix} 0.2763388 & 0.1454371 & 0.0 & 0.0 \\ 0.1454371 & 0.4543713 & 0.1020836 & 0.0 \\ 0.0 & 0.1020836 & 1.174446 & -4.36 \times 10^{-5} \\ 0.0 & 0.0 & -4.36 \times 10^{-5} & 0.9948441 \end{bmatrix}$

3. The matrices in Exercise 1 have the following eigenvalues, accurate to within 10^{-5}.

a. 3.414214, 2.000000, 0.58578644
b. −0.06870782, 5.346462, 2.722246
c. 1.267949, 4.732051, 3.000000
d. 4.745281, 3.177283, 1.822717, 0.2547188
e. 3.438803, 0.8275517, −1.488068, −3.778287
f. 0.9948440, 1.189091, 0.5238224, 0.1922421

5. **a.** Let

$$P = \begin{bmatrix} \cos\theta & -\sin\theta \\ \sin\theta & \cos\theta \end{bmatrix}$$

and $\mathbf{y} = P\mathbf{x}$. Show that $\|\mathbf{x}\|_2 = \|\mathbf{y}\|_2$. Then use the relationship $x_1 + ix_2 = re^{i\alpha}$, where $r = \|\mathbf{x}\|_2$ and $\alpha = \tan^{-1}(x_2/x_1)$, and $y_1 + iy_2 = re^{i(\alpha+\theta)}$.

b. Let $\mathbf{x} = (1, 0)^t$ and $\theta = \pi/4$.

CHAPTER 10

EXERCISE SET 10.2 (PAGE 408)

1. One example is $f(x_1, x_2) = \left(1, \dfrac{1}{|x_1 - 1| + |x_2|}\right)^t$.

3. **a.** $(-1, 3.5)^t$ and $(2.5, 4)^t$
b. $(0.11, 0.27)^t$ and $(-0.11, 0.23)^t$
c. $(1, 1, 1)^t$
d. $(1, -1, 1)^t$ and $(1, 1, -1)^t$

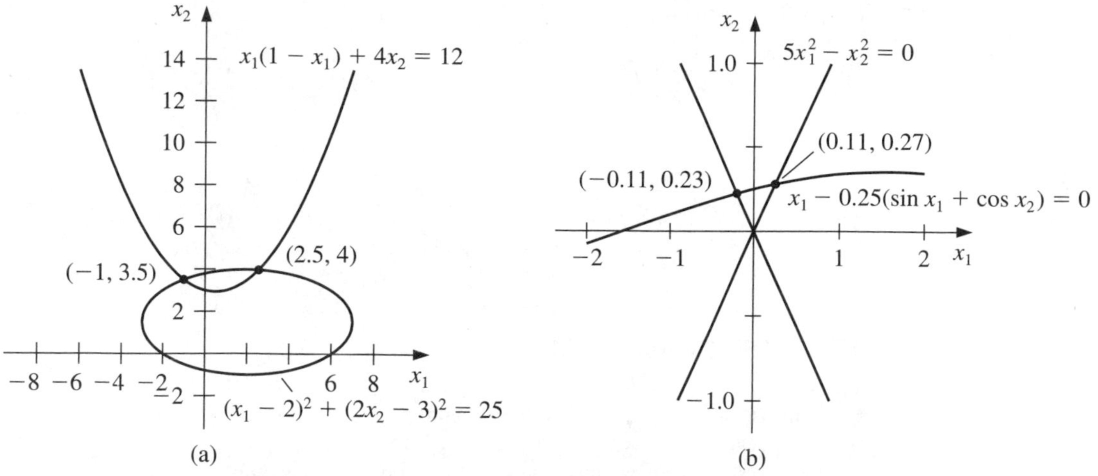

The graphs for parts (a) and (b) are shown with the approximate intersections. The three-dimensional graphs for parts (c) and (d) are not given since experimentation is needed in Maple to determine the approximate intersections.

5. **a.** With $\mathbf{x}^{(0)} = (-1, 3.5)^t$, $\mathbf{x}^{(1)} = (-1, 3.5)^t$, so $(-1, 3.5)^t$ is a solution. With $\mathbf{x}^{(0)} = (2.5, 4)^t$, $\mathbf{x}^{(3)} = (2.546947, 3.984998)^t$.

b. With $\mathbf{x}^{(0)} = (0.11, 0.27)^t$, $\mathbf{x}^{(6)} = (0.1212419, 0.2711051)^t$. With $\mathbf{x}^{(0)} = (-0.11, 0.23)^t$, $\mathbf{x}^{(4)} = (-0.09816344, 0.21950013)^t$.

c. With $\mathbf{x}^{(0)} = (1, 1, 1)^t$, $\mathbf{x}^{(3)} = (1.036401, 1.085707, 0.9311914)^t$.

d. With $\mathbf{x}^{(0)} = (1, -1, 1)^t$, $\mathbf{x}^{(5)} = (0.9, -1, 0.5)^t$, and with $\mathbf{x}^{(0)} = (1, -1, 1)^t$, $\mathbf{x}^{(5)} = (0.5, 1, -0.5)^t$.

7. With $\mathbf{x}^{(0)} = (1, 1, 1, 1)^t$, $\mathbf{x}^{(6)} = (0, 0.7071068, 0.7071068, 1)^t$.
With $\mathbf{x}^{(0)} = (1, 0, 0, 0)^t$, $\mathbf{x}^{(6)} = (0.8164966, 0.4082483, -0.4082483, 3)^t$.
With $\mathbf{x}^{(0)} = (-1, 0, 0, 0)^t$, $\mathbf{x}^{(6)} = (-0.8164966, -0.4082483, 0.4082483, 3)^t$.

9. **a.** $k_1 = 8.77125$, $k_2 = 0.259690$, $k_3 = -1.37217$

b. Solving the equation $500/\pi r^2 = k_1 e^{k_2 r} + k_3 r$ numerically gives $r = 3.18517$.

EXERCISE SET 10.3 (PAGE 415)

1. **a.** $\mathbf{x}^{(2)} = (0.4777920, 1.927557)^t$

b. $\mathbf{x}^{(2)} = (-0.3250070, -0.1386967)^t$

c. $\mathbf{x}^{(2)} = (0.5115893, -78.72872, -0.5120771)^t$

d. $\mathbf{x}^{(2)} = (-67.00583, 38.31494, 31.69089)^t$

3. **a.** $\mathbf{x}^{(9)} = (0.5, 0.8660254)^t$

b. $\mathbf{x}^{(8)} = (1.772454, 1.772454)^t$

c. $\mathbf{x}^{(9)} = (-1.456043, -1.664231, 0.4224934)^t$

d. $\mathbf{x}^{(5)} = (0.4981447, -0.1996059, -0.5288260)^t$

5. With $\mathbf{x}^{(0)} = (1, 1, 1, 1)^t$, we have $\mathbf{x}^{(6)} = (0, 0.7071068, 0.7071068, 1)^t$.
With $\mathbf{x}^{(0)} = (1, 0, 0, 2)^t$, we have $\mathbf{x}^{(12)} = (0.8164965, 0.4082484, -0.4082483, 3)^t$.
With $\mathbf{x}^{(0)} = (-1, 0, 0, 2)^t$, we have $\mathbf{x}^{(12)} = (-0.8164965, -0.4082484, 0.4082483, 3)^t$.

7. We have

$$\left[A^{-1} - \frac{A^{-1}\mathbf{x}\mathbf{y}^t A^{-1}}{1+\mathbf{y}^t A^{-1}\mathbf{x}}\right](A+\mathbf{x}\mathbf{y}^t) = A^{-1}A - \frac{A^{-1}\mathbf{x}\mathbf{y}^t A^{-1}A}{1+\mathbf{y}^t A^{-1}\mathbf{x}} + A^{-1}\mathbf{x}\mathbf{y}^t - \frac{A^{-1}\mathbf{x}\mathbf{y}^t A^{-1}\mathbf{x}\mathbf{y}^t}{1+\mathbf{y}^t A^{-1}\mathbf{x}}$$

$$= I - \frac{A^{-1}\mathbf{x}\mathbf{y}^t}{1+\mathbf{y}^t A^{-1}\mathbf{x}} + A^{-1}\mathbf{x}\mathbf{y}^t - \frac{A^{-1}\mathbf{x}\mathbf{y}^t A^{-1}\mathbf{x}\mathbf{y}^t}{1+\mathbf{y}^t A^{-1}\mathbf{x}}$$

$$= I - \frac{A^{-1}\mathbf{x}\mathbf{y}^t - A^{-1}\mathbf{x}\mathbf{y}^t - \mathbf{y}^t A^{-1}\mathbf{x}A^{-1}\mathbf{x}\mathbf{y}^t + A^{-1}\mathbf{x}\mathbf{y}^t A^{-1}\mathbf{x}\mathbf{y}^t}{1+\mathbf{y}^t A^{-1}\mathbf{x}}$$

$$= I + \frac{\mathbf{y}^t A^{-1}\mathbf{x}A^{-1}\mathbf{x}\mathbf{y}^t - \mathbf{y}^t A^{-1}\mathbf{x}(A^{-1}\mathbf{x}\mathbf{y}^t)}{1+\mathbf{y}^t A^{-1}\mathbf{x}} = I.$$

EXERCISE SET 10.4 (PAGE 421)

1. **a.** With $\mathbf{x}^{(0)} = (0,0)^t$, we have $\mathbf{x}^{(11)} = (0.4943541, 1.948040)^t$.
b. With $\mathbf{x}^{(0)} = (1,1)^t$, we have $\mathbf{x}^{(2)} = (0.4970073, 0.8644143)^t$.
c. With $\mathbf{x}^{(0)} = (2,2)^t$, we have $\mathbf{x}^{(1)} = (1.736083, 1.804428)^t$.
d. With $\mathbf{x}^{(0)} = (0,0)^t$, we have $\mathbf{x}^{(2)} = (-0.3610092, 0.05788368)^t$.

3. **a.** With $\mathbf{x}^{(0)} = (0,0,0)^t$, we have $\mathbf{x}^{(14)} = (1.043605, 1.064058, 0.9246118)^t$.
b. With $\mathbf{x}^{(0)} = (0,0,0)^t$, we have $\mathbf{x}^{(9)} = (0.4932739, 0.9863888, -0.5175964)^t$.
c. With $\mathbf{x}^{(0)} = (0,0,0)^t$, we have $\mathbf{x}^{(11)} = (-1.608296, -1.192750, 0.7205642)^t$.
d. With $\mathbf{x}^{(0)} = (0,0,0)^t$, we have $\mathbf{x}^{(1)} = (0, 0.00989056, 0.9890556)^t$.

5. **a.** With $\mathbf{x}^{(0)} = (0,0)^t$, we have $\mathbf{x}^{(8)} = (3.136548, 0)^t$ and $\mathbf{g}(\mathbf{x}^{(8)}) = 0.005057848$.
b. With $\mathbf{x}^{(0)} = (0,0)^t$, we have $\mathbf{x}^{(13)} = (0.6157412, 0.3768953)^t$ and $\mathbf{g}(\mathbf{x}^{(13)}) = 0.1481574$.
c. With $\mathbf{x}^{(0)} = (0,0,0)^t$, we have $\mathbf{x}^{(5)} = (-0.6633785, 0.3145720, 0.5000740)^t$ and $\mathbf{g}(\mathbf{x}^{(5)}) = 0.6921548$.
d. With $\mathbf{x}^{(0)} = (1,1,1)^t$, we have $\mathbf{x}^{(4)} = (0.04022273, 0.01592477, 0.01594401)^t$ and $\mathbf{g}(\mathbf{x}^{(4)}) = 1.010003$.

CHAPTER 11

EXERCISE SET 11.2 (PAGE 430)

1. The Linear Shooting method gives the results in the following tables.

a.

i	x_i	w_{1i}	$y(x_i)$
1	0.333333	0.5311664	0.5310687
2	0.666667	1.153515	1.153323

b.

i	x_i	w_{1i}	$y(x_i)$
1	0.25	0.3937095	0.3936767
2	0.50	0.8240948	0.8240271
3	0.75	1.337160	1.337086

3. The Linear Shooting method gives the results in the following tables.

a.

i	x_i	w_{1i}	$y(x_i)$
3	0.3	0.7833204	0.7831923
6	0.6	0.6023521	0.6022801
9	0.9	0.8568906	0.8568760

b.

i	x_i	w_{1i}	$y(x_i)$
5	1.25	0.1676179	0.1676243
10	1.50	0.4581901	0.4581935
15	1.75	0.6077718	0.6077740

c.

i	x_i	w_{1i}	$y(x_i)$
3	0.3	−0.5185754	−0.5185728
6	0.6	−0.2195271	−0.2195247
9	0.9	−0.0406577	−0.0406570

d.

i	x_i	w_{1i}	$y(x_i)$
3	1.3	0.0655336	0.06553420
6	1.6	0.0774590	0.07745947
9	1.9	0.0305619	0.03056208

5. The Linear Shooting method with $h = 0.1$ gives the following results.

i	x_i	w_{1i}	$y(x_i)$
3	0.3	0.05273437	0.04978707
5	0.5	0.00741571	0.00248825
8	0.8	0.00038976	0.00012341

The Linear Shooting method with $h = 0.05$ gives the following results.

i	x_i	w_{1i}
6	0.3	0.04990547
10	0.5	0.00673795
16	0.8	0.00033755

7. **a.** The approximate potential is $u(3) \approx 36.66702$ using $h = 0.1$.
b. The actual potential is $u(3) = 36.66667$.

9. **a.** There are no solutions if b is an integer multiple of π and $B \neq 0$.
b. A unique solution exists whenever b is not an integer multiple of π.
c. There are infinitely many solutions if b is an multiple integer of π and $B = 0$.

EXERCISE SET 11.3 (PAGE 437)

1. The Linear Finite-Difference method gives the results in the following tables.

a.

i	x_i	w_{1i}	$y(x_i)$
1	0.333333	0.5343259	0.5310687
2	0.666667	1.1579818	1.1533232

b.

i	x_i	w_{1i}	$y(x_i)$
1	0.25	0.3951247	0.3936767
2	0.50	0.8265306	0.8240271
3	0.75	1.3395692	1.3370861

3. The Linear Finite-Difference method gives the results in the following tables.

a.

i	x_i	w_i	$y(x_i)$
2	0.2	1.018096	1.0221404
5	0.5	0.5942743	0.59713617
7	0.7	0.6514520	0.65290384

b.

i	x_i	w_i	$y(x_i)$
5	1.25	0.16797186	0.16762427
10	1.50	0.45842388	0.45819349
15	1.75	0.60787334	0.60777401

c.

i	x_i	w_{1i}	$y(x_i)$
3	0.3	−0.5183084	−0.5185728
6	0.6	−0.2192657	−0.2195247
9	0.9	−0.0405748	−0.04065697

d.

i	x_i	w_{1i}	$y(x_i)$
3	1.3	0.0654387	0.0655342
6	1.6	0.0773936	0.0774595
9	1.9	0.0305465	0.0305621

5. The Linear Finite-Difference method gives the results in the following tables.

i	x_i	$w_i(h = 0.1)$	$y(x_i)$
3	0.3	0.05572807	0.04978707
6	0.6	0.00310518	0.00248825
9	0.9	0.00016516	0.00012341

i	x_i	$w_i(h = 0.05)$
6	0.3	0.05132396
12	0.6	0.00263406
18	0.9	0.00013340

7.

i	x_i	w_i
10	10.0	0.1098549
20	20.0	0.1761424
25	25.0	0.1849608
30	30.0	0.1761424
40	40.0	0.1098549

EXERCISE SET 11.4 (PAGE 445)

1. The Nonlinear Shooting method gives $w_1 = 0.405505 \approx \ln 1.5 = 0.405465$.

3. The Nonlinear Shooting Algorithm gives the results in the following tables.

a. 4 iterations required:

i	x_i	w_{1i}	$y(x_i)$
3	1.3	0.4347934	0.4347826
6	1.6	0.3846363	0.3846154
9	1.9	0.3448586	0.3448276

b. 6 iterations required:

i	x_i	w_{1i}	$y(x_i)$
3	1.3	2.069249	2.069231
6	1.6	2.225013	2.225000
9	1.9	2.426317	2.426316

c. 4 iterations required:

i	x_i	w_{1i}	$y(x_i)$
3	1.3	1.031597	1.031595
6	1.6	1.095007	1.095004
9	1.9	1.168174	1.168170

d. 13 iterations required:

i	x_i	w_{1i}	$y(x_i)$
6	1.6	1.92512377	1.92513487
12	2.2	8.39546972	8.39549397
18	2.8	22.60215126	22.60220545

5.

i	x_i	$w_{1i} \approx y(t_i)$	w_{2i}
3	0.6	0.71682963	0.92122169
5	1.0	1.00884285	0.53467944
8	1.6	1.13844628	−0.11915193

EXERCISE SET 11.5 (PAGE 449)

1. The Nonlinear Finite-Difference method gives $w_1 = 0.4067967 \approx \ln 1.5 = 0.4054651$.

3. The Nonlinear Finite-Difference method gives the results in the following tables.

a.

i	x_i	w_{1i}	$y(x_i)$
3	1.3	0.4347972	0.4347826
6	1.6	0.3846286	0.3846154
9	1.9	0.3448316	0.3448276

b.

i	x_i	w_{1i}	$y(x_i)$
3	1.3	2.0694081	2.0692308
6	1.6	2.2250937	2.2250000
9	1.9	2.4263387	2.4263158

c.

i	x_i	w_{1i}	$y(x_i)$
3	1.3	1.031970	1.031595
6	1.6	1.095321	1.095004
9	1.9	1.168271	1.168170

d.

i	x_i	w_{1i}	$y(x_i)$
6	1.6	1.89217972	1.92513487
12	2.2	8.32080768	8.39549397
18	2.8	22.54870722	22.60220545

5.

i	x_i	w_i
5	30	0.01028080
10	60	0.01442767
15	90	0.01028080

EXERCISE SET 11.6 (PAGE 463)

1. The Piecewise Linear method gives $\phi(x) = -0.07713274\phi_1(x) - 0.07442678\phi_2(x)$. This gives $\phi(x_1) = -0.07713274$ and $\phi(x_2) = -0.07442678$. The actual values are $y(x_1) = -0.07988545$ and $y(x_2) = -0.07712903$.

3. The Piecewise Linear method gives the results in the following tables.

a.

i	x_i	$\phi(x_i)$	$y(x_i)$
3	0.3	−0.212333	−0.21
6	0.6	−0.241333	−0.24
9	0.9	−0.090333	−0.09

b.

i	x_i	$\phi(x_i)$	$y(x_i)$
3	0.3	0.1815138	0.1814273
6	0.6	0.1805502	0.1804754
9	0.9	0.05936468	0.05934303

c.

i	x_i	$\phi(x_i)$	$y(x_i)$
5	0.25	−0.3585989	−0.3585641
10	0.50	−0.5348383	−0.5347803
15	0.75	−0.4510165	−0.4509614

d.

i	x_i	$\phi(x_i)$	$y(x_i)$
5	0.25	−0.1846134	−0.1845204
10	0.50	−0.2737099	−0.2735857
15	0.75	−0.2285169	−0.2284204

5. The Cubic Spline method gives the results in the following tables.

a.

i	x_i	$\phi(x_i)$	$y(x_i)$
3	0.3	−0.2100000	−0.21
6	0.6	−0.2400000	−0.24
9	0.9	−0.0900000	−0.09

b.

i	x_i	$\phi(x_i)$	$y(x_i)$
3	0.3	0.1814269	0.1814273
6	0.6	0.1804753	0.1804754
9	0.9	0.05934321	0.05934303

c.

i	x_i	$\phi(x_i)$	$y(x_i)$
5	0.25	−0.3585639	−0.3585641
10	0.50	−0.5347779	−0.5347803
15	0.75	−0.4509109	−0.4509614

d.

i	x_i	$\phi(x_i)$	$y(x_i)$
5	0.25	−0.1845191	−0.1845204
10	0.50	−0.2735833	−0.2735857
15	0.75	−0.2284186	−0.2284204

7.

i	x_i	$\phi(x_i)$	$y(x_i)$
3	0.3	1.0408182	1.0408182
6	0.6	1.1065307	1.1065306
9	0.9	1.3065697	1.3065697

9. A change in variable $w = (x - a)/(b - a)$ gives the boundary-value problem

$$-\frac{d}{dw}\left(p((b-a)w + a)\frac{dy}{dw}\right) + (b-a)^2 q((b-a)w + a)y$$
$$= (b-a)^2 f((b-a)w + a),$$

where $0 < w < 1$, $y(0) = \alpha$, and $y(1) = \beta$. Then Exercise 6 can be used.

11. Let $\mathbf{c} = (c_1, \ldots, c_n)^t$ be any vector and let $\phi(x) = \sum_{j=1}^{n} c_j\phi_j(x)$. Then

$$\mathbf{c}^t A\mathbf{c} = \sum_{i=1}^{n}\sum_{j=1}^{n} a_{ij}c_ic_j = \sum_{i=1}^{n}\sum_{j=i-1}^{i+1} a_{ij}c_ic_j$$
$$= \sum_{i=1}^{n}\left[\int_0^1 \{p(x)c_i\phi_i'(x)c_{i-1}\phi_{i-1}'(x) + q(x)c_i\phi_i(x)c_{i-1}\phi_{i-1}(x)\}\,dx\right.$$
$$+ \int_0^1 \{p(x)c_i^2[\phi_i'(x)]^2 + q(x)c_i^2[\phi_i'(x)]^2\}\,dx$$
$$\left.+ \int_0^1 \{p(x)c_i\phi_i'(x)c_{i+1}\phi_{i+1}'(x) + q(x)c_i\phi_i(x)c_{i+1}\phi_{i+1}(x)\}\,dx\right]$$
$$= \int_0^1 \{p(x)[\phi'(x)]^2 + q(x)[\phi(x)]^2\}\,dx.$$

So $\mathbf{c}^t A\mathbf{c} \geq 0$ with equality only if $\mathbf{c} = \mathbf{0}$. Since A is also symmetric, A is positive definite.

CHAPTER 12

EXERCISE SET 12.2 (PAGE 476)

1. The Poisson Equation Finite-Difference method gives the following results.

i	j	x_i	y_j	$w_{i,j}$	$u(x_i, y_j)$
1	1	0.5	0.5	0.0	0
1	2	0.5	1.0	0.25	0.25
1	3	0.5	1.5	1.0	1

3. The Poisson Equation Finite-Difference method gives the following results.

a. 30 iterations required:

i	j	x_i	y_j	$w_{i,j}$	$u(x_i, y_j)$
2	2	0.4	0.4	0.1599988	0.16
2	4	0.4	0.8	0.3199988	0.32
4	2	0.8	0.4	0.3199995	0.32
4	4	0.8	0.8	0.6399996	0.64

b. 29 iterations required:

i	j	x_i	y_j	$w_{i,j}$	$u(x_i, y_j)$
2	1	1.256637	0.3141593	0.2951855	0.2938926
2	3	1.256637	0.9424778	0.1830822	0.1816356
4	1	2.513274	0.3141593	−0.7721948	−0.7694209
4	3	2.513274	0.9424778	−0.4785169	−0.4755283

c. 126 iterations required:

i	j	x_i	y_j	$w_{i,j}$	$u(x_i, y_j)$
4	3	0.8	0.3	1.2714468	1.2712492
4	7	0.8	0.7	1.7509419	1.7506725
8	3	1.6	0.3	1.6167917	1.6160744
8	7	1.6	0.7	3.0659184	3.0648542

d. 127 iterations required:

i	j	x_i	y_j	$w_{i,j}$	$u(x_i, y_j)$
2	2	1.2	1.2	0.5251533	0.5250861
4	4	1.4	1.4	1.3190830	1.3189712
6	6	1.6	1.6	2.4065150	2.4064186
8	8	1.8	1.8	3.8088995	3.8088576

5. The approximate potential at some typical points is given in the following table.

i	j	x_i	y_j	$w_{i,j}$
1	4	0.1	0.4	88
2	1	0.2	0.1	66
4	2	0.4	0.2	66

EXERCISE SET 12.3 (PAGE 489)

1. The Heat Equation Backward-Difference method gives the following results.

a.

i	j	x_i	t_j	w_{ij}	$u(x_i, t_j)$
1	1	0.5	0.05	0.632952	0.652037
2	1	1.0	0.05	0.895129	0.883937
3	1	1.5	0.05	0.632952	0.625037
1	2	0.5	0.1	0.566574	0.552493
2	2	1.0	0.1	0.801256	0.781344
3	2	1.5	0.1	0.566574	0.552493

b.

i	j	x_i	t_j	w_{ij}	$u(x_i,t_j)$
1	1	1/3	0.05	1.59728	1.53102
2	1	2/3	0.05	−1.59728	−1.53102
1	2	1/3	0.1	1.47300	1.35333
2	2	2/3	0.1	−1.47300	−1.35333

3. The Forward-Difference method gives the following results.

a. For $h = 0.4$ and $k = 0.1$:

i	j	x_i	t_j	$w_{i,j}$	$u(x_i,t_j)$
2	5	0.8	0.5	3.035630	-2.5×10^{-9}
3	5	1.2	0.5	−3.035630	-1.6×10^{-9}
4	5	1.6	0.5	1.876122	-2.5×10^{-9}

For $h = 0.4$ and $k = 0.05$:

i	j	x_i	t_j	$w_{i,j}$	$u(x_i,t_j)$
2	10	0.8	0.5	0	-2.5×10^{-9}
3	10	1.2	0.5	0	-1.6×10^{-9}
4	10	1.6	0.5	0	-2.5×10^{-9}

b. For $h = \pi/10$ and $k = 0.05$:

i	j	x_i	t_j	$w_{i,j}$	$u(x_i,t_j)$
3	10	0.94247780	0.5	0.4864823	0.4906936
6	10	1.88495559	0.5	0.5718943	0.5768449
9	10	2.82743339	0.5	0.1858197	0.1874283

c. For $h = 0.2$ and $k = 0.04$:

i	j	x_i	t_j	$w_{i,j}$	$u(x_i,t_j)$
4	10	0.8	0.4	1.166149	1.169362
8	10	1.6	0.4	1.252413	1.254556
12	10	2.4	0.4	0.4681813	0.4665473
16	10	3.2	0.4	−0.1027637	−0.1056622

d. For $h = 0.1$ and $k = 0.04$:

i	j	x_i	t_j	$w_{i,j}$	$u(x_i,t_j)$
3	10	0.3	0.4	0.5397009	0.5423003
6	10	0.6	0.4	0.6344565	0.6375122
9	10	0.9	0.4	0.2061474	0.2071403

5. The Crank-Nicolson method gives the following results.

a. For $h = 0.4$ and $k = 0.1$:

i	j	x_i	t_j	$w_{i,j}$	$u(x_i, t_j)$
2	5	0.8	0.5	8.2×10^{-7}	-2.5×10^{-9}
3	5	1.2	0.5	-8.2×10^{-7}	-1.6×10^{-9}
4	5	1.6	0.5	5.1×10^{-7}	-2.5×10^{-9}

For $h = 0.4$ and $k = 0.05$:

i	j	x_i	t_j	$w_{i,j}$	$u(x_i, t_j)$
2	10	0.8	0.5	-2.6×10^{-6}	-2.5×10^{-9}
3	10	1.2	0.5	2.6×10^{-6}	-1.6×10^{-9}
4	10	1.6	0.5	-1.6×10^{-6}	-2.5×10^{-9}

b. For $h = \pi/10$ and $k = 0.05$:

i	j	x_i	t_j	$w_{i,j}$	$u(x_i, t_j)$
3	10	0.94247780	0.5	0.4926589	0.4906936
6	10	1.88495559	0.5	0.5791553	0.5768449
9	10	2.82743339	0.5	0.1881790	0.1874283

c. For $h = 0.2$ and $k = 0.04$:

i	j	x_i	t_j	$w_{i,j}$	$u(x_i, t_j)$
4	10	0.8	0.4	1.171532	1.169362
8	10	1.6	0.4	1.256005	1.254556
12	10	2.4	0.4	0.4654499	0.4665473
16	10	3.2	0.4	−0.1076139	−0.1056622

d. For $h = 0.1$ and $k = 0.04$:

i	j	x_i	t_j	$w_{i,j}$	$u(x_i, t_j)$
3	10	0.3	0.4	0.5440532	0.5423003
6	10	0.6	0.4	0.6395728	0.6375122
9	10	0.9	0.4	0.2078098	0.2071403

7. For the Modified Backward-Difference method, we have

i	j	x_i	t_j	w_{ij}
3	25	0.3	0.25	0.2883460
5	25	0.5	0.25	0.3468410
8	25	0.8	0.25	0.2169217

9.

i	j	x_i	t_j	w_{ij} (Backward-Difference)
2	10	0.3	0.225	1.207730
5	10	0.75	0.225	1.836564
9	10	1.35	0.225	0.6928342

EXERCISE SET 12.4 (PAGE 498)

1. The Wave Equation Finite-Difference method gives the following results.

i	j	x_i	t_j	w_{ij}	$u(x_i, t_j)$
2	4	0.25	1.0	−0.7071068	−0.7071068
3	4	0.50	1.0	−1.0000000	−1.0000000
4	4	0.75	1.0	−0.7071068	−0.7071068

3. **a.** The Finite-Difference method with $h = \pi/10$ and $k = 0.05$ gives the following results.

i	j	x_i	t_j	w_{ij}	$u(x_i, t_j)$
2	10	$\pi/5$	0.5	0.5163933	0.5158301
5	10	$\pi/2$	0.5	0.8785407	0.8775826
8	10	$4\pi/5$	0.5	0.5163933	0.5158301

b. The Finite-Difference method with $h = \pi/20$ and $k = 0.1$ gives the following results.

i	j	x_i	t_j	w_{ij}
4	5	$\pi/5$	0.5	0.5159163
10	5	$\pi/2$	0.5	0.8777292
16	5	$4\pi/5$	0.5	0.5159163

c. The Finite-Difference method with $h = \pi/20$ and $k = 0.05$ gives the following results.

i	j	x_i	t_j	w_{ij}
4	10	$\pi/5$	0.5	0.5159602
10	10	$\pi/2$	0.5	0.8778039
16	10	$4\pi/5$	0.5	0.5159602

5. The Finite-Difference method gives the following results.

i	j	x_i	t_j	w_{ij}
2	5	0.2	0.5	−1
5	5	0.5	0.5	0
8	5	0.8	0.5	1

7. Approximate voltages and currents are given in the following table.

i	j	x_i	t_j	Voltage	Current
5	2	50	0.2	77.769	3.88845
12	2	120	0.2	104.60	−1.69931
18	2	180	0.2	33.986	−5.22995
5	5	50	0.5	77.702	3.88510
12	5	120	0.5	104.51	−1.69785
18	5	180	0.5	33.957	−5.22453

EXERCISE SET 12.5 (PAGE 514)

1. With $E_1 = (0.25, 0.75)$, $E_2 = (0, 1)$, $E_3 = (0.5, 0.5)$, and $E_4 = (0, 0.5)$, the basis functions are

$$\phi_1(x, y) = \begin{cases} 4x, & \text{on } T_1 \\ -2 + 4y, & \text{on } T_2 \end{cases}$$

$$\phi_2(x, y) = \begin{cases} -1 - 2x + 2y, & \text{on } T_1 \\ 0, & \text{on } T_2 \end{cases}$$

$$\phi_3(x, y) = \begin{cases} 0, & \text{on } T_1 \\ 1 + 2x - 2y, & \text{on } T_2 \end{cases}$$

$$\phi_4(x, y) = \begin{cases} 2 - 2x - 2y, & \text{on } T_1 \\ 2 - 2x - 2y, & \text{on } T_2 \end{cases}$$

and $\gamma_1 = 0.323825$, $\gamma_2 = 0$, $\gamma_3 = 1.0000$, and $\gamma_4 = 0$.

3. The Finite-Element method with $K = 8$, $N = 8$, $M = 32$, $n = 9$, $m = 25$, and $NL = 0$ gives the following results.

$\gamma_1 = 0.511023$
$\gamma_2 = 0.720476$
$\gamma_3 = 0.507899$
$\gamma_4 = 0.720476$
$\gamma_5 = 1.01885$
$\gamma_6 = 0.720476$
$\gamma_7 = 0.507896$
$\gamma_8 = 0.720476$
$\gamma_9 = 0.511023$
$\gamma_i = 0, \quad 10 \le i \le 25$

$u(0.125, 0.125) \approx 0.614187$
$u(0.125, 0.25) \approx 0.690343$
$u(0.25, 0.125) \approx 0.690343$
$u(0.25, 0.25) \approx 0.720476$

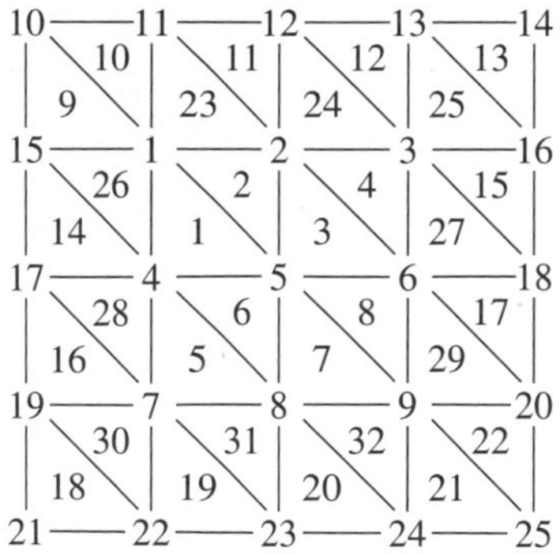

5. The Finite-Element Method with $K = 0$, $N = 12$, $M = 32$, $n = 20$, $m = 27$, and $NL = 14$ gives the following results.

$\gamma_1 = 21.40335$	$\gamma_8 = 24.19855$	$\gamma_{15} = 20.23334$	$\gamma_{22} = 15$
$\gamma_2 = 19.87372$	$\gamma_9 = 24.16799$	$\gamma_{16} = 20.50056$	$\gamma_{23} = 15$
$\gamma_3 = 19.10019$	$\gamma_{10} = 27.55237$	$\gamma_{17} = 21.35070$	$\gamma_{24} = 15$
$\gamma_4 = 18.85895$	$\gamma_{11} = 25.11508$	$\gamma_{18} = 22.84663$	$\gamma_{25} = 15$
$\gamma_5 = 19.08533$	$\gamma_{12} = 22.92824$	$\gamma_{19} = 24.98178$	$\gamma_{26} = 15$
$\gamma_6 = 19.84115$	$\gamma_{13} = 21.39741$	$\gamma_{20} = 27.41907$	$\gamma_{27} = 15$
$\gamma_7 = 21.34694$	$\gamma_{14} = 20.52179$	$\gamma_{21} = 15$	

$$u(1,0) \approx 22.92824$$
$$u(4,0) \approx 22.84663$$
$$u\left(\frac{5}{2}, \frac{\sqrt{3}}{2}\right) \approx 18.85895$$

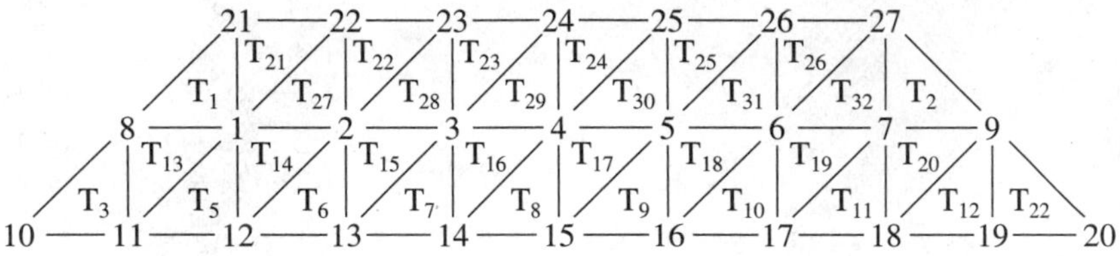

Index

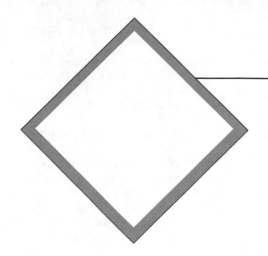

Trigonometry

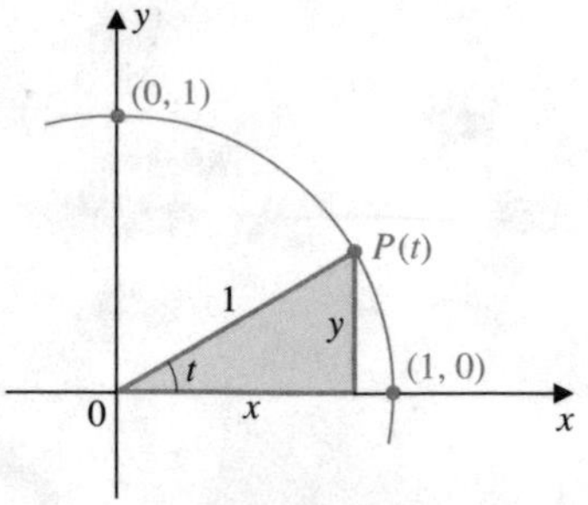

$$\sin t = y \qquad \cos t = x$$

$$\tan t = \frac{\sin t}{\cos t} \qquad \cot t = \frac{\cos t}{\sin t}$$

$$\sec t = \frac{1}{\cos t} \qquad \csc t = \frac{1}{\sin t}$$

$$(\sin t)^2 + (\cos t)^2 = 1$$

$$\sin(t_1 \pm t_2) = \sin t_1 \cos t_2 \pm \cos t_1 \sin t_2$$

$$\cos(t_1 \pm t_2) = \cos t_1 \cos t_2 \mp \sin t_1 \sin t_2$$

$$\sin t_1 \sin t_2 = \frac{1}{2}[\cos(t_1 - t_2) - \cos(t_1 + t_2)]$$

$$\cos t_1 \cos t_2 = \frac{1}{2}[\cos(t_1 - t_2) + \cos(t_1 + t_2)]$$

$$\sin t_1 \cos t_2 = \frac{1}{2}[\sin(t_1 - t_2) + \sin(t_1 + t_2)]$$

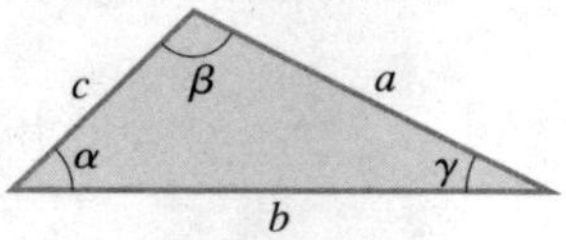

Law of Sines: $\dfrac{\sin\alpha}{\alpha} = \dfrac{\sin\beta}{\beta} = \dfrac{\sin\gamma}{\gamma}$

Law of Cosines: $c^2 = a^2 + b^2 - 2ab\cos\gamma$

Common Series

$$\sin t = \sum_{n=0}^{\infty} \frac{(-1)^n t^{2n+1}}{(2n+1)!} = t - \frac{t^3}{3!} + \frac{t^5}{5!} - \cdots$$

$$e^t = \sum_{n=0}^{\infty} \frac{t^n}{n!} = 1 + t + \frac{t^2}{2!} + \frac{t^3}{3!} + \cdots$$

$$\cos t = \sum_{n=0}^{\infty} \frac{(-1)^n t^{2n}}{(2n)!} = 1 - \frac{t^2}{2!} + \frac{t^4}{4!} - \cdots$$

$$\frac{1}{1-t} = \sum_{n=0}^{\infty} t^n = 1 + t + t^2 + \cdots, \qquad |t| < 1$$

The Greek Alphabet

Alpha	A	α	Eta	H	η	Nu	N	ν	Tau	T	τ
Beta	B	β	Theta	Θ	θ	Xi	Ξ	ξ	Upsilon	Υ	υ
Gamma	Γ	γ	Iota	I	ι	Omicron	O	o	Phi	Φ	ϕ
Delta	Δ	δ	Kappa	K	κ	Pi	Π	π	Chi	X	χ
Epsilon	E	ϵ	Lambda	Λ	λ	Rho	P	ρ	Psi	Ψ	ψ
Zeta	Z	ζ	Mu	M	μ	Sigma	Σ	σ	Omega	Ω	ω